Concepts and Applications in Environmental Geochemistry

Edited by

Dibyendu Sarkar
Rupali Datta

Department of Earth and Environmental Sciences
University of Texas at San Antonio
San Antonio, TX, USA

Robyn Hannigan

Department of Chemistry and Physics
Arkansas State University
Jonesboro, AR, USA

ELSEVIER

Amsterdam – Boston – Heidelberg – London – New York – Oxford
Paris – San Diego – San Francisco – Singapore – Sydney – Tokyo

Elsevier
The Boulevard, Langford Lane, Kidlington, Oxford, OX5 1GB, UK
Radarweg 29, PO Box 211, 1000 AE Amsterdam, The Netherlands

First edition 2007
Reprinted 2008

British Library Cataloguing in Publication Data
A catalogue record for this book is available from the British Library

Library of Congress Cataloging-in-Publication Data
A catalog record for this book is available from the Library of Congress

ISBN: 978-0-08-046522-7

For information on all Elsevier publications
visit our website at books.elsevier.com

Printed and bound by CPI Group (UK) Ltd, Croydon, CR0 4YY

Transferred to Digital Print 2011

Contents

List of Contributors

P.A. Abanda	Mid-South Community College, West Memphis, AR, USA
Hassan Y. Aboul-Enein	Pharmaceutical and Medicinal Chemistry Department, Pharmaceutical and Drug Industries Research Division, National Research Centre, Dokki, Cairo, Egypt
Honey Aggarwal	Department of Biotechnology & Environmental Sciences, Thapar Institute of Engineering & Technology (Deemed University), Patiala, Punjab, India
Imran Ali	Department of Chemistry, Jamia Millia Islamia (Central University), Jamia Nagar, New Delhi, India
C. Brannon Andersen	Department of Earth and Environmental Sciences, Furman University, Greenville, SC, USA
Andrea Anderson	Department of Earth and Environmental Sciences, University of Texas at San Antonio, San Antonio, TX, USA
S. Banerjee	Center for Study of Man and Environment, Sector II, Salt Lake City, Kolkata, India
M.L. Barrico	Department of Botany, University of Coimbra, Coimbra, Portugal
Tracy D. Branam	Indiana Geological Survey, Indiana University, Bloomington, IN, USA
B. Brocks	Towson University Geochemistry Laboratory, Towson University, Towson, MD, USA
Yong Cai	Department of Chemistry & Biochemistry and Southeast Environmental Research Center, Florida International University, University Park, Miami, FL, USA

Chinmoy Chakrabarti — Center for Study of Man and Environment, Sector II, Salt Lake City, Kolkata, India

Gargi Chatterjee — Center for Study of Man and Environment, Sector II, Salt Lake City, Kolkata, India

Tait Chirenje — Natural and Mathematical Science (NAMS), Richard Stockton College of New Jersey, Pomona, NJ, USA

Leonette Cox — Program for Environmental Sciences, Arkansas State University, State University, AR, USA

Rupali Datta — Department of Earth and Environmental Science, University of Texas at San Antonio, San Antonio, TX, USA

Sahadeb De — Earth Sciences and Resources Institute, University of South Carolina, Columbia, SC, USA

B.H.W.S. de Jong — Institute of Earth Sciences, Utrecht University, Budapestlaan, Utrecht, The Netherlands

M. Ding — Los Alamos National Laboratory, Earth and Environmental Science Division, Los Alamos, New Mexico, USA

L. Elango — Department of Geology, Anna University, Chennai, India

Claude Epstein — Natural and Mathematical Science (NAMS), Richard Stockton College of New Jersey, Pomona, NJ, USA

Jerry L. Farris — Environmental Sciences Program, Arkansas State University, Arkansas, USA

H. Freitas — Department of Botany, University of Coimbra, Coimbra, Portugal

J.L. Gardea-Torresdey — Department of Chemistry, University of Texas at El Paso, El Paso Texas, USA

Dinesh Goyal	Department of Biotechnology & Environmental Sciences, Thapar Institute of Engineering & Technology (Deemed University), Patiala, Punjab, India
Matthew Grunstra	Department of Earth and Environmental Science and Department of Biology, University of Texas at San Antonio, San Antonio, TX, USA
Sabine Grunwald	Soil and Water Science Department, University of Florida, Florida, USA
V.K. Gupta	Department of Chemistry, Indian Institute of Technology, Roorkee, India
Robyn E. Hannigan	Department of Chemistry and Physics, Arkansas State University, Arkansas, USA Environmental Sciences Program, Arkansas State University, Arkansas, USA
Syed E. Hasan	University of Missouri-Kansas City, Missouri, USA
Nancy R. Hasenmueller	Indiana Geological Survey, Indiana University, Bloomington, IN, USA
Curtis L. Hollabaugh	Department of Geosciences, University of West Georgia, Carrollton, GA, USA
Matthew Horton	Arkansas State University, Program for Environmental Sciences, State University, AR, USA
H. Hussein	Department of Earth and Environmental Science, The University of Texas at San Antonio, San Antonio, TX, USA
Atsushi Kamei	Interdisciplinary Faculty of Science and Engineering, Shimane University, Matsue, Shimane, Japan
R. Kannan	Department of Geology, Anna University, Chennai, India

Patrick Kinnicutt	Departments of Geology and Computer Science, Central Michigan University, Mount Pleasant, MI, USA
Gregory P. Lewis	Department of Biology, Furman University, Greenville, SC, USA
S.M. Lev	Towson University Geochemistry Laboratory, Towson University, Towson, MD, USA
Z.-Q. Lin	Environmental Sciences Program and Department of Biological Sciences, Southern Illinois University Edwardsville, Edwardsville, IL, USA
Guangliang Liu	Department of Chemistry & Biochemistry and Southeast Environmental Research Center, Florida International University, University Park, Miami, FL, USA
Konstantinos C. Makris	Department of Earth and Environmental Science, University of Texas at San Antonio, San Antonio, TX, USA
Raymond Mueller	Natural and Mathematical Science (NAMS), Richard Stockton College of New Jersey, Pomona, NJ, USA
Suresh Muthukrishnan	Department of Earth and Environmental Sciences, Furman University, Greenville, SC, USA
C. Nabais	Department of Botany, University of Coimbra, Coimbra, Portugal
Rachana Nagar	Department of Earth and Environmental Sciences, University of Texas at San Antonio, San Antonio, TX, USA
Kazuki Naito	Research Center for Deep Geological Environments, Advanced Industrial Science and Technology, Higashi, Tsukuba Ibaraki, Japan
G.A. O'Connor	Soil and Water Science Department, University of Florida, Gainesville, FL, USA

George M. Ogendi	Department of Chemistry and Physics, Arkansas State University, Arkansas, USA; Environmental Sciences Program, Arkansas State University, Arkansas, USA
Koichi Okuzawa	Research Center for Deep Geological Environments, Advanced Industrial Science and Technology, Higashi, Tsukuba Ibaraki, Japan
J.G. Parsons	Department of Chemistry, University of Texas at El Paso, El Paso Texas, USA
J.R. Peralta-Videa	Department of Chemistry, University of Texas at El Paso, El Paso Texas, USA
M.N.V. Prasad	Department of Plant Sciences, University of Hyderabad, Hyderabad, India
Shahida Quazi	University of Texas at San Antonio, San Antonio, TX, USA
K. Ramesh Reddy	Soil and Water Science Department, University of Florida, Florida, USA
Rosanna L. Rivero	Soil and Water Science Department, University of Florida, Florida, USA
Seth Rose	Department of Geosciences, Georgia State University, Atlanta, GA, USA
Aboubakar Sako	Department of Geology, Geography and Environmental Sciences, University of Stellenbosch, Matieland, South Africa
Dibyendu Sarkar	Department of Earth and Environmental Science, University of Texas at San Antonio, San Antonio, TX, USA
Yoji Seki	Research Center for Deep Geological Environments, Advanced Industrial Science and Technology, Higashi, Tsukuba Ibaraki, Japan

Saurabh Sharma	Department of Earth and Environmental Sciences, University of Texas at San Antonio, San Antonio, TX, USA
Shankar Sharma	University of California, SNRI, Merced, CA, USA
Jacqueline A. Shea	Washington State Department of Ecology, Richland, WA, USA
Richard O. Sleezer	Earth Science Department, Emporia State University, Emporia, KS, USA
Naoto Takeno	Research Center for Deep Geological Environments, Advanced Industrial Science and Technology, Higashi, Tsukuba Ibaraki, Japan
N. Terry	Department of Plant and Microbial Biology, University of California at Berkeley, Berkeley, CA, USA
Chacharee Therapong	Department of Earth and Environmental Science, University of Texas at San Antonio, San Antonio, TX, USA
Margaret A. Townsend	Kansas Geological Survey, University of Kansas, Lawrence, KS, USA
O.W. Van Auken	Department of Biology, University of Texas at San Antonio, San Antonio, TX, USA
George F. Vance	University of Wyoming, Laramie, WY, USA
Yoshio Watanabe	Research Center for Deep Geological Environments, Advanced Industrial Science and Technology, Higashi, Tsukuba Ibaraki, Japan
Z.H. Ye	School of Life Sciences, Zhongshan University, Guangzhou, China

Introduction to the Book Series

Environmental pollution has played a critical role in human lives since the early history of the nomadic tribes. During the last millennium, industrial revolution, increased population growth and urbanization have been the major determinants in shaping our environmental quality.

Initially, primary air pollutants such as sulfur dioxide and particulate matter were of concern. For example, the killer fog of London in 1952 resulted in significant numbers of human fatality leading to major air pollution control measures. During the 1950s, scientists also began to understand the cause and atmospheric mechanisms for the formation of the Los Angeles photochemical smog. We now know that surface level ozone and photochemical smog are a worldwide problem at regional and continental scales, with specific geographic areas of agriculture, forestry and natural resources, including their biological diversity at risk. As studies continue on the atmospheric photochemical processes, air pollutant transport, their atmospheric transformation and removal mechanisms, so is the effort to control the emissions of primary pollutants (sulfur dioxide, oxides of nitrogen, hydrocarbons and carbon monoxide), mainly produced by fossil fuel combustion.

During the mid-1970s, environmental concerns regarding the occurrence of "acidic precipitation" began to emerge to the forefront. Since then, our knowledge of the adverse effects of air pollutants on human health and welfare (terrestrial and aquatic ecosystems and materials) has begun to rise substantially. Similarly, studies have been directed to improve our understanding of the accumulation of persistent inorganic (heavy metals) and organic (polyaromatic hydrocarbons and polychlorinated biphenyls) chemicals in the environment and their impacts on sensitive receptors, including human beings. Use of fertilizers (excess nutrient loading) and herbicides and pesticides in both agriculture and forestry and the related aspects of their atmospheric transport, fate and deposition; their direct runoff through the soil and impacts on ground- and surface water quality and environmental toxicology have become issues of much concern.

In the recent times environmental literacy has become an increasingly important factor in our lives, particularly in the so-called developed nations. Currently the scientific, public and political communities are much concerned with the increasing global scale air pollution and the consequent global climate change. There are efforts being made to totally ban the use of chlorofluorocarbon and organo-bromine compounds at the global scale. However, during this millennium many developing nations will become major forces governing environmental health as their populations and industrialization grow at a rapid pace. There is an on-going international debate on policies and the mitigation strategies to be adopted to address the critical issue of climate change. Human health and environmental impacts and risk assessment and the associated cost-benefit analyses, including global economy are germane to this controversy.

An approach to understanding environmental issues in general and in most cases, mitigation of the related problems requires a systems analysis and a multi- and interdisciplinary philosophy. There is an increasing scientific awareness to integrate environmental processes and their products in evaluating the overall impacts on various receptors. As momentum is gained, this approach constitutes a challenging future direction for our scientific and technical efforts.

The objective of the book series *Developments in Environmental Science* is to facilitate the publication of scholarly works that address any of the described topics, as well as those that are related. In addition to edited or single- and multi-authored books, the series also considers conference proceedings and paperback computer-software packages for publication. The emphasis of the series is on the importance of the subject topic, the scientific and technical quality of the content and timeliness of the work.

Sagar V Krupa
Editor-in-Chief, Book Series

Concepts and Applications in Environmental Geochemistry

Developments in Environmental Science, Volume 5
D. Sarkar, R. Datta and R. Hannigan (Editors)

ISSN: 1474-8177/DOI:10.1016/S1474-8177(07)05001-2

Chapter 1

What goes around comes around: Today's environmental geochemistry

Robyn Hannigan

Abstract

What is environmental geochemistry? Is it simply the application of geochemical techniques to environmental media such as water or soil? Is it the study of contaminants and pollutants? Is it all low temperature geochemistry exclusive of the biosphere and natural weathering processes? For certain the definition is complex. The field is clearly diverse as is evidenced in this book. Indeed this book best represents the complexity of the field, a snapshot, if you will, of concepts and applications in the field. Bearing in mind that the field is ever changing, a dynamic and evolving discipline, we provide a brief historical perspective and an attempt to define today's environmental geochemistry.

1.1. Introduction

Many of us call ourselves environmental geochemists. If asked what that really means, it is difficult to explain. The boundaries between the scientific disciplines are significantly blurred. But were they always so blurry? Is there a place and time one can point to and say that "this" is the beginning of the field? First one has to define what environmental geochemistry is. Geochemistry is, most broadly, the study of the chemical composition of the earth and other planets, chemical processes, cycles, reactions that govern the composition of rocks and soils, and temporal and spatial changes in these controlling factors. While interactions with the other earth systems is implied they are not explicit. Biogeochemistry, the field of study centered on chemical, physical, geological, and biological processes, cycles, and reactions that govern the composition of the natural environment (including the biosphere, the hydrosphere, the pedosphere, the atmosphere, and the lithosphere), is more closely aligned with the concept of environmental geochemistry. However

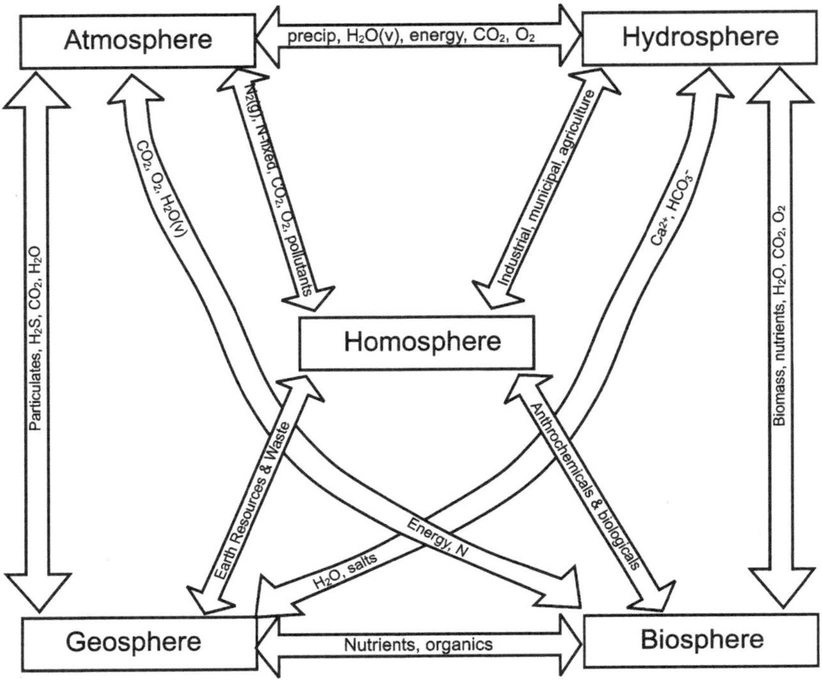

Figure 1.1. Illustration of the earth systems showing the linkages between the systems and the interactions of the homosphere with each.

today's environmental geochemistry is truly a hybrid between biogeochemistry and geochemistry. Environmental geochemistry is the field of study integrating the study of earth's chemical composition and earth system interactions with the interaction of humans with these systems. Therefore, a new "sphere" or dimension exists in environmental geochemistry that is mindful of the human interactions with the earth systems whether natural or man-made (Fig. 1.1).

1.1.1. What goes around ...

With environmental geochemistry defined as above it is possible to explore the history of this field and set this book in the context of where the field has been and where it is headed. Scientists have long explored the relations between mankind and the earth. As early as 1766 with Henry Cavendish's exploration of arsenic in soil, natural scientists understood that chemicals in the soil could have profound impacts on human life. Cavendish, in his fundamental study on the effect of lightening on soil (Cavendish, 1767), was able to duplicate the processes that produce nitrogen gas. Not long after Cavendish found that nitrogen gas could be generated from soil, Nicolas de Saussure showed that plants require nitrogen from the soil (Saussure, 1804). Connecting the observations of Cavendish and de Saussure, Thaer developed the humus theory for soil suggesting that "organic substances" are the primary sources of energy in soils (Thaer, 1846). It was Davy (1813) who extracted "gooey" substances from soils and in so doing unknowingly uncovered the role of carbon dioxide–carbonic acid in soil formation. One might consider Davy's work as the "moment" when the science of geochemistry and the environment were connected. But no one connected the two in the literature until Dumas stated "Everything that the plants take from the air they give to animals, the animals return it to the air; this is the eternal circle in which life revolves but where matter only changes place" (Dumas, 1841). Dumas followed this insightful comment with a seminal collaborative essay with Boussingault (Dumas and Boussingault, 1842) where they discussed how the "mysterious cycle of organic life on the globe" worked by plants drawing matter from the atmosphere to create organic substances (Davy's gooey substances), and how these substances are eaten by animals, who return these "materials to nature's grand reserve upon their death and ultimate decay". Here is the start of the field of biogeochemistry and the development of this fields focus on carbon, hydrogen, nitrogen, oxygen, phosphorus, and sulphur (CHNOPS) cycles. Where did the arsenic work of Cavendish go? How were humans, so fundamental to the studies of Thaer, the father of agronomy, so soon removed from the cycles of matter on the planet? Alas hindsight is 20/20. What goes around, thankfully, comes around.

Why focus on the history of soil chemistry and agronomy in an attempt to outline the history of environmental geochemistry? As the fields of geochemistry evolved and fundamental advances in our understanding of natural chemical processes on the planet were made, it is in soil science that the cycles and connections between the earth systems are elucidated with humans as actors in these cycles. For example, during the early and mid-1800s humans still had relatively little impact on the global cycling of nitrogen except in agricultural regions. In fact, Liebig (1840), or more correctly Sprengel (1837), attempted to capture the interaction of crops (anthropogenic) and nutrients (natural geochemical processes). This smacks of biogeochemistry except that humans plant the crops so it is still more closely aligned with our earlier definition of environmental geochemistry.

Geologist Eduard Suess (1875) tossed out the term "biosphere", but being a geologist, approached the earth systems from a traditional geochemical perspective. The current meaning of biosphere within an ecological context can be traced to Vladimir Vernadsky (1926) some 50 years later. Of course this seems to foreshadow Lovelock's (1979) Gaia Hypothesis, but I digress. Vernadksy's fundamentally important contribution to the understanding of the interaction of the living and non-living earth systems cannot be underestimated. Vernadsky's view of the biosphere was a new way of looking at our planet. His work led to major advances in ecology and ultimately blurred the lines between geochemistry and ecology to establish biogeochemistry as the field of science exploring chemical processes and interactions between geosphere–atmosphere–hydrosphere and living organisms. Humans, though certainly living organisms, were thought to have little influence on the earth as a whole and were basically end-users of the earth's abundant matter and energy.

In the years between Vernadsky's work and advancing the newly emerging field of biogeochemistry around from to World War II something changed: not an "earthquake" but field of a subtle shift occurred on the continents. Ecologists working in terrestrial aquatic systems began to make connections between natural geochemical processes, the influence of these processes on lake chemistry and availability of nutrients, and human interactions with the landscape.

Odum (1959, 1963) noted that the cycles of N, O, and water had a gaseous component, and that these cycles had self-regulating feedback mechanisms making them "perfect". However, scientists understood the imperfection of their "model" cycles when humans were considered in the biosphere as early as 1965 when concern abounded over disruption of the carbon cycle by burning of fossil fuels (Revelle et al., 1965) and nitrogen cycle through the introduction of pesticides (Cole, 1966). So here we have it, the beginnings of environmental geochemistry as a discipline and the introduction of humans as major players in the structure and the function of earth systems. A snapshot of the state of knowledge in 1960s can be found in Bormann and Likens (1967).

1.1.2. … comes around

As this introduction to concepts and applications in environmental geochemistry is not intended as a complete compilation of all research in the field, let us fast forward to Radojevic and Bashkin (1999) who summarized the analytical approaches to environmental chemistry. What is environmental chemistry? It is just environmental geochemistry, which, by any other name, would smell as sweet but makes chemists feel more at ease. Radojevic and Bashkin summarize the human influence on the cycles of C, S, N, and P as well as the consequences of human influence (Table 1.1).

Table 1.1. Relative contribution of human and natural sources of SO_2, NO_x, CO_2, and hydrocarbons to the atmosphere

Component	% of total emissions to the atmosphere	
	Natural	Human
SO_2	50	50
NO_x	50	50
CO_2	95	5
Hydrocarbons	84	16

Source: After Radojevic and Bashkin (1999).

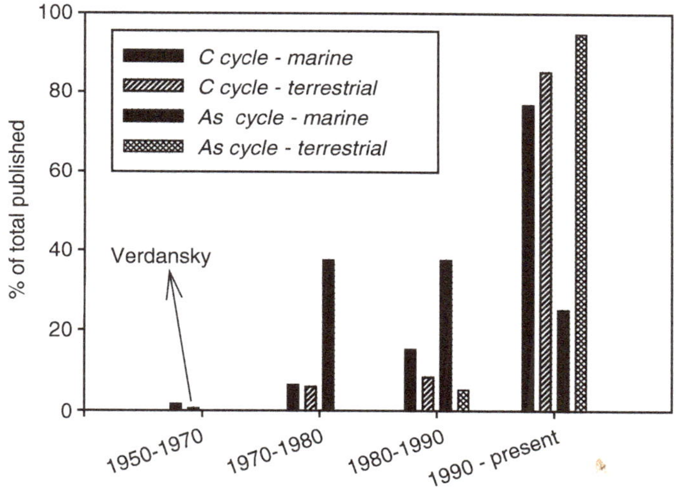

Figure 1.2. Relative percentage of publications derived from a GeoRef search using carbon cycle and arsenic cycle as keywords.

In 1999, humans contributed 16% of the hydrocarbon emissions to the atmosphere (5% of the CO_2), but up to 50% of the SO_2 and NO_x emissions. Human impact on the earth systems is no longer a subject of debate. Neither is the impact of these systems on humans. Witness the continued focus on arsenic in drinking water and the development of an intricate and relatively complete arsenic cycle based on a material's flow perspective (Ayres and Ayres, 1999).

So here we are in 2007. We are really not much farther along than Cavendish was in 1767 when he began his studies on arsenic. Much has changed over the intervening years between us and Cavendish. Publications in the field of environmental geochemistry are ever increasing (Fig. 1.2).

Although initially studies of the cycles of C and As were dominated by biogeochemists in the marine realm, terrestrial research has begun to dominate environmental geochemistry. As more and more research focuses on the relation of the homosphere to the other earth systems, disciplinary boundaries are obscured and the lines between once discrete disciplines such as geochemistry and biogeochemistry are now arbitrary, a vestige of the "old days".

Within this book you will find a great variety of environmental geochemistry topics ranging from mathematical approaches to exploring natural processes and their impact on the environment. Here we present a snapshot of environmental geochemistry as we know it today, much like it was in the beginning but with a larger vocabulary and few, if any, disciplinary boundaries. What goes around, comes around and we, as humans, are still in the thick of it.

REFERENCES

Ayres, R., Ayres, L.W., 1999. Accounting for Resources 2: The Life Cycle of Materials. Cheltenham, Edward Elgar.

Bormann, F.H., Likens, G.E., 1967. Nutrient cycling: Small watersheds can provide invaluable information about terrestrial ecosystems. Science, 55, 424–429.

Cavendish, H., 1767. Experiments on Rathbone-Cavendish Water: By the Hon. Henry Cavendish: F.R.S. Philosophical Transactions 57, 92–108.

Cole, L.C., 1966. Protect the friendly microbes. Saturday Review, pp. 46–47.

Davy, H., 1813. Elements of Agricultural Chemistry. W. Bulwer and Company, London.

Dumas, J.-B., 1841. Essai de Statique des Etres Organises. Fortin, Masson, and Ce, Librairies, Paris.

Dumas, J.-B., Boussingault, J.-B., 1842. The Chemical and Physiological Balance of Organic Nature, Essai de Statique des Etres Organises. Fortin, Masson, and Ce, Librairies, Paris.

Liebig, J., 1840. Die organische Chemie in ihrer Anwendung auf Agricultur und Physiologie. [*Organic Chemistry in Its Applications to Agriculture and Physiology*.] Friedrich Vieweg und Sohn Publishing Company, Braunschweig, Germany.

Lovelock, J.E., 1979. Gaia: A New Look at Life on Earth. Oxford University Press, Oxford.

Odum, E.P., 1959. Fundamentals of Ecology. Saunders, Philadelphia.

Odum, E.P., 1963. Ecology. Holt, Rinehart, and Winston, New York.

Radojevic, M., Bashkin, V.N., 1999. Practical Environmental Analysis. The Royal Society of Chemistry, Cambridge.

Revelle, R., Broecker, W., Craig, H., Keeling, C.D., 1965. Restoring the Quality of Our Environment. Government Printing Office, Washington, DC, pp. 111–133.

Saussure, N.T., 1804. Recherches chimiques sur la vegetation. Didot jeune, Paris.

Sprengel, C., 1837. Die Bodenkunde oder die Lehre vom Boden. [*Soil Science and Its Doctrines*.] Immanuel Müller Publishing Company, Leipzig, Germany.

Suess, E., 1875. Die Entstehung der Alpen. [*The Origin of Alps*.] W. Braunmüller, Vienna.

Thaer, A., 1809–1812 (1846). *The Principles of Agriculture*. Greeley and McElrath, New York.

Vernadsky, V.I., 1926. Biosphera. [*The Biosphere*.] Nauchnoe khimiko-technicheskoye izdatel'stvo (Scientific Chemico-Technical Publishing), Leningrad.

Section I: Today's Environmental Geochemistry—A Review of New Concepts and Innovative Practices

Developments in Environmental Science, Volume 5
D. Sarkar, R. Datta and R. Hannigan (Editors)

ISSN: 1474-8177/DOI:10.1016/S1474-8177(07)05002-4

Chapter 2

Modification of Goldschmidt's geochemical classification of the elements to include arsenic, mercury, and lead as biophile elements

Curtis L. Hollabaugh

Abstract

The geochemical distribution and classification of the elements was determined and described by Victor Moritz Goldschmidt. This paper describes proposed modifications of Goldschmidt's classification of the elements in the light of geochemical and environmental data obtained in the last 70 years. Arsenic, lead, and mercury have great affinity to concentrate in the biosphere. Under certain conditions they will reach relatively high concentrations in coal and humans. Arsenic and mercury form methyl compounds with carbon and can build up in the food chain. Lead, because it has the same charge and a similar ionic radius to calcium, builds up in the human body when ingested. This is especially true in young children (below 6 years old) that require high levels of calcium to grow. Toxic levels of As, Pb, and Hg can result in death to all ages. Less than toxic levels can cause birth defects and brain damage (Hg), cancer and skin warts (As), and brain damage and lower IQ (Pb). Research indicates that additional classification of these traditional chalcophile elements as biophile elements is indispensable to future proper understanding and utilization of arsenic, lead, and mercury in the modern industrial, developing, and impoverished world.

2.1. Introduction

In 1923 Victor Moritz Goldschmidt, the father of modern geochemistry, proposed the geochemical classification of the elements (Goldschmidt, 1923; Mason, 1992). The classification of the elements (Table 2.1) into lithophile (affinity for silicate), siderophile (affinity for iron–nickel), chalcophile (affinity for sulfide), and atmophile (affinity for the gaseous

Table 2.1. Goldschmidt's[a] geochemical classification of the elements

Siderophile	Chalcophile		Lithophile	Atmophile	Biophile (organisms)
	In meteorites	Terrestrial			
Fe, Ni, Co, P, C, Ru, Rh, Pd, Os, Ir, Pt, Au, Ge, Sn, Mo, Ta. Minor— As, W, Nb, Se, Te	S, Se, P, As, Cu, Ag, Zn, Cd, V, Cr, Mn, Fe. Minor— Te, Sb, Ti, Ca	S, Se, Te, As, Sb, Bi, Ga, In, Tl, Pb, Zn, Cd, Hg, Cu, Ag, Ni, Pd, Co, Fe, Ru, Mo. Minor— Ge, Sn, Au, Pt, Rh, Ir, Os	O, Si, Ti, Zr, Hf, Th, F, Cl, Br, I, B, Al, Sc, Y, La, Ce, Pr, Nd, Sm, Eu, Gd, Tb, Dy, Ho, Er, Tu, Yb, Lu, Li, Na, K, Rb, Cs, Be, Mg, Ca, Sr, Ba, V, Cr, Mn, Nb, Ta, W, U. Minor—S, P, H, Sn, Ga, Fe, Ni, Co, C	H, N, C, O, Cl, Br, I, He, Ne, Ar, Kr, Xe	C, H, O, N, P, S, Cl, I. Minor— B, Ca, Mg, K, Na, V, Mn, Fe, Cu

[a]Goldschmidt titled it "Geochemical Classification of the Elements, according to Distribution between Iron, Sulphides, Silicates, Atmosphere, and Organisms" (Goldschmidt, 1954).

state) has withstood the test of time (Cox, 1989; Emsley, 1998; Li, 2000; Railsback, 2003). Goldschmidt included in his classification the biophile elements. They are defined as "those elements that are the most typical in organisms and organic material, or are concentrated in and by living plants and animals" (Bates and Jackson, 1984). The biophile elements all occur as lithophile and/or chalcophile elements. Biophile elements include carbon, hydrogen, oxygen, nitrogen, phosphorus, sulfur, chlorine, iodine, bromine, calcium, magnesium, potassium, sodium, vanadium, iron, manganese, and copper. Sixty years after the publication many geochemistry textbooks had dropped the biophile elements from tables showing the geochemical classification of the elements (Table 2.2) (Mason and Moore, 1982; Krauskopf and Bird, 1995).

The purpose of this paper is to update the classification of the elements and to suggest that arsenic, lead, and mercury should be included in the biophile elements. Our research consists of determining the distribution

Table 2.2. Tables of the distribution of the elements in leading geochemistry textbooks in 1982 and 1995

Siderophile	Chalcophile	Lithophile	Atmophile
Mason and Moore (1982)			
Fe,[a] Co,[a] Ni,[a] Ru, Rh, Pd, Os, Ir, Pt, Au, Re,[b] Mo,[b] Ge,[a] Sn, W,[c] C,[c] Cu,[a] Ga,[a] Ge, As,[b] Sb[b]	Ag, Zn, Cd, Hg, Ga, In Tl, Pb, Bi, S, Se, Te, Mo. Minor—Cu, Ge, Sn, As, Sb, Fe, Os, Ru, Rh, Pd	Li, Na, K, Rb, Cs, Be, Mg, Ca, Sr, Ba, B, Al, Sc, Y, La, Ce, Pr, Nd, Sm, Eu, Gd, Tb, Dy, Ho, Er, Tm, Yb, Lu, Si, Ti, Zr, Hf, Th, P, V, Nb, Ta, O, Cr, U, H, F, Cl, Br, I, Mn. Minor—Fe, Zr, Ga	N, He, Ne, Ar, Kr, Xe. Minor—H, O
Krauskopf and Bird (1995)			
Fe, Co, Ni, Ru, Rh, Pd, Re, Os, Ir, Pt, Au, Mo, Ge, Sn, C, P. Minor[d]—Pb, As, W	Cu, Ag, Zn, Cd, Hg, Ga, In, Tl, Pb, As, Sb, Bi, S, Se, Te. Minor[d]—Fe, Mo, Re	Li, Na, K, Rb, Cs, Be, Mg, Ca, Sr, Ba, B, Al, Sc, Y, La, Ce, Pr, Nd, Sm, Eu, Gd, Tb, Dy, Ho, Er, Tm, Yb, Lu, Si, Ti, Zr, Hf, Th, V, Nb, Ta, O, Cr, W, U, Mn, F, Cl, Br, I. Minor[d]—Pb, C, P, Fe, H, Tl, Ga, Ge, N	H, N, He, Ne, Ar, Kr, Xe. Minor[d]—C, O, F, Cl, Br, I

[a]Chalcophile and lithophile in the earth's crust.
[b]Chalcophile in the earth's crust.
[c]Lithophile in the earth's crust.
[d]Minor elements belong for the most part in another group but do occur in this group.

ratio of arsenic, lead, and mercury in humans versus lithosphere and hydrosphere. This paper will discuss how humans have historically distributed arsenic, mercury, and lead around the world and the toxic effects of this distribution.

2.1.1. Biophile elements

The beauty of Goldschmidt's classification is that it is based on the distribution of the elements. The siderophile, chalcophile, lithophile, and atmophile elements are well described and long accepted (Garrett, 2005).

Goldschmidt's original list of major biophile elements was C, H, O, N, P, S, Cl, and I. His list of minor biophile elements was B, Ca, Mg, K, Na, V, Mn, Fe, and Cu. Biophile elements are enriched in the biosphere—life, organic waste, dead remains, and naturally processed materials from the biosphere (coal, oil, natural gas, and kerogen). The most significant biophile elements are carbon, hydrogen, oxygen, nitrogen, phosphorus, and sulfur. A complete listing of all biophile elements is compiled from:

1. Elements that form bonds with organic carbon and are the major elements of life (i.e., Goldschmidt lists the major biophile elements C, H, O, N, P, S, Cl, and I). Lindh (2005) discusses the major and minor elements in the human body. Eleven elements make up 99.8% of the atoms in humans. These are oxygen (65%), carbon (18%), hydrogen (10%), nitrogen (3%), calcium (1.4%), phosphorus (1.0%), magnesium (0.5%), potassium (0.34%), sulfur (0.26%), sodium (0.14%), and chlorine (0.14%). Combs (2005) lists as macronutrient elements Ca, P, Mg, Na, K, and Cl.
2. Elements that function as skeletal structure (calcium phosphate in bones, calcium carbonate in shells, silica in some exoskeletons, and fluorine for OH in bones and teeth). Calcium phosphate in human bones is similar to the mineral apatite, $Ca_5(PO_4)_3(OH)$. The calcium content of the human body is more than 1 kg with 99% of this in the skeleton. About 85% of the bodies' 700 g of phosphorus occur in bone. Human bones also contain magnesium with 60–65% of the 20–28 g of Mg in the bones (Combs, 2005).
3. Fluid mediums (i.e., K, Na, Cl in blood) (Combs, 2005).
4. The elements that are in minor or trace amounts play a necessary role in life. Combs (2005) lists ten elements as accepted essential micronutrient elements (Fe, Cu, Zn, I, Se, Mn, Mo, Cr, F, and Co) and six as suspected essential elements (Ni, Pb, As, B, V, and Si). Lindh (2005) lists 18 essential and possible essential trace elements (Li, V, Cr, Mn, Fe, Co, Ni, Cu, Zn, W, Mo, Si, Se, F, I, As, Br, and Sn).
5. Elements that because of functional similarities (size and charge) substitute for a biophile element (i.e., lead for calcium). Lead is the same charge as and is similar in size to calcium (Pb^{2+} radii is 1.19 Å and Ca^{2+} radii is 1.00 Å) and readily substitutes for calcium in bones. Skinner's (2005) proposed substitutes for calcium in bones and teeth are Pb, Cd, Mg, Sr, Ba, Mn, Na, and K. Substitutions for phosphate, PO_4^{-3}, in human bones can be AsO_4^{-3}, SiO_4^{-4}, VO_4^{-3}, SO_4^{-2}, SbO_4^{-3}, CO_3^{-2}, and AlO_4^{-5}. The hydroxyl site in apatite can have full substitution by fluorine and chlorine and limited substitution by Br and I. Carbonate can substitute for either phosphate or hydroxyl (Skinner, 2005).

6. Rare or trace elements that will complex with carbon compounds and build up in the food chain (methylmercury and methylarsenic). The mercury added to the environment can be added as methylmercury, $Hg(CH_3)_2$, or in the metallic, Hg, form. Methylmercury can form readily within living organisms from metallic or inorganic mercury. Methylmercury species include CH_3Hg^+, CH_3HgOH, CH_3HgCl, $(CH_3)_2Hg$, and $(CH_3Hg)_2S$ (Turner, 1987). Bacteria in soil, sediment, or water readily transforms mercury into $(CH_3)_2Hg$ and CH_3Hg^+ with the latter compound going into solution (Stober et al., 2001). In the form of CH_3Hg^+ plants pick up the mercury. This can then work its way up the food chain from plants to small fish to large fish to humans because it builds up faster than it can be expelled. Typical unpolluted surface waters contain 0.5–4 ng l^{-1} mercury. Movement of mercury as methylmercury up the food chain is so great that contaminated fish occur even in remote lakes. Fish at the top of the food chain, because of bioaccumulation of methylmercury, may have 1–10 million times greater concentration of mercury than surrounding water.
7. Elements that can be reduced and deposited by organic waste and remains (S, As, Pb, Hg, and U in peat, coal, and oil). For example, the organic-rich peat that is buried to form coal originates in wet, reducing environments. Under these conditions and the diagenesis that forms coal, chalcophile elements precipitate sulfide minerals. Pyrite in coal may contain arsenic, mercury, thallium, selenium, nickel, lead, and cobalt (Davies et al., 2005). Additional elements in coal are cadmium, chromium, copper, vanadium, fluorine, and uranium (Fuge, 2005).

Figure 2.1 shows that humans have high (>1) ratios of man to crust for H, C, O, N, P, S, Cl, Br, Cd, Se, Hg, and Te. Figure 2.2 shows that most elements have high (>1) ratios of man to ocean with Pb having a ratio less than that of C, N, and P but more than that of Ca, Zn, and Fe. Table 2.3 and Fig. 2.3 show the proposed list of biophile elements. Goldschmidt (1954) had listed eight major (C, H, O, N, P, S, Cl, and I) and nine minor (B, Ca, Mg, K, Na, V, Mn, Fe, and Cu) biophile elements. Based on the abundance of elements in humans (Lindh, 2005), Na, Mg, K, and Ca are included as major biophile elements and iodine is considered a minor biophile. Additional elements included as minor biophile are the essential micronutrient elements Fe, Cu, Zn, I, Se, Mn, Mo, Cr, F, and Co. Functional similarities in size and charge and its ease of substitution for calcium in human bones place lead as a minor biophile element. Other minor biophile elements that substitute into bone include Si, As, Br, Sr, and Cd. Another type of functional similarity is when an element located directly below an element on the periodic table readily

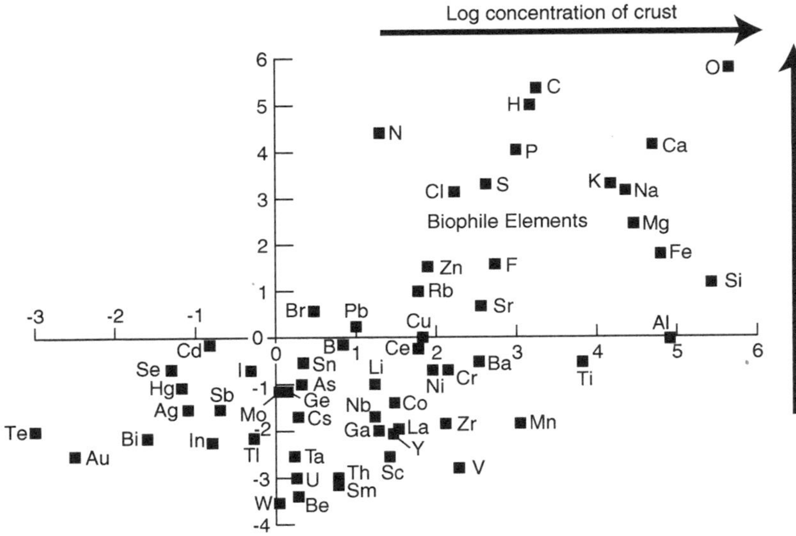

Figure 2.1. The abundance of elements in man versus the crust. Biophile elements are enriched in man relative to the crust.

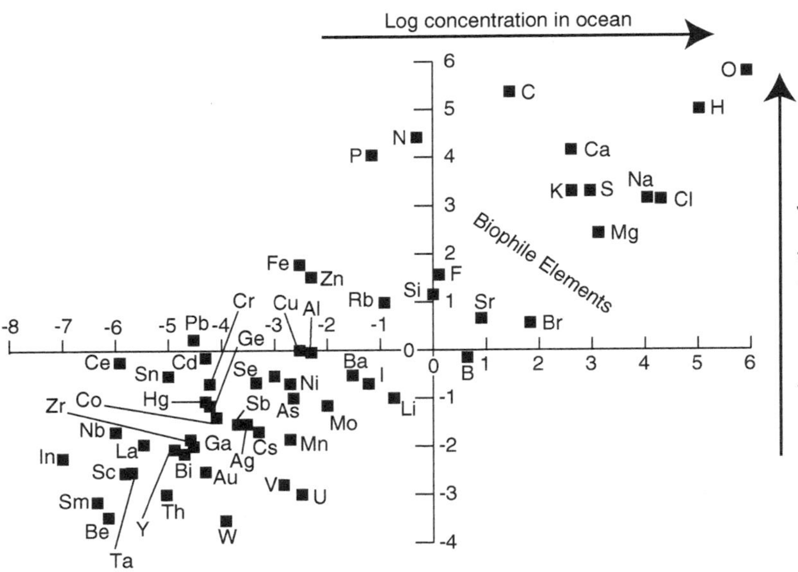

Figure 2.2. The abundance of elements in man versus the ocean. Some biophile elements (C, N, P) are enriched in man relative to the ocean while some biophile elements (Cl, Na, Mg, Br) are enriched in the ocean relative to man.

Table 2.3. The biophile elements ratio in man/crust and man/ocean

Element	Atomic number	Ratio	
		Man/crust	Man/ocean
Hydrogen, H	1	66.7	0.09
Boron, B	5	0.0805	0.158
Carbon, C	6	128	8,200
Nitrogen, N	7	1,300	52,000
Oxygen, O	8	1.35	0.7
Fluorine, F	9	0.0685	28.5
Sodium, Na	11	0.0622	0.1
Magnesium, Mg	12	0.00931	0.2
Silicon, Si	14	0.000011	14.3
Phosphorus, P	15	11.0	157,000
Sulfur, S	16	4.76	2.2
Chlorine, Cl	17	8.00	0.1
Potassium, K	19	1.33	4.8
Calcium, Ca	20	0.28	34.1
Vanadium, V	23	0.00000842	1.07
Chromium, Cr	24	0.00143	3,330
Manganese, Mn	25	0.0000127	7
Iron, Fe	26	0.000952	20,000
Cobalt, Co	27	0.00133	500
Copper, Cu	29	0.0147	333
Zinc, Zn	30	0.418	6,600
Arsenic, As	33	0.0476	43.5
Selenium, Se	34	4.2	467
Bromine, Br	35	1.23	0.055
Rubidium, Rb	37	0.162	80.8
Strontium, Sr	38	0.0128	0.568
Molybdenum, Mo	42	0.0636	7
Cadmium, Cd	48	4.67	14,000
Tellurium, Te	52	10	150,000
Iodine, I	53	0.408	3.3
Mercury, Hg	80	1.28	1,720
Lead, Pb	82	0.17	56,600

Note: Non-biophile elements with high ratio of human/ocean include: Al at 180, Sc at 1870, Ti at 290, Ni at 50, Ga at 333, Ge at 1170, Y at 662, Zr at 538, Nb at 20,000, Ag at 97, In at 57,000, Sn at 29,000, Sb at 145, Ba at 10, La at 3,235, Ce at 475,000, Sm at 1560, Ta at 1450, W at 2.41, Au at 60, Th at 109.

substitutes for it. For example, the chalcophile mineral sphalerite, ZnS, has as a common substitute for zinc the element cadmium (Greenwood and Earnshaw, 1997). Extending this type of functional similarity to biophile elements adds Rb, Sr, Cd, and Te as biophile elements because of their easy substitution for K, Ca, Zn, and Se, respectively.

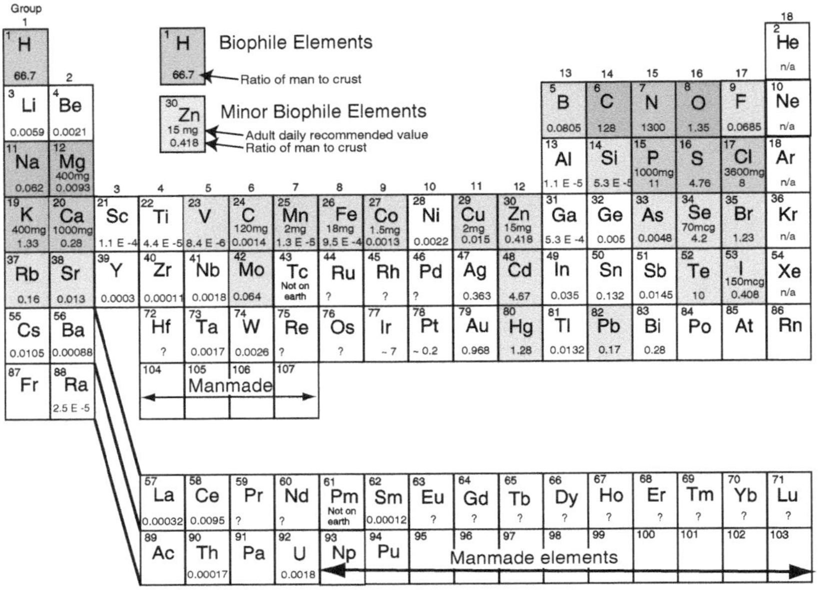

Figure 2.3. Periodic table showing biophile elements (shaded) and the ratio of man to crust.

2.2. Environmental geochemistry and anthropogenic dispersion of arsenic, lead, and mercury

2.2.1. Arsenic, As

At 1.8 ppm arsenic is the 53rd most abundant element in the earth's crust. Arsenic's typical concentration in soil is 5 ppm, while its average concentration in the United States coal is 22 ppm. Arsenic's abundance in seawater is 3 ppb. The oxidation states of Arsenic are As^{5+}, As^{3+}, As, As^{2-}, and As^{3-}. Most arsenic used in industry is obtained as As_2O_3 from flue dust at lead and copper smelters, though it is also obtained from arsenopyrite, FeAsS, and lollingite, $FeAs_2$. Figures 2.4 and 2.5 are plots of the human production and usage of arsenic (Table 2.4) that have spread arsenic around the world.

Arsenic is often considered the classic example of a substance that is very toxic (Fergusson, 1990; Thornton, 1996; ATSDR, 2000). The toxicity of ingested arsenic varies with oxidation state and whether it is organic or inorganic. Reduced forms are more toxic than oxidized forms and inorganic arsenic is more toxic than organic arsenic. The largest human intake of arsenic is from food (e.g., seafood). Organic arsenic is of

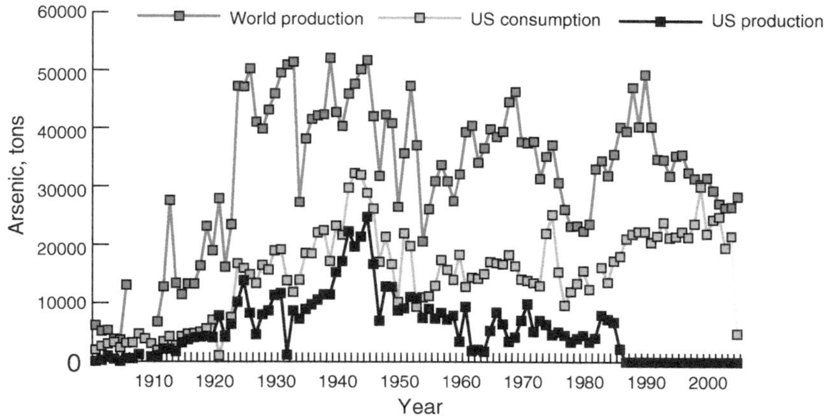

Figure 2.4. The world production and United States production and consumption of arsenic from 1900 to 2002. Arsenic production stopped by 1986 in the United States. Total world production of As from 1900 to 2002 was 3,250,000 tons. United States consumption was 1,500,000 tons. Data from USGS (2003, 2004).

low toxicity. As^{5+} is more soluble than As^{3+}. Inorganic forms of arsenic occurring in groundwater are most toxic.

Arsenic is a semimetal that has both high natural (12,000 tons per year) and manmade (19,000 tons per year) emissions into the atmosphere. These emission values are estimates and there exists a range in estimates that vary with time and the methods used to determine the estimate. Arsenic's occurrence in sulfide ores causes its high emissions (12,000 tons per year) from ore smelters (Loebenstein, 1994).

Total United States arsenic and arsenic compound emissions reported for 2002 by the EPA Toxic Release Inventory (TRI) are 89 tons. After emissions from electric utilities, metal mining and processing are the next major sources of arsenic emissions (EPA-TRI, 2004). Arsenic atmosphere emissions represent only 0.02% of total arsenic waste of 402,394 tons disposed of or released in 2002. Disposal of arsenic and arsenic compounds on land and on- and off-site were about 400,000 tons in 2002. Over 95% of this arsenic is from metal mining.

Arsenic in groundwater can be a risk because it is both reduced and inorganic and arsenic compounds readily dissolve in water. Drinking water has caused cases of chronic and acute arsenic poisoning in Taiwan, Argentina, China, and Mexico. A tragic modern epidemic of arsenic poisoning from groundwater is occurring in 17 nations including regions

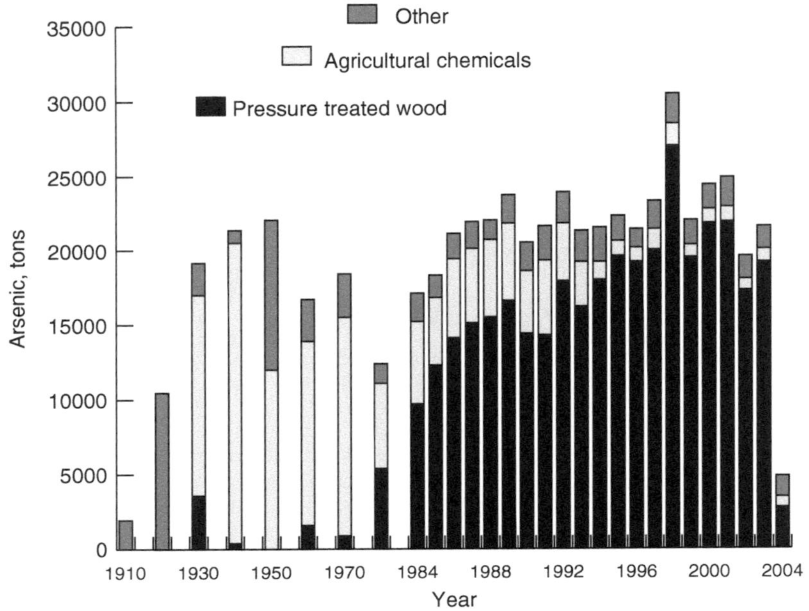

Figure 2.5. Use of arsenic in selected years in the United States after Bleiwas (2000) with additional data from USGS, Mineral Commodity Summaries and Loebenstein (1994). From 1960 to 2003 an estimated 420,000 tons of arsenic was used in the manufacture of pressure-treated wood and another 115,000 tons was used in agriculture. From 1910 to 1970 the largest usage of arsenic was in agriculture.

of Bangladesh, southern Nepal, China, Vietnam, and West Bengal and Bihar, India. Shallow tube wells were dug in these regions to provide drinking water free of the threats of diarrhea, dysentery, typhoid, cholera, and hepatitis common in surface water. In Bangladesh since 1971, 8–12 million shallow tube wells were dug, providing drinking water for 90% of a population of 140 million. In 1993 arsenic was discovered in the well water. Groundwater in one million of the wells has arsenic concentrations above 50 ppb with some wells containing over 1,000,000 ppb As. Chakraborti et al. (2003) in a study done in Bihar, India found 19.9% of tube wells had >300 ppb As. The study was done in a 4 km^2 area in a remote agricultural village of 5000 people. In some regions of south Asia, 50% of the wells have arsenic levels exceeding the USEPA maximum drinking water level of 10 ppb As (Frisbie et al., 2002). Arsenic threatens 40–60 million people in Bangladesh and has caused skin lesions (arsenical dermatitis) in over 100,000 people with tens of thousands of people

Table 2.4. Geochemical classification of the elements based on initial, primary, secondary, and tertiary geochemical differentiation[a]

Siderophile	Chalcophile	Lithophile	Atmophile	Biophile (organisms)
Fe, Ni, Co, P, C, Ru, Rh, Pd, Os, Ir, Pt, Au, Ge, Sn, Mo, Ta	**S**, Se, Te, As, Sb, Bi, Ga, In, Tl, Pb, Zn, Cd, Hg, Cu, Ag, Ni, Pd, Co, Fe, Ru, Mo	**O, Si**, H, Li, Be, B, C, N, F, Na, Mg, Al, Cl, K, Ca, Sc, Ti, V, Cr, Mn, Fe, Ga, Br, Rb, Sr, Y, Zr, Nb, Ag, Cd, In, Sn, Sb, I, Cs, Ba, La, Ce, Pr, Nd, Sm, Eu, Gd, Tb, Dy, Ho, Er, Tm, Yb, Lu, Hf, Ta, W, Hg, Tl, Bi, Th, U	**N, Ar**, Ne, Kr, Xe, O, C	**C, H, O**, N, Na, Mg, P, S, Cl, K, Ca
Minor—As, W, Nb, Se, Te	Minor—Ge, Sn, Au, Pt, Rh, Ir, Os	Minor—Cu, Ge, P, S, Se, He,[b] Ar,[b] Pb,[b] Kr, As	Minor—H,[c] He,[c] S,[d] Pb,[d] Hg[d]	Minor—B, F, Si, V, Cr, Mn, Fe, Co, Cu, Zn, As, Se, Br, Rb,[e] Sr,[e] Mo, Cd,[e] Te,[e] I, Hg, Pb[e]

Note: The key elements that characterize the group are in bold. Minor—elements belong for the most part in another group but do occur in this group.

[a]Initial differentiation is the formation of the earth and its separation from the solar system. Primary differentiation is the separation of earth into core and primitive mantle. Secondary differentiation is the formation of crust. Tertiary differentiation is the process of formation and development of hydrosphere and biosphere, and lithophile enrichment of crust by plume and plate tectonic activity.

[b]Radioactive decay of U and Th to He and Pb, and K to Ar places these elements into this group, He and Ar are trapped in silicate structures.

[c]H and He escape from the earth's atmosphere into space.

[d]Volcanic eruptions, forest fires, burning fossil fuels, smelting metal ores, and incineration of waste places these elements into this group. Mercury is in this group because of its high vapor pressure.

[e]Biophile elements because of easy substitution; Rb for K, Sr and Pb for Ca, Se and Te for S, and Cd for Zn.

displaying hyperkeratosis and chronic and acute arsenicosis (McArthur et al., 2001; WHO, 2001; Brooks, 2003; USGS International Program, 2004; Meharg, 2005). Children exposed to wells with >50 ppb As have reduced intellectual function relative to children with exposures to <5.5 ppb As (Wasserman et al., 2004). In Bangladesh an estimated 9000 deaths per year are arsenic related (Lokuge et al., 2004).

The high arsenic abundance in some coals is an important demonstration of the biophile nature of arsenic. Arsenic was concentrated in some coals during their formation; when the coal is burned As emissions are dispersed and can build up in humans. For example, high-arsenic lignite (500 ppm As) burned in Slovakia has resulted in arsenic emission from one power plant of 3000 metric tons since 1953 and nonmelanoma skin cancer (Ranft et al., 2003). An arsenic threat to human health is the burning of coal in the Guizhou Province of southwest China. Some coal samples from Guizhou Province contain up to 35,000 ppm As (3.5% As!), 50 ppm Hg, and 20 ppm Sb (USGS, 1998a; Liu et al., 2002). The coal is burned in un-vented ovens inside homes. The ovens can be used to dry crops such as chili peppers. The fresh chili peppers have less than 1 ppm arsenic whereas those dried over coal fires have more than 500 ppm As (USGS, 1998a; Schweinfurth, 2003). Over 3000 cases of arsenicosis have been reported. High levels of As, Hg, and F up to 181, 10.5 and 2259 ppm, respectively, occur in some locally used coals in China (Zhang et al., 2004).

A potential health problem of undefined proportions and dimensions is the several decade long usage of arsenic in pressure-treated wood (Fig. 2.5). Pressure-treated wood contains inorganic arsenic that protects it from decay and insects. The most common treatment is with chromated-copper-arsenate or CCA. One cubic foot (0.03 m^3) of CCA wood contains 1.15 oz (32.6 g) of arsenic. Such wood in an outdoor environment may resist termites and decay for 20–60 years. However, our research shows that in the warm humid southeastern United States, that framing timbers in contact with the ground may disintegrate after ten years (Hollabaugh, 2004, 2005). Over 420,000 tons of arsenic has been used in pressure-treated wood since 1960 in the United States. A voluntary agreement between the EPA and the manufacturers of pressure-treated wood removed arsenic in pressure-treated wood for residential uses in 2004 (Bleiwas, 2000; EPA, 2002; Brooks, 2003).

Another 350,000 tons of arsenic has been used since 1960 in the United States as agricultural chemicals (Fig. 2.5). In past decades arsenic-based agricultural chemicals were used for insecticides, selective plant killers, defoliants, rat poison, and parasitic control in chicken feed. Today arsenic as Roxarsone (4-hydroxy-3 nitrophenyl arsenic acid) is used as feed for broiler chickens to control intestinal parasites. This results in 3-fold greater arsenic concentrations in young chickens than in other meat and poultry (Lasky et al., 2004).

2.2.2. Lead, Pb

Lead is the 36th most abundant element in the earth's crust. Most geochemists report lead's crustal abundance to be between 12.5 and 16 ppm, although a detailed study by Taylor and McLennan (1985) determined that continental crust contains 8 ppm lead and ocean floor crust contains 0.8 ppm lead. Lead is the most abundant heavy metal in the earth's crust; some of this enhanced abundance is the product of the decay of extreme lithophile uranium and thorium into lead (Greenwood and Earnshaw, 1997). Oxidation states of lead are Pb, Pb^{2+}, Pb^{3+} (rare), and Pb^{4+} (rare). Common lead minerals are galena, PbS, cerussite, $PbCO_3$, and anglesite, $PbSO_4$. Lead concentration in seawater is 0.003 ppb. Figure 2.6 shows the United States production and consumption and world production of lead. Some human uses of lead shown in Table 2.5 and plotted in Fig. 2.7 have dispersed lead around the world.

Natural lead emissions into the atmosphere have been estimated from 2000 to 24,500 tons per year (Sannolo et al., 1995). During the middle of the 20th century, most of the lead in the atmosphere came from anthropogenic sources, with manmade emissions exceeding natural emissions by 346 times (Montgomery, 1989; Schlesinger, 1991). At that

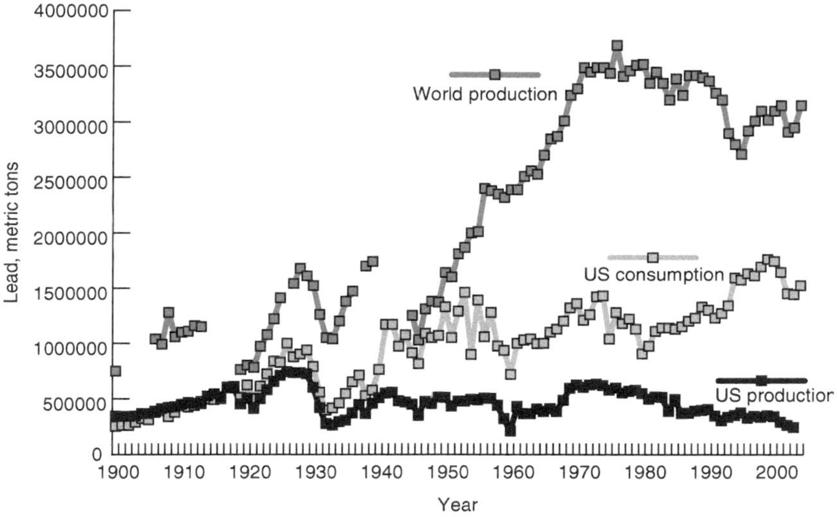

Figure 2.6. The world production and United States production and consumption of lead from 1900 to 2002. Total world production of Pb from 1900 to 2002 is estimated to be 214,000,000 tons. United States consumption was 97,000,000 tons. Data from USGS (2003, 2004).

Table 2.5. The major uses of arsenic, lead, and mercury during the last 140 years[a]

Metal	Usage
Arsenic Before 1900	Insecticides, herbicides, fungicides, glass and ceramics, folk remedies, metal alloys
1900–1985	Insecticides, herbicides, fungicides, leathering, tanning, paint pigments, treatment of syphilis and psoriasis until 1940s, glass making, metal alloys (in solder, with lead in ammunition, in bearings)
1985–present	Pressure-treated wood, metal alloys (additive to increase strength of posts on batteries), pesticide, desiccant on cotton fields, insecticides, herbicides (weed control), fungicides, fertilizers, semiconductors (computers, wireless phones, solar cells), light emitting diodes, animal feed (promote growth), glass making, paint pigment
Lead Before 1900	Ammunition, brass, burial vault liners, ceramic glazes, roofs, plumbing, soldering, paint and pigment, batteries, leaded glass and crystal, pewter, water lines and pipes
1900–1985	Batteries, gasoline additive, paint and pigment, ammunition, crystal glass, flint glass, insecticides, herbicides, fungicides, bearing metals, cable covering, caulking lead, solders, type metals, terne metal for gas tanks, radiation shielding, ballast and counter weights, brass, bronze, foil, wire
1985–present	Batteries (84% of US usage in 2003), ammunition (3%), glass and ceramics (3%), casting metals (2%), sheet lead (1%), radiation shielding, radiation shielding in medical analysis and video display equipment, covering for power and communication cables, building materials, solders for motor vehicles, metal containers
Mercury Before 1900	Gold and silver mining (amalgamation), medical, pharmaceutics, insecticides, herbicides, fungicides, electrical instruments, paint and pigment, wool and felt, antibacterial, batteries, mirrors, laboratory chemicals, thermometers, catalysts, explosives, dental amalgams
1900–1985	Batteries, gold mining, medical, electrical instruments, catalysts, insecticides, herbicides, fungicides, dental amalgams, paper industry for control of slime, laboratory chemicals, thermometers, chlorine and caustic soda production, explosives, preservative in cosmetics, antifungal agricultural products, pharmaceutics
1985–present	Chlorine and caustic soda production (33–50% of US usage in 2003), paint (anti-mildew marine paint, and fungicide and biocide additive), batteries, electrical instruments, fluorescent lights and Hg vapor lamps, dental amalgams, catalysts, laboratory chemicals, thermometers, barometers, thermostats, pigment, pharmaceutics, medical.[b] Globally—gold and silver mining, button-type batteries, cleansers, folk medicine, pesticides, skin-lightening cream and soap

[a]After Ayres and Rod (1986), Fergusson (1990), Jasinski (1994), ATSDR (2000), Bleiwas (2000), USGS (2005) and EPA (2004c and 2004d).
[b]See Appendix A and B of EPA (2004c).

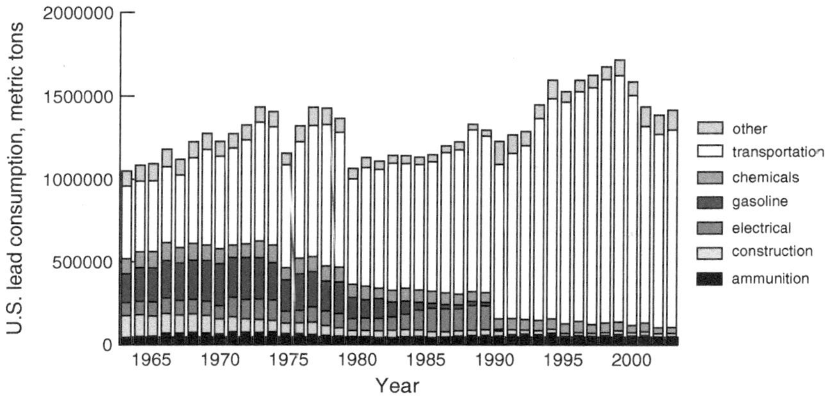

Figure 2.7. Use of lead in selected years in the United States. Data from USGS, Mineral Commodity Summaries (USGS, 1998a, 2004). Lead usage in gasoline in the United States ended in 1993. Transportation includes lead in car and truck batteries. Exports of ammunition and imports of glass, ceramics, and batteries cause an increase in lead consumption because of net imports of manufactured products (Biviano et al., 1999).

time atmosphere lead concentration was about 0.003 ppb with a residence time of two weeks to several years. The lead is removed from the atmosphere by both wet and dry precipitation.

Total United States lead and lead compound emissions reported for 2002 by the EPA Toxic Release Inventory (TRI) were 732 tons. TRI reports that metal mining and metal processing result in 351 tons of emissions (EPA-TRI, 2004). Lead atmosphere emissions are 0.17% of total Pb waste of 426,469 tons in 2002. Disposal of Pb and Pb compounds on land and on- and off-site were about 410,000 tons in 2002. Over 92% of this is from metal mining and primary metal processing. Industries in 2002 discharged 69 tons of lead into surface waters, 3589 tons into underground wells, 196,897 tons onto land, and 225,181 tons to on- and off-site disposal.

Lead is a classic siderophile element, occurring as galena in many sulfide deposits. Under oxidizing conditions the sulfur is removed or oxidized and lead behaves as a lithophile element forming cerussite and anglesite. Lead is a biophile element because of its close size and the same charge as calcium (Pb^{2+} radii is 1.19Å and Ca^{2+} is 1.00Å). Calcium in the brain plays a role in "converting an electrical neural signal into a chemical signal" however "Pb does not function as a neurotransmitter, effectively creating permanent neural differentiation defects resulting in mental retardation, learning disorders, and attention deficit hyper-activity disorder (ADHA)" (Filippelli et al., 2005).

Much of the lead in a human body is stored in the bone. Modern human bones contain 3–30 ppm Pb. Concentration of lead in soft tissue is about 0.18 ppm. Before the common and widespread burning of leaded gasoline, human bones contained from 0.05 to 0.08 ppm, and before heavy industrial development occurred after 1750 human bones contained 0.01 ppm Pb (Sannolo et al., 1995). However, during the peak of the Roman Empire lead production (80,000 tons Pb per year) when lead emissions into the atmosphere are estimated to have reached 4000 tons per year, analyses of 1st to 4th centuries AD Romano-Britons bones and teeth show 10 times more Pb than a 1960s British city-dweller (Keys, 2003). The body burden of an adult in the United States is 120–400 mg Pb/70 kg person (Sannolo et al., 1995; Emsley, 1998, 2001; Li, 2000).

Burning of leaded gasoline since 1924 has caused the entire biosphere to have lead levels at least 100 times natural background levels. Four million tons of lead used in gasoline remains in our dust and soil. Three million tons of lead is found in lead paint on 60 million private homes in the United States; about 10 million of these homes have families with children below seven years old. The 1990 publication of the results of a 11 year follow up study of children that had early lead poisoning indicates that childhood exposure to lead can cause long-term intellectual underdevelopment. There is also concern that even low-level exposure to lead may cause a lower intelligence quotient. If this study is correct millions of young children in the United States may be considered to have excessive lead levels. Lead poisoning is the number one environmental health problem to children (Needleman et al., 1990; Stapleton, 1994).

For adults and children in the United States, average blood lead levels have dropped as major uses of lead have declined—lead in paint, gasoline, drinking water systems, and food cans. The geometric mean blood lead level in children ages 1–5 has fallen from 14.9 $\mu g\,dl^{-1}$ in 1976–1980 to 2.0 $\mu g\,dl^{-1}$ in 1999 (Koller et al., 2004). In spite of good progress there are still areas of concern with lead. The first of these is well described by Filippelli et al. (2005). The number of children in the United States with lead poisoning has been lowered from 88% to 2.2%. However, in urban centers 15 to 29% of children can have unsafe blood lead levels ($>10\,\mu g\,dl^{-1}$). The 2.2% is still too high because it means 434,000 children in the United States have elevated blood lead levels (EPA, 2004a).

2.2.3. Mercury, Hg

Mercury is a classic chalcophile element occurring most commonly as cinnabar, HgS. Natural occurrences of high mercury concentrations are limited to a few areas in the crust where hydrothermal processes have concentrated cinnabar with lesser amounts of native mercury. Outside of these mineral deposits mercury is a rare substance; Hg is 68th in abundance at 80 ppb in the earth's crust, while the concentration in seawater is 0.03 ppb Hg. Oxidation states of mercury are Hg, Hg^{+}, and Hg^{2+}.

Mercury differs from other heavy toxic metals in many ways: (1) elemental mercury is a liquid with high volatility and readily vaporizes into the atmosphere (melting point = −38.9°C and boiling point = 357°C); (2) mercury vapor travels throughout the atmosphere; (3) mercury can undergo reduction with free mercury liquid produced in soils; (4) mercury is the most soluble metal; (5) mercury dissolves gold and silver and is dissolved in minor amounts in naturally occurring Au and Ag; (6) biomethylation of mercury produces very toxic and volatile compounds such as $(CH_3)_2Hg$; and (7) methylmercury bioaccumulates in the food chain (Fergusson, 1990; Sznopek and Goonan, 2000).

Figure 2.8 shows United States production and consumption and world production of mercury from 1900 to 2003. Before 1900 much usage of mercury was for extracting gold and silver from ore. The principle uses of mercury (Table 2.4) from 1941 to 2003 in the United States are shown in Fig. 2.9. Many of these uses have spread mercury throughout the

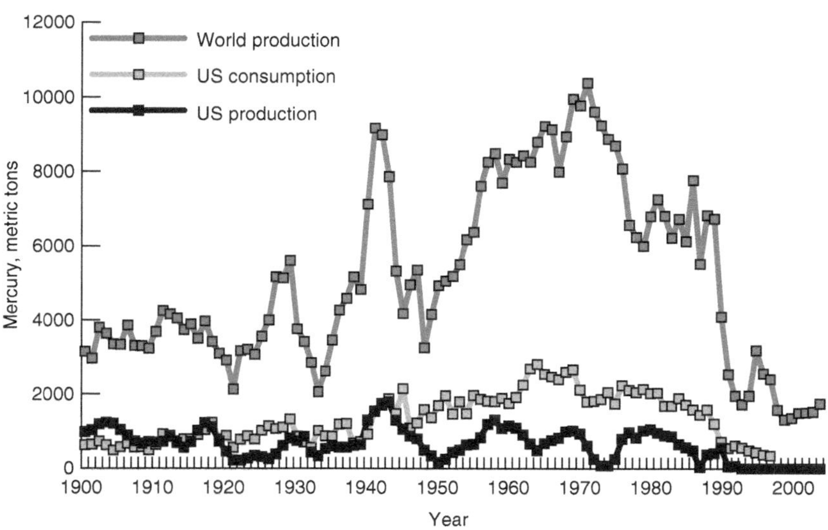

Figure 2.8. The world production and United States production and consumption of mercury from 1900 to 2002. Total world production of Hg from 1900 to 2002 is estimated to be 530,000 tons. United States consumption from 1900 to 1997 was 130,000 tons. There has been no Hg production in the United States since 1992. Data from USGS (2003, 2004).

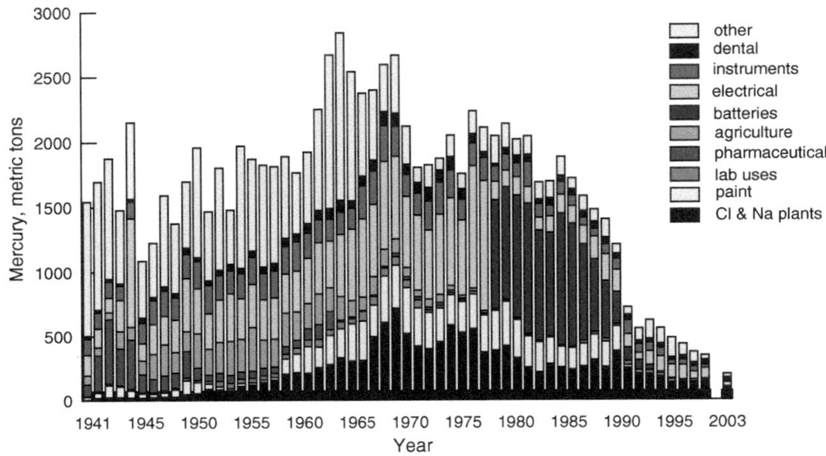

Figure 2.9. Use of mercury in selected years in the United States (Jasinski, 1994; Sznopek and Goonan, 2000). From 1941 to 1996 an estimated 93,000 tons of Hg was used in the United States. Major uses were for electrical (18,000 tons, including batteries, wiring devices and switches, and lighting), chlorine and caustic soda manufacture (12,500 tons), paint (8400 tons), laboratory and measuring and control instruments (8700 tons), pharmaceutical (3400 tons), agricultural chemicals (5500 tons) and dental (2700 tons). Recovery by amalgamation of gold and silver is part of "other". Before 1978 mercury in batteries was included within electrical.

environment—gold and silver mining, escape of mercury from chlorine and caustic soda production, batteries, paint, dental, insecticides, herbicides, and fungicides. Mercury usage in the United States has steadily declined since a high of 2800 tons in 1964 with environmental restrictions such as elimination of mercury from paint and most batteries (Sznopek and Goonan, 2000). In 2005 the principle uses of mercury in the United States are in the chlorine and caustic soda industry, fluorescent lights, Hg vapor lamps, and dental amalgams. Of these uses only dental amalgams appear to be safe for healthy working adults (Factor-Litvak et al., 2003). Worldwide, considerable mercury is still lost to the environment from gold mining.

Annual atmospheric discharges of mercury from natural (3750 tons) and manmade (3750 tons) sources combine for a total of 7500 tons per year (AEP, 2004). Mercury in the atmosphere can travel great distances and be deposited in remote regions of the world where it can build up in the food chain and contaminate fish. Manmade emissions occur from the burning of coal (coal contains 0.07 to 50 ppm Hg), mercury mining and smelting, municipal solid waste incinerators, medical waste incinerators, and chlor-alkali plants (chlorine and caustic soda production). The largest sources of mercury emissions in the United States are coal-fired electric power plants (40% or 50 tons per year), industrial boilers (10%), burning hazardous waste (5%), and chlorine production (5%) (EPA, 2004a).

Methylmercury is the most dangerous form of mercury. Fish and shellfish are the usual methods of mercury intake by humans. They contain 1000 to 10,000 times more methylmercury than other foods (cereals, potatoes, vegetables, fruits, meats, poultry, eggs, and milk). Most fish contain 0.001 to 0.5 ppm Hg and the average amount of methylmercury in commercial saltwater fish is less than 0.3 ppm. A few species of large predatory fish commonly have higher levels. Fish to be avoided by mothers and small children include shark, swordfish, king mackerel, and tilefish (EPA, 2004b). Avoiding consumption of certain top predatory freshwater fish is necessary; within the United States 48 states had issued 3089 fish advisories by 2003 with 76% of the advisories including mercury (EPA, 2004b). Large fish that may be on local consumption advisories (with mean and range of mercury in ppm) are eastern chain pickerel (0.61, 0.014–2.81), largemouth bass (0.52, 0.0005–8.94), walleye (0.43, 0.005–16), northern pike (0.36, 0.005–4.4), smallmouth bass (0.32, 0.005–3.34), lake trout (0.27, 0.005–2), yellow perch (0.25, 0.005–2.14) (EPA, 2001; Burger et al., 2005).

Mahaffey et al. (2004) discuss results of the 1999 and 2000 National Health and Nutrition Examination Survey that measured blood methylmercury concentration in 1709 women. The methylmercury concentration was 0.6 ppb at the 50th percentile. The concentration was 6.7 ppb at the 95th percentile. An estimated 7.8% of U.S. women have blood methylmercury levels greater that 5.8 ppb (EPA recommends whole blood levels be below 5 ppb), and a 1:1 ratio of mother to baby blood methylmercury concentrations. Mahaffey et al. (2004) conclude that over 300,000 newborns out of 4,058,814 births in the U.S. per year are at risk for problems in neurodevelopment. Hair mercury levels yielded similar results (McDowell et al., 2004).

2.3. Conclusion

Modifications of Goldschmidt's geochemical classification of the elements are made based on distribution and tendency of the elements. The biophile elements have tendencies that uniquely distinguish them. These tendencies are well demonstrated by arsenic, lead, and mercury. They are biophile because of substitutions (lead for calcium), bioaccumulation (arsenic and mercury), and methylation (mercury). In Bangladesh an estimated 9000 deaths per year are arsenic related and 40–60 million people are threatened by arsenic in drinking water. In the United States an estimated 434,000 children have their intellectual development threatened by elevated blood lead levels. Mercury from fish consumption in the United States threatens the neurodevelopment of an estimated 300,000 newborns each year. Today's world of rapid development necessitates the classification of As, Pb, and Hg as toxic biophile elements to avoid making future environmental mistakes.

ACKNOWLEDGEMENTS

The West Georgia Foundation funded research on lead in paint and drinking water and arsenic in pressure-treated wood. Helpful reviews were done by Randa Harris and two anonymous reviewers.

REFERENCES

AEP (American Electric Power Company), 2004. Environmental. Accessed on 2-21-2005 at http://www.aep.com/environmental/emissionscontrol/rtk/mercury.htm

ATSDR, 2000. Public health statement for arsenic. Agency for Toxic Substances and Disease Registry. Assessed on 7-18-2003 at http:astsdr1.cdc.gov/toxProfilesphs8802.html

Ayres, R.U., Rod, S.R., 1986. Patterns of pollution in the Hudson–Raritan Basin. Environment, 28(4), 14–20, 39–43.

Bates, R.L., Jackson, J.A., 1984. Dictionary of Geological Terms, third ed. American Geological Institute, Anchor Press, Garden City, New York, p. 571.

Biviano, M.B., Sullivan, D.E., Wagner, L.A., 1999. Total materials consumption: An estimation methodology and example using lead—a materials flow analysis. U.S. Geological Survey Circular 1183, p. 22. Accessed on 2-7-2005 at http://greenwood.cr.usgs.gov/pub/circulars/c1183/

Bleiwas, D.I., 2000. Arsenic and old waste. US Geological Survey. Accessed on 8-6-2004 at http://minerals.usgs.gov/minerals/mflow/d00-0195/

Brooks, W.E., 2003. Arsenic. U.S. Geological Survey Minerals Yearbook – 2003.

Burger, J., Stern, A.H., Gochfeld, M., 2005. Mercury in commercial fish: Optimizing individual choices to reduce risk. Environ. Health Perspect. 113, 1–6.

Chakraborti, D., Murkherjee, S.C., Pati, S., Sengupta, M.K., Rahman, M.M., Chowdhury, U.K., Lodh, A.D., Chanda, C.R., Chhakraborti, A.K., Gautam, 2003. Arsenic groundwater contamination in Middle Ganga Plain, Bihar, India: A future danger. Environ. Health Perspect. 111, 1194–1201.

Combs Jr., G.F., 2005. Geological impacts on nutrition. In: Selinus, O. (Ed.), Essentials of Medical Geology: Impacts of the Natural Environment on Public Health. Elsevier, Amsterdam, The Netherlands, pp. 161–178.

Cox, P.A., 1989. The Elements Their Origin, Abundance and Distribution. Oxford Science Publications, Oxford, England, p. 207.

Davies, B.E., Bowman, C., Davies, T.C., Selinus, O., 2005. Medical geology: Perspectives and prospects. In: Selinus, O. (Ed.), Essentials of Medical Geology: Impacts of the Natural Environment on Public Health. Elsevier, Amsterdam, The Netherlands, pp. 1–16.

Emsley, J., 1998. The Elements. third ed. Oxford University Press, p. 292.

Emsley, J., 2001. Nature's Building Blocks: An A–Z Guide to the Elements. Oxford University Press, Oxford, England, p. 538.

EPA, 2001. Fact Sheet. Mercury Uptake: Impact on fish advisories. EPA-823-F-01-011 p. 9.

EPA, 2002. Manufacturers to use new wood preservatives, replacing most residential uses of CCA. Accessed on 6-7-2004 at http://www.epa.gov/pesticides/factsheets/chemicals/cca_transition.htm

EPA, 2004a. Mercury Frequent Questions. Assessed on 9-3-2004 at http://www.epa.gov/air/mercury/information1.htm

EPA, 2004b. Fish Advisories: What you need to know about mercury in fish and shellfish. Assessed on 9-2-2004 at http://www.epa.gov/waterscience/fishadvice/advice.htm

EPA, 2004c. Reducing mercury use in healthcare promoting a healthier environment. Assessed on 9-3-2004 at http://www.epa.gov/getlakes/bnsdocs/merchealth/

EPA, 2004d. Lead in paint, dust, and soil. Assessed on 2-2-2005 at http://www.epa.gov/opptintr/lead/index.html

EPA-TRI, 2004. Data obtained from Toxic Release Inventory. Assessed on 3-1-2005 at http://www.epa.gov/triexplorer/trends/htm

Factor-Litvak, P., Hasselgren, G., Jacobs, D., Begg, M., Kline, J., Geier, J., Mervish, N., Schoenholtz, S., Graziano, J., 2003. Mercury derived from dental amalgams, and neuropsychologic function. Environ. Health Perspect. 111, 719–723.

Fergusson, J.E., 1990. The Heavy Elements: Chemistry, Environmental Impact and Health Effects. Pergamon Press, New York, p. 614.

Filippelli, G.M., Laidlaw, M.A.S., Latimer, J.C., Raftis, R., 2005. Urban lead poisoning and medical geology: An unfinished story. GSA Today 15, 4–11.

Frisbie, S.H., Ortega, R., Maynard, D.M., Sarkar, B., 2002. The concentrations of arsenic and other toxic elements in Bangladesh's drinking water. Environ. Health Perspect. 110, 1147–1153.

Fuge, R., 2005. Anthropogenic sources. In: Selinus, O. (Ed.), Essentials of Medical Geology: Impacts of the Natural Environment on Public Health. Elsevier, Amsterdam, The Netherlands, pp. 43–60.

Garrett, R.G., 2005. Natural distribution and abundance of elements. In: Selinus, O. (Ed.), Essentials of Medical Geology: Impacts of the Natural Environment on Public Health. Elsevier, Amsterdam, The Netherlands, pp. 17–42.

Goldschmidt, V.M., 1923. Geochemische Verteilungsgesetze der Elemente (1). Vidensk. skrifter. I. Mat.-naturv. klasse. No. 3, p. 17.

Goldschmidt, V.M., 1954. Geochemistry. Oxford University Press, Oxford, England, p. 730.

Greenwood, N.N., Earnshaw, A., 1997. Chemistry of the Elements. second ed. Butterworth-Heinemann, p. 1341.

Hollabaugh, C.L., 2004. The inclusion of arsenic, lead, and mercury as biophile elements. Geol. Soc. Am. Abstracts with Programs 36(5), 333.

Hollabaugh, C.L., 2005. Toxic biophile elements arsenic, lead, and mercury: From personal to professional undergraduate research in environmental geochemistry. Geol. Soc. Am. Abstracts with Programs 37(2), 8.

Jasinski, S.M., 1994. The materials flow of mercury in the United States. Bureau of Mines Information Circular 9412. p. 27.

Keys, D., 2003. How Rome polluted the world. Geographical 75, 45–48.
Koller, K., Brown, T., Spurgeon, A., Levy, L., 2004. Recent developments in low-level lead exposure and intellectual impairment in children. Environ. Health Perspect. 112, 987–994.
Krauskopf, K.B., Bird, D.K., 1995. Introduction to Geochemistry, third ed. McGraw-Hill, New York, p. 647.
Lasky, T., Sun, W., Kadry, A., Hoffman, M.K., 2004. Mean total arsenic concentrations in chicken 1989–2000 and estimated exposures for consumers of chicken. Environ. Health Perspect. 112, 18–21.
Li, Y-H., 2000. A Compendium of Geochemistry: From Solar Nebula to the Human Brain. Princeton University Press, Princeton, New Jersey, p. 475.
Lindh, U., 2005. Biological functions of the elements. In: Selinus, O. (Ed.), Essentials of Medical Geology: Impacts of the Natural Environment on Public Health. Elsevier, Amsterdam, The Netherlands, pp. 115–160.
Liu, J., Zheng, B., Aposhian, H.V., Zhou, Y., Chen, M.-L., Zhang, A., Waalkes, M.P., 2002. Chronic arsenic poisoning from burning high-arsenic-containing coal in Guizhou, China. Environ. Health Perspect. 110, 119–122.
Loebenstein, J.R., 1994. The materials flow of arsenic in the United States. Bureau of Mines Information Circular 9282, p. 12.
Lokuge, K.M., Smith, W., Caldwell, B., Dear, K., Milton, A., 2004. The effect of arsenic mitigation interventions on disease burden in Bangladesh. Environ. Health Perspect. 112, 1172–1177.
Mahaffey, K.R., Clickner, R.P., Bodurow, C.C., 2004. Blood organic mercury and dietary intake: National Health and Nutrition Examination Survey, 1999 and 2000. Environ. Health Perspect. 112, 562–570.
Mason, B., 1992. Victor Moritz Goldschmidt: Father of Modern Geochemistry. Special Publication No. 4. The Geochemical Society, New York, p. 184.
Mason, B., Moore, C.B., 1982. Principles of Geochemistry, fourth ed. John Wiley and Sons, New York, p. 344.
McArthur, J.M., Ravenscroft, P., Safiullah, S., Thirlwall, M.F., 2001. Arsenic in groundwater: Testing pollution mechanisms for sedimentary aquifers in Bangladesh. Water Resour. Res. 37, 109–117.
McDowell, M.A., Dillon, C.F., Osterloh, J., Bolger, P.M., Pellizzari, E., Fernando, R., de Oca, R.M., Schober, S.E., Sinks, T., Jones, R.L., Mahaffey, K.R., 2004. Hair mercury levels in U.S. children and women of childbearing age: Reference range data from NHANES 1999–2000. Environ. Health Perspect. 112, 1165–1171.
Meharg, A.A., 2005. Venomous Earth: How arsenic caused the world's worst mass poisoning. Macmillan, New York, p. 192.
Montgomery, C.W., 1989. Environmental Geology, second ed. Wm. C. Brown Publishers, Dubuque, Iowa, p. 476.
Needleman, H.L., Schell, A., Bellinger, D., Leviton, A., Allred, E.N., 1990. The long-term effects of exposure to low doses of lead in childhood. An 11-year follow-up report. N. Engl. J. Med. 322, 83–88.
Railsback, L.B., 2003. An Earth scientist's periodic table of the elements and their ions. Geology 31, 737–740.
Ranft, U., Miskovic, P., Pesch, B., Jakubis, P., Fabianova, E., Keegan, T., Hergemoller, A., Jakubis, M., Nieuwenhuijsen, M.J., 2003. Association between arsenic exposure from a coal-burning power plant urinary arsenic concentrations in Prievidza District, Slovakia. Environ. Health Perspect. 111, 889–894.

Sannolo, N., Carelli, G., De Lorenzo, G., Castellino, N., 1995. Environmental exposure. In: Castellino, N., Sannolo, N., Castellino, P. (Eds.), Inorganic Lead Exposure and Intoxications. CRC-Press, Boca Raton, Florida, p. 544.

Schlesinger, W.H., 1991. Biogeochemistry: An Analysis of Global Change. Academic Press, San Diego, p. 443.

Schweinfurth, S.P., 2003. Coal—A complex natural resource. U.S. Geological Survey Circular 1143, p. 39.

Skinner, H.C.W., 2005. Mineralogy of bone. In: Selinus, O. (Ed.), Essentials of Medical Geology: Impacts of the Natural Environment on Public Health. Elsevier, Amsterdam, The Netherlands, pp. 667–694.

Stapleton, R.M., 1994. Lead is a Silent Hazard. Walker and Company, New York, p. 224.

Stober, Q.J., Thornton, K., Jones, R., Richards, J., Ivey, C., Welch, R., Madden, M., Trexler, J., Gaiser, E., Scheidt, D., Rathbun, S., 2001. South Florida Ecosystem Assessment: Phase I/II—Everglades stressor interactions: Hydropatterns, eutrophication, habitat, alteration, and mercury contamination (Summary). EPA 904-R-01-002. p. 63.

Sznopek, J.L., Goonan, T.G., 2000. The materials flow of mercury in the economies of the United States and the world. U.S. Geological Survey Circular 1197, p. 28.

Taylor, S.R., McLennan, S.M., 1985. The Continental Crust: Its Composition and Evolution. Blackwell Scientific Publications, London, p. 312.

Thornton, I., 1996. Sources and pathways of arsenic in the geochemical environment: Health implications. In: Appleton, J.D., Fuge, R., McCall, G.J.H. (Eds.), Environmental Geochemistry and Health. Geological Society Special Publication No. 113, pp. 153–161.

Turner, D.R., 1987. Speciation and cycling of arsenic, cadmium, lead and mercury in natural waters. In: Hutchinson, T.C., Meema, K.M. (Eds.), Lead, Mercury, Cadmium and Arsenic in the Environment. SCOPE 31. John Wiley and Sons, Chichester, Great Britain, pp. 175–186.

United States Geological Survey, 1998a. Lead Statistical Compendium. Accessed on 2-7-2005 at http://minerals.usgs.gov/minerals/pubs/commodity/lead/stat/

United States Geological Survey, 2003. Mercury in the Environment Fact Sheet 146-00 (October, 2000). Accessed on 8-6-2003 at: http://www.usgs.gov/themes/factsheet/146-00/

United States Geological Survey, 2004. Minerals Database. Accessed on 2-2-2005 at: http://minerals.usgs.gov/minerals/pubs/commodity/lead/380303.pdf; http://minerals.usgs.gov/minerals/pubs/commodity/arsenic/160303.pdf; http://minerals.usgs.gov/minerals/pubs/commodity/mercury/430303.pdf

United States Geological Survey, 2005. Lead Statistics and Information. Accessed on 2-1-2005 at /minerals.usgs,gov/minerals/pubs/commodity/leadl/

USGS International Programs, 2004. Geologic, hydrologic, and geochemical characterization of the deep ground water aquifer system in the Bengal Delta of Bangladesh. Accessed on 1-5-2005 at http://international.usgs.gov/project/bg_arsenic.htm

Wasserman, G.A., Liu, X., Parvez, F., Ahsan, H., Factor-Livak, P., van Green, A., Slavkovich, V., Lolacono, N.J., Cheng, Z., Hussain, I., Momotaj, H., Graziano, J.H., 2004. Water arsenic exposure and children's intellectual function in Araihazar, Banfladesh. Environ. Health Perspect. 112, 1329–1333.

WHO, 2001. Arsenic in drinking water. Revised May 2001, World Health Organization. Accessed on 1-6-2005 at http://www.who.int/mediacentre/factsheets/fs210/en/

Zhang, J., Ren, D., Zhu, Y., Chou, C.-L., Zeng, R., Zheng, B., 2004. Mineral matter and potentially hazardous trace elements in coals from Qianxi Fault deposition area in southwestern Guizhou, China. J. Coal Geol. 57, 59–61.

Developments in Environmental Science, Volume 5
D. Sarkar, R. Datta and R. Hannigan (Editors)

ISSN: 1474-8177/DOI:10.1016/S1474-8177(07)05003-6

Chapter 3

Metal ions speciation in the environment: Distribution, toxicities and analyses

V.K. Gupta, Imran Ali and Hassan Y. Aboul-Enein

Abstract

Metal ions are the most toxic pollutants due to their acute toxicities, carcinogenicities and non-biodegradability. Speciation of metal ions is the most important area due to the different toxicities of the same metal ion species. This chapter describes the distribution of metal ions in water, sediment, soil, air, aquatic and terrestrial biota. Besides, discussion was also presented on metal ions toxicities and their analysis by chromatographic, capillary electrophoretic, spectroscopic and other methods. Some data of distribution and toxicities are also given in tabular form. A comparison of the activities of organic and inorganic metal ion species in biological systems is also included.

3.1. Introduction

Among various pollutants metal ions are very dangerous contaminants due to their acute toxicities, carcinogenicities and non-biodegradability. Transition metals, lanthanides and actinides exist in different oxidation states and form various species, e.g., arsenic (As) as As(III), As(V) and organic arsenic species (Caruso et al., 2000) and the concept is known as speciation. The different oxidation states of a particular metal ion possess different physico-chemical properties. Mainly, these oxidation states differ in their redox potential, complexation and hydration properties. It is very important to observe that these different oxidation states have different toxicities. The concept of the speciation dates back to 1954 when Goldberg (1954) introduced the term of speciation to improve the understanding of the bio-geochemical cycling of trace elements in seawater. Kinetic and thermodynamic information together with analytical data make it possible to differentiate between oxidized versus reduced, complexed or chelated versus free metal ions in dissolved form. The measurement of the total concentration of a metal ion cannot provide the information about the actual physico-chemical properties required for understanding its toxicity, bio-transformation, etc. Thus, in order to obtain information on toxicity and bio-transformation of a metal ion in the environment, speciation is needed of the individual element. Because of these points the knowledge of metal ion speciation is essential for the environmental, industrial and clinician point of views. Therefore, the scientific community and other regulatory authorities are asking data on speciation rather than the total metal ion concentrations. In view of these points, attempts have been made to discuss the distribution, toxicities and analyses of metal ions in this chapter.

3.2. Distribution

Originally, the concentrations of elemental species of anthropogenic origin was zero but today they are found widely in the ecosystem as they have been continued to be distributed in a manner that affects the life cycle. Total trace element concentration may be static and the species may be highly dynamic. Therefore, metal ions change continuously with respect to the changes in the environmental parameters such as pH, concentration of ligands for complex formation, the physiological state of a cell and state of health of a living body. The thermodynamic and kinetic stability of elemental species in the environment has also to be taken into account. Unstable species in the environment are predominant and this state requires some special attention. Basically, all metal ions are present in the earth's crust in different abundance and oxidation states. There are different ores containing various metal ions and these find their ways from these ores to the different components of the ecosystem. Some of them get dissolved into aquifers and contaminate groundwater while the others are extracted by human beings and used for different useful purposes. During all these activities, metal ions find their ways to the surface water, air, vegetables, other foodstuffs, etc. and finally enter the human bodies. Metals occur in the nature mostly in the combined forms but, sometimes, in the free state too. The active metal ions are found in the form of their salts while metals, which have little or no affinity for oxygen and water, are found in the free state. The distribution of metal ions is discussed herein in the different components of the ecosystem.

3.2.1. Distribution in water

Water is the main component of the ecosystem and unfortunately it is the most contaminated by toxic metal ions. All forms of water resources, i.e., ground, surface and waste are contaminated due to heavy toxic metal ions. The oxidation of the different metal species causes metal ions to become soluble and enter into the surrounding environment through the drainage water. The knowledge of the geographic distribution of metal ions species in the natural water system is important for the environmental consideration of the geochemical and biological cycling of the element. Furthermore, this also provides insight into the geochemical process responsible for elevated metal ion concentrations in the different geological environment.

Nowadays, arsenic disaster contamination is the worldwide problem and more than 20 countries are suffering from this (Chatterjee et al., 1995; Das et al., 1995a). The poisoning of groundwater due to mercury, cadmium, lead and chromium has been reported in some parts of the world. Timerbaev et al. (1999) reviewed the speciation of metal ions in tap water (groundwater) and reported the presence of arsenic, strontium, cadmium, chromium, copper, iron, etc. The status and the trend of river water pollution due to toxic metal ions in India has been discussed and described by Prebha and Selvapathy (1997). The Greenland snow and ice cap has been shown to contain universal archives of the large-scale atmospheric pollution of the Northern Hemisphere by lead and other metal ions (Adams et al., 1998). The first evidence of global contamination by lead was reported by Murozumi et al. (1969), who detected lead in snow samples covering the period from 1753 to 1965. This discovery had suffered from controversy but later recognized by the assessment of natural lead levels in old Greenland ice dated 5500 years BC (Ng and Patterson, 1981). Pleßow and Heinrichs (2000) presented the speciation of various trace elements in acidic pore waters from waste rock dumps. Pantsar-Kallio et al. (Pantsar-Kallio and Manninen, 1996, 1997) reported the speciation of chromium metal ions in wastewater. Similarly, Arar and Pfaff (1991) described the presence of hexavalent chromium in industrial wastewater. Carro and Mejuto (2000) reviewed the presence of organomercury species in different kinds of waters. Briefly, all types of water resources have been contaminated due to the heavy toxic metal ions.

3.2.2. Distribution in the sediment

A dynamic equilibrium is established between the sediment and water. Many reports have been published on the presence of heavy metal ions in the sediment. Due to the transfer process between different environmental components and the natural ability of the sediments; to accumulate metal ions, their concentrations are generally higher than those present in adjacent water and air. The enrichment of metal ions in the sediments by anthropogenic pollution is controlled by geological, mineralogical, hydrological and biological factors. Tack and Verloo (1995) reviewed the presence of metal ions on to the sediment and soil. Förstner and Wittmann (1983) described the presence of arsenic, selenium, zinc, lead, cadmium, chromium, copper, cobalt, nickel, mercury, silver, antimony, etc. metal ions on the sediment of lakes in North America and coastal areas of North America and British islands.

3.2.3. Distribution in the soil

The soils also get contaminated by the irrigation process by using the groundwater containing heavy metal ions. The contamination of soil also occurs through the use of fertilizers and pesticides containing metal ions. Besides, some metal ions present in the air also contaminate soil through rain. In addition to these, some man-made activities such as industrial and domestic discharges are also contributing toward the contamination of the soils. The humus layer of the soil plays a crucial role and this layer acts as a natural bio-geochemical barrier that suppresses the percolation of metal ions with seepage water, and thus, strongly accumulates the elements (Goldschmidt, 1937). Many metal ions have been reported on the soil at different parts of the world. Some reviews have been published on the distribution of metal ions on the soil (Das et al., 1995b). Matschullat (2000) reviewed the status of arsenic on the soil and the author reported a variable concentration of arsenic in different soils.

3.2.4. Distribution in the air

The atmospheric is also not untouched by the contamination of metal ions due to man-made and natural activities. Metal ions into atmosphere spreaded over a large area depending on the temperature, humidity, wind speed, etc. The determination of metal ions into ambient air, aerosol and dust is a subject of many environmental studies. Metal ions may absorb onto dust particles, fly ashes, etc. into the atmosphere. The determination of metal ions in aerosol is increasing continuously (Dabek-Zlotorzynska and Dlouhy, 1994, 1995a, 1995b). Matschullat (2000) reviewed the presence of arsenic in the atmosphere and the author reported 28,230–54,270 tons per year as arsenic flux released from different sources. Some toxic metal ions present in rain may be considered as the pollution into air as the rainwater dissolved these metal ions from atmosphere. Förstner and Wittmann (1983) reported aluminum, antimony, arsenic, cadmium, chromium, cobalt, copper, iron, lead, mercury, nickel, selenium, silver, vanadium, and zinc into the air.

3.2.5. Distribution in aquatic biota

Information of metal ions at high concentrations in animals and plants is an indication of the contamination of the environment. The role of heavy metals in aquatic organisms has been reported in the literature with emphasis on the toxicities of the individual metal ion (Doudoroff and Katz, 1953). Both essential (copper, zinc, iron, cobalt) and non-essential elements (arsenic, cadmium, lead, mercury) are found in the different aquatic organisms. Metal ions accumulation in aquatic organisms is controlled by biotic and abiotic factors. The biotic factors are physiological behavior, pattern of life cycle, seasonal variation in the organisms, species specific and individual variability, etc. while the abiotic factors include temperature and dissolved oxygen of water, hardness, pH, salinity, presence of several organic and inorganic compounds.

3.2.6. Distribution in the terrestrial biota

All the terrestrial animals and plants depend on water supply from the ground and surface waters and, hence, it is a common observation to find toxic metal ions in these organisms. The evolutionary studies of heavy metal tolerance have contributed for dealing with contaminated soils on several levels. Such types of studies have provided evidence for the toxic affects of heavy metal contamination on non-adapted genotypes. The evolutionary studies have also shown that the mechanism for metal tolerance is uptake, not exclusion, such that metal-tolerant genotypes are also metal accumulators. This last insight, in conjunction with physiological research on metal-tolerant plants, has led to a growing use of plants as a part of a clean-up technology for dealing with contaminated sites (Salt, 1995; Adler, 1996). Bradshaw and Trans (1991), in particular, have conducted extensive experiments on the evolutionary properties of plants growing in contaminated sites, such as mine spoils. Their findings, including growth of plants in contaminated sites, are genetically adapted to be tolerant of heavy metals. Metal-tolerant plants do not compete well in non-contaminated sites. This adaptation of plants to heavy metal contamination has been of particular interest because it is a character that appears to have evolved in part in response to human disturbance (Antonovics, 1971, 1975; Bradshaw and Trans, 1991). Dwivedi and Dey (2002) explored the possibility of translocation of heavy metals into humans and animals, the authors studied 28 commonly used medicinal plants and estimated their heavy metal content. Besides, some metal ions have been reported in the different kinds of food products. Basically, the type of metallic species and their concentrations in foodstuffs depend on the origin of the food products. For example, many metallic species have been detected in the seafood samples and the food products grown on the heavily contaminated soil and water resources. Aluminum is a ubiquitous element found in every food product. The other sources of aluminum are corn, yellow cheese, salt, herbs, spices, tea and tap water. Larsen et al. (1993) determined various arsenicals in the seafoods. Liang et al. (2003) reported mercury species in seafood samples. Benramdane et al. (1999) reviewed the presence of arsenic species in different foodstuffs. The distribution of metal ion species in the different components of the ecosystem is summarized in Table 3.1.

3.3. Toxicities

As discussed above different metal ions have different toxicities and therefore, the discussion on this issue is very important and discussed herein briefly. The changes in the degree of oxidation of an element have an important affect on the degree of bioavailability and toxicity (Stoeppler, 1992). The toxicities of metal ion species depend on age and sex of human beings. The biological activities of organic and inorganic metal ion species in living systems are given in Table 3.2. In oxidized environment As, Sb and Se appear mostly as oxyanions (Cutter, 1992). The valency state of a metal ion controls the behavior of metal ion in the aqueous system. For example, the toxicity of As(III) and Sb(III) is higher than that of their pentavalent species (Berman, 1980; Gesamp, 1986). The toxicity of the water-soluble species of arsenic varies in the order of arsenite > arsenate > monomethylarsonate (MMA) > dimethylarsinate (DMA) (Stugeron et al., 1989). Metals occur in the natural waters in different physico-chemical forms. Among these, simple hydrated metal ions are considered to be the most toxic forms while strong complexes and species associated with colloidal particles are usually assumed to be non-toxic ones. Organometallic compounds are more toxic than their corresponding inorganic species with the exception of arsenic.

Illmer and Mutschlechner (2004) studied the affect of pH on the toxicity of aluminum metal ion on *Arthrobacter* species. The authors reported EC_{100} indicating a complete inhibition of microbial growth with 185 μM aluminum for *Arthrobactor* 1 and 11 mM of aluminum for *Arthrobactor* 2. People drinking arsenic-contaminated water generally showed arsenical skin lesions, which are a late manifestation of arsenic toxicity. Long-term exposure of arsenic-contaminated water may lead to various diseases such as conjuctivitis, hyperkeratosis, hyperpigmentation, cardiovascular diseases, disturbance in the peripheral vascular and

Table 3.1. Distribution of metal ion species in the ecosystem

Metal ion species	Ecosystem component	Ref.
Water		
Cd, Cu, Pb & Zn	Vesdre river, Belgium	Houba et al. (1983)
Cd, Cu, Cr, Mn, Pb & Zn	Toyhira river, Japan	Sakai et al. (1986, 1993)
Cd, Cr, Cu, Pb & Ni	Pawtuxet river	Zhang and Huang (1993)
Fe, Mn, Cd, Pb, Ni & Zn	Ravi river, Pakistan	Mohammed and Hayat (1996)
Co, Cd & Zn	Kishon river, Israel	Barak and Kress (1997)
Cd, Co, Pb & Zn	Seine river, France	Alexandrine et al. (1998)
Cd, Cr, Cu, Fe, Ni, Pb & Zn	Yamuna river, India	Agarwal (1997)
Heavy metals	Rivers, USA & Europe	Förstner and Wittmann (1983)
Monobutyltin, dibutyltin and tributyltin	Sea water	Botana et al. (2002)
Co, Cu, Fe & Zn	Pond water	Regan et al. (1994)
Methyl-, ethyl- and methylethyl lead	Rain water	Zufiaurre et al. (1997)
Lead and other metal ions	Northern Hemisphere	Adams et al. (1998)
Lead	Snow samples	Murozumi et al. (1969)
Sediment		
Methylmercury	Sediment	Lai et al. (1998)
Butyltin(I) and butyltin(II)	Marine Sediment	Whang and Whang (1997)
Monto-, di- and tributyltin	Sediment Gulf of Gdansk, Baltic Sea	Flandysz et al. (2002)
Tributyltin and triphenyltin	Marine sediment	Feng and Narasaki (2002)
Trimethyltin, triethyltin and triphenyltin	Sediment	Yang, et al. (1995)
Monobutyltin, dibutyltin and tributyltin	Sediment	Ceulemans and Adams (1995)
Monobutyltin, dibutyltin and tributyltin	Sediment	Cai et al. (1993)
Soil		
Al species	Soil	Göttlein and Blasek (1996); Mitrovic and Milacic (2000)
As species	Soil	Prohaska et al. (1999); Yang et al. (1996)
Cu & Cr	Soil	Krajnc et al. (1995)
Hg	Soil	Hempel et al. (1995)
Me-Hg	Soil	Carro et al. (1998)
Pt	Soil	Michalke et al. (1997)
Organic lead, antimony, tin	Soil	Hirner et al. (2000)
Rare earth elements	Soil	Wang et al. (2001)
Air		
Al species	Air	Tripathi et al. (2002)
Mn, Sr & Cd	Aerosol	Dabek-Zlotorzynska et al. (1995b)
Zn & alkaline earth metal ion	Dust particles	Fung et al. (1998)

Table 3.1. (*Continued*)

Metal ion species	Ecosystem component	Ref.
Trimethylbutyllead, dimethylethylbutyllead, triethylbutyllead, methyldiethylbutyllead, dibutyldiethyllead and tetrabutyllead compounds	Dust	Nerin et al. (1995)
Alkali and alkaline earth metal ions	Atmosphere	Dabek-Zlotorzynska et al. (1995b); Fung et al. (1995)
Aquatic biota		
Cd, Cu, Mn, Pb & Zn	Diatoms, Monterey Bay, California	Knauer et al. (1973)
Mn	Phytoflagellate	Morris (1971)
Cu, Pb & Ag	Fucus vesiculosus	Lande (1977); Stenner and Nickless (1974)
Cr, Cu, Fe & Mn	Algae	Martin and Broenkow (1975)
As species	Algae	Bowen (1979)
As species	Crustacea, mollusca, porifera, chinodermata & coelentrata	Francesconi et al. (1994)
Hg species	Euphausiids, copepods, ctenophores & coelenterates	Knauer and Martin (1973)
Hg species	Zooplanktons	Skei et al. (1976)
Toxic metal ions	Crustaceans	Stenner and Nickless (1975)
Hg	Asellus aquaticus	Hasselrot and Göthberg (1974)
Zn	Platichthyes flesus	Papadopoulou (1978)
Hg & Se	Makaira indica	MacKay (1975)
MeHgCl & $HgCl_2$	Perca flavescens	Freitas et al. (1974)
Diethylmercury & ethylmethylmercury Specie	Dog fish	Logar et al. (2000)
Terrestrial biota		
Aluminum organic acid complexes	Plant sap	Bantan et al. (1999)
Fe species	Mugineic acid family plants	Weber, et al. (2002)
Se species	Terrestrial plants	Jung et al. (2002); Montes-Bayon et al. (2002); Zhang and Frankenberger (2001)
Cd	Pig	Crews et al. (1989)
Al & silicon	Human serum	Wrobel et al. (1995)
As species	Human serum & urine	He et al. (2000)
Alkyl lead	Human serum	Pons et al. (1998)
MeHg, EtHg, PhHg & Hg(II)	Human urine	Yin et al. (1998)
Organotin compounds	Human organs	Jiang et al. (2000)
Organo As, Ge, Se, Sn & Pb	Human urine	Kresimon et al. (2001)

Table 3.2. A comparison of the activities of organic and inorganic metal ion species in biological systems (Furst, 1987)

Biological action	Organic species	Inorganic species
Adsorption	Yes	Yes
Plasma protein binding	Albumin	Globulin
Target organs	Yes	Yes
Metabolized	Yes	Yes
Excretion (main route)	Bile	Urine
Induction	Enzyme (P-450)	Metallothiein
Oxidation states	Not important	Important
Biodegradation	Yes	No

nervous systems, cancer of the skin and gangrene, leucomelanosis, non-pitting swelling, hepatomegaly and splenomegaly (Kiping, 1977; Yamamura and Yamauchi, 1980; WHO, 1981; Pershagen, 1983). Arsenite is 20–100 times more toxic than arsenate in marine environment (Neff, 1997) and 60 times more toxic in human beings (Ferguson and Gavis, 1972).

Cadmium is supposed to be the most toxic metal ion with nausea and vomiting at 15 mg l^{-1} concentration. The kidneys are the critical target of cadmium contamination. The disorders associated with kidney are renal dysfunction, hypertension and anemia. The other side affects of cadmium are extreme restlessness and irritability, headache, chest pain, increased salivation, choking, abdominal pain, diarrhea, tenesmus, throat dryness, cough and pneumonitis. The exposure of cadmium chloride aerosol in rats has been reported of tumor formation (carcinogenesis). The different cadmium compounds found carcinogenic in nature are Cd, CdS, CdO, $CdSO_4$ and $CdCl_2$. WHO guideline for cadmium is 0.003 mg l^{-1}. Trivalent chromium [(Cr(III)] is essential for animal and human health and growth; at trace levels, with a safe and relatively innocuous level of 0.20 mg day^{-1} whereas hexavalent chromium [Cr(VI)] is a potent carcinogen and extremely toxic to the animals and humans. But recently, Mulyani et al. (2004) reported the oxidation of Cr(III) to Cr(VI) leading to toxicity of trivalent chromium in the biological systems. The main notorious affects of Cr(VI) are on liver, kidney, respiratory organs with hemorrhagic problem, dermatitis and ulceration of the skin for chronic and subchronic exposure.

Lead is the most significant toxin of the heavy metals and affects are of toxicological and neurotoxic in nature, which include irreversible brain damage. Inorganic forms of lead typically affect the central nervous system (CNS), peripheral nervous system (PNS), hematopoietic, renal, gastrointestinal, cardiovascular and reproductive systems. The various lead compounds reported to be carcinogens are PbO, $PbCrO_4$, $Pb_3(PO_4)_2$ and $PbAc_2$. Inorganic mercury compounds are toxic to kidney leading to neurological and renal disturbances while organic mercury compounds are very toxic to the CNS. Besides, discouragement, irritability, personality changes, learning disabilities, muscle tremors, jerky gait, spasms of extremities, inflammation of mouth and gums, swelling of salivary glands, excessive flow of saliva and loosening of teeth are the other hazardous affects of mercury poisoning. The mechanisms of toxicity and, particularly, the potential carcinogenicity of inorganic mercury are still under debate. The results of mutagenicity and genotoxicity testing with mercury have been inconsistent, i.e., mercury induces DNA single strand breaks at low concentrations in mammalian cells but has not proved mutagenic in several bacterial mutagenicity assays. Kungolos et al. (1989) studied the affects of six different forms of mercury on the growth of the yeast *Saccharomyces cerevisiae* using five kinds of strains of *S. cerevisiae*. In terms of EC_{50} toward the wild-type strain, the toxicity order for the inorganic forms was $Hg(NO_3)_2 > HgSO_4 > HgCl_2$. The different toxicities of various metal ion species are given in Table 3.3.

3.4. Analyses

Of course the analysis of the total concentration of metal ions is required and essential. Many reports are available in the literature on the analyses of metal ions but the data presented is not reliable. It is due to the different toxicities of various oxidation states of a particular metal ion, which makes the toxicity concept confused. The speciation of metal ions in the environment is a difficult job as the concentrations of heavy metals in the environment are generally very low. Moreover, sometimes, the physical and chemical properties of two or more species may be similar which makes the speciation again a challenging issue. Speciation strategy contains a careful plan or method or the art of developing or employing stratagems toward a goal. Scientists hope to learn everything about the elemental species, i.e., composition, mass, the bio- and environmental cycle, the stability of the species, its transformation and the interactions with inert or living matter. The work involved to achieve this goal is, however, challenging, if not impossible to complete. Therefore, a choice has to be made to identify the most important issues as elemental speciation studies are pursued. A first group of compounds to be studied very closely are those of anthropogenic in origin. In speciation studies, a lot of attention must to be paid to the stability. Species stability depends

Table 3.3. Toxicities of the different metallic species

Metallic species	Animals	LD_{50} (mg kg^{-1})
Arsenicals (Andreae, 1986; Kaise & Fukui, 1992; Penrose, 1974)		
Potassium arsenite	Rat	14
Calcium arsenate	Rat	20
MMA	Rat	700–1800
DMA	Rat	700–2600
Arsenobetaine	Rat	10,000
Lead compounds (Boyland et al., 1962; van Esch et al., 1962; Furst et al., 1976; Kanisawa and Schroeder, 1969; van Esch & Kroes, 1969)		
Pb-arsenite	Rat	100
Pb-naphthinate	Rat	5100
Pb-nitrate	Guinea pig	1300
Pb-chloride	Guinea pig	2000
Pb-fluoride	Guinea pig	4000
Pb-oleate	Guinea pig	8000
Mercury compounds (Cassidy & Furr, 1978; EPA, 1980)		
HgCl	Rat	210
HgI	Mouse	110
$HgNO_3$	Mouse	388
HgO	Rat	8.0
MeHg	—	10.0
EtHg	—	40.0
Organotin species (Merian, 1991)		
Bis(tributyltin) oxide	Rat	150–234
Trimethyltin hydroxide	Rat	540
Triphenyltin hydroxide	Rat	125
Trimethyltin acetate	Rat	9.1
Triethyltin acetate	Rat	4.0
Tributyltin acetate	Rat	4000
Antimony compounds (Furst, 1987)		
Sb(III) or Sb(V)	Rat	100
Sb_2O_3	Rat	3250
Sb_2S_3	Rat	1000
Sb_2O_5	Rat	4000
Sb_2S_5	Rat	1599

on the matrix and on the physical parameters, such as temperature, humidity, UV light, organic matter, etc. The isolation and purification of the species, the study of the possible transformation through the procedure, their characteristics and interactions are also important. New analytical procedures have to be devised, including appropriate quantification and calibration methodologies.

Various advanced techniques have been developed to overcome this problem. The different analytical methods used for metal ion speciation include spectroscopic, polarographic, radiometric, chromatographic and electrochemical. Various technologies used for metal ion speciation are shown in Fig. 3.1. Among these techniques, chromatographic and electrophoretic methods have been used frequently due to their ease of operation and reproducible results. The spectroscopic methods of the speciation involve the use of various spectroscopic techniques such as UV visible, flame atomic absorption, atomic emission, hydride generation, graphite furnace, etc. (Krull, 1991; Jain and Ali, 2000). Besides the spectroscopic methods of speciation, polarography, voltammetry and radiometry have also been used for the speciation of metal ions (Pretty et al., 1990, 1992, 1993).

3.4.1. Chromatographic methods

The chromatographic methods involve the use of gas or liquid separately as the mobile phases. Therefore, the former kind of chromatography is called gas chromatography (GC) while the later is termed as liquid chromatography (LC). Due to some problems associated with gas

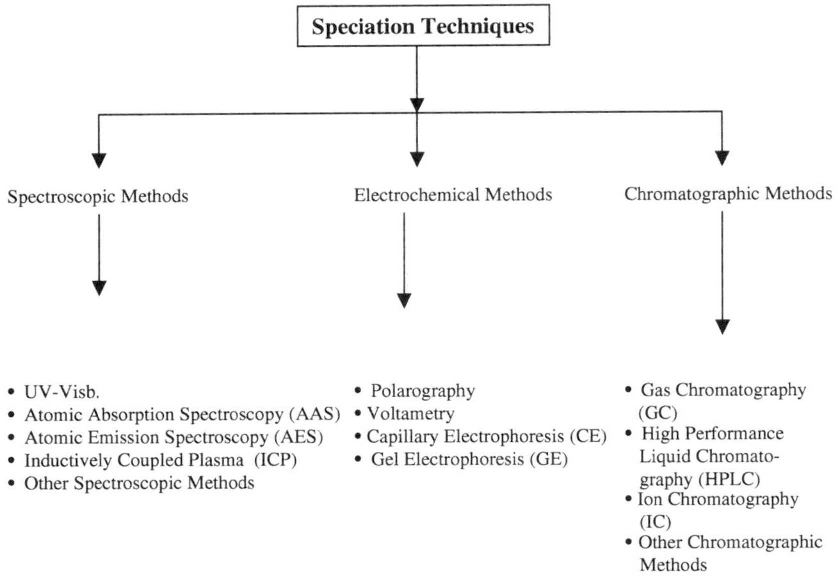

Figure 3.1. Different techniques used for meat ion speciation.

chromatography it could not be accepted as the method of the choice for the speciation of metal ions. The major disadvantage of GC is its requirement of the conversion of inorganic metal ions into organic and volatile metal ion derivatives, which is carried out by derivatization process. Contrarily, LC is the only and the best remaining technology for the speciation of a wide variety of metal ions. The main advantages of LC are its ability to speciate metallic species in the environmental samples directly. During the course of time various types of liquid chromatographic approaches were developed and used in this concern. Most important liquid chromatographic methods are high performance liquid chromatography (HPLC) and ion chromatography (IC). However, sub- and supercritical fluid chromatography (SFC), capillary electrochromatography (CEC) and thin layer chromatography (TLC) modalities can be used for this purpose.

Among the various liquid chromatographic techniques mentioned above, IC remains as the best modality due to its several advantages in comparison to others. The high speed, sensitivity and reproducible results make IC as the method of choice in the world. About 80% metal ion speciation has been carried out using IC mode of liquid chromatography. Various types of columns are available in IC, which make this technique popular in metal ion speciation. A variety of mobile phases including normal, reversed and new polar organic phases are used in IC. The composition of the mobile phases may be modified by the addition of various aqueous and non-aqueous solvents. Ion chromatography is a form of liquid chromatography that uses ion exchange resins to separate atomic or molecular ions based on their interaction with the resin. Its greatest utility is for analysis of ions for which there are no other rapid analytical methods. Most ion exchange separations are carried out with metallic pumps and columns. The column packing for ion chromatography consists of ion exchange resins bonded to inert polymeric particles. The optimization of metal ion speciation is carried out by a number of experimental parameters.

The use of supercritical fluids as the mobile phases for chromatographic separation was first reported more than 30 years ago, but most of the growth in SFC has occurred recently. In chromatographic systems, the solute diffusion coefficients are often of the order of higher magnitude in supercritical fluids than in tradition liquids. Contrarily, the viscosities are lower than those of liquid (Weast, 1973). At temperature below T_c and pressure above P_c, the fluid becomes a liquid. On the other hand, at temperature above T_c and pressure below P_c, the fluid behaves as a gas. Therefore, supercritical fluids can be used as a part of a mixture of liquid and gas (Berger, 1995). The commonly used supercritical fluids (SFCs) are carbon dioxide, nitrous oxide and trifluoromethane (Weast, 1973; Schoenmakers, 1988; Berger, 1995). The compatibility with most detectors, low critical temperature and pressure, low toxicity and environmental burden and low costs make carbon dioxide as the supercritical fluid of choice. The main drawback of supercritical carbon dioxide as mobile phase is to be seen in its inability to elute more polar organic derivatives of metal ions and other compounds. This can be improved by the addition of organic modifiers to the relatively apolar carbon dioxide.

Basically, capillary electro-chromatography (CEC) is a hybrid technique, which works on the basic principles of capillary electrophoresis and chromatography (Cronin and Pizarello, 1997). This mode of chromatography is used either on packed or tubular capillaries/columns. The packed column in CEC was first introduced by Pretorius et al. (1974), while the open tubular CEC was presented by Tsuda et al. (1982). In 1984, Terabe et al. (1984) introduced another modification in liquid chromatography, i.e. micellar electrokinetic capillary chromatography (MECC). Of course this mode also depends on the working principles of capillary electrophoresis and chromatography but also involves the formation of micelles. The high speed, sensitivity, lower limit of detection and reproducible results make CEC and MECC as the methods of choice in the analytical science. However, these methods could not be used frequently for speciation as these techniques are not fully developed and the research is underway on this subject. Many research papers and reports have appeared in the literature on metal ion speciation using chromatographic techniques, all of which cannot be summarized in this chapter. However, interested readers should consult some review articles on this issue (Robarts et al., 1991; Cagniant, 1992; Quevauviller, 1996; Sarzanini and Mentasti, 1997, 1999; Nielen et al., 1998; Liu and Lee, 1999; Pereiro and Diaz, 2002; Ponce de Leon et al., 2002).

3.4.2. Capillary electrophoretic method

In the last decade, a newly developed technique called capillary electrophoresis (CE) has emerged as one of the tools for the speciation of metal ions. CE has certain advantages in comparison to the other techniques. The advantages of CE include inexpensive running cost, simple, high speed of analysis, greater separation efficiency, unique selectivity and high degree of matrix independence, all of which make it ideal for the speciation of metal ions (Gareil, 1990; Tessier and Turner, 1996; Baraj et al., 2000). Among the electrophoretic methods of metal ion speciation various forms of capillary electrophoresis such as capillary zone electrophoresis (CZE), capillary isotachophoresis (CIF), capillary gel electrophoresis (CGE), capillary isoelectric focusing (CIEF), affinity capillary electrophoresis (ACE) and separation on microchips have been used. However, in contrast to others, CZE model was used frequently for this purpose (Chankvetadze, 1997). More than 100 research papers have been found in the literature on metal ion speciation using capillary electrophoresis but for quick search and information interested readers should consult some review articles on this subject (Dabek-Zlotorzynskaa et al., 1998; Liu and Lee, 1999; Timerbaev, 2002).

3.4.3. Spectroscopic and other methods

Spectroscopic methods of the speciation involve the use of UV-visible, flame atomic absorption, atomic emission, hydride generation, graphite furnace, etc. Some of them require the conversion of one form into another. Few metallic species form different colours with certain ligands and they can be speciated by UV-visible spectroscopy. For example, arsenic can be speciated by molybdenum blue and silver diethyldithiocarbamate methods. Even methods such as atomic emission and atomic absorption spectrometry coupled with flow injection analyser system (FIAS) give results that are more precise and exact when only one form of an element is present. Atomic absorption spectrometry with hydride generation is one of the most widely used methods for the speciation of some metal ions. For example, the speciation of arsenic is carried out by determining As(III) and the total arsenic, which is determined by the reduction of As(V) to As(III). The prior oxidation of arsenite to arsenate by KI has been reported (Bogdanova, 1984; Matsubara et al., 1987; Nasu and Kan, 1988; Tamari et al., 1989; Palanivelu et al., 1992). Besides, recently a method was developed for the determination of trace amounts of As(III) and total arsenic with L-cysteine as pre-reductant using flow injection hydride generation coupled with an in-house made non-dispersive AAS device (Yin et al., 1996). Inductively coupled plasma spectroscopy (ICP) involves the use of plasma formed at high temperature. This technique can be used for speciation after hyphenation with hydride generation unit. Lower limit of detection in ICP can be achieved by coupling it with mass spectrometer. Among electrochemical methods, potentiometry, polarography, cyclic, pulse and stripping voltammetric techniques are used for metal ion determination (Rajeshwar and Ibanez, 1997). The direct current polarography is able to detect different oxidation states of many metal ions. In addition to above, some reports are available on metal ion speciation by other techniques such as X-ray, X-ray fluorescence (XRF), EX-AFS, XA-NES spectroscopy, isotopic dilution, sensors, etc. Radiometric methods have been used for metal radioactive ion species. However, this method can be applied for speciation of non-radioactive metal ions by producing artificial radioactivity. Some reviews have been published on metal ion speciation using these techniques and readers should go through these articles (Florence, 1986; Luque de Castro, 1986; Cornelis, 1992; Taylor and van Staden, 1994; Williams and D'Silva, 1994; Rosenberg, 2003).

3.5. Conclusion

About 75,000 chemicals are being in use everyday having different species and some of them are in direct or indirect contact with human beings and, hence, speciation is an integral part of the specification criteria, because only certain defined chemical species are permitted as sources of the essential elements. In the lack of legislation governing the speciation concept people are consuming toxic metal ion species, which result into some serious side affects and ailments. The lack of awareness about the speciation may be a serious and challenging problem to the world community in the near future due to the continuous contamination of our environment. A little attention has been paid for regulating metal ion speciation at International level. In view of these facts, it is very important to determine the toxicities of all metallic species in order to facilitate the speciation regulators. The toxicological data should be capable to recognize the social awareness. The analytical techniques should be developed in such a way, which can speciate metal ions into the real-world sample with high selectivity, sensitivity and reproducibility. A social awareness of science is required recognizing the speciation issue among the communities. Briefly, there is a great need for a think of risk, safety and communication associated with metal ion speciation at International levels. The politician and the regulatory authorities should come forward to stipulate the regulations and legislation to control the speciation of metal ions into our environment.

REFERENCES

Adams, F.C., Heisterkamp, M., Candelone, J.P., Laturnus, F., van de Velde, K., Boutron, C.F., 1998. Speciation of organometal and organohalogen compounds in relation to global environmental pollution. Analyst 123, 767–772.

Adler, T., 1996. Botanical cleanup crews—Using plants to tackle polluted water and soil. Sci. News 150, 42–43.

Agarwal, A., 1997. Homicides by pesticides—What pollution does to our bodies. Center for Science & Environment, Excellent Printing Press, New Delhi.

Andreae, M.O., 1986. Organoarsenic compounds in the environment. In: Craig, P.J. (Ed.), Organometallic Compounds in the Environment, Principles and Reactions. Wiley, New York.

Antonovics, J., 1971. Heavy metal tolerance in plants. Adv. Ecol. Res. 7, 1–85.

Antonovics, J., 1975. Metal tolerance in plants: Perfecting an evolutionary paradigm. International conference on heavy metals in the environment, Toronto, Ontario, Canada, pp. 169–186.

Alexandrine, E., Jean-Marie, M., Thevenot, D.R., 1998. Urban runoff impacts on particulate metal concentrations in river Seine. Water Air Soil Pollut. 108, 83–105.

Arar, F.J., Pfaff, J.D., 1991. Determination of dissolved hexavalent chromium in industrial wastewater effluents by ion chromatography and post-column derivatization with diphenylcarbazide. J. Chromatogr. A 546, 335–340.

Bantan, T., Milacic, R., Mitrovic, B., Pihlar, B., 1999. Combination of various analytical techniques for speciation of low molecular weight aluminium complexes in plant sap. Fresen. J. Anal. Chem. 365, 545–552.

Baraj, B., Niencheski, L.F., Soares, J.A., Martinez, M., 2000. Comparison of chromium speciation by CZE and ion exchange followed by AAS. Fresen. J. Anal. Chem. 367, 12–16.

Barak, H., Kress, N., 1997. Particulate metals contamination in the Kishon river estuary. Mar. Pollut. Bull. 34, 706–711.

Benramdane, L., Bressolle, F., Vallon, J.J., 1999. Arsenic speciation in humans and food products: A review. J. Chromatogr. Sci. 37, 330–344.

Berger, T.A., 1995. In: Smith, R.M. (Ed.), Packed Column SFC. The Royal Society of Chemistry, Chromatography Monographs, Cambridge.

Berman, E., 1980. Toxic Metals and Their Analysis. Heyden & Sons, London.

Bogdanova, V.I., 1984. The choice of optimal conditions for the determination of arsenic in the form of molybdenum blue. Microkimica Acta II, 317–330.

Botana, J.C., Pereiro, I.R., Torrijos, R.C., 2002. Rapid determination of butyltin species in water samples by multicapillary gas chromatography with atomic emission detection following headspace solid-phase microextraction. J. Chromatogr. A 963, 195–203.

Bowen, H.J.M., 1979. Environmental Chemistry of the Elements. Academic Press, London, UK.

Boyland, E., Dukes, C.E., Grover, P.L., 1962. The introduction of renal tumors by feeding lead acetate to rats. Br. J. Cancer 16, 283–288.

Bradshaw, A.D., Trans, P., 1991. The croonian lecture, 1991. Genostasis and the limits to evolution. Royal Society (London), Biol. Sci. 333, 289–305.

Cagniant, D., 1992. Complexation chromatography. Marcel Dekker, New York.

Cai, Y., Rapsomanikis, S., Andreae, M.O., 1993. Determination of butyltin compounds in sediment using gas chromatography-atomic absorption spectrometry: Comparison of sodium tetrahydroborate and sodium tetraethylborate derivatization methods. Anal. Chim. Acta 274, 243–251.

Carro, A.M., Lorenzo, R.A., Cela, R., 1998. Comparative study of speciation techniques for the determination of methylmercury in biological and sediment samples. LC-GC 16, 926–936.

Carro, A.M., Mejuto, M.C., 2000. Application of chromatographic and electrophoretic methodology to the speciation of organomercury compounds in food analysis. J. Chromatogr. A 882, 283–307.

Caruso, J.A., Sutton, K.L., Ackley, K.L. (Eds.), 2000. Elemental speciation, comprehensive analytical chemistry, Vol. XXXIII, Elsevier, Amsterdam.

Cassidy, D.R., Furr, A., 1978. Toxicity of inorganic mercury compounds in animals. In: Oehme, F.W. (Ed.), Toxicity of Heavy Metals in the Environment – Part I. Marcel Dekker, New York.

Ceulemans, M., Adams, F.C., 1995. Evaluation of sample preparation methods for organotin speciation analysis in sediments—Focus on monobutyltin extraction. Anal. Chim. Acta 317, 161–170.

Chankvetadze, B., 1997. Capillary Electrophoresis in Chiral Analysis. Wiley, New York.

Chatterjee, A., Das, D., Mandal, B.K., Chowdhury, T.R., Samanta, G., Chakraborty, D., 1995. Arsenic in ground water in six districts of West Bengal, India: The biggest arsenic calamity in the world. Part I. Arsenic species in drinking water and urine of the affected people. Analyst 120, 643–650.

Cornelis, R., 1992. Use of radiochemical methods as tools for speciation purposes in environmental and biological sciences. Analyst 117, 583–588.

Crews, H.M., Dean, J.R., Ebdon, L., Massey, R.C., 1989. Use of radiochemical methods as tools for speciation purposes in environmental and biological sciences. Analyst 114, 895–900.

Cronin, J.R., Pizarello, S., 1997. Enantiometric excesses in meteoritic amino acids. Science 275, 951–955.

Cutter, G.A., 1992. Kinetic controls on the speciation of metalloids in seawater. Mar. Chem. 40, 65–80.

Dabek-Zlotorzynska, E., Dlouhy, J.F., 1994. Application of capillary electrophoresis in atmospheric aerosols analysis: Determination of inorganic and organic anions, J. Chromatogr. A 671, 389–395.

Dabek-Zlotorzynska, E., Dlouhy, J.F., 1995a. Application of capillary electrophoresis in atmospheric aerosol analysis: Determination of cations. J. Chromatogr. A 706, 527–534.

Dabek-Zlotorzynska, E., Dlouhy, J.F., Houle, N., Piechowski, M., Ritchie, S., 1995b. Comparison of capillary zone electrophoresis with ion chromatography and standard photometric methods for the determination of inorganic anions in atmospheric aerosols. J. Chromatogr. A 706, 469–478.

Dabek-Zlotorzynskaa, E., Lai, E.P.C., Timerbaev, A.R., 1998. Capillary electrophoresis. The state-of-the-art in metal speciation studies. Anal. Chim. Acta 359, 1–26.

Das, A.K., Chakraborty, R., Cervera, M.L., Delaguardia, M., 1995b. Metal speciation in solid matrices. Talanta 42, 1007–1030.

Das, D., Chatterjee, A., Mandal, B.K., Samanta, G., Chakroborty, D., Chanda, B., 1995a. Arsenic in ground water in six districts of West Bengal, India: The biggest arsenic calamity in the world. Part 2. Arsenic concentration in drinking water, hair, nails, urine, skin-scale and liver tissue (biopsy) of the affected people. Analyst 120, 917–924.

Doudoroff, P., Katz, M., 1953. Critical review of literature on the toxicity of industrial wastes and their components to fish. Sewage Ind. Wastes 25, 802–839.

Dwivedi, S.K., Dey, S., 2002. Medicinal herbs: A potential source of toxic metal exposure for man and animals in India. Arch. Environ. Health 57, 229–231.

EPA, Ambient water quality criteria for mercury, 1980. US EPA Washington, DC (EPA 440/5-80-058).

Feng, Y.L., Narasaki, H., 2002. Speciation of organotin compounds in marine sediments by capillary column gas chromatography-atomic absorption spectrometry coupled with hydride generation. Anal. Bioanal. Chem. 372, 382–386.

Ferguson, J.F., Gavis, J., 1972. A review of the arsenic cycle in natural water. Water Res. 6, 1259–1274.

Flandysz, J., Brzostowski, A., Szpunar, J., Rodriguez-Pereiro, I., 2002. Butyltins in sediments and three-spined stickleback (Gasterosteus aculeatus) from the marinas of the Gulf of Gdansk, Baltic Sea. J. Environ. Sci. Health A 37, 353–363.
Florence, T.M., 1986. Electrochemical approaches to trace element speciation in waters: A review. Analyst 111, 489–506.
Förstner, U., Wittmann, G.T.W. (Eds.), 1983. Metal Pollution in the Aquatic Environment. Springer, Berlin.
Francesconi, K.A., Edmonds, J.S., Morita, M., 1994. Biotransformation of arsenic in the marine environment. In: Nriagu, J.O. (Ed.), Arsenic in the Environment Part I, Cycling and Characterization. Wiley, New York.
Freitas, A.S.W., de Qadri, S.U., Case, B.E., 1974. Origins and fate of mercury compounds in fish. Proc. Int. Conf. Transp. Persist. Chem. Aquatic Ecosys, Ottawa, pp. 31–36.
Fung, Y.S., Lau, K.M., Tung, H.S., 1998. Analysis of leachable and total trace metals in air particulate matters by capillary electrophoresis. Talanta 45, 619–629.
Furst, A., 1987. Relationship of toxicological effects to chemical forms of inorganic compounds. In: Brown, S.S., Kodama, Y. (Eds.), Toxicology of Metals, Clinical and Experimental Research. Ellis Horwood, Chichester, p. 17.
Furst, A., Schlauder, M., Sasmore, D.P., 1976. Tumorigenic activity of lead chromate. Cancer Res. 36, 1779–1783.
Gareil, P., 1990. Capillary zone electrophoresis and electrokinetic chromatography II. Experimental aspects and applications. Analusis 18, 447–468.
Gesamp, 1986. (IMO/FAO/UNESCO/WMO/WHOIAEA/UN/UNEP Joint group of experts on the scientific aspects of marine pollution) review of potentially harmful substances—Arsenic, mercury and selenium. Rept. Stud. GESAMP 28.
Goldberg, 1954. Marine geochemistry 1. Chemical scavengers of the sea E.D. J. Geol. 62, 249–265.
Goldschmidt, V.M., 1937. Principles of distribution of chemical elements in minerals and rocks. J. Chem. Soc. (London), Part 1, 655–673.
Göttlein, A., Blasek, R., 1996. Analysis of small volumes of soil solution by capillary electrophoresis. Soil Sci. 161, 705–715.
Hasselrot, T.B., Göthberg, A., 1974. Ways of transport of mercury to fish. Proc. Int. Conf. Persist. Chem., Aquatic Ecosystem, Ottawa, Canada, pp. 37–47.
He, B., Jiang, G.B., Xu, X., 2000. Arsenic speciation based on ion exchange high-performance liquid chromatography hyphenated with hydride generation atomic fluorescence and on-line UV photo oxidation. Fresen. J. Anal. Chem. 368, 803–808.
Hempel, M., Wilken, R.D., Miess, R., Hertutich, J., Beyer, K., 1995. Mercury contaminated sites—Behaviour of mercury and its species in lysimeter experiments. Water Air Soil Pollut. 80, 1089–1098.
Hirner, A.V., Grüter, U.M., Kresimon, J., 2000. Metal(loid) organic compounds in contaminated soil. Fresen. J. Anal. Chem. 368, 263–267.
Houba, C., Remacle, J., Dubois, D., Thorez, J., 1983. Factors affecting the concentrations of cadmium, zinc, copper and lead in the sediments of the Vesdre. Water Res. 17, 1281–1286.
Illmer, P., Mutschlechner, W., 2004. Effect of temperature and pH on the toxicity of aluminium towards two new, soil born species of Arthrobacter sp. J. Basic Microbiol. 44, 98–105.
Jain, C.K., Ali, I., 2000. Arsenic: Occurrence, toxicity and speciation techniques. Water Res. 34, 4304–4312.
Jiang, G.J., Zhou, Q., He, B., 2000. Speciation of organotin compounds, total tin, and major trace metal elements in poisoned human organs by gas chromatography-flame

photometric detector and inductively coupled plasma-mass spectrometry. Environ. Sci. Technol. 34, 2697–2702.

Jung, M.C., Thornton, I., Chon, H.T., 2002. Arsenic, Sb and Bi contamination of soils, plants, waters and sediments in the vicinity of the Dalsung Cu-W mine in Korea. Sci. Total Environ. 259, 81–89.

Kaise, T., Fukui, S., 1992. The chemical form and acute toxicity of arsenic compounds in marine organisms. Appl. Organometal. Chem. 6, 155–160.

Kanisawa, M., Schroeder, H.A., 1969. Life term studies on the effect of trace elements on spontaneous tumors in mice and rats. Cancer Res. 29, 892–895.

Kiping, M.D., 1977. In: Lenihan, J., Fletcher, W.W. (Eds.), Arsenic, The Chemical Environment, Environment and Man, Vol. 6. Blackie, Glassgow.

Knauer, G.A., Martin, J.H., 1973. Seasonal variations of cadmium, copper, manganese, lead, and zinc in water and phytoplankton in Monterey Bay, California. Limnol. Oceanogr. 18, 597–604.

Krajnc, M., Stupar, J., Milicev, S., 1995. Characterization of chromium and copper complexes with fulvic acids isolated from soils in Slovenia. Sci. Total Environ. 159, 23–31.

Kresimon, J., Grüter, U.M., Hirner, A.V., 2001. HG/LT-GC/ICP-MS coupling for identification of metal(loid) species in human urine after fish consumption. Fresen. J. Anal. Chem. 371, 586–590.

Krull, I.S., 1991. Trace Metal Analysis and Speciation. Elsevier, Amsterdam.

Kungolos, A., Aoyma, I., Muramoto, S., 1989. Ecotoxicolo & Environ Safety 43, Speciation. Anal. Meth. Problems. CRC Press, London.

Lai, E.P.C., Zhang, W., Trier, X., Georgi, A., Kowalski, S., Kennedy, S., Muslim, T., Dabek-Zlotorzynska, E., 1998. Speciation of mercury at ng/ml concentration levels by capillary electrophoresis with amperometric detection. Anal. Chim. Acta 364, 63–74.

Lande, E., 1977. Heavy metal pollution in Trondheimsfjorden, Norway, and the recorded effects on the fauna and flora. Environ. Pollut. 12, 187–198.

Larsen, E.H., Pritz, G., Hansen, S.H., 1993. Arsenic speciation in seafood samples with emphasis on minor constituents: An investigation using high-performance liquid chromatography with detection by inductively coupled plasma mass spectrometry. J. Anal. Atom. Spectrom. 8, 1075–1084.

Liang, L.N., Jiang, G.B., Liu, J.F., Hu, J.T., 2003. Speciation analysis of mercury in seafood by using high-performance liquid chromatography on-line coupled with cold-vapor atomic fluorescence spectrometry via a post column microwave digestion. Anal. Chim. Acta 477, 131–137.

Liu, W., Lee, H.K., 1999. Chemical modification of analytes in speciation analysis by capillary electrophoresis, liquid chromatography and gas chromatography. J. Chromatogr. A 834, 45–63.

Logar, M., Horvat, M., Falnoga, I., Stibilj, V., 2000. A methodological study of mercury speciation using Dogfish liver CRM (DOLT-2). Fresen. J. Anal. Chem. 366, 453–460.

Luque de Castro, M.D., 1986. Speciation studies by flow-injection analysis. Talanta 33, 45–50.

MacKay, N.J., Kazacos, M.N., Williams, R.J., Leedowq, M.I., 1975. Selenium and heavy metals in black marlin. Mar. Pollut. Bull. 6, 57–61.

Martin, J.H., Broenkow, W.W., 1975. Cadmium in plankton. Elevated concentrations off Baja California. Science 190, 884–885.

Matschullat, J., 2000. Arsenic in the geosphere—a review. Sci. Total Environ. 249, 297–312.

Matsubara, C., Yamamoto, Y., Takamura, K., 1987. Rapid determination of trace amounts of phosphate and arsenate in water by spectrophotometric detection of their heteropoly

acid-malachite green aggregates following pre-concentration by membrane filtration. Analyst 112, 1257–1260.
Merian, E., 1991. Metal and their compounds in the environment. VCH, Verlagsgesellschaft, Weinheim.
Michalke, B., Lusting, S., Schramel, P., 1997. Analysis for the stability of platinum-containing species in soil samples using capillary electrophoresis interfaced online with inductively-coupled plasma mass spectrometry. Electrophoresis 18, 196–201.
Mitrovic, B., Milacic, M., 2000. Speciation of aluminum in forest soil extracts by size exclusion chromatography with UV and ICP-AES detection and cation exchange fast protein liquid chromatography with ETAAS detection. Sci. Total Environ. 258, 183–194.
Mohammed, J., Hayat, S., 1996. Planktonic productivity of river water as a bioindicator of freshwater contamination by metals. Proc. Pak. Conf. Zool. 16, 283–298.
Montes-Bayon, M., Yanes, E.G., Ponce, D., Jayasimhulu, K., Stalcup, A., Shann, J., Caruso, J.A., 2002. Initial studies of selenium speciation in brassica juncea by LC with ICP-MS and ES-MS detection: An approach for phytoremediation studies. Anal. Chem. 74, 107–113.
Morris, A.W., 1971. Trace metal variations in sea water of the Menai straits caused by a bloom of phaeocystis. Nature 233, 427–428.
Mulyani, I., Levina A., Lay, P.A., 2004. Biomimetic oxidation of chromium(III), does the antidiabetic activity of chromium(iii) involve carcinogenic chromium(VI)? Ang. Chem. Int. Ed. 43, 4504–4507.
Murozumi, M., Chow, T.J., Patterson, C.C., 1969. Chemical concentrations of pollutant lead aerosols, terrestrial dusts, and sea salts in Greenland and Antarctic snow strata. Geochim. Cosmochim. Acta 33, 1247–1294.
Nasu, T., Kan, R., 1988. Determination of phosphate, arsenate and arsenite in natural water by flotation-spectrophotometry and extraction-indirect atomic absorption spectrometry using Malachite Green as an ion-pair reagent. Analyst 113, 1683–1686.
Neff, J.M., 1997. Ecotoxicology of arsenic in marine environment. Environ.Toxicol. Chem. 16, 917–927.
Nerin, C., Pons, B., Zufiaurre, R., 1995. Extraction, clean-up and determination of ionic alkyllead compounds in road dust by gas chromatography–mass spectrometry. Analyst 120, 751–754.
Ng, A., Patterson, C.C., 1981. Natural concentrations of lead in ancient Arctic and Antarctic ice. Geochim. Cosmochim. Acta 45, 2109–2121.
Nielen, M.W.F., Frei, R.W., Brinkmann, U.A.Th., 1998. Selective Sample Handling and Detection in High Performance Liquid Chromatography. Elsevier, Amsterdam.
Palanivelu, K., Balasubramaniam, N., Ramakrishna, T.V., 1992. A chemical enhancement method for the spectrophotometric determination of trace amounts of arsenic. Talanta 39, 555–561.
Pantsar-Kallio, M., Manninen, P.K.G., 1996. Speciation of chromium in wastewaters by coupled column ion chromatography-inductively coupled plasma mass spectrometry. J. Chromatogr. A 750, 89–95.
Pantsar-Kallio, M., Manninen, P.K.G., 1997. Simultaneous determination of toxic arsenic and chromium species in water samples by ion chromatography-inductively coupled plasma mass spectrometry. J. Chromatogr. A 779, 139–146.
Papadopoulou, C., Kanias, G.D., Elli, M.K., 1978. Zinc content in otoliths of mackerel from the Aegean. Mar. Pollut. Bull. 9, 106–108.
Penrose, W.R., 1974. Arsenic in the marine and aquatic environments. Analysis, occurrence, and significance. CRC Crit. Rev. Environ. Control 4, 465–482.

Pereiro, I.R., Diaz, A.C., 2002. Speciation of mercury, tin, and lead compounds by gas chromatography with microwave-induced plasma and atomic-emission detection (GC-MIP-AED). Anal. Bioanal. Chem. 372, 74–90.

Pershagen, G., 1983. In: Fowler, B.A. (Ed.), The Epidemiology of Human Arsenic Exposure. Elsevier, Amsterdam.

Pleßow, A., Heinrichs, H., 2000. Speciation of trace elements in acidic pore waters from waste rock dumps by ultrafiltration and ion exchange combined with ICPMS and ICPOES. Aqua. Geochem. 6, 347–366.

Ponce de Leon, C., Montes-Bayon, M., Caruso, J.A., 2002. Elemental speciation by chromatographic separation with inductively coupled plasma mass spectrometry detection. J. Chromatogr. A 974, 1–21.

Pons, B., Carrera, A., Nerin, C., 1998. Speciation of ionic alkyllead compounds in human urine by gas chromatography-mass spectrometry after butylation through a Grignard reaction. J. Chromatogr. B 716, 139–145.

Prebha, S., Selvapathy, P., 1997. Heavy metal pollution in Indian rivers. Ind. J. Environ. Pollut. 17, 641–649.

Pretorius, V., Hopkins, B.J., Schieke, J.D., 1974. Electroosmosis. New concept for high speed liquid chromatography. J. Chromatogr. 99, 23–30.

Pretty, J.R., Blubaugh, E.A., Caruso, J.A., 1993. Determination of arsenic(III) and selenium(IV) using an on-line anodic stripping voltammetry flow cell with detection by inductively coupled plasma atomic emission spectrometry and inductively coupled plasma mass spectrometry. Anal. Chem. 65, 3396–3406.

Pretty, J.R., Blubaugh, E.A., Evans, E.H., Caruso, J.A., Davidson, T.M., 1992. Determination of copper and cadmium using an online anodic stripping voltammetry flow cell with detection by inductively coupled plasma mass spectrometry. J. Anal. Atom. Spectrom. 7, 1131–1137.

Pretty, J.R., Evans, E.H., Blubaugh, E.A., Shen, W.L., Caruso, J.A., Davidson, T.M., 1990. Minimization of sample matrix effects and signal enhancement for trace analytes using anodic stripping voltammetry with detection by inductively coupled plasma atomic emission spectrometry and inductively coupled plasma mass spectrometry. J. Anal. Atom. Spectrom. 5, 437–443.

Prohaska, T., Pfeffer, M., Tulipan, M., Stingeder, G., Mentler, A., Wenzel, W.W., 1999. Speciation of arsenic of liquid and gaseous emissions from soil in a microcosmos experiment by liquid and gas chromatography with inductively coupled plasma mass spectrometer (ICP-MS) detection. Fresen. J. Anal. Chem. 364, 467–470.

Quevauviller, Ph., 1996. Improvement of quality control of speciation analysis using hyphenated techniques: A decade of progress within the European Community. J. Chromatogr. A 750, 25–33.

Rajeshwar, K., Ibanez, J.G., 1997. Environmental Electrochemistry, Fundamentals and Application is Pollution Abatement. Academic Press, New York.

Regan, F.B., Meaney, M.P., Lunte, S.M., 1994. Determination of metal ions by capillary electrophoresis using on-column complexation with 4-(2-pyridylazo)resorcinol following trace enrichment by peak stacking. J. Chromatogr. B 657, 409–417.

Robarts, K., Starr, P., Patsalides, E., 1991. Metal determination and metal speciation by liquid chromatography: A review. Analyst 116, 1247–1274.

Rosenberg, E., 2003. The potential of organic (electrospray- and atmospheric pressure chemical ionization) mass spectrometric techniques coupled to liquid-phase separation for speciation analysis. J. Chromatogr. A 1000, 841–889.

Sakai, H., Kojima, Y., Saito, K., 1986. Distribution of heavy metals in water and sieved sediments in the Toyohira river. Water Res. 20, 559–567.

Sakai, K., Okazaki, M., Matsumoto, S., 1993. The chemical phase changes in heavy metals with drying and oxidation of the lake sediments. Water Res. 27, 1243–1251.
Salt, D.E., 1995. Phytoremediation: A novel strategy for the removal of toxic metals from the environment using plants. Biol. Technol. 13, 468–474.
Sarzanini, C., Mentasti, E., 1997. Determination and speciation of metals by liquid chromatography. J. Chromatogr. A 789, 301–321.
Sarzanini, C., Mentasti, E., 1999. Liquid chromatography: A tool for the analysis of metal species. J. Chromatogr. A 850, 213–228.
Schoenmakers, P.J., 1988. In: Smith, R.M. (Ed.), Packed Column SFC. The Royal Society of Chemistry, Chromatography Monographs, Cambridge.
Skei, J.M., Saunders, M., Price, N.B., 1976. Mercury in plankton from a polluted Norwegian Fjord. Mar. Pollut. Bull. 7, 34–36.
Stenner, R.D., Nickless, G., 1974. Absorption of cadmium, copper, and zinc by dog whelks in the Bristol Channel. Nature 247, 198–199.
Stenner, R.D., Nickless, G., 1975. Heavy metals in organisms of the Atlantic coast of S.W. Spain and Portugal. Mar. Pollut. Bull. 6, 89–92.
Stoeppler, M., 1992. Hazardous Metals in the Environment. Elsevier, Amsterdam.
Stugeron, R.E., Siu, K.W.M., Willie, S.N., Bermann, S., 1989. Quantification of arsenic species in a river water reference material for trace metals by graphite furnace atomic absorption spectrometric techniques. Analyst 114, 1393–1396.
Tack, F.M.G., Verloo, M.G., 1995. Chemical speciation and fractionation in soil and sediment heavy metal analysis: A review. Int. J. Environ. Anal. Chem. 59, 225–238.
Tamari, Y., Yamamoto, N., Tsuji, H., Kasuka, Y., 1989. Thorium co-precipitation method for spectrophotometric determination of arsenic(III) and arsenic(V) in groundwaters. Anal. Sci. 5, 481–484.
Taylor, M.J.C., van Staden, J.F., 1994. Spectrophotometric determination of vanadium(IV) and vanadium(V) in each other's presence. Rev. Anal. 119, 1263–1276.
Terabe, S., Otsuka, K., Ichikawa, K., Tsuchiya, A., Ando, T., 1984. Electrokinetic separations with micellar solutions and open-tubular capillaries. Anal. Chem. 56, 111–113.
Tessier, A., Turner, D.R. (Eds.), 1996. Metal Speciation and Bioavailabilty in Aquatic Systems. Wiley, New York.
Timerbaev, A.R., 2002. Recent advances and trends in capillary electrophoresis of inorganic ions. Elerctrophoresis 23, 3884–3906.
Timerbaev, A.R., Dabek-Zlotorzynska, E., van den Hoop, M.A.G.T., 1999. Inorganic environmental analysis by capillary electrophoresis. Analyst 124, 811–826.
Tripathi, R.M., Mahapatra, S., Raghunath, R., Kumar, A.V., Sadasivan, S., 2002. Daily intake of aluminium by adult population of Mumbai, India. Sci.Total Environ. 299, 73–77.
Tsuda, T., Nomura, K., Nagakawa, G., 1982. Open-tubular microcapillary liquid chromatography with electroosmosis flow using a UV detector. J. Chromatogr. 248, 241–247.
van Esch, G.J., Kroes, R., 1969. The induction of renal tumours by feeding basic lead acetate to mice and hamsters. Br. J. Cancer 23, 765–771.
van Esch, G.J., van Genderen, H., Vivik, H.H., 1962. The induction of renal tumours by feeding of basic lead acetate to rats. Br. J. Cancer 16, 289–297.
Wang, Q., Huang, B., Guan, Z., Yang, L., Li, B., 2001. Speciation of rare earth elements in soil by sequential extraction then HPLC coupled with visible and ICP-MS detection. Fresen. J. Anal. Chem. 370, 1041–1047.
Whang, K.S., Whang, C.W., 1997. Capillary electrophoresis of organotin compounds with indirect UV detection. Electrophoresis 18, 241–246.
Weast, R.C. (Ed.), 1973. Handbook of Chemistry and Physics. 54th ed. CRC Press, Cleveland.

Weber, G., Neumann, G., Romheld, V., 2002. Speciation of iron coordinated by phytosiderophores by use of HPLC with pulsed amperometric detection and AAS. Anal. Bioanal. Chem. 373, 767–771.

Williams, G., D'Silva, C., 1994. Field-based heavy metal analyser for the simultaneous determination of multiple cations on-site. Analyst 119, 2337–2342.

World Health Organisation, 1981. Environmental health criteria, 18, Arsenic. World Health Organisation, Geneva.

Wrobel, K., Gonzalez, E.B., Wrobel, K., Sanz-Medel, A., 1995. Aluminum and silicon speciation in human serum by ion-exchange high-performance liquid chromatography–electrothermal atomic absorption spectrometry and gel electrophoresis. Analyst 120, 809–816.

Yamamura, Y., Yamauchi, H., 1980. Arsenic metabolites in hair, blood and urine in workers exposed to arsenic trioxide. Industr. Health 18, 203–210.

Yang, H.J., Jiang, S.J., Yang, Y.J., Hwang, C.J., 1995. Speciation of tin by reversed phase liquid chromatography with inductively coupled plasma mass spectrometric detection. Anal. Chim. Acta 312, 141–148.

Yang, Q., Hartmann, C., Smeyers-Verbeke, J., Massart, D.L., 1995. Method development and validation for the determination of mineral elements in food and botanical materials by capillary electrophoresis. J. Chromatogr. A 717, 415–425.

Yin, X., Frech, W., Hoffmann, E., Lüdke, C., Skole, J., 1998. Mercury speciation by coupling cold vapour atomic absorption spectrometry with flow injection on-line pre-concentration and liquid chromatographic separation. Br. J. Cancer 361, 761–766.

Yin, X., Hoffmann, E., Ludke, C., 1996. Differential determination of arsenic(III) and total arsenic with L-cysteine as prereductant using a flow injection non-dispersive atomic absorption device. Fresen. J. Anal. Chem. 355, 324–326.

Zhang, J., Huang, W.W., 1993. Dissolved trace metals in the Huanghe: The most turbid large river in the world. Water Res. 27, 1–8.

Zhang, Y., Frankenberger, W.T. Jr., 2001. Speciation of selenium in plant water extracts by ion exchange chromatography-hydride generation atomic absorption spectrometry. Sci. Total Environ. 26, 39–47.

Zufiaurre, R., Pons, B., Nerin, C., 1997. Speciation of trimethyllead in rainwater by gas chromatography-mass spectrometry after ethylation with sodium tetraethylborate comparison with other alkylation methods. J. Chromatogr. A 779, 299–306.

Developments in Environmental Science, Volume 5
D. Sarkar, R. Datta and R. Hannigan (Editors)

ISSN: 1474-8177/DOI:10.1016/S1474-8177(07)05004-8

Chapter 4

International practice in high-level nuclear waste management

Syed E. Hasan

Abstract

Over the past five decades a huge quantity of high-level nuclear (radioactive) waste has accumulated worldwide in 30 countries. Despite the fact that high-level waste presents serious hazard to human health and the environment, not a single facility for disposal of high-level waste exits anywhere in the world. This paper presents a discussion of the utilization of nuclear energy on a global basis as well as the current practice of temporary storage and planned management of high-level waste. Progress made by Finland, Sweden, and the U.S.A. in finalizing permanent deep geological repositories is reviewed. Details of the Yucca Mountain Project, U.S.A., the Olkiluoto site in Finland, and Oskarshamn and Östhammar sites in Sweden are presented in the paper. Status of plans for permanent disposal of high-level waste by other countries is summarized. The recently proposed concept of shared regional/international deep geological repository is reviewed.

4.1. Introduction

Ever since the discovery of nuclear fission and development of technology to unleash the tremendous energy locked up in the nucleus of certain radioactive elements, many countries opted to use nuclear energy to meet their power needs. For about 60 years huge quantity of fissionable materials has been used for electric power generation and making atomic/nuclear bombs. One outcome of this technological innovation has been generation of a large quantity of very dangerous waste that requires careful handling, storage, transportation, and disposal.

According to the International Atomic Energy Agency (IAEA, 2007) as of April 18, 2007, 436 nuclear power plants were in operation worldwide, generating about 368.9 GWe of electric power (Figs. 4.1a, 4.1b). These power plants provide 16% of world electricity and produce 10,500

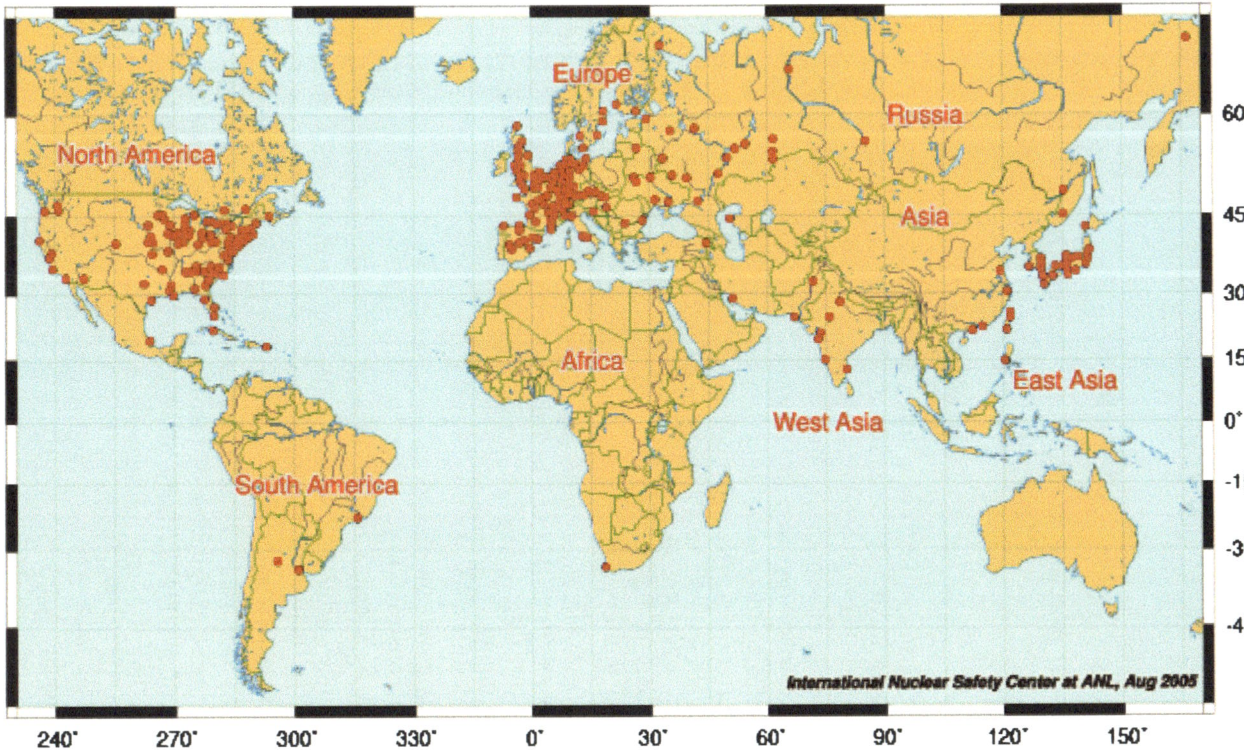

Figure 4.1a. Location of operating nuclear power plants in the world (Adapted from: International Nuclear Safety Center, Argonne National Laboratory, 2005).

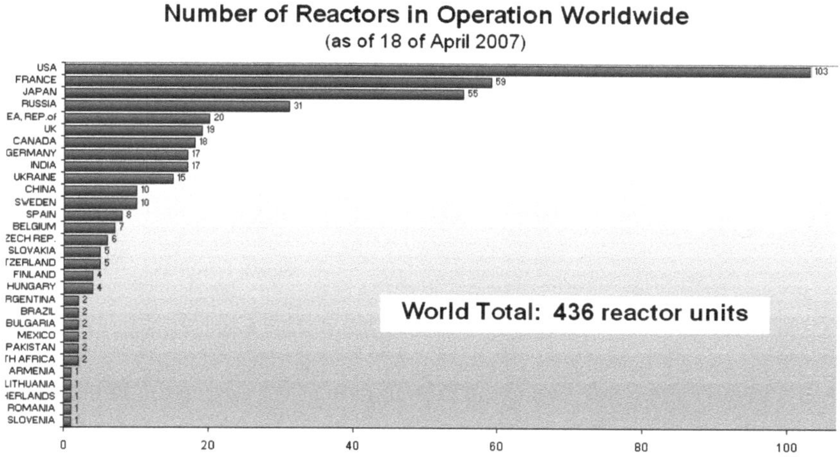

Figure 4.1b. Number of nuclear power plants worldwide as of April 18, 2007 (Adapted from: IAEA, 2007).

tonnes (metric) of high-level nuclear waste (HLW) every year. This paper presents an overview of the global status of management of HLW generated only in the civilian sector as information on use of nuclear materials in military programs of various countries is classified and is generally not available.

After the April 26, 1986 Chernobyl tragedy, interest in using nuclear energy for electricity plummeted and the nuclear industry remained in a state of dormancy for nearly 20 years. It is only in recent years that it has revived, and quite a few countries are moving ahead with construction of new nuclear power plants. As of November 2005, 30 new power plants were under construction in Argentina, Bulgaria, China, Finland, India, Iran, Japan, S. Korea, Pakistan, Romania, Russia, and Ukraine (IAEA, 2007) (Fig. 4.2) with 79% of them being located in Asian countries (IAEA, 2007). In addition to meeting the exploding energy need, another reason for the come back is the concern for global warming and adoption of the Kyoto Protocol by nearly 150 countries. Besides eliminating output of CO_2, nuclear power plants have several other advantages that might be the reason for its revival (Table 4.1).

A large volume of information on civilian use of nuclear energy is available in published books, articles, and reports, and also at websites maintained by many agencies, notably the IAEA (http://www.iaea.org/), World Nuclear Association (http://www.world-nuclear.org/), Nuclear Energy Institute (http://www.nei.org/), and the U.S. Department of Energy (http://www.eia.doe.gov/). Detailed technical information on the

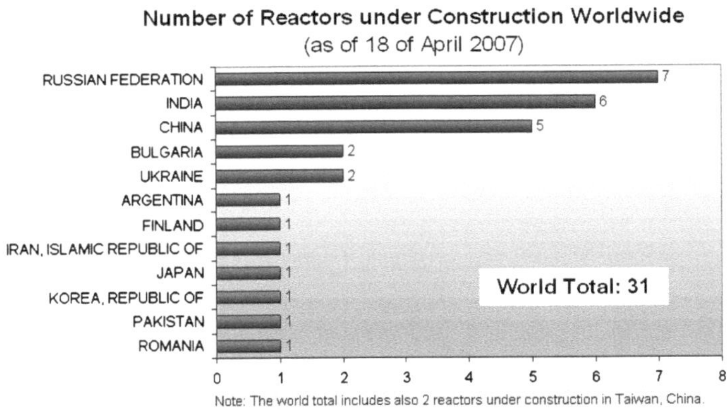

Figure 4.2. Nuclear power plants under construction, April 2007 (Adapted from: IAEA, 2007).

Table 4.1. Economic and environmental comparison for nuclear and coal-fired power plants (Data from International Atomic Energy Agency, World Energy Organization, and other)

Items	Nuclear power plant*	Coal-fired power plant*
Electricity cost per kWh (current)	\$0.02 (IAEA, 2004)	\$0.07–0.09
Projected cost after 2% rise in fuel cost	\$0.04–0.08 (200–400% increase)	\$0.42–0.63 (600–700% increase)
Capital cost	\$1500–2000 kW^{-1}	\$1250–1775 kW^{-1}
Construction time	Very long: several years to decade	<few years
Carbon emission	2–6 $g\,kW^{-1}\,h^{-1}$ (same as wind and solar)	20–60 $g\,kW^{-1}\,h^{-1}$
Waste generated/year	30 tonnes of HLW and 300 m^3 of LLW	300,000 tonnes of fly ash
Geographic distribution of fuel	Relatively widespread (U.S; Australia and Canada: > 50% of all uranium mined in 2003)	Concentrated in three countries (U.S; Russia and China account for 58% of world's reserve)
Land requirements	1–4 km^2	20–50 km^2 (solar); 50–150 km^2 (wind)
Environmental contamination	None—no harmful metal or gas released; compact waste sequestered at safe locations	Widespread—As, Hg, SO_x, NO_x, CO_2 disperse over large areas
Environmental issues	Does not contribute to acid rain or global warming	Major contributor to acid rain and global warming
Fuel transport	Concentrated energy form does not require extensive transportation system	Requires extensive transportation lines

*1000 MW installed capacity.

proposed deep geological repositories in the United States, Finland, and Sweden are available at http://www.ocrwm.doe.gov/, http://www.posiva.fi/englanti/, and http://www.skb.se/default____8563.aspx respectively.

The paper addresses the current status of management of high-level waste on a global basis. It does not include discussion of intermediate and low-level wastes because these are being managed in an acceptable manner in many countries and, unlike the HLW, their disposal is not a major concern.

4.2. Definition and classification

4.2.1. Definition

Nuclear or radioactive waste is a special kind of waste containing radioactive materials that emit harmful ionizing radiation along with attendant heat.

4.2.2. Classification

A number of classification schemes are in use internationally. Almost all of them essentially group the nuclear waste into three major categories

1. High-level nuclear wastes (HLW): includes solid, liquid, and gases and possess the following characteristics:
   - are highly radioactive;
   - come from nuclear power plants, defense research facilities, spent fuel reprocessing, and weapons production plants;
   - some have very long half-life, e.g. ^{239}Pu with a half-life of 24,000 years; and
   - continuously give off heat and radiation as they decay: therefore, need to be cooled and contained.

   HLW includes: (i) Spent nuclear fuel (SNF) which are fuel assemblies, about 4 m long and weighing about 410 kg, containing nearly 200 fuel rods that are removed from nuclear power plants every 4–6 years and (ii) wastes produced at military nuclear research facilities and weapons production plants. HLW from military facilities are mostly in liquid form and stored in tanks. Nuclear vessels generate HLW in the form of SNFs (Bodansky, 1996).
2. Transuranic waste (TRU): Comes mainly from reprocessing of spent nuclear fuel and fabrication of nuclear weapons. It consists of clothing, tools, rags, debris, soil and sludge, residues, and other disposable

Table 4.2. U.S. DOE's classification of nuclear waste (Modified from USNAS, 2003)

Category	Radioactivity level in liquid waste	Radiation dose on surface of solid waste package
HLW	$\geq 1\,Ci\,l^{-1}$	$\geq 1\,rad\,h^{-1}$
ILW	$1–10^{-5}\,Ci\,l^{-1}$	$1–>300\,mrad\,h^{-1}$
MLW	—	$300–30\,mrad\,h^{-1}$
LLW	$<10^{-5}\,Ci\,l^{-1}$	$<30\,mrad\,h^{-1}$

items (tarpaulins, plastic sheeting) contaminated with radioactive elements. These elements have atomic number greater than that of uranium, i.e. >92 and include Pu (94), Am (95), Cm (96), and Np (93)—man-made elements that are created during nuclear reactor operations. In the U.S.A. TRU waste is being disposed off at the WIPP site in New Mexico since March, 1999 in openings excavated in 600-m thick salt beds of the Permian age (250–290 Ma) Salado Formation at a depth of 654 m.

The IAEA subdivides TRU wastes into intermediate-level waste (ILW) and medium-level waste (MLW).

3. Low-level nuclear waste (LLW): Could be in solid, liquid, or gas form and is characterized by short half-life and low level of radiation.

The U.S. Department of Energy also categorizes nuclear waste into four categories based on its radioactivity levels and radiation dose (Table 4.2).

4.3. Scope of the problem

Nuclear fission was discovered in 1939 and the first successful chain reaction took place in a research reactor that was built in 1942 at the University of Chicago as part of the *Manhattan Project*. Electricity was first generated in a nuclear reactor at the National Reactor Testing Station in Idaho in December, 1951. The first commercial production of electricity from a nuclear reactor occurred in June, 1954 when the town of Obninsk near Moscow in the former USSR was connected to an electric power grid, providing 5000 kW of electricity to residences and businesses. For nearly 50 years since nuclear power has been used in the civilian sector, a huge quantity of high-level waste has accumulated. This waste, in the form of spent fuel assemblies, referred to as SNF rods, is much more radioactive than fresh fuel rods. The SNF assemblies have been sitting at temporary storage at nuclear power plant sites and military weapons facilities for half-a-century.

Spent nuclear fuel rods give off intense radiation and heat. They are kept under steel-lined concrete pools where water is force-circulated to allow it to cool off. With time, SNF radiation drops to levels much lower than newly discharged SNF. When the SNF undergoes sufficient cooling, it may be transported to a reprocessing plant to recover Pu or it may be stored at "dry storage" sites awaiting ultimate disposal in a geological repository.

In the U.S. alone, a total of 47,000 tonnes of high-level waste had accumulated by 2003. By 2005 the quantity was estimated to be over 52,000 tonnes; and by the year 2035 this amount will increase to an estimated 105,000 tonnes (USDOE, 2005). These figures do not include HLW placed in temporary storage at military weapons facilities.

Globally, as of December 2004, about 270,000 tonnes of HLW were in temporary storage, mainly at reactor sites across the world. The World Nuclear Association (2007) estimates that on an annual basis, 12,000 tonnes of HLW is generated, of which 3000 tonnes are reprocessed, leaving 9000 tonnes for disposal.

4.4. Management of high-level nuclear waste

High-level nuclear waste is being stored at temporary locations at 440 nuclear power plants (NPP) in 30 countries across the world. While eight countries reprocess the spent fuel rods to recover Pu and U to produce mixed-oxide (MOX) fuel, others have opted for direct disposal of the HLW (Table 4.3).

It is worth noting that as of May, 2007, not a single permanent disposal facility for HLW existed *anywhere* in the world. However, Sweden and Finland have received approvals from their governments to construct the deep geological repositories, planned to begin accepting waste in 2017 and 2020, respectively. In the U.S., the Yucca Mountain Site in Nevada was designated as the HLW repository by the President in 2002.

4.4.1. Geological and environmental considerations for locating HLW repository

A geological repository for safe disposal of HLW should, at a minimum, meet the following basic criteria:

- minimal or no possibility of exposure to the environment for 1000 s of years;
- stable geologic environment: no significant erosion, or possibility of major earthquakes, or volcanic activities; and
- fail-safe mechanisms for handling and transportation of HLW.

Table 4.3. Status of HLW management in various countries (Adapted from World Nuclear Association, April, 2007)

Country	Policy	Facilities and progress towards HLW repositories
Belgium	Reprocessing	Investigations underway at underground laboratory. Construction of repository to begin about 2035
Canada	Direct disposal	Underground repository laboratory established. Repository planned in crystalline rocks at 500–1000 m depth for use in 2025
China	Reprocessing	Feasibility studies planned for 2010–2020; repository operations after 2040
Finland	Direct disposal	Site near Olkiluoto selected for deep repository for spent fuel; construction commences 2010; waste emplacement planned for 2020
France	Reprocessing	Site selection studies underway for deep repository for commissioning 2020. Sites in clay and granite being studied
Germany	Reprocessing and direct disposal	Spent fuel storage at Ahaus and Gorleben. Sate dome at Gorleban studied from 1986 to 1999. Plans for HLW repository put on hold
India	Reprocessing	Research on deep geological disposal for HLW underway
Japan	Reprocessing	Nuclear waste Management Organization established in October 2002. Construction of repository in granite or sedimentary rock planned for the 2030 s
Russia	Reprocessing	Sites for final disposal in salt, granite, clay and/or basalt under early stage of investigation
South Korea	Direct disposal	Investigating deep HLW repository sites
Spain	Direct disposal	Preliminary design developed from studies at three candidate sites in clay, granite and salt. Final decision not until 2010
Sweden	Direct disposal	Site selection for repository in two volunteered locations in granitic rock completed; waste emplacement planned to begin in 2017
Switzerland	Reprocessing	Investigations being conducted at underground research laboratory for HLW repository. Also considering shared HLW repository
United Kingdom	Reprocessing	HLW is vitrified and stored at Sellafiled. Underground HLW repository planned; date for construction not determined
USA	Direct disposal	2002 decision to proceed with geological repository at Yucca Mountain; waste emplacement planned for 2010 (pending licensing)

The IAEA (1995) has developed a set of guidelines to ensure protection of human health and the environment. These include

- *Protection of human health and the environment.* HLW should be managed in such a way as to ensure an acceptable level of protection for human health and the environment.
- *Protection beyond national borders.* HLW management should include consideration of possible effects on human health and the environment beyond national borders.
- *Protection of future generations.* HLW should be managed in such a way that predicted impacts on the health of future generations will not be greater than relevant levels of impact that are acceptable today; no undue burdens should be placed on future generations.
- *National legal framework.* HLW should be managed within an appropriate national legal framework including clear allocation of responsibilities and provision for independent regulatory functions.
- *Control of radioactive waste generation.* Waste minimization techniques should be given due consideration to assure minimization of the volume of waste generated.
- *Safety of facilities.* The safety of facilities for radioactive waste management should be appropriately assured during their lifetime.

It is generally believed that allowing the SNF to sit at temporary storage sites under water is a good practice—provided its safety is assured—as it allows for significant reduction of radioactivity and heat content. Figure 4.3 shows that 1 tonne of SNF undergoes rapid loss of heat and radioactivity such that in about 30 years it loses 90% of its original radioactivity. Sweden allows the SNFs to be kept under water for up to 40 years after which the reduced levels of heat and radiation would make permanent disposal much safer. Most other countries, such as Belgium, Japan, U.K., and others, vitrify the solid HLW and plan to store their waste at temporary locations for periods ranging from 30 to 50 years.

4.4.2. Permanent disposal of HLW

Three countries, Finland, Sweden, and the United States, have been actively pursuing plans for construction of deep geological repositories for many years. In May 2001, Finland achieved the distinction of being the first country to approve plans to build a repository near Olkiluoto with construction beginning in 2011 and waste emplacement planned for 2020 (Posiva, 2003). Sweden also received a favorable referendum from the citizens in the municipalities of Oskarshamn and

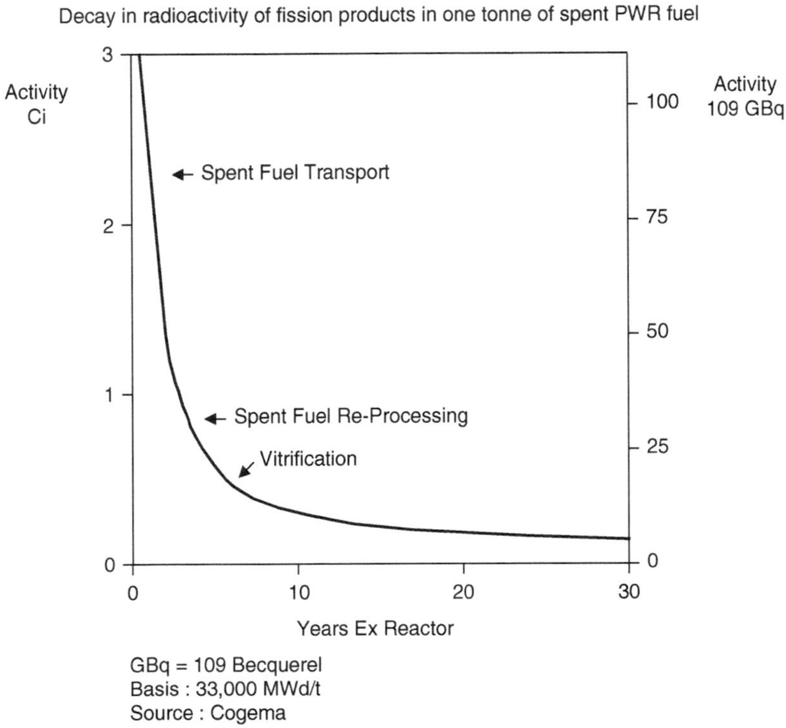

Figure 4.3. Reduction of radioactivity in SNF (Adapted from World Nuclear Association, April, 2007).

Östhammar where detailed site investigations have been conducted since 2001 with the goal of emplacing HLW in 2017 (Thegerström, 2004). In July 2002, the U.S. government approved a repository site at Yucca Mountain, but the license for its construction has not been submitted as of this writing. Worldwide, 20 more repositories are planned by the year 2030 (IAEA, 2007). Detailed research and feasibility studies at underground locations are in progress in Belgium, Canada, Switzerland, and the U.K.

4.4.3. Rock types being considered

All three main classes of rocks, igneous, sedimentary and metamorphic, are being investigated for disposal of HLW. Table 4.4 is a summary of the various rock types along with their gross mechanical properties that are being investigated for underground geological repository.

Table 4.4. Rock types being investigated for HLW repository (IAEA, 2003)

Rock type	Country	Mechanical property/limitation
Granite	Canada, China, Finland, Sweden, Switzerland	High strength, no lining required/fractures may allow groundwater movement causing radionuclide transport by advection and/or diffusion
Claystone, shale	France, Switzerland, Belgium	Low to medium strength, lining required. High porosity, negligible permeability unless fractured, may develop high pore water pressure/Radionuclides transport by diffusion
Salt domes	Germany	Plastic/no pores; no lining required, self sealing fractures/no transport of radionuclides
Volcanic tuff	U.S.A	Medium strength above groundwater table: pores and fractures unsaturated. Light lining at some places/some radionuclide transport possible with percolating water

4.4.4. Regional HLW repository

Recently, the idea of shared repositories for disposal of HLW has been proposed that has received favorable reaction from IAEA and some European countries. The concept is that for small countries situated in a particular geographic region, such as eastern Europe, it would be more economical and expedient to agree upon constructing a geological repository in one of the countries in the region where each participating country can send its HLW. This concept is especially attractive to countries with unfavorable geologic conditions who will greatly benefit from such regional repository. McCombie and Boutellier (2004) have argued in favor of such international collaboration, emphasizing the following advantages:

- ensuring common standards for safety, handling, and disposal;
- affordability by economically disadvantaged countries and possibility of fair compensation if the international repository is located in such country;
- ethical and political benefit of shared responsibility; and
- better security from potential terrorist activities.

A strong argument in favor of a shared international repository in Europe, according to McCombie and Boutellier (2004), is that the U.S.A.—a single country with a far greater area than the combined area of HLW-generating countries in Europe—is having serious problems in finalizing its first HLW repository; the problems would be far more complex and time consuming for the 17 eastern and western European countries if each decides to build its own repository.

4.5. HLW disposal in various countries

This section presents an overview of the current status of efforts being made by various countries for disposal of HLW. Detailed information is included for the Yucca Mountain project and a summarized discussion is provided for Sweden and Finland. Status of HLW disposal in other countries is summarized at the end of this section.

4.5.1. U.S.A.

The United States has 104 operating nuclear power plants that provide 20% of its electricity. The proposed site for disposal of HLW is located at Yucca Mountain in Nye County in the State of Nevada, 100 miles NW of Las Vegas (Fig. 4.4a, 4.4b). After spending over 20 years in intense scientific and engineering investigations and $6.6 billion, President Bush, on July 23, 2002, approved the site for a deep geological repository. If the Nuclear Regulatory Commission grants a license to the U.S. Department

Figure 4.4a. Location map of the proposed Yucca Mountain repository site (Adapted from: USDOE, 2004).

Figure 4.4b. Photograph of the Yucca Mountain in Nevada (Adapted from: USDOE, 2004).

of Energy, which would construct and operate the repository, the Yucca Mountain site is expected to begin receiving HLW in 2010. About 2,725 tonnes of HLW will be placed in the repository each year. By 2041, the emplacement of 70,000 tonnes of HLW will be complete; site closure operations would begin in 2110, and the program would end in 2119.

- The Yucca Mountain site has the following favorable conditions (US-DOE, 2002):
- Land owned by the federal government; no one lives within 22 km of the site.
- The area has a very dry climate—receiving a combined average of about 19.1 cm of precipitation per year. Approximately 95% of this either runs off, evaporates, or is taken up by the desert vegetation, with about 1 cm of infiltration. The dry climate is an important factor because water is the primary way by which radionuclides could move from a repository.
- The groundwater table is very deep, about 610 m. If a repository is built at Yucca Mountain, it would be located about 366 m below the surface and nearly 234 m above the groundwater table. So, any water that does not run off or evaporate at the surface would have to move down nearly 366 m before reaching the repository and then another 308 m before it reached the groundwater table.
- Insignificant to very low rate of erosion, 0.1–0.5 cm per thousand years in the Yucca Mountain area.

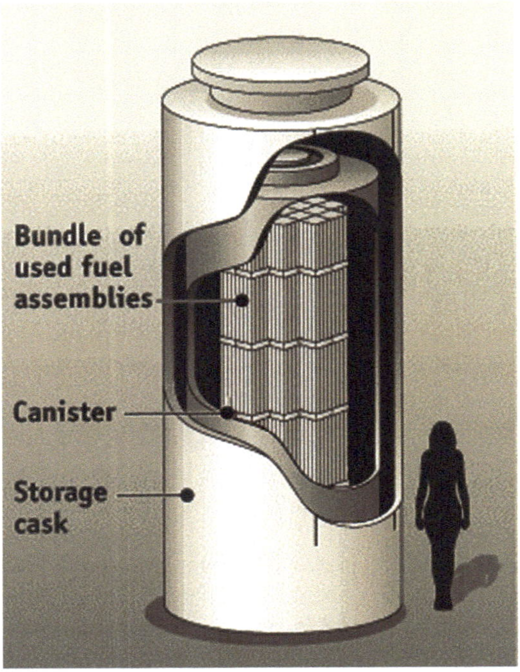

Figure 4.5. Multiple engineering barriers (Adapted from: USDOE, 2004).

The Yucca Mountain repository is being designed on the principle of multiple barriers. It is expected that a combination of suitable geologic medium and engineered barriers will prevent uncontrolled release of radionuclides into the environment. The first level of containment will come from the zircalloy metal cladding that encloses pellets in the fuel rod. Spent fuel assemblies will then be placed in a specially designed canister from which all water and air will be removed. After filling the canister with an inert gas, it will be welded shut (Fig. 4.5) and placed in a metal barrel ("cask") to prepare it for dry storage or transportation.

Figure 4.6 shows the essential features of the proposed repository at the Yucca Mountain. According to current plans, HLW will be shipped to the site by road and rail and stored at an above-ground facility. Waste emplacement is projected to begin in 2010 and the repository will become full in 2041. After 70 years of post-closure monitoring the site will be finally closed in 2119. The total cost of the project from its beginning in 1981 until 2119 is estimated at $49.3 billion in constant 2000 U.S. dollars (USDOE, 2002).

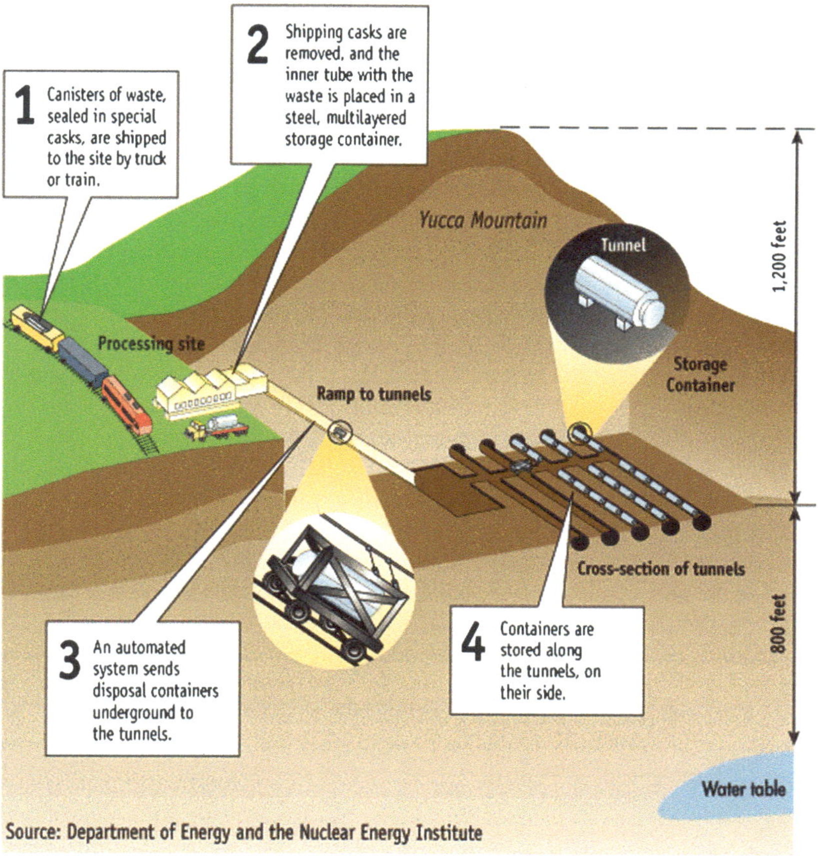

Figure 4.6. Layout of the proposed Yucca Mountain HLW repository (Adapted from: USDOE, 2004).

4.5.2. Sweden

Sweden currently has 11 operating nuclear power reactors that collectively meet 47% of the country's electric power need. Sweden's HLW management policy is based on the following three-step process (SKB, 2005a):

(1) The SNF rods are allowed to cool for one year at the nuclear power plant.
(2) The waste is then packaged and the casks are sent to Sweden's Central Interim Storage Facility, known as CLAB, to Oskarshamn in southeast

Sweden where pools are built in an underground rock cavern to shield the radiation. Thirty years of cooling will result in substantial lowering of temperature and 90% reduction of radioactivity (SKB, 2005b).

(3) Finally, the waste will be transferred to the deep geological repository in 2017.

Sweden is actively engaged in site investigations to build a deep geological repository that will isolate the HLW for 100,000 years. It also utilizes the concept of multiple engineered barriers. Spent fuel rods will be placed in a cast iron–copper canister. The canister will be entombed in a layer of bentonite before being deposited in the repository to be constructed in granitic rocks.

After preliminary site screening studies, conducted in eight municipalities between 1993 and 2000, two candidate sites were selected for further studies—in the municipalities of Östhammar and Oskarshamn. Detailed investigations commenced in 2002 and by July 2005, Forsmark in the former and Simpevarp and Laxemar in the latter were selected as potential repository locations.

The proposed repository will be located at a depth of 500 m (Fig. 4.7). The bedrock at potential sites is Precambrian-fractured granite, garnodiorite and diorite. Detailed investigations to characterize the hydrogeology of fractured granitic rocks, thermal effects from heat emanating from the SNF, along with data analyses and system modeling studies are underway. The contractor, Svensk Kärnbränslehantering AB (SKB) is planning to submit the permitting application in 2008 and expects to have the deep repository ready for waste emplacement in 2017.

4.5.3. Finland

Finland has four nuclear power plants that provide 33% of electric power and generate 70 tonnes of SNF each year. Finland's is a model case study in the sense that it illustrates the importance of thorough planning, public involvement, and political will of the government. Finland began planning for permanent disposal of HLW as early as the 1970s when its first two nuclear power plants were still under construction. Pursuant to its Nuclear Energy Act, the Finnish government in 1993 set the target date of 2010 for construction of the deep geological repository with the goal of opening it for waste emplacement by 2020. Finland aggressively moved ahead with site selection, detailed characterization, and environmental impact studies and selected a site for the deep geological repository on the island of Olkiluoto in Eurajoki municipality, in the southwestern part of the country (Fig. 4.8a). Selection of the Olkiluoto

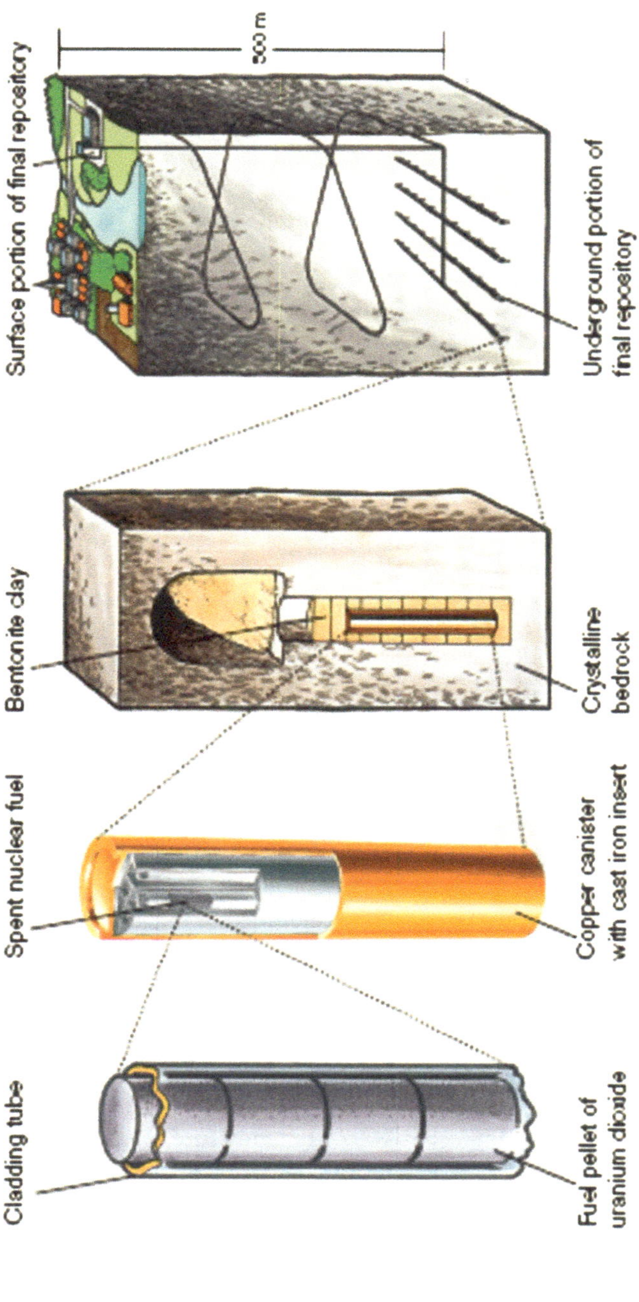

Figure 4.7. Multiple barrier approach and layout of the proposed HLW repository, Sweden (Adapted from: SKB, 2005a).

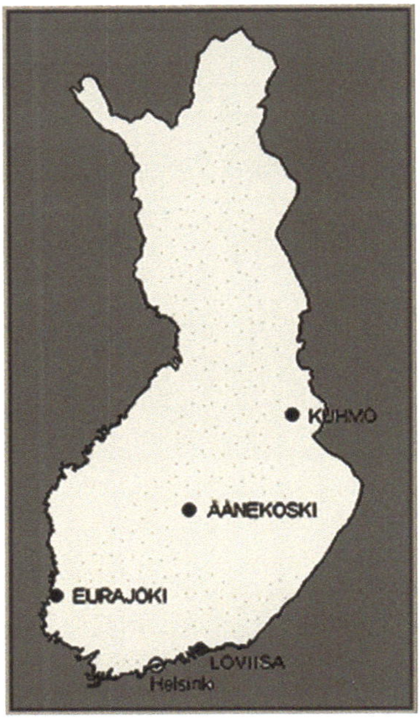

Figure 4.8a. Location map of the Olkiluoto site in the Eurajoki Municipality (Adapted from: Posiva. Oy, 2003).

site was based on: geological suitability, acceptance by local population, environmental considerations, and technical feasibility. Finland's example is instructive because it placed great importance on public acceptance and took steps to settle this critical issue in the very beginning. The public was involved in the decision-making process and the deep geological repository site was selected after the respective municipal council had voted in favor of the location. This is a most desirable approach as it saves later delays due to potential legal and political reasons.

The geological repository will be at 500 m depth in granitic rock (Fig. 4.8b) and is designed to receive 6,500 tonnes of SNF that will have accumulated by the 30–40 life of the existing nuclear power plants. Total cost, including construction, waste packaging, shipment, and final disposal of 6,500 tonnes of SNF is estimated at €845 million at December 2002 price level (1.14 billion at 2007 exchange rate).

Finland has collaborated with Sweden in the design of the copper canister for packaging spent fuel assemblies. Using the multiple barrier approach, fuel assemblies (after 30–40 years of cooling in pools) will be

Figure 4.8b. Conceptual layout of the proposed Olkiluoto repository, Finland (Adapted from: Posiva. Oy, 2003).

placed in a canister made of cast iron with a copper overpack layer, each holding between 9 and 12 fuel assemblies and weighing 16–21 tonnes. The canister will be transported to the repository site where it will be placed in the excavated rock, surrounded by bentonite before being backfilled by concrete or a similar material. These barriers, combined with 500-m deep disposal environment, would assure that the waste will remain isolated for a very long period of time and pose no danger to humans and/or the environment.

4.5.4. HLW disposal in other countries

The status of management of HLW in other countries ranges widely from very preliminary ideas of deep geological repositories to advanced research and testing being carried out in underground geological locations. Table 4.3 also includes a summary of the HLW management efforts in these countries.

4.6. Summary and conclusions

After two decades of inactivity in construction of new nuclear power plants, there is resurgence in nuclear industry and currently 31 nuclear power plants are under construction worldwide, mostly in Asia. With the passage of the Kyoto Treaty to combat greenhouse effect, many countries are leaning toward nuclear power plants to meet their future energy needs because of the distinct advantage of nuclear power plants in eliminating production of CO_2.

Ever since the coming-on-line of the first commercial nuclear power plant in Russia in 1954, a very large volume of extremely hazardous waste has accumulated in 30 countries where nuclear power plants and/or spent fuel reprocessing facilities are in operation. The concerning fact is that despite more than five decades of utilization of nuclear energy, there is not a single facility anywhere in the world for permanent disposal of HLW. Storage at temporary locations is not desirable from many considerations, including safety and security issues.

Three countries, Finland, Sweden, and the United States, have made impressive progress in their plans to construct deep geological repositories with planned dates for emplacement of HLW set at 2020, 2017, and 2010, respectively. Worldwide 20 more repositories are planned by the year 2030.

The IAEA, along with some other countries, is supporting the new concept of regional and international repositories that will be based on scientific, technical, economic, and security considerations. It is likely that some small European countries, sharing borders with each other within distances of 300–500 km, may seriously consider joining each other to construct a deep geological repository.

ACKNOWLEDGMENTS

The author is thankful to the series editor, Dr. Dibyendu Sarkar, for his invitation to contribute this paper. Dr. David A. Pickett, Southwest Research Institute, San Antonio, Texas, reviewed the paper and offered valuable suggestions. The author wishes to thank Dr. Pickett and an anonymous reviewer for their valuable suggestions. Word processing assistance from Rita Messina is gratefully acknowledged.

REFERENCES

Bodansky, D., 1996. Nuclear Energy: Principles, Practices, and Prospects. American Institute of Physics Press, Woodbury, NY, p. 123, 125.

International Atomic Energy Agency (IAEA), 1995. The Principles of Radioactive Waste Management, Safety Series 111-F, Vienna, p. 24 Available at: http://www-pub.iaea.org/MTCD/Publications/PDF/Pub989e_scr.pdf Accessed November 24, 2005.

International Atomic Energy Agency (IAEA), 2003. Scientific and Technical Basis for Geological Disposal of Radioactive Wastes. Technical Report Series No. 413, Vienna, p. 80.

International Atomic Energy Agency (IAEA), 2004. Nuclear Power's Changing Future (Press Release). Available at http://www.iaea.org/NewsCenter/PressReleases/2004/prn200405.html Accessed November 27, 2005.

International Atomic Energy Agency (IAEA), 2007. Nuclear Power Plant Information (PRIS Database). Available at: http://www.iaea.org/programmes/a2 Accessed May 13, 2007.

International Nuclear Safety Center, Argonne National Laboratory, 2005. Maps of Nuclear Power Reactor: World Map. Available at http://www.insc.anl.gov/pwrmaps/map/world_map.php. Accessed May 13, 2007.

McCombie, C., Boutellier C., 2004. Problems of an international repository for radioactive waste: political and legal aspects of international repositories. Available at: http://www.arius-world.org/pages/pdfs_pub/INLA-paper%2027-09-04.pdf Accessed July 7, 2005.

Posiva, O., 2003. Into Olkiluoto Bedrock—Final Disposal of Spent Nuclear Fuel in Finland. Available at: http://www.posiva.fi/esitteet/Bedrock_201102.pdf Accessed November 16, 2005.

SKB, 2005a. Encapsulation Plant and Final Repository for Spent Nuclear Fuel. Available at: http://www.skb.se/upload/publications/pdf/Ebso_InakapslTryckE.pdf Accessed November 27, 2005.

SKB, 2005b. Activities 2004. Available at: http://www.skb.se/upload/publications/pdf/SKB_ENGVerksam_2004.11.5.pdf Accessed July 15, 2005.

Thegerström, C., 2004. Down to Earth and Below: Sweden's Plans for Nuclear Waste. IAEA Bull 46(1), 36–38.

U.S. Department of Energy (USDOE), 2004. Information Digest, U.S. Nuclear Regulatory Commission, NUREG-1350, vol. 16, p. 151 Available at: http://www.nrc.gov/waste/hlw-disposal/photo-loc.html Accessed July 15, 2005.

U.S. Department of Energy (USDOE), Office of Civilian Radioactive Waste Management, Yucca Mountain Project: Now Much Nuclear Waste is in the United States? Available at: http://www.ocrwm.doe.gov/ymp/about/howmuch.shtml Accessed July 1, 2005.

U.S. Department of Energy (USDOE), Office of Civilian Radioactive Waste Management, 2002. Science and Engineering Report: Technical Information Supporting Site Recommendation Consideration, Revision 1. Report No. DOE/RW-0539-1, unpaginated. Available at: http://www.ocrwm.doe.gov/documents/ser_b/index.htm Accessed November 27, 2005.

U.S. National Academy of Sciences (USNAS), 2003. End Points for Spent Nuclear Fuel and High-Level Radioactive Waste in Russia and the United States. National Academy Press, Washington, D.C, p. 137.

World Energy Organization, Proved Coal Reserve at 1999 end. Available at: www.worldenery.org/wec-geis/publications/reports/ser/coal/coal.asp Accessed November 1, 2005.

World Nuclear Association, April, 2007. Waste Management in the Nuclear Fuel Cycle. Available at: http://www.world-nuclear.org/info/inf04.htm Accessed May 13, 2007.

Developments in Environmental Science, Volume 5
D. Sarkar, R. Datta and R. Hannigan (Editors)

ISSN: 1474-8177/DOI:10.1016/S1474-8177(07)05005-X

Chapter 5
Phytoremediation of some heavy metals by agronomic crops

Honey Aggarwal and Dinesh Goyal

Abstract

Remediation of metal contaminated soil faces challenges, as unlike organic contaminants metals cannot be degraded. Phytoremediation is an emerging technology that employs the use of higher plants for the cleanup of contaminated environments. Plant-based approach for remediation is cost effective and takes advantage of their remarkable ability to grow and uptake heavy metals from severely polluted sites. Among several crops, *Brassica juncea*, *Cucurbita pepo*, *Amaranthus sp.*, *Raphanus sativus oleiformis*, *Zea mays*, etc., has been demonstrated for remediation of cadmium (Cd), nickel (Ni), lead (Pb), zinc (Zn), selenium (Se), chromium (Cr), copper (Cu) and uranium (U). This review compiles the relevant information on possible uses of agronomic crops for metal extraction as an alternative for the removal of heavy metals excess from soil.

5.1. Introduction

A large portion of biosphere is contaminated by heavy metals as a result of human activities. Conventional solutions such as disposal of contaminated soil in landfills, which relies heavily on 'dig and dump' or encapsulation, neither of which takes into consideration the issue of decontamination of the soil, account for a large proportion of the remediation operations at present (Pulford and Watson, 2003). Remediation techniques like immobilization or extraction by physico-chemical techniques are expensive and are often appropriate only for small areas. On the other hand, for the decontamination of polluted sites phytoremediation seems attractive as it offers site restoration, partial decontamination, maintenance of biological activity and biorecovery of metals (Baker et al., 1991, 1994; Gardea-Torresdey et al., 2005a). It has attracted attention for its low cost of implementation and environmental benefits.

Phytoremediation is the use of green plants as well as associated rhizospheric microbes to remove pollutants from the environment or to render them harmless (Raskin et al., 1994; Salt et al., 1998). It is an innovative biological technique, which can be applied for the cleanup of severely polluted soil (Salt et al., 1995, 1998; Chaney et al., 1997; Chaudhry et al., 1998; Meagher, 2000; Lasat, 2002; Prasad and Freitas, 2003; Pulford and Watson, 2003; Alkorta et al., 2004; Gardea-Torresdey et al., 2005a).

Phytoremediation of metals occurs due to following activities:

- Phytoextraction: Pollutant accumulating plants remove metals from the soil and concentrate them in the harvestable part of plants (Kumar et al., 1995a).
- Rhizofiltration: Removal of contaminants from aqueous waste streams by absorption onto plant roots (Dushenkov et al., 1995).
- Phytostabilization: Immobilization or prevention of migration of contaminants in the environment by plant exudates, leading to the reduction in the mobility and bioavailability of the contaminants (Vangronsveld et al., 1995).
- Phytovolatilization: Volatilization of pollutants into the atmosphere via plants (Burken and Schnoor, 1997; Banuelos et al., 1997a).
- Phytomining: Capability of plants to extract large amount of metals from soils that can be exploited to recover metals of high economic value from ore deposits and other soils (Glass, 2000; Gardea-Torresdey et al., 2005a).

Governments worldwide are establishing research and decontamination programs to use this potential. Environment Canada has developed a database (PHYTOREM) of 775 plants with capabilities to accumulate or hyperaccumulate one or several of 19 key metallic elements (McIntyre, 2003). Phytoremediation is easier to manage because it is an autotrophic system of large biomass that requires little nutrient input (Evans and Furlong, 2003). Moreover, plants offer protection against water and wind erosion and in preventing spreading of contaminants (Pulford and Watson, 2003).

The success of phytoremediation as an environmental-cleanup technology depends upon number of factors including the extent of soil contamination, metal availability for uptake into roots and plant ability to intercept, absorb and accumulate metals in shoots (Ernst, 1996). This review aims to give a broad overview of the various phytoremediation technologies and their potential role in clean up of pollutants especially heavy metals.

5.2. Phytoextraction

The concept of using plants to cleanup contaminated environment is very old and cannot be traced to any particular source (Blaylock and Huang, 2000). About 300 years ago, plants were proposed for use in the treatment of wastewater (Hartman, 1975). At the end of the 19th century, *Thlaspi caerulescens* and *Viola calaminaria* were the first plant species documented to accumulate high levels of metals in leaves (Baumann, 1885). Plants able to accumulate upto 1% Ni in shoots were identified (Minguzzi and Vergnano, 1948). The idea of using plants to extract metals from contaminated soil was reintroduced and developed by Utsunamyia (1980) and Chaney (1983). At present, there are two basic strategies of metal phytoextraction; natural phytoextraction and induced or chemically assisted phytoextraction (Salt et al., 1998).

Phytoextraction is based on the use of pollutant accumulating plants for the removal of metals and organics from soil by concentrating them in the harvestable parts (Brooks, 1977; Salt et al., 1998; Reeves and Baker, 2000; Vassilev et al., 2004). Metal phytoextraction is not as extreme as conventional metal removal methods but still involves considerable alterations in the environment, which includes elimination of the existing vegetation cover and application of fertilizers and various soil amendments to increase metal availability to plants (Tichy et al., 1997; Ebbs et al., 1998; Huang et al., 1998; Pawlowska et al., 2000). Plants for phytoextraction, i.e., metal removal from soil, should have the following characteristics: (i) tolerant to high levels of the metal, (ii) accumulate reasonably high levels of the metal, (iii) rapid growth rate, (iv) produce reasonably high biomass in the field and (v) profuse root system (Garbisu et al., 2002).

The roots of the established hyperaccumulators absorb metal elements from the soil and translocate them to the above ground shoots, where they get accumulated in high concentration (Prasad and Freitas, 2003). It is also based on high biomass-producing plants used together with chemical agents enhancing both metal solubility and uptake by plants (Blaylock et al., 1997; Huang et al., 1997). After sufficient plant growth and metal accumulation, the above ground portions of the plants are harvested and removed, which results in the permanent removal of the metals from site. After removal of heavy metals from the soil, the disposal of the contaminated material is an environmental concern. Some researchers suggested incineration (Kumar et al., 1995a), while others suggested about the extraction of valuable metals from the metal-rich ash (Comis, 1996; Cunningham and Ow, 1996).

In general, the reports assessing metal phytoextraction potential are based on pot experiments, when compared to field experiments higher metal-extracting values have been observed, which might be due to higher solubility of metals and the effect of amendments aiming at mobilizing the metals etc. (Vassilev et al., 2004). The selection of heavy metal tolerant species is a reliable tool to achieve success in phytoremediation. One hundred and sixty three plant taxa belonging to 45 families have been found to be metal tolerant and are capable of growing on elevated concentration of toxic metals (Prasad and Freitas, 2003). The use of metal tolerant species and their metal indication and accumulation is a function of immense use for biogeochemical prospecting (Brooks, 1983; Badri and Springuel, 1994; Mcinnes et al., 1996). Chemically assisted phytoextraction is based on the use of non-accumulator plants with metal accumulation levels far below those of hyperaccumulators but with high biomass potential (Vassilev et al., 2004). This is aimed to overcome the main limitations of natural phytoextraction, i.e., a limited number of suitable hyperaccumulators for some important metal pollutants such as Pb (Huang et al., 1997; Lasat, 2000), as well as their low biomass production. Lopez et al. (2005) has shown the combined effects of ethylenediaminetetraacetic acid (EDTA) and phytohormone indole-3-acetic acid (IAA) on Pb uptake by *Medicago sativa.* After 10 days of treatment with 0.2 mM Pb and different combinations of EDTA and IAA, the quantification of Pb content in plant tissues using an inductively coupled plasma optical emission spectrometer (ICP/OES) showed that Pb accumulation in leaves was increased by about 2800% in plants cultivated with Pb/EDTA as compared to plants exposed to Pb alone respectively, where it was increased by only 600%.

The first field trial on natural phytoextraction was conducted in 1991–1992 in sewage sludge treated plot at Woburn, England (McGrath et al., 1993). The maximum Zn uptake was found in *Thlaspi caerulescens* accumulating 2000–8000 mg Zn/kg. It was also shown to accumulate 1000–4000 mg kg^{-1} Cd (Brooks, 1998). *Sedum alfredii* Hance has been identified as a new Zn and Cd hyperaccumulating plant species (Yang et al., 2004). Zn concentration in its shoot can reach over 20 g kg^{-1} when grown at 80 mg Zn/l in nutrient solution without showing any toxic symptoms (Yang et al., 2002). Cd concentration in leaves and stem of *S. alfredii* increased with increasing Cd supply levels, and reached a maximum upto 9000—65,000 mg kg^{-1} dry weight (Yang et al., 2004). Seedlings of *Sesbania drummondii* can hyperaccumulate Pb in a controlled hydroponic environment (Sahi et al., 2002). Green bean, beetroot, green cabbage, lettuce, onion, pea, radish, spinach, tomato, turnip, watercress and Iberis were grown on a thallium-contaminated soil and *Iberis intermedia* was found to hyperaccumulate thallium (LaCoste et al., 2001).

Maximum thallium levels ranged from nearly 400 mg kg^{-1} in *Iberis intermedia* down to just over 1 mg kg^{-1} in green bean. High Cu concentration has been found in *Betula* roots (Kozlov et al., 1995; Maurice and Lagerkvist, 2000) as well as in *Salix* roots (Punshon and Dickinson, 1997). The Pb, Zn and Cd phytoextraction potential of 14 different plants was assessed in a chelate induced phytoextraction experiment. EDTA and EDDS (ethylenediamine disuccinic acid) were used as chelates. The addition of these chelates increased the proportion of phytoavailable Pb, Zn and Cd in the soil and also their uptake by tested plants upto 48 times by *Sinapsis alba*, 4.6 times by *Raphanus sativus oleiformis* and 3.3 times by *Amaranthus sp.*, respectively. *Cannabis sativa* hyperaccumulated 105 times Pb, 2.3 times Zn and 31.7 times Cd higher than control (Kos et al., 2003). Blaylock et al. (1997) and Huang et al. (1997) found that application of EDTA at 2 g kg^{-1}soil resulted in a concentration of more than 1.5% Pb in the shoots of *Brassica juncea* and 1% in maize and pea plants. It was also shown that other chelators such as EGTA (ethylene–bis [oxy ethyletrinitrilo] tetracetic acid) had high affinity to Cd, while DTPA (diethylene triamine pentacetic acid) showed high affinity to Zn (Blaylock et al., 1997). Turgut et al. (2004) used two cultivars of *Helianthus annuus* in conjunction with EDTA and CA (citric acid) as chelators for phytoremediation of heavy metal contaminated soil. Results showed that EDTA at a concentration of 0.1 g kg^{-1} yielded the best results for both cultivars achieving a total metal (Cd, Cr and Ni) uptake of approximately 0.73 mg.

Restrictions apply, however, to both use of complexing agents and artificial soil acidification. It was found that EDTA and EDTA-heavy metal complexes are toxic for some plants and high doses of EDTA inhibited the development of arbuscular mycorrhiza (Dirilgen, 1998; Creman et al., 2001; Geebelen et al., 2002). *In-situ* application of chelating agents can cause groundwater pollution by uncontrolled metal dissolution and leaching (Creman et al., 2001; Sun et al., 2001).

The ideal plant species for metal phytoextraction has to be highly productive in biomass and to uptake and translocate a significant part of metals to its shoots (Vassilev et al., 2004). Some tree species mainly *Salix* sps. and *Populus* sps. exhibit these traits and are already used in phytoremediation programmes and for Cd phytoextraction from lightly polluted agricultural soils (Landberg and Greger, 1994). In fact, *Salix* sps. are not metal hyperaccumulators, but it was shown that among different clones there are some species which are hyperaccumulators of Cd and Zn. About 150 clones of *Salix* sps. have been screened for uptake, transport of metals to shoots and tolerance to Cd, Zn and Cu (Landberg and

Greger, 1994; Landberg and Greger, 1996). Some reports by Grant and Bailey (1997), Yankov et al. (2000) and Yankov and Tashin (2001), Griga et al. (2003) showed that crops for fibre or oil production could be used for profitable crop production accompanied by phytoextraction of metal from polluted soils. Brake fern (*Pteris vittata*) was reported to tolerate soils contaminated with arsenic as much as 1500 ppm and its fronds were found to concentrate up to 22,630 ppm in six weeks (Ma et al., 2001). Wang et al. (2002) showed that among maize, wheat, rapeseed, field pea and fodder vetch grown on a multiple metal contaminated site, maize had the highest concentrations of Mn, Zn and Cd, rapeseed had the highest concentration of Cr, Cu concentration was highest in fodder vetch and Pb was highest in wheat, but heavy metal accumulation was there in grain of wheat. The results suggested that on sites with multiple metal contaminations, growing maize and rapeseed would be safer than growing wheat or legumes and could be used for phytoremediation of lightly contaminated soils. A study by Peralta-Videa and Gardea-Torresdey (2002) showed that Alfalfa (*Medicago sativa*) plants were able to take up metals from a mixture of Cd(II), Cu(II), Ni(II) and Zn(II) in soils. The maximum relative uptakes in comparison to control were found to be 36 times for Ni, 23 times for Cd, 12 times for Zn and 6 times for Cu. In another field trial the possibility of using *Beta vulgaris*, *Cichorium intybus*, *Cucurbita pepo*, *Phaseolus vulgaris*, *Hordeum vulgare*, *Brassica oleracea*, *Zea mays*, *Medicago sativa* and *Pastinaca sativa* for removing heavy metals from soil was investigated. Results showed that the most effective crop for phytoextraction of Cd, Cu, Ni, Pb and Zn was *Cucurbita pepo* and for Cr – *Zea mays* (Ciura et al., 2005).

The main advantage of this technology is its lower cost as compared to other known remediation technologies (EPA, 2000; Glass, 2000). The possible metal recycling should provide further economic advantage as the ash of some hyperaccumulators consists of significant amount of metals (20–40% Zn for *T. caerulescens*) and there is no need to pay for safe disposal (Chaney et al., 1997), it can work without further disturbing the site, which is of great importance for its public acceptance (Vassilev et al., 2004). Besides all its advantages, it has certain limitations also, the major limitation is that it can only be used for low to moderately contaminated soils and it is applicable only to surface soils with few exceptions and is a time consuming process (Robinson et al., 1998; Blaylock and Huang, 2000; Vassilev et al., 2004). It is still at developmental stage, small companies and universities are driving much of its innovation and research, whereas, environmental engineering firms are involved in application projects.

5.3. Rhizofiltration

Rhizofiltration refers to the use of plant roots to sorb, concentrate and precipitate metal contaminants from surface or groundwater (Dushenkov et al., 1995). It is effective in cases where wetlands can be created and all of the contaminated water is allowed to come into contact with roots. Contaminants should be those that sorb strongly to roots such as hydroponic organics, Pb, Cr(III), etc. Rhizofiltration is primarily used to remediate extracted groundwater, surface water and wastewater with low concentration of contaminant and it can be very cost effective (Salt et al., 1995). It can be used for Pb, Cd, Cu, Ni, Zn and Cr, which are primarily retained within the roots (USEPA, 2000). An ideal plant for rhizofiltration should have rapidly growing roots with the ability to remove toxic metals from solution over extended periods of time. Dushenkov et al. (1995) demonstrated that many 'large root' species have the ability to absorb and precipitate heavy metals from solution, such as sunflower (*Helianthus sp.*), rye (*Elymus sp.*), corn (*Zea mays*) and Indian mustard (*Brassica juncea*). The mechanisms of toxic metal removal by plant roots are not necessarily similar for different metals.

In rhizofiltration, plants used have the ability to remove upto 60% of their dry weights as toxic metals (Salt et al., 1995). The process involves raising plants hydroponically and transplanting them into metal polluted waters where plants absorb and concentrate the metals in their roots and shoots (Dushenkov et al., 1995; Salt et al., 1995; Flathman and Lanza, 1998; Zhu et al., 1999). Root exudates and changes in the pH of rhizosphere soil may also cause metals to precipitate onto root surfaces. Plants for rhizofiltration should be able to accumulate and tolerate significant amounts of the target metals alongwith easy handling, low maintenance cost and a minimum of secondary waste requiring disposal (Dushenkov and Kapulnik, 2000). Several aquatic species have the ability to remove heavy metals from water, including water hyacinth (*Eichhornia crassipes*; Kay et al., 1984; Zhu et al., 1999), pennywort (*Hydrocotyle umbellate* L.; Dierberg et al., 1987) and duckweed (*Lemna minor* L.; Mo et al., 1989). As a result of their small, slow growing roots these plants are not much efficient at metal removal and have limited potential for rhizofiltration (Dushenkov et al., 1995). Sunflower (*Helianthus sp.*), Indian mustard (*Brassica juncea*), tobacco (*Nicotiana tabacum*), rye (*Elymus sp.*), spinach (*Spinacea oleracea*) and corn (*Zea mays*) have been studied for their ability to remove Pb from water, with sunflower having the greatest ability. In a study, after only 1 h of treatment sunflower reduced Pb concentration significantly (Raskin and Ensley, 2000). Indian mustard had bioaccumulation coefficient of 563 for Pb and had proven to be effective in removing a wide concentration range of Pb (4–500 mg l^{-1}) (Raskin and Ensley, 2000; USEPA, 2000). The hairy root cultures of *Brassica napus* were used to study the removal of 2,4-dichlorophenol (2,4-DCP), a common contaminant in industrial effluents that is highly toxic for human and aquatic life. High removal efficiencies (93–95%) were observed in a broad pH range (pH 3–9), reaching 98–99% in the pH range 4–8 (Agostini et al., 2003). The hairy root systems of *Brassica napus* and *Chenopodium amaranticolor* were used for removal of uranium from the solution of concentration up to 5000 μM. The results indicated that the hairy roots could remove uranium from the aqueous solution within a short period of incubation. *Brassica juncea* could take up 20–23% of uranium on dry weight basis and *Chenopodium amaranticolor* showed a slow and steady uptake of uranium upto 13% (Eapen et al., 2003). In pot culture experiment, palak (*Beta vulgaris* L.), coriander (*Coriandrum sativum*) and fenugreek (*Trigonella foenum-graecum*) grown in soil contaminated with heavy metals (Fe, Zn, Cu, Pb, Cr, Ni, Cd) showed elevated levels of heavy metals in roots as compared to shoot in the order of Fe > ZnNi > Cr > Cd > CuPb (Jaj, 2005).

The advantages associated with the rhizofiltration are the ability to use both terrestrial and aquatic plants for either *in-situ* or *ex-situ* applications, applicability to many problem metals, ability to treat high volumes, lesser need for toxic chemicals, reduced volume of secondary waste, possibility of recycling and likelihood of public acceptance (Dushenkov et al., 1995; Kumar et al., 1995b; Raskin and Ensley, 2000). The disadvantages include the constant need to maintain pH, plants may first need to be grown in a nursery and then transplantation and maintenance of successful hydroponic systems in the field would require expertise of qualified personnel, periodic harvesting and plant disposal and a good understanding of the chemical speciation (USEPA, 2000).

5.4. Phytostabilization

Phytostabilization is defined as immobilization of a contaminant in soil through absorption and accumulation by roots, adsorption onto roots or precipitation within the root zone of plants. It is also known as phytorestoration. It is a plant-based remediation technology that stabilizes wastes and prevents exposure pathways via wind and water erosion; provides hydraulic control, which suppresses the vertical migration of contaminants into groundwater and physically and chemically immobilizes contaminants by root sorption and by chemical fixation with various soil amendments (Cunningham et al., 1995; Salt et al., 1995; Flathman and Lanza, 1998; Berti and Cunningham, 2000; Schnoor, 2000). Phytostabilization of organic pollutants that are foreign to living organisms is based on sequestration processes, such as humification (McCutcheon and Schnoor, 2003). Phyostabilization involves root zone microbial and chemical processes. It can change metal solubility and mobility or impact the dissociation of organic compounds. The plant affected soil environment can convert metals from a soluble to an insoluble oxidation state (Salt et al., 1995). Phytostabilization can occur through sorption, precipitation, complexation or metal valence reduction (EPA, 1997). In a vegetative cap for phytostabilization, a combination of trees and grasses may be used. Fast transpiring trees such as poplar maintain an upward flow to prevent downward leaching, while grasses prevent wind erosion and lateral runoff with their dense root systems (Bennet et al., 2003; McCutcheon and Schnoor, 2003).

Sometimes there is no immediate effort to clean metal polluted sites, either because the responsible companies no longer exist or because the sites are of no high priority on a remediation agenda (Berti and Cunningham, 2000). Metal tolerant plants are required for heavy metal contaminated soils. Plants chosen for phytostabilization should be poor translocaters of metal contaminants to above ground plant tissues that could be consumed by humans and animals. The plants selected should be easy to establish and care for, grow quickly, have dense canopies and root systems and be tolerant to metal contaminants and other site conditions that may limit plant growth. *Brassica juncea* has been shown to reduce leaching of metals from soil by over 98% (Raskin et al., 1994). Arsenic might be taken up by plants because it is similar to the plant nutrient phosphate, although poplar leaves in a field study did not accumulate amounts of As. Poplars were grown in soil containing an average of $1250\,mg\,kg^{-1}$ As (Pierzynski et al., 2002). Grasses were used to stabilize mine wastes containing Cu (Salt et al., 1995). The research of Smith and Bradshaw (1992) led to the development of two cultivars of *Agrostis tenuis* and one *Festuca rubra*, which are now commercially available for phytostabilization of Pb, Zn and Cu contaminated soils. Soil amendments can also be used to stabilize metals in soils. Amendments should be selected that will maximize the growth of vegetation, which then also helps to phytostabilize the soil (Berti and Cunningham, 2000).

It has advantages over other soil remediation practices in that it has a lower cost and is less disruptive than other more vigorous soil remedial technologies (EPA, 2000), easy to implement (Schnoor, 2000). Revegetation offers aesthetic value and enhances ecosystem restoration. The lack of appreciable metals in shoot tissue also eliminates the necessity of treating harvested shoot residue as hazardous waste (Flathman and

Lanza, 1998). The main disadvantage is that the contaminants remain in place. The vegetation and soil may require long-term maintenance to prevent release of the contaminants and future leaching (EPA, 2000). Highly contaminated sites are not suitable for phytostabilization, because plant growth and survival is not possible (Berti and Cunningham, 2000).

5.5. Phytovolatilization

Phytovolatilization is a multimedia transfer of contaminants from water or soil to the atmosphere. Volatile organic compounds are taken up and transpired with water vapour or diffused out of the leaves, stems and roots (McCutcheon and Schnoor, 2003). Plants normally transpire water as vapour, but volatile compounds can be transpired as well. It occurs via diffusion from tree's xylem through its bark or leaves. In recent years, researchers have searched for plants that are capable of absorbing elemental forms of these metals from the soil, biologically converting them to gaseous species within the plant and releasing them into the atmosphere (Prasad and Freitas, 2003). Phytovolatilization has mainly been applied to groundwater, but it can be applied to soil, sediments and sludge. Plants may serve as effective pump and treat systems for mobile contaminants. Selenium was taken up and transpired from groundwater with concentrations of 100–500 $\mu g\,l^{-1}$ Se (Banuelos et al., 1997a) and at soil concentrations of 40 $mg\,l^{-1}$ (Banuelos et al., 1997b). Indian mustard (*Brassica juncea*) and canola (*Brassica napus*) have been used in the phytovolatilization of Se (Banuelos et al., 1997b). Lewis et al. (1966) first showed that both selenium non-accumulator and accumulator species volatilize selenium. Selenium (as selenate) was converted to less toxic dimethyl selenite gas and released to the atmosphere (Adler, 1996). Kenaf (*Hibiscus cannabinus* L. cv. Indian) and tall fescue (*Festuca arundinacea Schreb* cv. Alta) have also been used to take up Se, but to a lesser degree than canola (Banuelos et al., 1997b). Some aquatic plants, such as cattail (*Typha latifolia* L.) are also good for Se phytoremediation (Pilon-Smits et al., 1999). Plants that volatilize Hg were genetically modified, for example, *Arabidopsis thaliana* L. and *Nicotiana tabaccum* L. with bacterial organomercurial lyase (*MerB*) and mercuric reductase (*MerA*) genes (Heaton et al., 1998; Rugh et al., 1998). These plants absorb elemental Hg(II) and methylmercury (MeHg) from the soil and release volatile Hg(0) from the leaves into the atmosphere (Heaton et al., 1998). Heaton et al. (1998) suggested that the addition of Hg(0) into the atmosphere would not contribute significantly to the atmospheric pool.

Advantage of phytovolatilization is that the contaminants could be transformed to less-toxic forms, such as elemental mercury and dimethyl selenite gas, without need for plant harvesting and disposal. Contaminants or metabolites released to the atmosphere might be subjected to more effective or rapid natural degradation processes such as photodegradation. The only disadvantage is that the contaminant or a hazardous metabolite might be released into the atmosphere or get accumulated in vegetation (Newman et al., 1997).

5.6. Phytomining

Phytomining is defined as the production of crop of metal by growing high biomass plants that accumulate high metal concentration. Some of these plants are natural metal hyperaccumulators, while in others the property can be induced (Brooks, 1998). Plant ability to extract large amount of metals from soils can be exploited to recover metals of economic value from ore deposits and other soils (Glass, 2000). Studies have shown that using certain plants to extract metals from soil is commercially feasible. It may be a green alternative to destructive, opencast mining practices. It could be used to mine metals that are uneconomic by conventional methods (Brooks, 1998).

Research conducted at the University of Texas in El Paso, USA, has shown that gold accumulated by alfalfa plants and stored in leaf and stem biomass can be present as discrete nanoparticles of pure metal (Gardea-Torresdey et al., 2002). This discovery was made after alfalfa sprouts germinated on gold chloride enriched agar ($320\,mg\,kg^{-1}$ Au) were analysed using X-ray absorption spectroscopy (XAS) and transmission electron microscopy (TEM). The gold recovery rate in plants has been observed for many tested artificial and real substrates (Anderson et al., 1998). Preliminary studies by Gardea-Torresdey et al. (2005b) have shown that desert willow (*Chilopsis linearis*) is able to extract gold from a gold enriched medium. It grew well in the presence of NH_4SCN lower than $1 \times 10^{-4}\,mol\,l^{-1}$. After two weeks the effect on plant growth and gold content was determined by inductively coupled plasma-optical emission spectroscopy (ICP-OES). The results showed that addition of $10^{-4}\,mol\,l^{-1}$ NH_4SCN increased the concentration of Au by about 595, 396 and 467% in roots, stems and leaves, respectively. It was found that shoot elongation was not affected by thiocyanate. XAS studies showed that the oxidation state of gold was Au(0) and Au nanoparticles were formed inside the plants and *C. linearis* is a potential plant for Au phytomining (Gardea-Torresdey et al., 2005b). XAS has provided important information on the coordination chemistry of metals and toxic element interactions with phytoremediation systems. It has provided information in terms of the coordination environment of metals absorbed by plants and the bioreduction of metals within phytoremediation systems. It also provided information about the production of Au and Ag nanoparticles by metal interaction with the plants on phytomining systems (Gardea-Torresdey et al., 2005a). Broadhurst et al. (2004) had developed commercially viable phytomining technologies employing *Alyssum bertolonii* Ni-hyperaccumulator species, where the majority of Ni was stored either in the leaf epidermal cell vacuoles, or in the basal portions of the numerous stellate trichomes. The metal concentration in the trichome basal compartment was the highest ever reported for healthy vascular plant tissue, approximately 15–20% dry weight (Broadhurst et al., 2004).

Keeling et al. (2003) investigated the potential of South African high biomass Ni hyperaccumulator *Berkheya coddii* to phytoextract Co and Ni from artificial metalliferous media. Plant accumulation of both metals from single element substrates indicated that the bioaccumulation coefficient increased as total metal concentration increased. An important step in the phytomining operations is the recovery of metals from harvested plant material. In this work, a laboratory scale horizontal tube furnace was used to generate Ni-enriched bio-ore from the dried biomass of Ni-hyperaccumulator plants. Prior to furnace treatment, hairy roots of *Alyssum bertolonii* were exposed to Ni in liquid medium to give biomass having Ni concentration of 1.9–7.7% dry weight; whole plants of *Berkheya coddii* biomass was about 15 times greater than in *A. bertolonii.* After furnace treatment at 1200°C under air, Ni-bearing residues with crystalline morphology and containing upto 82% Ni were generated from *A. bertolonii* (Boominathan et al., 2004).

5.7. Conclusion

In recent years, phytoremediation has emerged as a promising low cost and environment friendly remediation technology, especially relevant for moderately polluted areas which is gradually approaching commercialization. Plants and associated microorganisms can remediate heavy metal contaminated soil via phytoextraction, rhizofiltration, phytostabilization, phytovolatilization and phytomining. Phytoremediation works effectively for a wide range of inorganic pollutants, the underlying biological processes are still largely unknown in most of the cases (Pilon-Smits, 2005). It is advantageous to use commonly cultivated agronomic crops such as

Table 5.1. Agronomic crops for phytoremediation of heavy metals from soil

Plant species	Heavy metals	Method/activity	Reference
Agrostis tenuis	Pb, Zn, Cu	Phytostabilization	Smith and Bradshaw (1992)
Alyssum bertolonii	Ni	Phytomining	Boominathan et al. (2004); Broadhurst et al. (2004)
Alyssum murale	Ni, Zn	Phytoextraction	Whiting et al. (2003)
Amaranthus sp.	Cd, Pb, Zn,	Phytoextraction	Kos et al. (2003)
Arabidopsis thaliana L.	Hg	Rhizofiltration	Heaton et al. (1998)
Avena sativa	Zn	Phytoextraction	Ebbs et al. (1997)
Berkheya coddii	Ni, Co	Phytoextraction	Keeling et al. (2003)
Brassica juncea	Cd, Cu	Phytoextraction	Kumar et al. (1995a)
Brassica juncea	Cd, Pb, Ni, Zn, Cr, Cu	Rhizofiltration	Dushenkov et al. (1995)
Brassica juncea	Ni, Cu	Phytoextraction	Ebbs et al. (1997)
Brassica juncea	Se	Phytovolatilization	Banuelos et al. (1997b)
Brassica napus	Zn	Phytoextraction	Ebbs et al. (1997)
Brassica napus	Cd	Phytoextraction	Wang et al. (2002)
Brassica napus	Se	Phytovolatilization	Banuelos et al. (1997b)
Brassica napus	U	Rhizofiltration	Eapen et al. (2003)
Cannabis sativa	Cd, Ni, Cr	Phytoextraction	Citterio et al. (2003)
Chenopodium amaranticolor	U	Rhizofiltration	Eapen et al. (2003)
Chilopsis linearis	Au	Phytomining	Gardea-Torresdey et al. (2005b)
Cucurbita pepo	Pb, Ni, Zn, Cu	Phytoextraction	Ciura et al. (2005)
Festuca rubra	Pb, Zn, Cu	Phytostabilisation	Smith and Bradshaw (1992)
Helianthus annuus	Cd, Ni, Cr	Phytoextraction	Turgut et al. (2004)
Iberis intermedia	Th	Phytoextraction	LaCoste et al. (2001)
Medicago sativa	Cd, Ni, Zn, Cu	Phytoextraction	Peralta-Videa and Gardea-Torresdey (2002)
Medicago sativa	Pb	Phytoextraction	Lopez et al. (2005)
Medicago sativa	Au	Phytomining	Gardea-Torresdey et al. (2002)
Nicotiana tabacum L.	Hg	Rhizofiltration	Heaton et al. (1998)
Pteris vittata	As	Phytoextraction	Ma et al. (2001)
Raphanus sativus oleiformis	Cd, Pb, Zn,	Phytoextraction	Kos et al. (2003)
Sedum alfredii	Cd, Zn	Phytoextraction	Yang et al. (2004)
Sesbania drummondii	Pb	Phytoextraction	Sahi et al. (2002)
Sinapis alba	Cd, Pb, Zn,	Phytoextraction	Kos et al. (2003)
Thlaspi caerulescens	Zn	Phytoextraction	Baker and Walker (1990); Brooks (1998)
Thlaspi caerulescens	Ni	Phytoextraction	Baker et al. (1991)
Thlaspi caerulescens	Cd	Phytoextraction	Brown et al. (1995)
Typha latifolia L.	Se	Phytovolatilization	Pilon-Smits et al. (1999)
Zea mays	Cd, Zn	Phytoextraction	Wang et al. (2002)
Zea mays	Cr	Phytoextraction	Ciura et al. (2005)

Brassica juncea, *Medicago sativa*, *Cucurbita pepo*, *Brassica napus* and *Zea mays* (Table 5.1), which are reported to accumulate many toxic metals. Agronomic crops can achieve dual purpose of treating contaminated sites alongwith farm produce and demonstration of application of phytoremediation. Identification and selection of more efficient plant varieties for phytoremediation, optimized doses of soil amendments and agronomic practices and co-ordination with developments in environmental and agricultural engineering can increase the efficiency of the phytoremediation (Salt et al., 1998). More fundamental research is required to better exploit the metabolic diversity of the plants for phytoremediation and to understand the complex interactions between metals, soil, plant roots and rhizosphere microorganisms. However major disadvantage of this technology is in requiring longer period for remediation. Well-designed and well-documented demonstration experiments and field trials are needed for public acceptance and policy makers to promote the use of agronomic crops for phytoremediation of toxic heavy metals.

REFERENCES

Adler, T., 1996. Botanical cleanup crews. Sci. News 150, 42–43.

Agostini, E., Coniglio, M.S., Milrad, S.R., Tigier, H.A., Giulietti, A.M., 2003. Phytoremediation of 2,4-dichlorophenol by *Brassica napus* hairy root cultures. Biotechnol. Appl. Biochem. 37, 139–144.

Alkorta, I., Allica, J.H., Becerril, J.M., Amezaga, I., Albizu, I., Onaindia, M., Garbisu, C., 2004. Chelate-enhanced phytoremediation of soils polluted with heavy metals. Rev. Environ. Sci. Biotechnol. 3, 55–70.

Anderson, C.W.N., Brooks, R.R., Stewart, R.B., Simcock, R., 1998. Harvesting a crop of gold in plants. Nature 395, 553–554.

Badri, M., Springuel, I., 1994. Biogeochemical prospecting in the south-easter desert of Egypt. J. Arid Environ. 28, 257–264.

Baker, A.J.M., McGrath, S.P., Sidoli, C.M.D., Reeves, R.D., 1994. The possibility of in situ heavy metal decontamination of polluted soils using crops of metal accumulating plants. Resour. Conserv. Recycl. 11, 41–49.

Baker, A.J.M., Reeves, R.D., McGrath, S.P., 1991. In situ decontamination of heavy metal polluted soils using crops of metal-accumulating plants – a feasibility study. In: Hinchee, R.E., Olfenbuttel, R.F. (Eds.), In situ Bioreclamation. Butterworth-Heinemann, Boston, pp. 600–605.

Baker, A.J.M., Walker, P.L., 1990. Ecophysiology of metal uptake by tolerant plants. In: Shaw, A.J. (Ed.), Heavy Metal Tolerance in Plants: Evolutionary Aspects. CRC Press, Boca-Raton, FL, pp. 155–177.

Banuelos, G.S., Ajwa, H.A., Mackey, L.L., Wu, C., Cook, S., Akohoue, S., 1997b. Evaluation of different plant species used for phytoremediation of high soil selenium. Environ. Qual. 26, 639–646.

Banuelos, G. S., Ajwa, H. A., Terry, N., Downey, S., 1997a. Phytoremediation of Selenium laden effluent. Fourth International in situ and on-site bioremediation symposium, April 28–May 1, New Orleans, LA. 3:303.

Baumann, A., 1885. Das Verhalten von Zinksatzen gegen Pflanzen und im Boden. Landwirtsch. Vers.-Statn. 31, 1–53.

Bennet, L.E., Burkhead, J.L., Hale, K.L., Terry, N., Pilon, M., Pilon-Smits, E.A.H., 2003. Analysis of transgenic Indian mustard plants for phytoremediation of metal-contaminated mine tailings. J. Environ. Qual. 32, 432–440.

Berti, W.R., Cunningham, S.D., 2000. Phytostabilization of metals. In: Raskin, I., Ensley, B.D. (Eds.), Phytoremediation of Toxic Metals: Using Plants to Clean up the Environment. Wiley, New York, pp. 71–88.

Blaylock, M.J., Huang, J.W., 2000. Phytoextraction of metals. In: Raskin, I., Ensley, B.D. (Eds.), Phytoremediation of Toxic Metals: Using Plants to Clean up the Environment. Wiley, New York, pp. 53–70.

Blaylock, M.J., Salt, D.E., Dushenkov, S., Zakharova, O., Gussman, C., Kapulnik, Y., Ensley, B.D., Raskin, I., 1997. Enhanced accumulation of Pb in Indian mustard by soil-applied chelating agents. Environ. Sci. Technol. 31, 860–865.

Boominathan, R., Saha-Chaudhury, N.M., Sahajwalla, V., Doran, P.M., 2004. Production of nickel bio-ore from hyperaccumulator plant biomass: Applications in phytomining. Biotechnol. Bioeng. 86, 243–250.

Broadhurst, C.L., Chaney, R.L., Angle, J.S., Maugel, T.K., Erbe, E.F., Murphy, C.A., 2004. Simultaneous hyperaccumulation of nickel, manganese, and calcium in *Alyssum* leaf trichomes. Environ. Sci. Technol. 38, 5797–5802.

Brooks, R.R., 1977. Copper and cobalt uptake by *Haumaniastrum* species. Plant Soil 48, 541–544.

Brooks, R.R., 1983. Biological methods of prospecting for minerals. Wiley-Interscience, New York, p. 313.

Brooks, R.R., 1998. Plants that hyperaccumulate heavy metals. Wallingford CAB International, p. 380.

Brown, S.L., Chaney, R.L., Angle, J.S., Baker, A.J.M., 1995. Zinc and cadmium uptake by hyperaccumulator *Thlaspi caerulescens* and metal tolerant *Silene vulgaris* grown on sludge amended soils. Environ. Sci. Technol. 29, 1581–1585.

Burken, J.G., Schnoor, J.L., 1997. Uptake and metabolism of atrazine by poplar trees. Environ. Sci. Technol. 31, 1399–1406.

Chaney, R.L., 1983. Plant uptake of inorganic waste constitutes. In: Parr, J.F., Marsh, P.B., Kla, J.M. (Eds.), Land treatment of hazardous wastes. Noyes Data Corp., Park Ridge, NJ, pp. 50–76.

Chaney, R.L., Malik, K.M., Li, Y.M., Brown, S.L., Brewer, E.P., Angle, J.S., 1997. Phytoremediation of soil metals. Curr. Opin. Biotechnol. 8, 279–284.

Chaudhry, T.M., Hayes, W.J., Khan, A.G., Khoo, C.S., 1998. Phytoremediation-Focusing on accumulator plants that remediate metal-contaminated soils. Aust. J. Ecotoxicol. 4, 37–51.

Citterio, S., Santagostino, A., Fumagalli, P., Prato, N., Ranalli, P., Sgorbati, S., 2003. Heavy metal tolerance and accumulation of Cd, Cr and Ni by *Cannabis sativa* L. Plant Soil 256, 243–252.

Ciura, J., Poniedzialek, M., Sekara, A., Jedrszczyk, E., 2005. The possibility of using crops as metal phytoremediants. Pol. J. Environ. Stud. 14, 17–22.

Comis, D., 1996. Green remediation: Using plants to clean the soil. J. Soil Water Conserv. 51, 184–187.

Creman, H., Velikonja-Bolta, S., Vodnik, D., Kos, B., Lestan, D., 2001. EDTA enhanced heavy metal phytoextraction: Metal accumulation, leaching and toxicity. Plant Soil 235, 105–114.

Cunningham, S.D., Berti, W.R., Huang, J.W., 1995. Phytoremediation of contaminated soils. Trends Biotechnol. 13, 393–397.

Cunningham, S.D., Ow, D.W., 1996. Promises and prospects of phytoremediation. Plant Physiol. 110, 715–719.

Dierberg, F.E., Debusk, T.A., Goulet, J.R., 1987. Removal of copper and lead using a thin-film technique. In: Reddy, K.B., Smith, W.H. (Eds.), Aquatic Plants for Water Treatment and Resource Recovery. Mangolia Publisher, pp. 497–504.

Dirilgen, N., 1998. Effects of pH and chelator EDTA on Cr toxicity and accumulation in *Lemna minor*. Chemosphere 37, 771–783.

Dushenkov, S., Kapulnik, Y., 2000. Phytofiltration of metals. In: Raskin, I., Ensley, B.D. (Eds.), Phytoremediation of Toxic Metals: Using Plants to Clean up the Environment. Wiley, New York, pp. 89–106.

Dushenkov, V., Kumar, P.B.A.N., Motto, H., Raskin, I., 1995. Rhizofiltration: The use of plants to remove heavy metals from aqueous streams. Environ. Sci. Technol. 29, 1239–1245.

Eapen, S., Suseelan, K.N., Tivarekar, S., Kotwal, S.A., Mitra, R., 2003. Potential for rhizofiltration of uranium using hairy root cultures of *Brassica juncea* and *Chenopodium amaranticolor*. Environ. Res. 91, 127–133.

Ebbs, S., Brady, D., Kochian, L., 1998. Role of uranium speciation in the uptake and translocation of uranium by plants. J. Exp. Bot. 49, 1183–1190.

Ebbs, S.D., Lasat, M.M., Brandy, D.J., Cornish, J., Gordon, R., Kochian, L.V., 1997. Heavy metals in the environment: Phytoextraction of cadmium and zinc from a contaminated soil. J. Environ. Qual. 26, 1424–1430.

EPA, 1997. Electrokinetic laboratory and field processes applicable to radioactive and hazardous mixed waste in soil and groundwater. EPA/402/R-97/006. Environmental Protection Agency, Washington, DC.

EPA, 2000. Introduction to phytoremediation. EPA/600/R-99/107. Environmental Protection Agency, Washington, DC.

Ernst, W.H.O., 1996. Bioavailability of heavy metals and decontamination of soils by plants. Appl. Geochem. 11, 163–167.

Evans, G.M., Furlong, J.C., 2003. Environmental Biotechnology: Theory and applications. Wiley, West Sussex, UK.

Flathman, P.E., Lanza, G.R., 1998. Phytoremediation: Current views on an emerging green technology. J. Soil Contam. 7, 415–432.

Garbisu, C., Hernandez-Allica, J., Barrutia, O., Alkorta, I., Becerril, J.M., 2002. Phytoremediation: A technology using green plants to remove contaminants from polluted areas. Rev. Environ. Health 17, 75–90.

Gardea-Torresdey, J.L., Parsons, J.G., Gomez, E., Peralta-Videa, J., Troiani, H.E., Santiago, P., Jose, Y.M., 2002. Formation and growth of Au nanoparticles inside live alfalfa plants. Nano Letters 2, 397–401.

Gardea-Torresdey, J.L., Peralta-Videa, J.R., Rosa, G., Parsons, J.G., 2005a. Phytoremediation of heavy metals and study of the metal coordination by X-ray absorption spectroscopy. Coordin. Chem. Rev. 249, 1797–1810.

Gardea-Torresdey, J.L., Rodriguez, E., Parsons, J.G., Peralta-Videa, J.R., Meitzner, G., Cruz-Jimenez, G., 2005b. Use of ICP and XAS to determine the enhancement of gold phytoextraction by *Chilopsis linearis* using thiocyanate as a complexing agent. Anal. Bioanal. Chem. 382, 347–352.

Geebelen, W., Vangronsveld, J., Adriano, D.C., Van Poucke, L.C., Clijsters, H., 2002. Effects of Pb-EDTA and EDTA on oxidative stress reactions and mineral uptake in *Phaseolus vulgaris*. Plant Physiol. 115, 377–384.

Glass, D.J., 2000. Economic potential of phytoremediation. In: Raskin, I., Ensley, B.D. (Eds.), Phytoremediation of Toxic Metals: Using Plants to Clean up the Environment. Wiley, New York, pp. 15–32.

Grant, C.A., Bailey, L.D., 1997. Effects of phosphorus and zinc fertilizer management on cadmium accumulation in flaxseed. J. Sci. Food Agric. 73, 307–314.

Griga, M., Bjelkova, M., Tejklova, E., 2003. Potential of flax (*Linum usitatissimum*) for heavy metal extraction and industrial processing of contaminated biomass – a review. In: Mench, M., Mocquot, B. (Eds.), Risk Assessment and Sustainable Land Management Using Plants in Trace Element-Contaminated Soils. COST Action 837, 4th WG2 Workshop, Bordeaux 2002. INRA Centre Bordeaux-Aquitaine, Villenave d, Ornon, pp. 173–179.

Hartman, W. J., Jr., 1975. An evaluation of land treatment of municipal wastewater and physical siting of facility installations. US Department of Army, Washington, DC.

Heaton, A.C.P., Rugh, C.L., Wang, N., Meagher, R.B., 1998. Phytoremediation of mercury and methylmercury – polluted soils using genetically engineered plants. Soil Contamin. 7, 497–510.

Huang, J., Blaylock, M., Kapulnik, Y., Ensley, B., 1998. Phytoremediation of uranium contaminated soils: Role of organic acids in triggering uranium hyperaccumulation in plants. Environ. Sci. Technol. 32, 2004–2008.

Huang, J.W., Chen, R., Berti, W.R., Cunningham, S.D., 1997. Phytoremediation of lead contaminated soils: Role of synthetic chelators in lead phytoextraction. Environ. Sci. Technol. 31, 800–805.

Jaj, N., 2005. Heavy metals and pathogens in leafy vegetables. M.Sc. Thesis, TIET, Patiala, India.

Kay, S.H., Haller, W.T., Garrard, L.A., 1984. Effect of heavy metals on water hyacinth [*Eichhornia crassipes* (Mart.) Solms]. Aqua. Toxicol. 5, 117–128.

Keeling, S.M., Stewart, R.B., Anderson, C.W., Robinson, B.H., 2003. Nickel and cobalt phytoextraction by the hyperaccumulator *Berkheya coddii*: Implications for polymetallic phytomining and phytoremediation. Int. J. Phytoremediat. 5, 235–244.

Kos, B., Greman, H., Lestan, D., 2003. Phytoextraction of lead, zinc and cadmium from soil by selected plants. Plant Soil Environ. 49, 548–553.

Kozlov, M.V., Haukioja, E., Bakhtiarov, A.V., Stroganov, D.N., 1995. Heavy metals in Birch leaves around a nickel–copper smelter at Monchegorsk, north western Russia. Environ. Pollut. 90, 291–299.

Kumar, P.B.A.N., Dushenkov, V., Motto, H., Raskin, I., 1995a. Phytoextraction: The use of plants to remove heavy metals from soils. Environ. Sci. Technol. 29, 1232–1238.

Kumar, P.B.A.N., Motto, H., Raskin, I., 1995b. Rhizofiltration: The use of plants to remove heavy metals from aqueous streams. Environ. Sci. Technol. 29, 1239–1245.

LaCoste, C., Robinson, B., Brooks, R., 2001. Uptake of thallium by vegetables: Its significance for human health, phytoremediation and phytomining. J. Plant Nutr. 24, 1205–1215.

Landberg, T., Greger, M., 1994. Can heavy metal tolerant clones of Salix be used as vegetation filters on heavy metal contaminated land? In: Aronsson, P., Perttu, K. (Eds.), *Willow* Vegetation Filters for Municipal Wastewaters and Sludges. A Biological Purification System. Report 50, Swedish University of Agricultural Science, Uppsala, pp. 133–144.

Landberg, T., Greger, M., 1996. Differences in uptake and tolerance to heavy metals in *Salix* from unpolluted and polluted areas. Appl. Geochem. 11, 175–180.
Lasat, M.M., 2000. Phytoextraction of metals from contaminated soil: A review of plant/soil/metal interaction and assessment of pertinent agronomic issues. J. Hazar. Subst. Res. 2, 5–25.
Lasat, M.M., 2002. Phytoextraction of toxic metals: A review of biological mechanisms. J. Environ. Qual. 31, 109–120.
Lewis, B.G., Johnson, C.M., Delwiche, C.C., 1966. Release of volatile selenium compounds by plants. Collection procedures and preliminary observations. J. Agric. Food Chem. 14, 638–640.
Lopez, M.L., Peralta-Videa, J.R., Benitez, T., Gardea-Torresdey, J.L., 2005. Enhancement of lead uptake by alfalfa (*Medicago sativa*) using EDTA and a plant growth promoter. Chemosphere 61, 595–598.
Ma, L., Komar, K., Tu, C., Zhang, W., Cai, Y., Kennelley, E., 2001. A fern that hyperaccumulates arsenic. Nature 409, 579.
Maurice, C., Lagerkvist, A., 2000. Using *Betula pendula* and *Telephora caryophyllea* for soil pollution assessment. J. Soil Contam. 9, 31–50.
McCutcheon, S.C., Schnoor, J.L., 2003. Overview of phytotransformation and control of wastes. In: McCutcheon, S.C., Schnoor, J.L. (Eds.), Phytoremediation-Transformation and Control of Contaminants. Wiley, Hoboken, NJ, pp. 3–58.
McGrath, S., Sidoli, C., Baker, A.J.M., Reeves, R., 1993. The potential for the use of metal-accumulating plants for the in situ decontamination of metal-polluted soils. In: Eijsackers, H., Hamers, T. (Eds.), Integrated Soil and Sediment Research: A Basis for Proper Protection. Kluwer, Dordrecht, pp. 673–676.
Mcinnes, B.I.A., Dunn, C., Cameron, E.M., Kameko, L., 1996. Biogeochemical exploration for gold in tropical rain forest regions of Papua New Guinea. J. Geochem. Explor. 57, 227–243.
McIntyre, T.C., 2003. Databases and protocol for plant and microorganism selection: Hydrocarbons and metals. In: McCutcheon, S.C., Schnoor, J.L. (Eds.), Phytoremediation: Transformation and Control of Contaminants. Wiley, New York, pp. 887–904.
Meagher, R.B., 2000. Phytoremediation of toxic elemental and organic pllutants. Curr. Opin. Plant Biol. 3, 153–162.
Minguzzi, C., Vergnano, O., 1948. Il contento di nichel nelli ceneri di Alyssum bertlonii Desv. Atti della Societa Toscana di Science Naturali, Mem. Ser. A 55, 49–77.
Mo, S.C., Choi, D.S., Robinson, J.W., 1989. Uptake of mercury from aqueous solution by duckweed: The effect of pH, copper and humic acid. J. Environ. Health 24, 135–146.
Newman, L.A., Strand, S.E., Choe, N., Duffy, J., Ekuan, G., 1997. Uptake and biotransformation of trichloroethylene by hybrid poplars. Environ. Sci. Technol. 31, 1062–1067.
Pawlowska, T.E., Chaney, R.L., Chin, M., Charvat, I., 2000. Effects of metal phytoextraction practices on the indigenous community of arbuscular mycorrhizal fungi at a metal contaminated landfill. Appl. Environ. Microb. 66, 2526–2530.
Peralta-Videa, J.R., Gardea-Torresdey, J.L., 2002. Effect of mixed cadmium, copper, nickel and zinc at different pHs upon alfalfa growth and heavy metal uptake. Environ. Pollut. 119, 291–301.
Pierzynski, G.M., Schnoor, J.L., Youngman, A., Licht, L., Erickson, L.E., 2002. Poplar trees for phytostabilization of abandoned zinc–lead smelter. Prac. Period. Hazar. Toxic Radioact. Waste Manage. 6, 177–183.
Pilon-Smits, E., 2005. Phytoremediation. Annu. Rev. Plant Biol. 56, 15–39.

Pilon-Smits, E., de Souza, M.P., Hong, G., Amini, A., Bravo, R.C., 1999. Selenium volatilization and accumulation by twenty aquatic plant species. J. Environ. Qual. 28, 1011–1017.

Prasad, M.N.V., Freitas, H.M.O., 2003. Metal hyperaccumulation in plants – Biodiversity prospecting for phytoremediation technology. Electron. J. biotechnol. 6, 285–321.

Pulford, I.D., Watson, C., 2003. Phytoremediation of heavy metal-contaminated land by trees – a review. Environ. Int. 29, 529–540.

Punshon, T., Dickinson, N.M., 1997. Acclimation of *Salix* to metal stress. New Phytol. 137, 303–314.

Raskin, I., Ensley, B.D., 2000. Phytoremediation of toxic metals using plants to clean up the environment. Wiley, New York, p. 316.

Raskin, I., Kumar, P.B.A.N., Dushenkov, S., Salt, D.E., 1994. Bioconcentration of heavy metals by plants. Curr. Opin. Biotech. 5, 285–290.

Reeves, R.D., Baker, A.J.M., 2000. Metal-accumulating plants. In: Raskin, I., Ensley, B.D. (Eds.), Phytoremediation of Toxic Metals: Using Plants to Clean up the Environment. Wiley, New York, pp. 193–230.

Robinson, B.H., Leblanc, M., Petit, D., Brooks, R.R., Kirkman, J.H., Gregg, P.E.H., 1998. The potential of *Thlaspi caerulescenss* for phytoremediation of contaminated soils. Plant Soil 203, 47–56.

Rugh, C.L., Gragson, G.M., Meagher, R.B., Merkle, S.A., 1998. Toxic mercury reduction and remediation using transgenic plants with a modified bacterial gene. Hortsicence 33, 618–621.

Sahi, S.V., Bryant, N.L., Sharma, N.C., Singh, S.R., 2002. Characterization of a lead hyperaccumulator shrub, Sesbania drummondii. Environ. Sci. Technol. 36, 4676–4680.

Salt, D.E., Blaylock, M., Kumar, P.B.A.N., Dushenkov, V., Ensley, B.D., Chet, I., Raskin, I., 1995. Phytoremediation: A novel strategy for the removal of toxic metals from the environment using plants. Biotechnology 13, 468–474.

Salt, D.E., Smith, R.D., Raskin, I., 1998. Phytoremediation. Annu. Rev. Plant Physiol. 49, 643–668.

Schnoor, J.L., 2000. Phytostabilization of metals using hybrid poplar trees. In: Raskin, I., Ensley, B.D. (Eds.), Phytoremediation of Toxic Metals: Using Plants to Clean up the Environment. Wiley, New York, pp. 133–150.

Smith, R.A.H., Bradshaw, A.D., 1992. Stabilization of toxic mine wastes by the use of tolerant plant populations. Trans. Inst. Mining Metall. 81, A230–A237.

Sun, B., Zhao, F.J., Lombi, E., McGrath, S.P., 2001. Leaching of heavy metals in contaminated soils using EDTA. Environ. Pollut. 113, 111–120.

Tichy, R., Fajtl, J., Kuzel, S., Kolar, L., 1997. Use of elemental sulphur to enhance a cadmium solubilization and its vegetative removal from contaminated soil. Nutr. Cycl. Agroecosys. 46, 249–255.

Turgut, C., Katie, P.M., Cutright, T.J., 2004. The effect of EDTA and citric acid on phytoremediation of Cd, Cr and Ni from soil using *Helianthus annuus*. Environ. Pollut. 131, 147–154.

USEPA, 2000. Introduction to Phytoremediation. EPA/600/R-99/107. Office of Research and Development, Washington, DC.

Utsunamyia, T., 1980. Japanese Patent Application No. 55-72959.

Vangronsveld, J., van Assche, F., Clijsters, H., 1995. Reclamation of a bare industrial area contaminated by non-ferrous metals: In situ metal immobilization. Environ. Pollut. 87, 51–59.

Vassilev, A., Schwitzguebel, J.P., Thewys, T., van der Lelie, D., Vangronsveld, J., 2004. The use of plants for remediation of metal-contaminated soils. Sci. World J. 4, 9–34.

Wang, Q., Liu, X., Cui, Y., Dong, Y., Christie, P., 2002. Responses of legume and non-legume crop species to heavy metals in soils with multiple metal contamination. J. Environ. Sci. Health A 37, 611–621.

Whiting, S.N., Neumann, P.M., Baker, A.J.M., 2003. Nickel and zinc hyperaccumulation by *Alyssum murale* and *Thlaspi caerulescens* (Brassicaceae) do not enhance survival and whole plant growth under drought stress. Plant Cell Environ. 26, 351–360.

Yang, X.E., Long, X.X., Ni, W.Z., Fu, C.X., 2002. *Sedum alfredii* H: A new Zn hyperaccumulating plant first found in China. Chin. Sci. Bull. 47, 1634–1637.

Yang, X.E., Long, X.X., Ye, H.B., He, Z.L., Calvert, D.V., Stoffella, P.J., 2004. Cadmium tolerance and hyperaccumulation in a new Zn-hyperaccumulating plant species (*Sedum alfreii* Hance). Plant Soil 259, 181–189.

Yankov, B., Delibaltova, V., Bojinov, M., 2000. Content of Cu, Zn, Cd and Pb in the vegetative organs of cotton cultivars grown in industrially polluted regions. Plant Sci. 37, 525–531.

Yankov, B., Tashin, N., 2001. Accumulation and distribution of Pb, Cu, Zn and Cd in sunflower (*Helianhus annuus* L.) grown in an industrially polluted region. Helia 24, 131–136.

Zhu, Y.L., Zayed, A.M., Quian, J.H., De Souza, M., Terry, N., 1999. Phytoaccumulation of trace elements by wetland plants: II. Water hyacinth. J. Environ. Qual. 28, 339–344.

Developments in Environmental Science, Volume 5
D. Sarkar, R. Datta and R. Hannigan (Editors)

ISSN: 1474-8177/DOI:10.1016/S1474-8177(07)05006-1

Chapter 6

Environmental geochemistry of trace metal pollution in urban watersheds

Seth Rose and Jacqueline A. Shea

Abstract

Trace metal concentrations (notably lead, zinc, cadmium, chromium, and nickel) within urban watersheds on all continents remain higher than background levels primarily as the result of "non-point source" (i.e., diffuse) pollution. Metal contamination can be generated within a given urban watershed (*endogenous* pollution) or originate from and spread beyond its watershed boundaries (*exogenous* pollution). Zinc is likely the most prevalent urban metal contaminant today and the most important source of this trace metal is automotive tire wear. Other sources of urban metal pollution include road surfaces, roofs, paint, waste incineration, and untreated sewage effluent.

The largest flux of metals in most urban watersheds likely comes with the "first flush" (rising limb of the hydrograph). The dominant geochemical process controlling the fate of metals is adsorption in that most metal ions partition onto suspended sediment (operationally defined as particulate matter $>0.45\,\mu m$). The principal concern is that a large percentage of sediment-bound metal is "bioavailable" and therefore can be harmful to aquatic ecosystems (notably estuaries downstream of major urban centers) even at low concentrations.

The utilization of "cleaner" sampling methodologies and improved analytical capabilities indicate that metal concentrations in many urban and non-urban stream systems today are at low-trace levels (i.e., $<1\,\mu g\,l^{-1}$). Sediment core data from numerous urban lakes and reservoirs suggest that the Clean Water Act of 1972 and other environmental legislation resulted in reduced lead loading to our streams. Zinc concentrations, however, may be increasing as the result of increased urbanization and automotive traffic. Metal concentrations within sediment cores from many urban localities are above environmental quality guidelines indicating that more work has to be done in reducing metal influx to our stream basins.

6.1. Introduction and Background

The horrendous occurrences of Itai–itai and Minamata diseases in Japan during the late 1950s and early 1960s made the modern world painfully aware of the health dangers posed by cadmium and methyl mercury pollution in natural waters. In high concentrations, heavy metals, including mercury, chromium, nickel, cadmium, and lead, can be toxic, carcinogenic, and mutagenic (Siegel, 2002). Lead has been implicated as an agent of mental retardation for two millennia (at least since Roman times) and chromium and cadmium are known to cause immunosuppressant effects in humans (McCabe, 1998). A few of the health problems as well as a few of the common sources of metals in drinking water are given in Table 6.1.

In addition to medical problems, elevated concentrations of trace metals entering streams and rivers from the urban environment may result in significant damage to ecosystems (Dunne and Leopold, 1978).

Table 6.1. US EPA National Primary Drinking Water Quality Standards (MCL), sources, and common medical effects for trace metals (from EPA, 2005)

Metal	MCL ($\mu g\, l^{-1}$)[a]	Potential health effects from water ingestion[a]	Common industrial or urban sources of contaminant in drinking water[a,b]
Cadmium	5	Kidney damage	Galvanized pipes, refineries, waste batteries and paints; anti-corrosive coatings, coal combustion
Chromium	100	Allergic dermatitis	Steel and pulp mills, erosion of natural deposits, steel alloys, plating operations, wood treatment
Copper	1300	Gastrointestinal distress, liver, kidney damage	Corrosion of household plumbing systems, paint pigments, alloys
Lead	15	Delays of physical and mental development, kidney damage, high blood pressure	Corrosion of household plumbing systems, anti-knock fuel agents, tetramethyl lead, lead–acid batteries, paint pigments, plastics, alloys, pipes

MCL, Maximum concentration limit.
[a]EPA (2005).
[b]Siegel (2002).

Cadmium concentrations as low 1–10 μg l^{-1} can have lethal effects upon aquatic organisms (Eisler, 2000) and metal poisoning can occur via direct intake from water or indirectly from contaminated sediment or nutrients. Many of the so-called *urban metals* including lead, cadmium, zinc, and chromium are to some extent available to aquatic organisms. Microorganisms feeding off sediment have the ability to oxidize, reduce, methylate, or demethylate metals that affect their "bioavailability" for higher organisms (Amiard, 1992). In most modern urban environments, metal pollution likely poses a greater direct threat to ecosystems than it does to drinking water.

The Clean Water Act established in 1972 has greatly reduced the impact of metal pollution from industrial and municipal sources within urban watersheds in the United States. Similar environmental legislation during the past 30–40 years has had beneficial effects in many other nations as well. To cite a few examples of improving conditions, median concentrations of copper, cadmium, nickel, and zinc within the Hudson River estuary declined between 53% and 90% since the mid-1970s with the reduction of wastewater discharge (Sañudo-Wilhelmy and Gill, 1999). Lead importation to the Gulf of Mexico from the Mississippi River has declined by about 40% between 1974 and 1984 as a result of reduced lead consumption in gasoline within the United States (Treyfrey et al., 1985). Nonetheless, urbanization and the advent of the *megalopolis* is a worldwide phenomenon and continues to produce higher-than-background metal loadings in river networks throughout the world. For example, concentrations of the "urban metals" including lead, cadmium, copper, and zinc vary by three orders of magnitude within the Seine River basin and are four times higher than upstream in the Paris region where they approach background levels (Horowitz et al., 1999). Furthermore, the exceptional metal contamination of the urbanized regions within the Seine basin remains high following 20 years of generally improved conditions (Meybeck et al., 2004).

Urban watersheds are complex systems that potentially sequester and remobilize both historical and contemporary metal pollutant inputs. Therefore, the long-standing health and environmental problems associated with trace metal pollution persist as a result of the contamination of sediment deposited within urban stream basins and estuaries downstream of our cities. Flood-plain soils and other urban sediment can remain polluted in a similar manner long after point-source pollution problems have been mitigated. In a recent study documenting this persistence, Filippelli et al. (2005) have measured lead concentrations between 200 mg kg^{-1} and greater than 1000 mg kg^{-1} within inner city soils of Indianapolis, Indiana and claim that 400,000 children within the

United States (mostly in the inner cities) have unsafe lead levels in their blood. Rice (1999) estimated that 49% of the stream sediments in U.S. urban settings are characterized by concentrations of trace metals that exceed levels at which adverse biological effects could occur in aquatic biota.

The problem of returning trace metal concentrations to *background levels* within urban watersheds is particularly difficult in that many of the remaining sources are diffuse (i.e., non-point sources). Our urban and suburban way of life is presently dependent upon automotive transportation, which with mining and waste incineration comprise the dominant sources of global metal contamination. Vink and Peters (2003), in a study of metal contamination within the Elbe River basin (which drains Berlin, Hamburg, Prague, and Leipzig), concluded that 84% of the input of lead and 64% of the input of mercury originate from diffuse sources and more metal is imported to this basin than is exported.

Although the literature is heavily weighted to studies of stream sediment within the United States, trace metal contamination is a global problem as documented by studies of urban watersheds in India (Singh, 2001), England (Dawson and Macklin, 1998), Sweden (Palm and Östlund, 1996), southern Africa (Zhai et al., 2003), Switzerland (von Gunten et al., 1997), Spain (de Miguel et al., 1998), Italy (Renoldi et al., 1997), Germany (Behrendt, 1993), and New Zealand (Abrahim and Parker, 2002). These studies suggest that metal concentrations within urban stream sediment are elevated many times above "background" levels. For example, Lindström (2000) observed that cadmium, copper, mercury, and lead in the central areas of Stockholm are approximately five times above the surrounding regions. Widianarko et al. (2000) noted that while urban stream sediments in Semarang (Indonesia's fifth largest city) are not universally contaminated; there are "hot spots" in which zinc, lead, and copper concentrations were hundreds of times above background levels.

In this communication, we summarize our present understanding of the trace metal problem within urban watersheds by discussing the most common sources of metals, the dominant hydrogeochemical processes affecting the partitioning of metals between water and stream sediment, the developing methodologies used to measure and monitor trace metals, and historical trends for metals provided by sediment core data recovered from urban reservoirs. We focus primarily upon lead (Pb), zinc (Zn), and copper (Cu) in that these metals are readily associated with the urban environment and have been identified as priority pollutants in urban runoff (EPA, 1986). We also include some data for cadmium (Cd), nickel (Ni), vanadium (V), chromium (Cr), and cobalt (Co) in that these trace elements are often coanalyzed and associated with the three major priority contaminants. Other metals, notably mercury (Hg), silver (Ag), aluminum (Al), barium (Ba), platinum (Pt), palladium (Pd), tin (Sb), and metalloids, such as arsenic (As) and selenium (Se), are potentially problematic contaminants within the urban environment but are not emphasized herein.

6.2. Sources and sinks of trace metals in the urban environment

All watersheds are potentially *open* systems in that metals can be transported from atmospheric, industrial, municipal, upstream riverine, and other sources such as mining operations that are not indigenous to the specific watershed under consideration (see Fig. 6.1). The urban watershed in turn has numerous sources of trace metals within its own boundaries that include automotive pollution, rooftops, galvanized pipe, paint, sewage effluent, industrial wastes, and contaminated soils (Table 6.1). Urban air is often contaminated to some degree with metals and both wet and dry deposition account for significant metal flux. Larger particulate

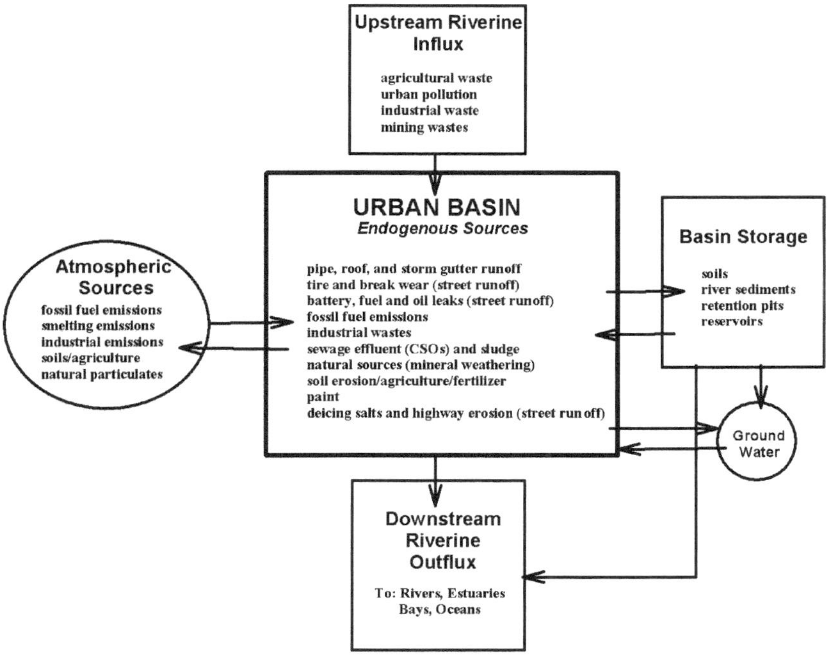

Figure 6.1. Common metal fluxes to and from upstream urban stream basins.

matter delivers most of the metals in dry precipitation or fallout (Sweet et al., 1998). Dusts emitted from the burning of fossil fuels in industrial regions are enriched in V, Ni, and Co (Norra and Stüben, 2004). Municipal waste incineration (Chillrud et al., 1999) along with fossil fuel emissions are potentially important sources of metal contamination within watersheds (*endogenous* pollution) and beyond the divides of urban watersheds (*exogenous* pollution).

Regional studies provide convincing evidence that the most salient feature of metal pollution is its diffuse (non-point source) nature. For example, in the vast Great Lakes Basin in Canada and the United States, urban metal pollution ranks second only behind agriculture as a non-point source of regional pollution (Marsalek, 1990). Aerosols in the Russian Arctic atmosphere are polluted with heavy metals originating from Europe, the Kola Peninsula, and North America (Shevchenko et al., 2003). Pb, Cu, and Zn concentrations within dry air were higher in Chicago than in the surrounding Lake Michigan region; however, Lake Michigan was strongly influenced by Chicago's metal emissions (Paode et al., 1998). Undisturbed urban soils in Madrid, Spain exceed background levels of Cu, Zn, and Pb by a factor of 2–4, likely due to the atmospheric fallout of metals from its own traffic (de Miguel et al., 1998). The extensive interstate highway system within the United States and similar road networks throughout much of the world serve not only as arteries for travel but also as conduits for automotive pollution. In short, the urban watershed should be viewed as both a source and a sink for metals and these metal contaminants are generated both endogenously and exogenously.

The urban watershed can store metals that have become sorbed to lake, reservoir, and stream basin sediment and soils. These watersheds can also export contaminated water and sediment downstream to estuaries, bays, and the oceans (Fig. 6.1). Groundwater–surfacewater interaction (both in terms of "gaining" and "losing" streams) is also a factor that in some cases need to be considered in understanding the total hydrogeochemical flux of metals within a watershed. We will now elaborate upon various sources of metal pollution, emphasizing automotive and roadway contamination in that these are often the most important non-point sources of trace metal pollution within the urban setting.

6.2.1. Natural sources of metals

As the term "trace metals" imply, natural concentrations of metals within the earth's crust and soils are relatively low, typically less than 100 $\mu g\ g^{-1}$ (Li, 2000; Callender, 2004). Natural concentrations of V and Zn approach 100 $\mu g\,g^{-1}$ while Cd concentrations are lower than 1 $\mu g\,g^{-1}$. Trace metal concentrations in river water reflect crustal concentrations and are often exceedingly low (typically <1 $\mu g\,l^{-1}$; Gaillardet et al., 2004; Table 6.2). In that many of the metals are *refractory* they are concentrated in mafic rocks (Li, 2000) and trace metals also tend to preferentially concentrate within shales (Wedepohl, 1991). Rock weathering and volcanic emissions account for an estimated 80% of the natural source of metals and are responsible for the release of between 1400 metric tonnes/year Cd and 45,000 metric tonnes/year Zn (Callender, 2004). In many locations, lithological factors pose only minor environmental controls in that the anthropomorphic flux of trace metals including Cd, Cu, Hg, Zn, and V exceed natural fluxes throughout the earth's surface by factors between 3 and 11 (Nriagu and Pacyna, 1988). However, lithological factors cannot be ignored, particularly in those watersheds where metal concentrations approach background levels.

6.2.2. Automotive sources of metals

In many urban areas, the industrial emission of atmospheric contaminants has been greatly reduced during the past three decades. In these regions the most significant source of metal pollution today is related to automotive use (Novotny and Olem, 1994 and references therein). Vehicular traffic generates waste products from exhaust emissions, lubrication fluid losses, tire wear, break pad linings, catalytic converters, deicing compounds, and road wear (Hares and Ward, 1999). The metals most commonly attributed to urban environments, and specifically to automobile pollution, are the priority pollutants Pb, Zn, and Cu; however, accumulations of Cd, Sb, Ba, Pt, and Pd have also been identified in sediment proximal to roadways (EPA, 1986). Zn flux is more significant than any of the other metals resulting from automotive wear and Pb is the dominant metal present within the automotive emissions (Legret and Pagotto, 1999; Table 6.3). There have been relatively few studies that have quantified inputs of metals from various sources within urban watersheds. However, in one such study Novotny and Olem (1994) estimate that ~45% of Pb in Milwaukee, Wisconsin watersheds originated from freeway and parking lot runoff and these sources account for approximately twice as much Pb as industrial, commercial, and residential runoff (Fig. 6.2).

Several studies have correlated traffic density or fuel consumption with Zn and Pb sediment concentrations in urban areas (Novotny and Olem, 1994; Weisner et al., 1998; Callender and Rice, 2000). Topsoils immediately adjacent to highways with high-traffic densities are typically

Table 6.2. Trace metal concentrations in urban sediments and other sources

	Cu	V	Cd	Cr	Co	Ni	Pb	Zn	Reference
Global reference materials									
Upper crust ($\mu g\, g^{-1}$)	25	86	0.1	69	18	55	17	67	Li (2000) and Callender (2004)
Average soil ($\mu g\, g^{-1}$)	39	82	0.1	61	13	27	26	74	Li (2000) and Callender (2004)
Average shale ($\mu g\, g^{-1}$)	45	130	0.3	90	14	50	20	95	Li (2000)
River water ($\mu g\, l^{-1}$)	1.5	NR	0.08	0.7	NR	0.8	0.08	0.60	Gaillardet et al. (2003)
Environmental quality guidelines ($\mu g\, g^{-1}$)									
Agency or organization									
Georgia Environmental Protection Division[1]	100	100	NR	100	20	50	75	100	Leigh and Gamble (1997)
Canadian Environmental Quality Guideline[2]	36	NR	NR	37	NR	NR	35	123	CCME (1995)
[1]Soil Criteria; [2]Threshold Effect Levels, values greater than guideline indicate possible adverse effects on aquatic life; NR = not reported									
Urban soils, suspended and Benthic sediments (from streams, lakes, canals, reservoirs; in $\mu g\, g^{-1}$)									
Source of metal/study area									
Estuary sediment, Lower Hudson Valley, New York (average of 31 locations)	76	NR	1.2	NR	NR	NR	81	182	Feng et al. (1988)
Ala Wai Urban Canal sediment, Honolulu Hawaii (at depth of 49 cm below water)	16	245	1.1	398	55	282	276	456	De Carlo and Anthony (2002)
Suspended sediment, Yamuna River, Delhi, India (median values)	171	NR	3.0	303	18	69	37	232	Singh (2001)
Suspended sediments, River Manzanares, Madrid, Spain (mean "urban" samples, eight locations)	101	34	NR	48	6.4	22	146	294	de Miguel et al. (2004)

Stream-bed sediment from the US (median values for 540 sites)	27	NR	0.4	64	NR	27	27	110	Rice (1999)
Suspended sediment, Seine Mouth, downstream of Paris	130	NR	3.3	NR	NR	NR	147	510	Horowitz et al. (1999)
Suspended sediment, Seine background, upstream of Paris	20	NR	0.35	NR	NR	NR	19	60	Horowitz et al. (1999)
Roadside topsoil adjacent to highways in Osbgobo, Nigeria (mean values)	21.2	NR	0.60	NR	NR	8.4	68.7	42.4	Fakayode and Olu-Owolabi (2003)
Semerang Central Java, Indonesia (median values, 101 sites)	42	NR	NR	NR	NR	NR	18	105	Widianarko et al. (2000)
Gaborone, Boswana (average value of for nine sites in the city center)	36	NR	1.6	72	17	48	112	248	Zhai et al. (2003)
Post 1750 floodplain sediment, River Swale, northeastern England (mean concentrations)	18	NR	NR	33	NR	NR	1660	637	Hudson-Edwards et al. (1999)
Core from Burlington harbor, Vermont (maximum values)	110	NR	1.3	31	NR	NR	272	404	Mecray et al. (2001)
Stream and drainage system sediments, Ganga Plain, India, mean values, 1994	132	NR	1.5	2,685	55	102	197	857	Ansari et al. (1998)
			Urban stream water ($\mu g\, l^{-1}$)						
River Thames estuary, England (1995)	10.7	NR	0.32	NR	NR	6.3	9.9	29.1	Power et al. (1999)
Peachtree Creek, urban Atlanta, storm runoff (median storm values)	1.9	0.9	NR	NR	NR	2.0	1.0	60	Rose et al. (2001)
Calder river in (urbanized stream in NE England, mean values)	5.8	NR	NR	4.82	1.86	7.6	1.25	35.9	Neal et al. (1996)
14 non-urbanized Wisconsin streams (mean concentrations)	0.6	NR	0.01	NR	NR	NR	0.08	0.46	Shafer et al. (1999)
Small urban watercourse, Dagenham, UK	11.3	NA	3.3	3.7	NR	70	7.7	65	Scholes et al. (1998)

Table 6.2. (*Continuied*)

	Cu	V	Cd	Cr	Co	Ni	Pb	Zn	Reference
			Large river systems ($\mu g\,l^{-1}$)						
Ottawa River, Canada	1.14	0.35	0.02	NR	0.07	0.83	0.11	3.53	Gaillardet et al. (2004)
Average of 21 Amazon River tributaries (sampled twice)	1.0	0.48	0.08	0.48	0.08	0.55	NA	1.2	Seyler and Boaventura (2003)
Average Mississippi River water	2.1	<3.0	<5.0	1.1	<0.5	1.8	<0.5	2.2	Horowitz et al. (2001b)
17 east coast rivers, United States (average values)	1.1	NR	0.01	NR	NR	NR	0.02	0.85	Windom et al. (1991)
Niger River, Africa	0.63	0.59	NR	0.45	0.04	0.29	0.04	0.89	Gaillardet et al. (2004)
Ohio River water, Olmstead, Illinois, 1990–1992	1.76	0.43	NR	NR	NR	1.21	11.74	0.33	Schiller (1997)
			Street or highway runoff ($\mu g\,l^{-1}$)						
Svaneburg, Sweden, Highway E20 (summer)	13.3	NR	0.05	NR	1.4	NR	7.0	89.3	Bäckström et al. (2003)
Norsholm, Sweden, Highway E4 (winter)	126	NR	0.53	NR	13.3	NR	21.0	287	Bäckström et al. (2003)
Urban street runoff Atlanta, USA (median values)	16.9	3.2	NR	NR	NR	4.0	11.4	905	Rose et al. (2001)
Highway runoff, inflow to a biofiltration pond, London	274	29	14.1	105	6.7	93	81	208	Hares and Ward (1999)
IP 4 mountain highway runoff, Portugal (average values)	10.7	NR	NR	NR	NR	NR	10.8	172	Barbosa and Hvitved-Jacobsen (1999)

Table 6.3. Emission fluxes (kg km^{-1} $year^{-1}$) on roads, estimated by measuring mass fluxes of particles and metals over 10 months, for 1 km of road

Sources	Pb	Cu	Cd	Zn
Vehicles				
Tire wear	0.002	0.0006	0.0008	3.22
Brake linings	0.390	14.2	0.0003	2.17
Gasoline	13.0			
Road				
Safety barrier	0.002	0.0002	0.0002	0.95
Deicing agents	0.015	0.002	0.0007	0.002
Vehicles				
Leaded gasoline	200			
Unleaded gasoline	17			
Brake linings	3900	142,000	2.7	21,800
Tire rubber	6.3	1.8	2.6	10,250
Road				
Deicing agent	3.3	0.5	0.2	0.5

Note: Metal content (mg kg^{-1}) in vehicles and road materials (from Legret and Pagotto, 1999).

contaminated with metals (e.g., high Pb, Cd, Cu, Ni, and Zn concentrations adjacent to highways in Osobogo, Nigeria; Fakoyde and Olu-Owolabi, 2003). Lead contamination from street and roadside sources is of particular concern because Pb loads within surface water ranks second only to Zn (Weisner et al., 1998). Kelly and Thornton (1996) showed that Pb concentrations in roadside soils generally decrease exponentially with distance from the road. Elevated Pb concentrations are common in areas where traffic density is high and vehicles are required to start and stop regularly (Lyons et al., 1990; Kelly and Thornton, 1996). Shea et al. (2004) found a convincing relationship between population and road densities and exchangeable Pb and Zn in stream sediment from the Atlanta, Georgia metropolitan region. However, there was far-less correlation between other trace metals and the degree of urbanization in the same basins. The maturation of global information system (GIS) analysis has added a needed spatial dimension for the correlation of trace metal distribution patterns with population and road density.

Trace metal discharges from gasoline combustion and other petroleum-related sources are, no doubt, important sources of metals; however, vehicle-related metal inputs are largely a product of wear, rather than derivatives of fuel combustion (Sternbeck et al., 2002; Harrison et al., 2003). Tire erosion constitutes the most substantial source of

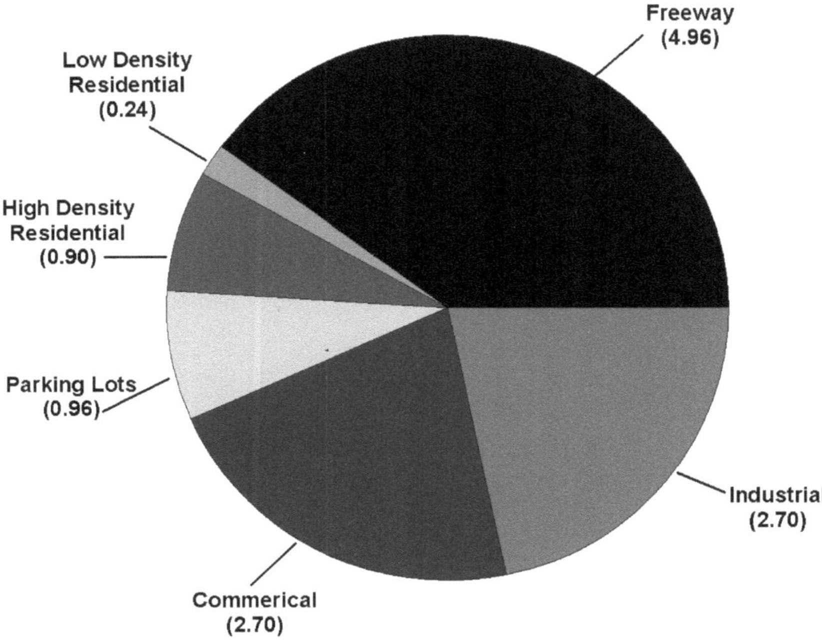

Figure 6.2. Lead sources to urban watersheds in Milwaukee, Wisconsin region. Numbers in parentheses indicate lead totals in kg/hectare/year (data from Novotny and Olem, 1994).

anthropogenic Zn in urban watersheds and also contributes smaller amounts of other trace metals including Cd, Co, Ni, Cu, and Pb (Adachi and Tainosho, 2004). Brake pad wear has been identified as a dominant source of Cu in urban runoff (Legret and Pagotto, 1999). In a study of urban pollution in the Santa Clara Valley region in California, break pads accounted for 80% of Cu within the runoff discharged into the South San Francisco Bay (SCVURPPP, 1994). Legret and Pagotto (1999) estimated Cu emissions from brake linings between 20 and 47 mg/vehicle/km. Catalytic converters unquestionably have improved air quality; however, they are also sources of Pt and Pd in urban watersheds. Car batteries must also be considered as a potential source of Zn and Pb in the urban environment (Palm and Östlund, 1996).

Ranking only behind mining and waste incineration, tire wear is the third largest source of Zn to the environment (Callender, 2004). Automotive tires have a zinc content of approximately 1% (ZnO is added to tires as part of the "vulcanization" or hardening process) and an estimated 10,000 metric tonnes of Zn is released annually within the

United States. Typically, a tire will contain 25–40 g of Zn and wear at a rate of 0.05 g km^{-1}; hence, there is a very strong correlation between traffic density and Zn accumulation in stream sediment within urban areas (Councell et al., 2004). Storage of used tires can also pose a potential hazard as a result of leaching (Smolders and Degryse, 2002).

Callender and Rice (2000) defined an "urban environmental gradient" where Zn and Pb concentrations in reservoir sediment collected within the Chattahoochee River basin varied directly with population density in the Atlanta, Georgia region. Filippelli et al. (2005) have identified a pronounced urban–suburban gradient for Pb brought about by the many decades of leaded gasoline use in the United States with the primary impact zones located in the inner cities. Automotive use, rather than population density, is likely a better indicator of metal flux as indicated by Zn concentrations in sediment from the Houston, Texas area. Sediment from this "automotive city" were eight times greater than more contaminated with Zn than similar sediment from other major cities such as Tokyo, Paris, Venice, and Hong Kong (Weisner et al., 1998).

The urban street surface is both a receptacle and means of conveyance for metals and other contaminants, notably motor oil, fuel, and their derivatives. Street runoff is often contaminated with organics resulting in high total organic carbon (TOC) and biological oxygen demand (BOD) levels in stream water. The initial portion of rain events ("first flush" phenomenon) can overload treatment systems with organic and metal contaminants from street runoff (Lazaro, 1990). For example, maximum concentrations of Cr, Pb, and Cu in road runoff from the Bayside Bridge in Tampa, Florida were 22, 440, and 1100 μg l^{-1}, respectively (Forman et al., 2003). Concentrations of metals in *street* runoff are often many times greater than *stream* runoff during storm events. In a study within the Atlanta metropolitan region, Rose et al. (2001) found that median Zn concentrations in street runoff were ~15 times greater than within storm runoff in urbanized streams. Total metal loads (concentration × discharge) are also greater in storm runoff as a result of the prolific discharge generated by impermeable roads.

Particulate matter on road surfaces can be an important source of metals. In cold climates such as Sweden's, the export of metals from urban streams occurs mainly in the winter as the result of the erosion of roads by tire studs in combination with the use of deicing salts (Bäckström et al., 2003). The abrasion of asphalt that comes about as roads are eroded by vehicular traffic releases Ni and V to waterways (Ozaki et al., 2004). Most of the metals from the roads are derived from filler rock bound to the asphalt, rather than the asphalt itself (Lindgren, 1996). Metals that accumulate on road surfaces are most commonly associated with relatively insoluble finer particulates (often iron oxides) (Wilber and Hunter, 1979) and therefore street sweeping can be ineffective in that it selectively removes mostly coarser particles (German and Svensson, 2002). Permeable pavement offers a viable alternative to impervious asphalt-based surfaces and has been shown to lower Cu and Zn concentrations in infiltration water as well as lower motor oil and diesel fuel in street runoff (Brattebo and Booth, 2003).

6.2.3. Some other sources of metals in the urban environment

High-intensity rainfall can release Pb and other metals from painted wood and brick urban structures, particularly from older buildings (Davis and Burns, 1999). The most common source of Zn in urban structures is roofs, ranking only behind automobiles as a source of metal pollution in Stockholm, Sweden (Palm and Östlund, 1996). Metallic roofs and their drainage systems contribute Pb, Zn, Cu, Ni, Cd, Hg, Cr, and As to runoff and they are an important pathway from the atmosphere to water and soils (Förster, 1999; Davis et al., 2001). Roofs are both a temporary sink for atmospheric dusts and a source of urban metal pollution. Concentrations of Zn as high as 11900 $\mu g\, l^{-1}$ were measured form the first flush emanating from galvanized roof surfaces (Good, 1993). Roof runoff accounted for an estimated 20% of Zn and Pb to watersheds in an Austin, Texas watershed and also contributed polyaromatic hydrocarbon (PAH) pollution (Van Metre and Mahler, 2003). Metal roofing materials constituted the main source of Cd, Pb, and Zn in an experimental catchment within the LeMarais basin in the center of Paris (Chebbo and Gromaire, 2004).

A principal source of metals in many urban locations is the illegal or accidental disposal of sewage wastes. In the river Manzanares watershed in Madrid, urban metals (Ag, Cr, Cu, Pb, and Zn) were enriched in sediment by factors of 200–400% as the result of uncontrolled sewage disposal (de Miguel et al., 2005). Sewage sludge is a potential source of metal and these composts were found to exceed "background" concentrations of Cu, Zn, and Pb by factors of 5–8 in Madrid (de Miguel et al., 2005). Sewage disposal problems and related metal contamination are not restricted to the expanding urban centers of the developing nations; however, untreated effluent is particularly problematic in these locations (e.g., the Ganga alluvial plain in north central India; Ansari et al., 1999). Numerous cities such as older sections of Atlanta convey both their street runoff and sewage through combined sewage overflow (CSO) systems that are unfortunately designed to pollute river basins during storm periods. CSOs in Paris convey a large urban metal (Cd, Cu, Pb, and

Zn) load with suspended sediment that rapidly settles in the Seine River channel. These metals are for the most part "bioavailable" and provide a chronic flux to the Seine basin downstream of Paris (Estébe et al., 1997).

6.2.4. Sinks

The sediment deposited in lakes, canals, reservoirs, and estuaries are local (and often only temporary) "sinks" for many of the contaminants present in urban effluents. Estuaries are the most problematic in that these are complex and dynamic natural features where metal-sensitive ecosystems have evolved. Concentrations of Pb, Cu, Zn, and Cd were enriched by factors of 3–25 in the tops of cores taken from the Tamaki estuary in the Auckland, New Zealand metropolitan region (Abrahim and Parker, 2002). The Hudson River provides another well-studied example of how municipalities (New York City and other cities) can pollute an estuary with metals (e.g., Ag, Cu, and Pb) present in urban runoff and wastewater discharges (Feng et al., 1998). Estuary sediments record historical trends with respect to upstream urbanization and related pollution inputs. For example, sediment cores from the River Thames estuary downstream of London show a recent marked decline with respect to most of the urban metals (Cd, Cu, Hg, Ni, and Zn). This is partially the result of the reduction of sewage plant discharge into the river system from upstream municipalities, notably London and its surroundings (Power et al., 1999).

In one of the few quantitative studies of metal fluxes, watershed yields of various urban metals (notably Cd, Zn, and Pb) were found to represent only between 1% and 2% of the atmospheric input in the Tampa Bay region (Forman et al., 2003). This indicates that the watershed and bay were sinks for most of the metal input to this urban basin. Zn, Ni, Cu, and Cd readily form complexes with chlorides, sulfates, and organic matter, and are therefore relatively mobile and bioavailable within an estuary (Church and Scudlark, 1998). Most metals within the major Humber estuary in England are in the "acid soluble" particulate phase (i.e., can be removed in nitric acid; Neal et al., 1996) and are potentially exchangeable with water. In short, the estuary is the most important "sink" for metals generated by urban activities; however, estuary sediment often only provides a temporary repository for these contaminants. Metals within estuary sediment deposits are "bioavailable" to a variable degree and subject to remobilization within the water column, aquatic ecosystem, and food chain.

6.3. Methodological considerations

Improving methodologies for the sampling, storage, laboratory handling, and analysis of trace metals have greatly enhanced our ability to accurately assess the effects of urban pollution upon stream systems. However, as a result of these new "clean" techniques, a controversy has arisen regarding the validity of trace metal concentrations that were measured in river water decades ago. (The term *clean* usually refers to "metal-free" or an environment that is as free from contact with extraneous metals as possible.) Citing a representative example of recent research, clean room technology produced trace metal values from highway runoff in the Olentangy River near Columbus, Ohio that were ~1/10 of those previously reported (Gardner and Carey, 2004). In a landmark study involving one of the largest historical sets of water quality data within the United States (the United States Geological Survey's "NASQAN" database), Windom et al. (1991) cast doubt upon the relatively high-stream water metal concentrations measured decades ago (refer also to Horowitz et al., 2001a). Shiller and Boyle (1985) reached similar conclusions regarding data from the 1960s and 1970s and noted that clean-techniques produced Zn concentrations in Mississippi River water that were often less than $1\,\mu g\,l^{-1}$ as opposed to earlier studies in which Zn concentrations were as high as $30\,\mu g\,l^{-1}$.

Today's U.S. government (NASQAN) quality assurance/quality control (QA/QC) protocols call for field blanks, split samples (to be sent to different laboratories), sample replicates, "clean hands", clean sampling equipment, and clean analytical facilities. Recent results have shown that metal concentrations (Pb, Co, V, and other metals) in major river systems within the United States are often below detection limits and in most cases certainly well below $1\,\mu g\,l^{-1}$ (Horowitz et al., 2001b). In addition to improved sampling techniques, there has been a recent emphasis upon sampling stream water and stream sediment at multiple locations along a given stream transect (width- and depth-integrated sampling; Hart and Hines, 1995). Integrated sampling methods will be helpful in determining the effects of current velocity upon metal transport and will potentially provide a better estimate of total metal loads than single-point sampling.

Metal partitioning into the *dissolved*, *colloidal*, and *particulate* states in stream basins and estuaries poses a strong control upon metal concentrations both in solution and upon sediment. There is no universal agreement as to what "size" constitutes the upper and lower limits for these gradations although we often characterize metal distributions in the environment using these size classes. In a recent summary, Gaillardet et al. (2004) characterize the *dissolved* state as below 1 nm, *colloids* are operationally defined as greater than 1 or several nanometers and less than 0.45–0.2 μm, and *particulates* or suspended matter as >0.45 μm. Horowitz et al. (1996) imply that the once-typical characterization of "dissolved metals" as those which pass through a 0.45 μm membrane is inadequate for determining metal concentrations because it does not accurately account for the significant loads of colloidal phase metals. The filtration of samples is an extremely important matter that will influence all subsequent characterization of metal distributions in natural systems. The most important consideration is the separation of colloidal material that involves removing material ~90 times finer than what is commonly removed using conventional filtering methods (Horowitz et al., 1996). Removing this colloidal material would involve the utilization of ultra-centrifugation techniques or the repeated filtering of samples in order to clog and thereby reduce the pore size of ordinary membranes (Shiller and Boyle, 1987).

Inductively coupled plasma emission spectrometry (ICP-AES) and mass spectrometry (ICP-MS) are now the most commonly used analytical techniques for the measurement of trace metal concentrations in natural waters. ICP methodologies have largely supplanted (but not completely replaced) earlier atomic adsorption and emission techniques in that the concentrations of a suite of metals (~20), rather than a single metal, can be simultaneously determined at high excitation temperatures (7000–8000 K) within the plasma state (Ortega, 2002). Detection limits using conventional ICP-AES techniques are between 1 and 50 $\mu g\,l^{-1}$ while newer ICP-MS technologies (quadrupole and mass sector analyzers) can measure metal concentrations in the low $ng\,l^{-1}$ and high $pg\,l^{-1}$ range. Furthermore, the mass analyses for many of the metals are linear over many orders of magnitudes (Ortega, 2002). Anodic stripping voltammetry (ASV) methods can be used to analyze selected labile organic and inorganic metal–ligand complexes of environmentally important metals in natural waters (Salomons and Förstner, 1984); however, this is a seldom-used technique.

Horowitz (1997) cautions that our improved analytical techniques for measuring low concentrations of metals have dwarfed our ability to collect uncontaminated (i.e., "meaningful") environmental samples. In summary, the recent advances with regard to analytical technology, filtering methodologies, and improved clean room and sampling techniques have brought the state-of-the-science to new levels within the last decade. These expanded methodologies based upon lower detection limits will continue to stimulate new research within the area of metal geochemistry in urban and natural watersheds. Lower detection limits might also conceivably promulgate lower concentration limit standards for metals in drinking water.

6.4. Hydrological and geochemical processes

6.4.1. Storm runoff in the urban environment

A complex interactive set of hydrological and geochemical processes control the transport and fate of metals in urban and natural watersheds (Fig. 6.3). Probably, rainfall intensity is the most important hydrological factor determining metal concentrations and metals loads for any given period of sampling. High-intensity rainfall often flushes street wastes and fined grained sediment into streams and also directly transports metals from the atmosphere into watersheds; however, the control upon metal flux is often the amount of precipitation, rather than the metal concentration within wet precipitation (Sweet et al., 1998).

The most distinctive feature of urban surface water hydrology is the linear and grid-like drainage networks (both streets and storm water

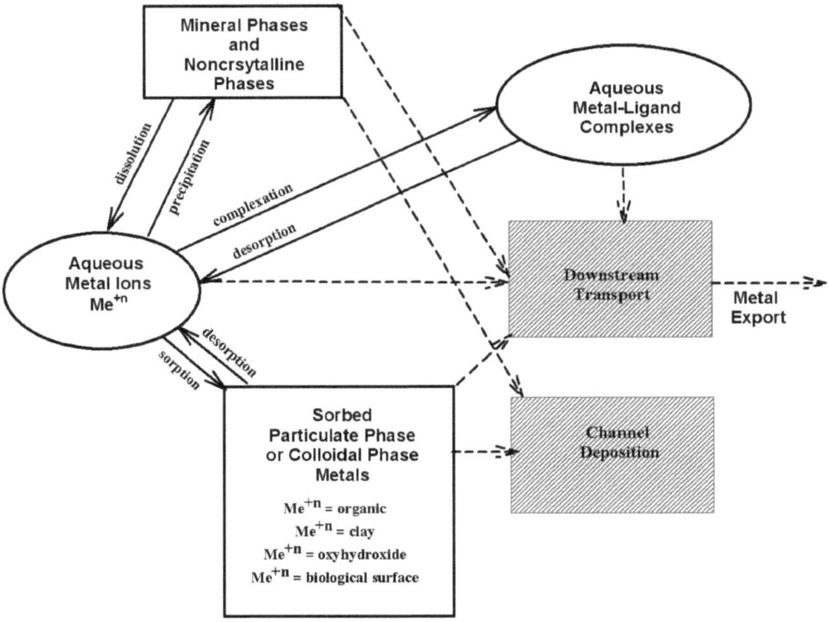

Figure 6.3. Some common geochemical processes (solid lines) and physical processes (dashed lines) affecting the fate of metals in stream water.

collection pipes) that result in higher peak flows and reduced lag time between rainfall and runoff (Lazaro, 1990). Relatively small changes in the percentage of impervious road cover will cause significant changes in the amount of runoff. For example, when the percentage of impervious street cover increases from 30% to 50%, the runoff coefficient (runoff/rainfall ratio) nearly doubles from 22% to 42% (Novotny and Olem, 1994). Higher metal concentrations and loads typically accompany the increased urban storm runoff. Generally, the highest concentrations of urban metals (notably Pb, Cu, and Zn) come with the first flush, coincident with the *rising limb* of the stream hydrograph (Rose et al., 2001; Blake et al., 2003). Mason et al. (1999) noted that the first flush of Cd, Cr, and Cu from rooftops quickly forms metal–ligand complexes that are often mobile in surface water.

6.4.2. Surface water–groundwater interactions and hydrograph "hysteresis"

Groundwater often dominates the *falling limb* of the hydrograph and in most cases is characterized by relatively low-metal concentrations. Therefore, the solute (metal)–discharge relationship in most urban basins will exhibit a *clockwise* hysteresis loop (Blake et al., 2003; Mishra et al., 2003) with higher-metal concentrations coincident with the earlier portion of the storm cycle or rising limb of the storm hydrograph (Fig. 6.4). There has been relatively little attention paid to metal concentrations in base flow (groundwater and soil water) which often is the dominant

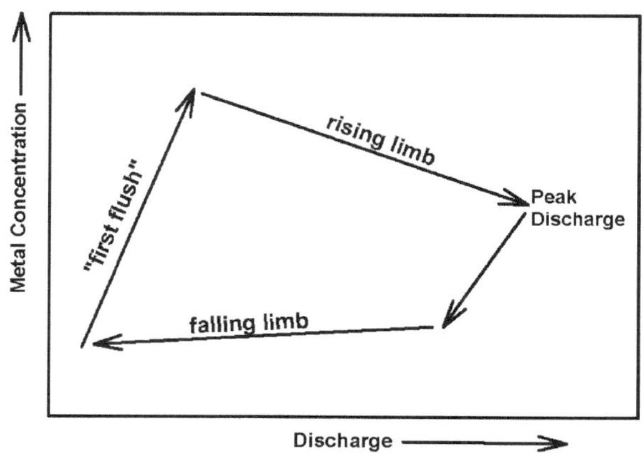

Figure 6.4. Metal concentration–discharge relationship for a typical urban stream showing clockwise hysteresis and "first flush" phenomenon.

year-round water resource in many stream basins in humid regions. Concentrations of common metals in groundwater are often exceedingly low (i.e., below detection limits) principally as the result of metal adsorption. Urban soils in which clay, iron oxyhydroxide and aluminum oxide coatings are abundant will provide the best sorbents for metals (see next section) and groundwater infiltrating through these soils should have the lowest metal concentrations. Metal concentrations within unpolluted base flow may in some cases provide a reference value for the contribution of lithological sources to total metal loads within a watershed.

Although there are relatively few sources for metal pollution within the subsurface, sewage effluent from leaky conveyance pipes often impact metal loading within urban basins. Counterclockwise hysteresis loops would most likely indicate that metal concentrations are higher within groundwater than in surface runoff and such a solute–discharge relationship would probably result form some subsurface source of metal pollution. The movement of stream water through pore waters adjacent to the stream and beneath the stream channel (i.e., the *hyporheic zone*) can result in the exchange of metals between stream water and sediment (Ren and Packman, 2004). Hyporheic exchange is a critical process, particularly in contaminated streams, in that under oxic conditions a significant fraction of exchangeable metals are bound to iron and manganese oxyhydroxides that precipitate in the near-stream zone (Fuller and Harvey, 2000).

6.4.3. Metal–ligand complexation

The simplest of the many chemical partitioning processes that occur in stream water (Fig. 6.3) is the formation of metal–ligand complexes. The most common organic ligands are phenols and carboxylic groups that are derived from the breakdown of humic substances. Organic-metal complexes are often chemically unstable and can break down in sunlight (Martino et al., 2003). The complexation of metals within naturally occurring organic matter is kinetically controlled by the metal-to-ligand ratio, the ionic potential (charge/ionic radius) of the metal and the ligand field stabilization energy of the metal within the complex (Fasfous et al., 2004). Typically chloride, sulfate and bicarbonate are the dominant inorganic ligands in most natural waters. Many of the trace metals form fairly stable complexes with oxygen and hydroxide ligands; common inorganic ionic complexes that form at near neutral pH under these oxidizing conditions include $CuOH^+$, $CuCl^+$, $CuHCO_3^+$, $Zn(OH)_2^0$ $PbSO_4^0$ and $PbOH^+$ (van Loon and Duffy, 2000). To illustrate the relative significance of inorganic complexes, Hermann et al. (1994) found that greater than 70% of the Pb from roof runoff was bound in complex form ($PbSO_4^0$ and $PbOH^+$) while 30–80% of Cu and virtually all Cd were in free-ion form. The portion of the total metal load that is partitioned into complexes cannot precipitate to form solids; however, the formation of metal–ligand complexes may significantly enhance the adsorption of some trace metals onto iron oxyhydroxide surfaces (Davis and Leckie, 1978).

6.4.4. Partitioning with colloidal matter

Much of the aqueous trace metal load is associated with *colloidal* matter, which has a very large surface area and hence has a high affinity for trace metals. The colloidal state is between the particulate and dissolved states and has been operationally defined as between 0.006 and 0.2 μm (Hart and Hines, 1995) or 0.001–0.2 μm (Gaillardet et al., 2004). Most organic colloids are made up of humic and fulvic substances and the inorganic colloids are typically oxyhydroxides of aluminum, iron, and manganese. At the pH of most natural waters, colloids are negatively charged and hence have an electrostatic attraction for trace metals and positively charged metal complexes. Cr, Pb, Zn, Cd, Cu, and Pb are either divalent or trivalent and therefore at a pH of 6 or above should be nearly 100% partitioned onto negatively charge colloidal (or particulate) surfaces (Gaillardet et al., 2004). To determine the proportion of the metal bound to colloids, ultracentrifugation or multiple filtration methods must be employed (refer to previous discussion). Grout et al. (1999), using ultracentrifugation techniques, found that ~43% of the total Cr and ~29% of the total Zn were colloidal in urban storm water runoff from the Houston, Texas area.

6.4.5. Chemical precipitation and redox considerations

In most cases, the pH of stream water is a master variable controlling both the adsorption of metals onto colloidal and particulate surfaces and the precipitation of metals. In stream water, Cu, Zn, Pb, and Cd are all "alkaline" type metals that precipitate with increasing pH (Salomons and Förstner, 1984). However, the precipitation of trace metals to form crystalline solids is often only a minor geochemical process in that metal concentrations in most river systems are usually exceedingly low and therefore surface waters remain grossly under saturated with respect to most metallic mineral phases. The oxidation state of stream water is an another key variable controlling the precipitation of metals. Typically, the non-stagnant river water is oxic; however, in some urban streams (e.g., the River Meuse in the Netherlands (Van Den Berg et al., 1999)) the water–sediment interface is anoxic due to high-organic loads. Under these "micro-reducing" conditions, metals can precipitate to form sulfide minerals and then possibly become remobilized as reduced sediments subsequently dissolve under oxidizing conditions (von Gunten et al., 1991).

The single most important control upon the oxidation state of most urban streams is the degree to which they are polluted and thereby electrochemically reduced with organic matter from sewage effluent and other sources. Sunlight also plays a role in controlling metal concentrations in that it catalyzes photosynthetically enhanced redox reactions (Scott et al., 2002). In a study of the non-urbanized Clark Fork River in southwestern Montana, Brick and Moore (1996) found that metal (Fe and Al) and trace metal concentrations increased by factors of two or three at night (i.e., "diel variation") when the absence of photosynthesis dictates that dissolved oxygen concentrations and pH are at a minimum and CO_2 is at a maximum. Probably, the most important redox consideration involves the precipitation of iron oxyhydroxides (which occurs at relatively high-redox potentials) in that these phases provide the primary surface area for metal adsorption. Metals can be released to water when the oxyhydroxides dissolve under reducing conditions, such as those that can occur on lake bottoms or in stagnant streams, mires, and bogs (Shiller, 1997).

6.4.6. Adsorption

The most important mechanism of metal transport within most river systems is adsorption upon suspended sediment or particulates. The diffusion of trace metals within microporous amorphous iron and aluminum oxyhydroxides and oxides is another physiochemical mechanism for sequestering metals; however, it would be difficult to distinguish from adsorption. Greater than 70% of the Cu, Zn, Cr, and Ni within many of the major river systems throughout the United States are transported in association with suspended sediment (Horowitz et al., 2001b). Adsorption is the major reason why metal concentrations in relatively uncontaminated rivers such as the Amazon are often less than $1\,\mu g\,l^{-1}$ (Seyler and Boaventura, 2003). In highly polluted waters such as the Guadaira River (southwest Spain), Pb, Zn, Cr, Cu, and Cd are 50–7500 times more concentrated in sediment than in river water (Gonzalez et al., 2000). Marine sediments contain very little exchangeable metals; however, 50–70% of the Pb and Zn in terrestrial sediments are exchangeable (Callender, 2004).

One of the most commonly referenced schemes for the extraction of metals from sediment was devised by Tessier et al. (1979) in order to differentiate the "exchangeable" fraction of the total metals from that associated with carbonates, organic matter, Mn and Fe oxides, and sulfides. Some other investigators (e.g., Shea, 2003) simply approximate "exchangeable" or "adsorbed" metals as those liberated in contact with strong ultra-clean nitric acid (i.e., at pH $<$2–3). The utilization of different treatment techniques to quantify sorbed metal concentrations makes it extremely difficult to meaningfully compare the results between different studies and different study basins.

Nonetheless, *partition coefficients* ((K values) where K is the equilibrium ratio of particulate metal concentrations to dissolved metal concentrations) for metals onto natural particulate matter in rivers throughout the world are similar and are in the range of 10^4–10^5 (Sigg, 1998). Sediment grainsize is the primary factor controlling the adsorption of metals (Horowitz and Elrick, 1987) in that surface area increases within decreasing grain size and adsorption is proportional to surface area. The electrostatic affinity of a surface for a given metal varies primarily as a function of the metal's ionic potential. However, trace metals, regardless of their size and charge are almost always preferentially concentrated within the $<63\,\mu m$ or $125\,\mu m$ size fraction (Horowitz and Elrick, 1987).

Sediments are not merely "passive recipients" of metals; rather they act as "biogeochemical reactors" where numerous processes including redox-mediated microbiological transformations occur (Van Cappellen and Wang, 1995). In most cases, adsorbed metals are considered to be "bioavailable" and hence can potentially work their way up the food chain. However, biological activity is probably not as an important control upon trace metal concentrations in natural waters as adsorption (Hart and Hines, 1995). Probably all particulate matter including bacterial surfaces, clay minerals, organic substances, crystalline mineral phases, and amorphous iron oxyhydroxides can adsorb metals to some degree, depending upon their specific surface area and surface charge. Iron oxyhydroxides have received the greatest attention in that they are ubiquitous in oxic waters and often mask the effects of other sorbents such as the clay minerals that they coat. Naturally occurring aluminum oxides in soils and stream sediments also are effective sorbents of trace metals.

The chemisorption of metals upon oxyhydroxide surfaces is dependent upon numerous factors including the ionic potential and concentration of the metal, solution pH, the type and concentration of sorbent, ligand concentrations (Benjamin and Leckie, 1981), and redox conditions. At relatively low concentrations there might not be significant competition among the metals for exchange sites on hydrous ferric oxides

(Swallow et al., 1980); however, the major cations (notably Ca^{2+} and Mg^{2+}) in stream water will successfully compete with trace metals for these sites. The amorphous iron oxyhydroxide surface is composed of many groups of binding sites and the strength of bonding may vary by an order of magnitude or more (Benjamin and Leckie, 1981). The more electronegative metals form the strongest covalent bonds with oxygen atoms on mineral surfaces and therefore Cu and Zn tend to be more effectively bound than Co and Pb (McBride, 1994). The same metal can bond to multiple sites to form *monodentate*, *bidentate*, or *polydentate* bonds, all with different free energies or strengths (Davis and Leckie, 1978). Interestingly, the adsorption of metals tend to form *binuclear complexes* which increases the stability of the oxyhydroxides, thereby lowering their solubility and the solubility of adsorbed metals (Bondietti et al., 1993).

Probably, the most important environmental factor controlling metal adsorption onto the oxyhydroxide surface is the pH of stream water. In all cases, Pb, Cu, Zn, and Cd adsorption decreases with decreasing pH (Benjamin and Leckie, 1981), and at a pH <4.5 Cu, Pb, Ni, Zn, Cd, and Cr are effectively desorbed from oxyhydroxide and ferrihydrite surfaces (Schultz et al., 1987). Conversely, as the pH of the solution approaches neutral, the percentage of metals (Pb, Cu, Zn, and Cd) that are sorbed sharply increases as shown by the "adsorption edges" in Fig. 6.5 (Benjamin and Leckie, 1981). Urban streams often maintain near neutral

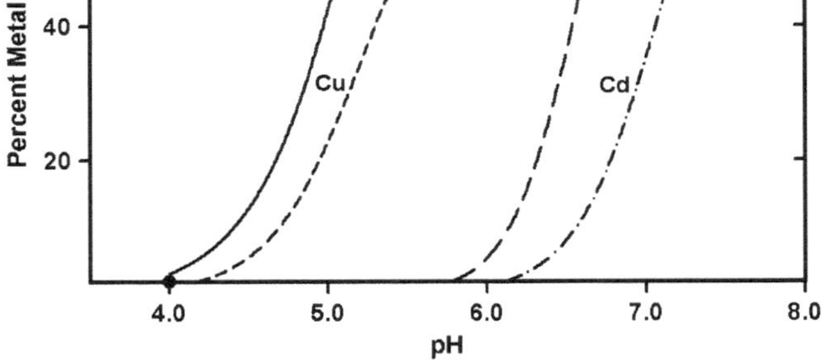

Figure 6.5. Metal adsorption as a function of pH for four common trace metals. These plots are known as "adsorption edges" (after Benjamin and Leckie, 1981).

pH values and adsorption of metals onto iron oxyhydroxides (and clay minerals making up the suspended sediment) is a very important factor in keeping the concentration of metal contaminants within stream water in the low parts per billion range or less. Development and construction activities within urban watersheds often supply streams with large quantities of fine-grained suspended sediment (i.e., turbidity), which provide abundant surface area for the adsorption of trace metals. Fast-acting adsorption processes may be an important factor in explaining why metal concentrations in urban stream runoff are often so much lower than metal concentrations in street runoff.

Bacterial surface coatings typically have a negative charge and therefore represent potential binding sites for metals and positively charged complexes. *Escherichia coli*, a common microbe in sewage waste, has a remarkable affinity for metals and the biofilms that form stable coats on algae, fungi, and bacteria are also potentially important metal sorbents (Schultze-Lam et al., 1995). Metals bind to numerous anthropogenic compounds such as NTA in detergents, phosphates (in detergents and fertilizers), and EDTA (in cleaning agents) (van Loon and Duffy, 2000); many of which are present in urban stream water.

The most important of the organic sorbents are the humic and phenolic compounds found in naturally occurring organic matter (NOM) and are widely distributed in aquatic systems (Kerndorff and Schnitzer, 1980). Sorption of trace metals on organic substances (either colloidal or particulate) varies directly with pH, similar to the relationship between metals and inorganic surface coatings. At $pH > 6$, more than 40% of the Ni, Zn, Cr, Cu, Pb, and Hg were found under laboratory conditions to be sorbed upon humic acids (Kerndorff and Schnitzer, 1980). Organic acids not only provide direct surface sites for sorption but also enhance the sorption of Cu^{+2} on ferric oxyhydroxide suspensions and shift the "adsorption edge" to lower pH values (Ali and Dzombak, 1996).

6.5. Summary of metal pollution trends

Core analysis of lake and flood plain sediment provides convincing evidence that the anthropogenic input of metals to stream systems did not begin within the modern industrial and urbanized era. The lead isotope record in high latitude Swedish lake sediment reveals a 3000-year history of pollution input and a prominent peak during the medieval period between AD 900 and 1600 (Brännvall et al., 1999). The Rio Tinto estuary in southwestern Spain was mined for perhaps 5000 years and has been heavily polluted by V, Zn, Ni, Co, and Cu since pre-Roman times.

Metal concentrations at this site are to this day 3–6 orders or magnitude greater than contaminant levels found in most other watersheds (Davis et al., 2000). Lake cores in northeastern England flood plain sediment reveal similar high levels of Pb and Zn ($>500,000\,\mu g\,kg^{-1}$) since AD 1750 as a result of mining in the Yorkshire district (Hudson-Edwards et al., 1999).

In recent decades, the flux of many trace metals into surface waters has apparently declined as the result of conservation, replacement, and improved disposal technologies (van der Voet et al., 2000). Interpreting recent historical trends from archival water data is problematic due to the inaccurate measurements made prior to the advent of the cleaner sampling methods used today (see proceeding section). Nonetheless, sediment core data suggest that metal concentrations (notably Cu, Cr, Pb, and Ag) within urban streams have generally declined during the past two decades in many locations on different continents (von Gunten et al., 1997; Power et al., 1999; Mecray, 2001).

Lead concentrations in the recent sedimentary record for 30 U.S. reservoirs (many of which are urban) have undergone the most unequivocal decline while Cu and Ni trends are much more ambiguous (Callender, 2004). Lead concentrations in sediment have declined the most notably since the 1970s peak. This is the result of both reduced automotive emissions and better sewage treatment disposal within the United States and elsewhere. Core analysis from lakes in Boston and Forth Worth, Texas indicates that pesticides, PCBs, and PAHs have declined somewhat with Pb during the past three decades (Van Metre and Mahler, 2004). Unfortunately, Zn concentrations during this period of time have increased in numerous reservoir cores as the result of diffuse non-point source urban contamination (mostly tire wear). Despite reductions in contaminant inputs, concentrations of metals in numerous lakes within densely populated urban watersheds remain greater than sediment quality guidelines (Callender, 2004; Van Metre and Mahler, 2004).

6.6. Concluding statement

The prevalence of Pb in gasoline used in many developing countries, the continued high rates of atmospheric release of metals from mining and industrial sources, increasing automobile traffic with the concomitant paving of urban watershed surfaces strongly suggest that metal pollution will continue to be part of the "urban hydrology problem". The reduction of point source pollution has undoubtedly lowered metal loading within urbanized and undeveloped watersheds within the United States and elsewhere. However, non-point source pollution, particularly that associated with automotive traffic continues to present a risk to urban streams and major estuary ecosystems downstream of our cities. Even if all sources of metal pollution were eliminated today, past inputs of metals can become remobilized through natural chemical and physical weathering processes. Hence, efforts to lower metal loads including conservation, reduction of automotive traffic, improvement of wastewater treatment and conveyance infrastructure, design of phreatophyte remediation schemes, and construction of artificial wetlands or other buffers should be encouraged.

REFERENCES

Abrahim, G., Parker, R., 2002. Heavy metal contaminants in Tamaki estuary: Impact of city development and growth, Auckland, New Zealand. Environ. Geol. 42, 883–890.

Adachi, K., Tainosho, Y., 2004. Characterization of heavy metal particles embedded in tire dust. Environ. Int. 30, 1009–1017.

Ali, M.A., Dzombak, D.A., 1996. Effects of simple organic acids on sorption of Cu^{2+} and Ca^{2+} on goethite. Geochimica et Cosmochimica Acta 60, 291–304.

Amiard, J.-C., 1992. Bio-availability of sediment-bound metals for benthic aquatic organisms. In: Vernet, J.-P. (Ed.), Impact of Heavy Metals on the Environment. Elsevier, Amsterdam, pp. 183–202.

Ansari, A.A., Singh, I.B., Tobschall, H.J., 1999. Status of anthropogenically induced metal pollution in the Kanpur–Unnao industrial region of the Ganga Plain, India. Environ. Geol. 38, 25–33.

Bäckström, M., Nillsson, U., Håkansson, K., Allard, B., Karlsson, S., 2003. Speciation of heavy metals in road runoff and roadside total deposition. Water Air Soil Poll. 147, 343–366.

Barbosa, A.E., Hvitved-Jacobsen, T., 1999. Highway runoff and potential for removal of heavy metals in an infiltration pond in Portugal. Sci. Total Environ. 235, 151–159.

Behrendt, H., 1993. Separation of point and diffuse loads of pollutants using monitoring data of rivers. Water Sci. Technol. 28, 165–175.

Benjamin, M.M., Leckie, J.O., 1981. Multiple-site adsorption of Cd, Cu, Zn, and Pb on amorphous iron oxyhydroxide. J. Colloid Interf. Sci. 79, 209–221.

Bondietti, G., Sinniger, J., Stumm, W., 1993. The reactivity of Fe(II) (hydr)oxides: Effects of ligands in inhibiting the dissolution. Colloid. Surface. A 79, 157–167.

Blake, W.H., Walsh, R.P.D., Barnsley, M.J., Palmer, G., Dyrynda, P., James, J.G., 2003. Heavy metal concentrations during storm events in a rehabilitated industrialized catchment. Hydrol. Proc. 17, 1923–1939.

Brännvall, M.J., Bindler, R., Renberg, I., Emteryd, O., Bartnicki, J., Billström, K., 1999. The medieval metal industry was the cradle of modern large-scale atmospheric lead pollution in northern Europe. Environ. Sci. Technol. 33, 4391–4395.

Brattebo, B.O., Booth, D.B., 2003. Long-term stormwater quantity and quality performance of permeable pavement systems. Water Research 37, 4369–4376.

Brick, C.M., Moore, J.N., 1996. Diel variation of trace metals in the Upper Clark Fork River, Montana. Environ. Sci. Technol. 30, 1953–1960.

Callender, E., 2004. Heavy metals in the environment – Historical trends. In: Holland, H.D., Turekian, K.K. (Eds.), Treatise on Geochemistry. Elsevier, Amsterdam, pp. 67–105.

Callender, E., Rice, K.C., 2000. The urban environmental gradient: Anthropogenic influences on the spatial and temporal distributions of lead and zinc in sediments. Environ. Sci. Technol. 34, 232–238.

CCME, 1995. Protocol for the derivation of Canadian sediment quality guidelines for the protection of aquatic life: Canadian Council of Ministers of the Environment, CCME EPC-98E.

Chebbo, G., Gromaire, M.C., 2004. The experimental urban catchment "Le Marais" in Paris: What lessons can be learned from it? J. Hydrol. 299, 312–323.

Chillrud, S.N., Bopp, R.F., Simpson, H.J., Ross, J.M., Shuster, E.L., Chaky, D.A., Walsh, D.C., Choy, C.C., Tolley, L.R., Yarme, A., 1999. Twentieth century atmospheric metal fluxes into Central Park Lake, New York City. Environ. Sci. Technol. 33, 657–662.

Church, T.M., Scudlark, J.R., 1998. Trace metals in estuaries: A Delaware Bay synthesis. In: Allen, H.E., Garrison, A.W., Luther, G.W. (Eds.), Metals in Surface Waters. Ann Arbor Press, Chelsea, MI, pp. 1–22.

Councell, T.B., Duckenfield, K.U., Landa, E.R., Callender, E., 2004. Tire-wear as a source of zinc to the environment. Environ. Sci. Technol. 38, 4206–4214.

Davis, A.P., Burns, M., 1999. Evaluation of lead concentrations in runoff from painted structures. Water Res. 33, 2949–2958.

Davis, A., Shokouhian, M., Ni, S., 2001. Loading estimates of lead, copper, cadmium, and zinc in urban runoff from specific sources. Chemosphere 44, 997–1009.

Davis, J.A., Leckie, J.O., 1978. Effect of adsorbed complexing ligands on trace metal uptake by hydrous oxides. Environ. Sci. Technol. 12, 1309–1315.

Davis, R.A., Welty, A.T., Borrego, J., Morales, J.A., Pendon, J.G., Ryan, J.G., 2000. Rio Tinto estuary (Spain): 5000 years of pollution. Environ. Geol. 39, 1107–1116.

Dawson, E.J., Macklin, M.G., 1998. Speciation of heavy metals on suspended sediment under high flow conditions in the River Aire, West Yorkshire, UK. Hydrol. Proc. 12, 1483–1494.

De Carlo, E.H., Anthony, S.S., 2002. Spatial and temporal variability of trace element concentrations in an urban subtropical watershed, Honolulu, Hawaii. Appl. Geochem. 17, 475–492.

de Miguel, E., Jiménez de Grado, M., Llamas, J.F., Martín-Dorado, A., Mazadiege, L.F., 1998. The overlooked contribution of compost application to the trace element load in the urban soil of Madrid (Spain). Sci. Total Environ. 215, 113–122.

de Miguel, E., Charlesworth, S., Ordóñez, A., Siejas, E., 2005. Geochemical fingerprints and controls in the sediments of an urban river: River Manzanares, Madrid (Spain). Sci. Total Environ. 340, 137–148.

Dunne, T., Leopold, L., 1978. Water in Environmental Planning. W.H. Freeman and Company, New York.

Eisler, R., 2000. Handbook of Chemical Risk AssessmentVol. 1. Lewis Publishers, Boca Raton, p. 738.

EPA, 1986. Quality criteria for water: US Environmental Protection Agency Update 1 and 2, EPA-440/5-86-001.

EPA, 2005. National primary drinking water quality standards (MCL). Available at http://www.epa.gov/safewater/mcl.html

Estébe, A., Mouchel, J.M., Thévenot, D.R., 1997. Urban runoff on particulate metal concentrations in River Seine. Water Air Soil Poll. 108, 83–105.

Fakoyde, S.O., Olu-Owolabi, B.I., 2003. Heavy metal contamination of roadside topsoil in Osogbo, Nigeria: Its relationship to traffic density and proximity to highways. Environ. Geol. 44, 150–157.

Fasfous, I.I., Yapachi, T., Murimboh, J., Hassan, N.M., Chakrabati, C.L., Back, M.H., 2004. Kinetics of trace metal composition in the freshwater environment: Some fundamental considerations. Environ. Sci. Technol. 38, 4979–4986.

Feng, H., Cochran, J.K., Lwiza, H., Brownawall, B.J., Hirschberg, D.J., 1998. Distribution of heavy metal and PCB contaminants in the sediments of an urban estuary: The Hudson River. Mar. Environ. Res. 45, 69–88.

Filippelli, G.M., Mark, A.S., Latimer, J., 2005. Urban lead poisoning and medical geology: An unfinished story. GSA Today 15, 4–11.

Forman, R.T.T., Sperling, D., Bissonette, J.A., Clevenger, A.P., Cutshall, C.D., Dale, V.H., Fahrig, L., France, R., Goldman, C.R., Heanue, K., Jones, J.A., Swanson, F.J., Turrentine, T., Winter, T.C., 2003. Roadside Ecology. Island Press, Washington, DC, p. 481.

Förster, J., 1999. Variability of roof runoff quality. Water Sci. Technol. 39, 137–144.

Fuller, C.C., Harvey, J.W., 2000. Reactive uptake of trace metals in the hyporheic zone of a mining-contaminated stream, Pinal Creek, Arizona. Environ. Sci. Technol. 34, 1150–1155.

Gaillardet, J., Viers, J., Dupré, B., 2004. Trace elements in river waters. In: Holland, H.D., Turekian, K.K. (Eds.), Treatise on Geochemistry. Elsevier, Amsterdam, pp. 225–272.

Gardner, C.B., Carey, A.E., 2004. Trace metal and major ion inputs into the Olentangy River from an urban storm sewer. Environ. Sci. Technol. 38, 5319–5326.

German, J., Svensson, J.G., 2002. Metal content and particle size distribution of street sediments and sweeping waste. Water Sci. Technol. 46, 191–198.

Gonzalez, A.E., Rodriguez, M.T., Sa, J.C.J., Espinosa, A.J.F., DeLaRosa, F.J.B., 2000. Assessment of metals in sediments in a tributary of Guadalquivir River (Spain): Heavy metal partitioning and relation between the water and sediment system. Water Air Soil Poll. 121, 11–29.

Good, J.C., 1993. Roof runoff as a diffuse source of metals and aquatic toxicity of storm water. Water Sci. Technol. 28, 317–321.

Grout, H., Weisner, M.R., Bottero, J.Y., 1999. Analysis of colloidal phases in urban stormwater runoff. Environ. Sci. Technol. 33, 831–839.

Hares, R.J., Ward, N.I., 1999. Comparison of the heavy metal content of motorway stormwater following discharge into wet biofiltration and dry detention ponds along the London Orbital (M25) motorway. Sci. Total Environ. 235, 169–178.

Harrison, R., Tilling, R., Callèn-Romero, M., Harrad, S., Jarvis, K., 2003. A study of trace metals and polycyclic aromatic hydrocarbons in the roadside environment. Atmos. Environ. 37, 2391–2402.

Hart, B.T., Hines, T., 1995. Trace elements in rivers. In: Salbu, B., Steines, E. (Eds.), Trace Elements in Natural Waters. CRC Press, Boca Raton, pp. 203–222.

Hermann, R., Daub, J., Föster, J., Streibel, T., 1994. Chemodynamics of pollutants during roof and street runoff. Water Sci. Technol. 29, 73–82.

Horowitz, A.J., Elrick, K., 1987. The relation of stream sediment surface area, grain size, and composition to trace element chemistry. Appl. Geochem. 2, 437–451.

Horowitz, A.J., Lum, K.R., Garbarino, J.R., Hall, G.E.M., Lemieux, C., Demas, C.R., 1996. Problems associated with using filtration to define dissolved trace element concentrations in natural waters. Environ. Sci. Technol. 30, 954–963.

Horowitz, A.J., 1997. Some thoughts on problems associated with various sampling media used for environmental monitoring. Analyst 122, 1193–1200.

Horowitz, A.J., Meybeck, M., Idlafkih, Z., Biger, E., 1999. Variations in trace element geochemistry in the Seine River basin based on floodplain deposits and bed sediments. Hydrol. Proc. 13, 1329–1340.

Horowitz, A.J., Elrick, K.A., Smith, J.J., 2001a. Estimating suspended sediment and trace element fluxes in large river basins: Methodological considerations as applied to the NASQAN programme. Hydrol. Proc. 15, 1107–1132.

Horowitz, A.J., Elrick, K.A., Smith, J.J., 2001b. Annual suspended sediment and trace element fluxes in the Mississippi, Columbia, Colorado, and Rio Grande drainage basins. Hydrol. Proc. 15, 1169–1207.

Hudson-Edwards, K.A., Macklin, M.G., Taylor, M.P., 1999. 2000 years of sediment-borne heavy metal storage in the Yorkshire Ouse basin, NE England, UK. Hydrol. Proc. 13, 1087–1102.

Kelly, J., Thornton, I., 1996. Urban geochemistry: A study of the influence of anthropogenic activity on the heavy metal content of soils in traditionally industrial and non-industrial areas of Britain. Appl. Geochem. 11, 363–370.

Kerndorff, H., Schnitzer, M., 1980. Sorption of metals on humic acid. Geochimica et Cosmochimica Acta 44, 1701–1708.

Lazaro, T.R., 1990. Urban Hydrology. Technology Publishing Co., Lancaster, Pennsylvania, p. 243.

Legret, M., Pagotto, C., 1999. Evaluation of pollutant loadings in the runoff waters from a major rural highway. Sci. Total Environ. 235, 143–150.

Leigh, D.S., Gamble, D.W., 1997. Survey of nonpoint trace metal inputs to Lake Lanier. In: Hatcher, K. (Ed.), Proceedings of the 1997 Georgia Water Resources Conference, Athens, Georgia, pp. 170–173.

Li, Y.H., 2000. A Compendium of Geochemistry. Princeton University Press, Princeton, New Jersey, p. 475.

Lindgren, Å., 1996. Asphalt wear and pollution transport. Sci. Total Environ. 189, 281–286.

Lindström, M., 2000. Predictive Modelling in Urban Lakes. Comprehensive Summaries of Upssala Dissertations from the Faculty of Science and Technology 584. Ph.D. Dissertation, University of Upsalla, Sweden.

Lyons, T., Kenworth, J., Newman, P., 1990. Urban structure and air pollution. Atmos. Environ. 24B, 43–48.

Marsalek, J., 1990. Evaluation of pollutant loads from urban nonpoint sources. Water Sci. Technol. 22, 23–30.

Martino, M., Turner, A., Millward, G.E., 2003. Influence of organic complexation on the adsorption kinetics of nickel in river waters. Environ. Sci. Technol. 37, 2383–2388.

Mason, Y., Ammann, A.A., Ulrich, A., Sigg, L., 1999. Behavior of heavy metals, nutrients, and major runoff components during roof runoff infiltration. Environ. Sci. Technol. 33, 1588–1597.

McBride, M.B., 1994. Environmental Chemistry of Soils. Oxford University Press, New York, p. 406.

McCabe, M.J., 1998. Lead. In: Zelikoff, J.T., Thomas, P.T. (Eds.), Immunotoxicology of Environmental and Occupational Metals. Taylor & France Ltd, London, pp. 111–130.

Mecray, E.L., King, J.W., Appleby, P.G., Hunt, A.S., 2001. Historical trace metal accumulation in the sediments of an urbanized region of the Lake Champlain watershed, Burlington, Vermont. Water Air Soil Poll. 125, 201–230.

Meybeck, M., Horowitz, A.J., Gosbois, C., 2004. The geochemistry of Seine River Basin particulate matter: Distribution of an integrated metal pollution index. Sci. Total Environ. 328, 219–236.

Mishra, S.K., Sansalone, J.J., Singh, V.P., 2003. Hysteresis-based analysis of overland metal transport. Hydrol. Proc. 17, 1579–1606.
Neal, C., Smith, C.J., Jeffey, H.A., Jarvie, H.P., Robson, A.J., 1996. Trace element concentrations in the major rivers entering Humber estuary, NE England. J. Hydrol. 182, 37–64.
Norra, S., Stüben, D., 2004. Trace element patterns and seasonal variability of dust precipitation in a low polluted city – The example of Karlsruhe/Germany. Environ. Monit. Assess. 93, 203–228.
Nriagu, J.O., Pacyna, J.M., 1988. Quantitative assessment of worldwide contamination of air, water, soils by trace metals. Nature 333, 134–139.
Novotny, V., Olem, H., 1994. Water Quality Prevention, Identification and Managment of Diffuse Pollution. Van Nostrand Reinholt, New York, p. 1054.
Ortega, R., 2002. Analytical methods for heavy metals in the environment: Quantitative determination, speciation, and microscopic analysis. In: Sarkar, B. (Ed.), Heavy Metals in the Environment. Marcel Dekker, Inc., New York, pp. 35–68.
Ozaki, H., Watanabe, I., Kuno, K., 2004. Investigation of the heavy metal sources in relation to automobiles. Water Air Soil Poll. 157, 209–223.
Palm, V., Östlund, C., 1996. Lead and zinc flows from technosphere to biosphere in a city region. Sci. Total Environ. 192, 95–109.
Paode, R.D., Sofuoglu, S.C., Sivadechathep, J., Noll, K.E., Holsen, T.M., 1998. Dry deposition fluxes and mass size distributions of Pb, Cu, and Zn measured in southern Lake Michigan during AEOLOS. Environ. Sci. Technol. 32, 1629–1635.
Power, M., Atrill, M.J., Thomas, R.M., 1999. Heavy metal concentration trends in the Thames estuary. Water Res. 33, 1672–1680.
Ren, J., Packman, A.I., 2004. Stream-subsurface exchange of zinc in the presence of silica and kaolinite colloids. Environ. Sci. Technol. 38, 6571–6581.
Renoldi, M., Cmusso, M., Tartari, G., 1997. The highly polluted Lambro River (N. Italy): Dissolved and solid transport of Cu, Cr, and Fe. Water Air Soil Pollut. 95, 99–118.
Rice, K.C., 1999. Trace-element concentrations in streambed sediment across the conterminous United States. Environ. Sci. Technol. 33, 2499–2504.
Rose, S., Crean, M.S., Sheheen, D.K., Ghazi, A.M., 2001. Comparative zinc dynamics in Atlanta metropolitan region stream and street runoff. Environ. Geol. 40, 983–992.
Salomons, W., Förstner, U., 1984. Metals in the Hydrocycle. Springer-Verlag, Berlin, p. 349.
Sañudo-Wilhelmy, S.A., Gill, G., 1999. Impact of the Clean Water Act on the levels of toxic metals in urban estuaries: The Hudson River estuary revisited. Environ. Sci. Technol. 33, 3477–3481.
SCVURPPP, 1994. Contribution of heavy metals to storm runoff from automotive disc break wear: Santa Clara Valley Urban Pollution Prevention Program. Woodward-Clyde Consulting Report.
Scholes, L., Shutes, R.B.E., Revitt, D.M., Forshaw, M., Purchase, D., 1998. The treatment of metals in urban runoff by constructed wetlands. Sci. Total Environ. 214, 211–219.
Schultze-Lam, S., Urrutia-Mera, M., Beveridge, T.J., 1995. Metal and silicate sorption and subsequent mineral formation on bacterial surfaces: Subsurface implications. In: Allen, H.E. (Ed.), Metal Contaminated Aquatic Sediments. Ann Arbor Press, Chelsea, MI, pp. 111–148.
Schultz, M.F., Benjamin, M.M., Ferguson, J.F., 1987. Adsorption and desorption of metals on ferrihydrite: Reversibility of the reaction and sorption properties on the regenerated surface. Environ. Sci. Technol. 21, 863–869.
Scott, D., McKnight, D.M., Voelker, B.M., Hrncir, D.C., 2002. Redox processes controlling manganese fate and transport in a mountain stream. Environ. Sci. Technol. 36, 453–459.

Seyler, P.T., Boaventura, G.R., 2003. Distribution and partition of trace metals in the Amazon basin. Hydrol. Proc. 17, 1345–1361.

Shafer, M.M., Overdier, J.T., Phillips, H., Webb, D., Sullivan, J.R., Armstrong, D.E., 1999. Trace metal levels and partitioning in Wisconsin rivers. Water Air Soil Pollut. 110, 273–311.

Shea, J.A., 2003. Trace metal adsorption on suspended sediments in surface water: A comparison of urban and less-urbanized stream basins in the Piedmont Province of Georgia. M.S. Thesis, Georgia State University, Atlanta, GA.

Shea, J.A., Rose, S.E., Hallisey, E.J., 2004. A GIS investigation of urbanization and trace metal variability in stream sediments from the Atlanta metropolitan region. Geological Society of America Abstracts with Program 36, CD-ROM.

Shevchenko, V., Lisitzin, A., Vinogradova, A., Stein, R., 2003. Heavy metal aerosols over the seas of the Russia Arctic. Sci. Total Environ. 306, 11–25.

Shiller, A.M., 1997. Dissolved trace elements in the Mississippi River: Seasonal, interannual, and decadal variability. Geochimica et Cosmochimica Acta 61, 4321–4330.

Shiller, A.M., Boyle, E., 1985. Dissolved zinc in rivers. Nature 317, 49–52.

Shiller, A.M., Boyle, E., 1987. Variability of dissolved trace metals in the Mississippi River. Geochimica et Cosmochimica Acta 51, 3273–3277.

Siegel, F.R., 2002. Environmental Geochemistry of Potentially Toxic Metals. Springer, Berlin, p. 218.

Sigg, L., 1998. Partitioning of metal to suspended particles. In: Allen, H., Garrison, A., Luther, G. (Eds.), Metals in Surface Waters. Ann Arbor Press, Chelsea, MI, pp. 217–240.

Singh, M., 2001. Heavy metal pollution in freshly deposited sediments of the Yamuna River (the Ganges River tributary): A case study from Delhi and Agra urban centers, India. Environ. Geol. 40, 664–671.

Smolders, E., Degryse, F., 2002. Fate and effect of zinc from tire debris in soil. Environ. Sci. Technol. 36, 3706–3710.

Sternbeck, J., Sjödin, A., Andrèasson, K., 2002. Metal emissions from road traffic and the influence of resuspension – Results from two tunnel studies. Atmos. Environ. 36, 4735–4744.

Swallow, K.C., Hume, D.N., Morel, F.M.M., 1980. Sorption of copper and lead by hydrous ferric oxide. Environ. Sci. Technol. 14, 1326–1331.

Sweet, C.W., Weiss, A., Vernette, S.J., 1998. Atmospheric deposition of trace metals at three sites near the Great Lakes. Water Air Soil Pollut. 103, 423–439.

Tessier, A., Campbell, P.G.C., Bisso, M., 1979. Sequential extraction procedure for the speciation of particulate metals. Anal. Chem. 51, 844–851.

Treyfrey, J.H., Metz, S., Trocine, R.P., Nelsen, T.A., 1985. A decline in lead transport by the Mississippi River. Science 230, 439–441.

Van Cappellen, P., Wang, Y., 1995. Metal cycling in surface sediments: Modelling the interplay of transport and reaction. In: Allen, H.E. (Ed.), Metal Contaminated Aquatic Sediments. Ann Arbor Press, Chelsea, MI, pp. 21–64.

Van Den Berg, G.A., Loch, J.P.G., Van Der Heijdt, L.M., Zwolsman, J.J.G., 1999. Mobilisation of heavy metals in contaminated sediments in the River Meuse, The Netherlands. Water Air Soil Pollut. 116, 567–586.

van der Voet, E., Udo de Haes, H.A., Guineé, J.B., 2000. General introduction. In: Van der Voet, E., Udo de Haes, H.A., Guineé, J.B. (Eds.), Heavy Metals: A Problem Solved. Kluwer Academic Publishers, Dordrecht, pp. 3–10.

van Loon, G.W., Duffy, S.J., 2000. Environmental Geochemistry: A Global Perspective. Oxford University Press, New York, p. 492.

Van Metre, P.C., Mahler, B.J., 2003. The contribution of particles washed from rooftops to contaminant loading to urban streams. Chemosphere 52, 1727–1741.

Van Metre, P.C., Mahler, B.J., 2004. Contaminant trends in reservoir sediment cores as records of influent stream quality. Environ. Sci. Technol. 38, 2978–2986.

Vink, R., Peters, S., 2003. Modelling point and diffuse heavy metal emissions and loads in the Elbe basin. Hydrol. Proc. 17, 1307–1328.

von Gunten, H.R., Karamtaxas, G., Krähenbühl, U., Kuslys, M., Giovanoli, R., Hoen, E., Keil, R., 1991. Seasonal biogeochemical cycles in riverborne groundwater. Geochimica et Cosmochimica Acta 55, 3597–3609.

von Gunten, H.R., Sturm, M., Moser, R.N., 1997. 200-year record of metals in lake sediments and natural background concentrations. Environ. Sci. Technol. 31, 2193–2197.

Wedepohl, K.H., 1991. The composition of the upper earth's crust and the natural cycles of selected metals. Metals in raw materials. In: Merian, M. (Ed.), Metals and Their Compounds in the Environment. VCH Publishers, Weinheim, pp. 3–18.

Weisner, M., Charaddis, G., Brejchova, D., 1998. Metals and colloids in urban runoff. In: Allen, H., Garrison, A., Luther, G. (Eds.), Metals in Surface Waters. Ann Arbor Press, Chelsea, Milwaukee, pp. 23–65.

Widianarko, B., Verweij, R.A., Van Gretsel, C.A.M., Van Straalen, N.M., 2000. Spatial distribution of trace metals in sediments from urban streams of Seamarang, Central Java, Indonesia. Ecotoxicology and Environmental Safety 46, 95–100.

Wilber, W.G., Hunter, J.V., 1979. Distribution of metals in street sweepings, stormwater solids, and urban aquatic sediments. J. Water Pollut. Con. F. 51, 2810–2822.

Windom, H.L., Byrd, J.T., Smith, R.G., Huan, F., 1991. Inadequacy of NASQAN data for assessing metal trends in the nation's rivers. Environ. Sci. Technol. 25, 1137–1142.

Zhai, M., Kampunzu, H.A.B., Modisi, M.P., Totolo, O., 2003. Distribution of heavy metals in Gaborone urban soils (Botswana) and its relationship to soil pollution and bedrock composition. Environ. Geol. 45, 171–180.

Section II:
Geochemistry in Surface- and Groundwater Research

Developments in Environmental Science, Volume 5
D. Sarkar, R. Datta and R. Hannigan (Editors)

ISSN: 1474-8177/DOI:10.1016/S1474-8177(07)05007-3

Chapter 7

Geochemical cycling of trace and rare earth elements in Lake Tanganyika and its major tributaries

Aboubakar Sako

Abstract

Fe and Mn and light rare earth elements (LREE) were used to evaluate the biogeochemical factors affecting trace element distribution and cycling in Lake Tanganyika, East Africa. Dissolved trace elements were measured in samples from the lake and its three major tributaries (the Ruzizi, Malagarasi and Lufubu rivers) by ICP-MS. The river waters were relatively trace element—and REE—enriched compared to the lake waters. Ce-anomalies varied among the rivers and within the lake. Normalized REE patterns of the lake waters to the average upper continental crust were more similar to those of the Ruzizi River, which might be due to the relatively high pH values in the two systems. Hydrodynamics, Fe, and Mn coating and biogenic activity appeared to be the major biogeochemical factors that affect horizontal distribution of trace elements in the lake. Vertical fractionation of LREE was greater in the Southern Basin than the Northern Basin. The high-trace element and REE concentrations near the surface in the Southern Basin might be due to upwelling in this basin during the dry season. The coincidence of Ce/Ce* and Sm_{UCC}/Yb_{UCC} maxima with those of Fe and Mn (60 m) suggested that Fe- and Mn-oxyhydroxides were responsible for the vertical distribution of trace elements in Lake Tanganyika. The occurrence of nutrient profile minima at 60 m also suggested that trace elements and LREE cycling may be enhanced by bacterial reduction process. This study highlighted the importance of riverine inputs, Fe, Mn and LREE geochemistry and hydrodynamics in spatial abundance, and cycling of trace elements across Lake Tanganyika.

7.1. Introduction

The cycling and distribution of trace elements in lacustrine systems are a function of various geochemical and biological processes, which are governed by environmental factors such as pH, dissolved oxygen contents (DO), nutrient availability and the extent of vertical mixing (Heaney et al., 1986). Because of the disproportional distribution of nutrients, DO and spatial variations in hydrodynamics (e.g., wind strength), the distribution of certain trace elements is likely to vary across large and semi-productive tropical lakes such as Lake Tanganyika. Although a large dataset of dissolved major element geochemistry of the lake and its main tributaries is available (e.g., Plisnier, 1996; Langenberg et al., 2003), those on dissolved trace elements are scarce. The data presented in this paper may therefore contribute to the assessment of the factors that govern dissolved trace element cycling and distribution in Lake Tanganyika.

The geochemical composition of rivers is the primary source of trace element loadings into large waterbodies, such as lakes and oceans (e.g., Garrels and Mackenzie, 1971; Meybeck, 1987). To better understand the processes that control the chemical composition of lakes, an investigation of the geochemical continuity between lakes and rivers is therefore needed. Fe and Mn play a crucial role in controlling trace element distribution in freshwater systems, as riverine dissolved trace elements are generally associated with organically stable Fe- and Mn-oxyhydroxides (Hoyle et al., 1984). Although at river mouths the behavior of Fe and Mn may be altered, following variations in DO and pH, Fe and Mn may allow an accurate evaluation of changes in trace element distribution and their fate after mixing with lake waters (Haberyan and Hecky, 1987; Green et al., 1989).

Furthermore, the vertical distribution of lacustrine trace elements is a function of the oxidative states of Fe and Mn (Haberyan and Hecky, 1987; Viollier et al., 1995; De Carlo and Green, 2002). Because of high DO contents at epilimnion, Fe and Mn may precipitate as oxyhydroxides upon their entry to lakes and sink. As a result, the surface water is generally depleted in dissolved Fe and Mn (e.g., Viollier et al., 1995; Coffey et al., 1997). A subsequent increase in organic matter decompositions with depth may follow, leading to a substantial decrease in pH and DO and reductions of Fe(III) to Fe(II) and Mn(IV) to Mn(II) (De Carlo and Green, 2002). Fe and Mn redox reactions are then accompanied by the release of previously coated trace elements into the water column, resulting in a considerable increase in dissolved Fe(II), Mn(II) and other trace elements with depth (Viollier et al., 1995; De Carlo and Green, 2002).

REE behave in a coherent and predictable manner in natural environments, and they have been used to investigate mixing between various water sources (e.g., Sholkovitz, 1993; Johannesson et al., 1997; Elbaz-Poulichet and Dupuy, 1999; Barrat et al., 2000). Like other trace elements, REE tend to be adsorbed onto the surface sites of clay and organic particles (Byrne and Kim, 1990; Atsuyuki and Kawabe, 2001; Shan et al., 2002). However, their affinity for these ligands is mainly dependent on the decreasing pattern of their ionic radii from LREE to HREE. Hence, HREE, with smaller ionic radii, and which are relatively compatible with organic and inorganic ligands, are less susceptible to adsorption onto oxyhydroxides and are more resistant to scavenging than LREE (Aagaard, 1974; Johannesson et al., 1996). Owing to these unique characteristics, the normalized REE patterns in natural waters compared to their average upper continental crust (UCC) (Taylor and McLennan, 1985) compositions commonly show HREE-enrichment relative to LREE. Dissolved Fe, Mn and LREE, which are non-conservative elements in most aquatic systems, can therefore be used to study the fate and behavior of riverine trace elements in an adjacent lacustrine environment (Webster, 1993; De Carlo and Green, 2002).

The objectives of this study are therefore to (1) use dissolved Fe, Mn and LREE concentrations to investigate the fate and behavior of trace elements from the Ruzizi, Malagarasi and Lufubu rivers coming into Lake Tanganyika; (2) evaluate surface distributions of trace elements in different sub-basins of the lake and (3) study the vertical cycling of trace elements in the Kigoma and Mpulungu sub-basins.

7.2. Hydrogeological setting

Lake Tanganyika, located in the Western Branch of the East African Rift Valley, is the second deepest lake (maximum depth of 1470 m) in the world after Lake Baïkal (Coulter and Spigel, 1991). The lake is 50 km wide and approximately 650 km long in a north–south direction. Lake Tanganyika has a total surface area of 34,000 km^2 and a water volume of 18,880 km^3. The bottom waters of the lake are divided into Northern and Southern basins by a naturally occurring sill of 600 m depth and several sub-basins by minor sills (Tiercelin and Mondeguer, 1991). The annual inflow of Lake Tanganyika is estimated to be 53 $cm\,cm^{-2}$ lake surface, whereas the difference between its annual evaporation and outflow is 127 $cm\,cm^{-2}$ lake surface (Haberyan and Hecky, 1987).

Although the thermal stratification varies seasonally from one basin to another, Lake Tanganyika has been classified as a meromictic lake with a thermocline depth varying between 50 m and 80 m in the Northern Basin and up to 150 m in the Southern Basin (Coulter and Spigel, 1991). The

Southern Basin is characterized by a deeper oxycline (*ca.* 250 m) than the Northern Basin (*ca.* 100 m; Tiercelin and Mondeguer, 1991).

Three major rivers, the Malagarasi, Ruzizi and the Lufubu, contribute to the lake recharge. The Lukuga River, which flows into the Congo River and later into the Atlantic Ocean, is the only outlet from Lake Tanganyika. The temperature of the Malagarasi and Lufubu river waters at their mouths is similar to the surface water of the lake (26.5–28.1°C), and thus their waters mix with the lake water. In contrast, the Ruzizi River water is generally denser (i.e., high-sediment loading) than the surface water of the lake, and does not immediately mix with the lake water (Hecky et al., 1991). The Malagarasi River has the greatest annual flow (*ca.* 300 $m^3 s^{-1}$; Langenberg et al., 2003), followed by the Ruzizi River (*ca.* 200 $m^3 s^{-1}$; Devroey, 1954; Cazenave-Piarrot, 1981) and Lufubu rivers (*ca.* 100 $m^3 s^{-1}$; Sichingabula, 1999).

7.3. Materials and methods

7.3.1. Water sampling

Fourteen surface water samples were collected at the mouths of the three rivers and upstream. Samples were collected once in the rainy and once in the dry season (2003). Because of the large size of the lake, twenty water samples were collected inshore and offshore of four sub-basins (Bujumbura, Kigoma, Moba and Mpulungu sub-basins). The Northern and Southern basins of the lake are characterized by different hydrodynamics, and thus replicate profile samples were collected in the Northern (the Kigoma Sub-basin) and Southern basins (the Mpulungu Sub-basin; Fig. 7.1), respectively.

Surface waters were directly filtered (0.45-μm membrane pore diameter) by a hand pump into pre-acid-washed polyethylene bottles. The tubing system was thoroughly flushed with lake or river waters between samples. Profile samples were collected by a conductivity–temperature–depth (CTD) meter. One sample from each profile was transferred into a clean, pre-acidified polyethylene container, and a portion was filtered and acidified (pH 2). The sampling was carried out under ultra-trace element conditions (Patterson and Settle, 1976). Thus, plastic gloves were worn through out the sampling, and contact with metallic tools was minimized. Samples were subsequently stored in double-zip loc bags.

7.3.2. Water analysis

Cadmium reduction and ascorbic acid methods were used to measure nitrate (NO_3^-) and orthophosphate (PO_4^{3-}) concentrations on the filtered

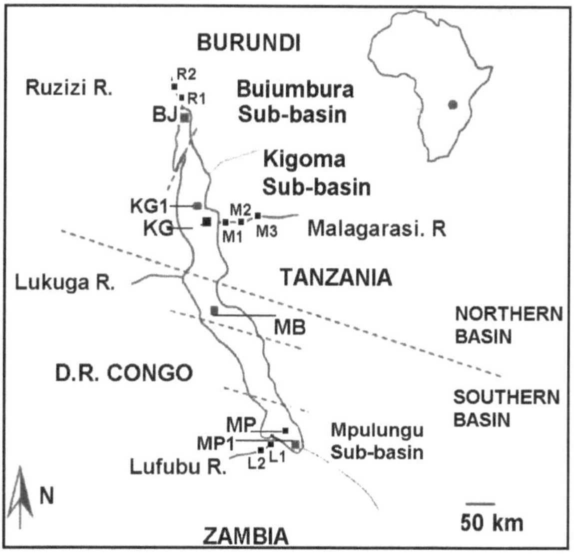

Figure 7.1. Schematic map of Lake Tanganyika, showing the sampling sites; BJ and MB stand for the Bujumbura and Moba sub-basins, whereas KG and MP are profile samples from the Kigoma and Mpulungu sub-basins, respectively.

water samples, using spectrophotometry (Hach DR/2000), whereas pH was measured by a portable Hach pH meter. The analytical precision for the methods was $\pm 0.01\,mg\,l^{-1}$.

The Finnigan Element 2 high-resolution inductively coupled plasma mass spectrometry HR-ICP-MS was used to analyze dissolved trace element and REE concentrations in the laboratory of Isotope and Trace Element Research (LITER) at Old Dominion University, Virginia, without further treatment. The following isotopes were used to control isobaric interferences on REE concentrations: ^{139}La, ^{140}Ce, ^{141}Pr, ^{146}Nd, ^{147}Sm, ^{157}Gd, ^{159}Tb, ^{163}Dy, ^{165}Ho, ^{167}Er, ^{169}Tm, ^{172}Yb and ^{175}Lu. A four-point calibration was used to determine concentrations of trace elements and REE ranging 1–10 ppb for Fe, Mn, Cu, Cd and Zn, 2–200 ppt for Pb and 2–20 ppt for Th, Y and REE. ^{115}In was used as an internal standard to mitigate instrument drift and analyte loss during the analysis (Graham et al., 1996). A blank (i.e., 1% HNO_3) and control standard was run every eight samples, and the blank data were used to calculate instrument detection limits (Vandecasteel and Block, 1993). Relative standard deviations (RSD), based on triplicate analysis, were better than 10% for most elements except for Fe (RSD = 17.49%) and Yb (RSD = 15.52%).

HR-ICP-MS was also successfully used to analyze dissolved riverine and lacustrine Na (RSD = 3.08%), Ca (5.12%) and Mg (4.63%) concentrations. DO, temperature, pH, conductivity and nutrient profiles, measured in the Kigoma Sub-basin in the dry season in 2002 (Sako, unpublished data), were also used in the present study to illustrate their decreasing behavior with depth.

7.3.3. Multivariate analysis

Factor analysis, one of the multivariate methods used in hydrogeochemical studies to investigate inter-relationships between several variables (elements), is a useful technique for reducing a dataset that contains several variables to a few interpretable factors (Reeder et al., 1972; Ruiz et al., 1990; Simeonov et al., 2000). To gain insight into the horizontal distribution of trace elements across the lake, samples collected from the Bujumbura and Kigoma sub-basins during the two sampling periods were categorized as the Northern Basin samples ($n = 10$), whereas those from Moba and Mpulungu sub-basins were treated as the Southern Basin ($n = 10$). Log_e-transformed concentrations of nine trace elements (Cu, Rb, Ba, V, Sr, Zn, Pb, Cd and U) were normally distributed, and consequently these elements were used in the factor analysis test. The test was performed using Minitab version 12.21 (Minitab Inc, 1998).

Varimax rotation, which maximizes the variance of the squared factor loadings, increased the data interpretability (Kaiser, 1958), and was selected to rotate the data; principle component analysis was chosen as the extraction method. The initial extraction yielded nine possible factors, which proved difficult to interpret. However, 96–97% of the total variability for both basins was explained by the first four factors. Although Eigen values for factors 3 and 4 were relatively low (< 1.0), each of these factors accounted for about 12% of the total variance, and were therefore used for further analysis. Furthermore, the communality for factors 1, 2, 3 and 4 was close to 100%, suggesting that the variance of each element is well described by these factors, and the elimination of the five factors did not cause a significant loss of information (Drever, 1997).

7.4. Results

7.4.1. Riverine trace elements and REE abundance

The Lufubu River represented the highest average dissolved Fe and the lowest Mn concentrations among the three rivers followed by the

Table 7.1. Dissolved major, trace element and REE concentrations in the Ruzizi River

Elements	DL	R1-R	R2-R	Mean	R1-D	R2-D	Mean
Na (ppm)	0.01	55.18	53.23	54.21	79.70	82.79	81.24
Mg (ppm)	0.00	40.41	40.27	40.34	57.27	62.35	40.34
Ca (ppm)	0.00	7.50	5.23	6.36	7.00	7.50	7.29
Cd (ppb)	0.00	0.02	0.02	0.02	0.01	0.01	0.01
V (ppt)	0.34	2470.96	685.96	1578.46	678.11	2743.18	1710.65
Mn (ppb)	0.00	1.94	6.69	4.31	9.68	0.52	5.10
Fe (ppb)	0.00	10.21	116.00	63.10	2.04	1.05	1.54
Cu (ppb)	0.00	14.94	1.21	8.07	1.07	1.05	1.06
Zn (ppb)	0.01	23.72	4.49	14.11	0.66	0.21	0.44
La (ppt)	0.00	117.32	24.40	70.86	34.39	43.48	38.94
Ce (ppt)	0.00	210.74	30.81	120.78	32.58	43.55	38.06
Pr (ppt)	0.01	16.87	4.09	10.48	5.71	7.19	6.45
Nd (ppt)	0.01	6.31	36.01	21.16	15.00	18.74	16.87
Sm (ppt)	0.09	5.42	6.02	5.72	4.23	5.34	4.78
Gd (ppt)	0.22	3.27	4.29	3.78	4.95	6.28	5.61
Tb (ppt)	0.02	1.52	1.64	1.58	2.37	2.88	2.62
Dy (ppt)	0.15	8.27	7.24	7.75	13.74	16.17	14.95
Ho (ppt)	0.01	2.21	1.59	1.90	3.76	5.00	4.38
Er (ppt)	0.15	7.30	4.37	5.83	12.59	14.39	13.49
Tm (ppt)	0.03	1.20	0.70	0.95	1.83	2.19	13.49
Yb (ppt)	0.25	8.32	3.69	6.01	10.74	12.81	11.77
Lu (ppt)	0.03	1.15	0.67	0.91	1.74	2.11	1.92
ΣREE		423.81	129.72	276.77	143.61	180.13	161.87
Y (ppt)	0.01	64.31	41.52	52.91	66.01	76.31	71.16
Ce/Ce*		1.05	0.69	0.87	0.52	0.55	0.54
$(La/Yb)_{UCC}$		1.03	0.48	0.87	0.52	0.55	0.54
$(La/Sm)_{UCC}$		2.06	0.29	1.18	0.40	0.40	0.40
$(Gd/Yb)_{UCC}$		0.23	0.67	0.45	0.27	0.28	0.27
$(Er/Nd)_{UCC}$		2.05	4.81	3.34	9.48	8.67	9.07
$(Sm/Yb)_{UCC}$		0.50	1.65	1.08	0.59	0.62	0.60

Note: DL, Detection limit.
R1 and R2 are samples collected at the Ruzizi River mouth and upstream, respectively.
R and D stand for rainy and dry seasons.
$Ce/Ce^* = (Ce_{wat}/Ce_{UCC})/[(La_{wat}/La_{UCC})^{0.5} X (Pr_{wat}/Pr_{UCC})^{0.5}]$.
REE ratios (e.g., $(La/Yb) = (La_{wat}/LaUCC)/(Yb_{wat}/Yb_{UCC})$ are normalized to upper continental crust (UCC, Taylor and McLennan, 1985).

Malagarasi and then the Ruzizi (Tables 7.1–7.3). The highest Mn concentrations were found in the Malagarasi and Ruzizi rivers. The average Cd concentrations were slightly greater in the Malagarasi River (Table 7.2) compared to the Lufubu and Ruzizi rivers. As with the other trace elements, the dissolved V concentrations showed significant variations among the three rivers, with the highest concentrations in the Ruzizi River (Tables 7.1–7.3).

Table 7.2. Dissolved major, trace element and REE concentrations in the Malagarasi River

Elements	M1-R	M2-R	M3-R	Mean	M1-D	M2-D	M3-D	Mean
Na (ppm)	24.80	8.67	8.41	13.96	8.41	12.75	12.93	11.336
Mg (ppm)	18.52	7.11	6.52	10.72	6.52	10.11	9.92	8.85
Ca (ppm)	7.26	10.27	9.08	8.87	9.08	11.95	12.65	11.23
Cd (ppb)	7.05	0.22	0.18	2.48	0.01	0.01	0.40	0.14
V (ppt)	571.48	602.21	631.18	601.62	2854.36	747.00	690.79	1430.72
Mn (ppb)	3.40	9.39	7.74	6.84	0.95	9.83	10.77	7.18
Fe (ppb)	62.56	86.07	123.31	90.65	1.37	2.00	1.52	1.63
Cu (ppb)	1.91	4.06	2.64	2.87	1.37	0.82	0.61	0.93
Zn (ppb)	16.43	39.53	29.56	28.51	0.25	1.07	0.23	0.52
SiO_2 (ppm)	7.02	6.97	6.89	6.96	n.d	n.d	n.d	n.d
T (°C)	26.90	27.70	27.20	27.27	n.d	n.d	n.d	n.d
pH	8.36	8.35	8.30	8.34	n.d	n.d	n.d	n.d
DO (ppm)	7.48	6.80	7.18	7.15	n.d	n.d	n.d	n.d
La (ppt)	22.87	39.65	50.16	37.56	10.19	9.86	5.96	8.67
Ce (ppt)	31.28	51.94	70.73	51.32	12.92	11.62	5.87	10.14
Pr (ppt)	2.73	5.28	6.53	4.85	1.27	1.23	0.72	1.07
Nd (ppt)	6.41	12.43	15.25	11.36	3.18	3.09	2.00	2.76
Sm (ppt)	1.82	3.07	3.96	2.95	0.84	0.96	0.59	0.80
Gd (ppt)	12.13	3.37	4.45	3.32	0.95	1.18	0.71	0.94
Tb (ppt)	0.77	1.37	1.57	1.24	0.34	0.36	0.27	0.32
Dy (ppt)	3.66	6.30	6.84	5.60	1.77	1.68	1.44	1.63
Ho (ppt)	1.55	1.40	2.30	1.75	0.41	0.36	0.27	0.32
Er (ppt)	6.52	3.94	7.92	6.13	1.20	1.21	1.14	1.19
Tm (ppt)	0.52	0.59	0.68	0.60	0.20	0.16	0.13	0.16
Yb (ppt)	9.11	3.29	9.67	7.36	0.99	1.06	0.85	0.97
Lu (ppt)	0.90	0.55	1.04	0.83	0.16	0.16	0.14	0.15
ΣREE	90.29	133.19	181.11	134.87	34.42	32.97	20.21	29.20
Y (ppt)	16.57	16.20	17.03	16.60	6.14	5.84	5.42	5.80
Ce/Ce*	0.85	0.79	0.85	0.83	0.78	0.72	0.61	0.70
$(La/Yb)_{UCC}$	0.18	0.88	0.38	0.48	0.76	0.68	0.51	0.65
$(La/Sm)_{UCC}$	0.72	0.61	0.64	0.66	0.59	0.54	0.50	0.54
$(Gd/Yb)_{UCC}$	0.14	0.59	0.27	0.33	0.55	0.64	0.48	0.56
$(Er/Nd)_{UCC}$	11.49	3.57	5.87	6.97	4.27	4.44	6.45	5.05
$(Sm/Yb)_{UCC}$	0.26	1.45	0.59	0.77	0.27	1.26	1.03	1.19

Note: M1, M2 and M3 are samples collected at the Malagarsi River mouth, 500m and 1000m upstream, respectively. n.d not determined.

The average total REE concentrations (ΣREE) were higher in the Lufubu River than the two other rivers (Tables 7.1–7.3). Although the Ruzizi and Malagarasi rivers showed the same seasonal variation patterns in their REE concentrations, the average seasonal ΣREE were higher in the Ruzizi River compared to the Malagarasi River. Whereas the average dissolved Y concentrations were comparable between the Ruzizi and

Table 7.3. Dissolved major, trace element and REE concentrations in the Lufubu River

Elements	L1-R	L2-R	Mean	L1-D	L2-D	Mean
Na (ppm)	3.31	3.21	3.26	3.03	2.95	2.99
Mg (ppm)	0.56	0.45	0.50	0.58	0.58	0.58
Ca (ppm)	1.04	1.30	1.17	1.34	1.34	1.34
Cd (ppb)	0.23	0.18	0.20	0.02	0.03	0.02
V (ppt)	238.08	195.41	216.75	158.69	159.52	159.11
Mn (ppb)	1.89	0.87	1.38	3.52	3.47	3.50
Fe (ppb)	129.98	94.75	112.36	100.09	97.74	98.91
Cu (ppb)	3.82	3.09	3.46	1.81	1.71	1.76
Zn (ppb)	44.80	31.82	38.31	7.98	8.56	8.27
SiO_2 (ppm)	6.22	0.95	3.58	1.081	1.53	1.31
pH	7.83	7.35	7.59	8.88	8.93	8.91
T (°C)	28.00	28.00	28.00	24.60	22.90	23.75
La (ppt)	419.12	371.06	395.09	416.41	440.53	428.47
Ce (ppt)	412.98	421.21	417.10	216.76	230.72	223.74
Pr (ppt)	41.22	42.53	41.87	20.70	21.89	21.29
Nd (ppt)	92.08	94.21	93.14	10.07	10.64	10.36
Sm (ppt)	21.04	17.79	19.41	10.07	10.64	10.36
Gd (ppt)	22.07	16.27	19.17	10.47	11.05	10.76
Tb (ppt)	7.47	5.40	6.44	3.54	3.76	3.65
Dy (ppt)	32.61	23.62	28.12	15.58	16.84	16.21
Ho (ppt)	7.46	5.40	6.43	3.42	3.67	3.55
Er (ppt)	21.49	15.53	18.51	9.46	10.03	9.74
Tm (ppt)	3.06	2.29	2.67	1.47	1.50	1.49
Yb (ppt)	17.46	12.71	15.09	8.44	9.05	8.74
Lu (ppt)	2.73	1.95	2.34	1.21	1.31	1.26
Sum REE	1100.79	1029.98	1065.39	762.26	808.31	785.29
Y (ppt)	84.10	61.26	72.68	36.07	38.08	37.07
Ce/Ce*	0.65	0.72	0.68	0.40	0.41	0.40
$(La/Yb)_{UCC}$	1.76	2.14	1.95	3.62	3.57	3.59
$(La/Sm)_{UCC}$	1.02	2.89	1.95	2.09	2.11	2.10
$(Gd/Yb)_{UCC}$	0.73	0.74	0.74	0.67	0.71	0.69
$(Er/Nd)_{UCC}$	2.63	1.86	2.24	2.38	2.39	2.38
$(Sm/Yb)_{UCC}$	1.72	1.75	1.74	1.74	1.69	1.71

Note: L1 and L2 are the samples collected at the Ruzizi mouth and 500m upstream. R and D stand for rainy and dry seasons, respectively.

Lufubu rivers for both sampling periods, they were considerably lower in the Malagarasi River (Tables 7.1–7.3).

A small positive Ce-anomaly ($Ce/Ce^* = 1.05$) was observed at the Ruzizi River mouth during the rainy season, whereas a substantial negative anomaly occurred upstream ($Ce/Ce^* = 0.69$). The Ce/Ce^* values were similar within the Ruzizi River during the dry season (0.52 and 0.55). The Malagarasi River exhibited the lowest negative Ce-anomalies

($0.60 < Ce/Ce^* < 0.85$), without a significant seasonal variation. The highest negative Ce-anomalies occurred in the Lufubu River, which decreased in the rainy season samples (Table 7.3).

With the exception of its mouth ($La_{UCC}/Yb_{UCC} = 1.05$), the Ruzizi River showed HREE-enrichment (La_{UCC}/Yb_{UCC} ratios ranged from 0.23 to 0.48). Likewise, the Malagarasi River samples were HREE-enriched relative to LREE (La_{UCC}/Yb_{UCC} ratios ranged from 0.18 to 0.88; Figs. 7.2 and 7.3). In contrast to the two northern rivers, the Lufubu River showed LREE-enrichment (La_{UCC}/Yb_{UCC} ratios ranged from 1.76 to 3.62). However, Er_{UCC}/Nd_{UCC} values for the three rivers were greater than unity, suggesting that La concentrations were exceptionally high in certain samples (particularly in the Lufubu River samples). All samples showed HREE-enrichment relative to the middle REE (MREE; Sm to Dy) with $Gd_{UCC}/Yb_{UCC} < 1$.

7.4.2. Surface trace elements and REE distribution in the Lake

Surface trace elements and REE concentrations of Lake Tanganyika are presented in Tables 7.4–7.7. Elements such as Fe, Mn, Th, Ba, Cd, Zn and Pb fluctuated widely across the lake, whereas Sr, Rb, to a lesser extent, U and V were constant. Finally, the average U/Sr values for different the sub-basins were: $3.18 \pm 0.21 \times 10^{-3}$ (Bujumbura), $2.93 \pm 0.19 \times 10^{-3}$ (Kigoma), $2.87 \pm 0.23 \times 10^{-3}$ (Moba) and $2.69 \pm 0.16 \times 10^{-3}$ (Mpulungu).

7.4.2.1. Factors controlling horizontal distribution of trace elements in the lake

The factor analysis test, performed on surface waters, identified four dominant factors that may explain the distribution and behavior of Cu, Rb, Ba, V, Sr, Zn, Pb, Cd and U across the lake (Tables 7.8 and 7.9). Factor 1, which accounted for 44% and 54% of the total variability in the Northern and Southern basins, respectively, was characterized by high-positive loading coefficients for Ba, V, Sr and U and a moderate negative loading for Zn (55%) in the Northern Basin, and high-positive loading coefficients for Rb, Ba, V, Sr and U in the Southern Basin. Factor 1 also showed moderate positive loadings for Pb (47%) in the Northern Basin and Cd (35%) in the Southern Basin. Factor 2 accounted for 26% and 19% of the total variability in the Northern and Southern basins, respectively, with high-negative loading coefficients for Cu, Rb and Zn in the Northern Basin and high-positive loading coefficients for Cu and Pb in the Southern Basin. Factor 3 had a significant influence on Pb concentrations in the Northern Basin, whereas Cd was most affected by factor 3 in the Southern Basin. Factor 4, which only accounted for 12%

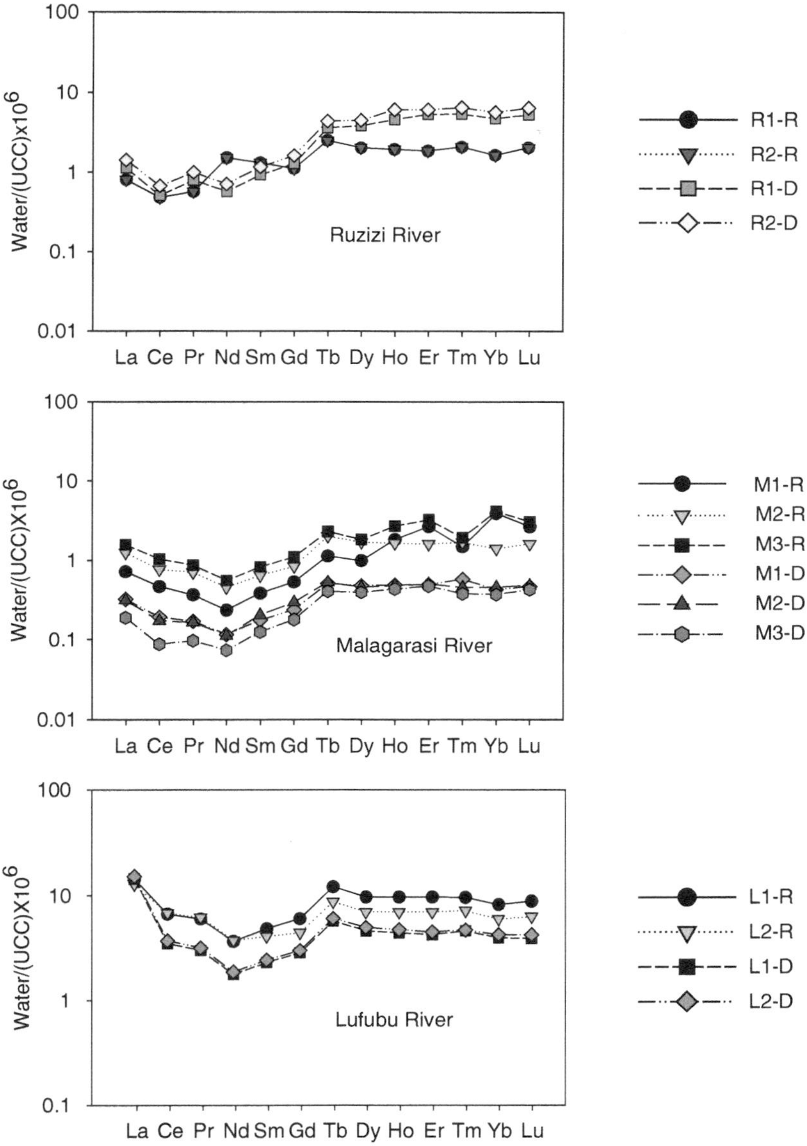

Figure 7.2. UCC-normalized riverine REE patterns. 1, 2 and 3 stand for river mouth, at 500 m and 1000 m upstream, respectively. R&D represents the rainy and dry seasons.

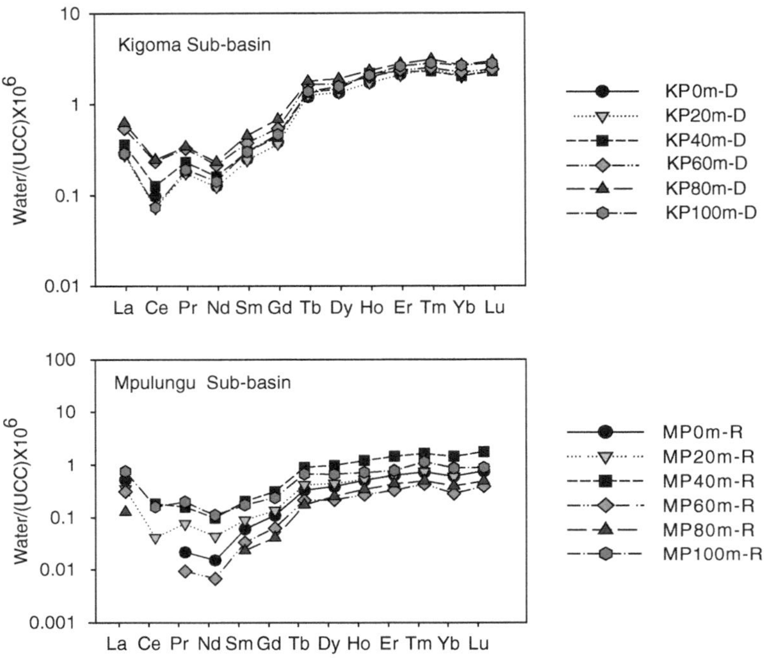

Figure 7.3. Representative UCC-normalized lacustrine REE patterns of the rainy and dry season profile samples from the Kigoma and Mpulungu sub-basins, respectively.

(Northern Basin) and 11% (Southern Basin) of the total variability, was characterized by high-negative loadings for Cd in the Northern Basin and Zn in the Southern Basin.

7.4.2.2. REE distribution and patterns

Average ΣREE concentrations showed large spatial and temporal variations across the lake with the highest values in the Mpulungu Sub-basin followed by the Bujumbura, Kigoma and Moba sub-basins (Tables 7.4–7.7). As with Fe and Mn, most of the LREE concentrations were below detection limits in the Kigoma and Moba sub-basins during the rainy season. The Mpulungu Sub-basin exhibited the highest negative Ce-anomalies, whereas the lowest negative values were recorded in the Bujumbura Sub-basin (Tables 7.4–7.6). The UCC-normalized REE patterns (Fig. 7.4) showed a predominantly HREE-enrichment over LREE with La_{UCC}/Yb_{UCC} ratios lower than unity, except for samples PM1-R

Table 7.4. Dissolved major, trace element and REE in Bujumbura Sub-basin (Littoral zone)

Elements	BL-R	BL-D	Mean	UP-R	UP-D	Mean
Na (ppm)	64.77	56.83	60.80	53.99	55.54	54.77
Mg (ppm)	15.32	41.08	28.20	38.62	39.66	39.14
Ca (ppm)	10.78	10.50	10.64	10.49	10.29	10.39
Rb (ppb)	8.82	24.40	16.61	22.70	22.74	22.72
Cd (ppb)	0.01	0.05	0.03	0.02	2.20	1.11
Ba (ppb)	50.04	73.75	61.90	124.06	70.25	97.16
Pb (ppt)	136.05	193.00	164.56	77.91	28.80	53.35
Th (ppt)	5.34	0.31	2.83	<DL	1.01	1.01
U (ppt)	33.38	352.47	192.93	325.47	345.82	335.65
V (ppt)	551.23	907.39	729.31	963.19	835.17	899.18
Mn (ppb)	58.46	1.65	30.05	5.5	0.21	2.85
Fe (ppb)	<DL	0.71	0.71	24.76	0.85	12.81
Cu (ppb)	9.55	0.68	5.12	3.04	0.41	1.72
Zn (ppb)	27.52	0.70	14.11	8.84	0.8	4.82
Sr (ppb)	98.06	119.64	108.85	99.69	103.40	101.54
DO (ppm)	n.d	7.50	7.50	8.00	n.d	8.00
pH	9.00	9.10	9.05	9.81	n.d	9.81
Y (ppt)	17.65	8.83	13.24	37.71	1.94	19.83
La (ppt)	54.02	7.33	30.67	23.29	27.93	25.61
Ce (ppt)	72.85	5.76	39.31	30.16	24.84	27.50
Pr (ppt)	6.26	1.13	3.67	8.36	10.66	9.51
Nd (ppt)	14.24	3.11	8.67	8.36	10.66	3.87
Sm (ppt)	3.50	0.98	2.24	2.59	5.14	3.87
Gd (ppt)	3.84	1.21	2.52	3.16	3.56	3.36
Tb (ppt)	1.26	0.66	0.96	1.31	1.48	1.39
Dy (ppt)	5.45	4.43	4.94	7.13	8.15	7.64
Ho (ppt)	1.31	1.62	1.47	2.11	2.36	2.23
Er (ppt)	3.24	4.59	3.92	6.11	6.82	6.46
Tm (ppt)	0.59	0.70	0.64	0.98	1.00	0.95
Yb (ppt)	3.35	3.91	3.63	6.93	5.85	6.39
Lu (ppt)	0.49	0.70	0.60	0.94	0.96	0.95
Sum REE	170.41	36.12	103.27	96.41	102.96	99.68
Ce/Ce*	0.85	0.45	0.65	0.76	0.51	0.63
$(La/Yb)_{UCC}$	1.18	0.14	0.66	0.25	0.35	0.30
$(La/Sm)_{UCC}$	0.79	0.37	0.58	0.48	0.47	0.48
$(Gd/Yb)_{UCC}$	0.66	0.18	0.42	0.26	0.35	0.31
$(Er/Nd)_{UCC}$	2.57	16.70	9.63	8.25	7.23	7.74
$(Sm/Yb)_{UCC}$	1.49	0.38	0.93	0.51	0.74	0.63

Note: BL, sample collected in the littoral zone in the Bujumbura area (Bujumbura Sub-basin); UP, sample collected in the pelagic zone in the Uvira area (Bujumbura Sub-basin); R and D stand for rainy and dry seasons, respectively.

Table 7.5. Dissolved major, trace element and REE concentrations in the Kigoma Sub-basin (littoral zone)

Elements	KG1-R	KG2-R	KG3-R	Mean	KG1-D	KG2-D	KG3-D	Mean
Na (ppm)	n.d	n.d	n.d	n.d	57.76	54.23	58.53	56.84
Mg (ppm)	n.d	n.d	n.d	n.d	43.01	38.34	41.22	40.86
Ca (ppm)	n.d	n.d	n.d	n.d	10.50	9.97	10.68	10.38
Rb (ppb)	5.93	1.38	3.39	3.57	22.39	21.56	22.56	22.17
Cd (ppb)	0.03	0.00	0.01	0.01	0.01	0.03	0.08	0.04
Ba (ppb)	35.90	7.38	18.14	20.47	70.70	68.93	70.37	70.00
Pb (ppt)	44.13	2.61	18.22	21.65	24.15	54.10	n.d	26.09
Th (ppt)	<DL	<DL	<DL		0.55	0.52	0.85	0.64
U (ppt)	103.59	24.77	45.47	57.95	309.27	296.89	308.49	304.88
V (ppt)	295.14	79.24	116.73	163.75	893.20	750.23	872.19	838.54
Mn (ppb)	0.01	<DL	<DL	0.01	0.06	0.20	0.29	0.18
Fe (ppb)	0.04	<DL	<DL	0.04	0.37	0.90	2.97	1.41
Cu (ppb)	0.23	0.08	0.07	0.13	0.08	0.23	0.37	0.23
Zn (ppb)	4.06	2.57	3.54	3.39	0.10	0.57	7.77	2.81
Sr (ppb)	35.77	7.60	16.15	19.84	108.37	96.54	113.02	105.98
SiO_2 (ppm)	0.24	0.24	0.20	0.23	n.d	n.d	n.d	n.d
NO_3^- (ppm)	0.15	0.10	0.03	0.23	n.d	n.d	n.d	n.d
PO_4^{-3} (ppm)	0.00	0.02	5.69	1.90	n.d	n.d	n.d	n.d
DO (ppm)	8.50	8.60	7.90	8.33	n.d	n.d	n.d	n.d
pH	9.17	9.09	9.23	9.16	n.d	n.d	n.d	n.d
Y (ppt)	3.70	23.66	28.27	18.54	2.78	45.61	34.26	27.55
La (ppt)	4.04	0.47	<DL	1.50	16.61	19.84		12.15
Ce (ppt)	1.80	<DL	<DL	0.60	6.63	8.84	19.36	11.61
Pr (ppt)	0.50	<DL	<DL	0.17	1.69	1.67	3.00	2.12
Nd (ppt)	1.36	<DL	<DL	0.45	4.60	4.47	8.54	5.87
Sm (ppt)	0.14	0.00	<DL	0.05	1.56	1.41	2.42	1.80
Gd (ppt)	0.23	<DL	<DL	0.08	1.96	1.73	2.85	2.18
Tb (ppt)	0.29	0.04	0.03	0.12	0.90	0.87	1.12	0.96
Dy (ppt)	1.58	0.34	0.52	0.81	5.61	5.29	6.70	5.87
Ho (ppt)	0.41	0.19	0.52	0.37	1.59	1.49	1.83	1.64
Er (ppt)	1.33	0.22	1.28	0.94	5.45	5.08	6.15	5.56
Tm (ppt)	0.20	0.04	0.06	0.10	0.88	0.83	0.93	0.88
Yb (ppt)	1.13	0.21	0.26	0.53	5.27	4.71	5.70	5.23
Lu (ppt)	0.16	0.01	0.05	0.07	0.86	0.82	0.91	0.86
ΣREE	13.17	1.52	2.72	5.80	53.60	57.06	59.51	56.72
Ce/Ce*	0.28			0.28	0.26	0.31		0.29
$(La/Yb)_{UCC}$	0.26	0.17		0.21	0.23	0.31		0.27
$(La/Sm)_{UCC}$	0.26	0.17		0.21	0.23	0.31		0.27
$(Gd/Yb)_{UCC}$	0.12			0.26	0.23	0.31		0.21
$(Er/Nd)_{UCC}$	10.99			10.99	13.39	12.84	8.13	11.45
$(Sm/Yb)_{UCC}$	0.58			0.58	0.14	0.15		0.15

Table 7.6. Dissolved major, trace element and REE concentrations in the Moba Sub-basin (littoral zone)

Elements	MB1-R	MB2-R	MB3-R	Mean	MB1-D	MB2-D	MB3-D	Mean
Na (ppm)	n.d	n.d	n.d	n.d	60.35	57.13	55.55	57.68
Mg (ppm)	n.d	n.d	n.d	n.d	42.14	40.89	40.22	41.08
Ca (ppm)	n.d	n.d	n.d	n.d	10.96	10.74	10.63	10.78
Rb (ppb)	11.02	1.42	6.83	6.42	22.73	22.47	22.82	22.67
Cd (ppb)	0.02	0.00	0.02	0.01	0.02	0.00	0.07	0.03
Ba (ppb)	33.41	5.11	18.62	19.05	72.04	71.02	70.63	71.10
Pb (ppt)	5324.48	37.16	85.11	1815.58	92.96	12.17	477.63	194.25
Th (ppt)	<DL	<DL	<DL		1.07	0.39	0.72	0.73
U (ppt)	168.13	22.59	95.20	95.30	301.43	306.69	315.28	307.80
V (ppt)	449.42	48.06	270.50	255.99	861.04	877.88	876.41	871.78
Mn (ppb)	0.05	<DL	<DL	0.05	0.17	0.11	0.44	0.24
Fe (ppb)	4.03	<DL	<DL	4.03	1.21	0.85	5.13	2.39
Cu (ppb)	4.34	0.10	0.44	1.63	0.46	0.12	1.56	0.71
Zn (ppb)	7.01	1.85	6.74	5.20	0.50	0.13	0.93	0.52
Sr (ppb)	57.39	6.96	31.98	32.11	116.15	115.24	110.49	113.96
Y (ppt)	10.20	1.89	2.78	4.96	29.54	31.92	33.17	24.90
La (ppt)	7.50	<DL	<DL	7.50	11.21	12.27	13.36	11.09
Ce (ppt)	9.14	<DL	<DL	9.14	8.33	8.93	9.92	9.08
Pr (ppt)	0.94	<DL	<DL	0.94	1.68	1.95	2.02	1.65
Nd (ppt)	2.24	<DL	<DL	2.24	4.40	5.20	5.34	4.29
Sm (ppt)	1.71	<DL	<DL	1.71	4.21	4.88	5.34	4.29
Gd (ppt)	0.55	<DL	<DL	0.55	1.58	1.94	2.21	1.57
Tb (ppt)	0.28	<DL	<DL	0.28	0.87	0.94	1.01	0.78
Dy (ppt)	1.94	0.12	0.26	0.77	5.35	5.87	6.01	4.50
Ho (ppt)	0.90	0.23	0.47	0.53	1.76	1.85	2.09	1.56
Er (ppt)	1.73	0.24	1.01	0.99	5.24	5.72	5.98	4.48
Tm (ppt)	0.26	0.02	0.07	0.12	0.87	0.92	0.90	0.70
Yb (ppt)	1.84	0.15	0.46	0.81	4.97	5.17	5.69	4.16
Lu (ppt)	0.30	0.01	0.09	0.13	0.85	0.86	0.95	0.70
Sum REE	29.34	0.76	2.36	10.82	51.31	56.50	60.83	48.59
Ce/Ce*	0.75				0.43	0.41	0.42	0.42
$(La/Yb)_{UCC}$					0.17	0.15	0.70	0.34
$(La/Sm)_{UCC}$					1.28	0.50	0.84	0.87
$(Gd/Yb)_{UCC}$					0.18	0.33	0.85	0.87
$(Er/Nd)_{UCC}$					13.47	12.42	12.99	12.96
$(Sm/Yb)_{UCC}$					0.13	0.30	0.83	0.42

and PM1-D in the Mpulungu Sub-basin and BL-R in the Bujumbura Sub-basin (Figs. 7.2 and 7.3). Er_{UCC}/Nd_{UCC} values also indicated HREE-enrichment relative to LREE. As was the case with the rivers, the lake waters showed HREE-enrichment relative to MREE with Gd_{UCC}/Yb_{UCC} ratios less than unity.

Table 7.7. Dissolved major, trace element and REE concentrations Mpulungu Sub-basin (littoral zone)

Elements	MP1-R	MP2-R	Mean	MP1-D	MP2-D	Mean
Na (ppm)	56.84	56.04	56.44	57.25	54.74	56.00
Mg (ppm)	41.19	40.85	41.02	42.70	41.64	42.17
Ca (ppm)	10.55	10.16	10.35	11.04	10.64	10.84
Rb (ppb)	22.93	22.73	22.83	23.29	22.86	23.07
Cd (ppb)	0.03	0.02	0.03	0.04	0.00	0.02
Ba (ppb)	74.29	73.27	73.78	101.22	75.65	88.43
Pb (ppt)	485.01	180.45	332.73	1879.12	1181.85	1530.48
Th (ppt)	2.92	1.57	2.25	0.58	0.34	0.46
U (ppt)	296.38	292.64	294.51	291.49	294.05	292.77
V (ppt)	792.36	755.72	774.04	840.76	773.72	807.24
Mn (ppb)	1.11	0.19	0.65	0.59	0.18	0.38
Fe (ppb)	8.77	0.90	4.84	2.79	1.41	2.10
Cu (ppb)	2.23	0.67	1.45	2.23	0.67	1.45
Zn (ppb)	10.18	0.99	5.58	12.98	2.67	7.82
Sr (ppb)	106.81	99.81	103.31	116.31	108.30	112.31
pH	8.71	8.72	8.71	9.02	9.02	9.02
Y (ppt)	45.61	34.26	39.93	42.02	33.82	37.92
La (ppt)	116.63	40.71	78.67	113.28	78.10	37.92
Ce (ppt)	84.88	20.11	52.49	47.73	21.00	34.37
Pr (ppt)	9.49	3.26	6.37	7.11	3.36	5.23
Nd (ppt)	21.45	8.22	14.83	15.76	7.84	11.80
Sm (ppt)	15.79	6.78	11.29	10.69	6.19	8.44
Gd (ppt)	6.17	2.87	4.52	4.46	2.89	3.67
Tb (ppt)	2.30	1.19	1.75	1.80	1.18	1.49
Dy (ppt)	11.38	6.85	9.11	9.33	6.67	8.00
Ho (ppt)	4.44	2.62	3.53	2.71	2.00	2.35
Er (ppt)	8.83	6.38	7.60	7.86	6.38	7.12
Tm (ppt)	1.39	0.97	1.18	1.19	0.94	1.07
Yb (ppt)	8.12	5.92	7.02	6.82	5.80	6.31
Lu (ppt)	1.35	1.01	1.18	1.10	0.94	1.02
ΣREE	337.82	141.13	239.47	271.86	177.12	224.49
Ce/Ce*	0.51	0.35	0.43	0.31	0.21	0.26
$(La/Yb)_{UCC}$	1.05	0.50	0.78	0.99	1.22	1.10
$(La/Sm)_{UCC}$	3.18	2.53	2.85	4.90	4.25	4.57
$(Gd/Yb)_{UCC}$	0.44	0.28	0.36	0.29	0.38	0.33
$(Er/Nd)_{UCC}$	8.77	9.19	8.98	31.96	5.63	18.79
$(Sm/Yb)_{UCC}$	0.33	0.20	0.27	0.20	0.29	0.24

7.4.2.3. Trace elements and REE distributions in the water column

Vertical distribution of trace elements in lakes is mainly a function of physico-chemical parameters such as temperature, DO, pH and nutrients (Webster, 1993; De Carlo and Green, 2002). The average Sr and Ba concentrations showed slight seasonal variations in the Kigoma

Table 7.8. Four factors, controlling trace element loading into the Northern Basin of Lake Tanganyika

Elements	Factor 1	Factor 2	Factor 3	Factor 4	Communality
Cu	0.177	**−0.911**	−0.222	−0.098	0.921
Rb	0.294	**−0.905**	−0.06	0.059	0.913
Ba	**0.884**	−0.0353	−0.178	−0.178	0.97
V	**0.909**	−0.285	−0.165	−0.227	0.986
Sr	**0.874**	−0.353	−0.178	−0.178	0.97
Zn	−0.557	**−0.612**	−0.534	0.015	0.97
Pb	0.472	−0.233	**−0.837**	−0.101	0.987
Cd	0.371	−0.011	−0.067	**−0.924**	0.996
U	**0.916**	0.118	−0.062	−0.269	0.928
Variance	3.999	2.412	1.147	1.08	8.64
% of variability	0.444	0.268	0.127	0.12	0.96

Note: Significant loading coefficients are in bold.

Table 7.9. Four factor, controlling trace element loading into the Southern Basin of Lake Tanganyika

Elements	Factor 1	Factor 2	Factor 3	Factor 4	Communality
Cu	0.22	**0.922**	0.058	−0.197	0.94
Rb	**0.965**	0.154	0.174	−0.111	0.998
Ba	**0.949**	0.188	0.139	−0.201	0.995
V	**0.963**	0.133	0.221	−0.038	0.995
Sr	**0.969**	0.174	0.153	−0.081	0.999
Zn	0.162	0.354	0.038	**−0.918**	0.996
Pb	0.161	**0.764**	0.413	−0.296	0.868
Cd	0.355	0.235	**0.889**	−0.027	0.972
U	**0.966**	0.155	0.183	−0.08	0.998
Variance	4.859	1.745	1.121	1.036	8.763
% of variability	0.54	0.194	0.125	0.115	0.974

Note: Significant loading coefficients are in bold.

Sub-basin (Table 7.10), whereas the Ba profile coincided with those of NO_3^- and PO_4^{3-} in the Mpulungu Sub-basin at 20 m (maxima) and 60 m (minima; Fig. 7.4). Additionally, Th, U, V, Mn and Fe exhibited relatively high concentrations in the dry season compared to the rainy season, whereas the opposite was observed for Pb, Zn and Cu (Tables 7.10 and 7.11). In contrast to Fe, Mn, Th, Ba and nutrient profiles, which showed non-conservative behavior in the two sub-basins, Sr (Fig. 7.4) and Rb (Tables 7.10 and 7.11) concentrations appeared to be less affected by changes in pH, DO, or biological uptake.

A striking similarity was found between the Fe and Mn distributions in the two locations, which is probably to be expected. Dissolved concentrations

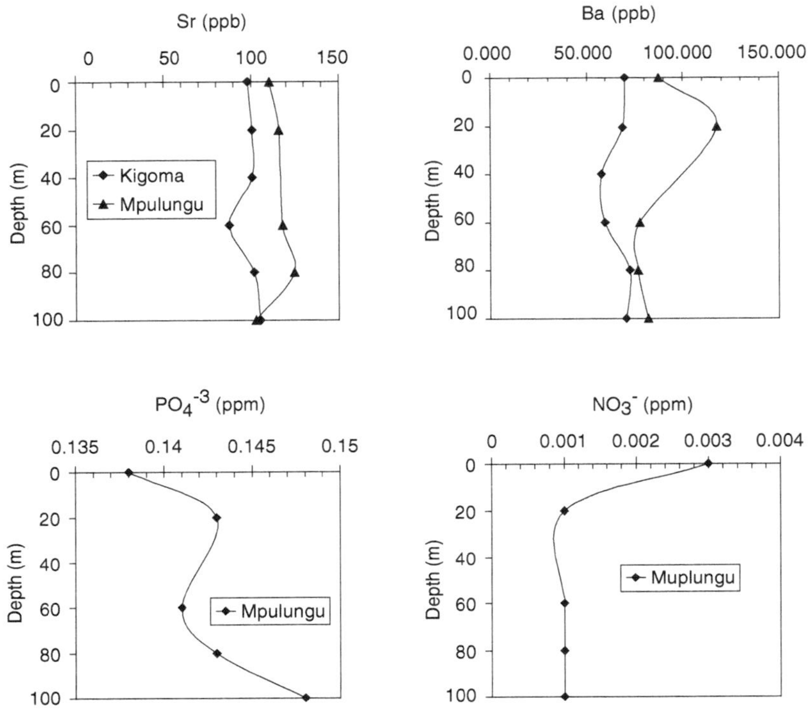

Figure 7.4. Sr, Ba, PO_4^{3-} and NO_3^- profiles in the Kigoma and Mpulungu sub-basins (dry season in 2003).

of these two elements reached their maxima at 60 m (Fig. 7.5). The maximum concentrations of Th were also observed at this depth.

REE showed decreasing concentrations from the surface to 60 m, where they reached their maxima, dropped at 80 m and increased at 100 m. The local REE maxima were observed at 40 m in the Mpulungu Sub-basin with a similar fluctuation across the profile, as seen in the Kigoma Sub-basin. The absolute concentrations of REE, in the water column, were higher in the dry season than the rainy season for the two sub-basins (Table 7.12).

7.4.2.4. Inter-element relationships

The PO_4^{-3} and Cd profiles collected in the Mpulungu Sub-basin during the dry season showed a substantial negative correlation ($r = -0.823$; $P = 0.087$). A strong positive correlation was also found between V and U ($r = 0.976$; $P < 0.001$; Fig. 7.6a) and between V and Rb ($r = 0.959$;

Table 7.10. Dissolved major and trace element profiles from the Kigoma Sub-basin

	Sample						Mean
	KP1-R	KP2-R	KP3-R	KP4-R	KP5-R	KP6-R	
	(0 m)	(20 m)	(40 m)	(60 m)	(80 m)	(100 m)	
Na (ppm)	43.47	53.33	28.33	54.01	27.51	55.36	43.67
Mg (ppm)	32.25	37.86	21.48	39.82	20.08	36.79	31.38
Ca (ppm)	n.d	n.d	n.d	9.67	4.97	9.62	8.09
Rb (ppb)	16.13	21.25	11.33	21.55	10.34	20.83	16.90
Cd (ppb)	0.03	0.24	0.01	0.03	0.02	0.02	0.06
Ba (ppb)	52.50	66.85	36.79	66.54	34.73	67.100	54.080
Pb (ppt)	51.43	207.20	15.84	125.53	29.08	104.81	88.98
Th (ppt)	<DL	<DL	<DL	1.54	0.58	2.24	0.73
U (ppt)	232.38	281.6	162.46	269.12	153.75	264.57	227.31
V (ppt)	639.24	775.96	437.05	700.74	323.39	575.24	575.27
Mn (ppb)	<DL	0.07	<DL	0.21	<DL	<DL	0.04
Fe (ppb)	2.03	<DL	<DL	0.27	<DL	<DL	0.38
Cu (ppb)	0.58	0.93	0.030	1.39	0.19	0.33	0.62
Zn (ppb)	1.59	9.94	0.56	1.67	0.57	1.16	2.56
Sr (ppb)	88.85	91.26	54.49	95.37	51.99	94.89	79.47
NO_3^- (ppm)	0.07	0.03	0.06	0.24	0.09	0.28	0.13
PO_4^{-3} (ppm)	0.00	0.08	0.006	0.006	0.01	0.08	0.03
Ω(mv)	657	657	668	676	677	678	668
DO (ppm)	5.60	5.32	5.20	5.32	5.42	2.50	4.89
T (°C)	27.40	27.30	26.20	24.90	24.60	24.40	25.80
pH	9.50	9.50	9.40	9.30	9.20	9.20	9.35
	KP-D	KP2-D	KP3-D	KP4-D	KP5-D	KP6-D	Mean
Na (ppm)	52.83	53.32	45.79	45.52	54.82	54.88	51.19
Mg (ppm)	37.2	38.39	34.83	33.20	41.10	40.57	37.55
Ca (ppm)	10.45	9.89	8.88	8.61	10.27	10.31	9.73
Rb (ppb)	22.32	21.90	18.37	18.76	22.97	22.21	21.09
Cd (ppb)	0.03	0.002	0.003	0.004	0.04	0.003	0.007
Ba (ppb)	69.96	68.68	57.85	59.46	72.32	70.43	66.45
Pb (ppt)	44.66	15.92	20.77	35.54	46.18	54.30	36.23
Th (ppt)	0.24	0.22	0.41	1.66	1.62	0.60	0.79
U (ppt)	294.63	297.25	256.48	256.46	312.01	297.25	285.69
V (ppt)	769.88	798.62	756.44	675.32	778.43	701.35	746.67
Mn (ppb)	0.20	0.08	0.12	0.30	0.14	0.17	0.17
Fe (ppb)	0.34	0.22	0.94	5.40	2.74	0.17	1.64
Cu (ppb)	0.59	0.20	0.20	0.23	0.30	0.18	0.28
Zn (ppb)	0.58	0.16	0.32	0.56	0.63	0.66	0.49
Sr (ppb)	98.00	100.46	100.59	87.24	101.49	104.90	98.78

Note: KP1, KP2, KP3, KP4, KP5 and KP6 correspond to depth 0 m, 20 m, 40 m, 60 m, 80 m and 100 m, respectively. R and D stand for rainy and dry seasons.

Table 7.11. Dissolved major and trace element profiles from the Mpulungu Sub-basin

	Samples						Mean
	MP1-R	MP2-R	MP3-R	MP4-R	MP5-R	MP6-P	
	(0 m)	(20 m)	(40 m)	(60 m)	(80 m)	(100 m)	
Rb (ppb)	6.37	5.22	12.88	3.25	4.21	7.03	6.49
Cd (ppb)	0.00	0.00	0.03	0.00	0.00	0.00	0.01
Ba (ppb)	21.69	19.09	43.25	13.87	14.66	29.51	23.68
Pb (ppt)	44.040	38.420	236.380	8.060	16.320	33.930	62.860
Th (ppt)	<DL	<DL	<DL	<DL	<DL	2.34	2.34
U (ppt)	95.59	86.45	169.56	49.33	60.91	98.90	93.46
V (ppt)	225.28	222.82	378.81	127.00	135.66	275.22	227.47
Mn (ppb)	<DL	<DL	<DL	<DL	<DL	<DL	<DL
Fe (ppb)	<DL	<DL	<DL	<DL	<DL	0.96	0.96
Cu (ppb)	7.65	0.35	0.37	0.04	0.10	0.24	1.46
Zn (ppb)	1.13	1.13	2.60	0.99	0.44	1.66	1.33
Sr (ppb)	35.10	29.02	55.92	15.88	22.13	35.85	32.31
SiO_2 (ppm)	1.57	1.63	1.65	1.71	1.88	2.02	1.74
NO_3^- (ppm)	0.04	0.01	0.19	0.35	0.36	0.58	0.25
PO_4^{-3} (ppm)	0.00	0.00	0.03	0.04	0.07	0.09	0.04
Ω(mv)	688	683	694	696	698	700	694
DO (ppm)	7.66	7.75	5.35	4.80	4.50	3.86	5.65
T (°C)	28.52	27.27	26.08	25.51	24.64	26.26	26.05
pH	9.38	13.78	13.70	13.67	13.63	13.58	12.96
	MP1-D	MP2-D	MP3-D	MP4-D	MP5-D	MP6-D	Mean
Na (ppm)	56.02	55.10		58.35	62.06	52.85	56.87
Mg (ppm)	40.30	41.69		42.60	43.96	39.63	41.66
Ca (ppm)	10.86	10.49		10.80	11.19	10.42	10.75
Rb (ppb)	23.44	21.94		22.95	22.89	22.05	13.23
Cd (ppb)	0.01	0.01		0.01	0.0	0.0	0.01
Ba (ppb)	87.24	117.61		77.68	76.62	82.02	50.58
Pb (ppt)	635.24	343.32		628.00	141.80	167.01	196.28
Th (ppt)	1.49	0.75		1.79	0.63	0.27	1.37
U (ppt)	276.55	266.47		277.68	278.67	266.48	168.34
V (ppt)	634.03	597.26		685.81	667.36	565.97	395.23
Mn (ppb)	1.04	1.19		1.42	0.55	0.63	0.97
Fe (ppb)	8.81	2.80		46.83	3.80	2.59	9.54
Cu (ppb)	7.49	10.93		9.18	2.46	8.67	4.08
Zn (ppb)	4.39	11.86		7.89	2.57	5.81	3.48
Sr (ppb)	110.27	115.84		117.75	124.48	102.53	66.42
SiO_2 (ppm)	1.57	1.72		1.70	1.73	1.77	1.72
NO_3^- (ppm)	0.00	0.00		0.00	0.15	0.15	0.15
PO_4^{-3} (ppm)	0.14	0.14		0.14	0.14	0.15	0.08
PH	8.80	8.37		8.95	8.96	8.92	8.80

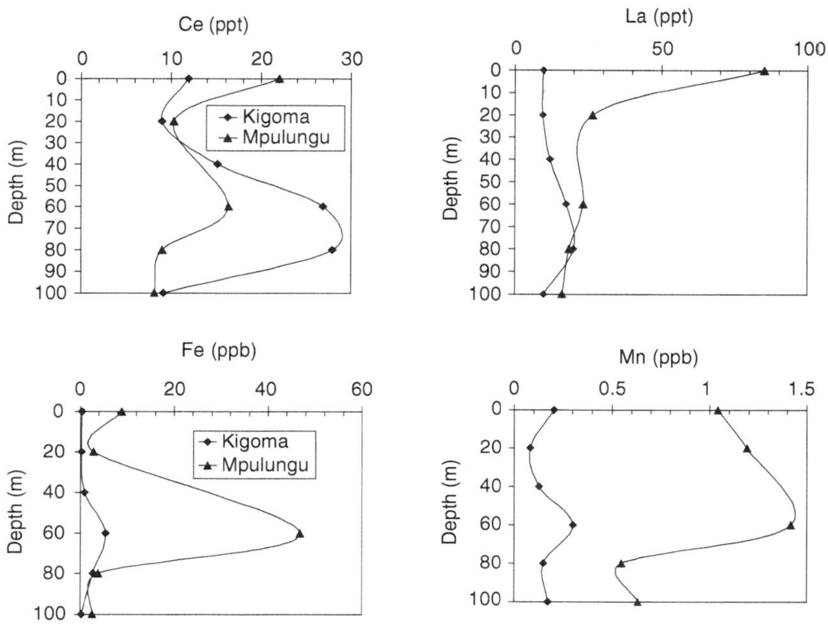

Figure 7.5. Dry season profiles of Ce, La, Fe and Mn from the Kigoma and Mpulungu sub-basins. Both Fe and Mn maxima were found at 60 m in the two sub-basins. Ce and La reached their maxima at the surface in the Mpulungu Sub-basin and at 80 m in the Kigoma Sub-basin.

$P = 0.02$; Fig. 7.6c) in the two sub-basins. Fe and Th profiles, collected during the dry season in the Kigoma Sub-basin, showed a strong positive correlation ($r = 0.89$; $P = 0.016$), whereas the relationship between the two elements was weak in the Mpulungu Sub-basin ($r = 0.79$; $P = 0.110$). The Ba and Sr profiles, however, showed strong positive correlations in the two sub-basins ($r = 0.965$; $P = 0.002$). There were no apparent relationships between Fe and U, Ba, Sr and V.

7.4.2.5. The relationships between REE, Fe and Mn profiles

The UCC-normalized REE of the two dry season profiles, collected from the Kigoma and Mpulungu sub-basins, resembled a typical seawater pattern, which is characterized by HREE-enrichment over LREE (Fig. 7.3). However, the absolute REE concentrations were substantially higher in the Southern Sub-basin compared to the Northern Sub-basin (Table 7.9). The higher REE concentrations were particularly for pronounced LREE (e.g., La and Nd).

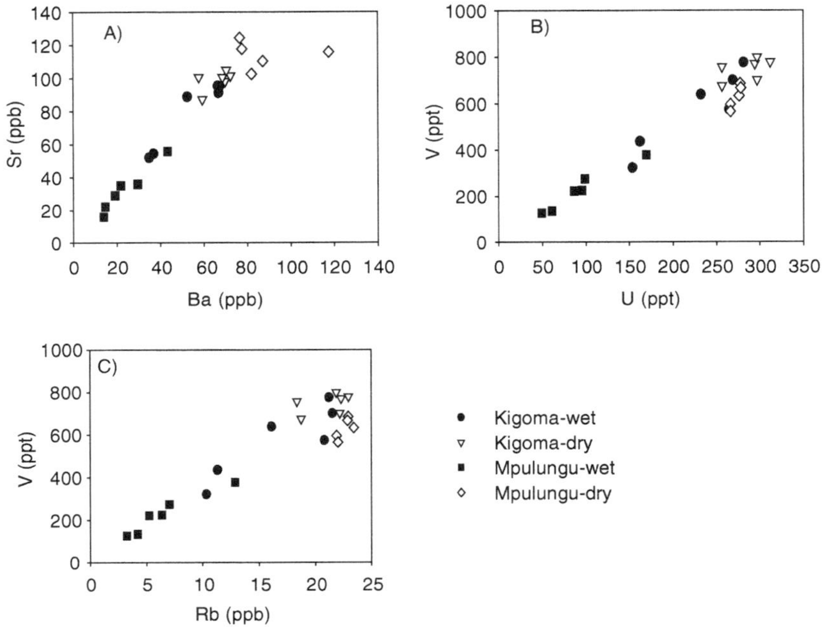

Figure 7.6. A–C relationship between Ba and Sr vertical profiles from the Kigoma and Mpulungu sub-basins.

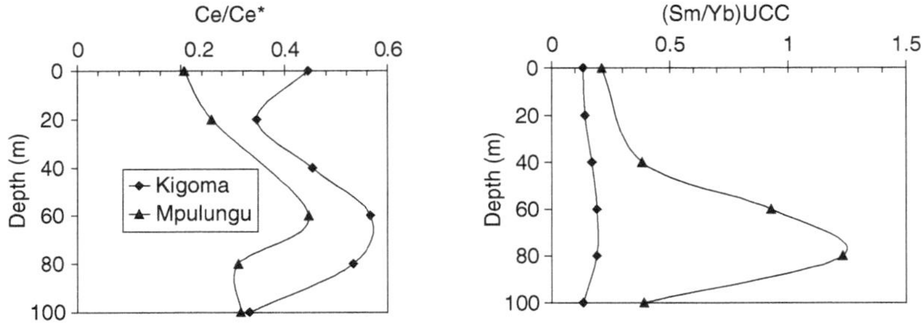

Figure 7.7. Profile samples from the Kigoma and Mpulungu sub-basins, showing LREE/HREE fractionation. According to these plots, the lowest deviation of Ce from the other trivalent species and the maximum Sm release into the water column were found at 60 m depth.

The release of LREE in the water column was similar in the two sub-basins (Fig. 7.7). However, the extent of LREE release was higher in the Mpulungu Sub-basin (Sm_{UCC}/Yb_{UCC} ranged from 0.21 to 1.23) compared to the Kigoma Sub-basin (Sm_{UCC}/Yb_{UCC} ranged from 0.13 to

Table 7.12. Seasonal Y and REE concentration profiles (ppt) from the Kigoma Sub-basin

	Samples						Total
	KP1-R	KP2-R	KP3-R	KP4-R	KP5-R	KP6-R	
	(0 m)	(20 m)	(40 m)	(60 m)	(80 m)	(100 m)	
Y	16.16	30.99	15.26	26.78	10.20	1.89	101.29
La	11.41	9.90	4.47	11.68	4.76	6.81	49.03
Ce	10.30	3.80	1.41	13.74	2.13	3.87	35.24
Pr	1.59	1.16	0.83	1.89	0.69	1.07	7.24
Nd	3.87	3.34	2.18	4.88	2.20	2.99	19.47
Sm	1.27	1.03	0.87	1.61	0.63	1.00	6.39
Gd	1.57	1.25	0.83	2.00	0.74	1.25	7.63
Tb	0.76	0.77	0.46	0.96	0.46	0.76	4.16
Dy	4.45	4.83	2.75	5.25	2.72	4.48	24.48
Ho	1.26	1.41	0.53	1.57	0.79	1.35	6.91
Er	4.33	4.94	2.81	5.57	2.70	4.83	25.17
Tm	0.66	0.80	0.43	0.84	0.40	0.72	3.83
Yb	3.95	4.50	2.43	4.87	2.40	4.00	22.15
Lu	0.65	0.78	0.42	0.87	0.42	0.77	3.90
ΣREE	46.08	38.48	20.41	55.72	21.04	33.89	
Ce/Ce*	0.53	0.24	0.17	0.65	0.26	0.32	
	KP1-D	KP2-D	KP3-D	KP4-D	KP5-D	KP6-D	
Y	9.80	9.48	18.87	5.60	6.39	12.12	62.25
La	9.65	9.36	11.96	17.45	19.91	9.75	78.08
Ce	7.61	5.72	9.65	17.14	17.82	5.83	63.77
Pr	1.50	1.45	1.88	2.58	2.50	1.58	11.68
Nd	3.93	3.83	4.93	6.42	6.91	4.37	30.39
Sm	1.29	1.24	1.51	1.86	2.22	1.52	9.63
Gd	1.60	1.53	1.83	2.21	2.74	1.92	11.83
Tb	0.78	0.81	0.87	1.05	1.13	0.89	5.53
Dy	4.70	4.74	5.14	5.76	6.55	5.47	32.33
Ho	1.36	1.35	1.70	1.57	1.81	1.62	9.41
Er	4.71	4.68	5.07	5.23	8.08	5.74	31.54
Tm	0.77	0.78	0.73	0.79	0.97	0.89	4.942
Yb	4.71	4.31	4.35	4.79	5.67	5.61	29.42
Lu	0.76	0.74	0.71	0.74	0.90	0.86	4.71
ΣREE	43.39	40.53	50.32	67.58	75.38	46.05	
Ce/Ce*	0.45	0.35	0.45	0.57	0.53	0.33	

0.191). Ce-anomalies (Ce/Ce^*) were relatively lower in the Kigoma Sub-basin ($Ce/Ce^* = 0.45$–0.57) compared to the Mpulungu Sub-basin ($Ce/Ce^* = 0.21$–0.45). Although the La and Ce maxima did not coincide with Fe and Mn, Ce/Ce^* maxima coincide with Fe and Mn at 60 m. Ce and La reached their maxima at 80 m in the Kigoma Sub-basin for the

Table 7.13. Seasonal Y and REE profiles (ppt) from the Mpulungu Sub-basin

	Samples						Total
	MP1-R	MP2-R	MP3-R	MP4-R	MP5-R	MP6-R	
	(0 m)	(20 m)	(40 m)	(60 m)	(80 m)	(100 m)	
Y	9.80	9.48	18.90	5.60	6.40	12.10	62.25
La	16.96	11.69	168.10	10.60	4.60	23.80	235.73
Ce		3.47	13.66	<DL	<DL	12.11	29.24
Pr	0.21	0.67	1.33	0.10	<DL	1.64	3.95
Nd	0.55	1.48	3.12	0.24	0.24	3.51	9.15
Sm	0.63	1.47	2.79	0.36	0.35	2.96	8.56
Gd	0.50	0.62	1.33	0.30	0.21	1.03	3.99
Tb	0.24	0.30	0.60	0.16	0.13	0.48	1.88
Dy	1.50	1.69	3.53	0.87	1.01	2.47	11.07
Ho	0.70	0.92	1.75	0.50	0.53	1.30	5.70
Er	1.55	1.51	3.30	0.84	1.09	1.88	10.17
Tm	0.25	0.26	0.53	0.16	0.18	0.38	1.76
Yb	1.43	1.40	3.18	0.71	0.96	1.99	9.66
Lu	0.25	0.26	0.55	0.14	0.17	0.29	1.67
ΣREE	24.77	25.74	203.73	14.96	9.46	53.85	
Ce/Ce*		0.22	0.07			0.37	
	MP1-D	MP2-D	MP3-D	MP4-D	MP5-D	MP6-D	
Y	33.63	30.55		33.16	31.27	29.25	157.86
La	85.00	26.26		23.29	18.50	16.03	169.08
Ce	22.04	10.32		16.38	9.07	8.19	65.96
Pr	3.66	2.27		2.61	2.07	1.96	12.58
Nd	8.55	5.79		6.50	4.88	4.33	27.63
Sm	7.97	4.98		5.47	4.88	4.33	27.69
Gd	3.14	2.44		2.35	1.90	2.20	12.03
Tb	1.22	1.02		1.04	0.93	0.97	5.17
Dy	6.98	6.09		6.32	5.94	5.60	30.94
Ho	2.39	1.85		1.99	1.82	1.73	9.77
Er	6.29	5.61		6.09	5.63	5.45	29.06
Tm	0.95	0.88		0.95	0.87	0.82	4.47
Yb	5.99	4.89		5.63	5.13	5.05	26.70
Lu	0.94	0.79		0.91	0.87	0.81	4.32
ΣREE	155.13	73.29		79.54	62.94	58.09	
Ce/Ce*	0.21	0.26		0.45	0.31	0.32	

dry season samples, whereas they reached their maxima at the surface in the Mpulungu Sub-basin (Tables 7.10 and 7.11). Ce/Ce* and Sm_{UCC}/Yb_{UCC} maxima (Fig. 7.5), however, coincided with those of Fe and Mn at 60 m in the two sub-basins, which also corresponded to the nutrient minima in the Mpulungu Sub-basin (Table 7.13).

7.5. Discussion

7.5.1. Rivers–lake interface

The relatively high concentrations of dissolved Fe, Mn and LREE in the rivers compared to the lake indicated that the epilimnion waters of the lake may contain more DO than those of the rivers. The high average ΣREE observed in the Ruzizi River compared to the Malagarasi River, despite the relatively high pH value in Ruzizi River, suggested that, in addition to pH, other factors may control REE behavior in the Ruzizi River. The lower average ΣREE (particularly LREE) in the Bujumbura, Kigoma and Mpulungu sub-basins compared to the Ruzizi, Malagarasi and Lufubu rivers may reflect their co-precipitation and scavenging by Fe-oxyhydroxides at river–lake interface. This observation is consistent with REE behavior at river–ocean interface (Hoyle et al., 1984; Elderfield et al., 1990; Sholkovitz, 1995).

The absolute HREE (0.5–6.32 ppt) concentrations in the lake were consistent with those reported by Barrat et al. (2000). However, LREE concentrations (particularly, Nd and Sm) were lower than those reported by Barrat et al. (2000), which may be due to the preferential removal of these two elements by Fe colloids (Protano and Riccobono, 2002). HREE concentrations in aquatic systems are generally associated with high-dissolved organic carbon and/or low pH (Dupré et al., 1996; Braun et al., 1997; Gaillardet et al., 1997). Thus, the relatively low-dissolved organic carbon contents (2.2–2.9 $mg\,l^{-1}$, Degens et al., 1971; Hecky et al., 1978; Jârvinen et al., 1996) and high pH (*ca.* 9) of Lake Tanganyika may explain the low-dissolved REE in the lake.

Following the removal of Fe through precipitation, the UCC-normalized patterns of lacustrine REE were more similar to seawater patterns with pronounced HREE-enrichment and LREE-depletion (e.g., Balistrieri et al., 1981; Elderfield and Greaves, 1982). This was clearly noticeable when comparing the UCC-normalized REE patterns of Lake Tanganyika and its three major tributaries. As a result of complexation with carbonate and other ligands, HREE are chiefly found in the dissolved phase, whereas LREE are more likely to be scavenged by Fe-oxyhydroxides (e.g., Sholkovitz and Elderfield, 1988; Elderfield et al., 1990; Sholkovitz and Szymczak, 2000; Quinn et al., 2004). The UCC-normalized REE patterns of the lake waters were more similar to those of the Ruzizi River than the Malagarasi and Lufubu rivers. The similar pattern between the lake and the Ruzizi River may be attributed to their identical pH values (*ca.* 9), or perhaps because this river is the dominant source of REEs for the lake. For example, HREE-enrichment in alkaline waters such as the

Tanganyika system is due to the complexation of REEs with bicarbonate ions (e.g., Johannesson and Lyons, 1994).

Ce-anomalies in the rivers were in the range of the world rivers (Elderfield et al., 1990). However, seasonal variations of Ce/Ce* and their lower values in the Lufubu Rivers with a relatively lower average annual pH of 7.8 compared to 8.7 and 9.0 for the Malagarasi and Ruzizi rivers (Langenberg et al., 2003) suggested that Ce-anomalies in these rivers were chiefly controlled by DO fluctuations and probably microbial reduction rather than pH (Moffett, 1990). Additionally, the spatial variations in Ce-anomalies may be attributed to hydrodynamics and seasonal changes in DO across the lake.

7.5.2. Geochemical control of trace element distribution

In oxic water systems, such as Lake Tanganyika and its tributaries, dissolved Fe and Mn concentrations are found at very low concentrations (Morel et al., 1991). Owing to its relatively low redox potential, the concentration of dissolved Mn tends to be higher than that of Fe in oxic waters (Hutchinson, 1957; Delfino and Otto, 1986). However, Mn is generally found as an inorganic free-ion, whereas Fe is associated with organically stable colloids, which can pass through filter pore of 0.45 μm (Nordstrom and Ball, 1989; Pokrovsky and Schott, 2002). Consequently, water samples filtered through 0.45 μm filter pore size may represent a mixture of dissolved and colloidal Fe phases. The filtration technique used in the present study may therefore explain the high-dissolved Fe contents compared to Mn. According to Sokolowiski et al. (2001), the relatively low Mn concentrations with respect to Fe in freshwater systems can also be attributed to the oxidation of dissolved Mn into less-soluble species, whereas under low DO conditions, Fe(III) may be rapidly reduced to Fe(II). This mechanism may also provide an additional explanation of the relatively low-dissolved Mn concentrations in the Tanganyika system.

High V concentrations in freshwater systems are generally attributed to silicate weathering in catchment basins, whereas high U concentrations are associated with carbonate or garanite bedrock (Shiller and Boyle, 1987). This geochemical process was supported by relatively high concentrations of dissolved V compared to U in Lake Tanganyika and its tributaries, as the lithology of the lake catchment basin is predominantly made of aluminosilicate minerals (Bellon and Pouclet, 1980; Cohen and Palacios-Fest, 1999).

Sr concentrations are consistent with those reported by Barrat et al. (2000), with narrow spatial and temporal variations across the lake. Furthermore, Sr appeared to be less affected by scavenging and biogenic activity (Broecker and Peng, 1982). In contrast to Sr, Ba and Rb, which are normally found in true dissolved phase, U and V can be transported in freshwater systems in various colloidal phases (Riotte et al., 2003). However, the high-loading coefficients for Sr, Ba, Rb, V and U suggested that these elements were less affected by colloidal complexes (organic and inorganic phases) in Lake Tanganyika, and thus Factor 1 may represent the soluble state. In dilute waters, low Fe and organic carbon concentrations (e.g., Lake Tanganyika), U is also thought to be in soluble form with low U/Sr values (1.9×10^{-3} to 6×10^{-3}; Riotte et al., 2003). In contrast to Sr and Rb, U may be associated with organic and inorganic colloids, and thus U/Sr and U/Rb ratios can be used as indices to assess U behavior in freshwater systems (Riotte et al., 2003). The relatively low U/Sr ratios in Lake Tanganyika suggested that U was chiefly transported in the dissolved phase in this system. The high solubility of U in the Tanganyika system can be attributed to its association with carbonate ions in solution (e.g., Langmuir, 1978; Waite et al., 1994). However, the spatial variations in U/Sr ratios across the lake confirmed the spatial variations in hydrodynamics. Hence, the Mpulungu Sub-basin, with the lowest U/Sr ratio, appeared to be the least affected by colloids (i.e., particulates). The lack of relationship between V, U and Rb, elements characterized by a true-dissolved phase, and Fe supported the suggestion that Factor 1 represents mainly the high degree of solubility, or the differences in concentration (Stetzenbach et al., 2001).

Factor 2, however, can be related to the inverse hydrodynamic variations in the two basins. For example, in oxic waters Cu is mainly present in organic complex forms and rarely in free ionic forms (e.g., $CuCl^{+}$; Calvert and Pedersen, 1993; Xue and Sund, 1997), whereas Pb is likely to exist in soluble carbonate form ($PbCO_3$) under high pH conditions, such as Lake Tanganyika (Drever, 1997). Pb may also be associated with organic ligands (Sokolowski et al., 2001). The discrepancy observed between the Cu and Pb loadings in the two basins may be attributed to the higher concentration of dissolved organic carbon in the Northern Basin compared to the Southern Basin, which may be indirectly linked to fluvial inputs, vertical mixing and primary productivity. For example, the Bujumbura (0.40 NTU) and Mpulungu sub-basins (0.35 NTU) are characterized by high seasonal turbidity compared to the Kigoma Sub-basin (0.25 NTU; Plisnier et al., 1996). However, the high-negative loading for Rb in the Northern Basin suggested that Factor 2 did not represent the same biogeochemical factor in the Southern Basin as was the case in the Northern Basin. In oxic waters, Zn tends to be in free ionic phases (e.g., Zn^{2+} and $ZnCl^{+}$; Calvert and Pedersen, 1993), or form strong organic complexes (e.g., Bruland, 1989; Ellwood and Van den Berg, 2000). However, the common factor that may explain the high-negative loadings of

Rb, Cu and Zn in the Northern Basin is their adsorption onto particulate biogenic material. Filtered water samples through 0.45 to 0.20-μm membrane pore sizes may contain fine-grained particulate fraction (Delfino and Otto, 1986). It can be suggested that the solubility of Cu, Rb and Zn was dramatically reduced in the Northern Basin due to high turbidity. Furthermore, Factor 3 may also represent another type of adsoption/desorption of Zn and Pb in the Northern Basin. Significant correlations between Fe and Th in the Kigoma and Mpulungu sub-basins indicated the impact of dissolved organic matter on Th geochemistry. A similar relationship between Fe and Th has been observed in an organic river system in Eastern Cameroon (Braun et al., 1997).

7.5.3. Biological control of trace element distribution in the lake

On the basis of the differences in their chemical, chlorophyll and bacterial concentrations, Lake Tanganyika waters are classified into Northern, Central and Southern waters (Hecky, 1978, 1981). This biological delimitation of the lake may explain the non-conservative behavior of certain trace elements such as Fe, Mn, Ba, Cd, and Cu and to a lesser extent Sr across Lake Tanganyika. For example, the behavior of Fe, Mn and Ba is generally governed by both geochemical and biogenic processes (e.g., Evans et al., 1977; Finlay et al., 1983; Hunt, 1983). Fe, in addition to its scavenging and desorption role, is also subject to active biological uptake (Kolber et al., 1994).

The coincidence of the Ba profile maximum with that of nutrients in the Mpulungu Sub-basin suggested that Ba might be subject to biological uptake. This was consistent with a previous study, which has demonstrated that the removal of Ba from estuarine surface waters is directly associated with seasonal phytoplankton blooms (Stecher and Kogut, 1999). Moreover, a strong negative correlation between Ba removal and protozoa abundance has also been observed (Finlay et al., 1983).

The spatial distribution of dissolved Cd in the two basins was consistent with biogenic activity, as a high-positive correlation between PO_4^{-3} and Cd profiles in the Mpulungu Sub-basin was observed. This observation was supported by Factor 3 with a high-positive loading for Cd in the Southern Basin and Factor 4 with a high-negative loading score in the Northern Basin. Thus, Factor 3 may reflect the breakdown of organic matter and a subsequent release of Cd into the water column, and Factor 4 may represent the uptake of dissolved Cd by organisms in the Northern Basin, and iron sulfide scavenging of Zn in the Southern Basin. For example, an inverse association between Cd and PO_4^{-3} has been observed in the Tay Estuary (Owens and Balls, 1997). According to Owens and

Balls (1997), the positive correlation between Cd and PO_4^{-3} is due to the breakdown of organic matter, and the release of both Cd and PO_4^{-3} into the water column. Co-precipitation of Zn by Fe sulfide has been reported in another lacustrine environment (Acterberg et al., 1997). The occurrence of hydrogen sulfide near the surface in the Mpulungu Sub-basin during the dry season may explain the high-negative loading of Zn in this sub-basin, which may co-precipitate with Fe as FeS.

In conclusion, Factors 2, 3 and 4 appeared to be the most important biogeochemical factors in Lake Tanganyika, as the scatter plots based on their loading scores grouped the samples into two different categories (Fig. 7.8). Because these factors are indirectly affected by primary productivity, it can be inferred from this study that trace element

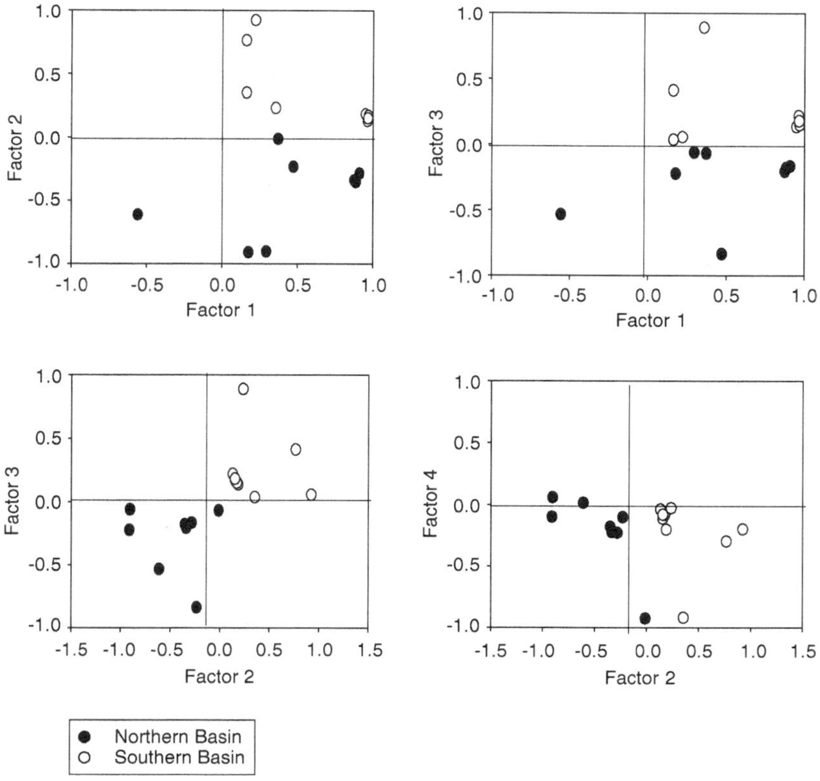

Figure 7.8. Scatter plots of varimax factor loading scores. Factor 2, 3 and 4 appeared to be the most important biogeochemical factors in the lake, as the plots of these three factor loading scores grouped the samples to two distinct categories (North and South).

distribution in Lake Tanganyika was in good agreement with the biological delineation.

7.5.4. Vertical cycling of REE

In permanently stratified lakes such as Lake Tanganyika, pH and DO decrease with depth following organic matter decomposition (Fig. 7.9). As a result, particulate Fe and Mn are broken down from anoxic sediments into the water column, and precipitate at the oxic–anoxic boundary (Haberyan and Hecky, 1987; De Carlo and Green, 2002). This phenomenon plays a fundamental role in trace element cycling in the water column (e.g., LREE; Elderfield et al., 1990). Thus, the coincidence of Fe and Mn maxima at 60 m in the two sub-basins may indicate a possible release

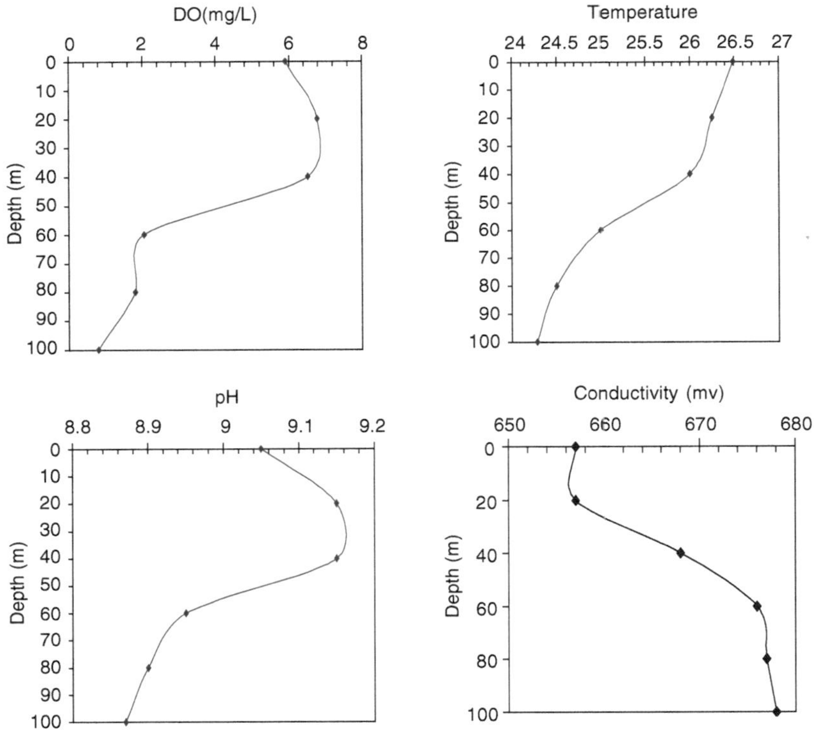

Figure 7.9. A contrast between dissolved oxygen, temperature, pH and conductivity depth in the Kigoma Sub-basin (dry season 2002; Sako, unpublished). A sharp drop of dissolved oxygen content and pH occurred around the thermocline depth (60 m).

of Fe(II) from the monimolimnion and its subsequent removal near mixolimnion through the oxidation process of Mn(IV) (Postma, 1985).

Although the maxima of La and Ce did not coincide with that of Fe, Ce/Ce* and Sm_{UCC}/Yb_{UCC} for both profiles reached their maxima at 60 m, which corresponded to Fe and Mn maxima in the two sub-basins, suggesting that LREE removal and cycling in the water column can be controlled by Fe and Mn geochemistry.

Another striking difference in trace element patterns between the Northern and Southern basins during the dry season was the relatively high LREE fractionation in the Mpulungu Sub-basin compared to the Kigoma Sub-basin. The high LREE release in the Mpulungu Sub-basin may be a combination of the changes in DO with depth and intensive vertical mixing during this period (Plisnier et al., 1996).

The drop in NO_3^- and PO_4^{3-} minima at 60 m can be attributed to a probable high planktonic density at this depth, resulting in a possible high-nutrient uptake, and a subsequent high bacterial activity may occur. Thus, the occurrence of Fe and Mn maxima at 60 m despite the presence of DO implied that, in addition to changes in pH, trace elements and LREE cycling in Lake Tanganyika was subject to bacteria-mediated reduction. This observation is consistent with previous work (Moffett, 1990; German et al., 1992; Sholkovitz, 1992; Moffett, 1994a, b).

Pronounced high concentrations of most elements at the surface in the Mpulungu Sub-basin compared to the Kigoma Sub-basin supported the influence of seasonal upwelling on water chemistry across the lake. In the Mpulungu Sub-basin, anoxic water is normally found at a depth of 200 m and is rich in hydrogen sulfide and nutrients, which may appear at the surface during the dry season (Coulter, 1963), and sulfide is known to be associated with HREE concentrations (Schijf, 1992; Schijf et al., 1995).

7.6. Conclusions

The three major rivers feeding Lake Tanganyika were Fe- and LREE-enriched compared to the lake, whereas HREE tended to be more conservative. The conservative nature of HREE may be due to the presence of a dominant inorganic ligand such as HCO_3^-. Ce/Ce* values varied between the rivers and within the lake, according to changes in pH and DO. Solubility, biogenic activity, and Fe- and Mn-oxyhydroxides appeared to be the major biogeochemical factors that control trace element behavior in Lake Tanganyika. The coincidence of Ce-anomalies and

Sm_{UCC}/Yb_{UCC} maxima with those of Fe and Mn in the Kigoma and Mpulungu sub-basins implied that LREE were preferentially scavenged from the water column by Fe- and Mn-oxyhydroxides.

The release of LREE into the water column, which seemed to be enhanced by microbial mediated reduction, was greater in the Mpulungu Sub-basin than in the Kigoma Sub-basin. High LREE fraction and trace element abundance at the near surface in the Mpulungu Sub-basin may result from the intensive upwelling in this sub-basin during the dry season. High-trace element concentrations, observed in the depth in the Mpulungu Sub-basin profile compared to the Kigoma Sub-basin, provided insights into understanding the effects of changes in hydrodynamics on trace element behavior in the two extreme basins of Lake Tanganyika.

This study has demonstrated that the spatial distribution of most trace elements in Lake Tanganyika was not conservative, but varied according to riverine input, scavenging by Fe-oxyhydroxides, sedimentation, biological uptake and reduction and remineralization.

ACKNOWLEDGEMENTS

I thank Willy Mbemba, Ismael Kimirei, Mukuli Mupape and Lawrence Makasa for their invaluable assistance in sample collections. I also thank Dr. Zhongxing Chen of the Old Dominion University for ICP-MS analysis. Logistic assistance from NSF funded REU the Nyanza Project, Tanzania Fisheries Research Institute and Zambian Fisheries Unit is greatly appreciated. I would also like to thank Drs Catherine M. O'Reilly and Roland Johnson for their insightful comments and suggestions for the early version of this manuscript. The paper benefited from constructive reviews by Dr. Karen Johannesson and an anonymous reviewer.

REFERENCES

Aagaard, P., 1974. Rare earth element complexation on clay minerals. Bull. Group France 26, 193–199.

Acterberg, E.P., Van den Berg, C.M.G., Boussemart, M., Davison, W., 1997. Speciation and cycling of trace metals in Estwaite water a productive English Lake with seasonal deep-water anoxia. Geochim. Cosmochim. Acta 54, 971–991.

Atsuyuki, O., Kawabe, I., 2001. REE (III) adsorption onto Mn dioxide (σ-MnO_2) and Fe oxyhydroxides: Ce (III) oxidation by σ-MnO_2. Geochim. Cosmochim. Acta 65(5), 695–703.

Balistrieri, L., Brewer, P.G., Murray, J.W., 1981. Scavenging residence times of trace metals and surface chemistry of sinking particles in Deep Ocean. Deep-Sea Res. 28A, 101–121.

Barrat, J.A., Boulègue, Tiercelin, J.J., Lesoud, M., 2000. Strontium isotopes and rare-earth element geochemistry of hydrothermal carbonate deposits from Lake Tanganyika, East Africa. Geochim. Cosmochim. Acta 64, 287–298.

Bellon, H., Pouclet, A., 1980. Datations K-Ar de quelques laves du rift ouest de l'Afrique Centrale: Implications sur l'évolution magmatique et structurales. Geologishes Rundschau 69(1), 49–62.

Braun, J.J., Viers, J., Dupré, B., Polve, M., Nsill, J., Muller, J.P., 1997. Solid/liquid REE fractionation in the lateritic system of Goyoum, East Cameroon: The implication for the present dynamics of the soil covers of the humid tropical regions. Geochim. Cosmochim. Acta 62(2), 273–299.

Broecker, W.S., Peng, T.-H., 1982. Tracers in the Sea. Eldigo Press, Palisades, NY, p. 690.

Bruland, K.W., 1989. Complexation of zinc by natural organic ligands in the Central North Pacific. Limnol. Oceanogr. 34, 269–285.

Byrne, R.H., Kim, K.-H., 1990. Rare earth scavenging in seawater. Geochim. Cosmochim. Acta 54, 2645–2656.

Calvert, S.E., Pedersen, T.F., 1993. Geochemistry of recent oxic and anoxic marine sediments: Implications for the geological records. Mar. Geol. 113, 67–88.

Cazenave-Piarrot, F., 1981. L'originalité des milieux natureles dans le grand rift occidental africain. L'exemple de l'Imbo au Burundi. Recherches sur les hautes terres d'Afrique centrale. Travaux et documents de géographie tropical, CEGET 42, p. 59.

Coffey, M., Dehairs, F., Collette, O., Luther, G., Church, T., Jickells, T., 1997. The behaviour of dissolved barium in estuaries. Estuar. Coast. Shelf Sci. 45, 113–121.

Cohen, S.A., Palacios-Fest, R.M., 1999. Paleolimnological investigations. Pollution control and other measures to protect biodiversity in Lake Tanganyika (RAF/92/G32).

Coulter, G.W., 1963. Thermal stratification in the deep hypolimnion of Lake Tanganyika. Limnol. Oceanogr. 13(2), 385–387.

Coulter, G.W., Spigel, H.R., 1991. Hydrodynamics. In: Coulter, G.W. (Ed.), Lake Tanganyika and Its Life. Natural History Museum Publication, Oxford University Press, London, pp. 49–75.

De Carlo, E.H., Green, J.W., 2002. Rare earth elements in the water column of Lake Vanda, McMurdo Dry Valleys, Antarctica. Geochim. Cosmochim. Acta 66, 1323–1333.

Degens, E.T., Von Herzen, R.P., Wong, H.-K., 1971. Lake Tanganyika: Water chemistry, sediments, geological structure. Naturwissenschaften 58(5), 229–241.

Delfino, J.J., Otto, G.R., 1986. Trace metal transport in two tributaries of the Upper Chesapeake Bay: The Susquehanna and Bush rivers. Mar. Chem. 20, 29–44.

Devroey, E., 1954. Annuailre hydrologique du C. B. et U. 1953. Public. Com. Hydro. Bass. Congo **7**, p. 326.

Drever, J.I., 1997. The Geochemistry of Natural Waters, third ed. Prentice-Hall, Upper Saddle River, NJ, p. 436.

Dupré, B., Gaillardet, J., Rousseau, D., Allègre, C.J., 1996. Major and trace elements of river-borne material: The Congo basin. Geochim. Cosmochim. Acta 60, 1301–1321.

Elbaz-Poulichet, F., Dupuy, C., 1999. Behaviour of rare earth elements at the freshwater–seawater interface of two acid mine rivers: The Tito and Odiel (Andalucia, Spain). Appl. Geochem. 14, 1063–1072.

Elderfield, H., Greaves, M.J., 1982. The rare earth elements in seawater. Nature 296, 214–219.

Elderfield, H., Upstill-Goddard, R., Sholkovitz, R.E., 1990. The rare earth elements in rivers, estuaries, and coastal seas and their significance to the composition of ocean waters. Geochim. Cosmochim. Acta 54, 971–991.

Ellwood, M.J., Van den Berg, C.M.G., 2000. Zinc speciation in the Northeastern Atlantic Ocean. Mar. Chem. 68, 295–306.
Evans, D.W., Cutshall, N.H., Goss, F.A., Wolfe, D.A., 1977. Manganese cycling in the Newport River estuary, North Carolina. Estuar. Coast. Mar. Sci. 5, 71–80.
Finlay, B.J., Hetherington, N.B., Davison, W., 1983. Active biological participation in lacustrine barium chemistry. Geochim. Cosmochim. Acta 47, 1325–1329.
Gaillardet, J., Dupré, B., Allègre, J.C., Négrel, P., 1997. Chemical and physical denudation in the Amazon River Basin. Chem. Geol. 142, 141–173.
Garrels, R.M., Mackenzie, F.T., 1971. Evolution of Sedimentary Rocks. Norton, New York.
German, C.R., Holliday, B.P., Elderfield, H., 1992. Reply to the comment by E.R. Sholkovitz on "redox cycling on rare earth elements in the suboxic zone of the Black Sea". Geochim. Cosmochim. Acta 56, 4309–4313.
Graham, E.Y., Ramsey, L.A., Lyons, W.B., Welch, K.A., 1996. Determination of rare earth elements in Antarctica lakes and streams of varying ionic strengths. In: Holland, G., Tanner, S.D. (Eds.), Plasma Source Mass Spectrometry: Developments and Applications. The Royal Society of Chemistry, London, pp. 253–262.
Green, W.J., Ferdelman, T.G., Canfield, D.E., 1989. Metal dynamics in Lake Vanda (Wright Valley, Antarctica). Chem. Geol. 76, 85–94.
Haberyan, K., Hecky, E.R., 1987. The late Pleistocene and Holocene stratigraphy and paleolimnology of lakes Kivu and Tanganyika. Palaeogeogr. Palaeoecol. 61, 169–197.
Heaney, S.L., Smyly, W.J.P., Talling, J.F., 1986. Interactions of physical, chemical and biological processes in depth and time within a productive English lake during summer stratification. Intern. Rev. Ges. Hydrol. 71, 441–494.
Hecky, R.E., 1978. The Kivu-Tanganyika basin: The last 14000 years. Polskie Archiwum Hydrobiologii 25(1–2), 159–165.
Hecky, R.E., 1981. East African lakes and their trophic efficiencies. AAAS 1981, Toronto Symposium.
Hecky, R.E., Fee, E.J., Kling, H.J., Rudd, J.W., 1978. Studies on the planktonic ecology of Lake Tanganyika. Can. Dept. Fish. Environm. Fisheries and Marine Services Technical Report 816, 1–51.
Hecky, R.E., Spigel, R.H., Coulter, G.W., 1991. The nutrient regime. In: Coulter, W.G. (Ed.), Lake Tanganyika and Its Life. Natural History Museum Publications, Oxford University Press, London, pp. 75–89.
Hoyle, J., Elderfield, E., Gledhill, A., Greave, M., 1984. The behaviour of rare earth elements during mixing of river and sea waters. Geochim. Cosmochim. Acta 48, 143–149.
Hunt, C.D., 1983. In corporation and deposition of Mn and other trace metals by flocculent organic matter in a controlled marine ecosystem. Limnol. Oceanogr. 28, 913–923.
Hutchinson, G.E., 1957. A treatise on limnology. Geography, Physics and Chemistry, Vol. I, Wiley, New York, p. 1015.
Jârvinen, M., Slonen, K., Sarvala, J., 1996. Experiments on phytoplankton and bacterial production ecology in Lake Tanganyika: The results of the first lake-wide research cruise on R/V Tanganyika Explorer. GCP/RAF/271/FIN–DT/44, p. 33.
Johannesson, K.H., Lyons, W.B., 1994. The rare earth element geochemistry of Mono Lake water and the importance of carbonate complexing. Limnol. Oceanogr. 39, 1141–1154.
Johannesson, K.H., Stetzenbach, K.J., Hodge, V.F., 1997. Rare earth elements as geochemical tracers of regional groundwater mixing. Geochim. Cosmochim. Acta 61, 3605–3618.
Johannesson, K.H., Stetzenbach, K.J., Hodge, V.F., Lyons, W.B., 1996. Rare earth element complexation behavior in circumneutral pH groundwaters: Assessing the role of carbonate and phosphate ions. Earth Planet. Sci. Lett. 139, 305–319.

Kaiser, H.F., 1958. The varimax criterion for analytic rotation in factor analysis. Psychometrika 23, 187–200.

Kolber, Z.S., Barber, R.T., Coale, K.H., Fitzwater, S.Z., Greene, R.M., Johnson, K.S., Lindley, S., Falkowski, P.G., 1994. Iron limitation of phytoplankton photosynthesis in equatorial Ocean. Nature 37, 145–149.

Langenberg, V.T.H., Nyamushahu, S., Roijackers, R., Koelmans, A., 2003. External nutrient sources for Lake Tanganyika. J. Great Lakes Res. 29(2), 169–180.

Langmuir, D., 1978. Uranium solution-mineral equilibria at low temperatures with applications to sedimentary ore deposits. Geochim. Cosmochim. Acta 42, 547–569.

Meybeck, M., 1987. Global chemical weathering of surficial rocks estimated from river dissolved loads. Am. J. Sci. 287, 401–428.

Moffett, J.W., 1990. Microbially mediated cerium oxidation in sean water. Nature 345, 421–423.

Moffett, J.W., 1994a. A radiotracer study of cerium and manganese uptake in Chesapeake Bay. Geochim. Cosmochim. Acta 58, 695–703.

Moffett, J.W., 1994b. The relationship between cerium and manganese oxidation in the marine environment. Limnol. Oceanogr. 39, 1309–1318.

Morel, F.M.M., Rueter, J.G., Price, N.M., 1991. Iron nutrition of phytoplankton and its possible importance in the ecology of ocean regions with high nutrient and low biomass. Oceanography 4(2), 56–61.

Nordstrom, D.K., Ball, J.W., 1989. Mineral saturation states in natural waters and their sensitivity to thermodynamic and analytical errors. Sci. Geol. Bull. 42, 269–280.

Owens, R.E., Balls, P.W., 1997. Dissolved trace metals in the Tay Estuary. Estuar. Coast. Shelf Sci. 44, 421–434.

Patterson, C.C., Settle, D.M., 1976. The reduction of orders of magnitude errors in lead analyses of biological materials and natural waters by evaluating and controlling the extent and source of industrial lead, handling, and analysis. Natl. Bureau Stand. Spec. Publ. 422, 321–351.

Plisnier, P.-D., 1996. Limnological sampling during a second annual cycle (1994–1995) and some comparisons with year one on Lake Tanganyika. FAO/FINNIDA Research for the management of the fisheries of Lake Tanganyika. GCP/RAF/271/FIN–TD/54, p. 48.

Plisnier, P.-D., Langenberg, V., Mwape, L., Chitamwembwa, D., Tshibangu, K., Coenen, E., 1996. Limnological sampling during an annual cycle at three stations on Lake Tanganyika (1993–1994). FAO/FINNIDA Research for the management of the fisheries of Lake Tanganyika. GCP/RAF/271/FIN–TD/46, p. 124.

Pokrovsky, O.S., Schott, J., 2002. Iron colloids/organic matter associated transport of major and trace elements in small boreal rivers and their estuaries (NW Russia). Chem. Geol. 190, 141–179.

Postma, D., 1985. Concentration of Mn and separation from Fe in sediments. I. kinetics and stoichiometry of the reaction between birnessite and dissolved Fe (II) at 10°C. Geochim. Cosmochim. Acta 49, 1023–1033.

Protano, G., Riccobono, F., 2002. High contents of rare earth elements (REEs) in stream waters of Cu–Pb–Zn mining area. Environ. Pollut. 117, 499–414.

Quinn, K.A., Byrne, R.H., Schijf, J., 2004. Comparative scavenging of yttrium and the rare earth elements in seawater: Competitive influences of solution and surface chemistry. Aquat. Geochem. 10(1/2), 59–80.

Reeder, S.W., Hitchon, B., Levinson, A.A., 1972. Hydrogeochemistry of the surface waters of the Mackenzie River drainage basin, Canada-I. Factors controlling inorganic composition. Geochim. Cosmochim. Acta 36(8), 825–865.

Riotte, J., Chabaux, F., Benedetti, M., Dia, A., Gérard, M., Boulègue, J., Etamé, J., 2003. Uranium colloidal transport and origin of the ^{234}U–^{238}U fractionation in surface waters: New insights from Mount Cameroon. Chem. Geol. 202, 365–381.
Ruiz, F., Gomis, V., Blasco, P., 1990. Application for factor analysis to the hydrogeochemical study of coastal aquifer. J. Hydrol. 119, 169–177.
Schijf, J., 1992. Aqueous geochemistry of rare earth elements in marine anoxic basins. PhD Dissertation, University of Utrecht, Utrecht (Geol. Utraiectina No. 85).
Schijf, J., de Baar, H.J.W., Millero, F.J., 1995. Vertical distributions and speciation of dissolved rare earth elements in the anoxic brines of the Bannock Basin, eastern Mediterranean Sea. Geochim. Cosmochim. Acta 59, 3285–3299.
Shan, X.-Q., Lian, J., Wen, B., 2002. Effect of organic acids on adsorption and desorption of rare earth elements. Chemosphere 47(7), 701–710.
Shiller, M.A., Boyle, E.A., 1987. Dissolved vanadium in rivers and estuaries. Earth Planet. Sci. Lett. 86, 214–224.
Sholkovitz, R.E., Elderfield, H., 1988. Cycling of dissolved rare earth elements in Chesapeake Bay. Global Biogeochem. Cy. 2(2), 157–176.
Sholkovitz, E.R., 1992. Comment on "Redox cycling of rare earth elements in the suboxic zone of the Black Sea" by C.R. German, B.P., Holliday, Elderfield, H. Geochim. Cosmochim. Acta 56, 4305–4307.
Sholkovitz, R.E., 1993. The geochemistry of rare earth elements in the Amazon River estuary. Geochim. Cosmochim. Acta 57, 2181–2190.
Sholkovitz, R.E., 1995. The aquatic chemistry of rare earth elements in rivers and estuaries. Aquat. Geochem. 1, 1–34.
Sholkovitz, E.R., Szymczak, R., 2000. The estuarine chemistry of rare earth elements: Comparison of the Amazon, Fly, Spik and the Gulf of Papua Systems. Earth Planet. Sci. Lett. 197, 299–309.
Sichingabula, H.M., 1999. Analysis and results of discharge and sediment monitoring activities in the southern Lake Tanganyika basin, Zambia. Special study on sediment discharge and its consequences. Pollution control and other measures to protect biodiversity in Lake Tanganyika (RAF/92/G32).
Simeonov, V., Stefanov, S., Tsakovski, S., 2000. Environmetrical treatment of water quality survey data from Yantra River, Bulgaria. Mikrochim. Acta 134(1/2), 15–21.
Sokolowski, A., Wolowicz, M., Hummel, H., 2001. Distribution of dissolved and labile particulate trace elements in the overlying bottom water in the Vistula River Plume (Southern Baltic Sea). Mar. Pollut. Bull. 42(10), 967–980.
Stecher, H.A., Kogut, M.B., 1999. Episodic barium removal in the Delaware estuary. Geochim. Cosmochim. Acta 63(7–8), 1003–1012.
Taylor, S.R., McLennan, S.M., 1985. The continental Crust: Its composition and evolution. Blackwell, Oxford, p. 312.
Tiercelin, J.J., Mondeguer, A., 1991. The geology of the Tanganyika trough. In: Coulter, G.W. (Ed.), Lake Tanganyika and Its Life. Natural History Museum Publication, Oxford University Press, London, pp. 6–48.
Vandecasteel, C., Block, C.B., 1993. Modern Methods for Trace Element Determination. Willey, Chichester.
Viollier, E., Jézéquel, D., Michard, G., Pèpe, M., Sarazin, G., Albéric, P., 1995. Geochemical study of a crater lake (Pavin Lake, France): Trace-element behaviour in the monimolimnion. Chem. Geol. 125, 61–72.
Waite, T.D., Davis, J.A., Payne, T.E., Waychunas, G.A., Xu, N., 1994. Uranium (VI) adsorption to ferrihydrite: Application of a surface complexation model. Geochim. Cosmochim. Acta 58, 5465–5478.

Webster, G.J., 1993. Trace-metal behaviour in oxic and anoxic Ca–Cl brines of the Wright Valley drainage, Antarctica. Chem. Geol. 112, 255–274.
Xue, H., Sund, W.G., 1997. Comparison of [Cu^{2+}] measurements in lake water determined by ligand exchange and Cathodic stripping voltammetry and by ion-selective electrode. Environ. Sci. Technol. 31, 1902–1909.

Developments in Environmental Science, Volume 5
D. Sarkar, R. Datta and R. Hannigan (Editors)

ISSN: 1474-8177/DOI:10.1016/S1474-8177(07)05008-5

Chapter 8

Baseline water chemistry, nitrate concentrations, and aquifer sensitivity of glacial sequences in LaGrange County, Indiana

Nancy R. Hasenmueller and Tracy D. Branam

Abstract

A project was initiated to evaluate the chemistry and relative age of groundwater in glacial sequences in LaGrange County in northeastern Indiana in 2002. The intent was to acquire a baseline dataset for selected ions, trace elements, and isotopes in water from 50 wells. In 2003, samples were collected from 18 additional wells that produce groundwater known to have high nitrate–nitrogen (NO_3–N) concentrations. The 68 wells draw water from sand and gravel deposits in the Topeka fan, Lima plain, interlobate, and Oliver Lake glacial sequences.

The prevalent chemical character of the groundwater from the 68 wells is dominated by calcium bicarbonate. The 2002 water samples average 63% calcium and 85% bicarbonate of the major ion distribution; the 2003 water samples average 61% calcium and 64% bicarbonate. The decrease in the percentage of bicarbonate is attributed to the addition of nitrate.

Nitrate–nitrogen concentrations in the water samples collected in 2002 are generally low; 84% of the samples have concentrations below the detection limit. Only one water sample has a NO_3–N concentration above the U.S. Environmental Protection Agency (EPA) Maximum Contaminant Level (MCL); that sample is from a well located in the Topeka fan. Nitrogen isotopes for four wells have ratios ranging from 7.60 to 19.3 parts per thousand (‰). The nitrogen isotope ratios are similar to those measured in shallow groundwater from wells in the following land-use settings in the Minnesota sand-plains aquifers: livestock feedlots, residential areas with septic systems, and nonirrigated-cultivated areas. The nitrate data suggest that the Topeka fan aquifers are sensitive to anthropogenic contamination.

Nitrate–nitrogen in the 2003 groundwater samples ranges from 7.0 to 30.4 mg l^{-1}; NO_3–N concentrations in 72% of the samples are above the EPA MCL. The isotope ratios for samples from the Lima plain and Oliver Lake sequences range from 4.9 to 9.6‰ and show a positive correlation with sodium concentration. This suggests that the source of the nitrogen in these sequences may include septic waste. The nitrogen isotope ratios for samples from the Topeka fan sequence range from 5.3 to 22.3‰. The nitrogen isotope ratios are similar to those measured in the following land-use settings in the Minnesota sand plains: livestock feedlots, residential areas with septic systems, and nonirrigated-cultivated areas.

The tritium values in the 2002 water samples range from below the detection limit (<0.8 TU) to 17.0 TU; 35 of the samples have values above the detection limit. Tritium is generally not detected in waters from wells 30.48 m or greater in depth. The $\delta^{18}O$ and $\delta^{2}H$ values in the water samples from LaGrange County are similar to those reported for groundwater samples from unconsolidated glacial deposits in northwestern Indiana.

The tritium data indicate that 62% of the water in wells sampled in the Topeka fan and Oliver Lake basin has been recharged from the surface since 1952. The groundwater in the remaining wells in the Topeka fan and Oliver Lake glacial sequences has been recharged between 10,000 years before present and 1953.

8.1. Background

LaGrange County is located in northeastern Indiana and lies in the St. Joseph River Basin; the estimated 2002 population of the county was 35,410 (U.S. Census Bureau, 2004). Groundwater is the source of 75% of the water withdrawn in the basin (Indiana Department of Natural Resources Division of Water, 1987). Agriculture is the primary land use in the area, and farms make up nearly 78% of the acreage in the county (1997 data from U.S. Department of Agriculture, 2004).

A number of these farms are small and employ older farming methods that rely on crop rotation and the use of manure as fertilizer. The farmsteads, several of the smaller towns, and lakeshore residences concentrated along the many natural lakes in the region rely on septic systems for sewage disposal. The towns of LaGrange (population: 2906 in 2002), Shipshewana (531 in 2002), Topeka (1155 in 2002), and Wolcottville (460 in 2002) (STATS Indiana, 2004) have sewage treatment facilities (W. F. Grant, LaGrange County Health Department). Wolcottville lies in both LaGrange County and Noble County. Only the population in

LaGrange County is reported here. The town of Howe in northern LaGrange County does not have a sewage treatment facility.

To evaluate the chemistry, potential contamination, and relative age of the groundwater in LaGrange County, the Indiana Geological Survey (IGS) at Indiana University began this project in 2001. The study was conducted to determine the baseline water chemistry and the levels of naturally occurring and man-made contaminants in the groundwater.

8.2. Aquifers

Water samples were collected from 50 public and private wells in LaGrange County during the summer of 2002 (Fig. 8.1). These wells, which are from 3.05 to 67.06 m deep, draw water from mixed deposits of gravel, sand, silt, and clay that were deposited during the Pleistocene.

Some sampled wells draw water from a glacial deposit composed of sand and gravel and known as the Topeka fan glacial sequence (Figs. 8.1 and 8.2a). Other sampled wells draw water from sand and gravel in the lacustrine Oliver Lake glacial sequence (Figs. 8.1 and 8.2b). The Oliver Lake sequence contains more clay and silt than the Topeka fan sequence.

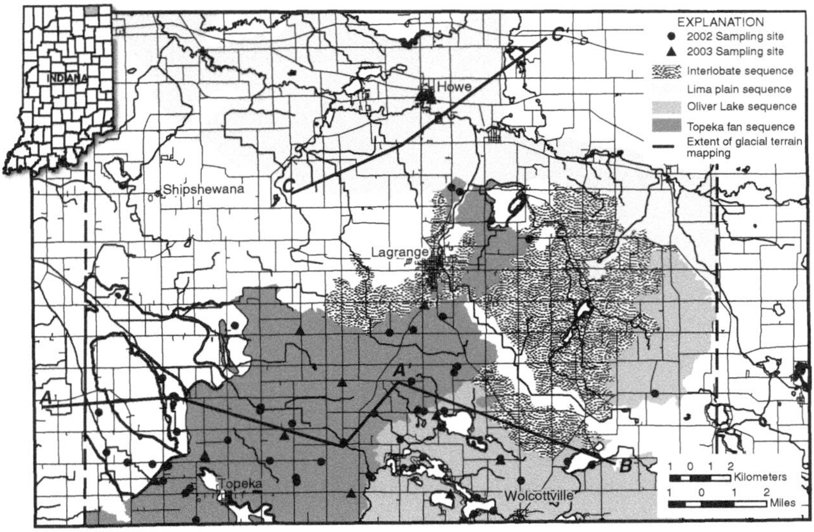

Figure 8.1. Map showing the areal extent of the Topeka fan, Oliver Lake, Lima plain, and interlobate glacial sequences and the 2002 and 2003 sampling sites. Modified from Fleming et al. (1997), Brown et al. (1998), and Brown and Jones (1999a–d).

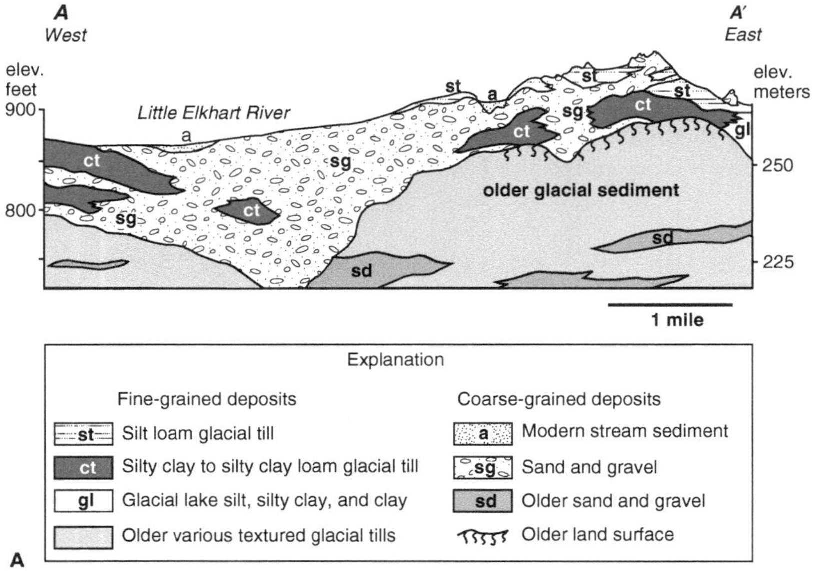

Figure 8.2a. West-to-east schematic cross section showing the Topeka fan glacial deposits. The Topeka fan sequence is characterized by coarser-grained deposits. Cross section was modified by Steven E. Brown from Brown (1999).

Much of the clay and silt was deposited in glacial lakes that once covered parts of the landscape.

The 2003 wells draw water from either sand and gravel deposits in the Topeka fan, Lima plain (Figs. 8.1 and 8.2c), and interlobate glacial sequences or localized sand and gravel deposits in the Oliver Lake sequence. The Topeka fan, Lima plain, and interlobate sequences are areas of groundwater recharge; the Oliver Lake basin is an area of groundwater discharge with areas of local recharge.

8.3. Site selection

The 2002 wells sampled in the Topeka Fan and Oliver Lake glacial sequences were selected from the IGS natural gamma-ray log database and the LaGrange County Health Department (LCHD) nitrate database. The majority (36) of the well sites were selected from the gamma-ray log database because it contains information for newer wells (1994–1998), which were drilled and completed professionally by water well drillers.

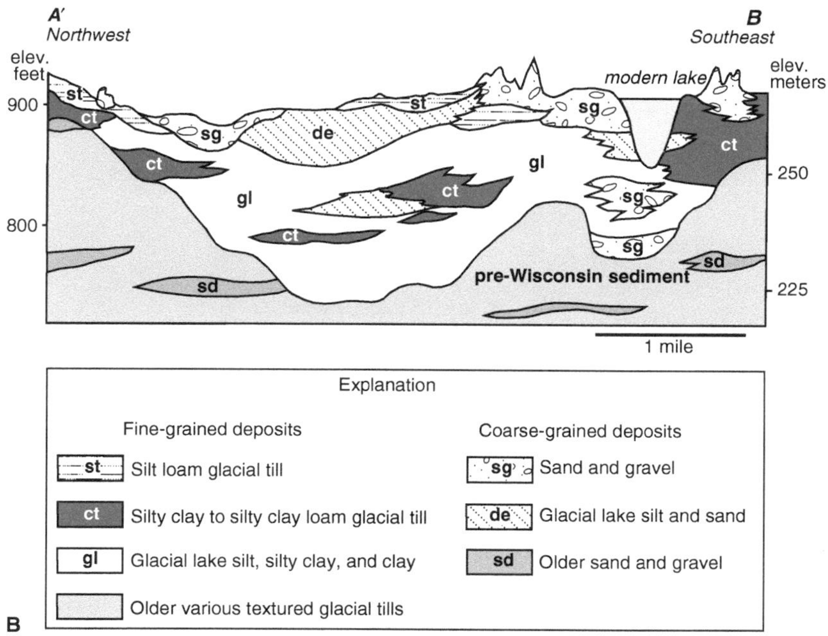

Figure 8.2b. Northwest-to-southeast schematic cross section showing the Oliver Lake glacial deposits. The Oliver Lake sequence is characterized by finer-grained deposits. Cross section was modified by Steven E. Brown from Brown (1999).

Also, gamma-ray geophysical logs and samples are available for many of these wells. The remaining 14 well sites were selected from the LCHD nitrate database, which contains data compiled from 1983 to 1998. The wells selected had reported nitrate values of less than $3\,mg\,l^{-1}$.

The 18 wells sampled in 2003 in the Topeka fan, Lima plain, interlobate, and Oliver Lake sequences were selected from the LCHD nitrate database, which contains information from 1983 to 1998, and the LCHD nitrate spreadsheets, which contain information from 1998 to 2003. The wells selected from the database and spreadsheet had reported nitrate values of $10\,mg\,l^{-1}$ or greater.

The Indiana Department of Natural Resources Division of Water Well Record Database was reviewed to obtain additional information pertaining to these water wells. During this review and discussions with the well owners, it was noted that some of the wells having high-nitrate values originally reported in the LCHD database had been abandoned and new water wells had been drilled to deeper aquifers.

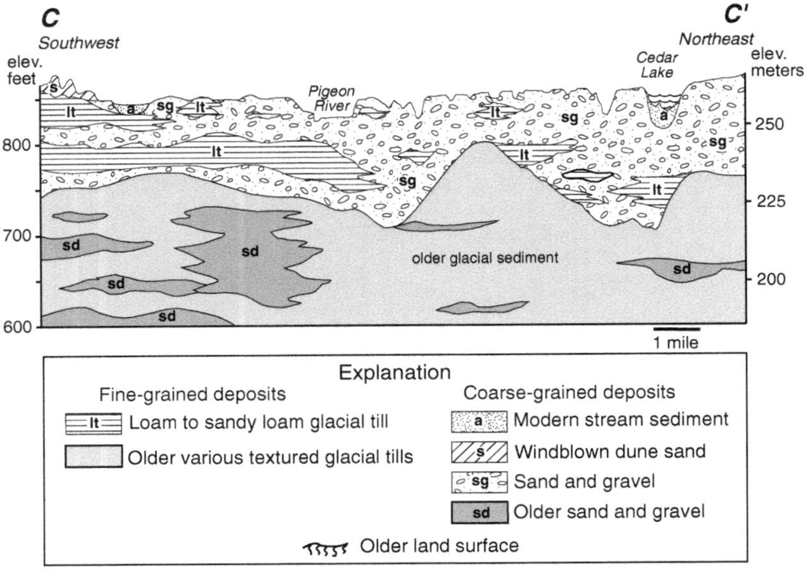

Figure 8.2c. Southwest-to-northeast schematic cross section showing the Lima plain glacial deposits. The Lima plain sequence is characterized by coarser-grained deposits. Cross section modified by Steven E. Brown from Fleming et al. (1997).

8.4. Groundwater chemistry

8.4.1. Introduction

The intent of the 2002 sampling was to acquire baseline data by analyzing for selected major and minor ions, trace elements, and isotopes in groundwater. The chemical analyses of the groundwater samples included pH and alkalinity (from which bicarbonate and carbonate concentrations were calculated); the major ions chloride, sulfate, calcium, magnesium, and sodium; the minor ions fluoride, nitrate–nitrogen, iron, manganese, silicon, strontium, and zinc; and the trace elements arsenic, barium, cadmium, chromium, lead, mercury, selenium, and silver. The total dissolved solids (TDS) were calculated from the major and minor ions. Total hardness for each sample was calculated from calcium and magnesium concentrations. Water samples were also analyzed to determine the concentrations of the radioactive isotope tritium and the stable isotope ratios of oxygen, deuterium (hydrogen), and nitrogen. The intent of the 2003 sampling was to collect water samples known to have high-nitrate values and to determine the NO_3–N concentrations and potential sources of the nitrogen by analyzing stable isotope ratios of nitrogen in nitrates. In addition to nitrogen isotopes, the field parameters and major ions noted above were determined for the 2003 samples.

8.4.2. Methods

Samples were collected from the water wells after receiving permission to sample from the well owners. The field crew flushed the well pressure tank by opening a spigot to which a flow-through cell with a multi-probe sonde was connected. When temperature and conductivity readings were observed to stabilize, indicating that fresh groundwater was being flushed through the system, field parameters were recorded and sample collection commenced. Samples were collected for chemical analyses by connecting the discharge hose to a positive-pressure stainless steel tripod filter stand containing a 0.45 μm cellulose nitrate filter and dispensing several aliquots for each sample into high-density polyethylene (HDPE) plastic bottles. One aliquot collected for selected anion analysis was kept refrigerated at 4°C. Two additional aliquots collected for determining major cations and trace elements were preserved with concentrated nitric acid so that pH was less than 2; these aliquots were transported on ice and refrigerated until used. Additional unfiltered aliquots were collected for isotope analyses and stored on ice until the field crew returned to the IGS, where the samples were frozen and shipped to the University of Waterloo Environmental Isotope Laboratory where isotopic analyses were performed.

At the IGS geochemistry laboratories, anions were analyzed using EPA method 300.0 (U.S. Environmental Protection Agency, 1993) on a Dionex DX100 suppressed ion chromatograph (IC) equipped with a conductivity detector; cations were determined by EPA method 200.7 (U.S. Environmental Protection Agency, 1994) employing simultaneous inductively coupled plasma-atomic emission spectroscopy (ICP-AES); trace element concentrations were measured using graphite furnace atomic absorption spectrophotometry (GFAA) following EPA method 200.9 (U.S. Environmental Protection Agency, 1994); and potassium levels were quantified by atomic emission spectroscopy (AES) as defined in Standard Methods section 3111B (American Public Health Association, 1995). At the University of Waterloo Environmental Isotope Laboratory, stable isotopes were determined by ratio mass spectrometry on processed samples. Deuterium ratios were measured for hydrogen gas produced from water samples reduced on hot manganese. Oxygen ratios were assessed from CO_2 equilibrated with water samples at constant temperature, and nitrogen ratios for dissolved nitrate were analyzed after sample concentration using an anion resin. Tritium values were determined by the University of Waterloo Environmental Isotope Laboratory on samples enriched by electrolysis and analyzed with a liquid scintillation counter.

All lab procedures used check standards, sample replicates, and sample spikes to monitor instrument performances. Ten percent of samples collected included duplicate aliquots to monitor precision. Data quality control also included charge balance calculations from major ion concentrations as the percentage of the difference between the anions and cations divided by the sum of the anions and cations. Any charge balance exceeding 10% was considered to have significant error in one or more major ion component and was not included in data evaluation.

8.4.3. Prevalent water chemistry

The composition of groundwater collected from 44 of the 50 wells in 2002 is dominated by calcium bicarbonate (Fig. 8.3a). Three of the remaining six wells contained no dominant cation while three others contained 100% sodium and are suspected of having been treated with an in-line water softener. Calcium averages 63% of the cation content, varying from 27 to 71% in the 47 wells known to be untreated with a water softener (Table 8.1). Bicarbonate is the dominant anion found in all wells, varying from 54 to 100% of the anion content and averaging 85% for the 47 nontreated wells. The high proportion of calcium and bicarbonate suggests that carbonate minerals dominate the water–rock interactions in the aquifer.

Data from the 18 wells sampled in 2003 contain a greater degree of mixture in the prevalent chemistry (Fig. 8.3b). Calcium remains the dominant cation for most of the samples averaging 61%, while sodium and potassium content increases from an average of 7% in 2002 to 14% in wells sampled in 2003 (Table 8.1). The relative percentage of calcium remains nearly unchanged in the cation distribution; however, bicarbonate does not maintain as dominant a proportion of the anion composition as in the 2002 dataset. The presence of nitrate in this set of wells exerts a significant enough influence on the anion distribution that the average bicarbonate proportion decreases from 85% in the 2002 wells to 64% in the 2003 wells while average nitrate + chloride increases from 5% in the 2002 wells to 27% in the 2003 wells (Table 8.1).

The calculated TDS and measured alkalinity suggest that the change in major ion proportions is primarily due to nitrate + chloride addition to the ion chemistry rather than exchange reactions. The TDS average for water sampled in 2002 is 367 $mg\,l^{-1}$; the TDS average for water sampled in 2003 is 501 $mg\,l^{-1}$ (Table 8.1). In contrast, the average alkalinity for 2002

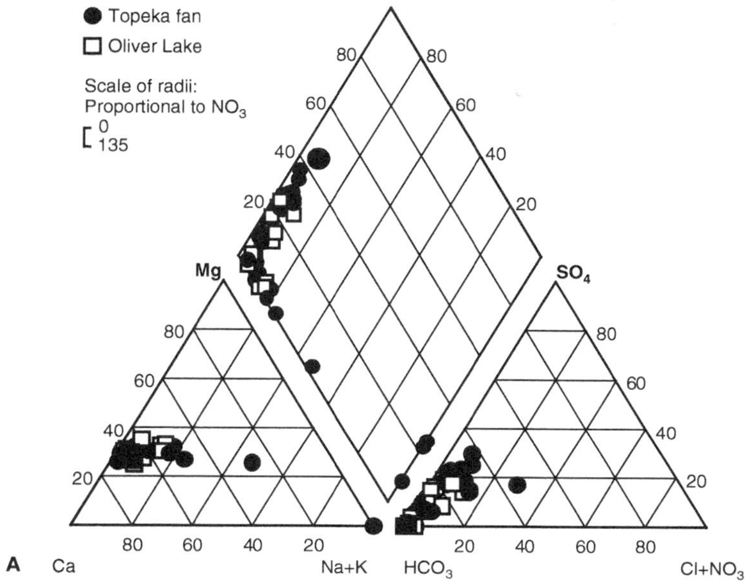

(a)

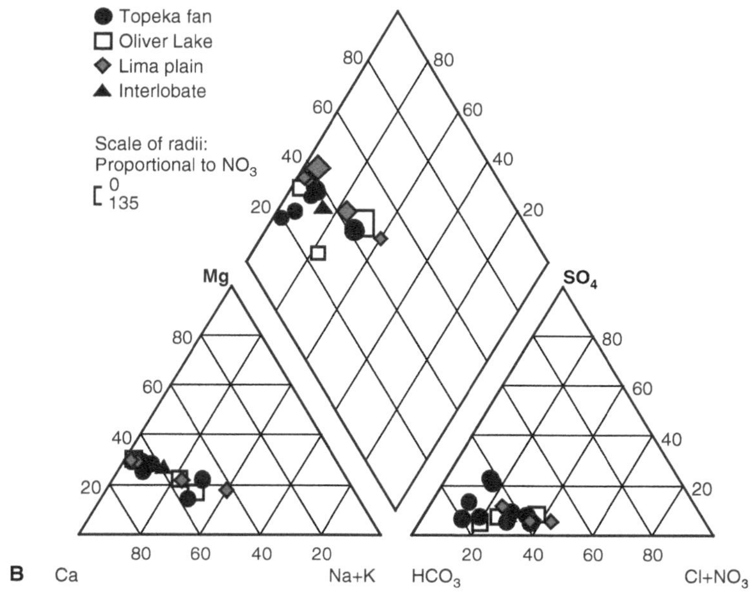

(b)

Figure 8.3. (a) Trilinear plot showing the chemistry of the groundwater samples collected in 2002. Modified from Hasenmueller and Branam (2004). (b) Trilinear plot showing the chemistry of the groundwater samples collected in 2003.

Table 8.1. Relative percentages of major ions

	2002				2003			
	Minimum	Maximum	**Average**	**Median**	Minimum	Maximum	**Average**	**Median**
Ca (%)	27	71	**63**	**65**	42	68	**61**	**65**
Mg (%)	25	34	**30**	**30**	15	31	**26**	**28**
Na + K (%)	2	46	**7**	**4**	2	40	**14**	**8**
HCO_3 (%)	54	100	**85**	**86**	51	80	**64**	**63**
SO_4 (%)	0	30	**10**	**10**	5	23	**9**	**8**
Cl + NO_3 (%)	0	29	**5**	**3**	12	44	**27**	**29**
TDS[a]	273	579	**367**	**344**	387	610	**501**	**518**
Alkalinity	207	382	**289**	**285**	223	358	**286**	**279**

[a]Calculated values.

wells is 289 mg l^{-1} as calcium carbonate equivalents and the average for 2003 wells is 286 mg l^{-1}. This indicates that while the percentage distribution of bicarbonate is less in the 2003 dataset, the actual average bicarbonate concentrations are virtually the same in both datasets and the shift in relative percentages is due to the addition of nitrate and chloride as confirmed by the TDS data.

8.4.4. Nitrate

Nitrogen is a biologically active element involved in chemical reactions that are important to life and that affect water quality. Decaying plant and animal materials release organic nitrogen, which is altered through transformation processes to produce reduced or oxidized species. Komor and Anderson (1993) identify the following transformation reactions: ammonification, nitrification, denitrification, and dissimilatory nitrate reduction. The first two reactions occur above the water table, while the last two, both nitrate-reducing reactions, occur below the water table under oxygen-deficient conditions (Freeze and Cherry, 1979). Other potential sources of nitrate–nitrogen in groundwater are inorganic nitrogen-based fertilizers, septic systems, animal waste, and fixations of atmospheric nitrogen. The EPA has established a MCL of 10 mg l^{-1} for NO_3–N in drinking water.

Specific sources of nitrate contamination can be identified by studying the proportions of stable nitrogen isotopes. The ability to identify specific sources is dependent on the number and types of sources, the distance from the sources, and the chemical or biological processes that can alter the isotopic signature of the nitrogen source. In the current study, adequate time and resources were not available to pursue the identification of the number and types of nitrogen sources.

Another factor affecting the concentration and isotopic signature of nitrate in groundwater is the variability of nitrate through time owing to changing source concentrations and isotope-altering processes, namely, denitrification with well depth. The nitrate-reducing reactions described by Komor and Anderson (1993) are catalyzed by microbial activity and have been characterized by $\delta^{15}N$ depletion in products and $\delta^{15}N$ enrichment in the residual nitrate reactant (Kaplan, 1983; Heaton, 1986). In a study on water quality in private rural wells in the Midwest, Richards et al. (1996) noted that nitrate contamination in rural wells varies from region to region owing to geologic and hydrologic factors and land use. Because of these variables, a long-term site-specific sampling program might result in a more definitive dataset to determine the potential nitrate sources.

The 2002 nitrate data represent NO_3–N concentrations and potential sources at the specific time of the field sampling of the water wells. The 2003 nitrate data also represent NO_3–N concentrations and potential sources at the specific time of field sampling; however, nitrate values for these wells are known to have been high in the past from the data in the LCHD nitrate database and spreadsheets.

The NO_3–N concentrations in groundwater sampled and analyzed in 2002 were generally low; 42 of the samples had concentrations below the detection limit ($<0.2\,mg\,l^{-1}$). Only one sample had a NO_3–N concentration above the EPA MCL (Fig. 8.4). Seven of the eight samples

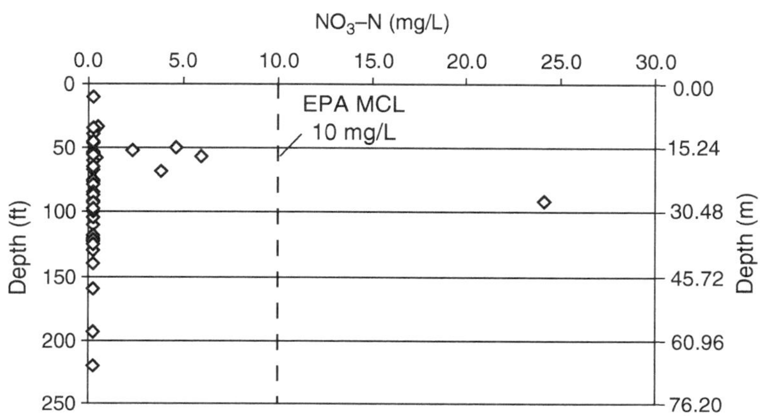

Figure 8.4. Graph showing well depth versus NO_3–N concentration for the 2002 groundwater samples. Modified from Hasenmueller and Branam (2004).

containing measurable NO_3–N were from wells in the Topeka fan glacial sequence.

In a report on the quality of well water and its vulnerability to agricultural contamination in the midwestern United States, Richards et al. (1996) reported that shallower and older wells are more likely to be contaminated than deeper and newer wells. In the LaGrange County project, total well depths were available for seven of the eight 2002 sampling sites where the water samples contained measurable NO_3–N. The seven water samples containing NO_3–N were from wells less than 30.48 m in depth (Fig. 8.4).

Nitrogen isotope ratios were used to aid in the identification of the sources of nitrates in wells where sufficient concentrations allowed the determination of the isotope ratios. Stable nitrogen isotope ratios for four 2002 Topeka fan samples ranged from 7.60 to 19.3‰. The pattern of the distribution of the isotope ratios for the 2002 and the 2003 Topeka fan samples are similar to the pattern of the distribution of nitrogen isotope ratios reported by Komor and Anderson (1993) for groundwater from shallow wells in the central Minnesota sand-plain aquifers (Fig. 8.5). The

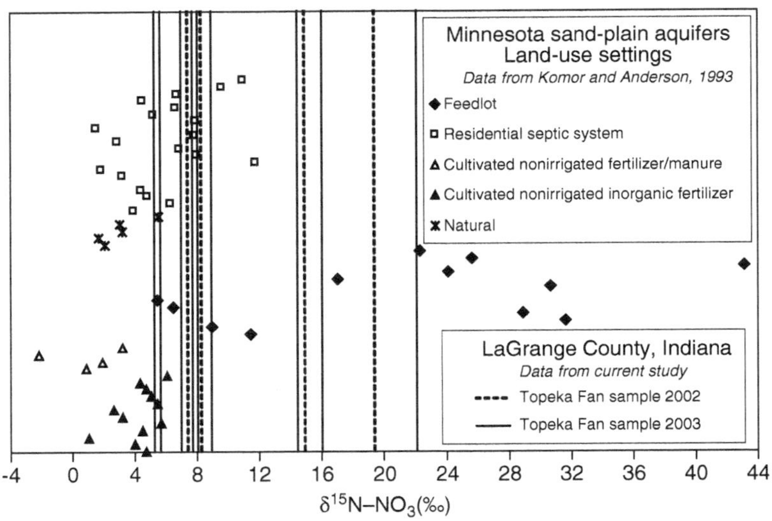

Figure 8.5. Graph showing the range of nitrogen isotope ratios for the 2002 and 2003 groundwater samples from the Topeka fan glacial sequence in LaGrange County, Indiana. The isotope ratios from the Topeka fan samples cluster into two subsets on the graph. The LaGrange County isotope data are overlain on data from the Minnesota sand-plain aquifers (Komor and Anderson, 1993, Table 3).

Minnesota sand plains are composed of glacial outwash of Quaternary age (Wright and Ruhe, 1965). The 2002 nitrogen isotope ratios are similar to those measured in the following land-use settings in Minnesota: livestock feedlots, residential areas with septic systems, and nonirrigated-cultivated areas (Fig. 8.5).

The 18 wells sampled in 2003 were selected because the previously reported nitrate values for the groundwater from these wells were high. The NO_3–N concentrations in the 2003 groundwater samples ranged from 7.0 to 30.4 $mg\,l^{-1}$. Seventy-two percent of the 2003 water samples had NO_3–N concentrations above the EPA MCL. The NO_3–N concentration in the water sample from the interlobate glacial sequence was 19.7 $mg\,l^{-1}$. Nitrate–nitrogen concentrations in the water samples in the other glacial sequences ranged from 8.3 to 30.2 $mg\,l^{-1}$ in the Lima plain sequence, 12.5 to 30.4 $mg\,l^{-1}$ in the Oliver Lake sequence, and 7 to 20.9 $mg\,l^{-1}$ in the Topeka fan sequence. The water samples from the Lima plain sequence are from home-use wells located in the town of Howe, which has no wastewater treatment facility. In regard to the NO_3–N in wells sampled in a study of private rural wells in the midwestern United States, Richards et al. (1996) indicated that consistent incidences of contamination may indicate contamination of the aquifer locally. This may be the case in the vicinity of the town of Howe. Several of the water samples from the Oliver Lake sequence are from lakeshore residences concentrated along natural lakes; these sites have no wastewater treatment facilities. The water samples from the Topeka fan are predominantly from rural sites.

Nitrogen isotope ratios for the 2003 groundwater samples range from 4.9 to 22.3‰. The isotope ratios for the samples from wells in the Lima plain and Oliver Lake sequences range from 4.9 to 9.6‰ (Fig. 8.6). These $\delta^{15}N$ values indicate multiple sources when compared to the data from Komor and Anderson (1993); the positive correlation of the $\delta^{15}N$ with sodium concentrations (Fig. 8.7) suggests that septic waste is a primary source of the nitrogen in these two sequences (Bleifuss et al., 2004).

The nitrogen isotope ratios for the 2003 water samples from the Topeka fan sequence range from 5.3 to 22.3‰ (Fig. 8.5). In the Topeka fan glacial deposits, one group of isotope data ranges from 5.32 to 9.14‰; the other group has heavier isotopic signatures and ranges from 14.5 to 22.32‰ (Fig. 8.5). The group of lighter nitrogen isotope ratios is similar to ratios reported for livestock feedlots, residential areas with septic systems, and nonirrigated-cultivated areas by Komor and Anderson (1993). The group of heavier nitrogen isotope ratios is similar to the ratios reported for feedlots by Komor and Anderson (1993).

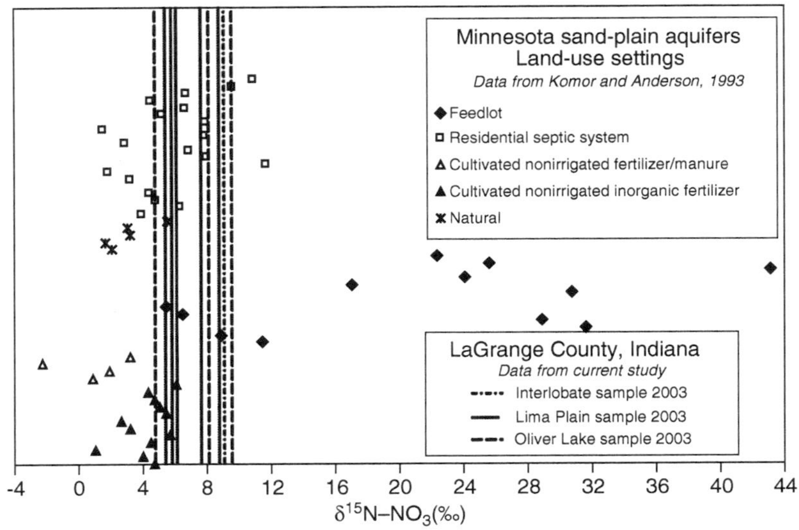

Figure 8.6. Graph showing the range of nitrogen isotope ratios for the 2003 groundwater samples from the interlobate, Lima plain, and Oliver Lake glacial sequences in LaGrange County, Indiana. The LaGrange County isotope data are overlain on data from the Minnesota sand-plain aquifers (Komor and Anderson, 1993, Table 3).

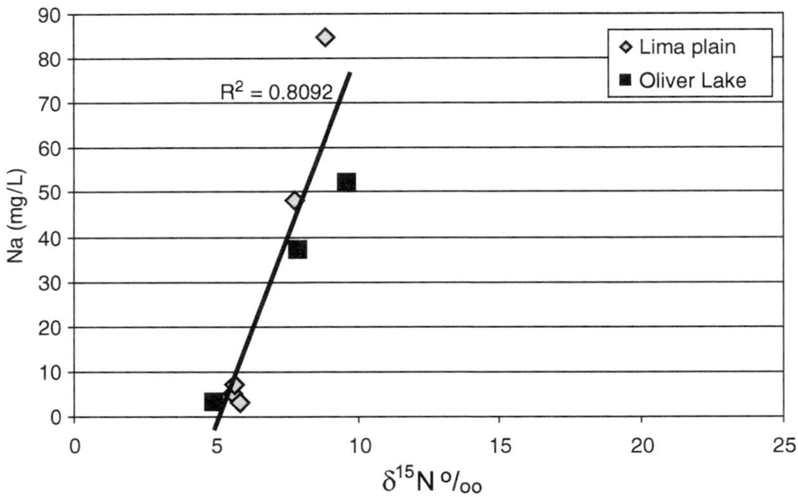

Figure 8.7. Graph showing the relationship of sodium (mg l^{-1}) to nitrogen isotope ratios for the 2003 groundwater samples from the Lima plain and Oliver Lake glacial sequences.

8.4.5. Tritium, ^{18}O, and Deuterium

Tritium (^3H), ^{18}O, and deuterium (^2H) can be used to determine the locations of groundwater recharge areas, the circulation patterns in aquifers, the sources of dissolved solids in groundwater, and the age of groundwater (Fetter, 1988). Knowledge of relative ages of groundwater contributes to the understanding of groundwater flow and sensitivity to contamination. Only the water samples collected in 2002 were analyzed for tritium, ^{18}O, and deuterium.

Tritium is an unstable radioactive isotope of hydrogen with a half-life of 12.43 years. It is produced naturally in low concentrations by the interaction of cosmic rays with nitrogen and oxygen in the atmosphere. The most significant source of tritium, however, is anthropogenic from the atmospheric testing of nuclear weapons from 1952 to 1969. Tritium in the atmosphere is directly incorporated into the water molecule and is introduced to groundwater through rainfall. Tritium is used to identify modern recharge to groundwater systems.

In the 2002 water samples, tritium concentrations ranged from below the detection limit (<0.8 tritium units [TU]) to 17.0 TU. Of the 50 samples, 35 had values above the detection limit. A tritium concentration greater than 2 TU is interpreted as indicative of the presence of post-1952 recharge water (Fontes, 1980). A similar range of values was observed in aquifers of unconsolidated glacial deposits in northwestern Indiana, where tritium values from the detection limit (0.8 TU) to 28.2 TU were reported by Hasenmueller et al. (2001).

In the LaGrange County study, the presence or absence of tritium correlated with the depth of the well. A total well depth was available for 44 of the 2002 sampling sites. Tritium concentrations greater than 2.0 TU were detected in 83% of the wells that were less than 30.48 m in depth (Fig. 8.8). Concentrations greater than 2.0 TU were detected in only 20% of the wells that were 30.48 m or greater in depth (Fig. 8.8).

In the water samples collected and analyzed in 2002, the $\delta^{18}O$ values ranged from −9.07 to −5.47‰ and the δ^2H values ranged from −60.11 to −40.76‰. The $\delta^{18}O$ and δ^2H values for the 19 water samples with tritium concentrations of 2 TU or less exhibited a smaller range −8.99 to −8.35‰ (Fig. 8.9a) and −58.62 to −54.97‰ than the $\delta^{18}O$ and δ^2H values for the 31 water samples with tritium concentrations greater than 2 TU (Fig. 8.9b).

The $\delta^{18}O$ and δ^2H values in the water samples from LaGrange County are slightly more negative than the current mean average annual values calculated by Drever (1988) for this area. The average $\delta^{18}O$ and δ^2H values for the 2002 LaGrange County water samples were −8.59‰ and −55.83‰. Similar values were reported for $\delta^{18}O$ and δ^2H for

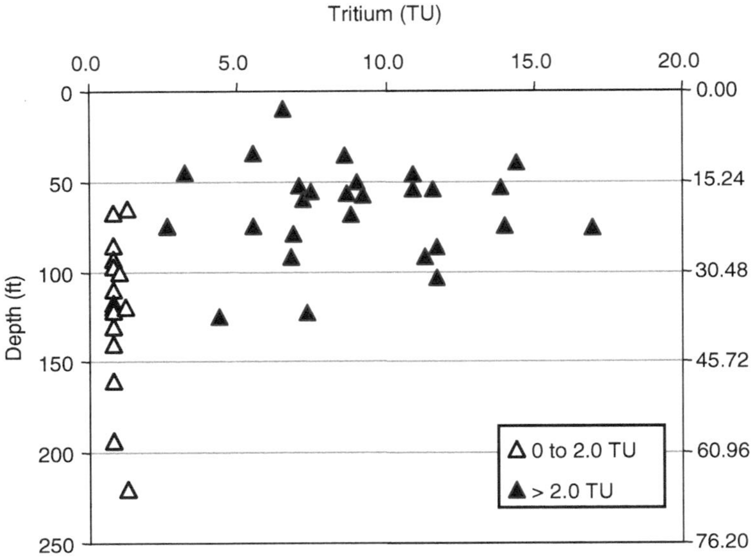

Figure 8.8. Graph showing well depth versus tritium concentration for the 2002 groundwater samples. Modified from Hasenmueller and Branam (2004).

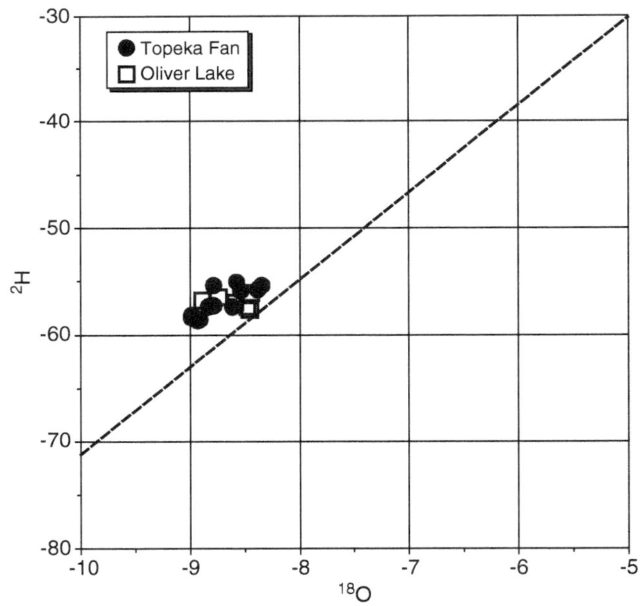

Figure 8.9a. Graph showing the $\delta^{18}O$ and $\delta^{2}H$ values for the 19 water samples with tritium concentrations of 2 TU or less.

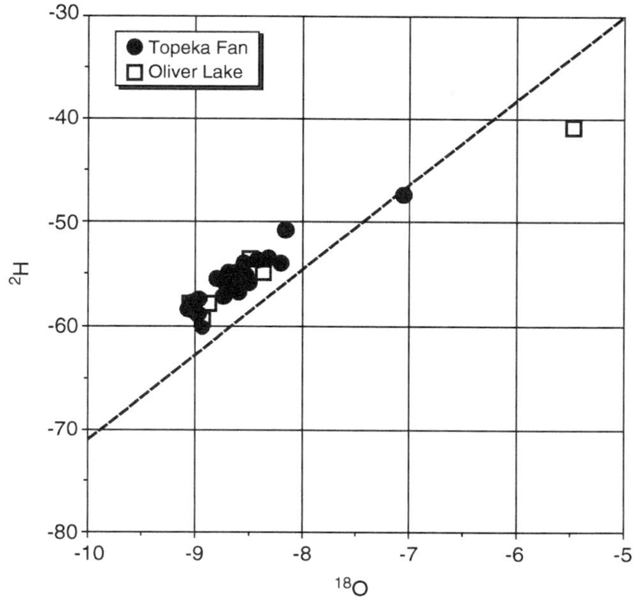

Figure 8.9b. Graph showing the $\delta^{18}O$ and δ^2H values for the 31 water samples with tritium concentrations greater than 2 TU.

groundwater samples from unconsolidated glacial deposits in northwestern Indiana (Hasenmueller et al., 2001). In that study, the $\delta^{18}O$ values ranged from −9.3 to −6.66‰ and averaged −8.35‰; δ^2H values ranged from −61.84 to −51.83 ‰ and averaged 55.52‰. The water isotope data from the unconsolidated deposits in LaGrange County and northwestern Indiana plot close to the meteoric line (Fig. 8.10).

8.5. Conclusions and recommendations

The prevalent chemical character of the groundwater in LaGrange County is calcium-bicarbonate. The primary difference between the 2002 and 2003 datasets is (1) the addition of nitrate that caused an increase in the TDS content and (2) a moderate shift in the relative percentages of major anion species, which was not sufficient enough to modify the prevalent chemical character.

The 2002 and 2003 nitrate data indicate that groundwater in the Topeka fan, Lima plain, and interlobate glacial sequences and in localized sand and gravel deposits in the Oliver Lake sequence, is sensitive to

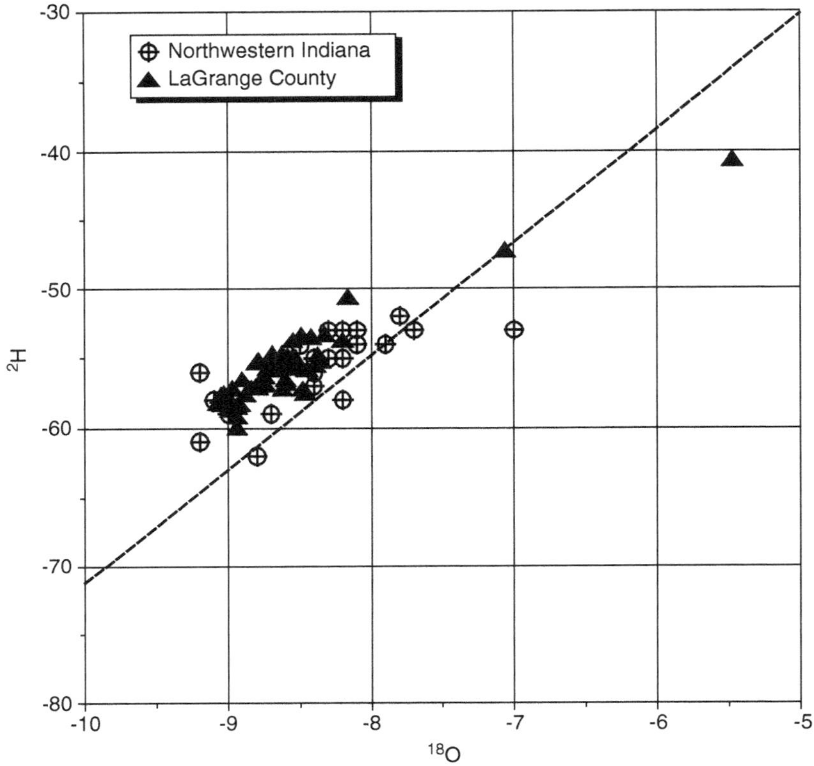

Figure 8.10. Graph showing stable isotopes correlations to meteoric water line for water samples from unconsolidated glacial materials in northern Indiana.

anthropogenic contamination from the surface. The nitrogen isotope ratios from groundwater in wells in LaGrange are similar to those measured in the following land-use settings in Minnesota sand plains: livestock feedlots, residential areas with septic systems, and nonirrigated-cultivated areas. Poorly performing septic systems may be a source of some of the nitrate and may be more readily identified as nitrogen sources when positive correlations occur between nitrogen isotope values and other component concentrations, such as sodium. Septic failure often occurs as the result of the absorption field being plugged; however, this type of failure is not as common in LaGrange County. In this county, most failures occur when septic systems are completed in coarse-grained deposits, which are permeable and allow sewage to flow into the groundwater aquifer system. Such failures can cause groundwater contamination.

The 2002 tritium data indicate that the groundwater in 62% of the wells sampled in the Topeka fan and Oliver Lake areas has been recharged since 1952. The groundwater in the remaining wells in the Topeka fan and Oliver Lake glacial sequences has been recharged between 10,000 years before present and 1953.

The coarse sand and gravel aquifers in the area facilitate the rapid flow of groundwater. To prevent contamination of the groundwater in the county, property owners should consider nearby sources of potential contamination, such as septic systems or farm-animal storage areas, before drilling new water wells. Farmers should be encouraged to participate in the voluntary manure management program currently available. And when feasible, additional wastewater treatment facilities should be built to treat waste from the smaller communities and lakeshore properties.

To more definitively determine the nitrate sources contributing to contamination, long-term site-specific water monitoring and site evaluation are necessary. This should include water sampling at various distances from identified sources and at various depths.

ACKNOWLEDGMENTS

This project was funded in part by the Indiana Department of Environmental Management (IDEM) (ARN# A305-1-01-280). Greg Overtoom of IDEM was the project manager. The project participants worked cooperatively with William F. Grant, Director of the LaGrange County Health Department. We made extensive use of the glacial terrain data for LaGrange County compiled by Anthony H. Fleming, Steven E. Brown, Holly Jones, Angela J. Smous, and Tara L. Schrader. Steven E. Brown and Anthony H. Fleming served as advisors to the project. We appreciate the work of the following IGS staff members, who collected and analyzed the water samples and participated in the preparation of the report and geographic information system database: Sarah M. Butler, Margaret V. Ennis, Barbara T. Hill, Karen K. Like, Jeffery J. Olyphant, Ronald T. Smith, and Kimberly H. Sowder. Finally, we wish to thank many property owners who allowed us to sample and analyze water from their wells.

REFERENCES

American Public Health Association, 1995. *Standard Methods for the Examination of Water and Wastewater*, 19th ed. American Public Health Association, American Water Works Association, Water Pollution Control Federation, Washington, D.C.

Bleifuss, P.S., Hanson, G.N., Schoonen, M.A.A., 2004. Tracing sources of nitrate in the Long Island aquifer system. Available at http://pbisotopes.ess.sunysb.edu/reports/bleifuss/ accessed June 11, 2004.

Brown, S.E., 1999. Part 2—Ice marginal environments of Huron-Erie and Saginaw Lobes, north central Indiana. In: Brown, S.E., Fisher, T.G., Kehew, A.E., Taylor, L.D. (Eds.), *Pleistocene Geology of North-Central Indiana and South-Central Michigan*. Guidebook for the 45th Midwest Friends of the Pleistocene Field Conference. Indiana Geological Survey Open-File Study 99-1, p. 77.

Brown, S.E., Fleming, A.H., Jones, H., Schrader, T.L., 1998. Glacial terrains of the Mongo, Wolcottville, and the Indiana part of the Burr Oak 7.5-minute quadrangles, LaGrange and Noble Counties, Indiana. Indiana Geological Survey Open-File Study 98-06, p. 24.

Brown, S.E., Jones, H., 1999a. Glacial terrains of the Middlebury 7.5-minute quadrangle. Indiana Geological Survey Open-File Study 99-15, 1 pl.

Brown, S.E., Jones, H., 1999b. Glacial terrains of the Millersburg 7.5-minute quadrangle. Indiana Geological Survey Open-File Study 99-14, 1 pl.

Brown, S.E., Jones, H., 1999c. Glacial terrains of the Orland and Indiana part of the Bronson South 7.5-minute quadrangles. Indiana Geological Survey Open-File Study 99-13, 3 pls.

Brown, S.E., Jones, H., 1999d. Glacial terrains of the Stroh 7.5-minute quadrangle. Indiana Geological Survey Open-File Study 99-12, 3 pls.

Drever, J.I., 1998. The Geochemistry of Natural Waters, second ed. Prentice Hall, Englewood Cliffs, NJ, p. 437.

Fetter, C.W., 1988. Applied Hydrogeology, second ed. Merrill Publishing Company, Columbus, OH, p. 592.

Fleming, A.H., Brown, S.E., Smous, A.J., Schrader, T.L., 1997. Glacial terrains of the Topeka, Shipshewana, Oliver Lake, LaGrange and Sturgis 7.5-minute quadrangles, LaGrange and Noble Counties, Indiana. Indiana Geological Survey Open-File Study 97-14, Expanded Explanation, p. 26.

Fontes, J.Ch., 1980. Environmental isotopes in groundwater hydrology. In: Fritz, P., Fontes, J.Ch. (Eds.), Handbook of Environmental Isotope Geochemistry. Elsevier, Amsterdam, Vol. 1, pp. 75–140.

Freeze, R.A., Cherry, J.A., 1979. Groundwater. Prentice Hall, Englewood Cliffs, NJ, p. 604.

Hasenmueller, N.R., Branam, T.D., 2004. Evaluation of ground-water chemistry and age for aquifer systems in LaGrange County, Indiana. Proceedings of the 25th Annual Water Resources Symposium, June 23–25, 2004, Valparaiso, Indiana, p. 9.

Hasenmueller, N.R., Branam, T.D., Bleuer, N.K., 2001. Evaluation of ground-water age and chemistry relations in aquifer systems in Lake, Porter, and La Porte Counties of Indiana. Indiana Geological Survey Open-File Study 01-21, p. 24.

Heaton, T.H.E., 1986. Isotopic studies of nitrogen pollution in the hydrosphere and atmosphere: A review. Chem. Geol. 59, 87–102.

Indiana Department of Natural Resources Division of Water, 1987. Water resource availability in the St. Joseph River Basin, Indiana. Indiana Department of Natural Resources Division of Water, Water Resource Assessment, 87-1, p 139.

Kaplan, I.R., 1983. Stable isotopes of sulfur, nitrogen, and deuterium in Recent marine environments. In: Arthur, M.A. (organizer), Stable Isotopes in Sedimentary Geology, SEPM Short Course No. 10, pp. 2-1 to 2-108.

Komor, S.C., Anderson, H.W. Jr., 1993. Nitrogen isotopes as indicators of nitrate sources in Minnesota sand-plan aquifers. Ground Water 31, 260–270.

Richards, R.P., Baker, D.B., Creamer, N.L., Kramer, J.W., Ewing, D.E., Merryfield, B.J., Wallrabenstein, L.K., 1996. Well water quality, well vulnerability, and agricultural contamination in the midwestern United States. J. Environ. Qual. 25, 389–402.

STATS Indiana, 2004. LaGrange County in depth profile. Indiana Business Research Center, Indiana University's Kelley School of Business Web page, http://www.stats.indiana.edu/profiles/pr18087.html accessed July 19, 2004.

U.S. Census Bureau, 2004. U.S. Census Bureau state and county *QuickFacts*—LaGrange County. U.S. Census Bureau Web page, http://quickfacts.census.gov/qfd/states/18/18087.html accessed May 28, 2004.

U.S. Department of Agriculture, 2004. Indiana agricultural statistics—Historical data Indiana farm land use history, LaGrange County, Indiana. U.S. Department of Agriculture Web page, http://www.nass.usda.gov/in/historic/h97lagra.pdf accessed July 19, 2004.

U.S. Environmental Protection Agency, 1993. Methods for the Determination of Inorganic Substances in Environmental Samples. U.S. Environmental Protection Agency EPA/600/R-93/100, CD-ROM.

U.S. Environmental Protection Agency, 1994. Methods for the Determination of Metals in Environmental Samples, Supplement 1 to EPA/600-4-91-010. U.S. Environmental Protection Agency EPA/600/R-94/111, CD-ROM.

Wright Jr., H.E., Ruhe, R.V., 1965. Glaciation of Minnesota and Iowa. In: Wright, Jr., H.E., Frey, D.G. (Eds.), The Quaternary of the United States. Princeton University Press, Princeton, NJ, pp. 29–41.

Developments in Environmental Science, Volume 5
D. Sarkar, R. Datta and R. Hannigan (Editors)

ISSN: 1474-8177/DOI:10.1016/S1474-8177(07)05009-7

Chapter 9

Agriculture-induced contamination of surface water and groundwater in Portugal

C. Nabais, M.L. Barrico, H. Freitas and M.N.V. Prasad

Abstract

From the end of the 1980s, many studies have drawn the attention to the environmental consequences of "industrial agriculture", namely significant soil erosion and a decrease in water quality. European citizens have demonstrated a growing interest in understanding how their tax money is used in efforts to guarantee the safety and quality of consumer goods and to protect the environment. When Portugal joined the EC in 1986, its main agricultural indices contrasted significantly with those of the other member states. Agriculture made up 12% of gross domestic product, and provided employment to roughly 20% of the active population. In Portugal, the adoption of new technologies, namely improved seeds, higher inputs of fertilizers and pesticides, and mechanization, were introduced later than in other European countries. Sparingly used fertilizers and pesticides partially avoided the negative impact in terms of soil and underground water contamination that took place in many European countries. However, maintenance of the traditional pattern of land allocation to crops had negative impacts in some regions. Extensive cattle breeding in silvo-pastoral systems, low intensity of crop rotation, and excessive tilling resulted in erosion and degradation of soil organic matter with consequent loss of fertility.

The adoption of less-polluting farming practices can only be made with the full cooperation of the agriculture community and largely depends on demonstrating both the environmental and economic benefits of such practices to land managers.

9.1. Agriculture, water quality and ecodevelopment

Agricultural research in 20th century was primarily focused on increased production of food ignoring the impact of agrochemicals on ecosystems.

Agriculture worldwide has several positive effects such as open space and landscape amenities as well as habitat preservation for many wildlife species (Vitousek et al., 1997). Contrarily, negative effects associated with agriculture include loss of biodiversity from conversion of forests and wetlands to agricultural land, degradation of soil and water quality, and overuse of surface water and groundwater for irrigation (Abler et al., 2002). Water quality in rivers is generally linked with land-use in the catchment areas that can affect the amount and quality of run-off during and following rainfall (Ngoye and Machiwa, 2004). Forestry, agriculture, industrialization and urbanization modify watershed management and consequently influence run-off quality and quantity. Additionally, with many rivers experiencing severely reduced flow, wetlands diminishing, and major groundwater aquifer depletion at unprecedented rates, we are moving towards significant scarcity of water resources. The projected need to double food production "must largely take place on the same land area and using less water" (Task Force on Building a Science Roadmap for Agriculture, 2001).

In the European Community (EC), between 1960 and 1980, employment in the primary sector fell from 18.2% to 7.8%, while income from agriculture fell from 11% of total income in 1958 to 3.4% in 1980 (Freire and Parkhurst, 2002). When Portugal joined the EC in 1986, its main agricultural indices contrasted significantly with those of the other member states. Agriculture made up 12% of gross domestic product, and provided employment to roughly 20% of the active population (Freire and Parkhurst, 2002). In Portugal, the adoption of new technologies, namely improved seeds, higher inputs of fertilizers and pesticides, and mechanization, were introduced later than in other European countries, mainly after the entry of Portugal into the European Economic Community. Sparingly used fertilizers and pesticides partially avoided the negative impact in terms of soil and underground water contamination that took place in many European countries. Therefore, Portuguese agro-ecosystems supported an important agro-biodiversity (crop varieties and livestock breeds) and wild biodiversity (non-domesticated animals and plants). On the other hand, maintenance of the traditional pattern of land allocation to crops had negative impacts in some regions. For instance, the soil erosion that occurred in the basin of the Guadiana River resulted from the use of marginal land (with low-quality soils and excessive slopes for tillage) for the production of wheat under the Wheat Campaign of the 1930s, aiming to ensure the country's self-sufficiency in wheat production (Giordano et al., 1992). Extensive cattle breeding in silvo-pastoral systems, low intensity of crop rotation, and excessive tilling resulted in erosion and degradation of soil organic matter with consequent loss of fertility.

Agricultural intensification in certain regions (mainly in Algarve, Ribatejo and Oeste, coastal Alentejo and coastal Beira Litoral), has led to the overexploitation of underground water. Leaches from pesticides and fertilizers not only polluted groundwater but also surface water with farm run-off, particularly phosphorus (INAG, 1997). In Portuguese agro-ecosystems, the run-off is extremely variable due to intra- and inter-annual variations in precipitation. The highest run-off occurs during the winter months followed by prolonged periods of low run-off. Mean annual rainfall is 922 mm, with 577 mm of the water budget in evapo-transpiration and 385 mm in run-off (INAG, 2001). Total annual water consumption in Portugal is 8754 hm^3, with about 45% from surface run-off and the remainder from underground sources (INAG, 2001). About 74% of water consumption is associated with agricultural activity.

Diffuse pollution is recognized as the main cause of pollution in Portuguese surface water and groundwater, where most of the detected compounds (pesticides and chlorophenols) are present at concentration levels $< 1\,\mu g\,l^{-1}$. Higher values (up to $12\,\mu g\,l^{-1}$) are due to specific and punctual pollution episodes, local treatments or accidental spills (Lacorte et al., 2001).

The socio-economic development of Portugal is characterized by a concentration of the population and economic activities near the coast, causing an imbalance in managing water resources. Over the last few years, water resource management has been implemented through the EU Water Framework Directive (2000/60/EC). It requires all inland and coastal waters to reach "good status" by 2015. It will do this by establishing a river basin district structure within which demanding environmental objectives will be set, including ecological targets for surface waters.

9.2. Nitrate pollution

Increased agricultural production over the last 50 years has resulted in increased nitrate (NO_3^-) concentrations in rivers, lakes and underground aquifers. This loss of nitrogen from agriculture represents an economic shortfall, in that the applied nitrogen is not being utilized for food production (Merrington et al., 2002). Nitrate pollution has major impacts on the quality of drinking water and public health, and the eutrophication of surface waters.

The European Community maximum admissible content of nitrate in water for human consumption is $50\,mg\,l^{-1}$. The Alentejo region, in the South of Portugal, is subjected to periods of acute water shortages, as it is common in a Mediterranean climate. This region is extremely dependent on groundwater resources for public supply and agriculture. Beja is a major town of the Alentejo region (*ca.* 20,000 inhabitants) and its public water supply depends partially on groundwater resources of the gabbro-dioritic aquifer surrounding the city and surface reservoir of the Roxo dam (Paralta and Ribeiro, 2001). Increasing concentrations of nitrate in groundwater supplies in Beja were recognized as early as the 1940s. Nitrate monitoring of Beja Aquifer System, performed between 1997 and 2000, revealed median values between 53 and 86 $mg\,l^{-1}$, and maximum nitrate content between 126 to 225 $mg\,l^{-1}$ (Paralta and Ribeiro, 2001).

According to a report of the Portuguese Ministry of Environment and Ministry of Agriculture (MAOT, 2000), three vulnerable areas of nitrate pollution were recognized, based on groundwater analysis: Esposende/Vila do Conde (North), Aveiro (North) and Campina de Faro (Algarve, South). To reduce nitrate pollution of groundwater, some agricultural measurements were implemented in these three areas, namely reduction of organic fertilizers applied per hectare. The simulation model Root Zone Water Quality Model (RZWQM), developed by the USDA-ARS (Ma et al., 2001), was used to simulate/evaluate the effectiveness of the agricultural measurements to reduce nitrate leachates. According to the model, the three vulnerable areas would reduce by 50% the nitrate leachates within a nine-year period (MAOT, 2000).

9.3. Pesticides contamination and pollution

Pesticides have become an integral part of modern farming, with most crops receiving several applications per year. Pesticides reduce attack by pests, diseases and weeds and contribute to higher yields and higher economic return. An ideal pesticide would only affect its target organism, be non-persistent and have no harmful environmental effects. However, most pesticides do not have these traits. Pesticides may cause pollution by direct contamination of ground and surface waters, by soil contamination and subsequent leaching into ground and surface waters, and by contamination of non-target organisms (Merrington et al., 2002). In a review work, Pimentel (1995) concluded that less than 0.1% of pesticides reach their target organism.

The amount of plant protection products sold in Portugal, during 2003 was 17,030,910 kg, expressed as active ingredients (Table 9.1). The evaluation of exposure of Portuguese groundwater to pesticides started in the early 1990s (Batista et al., 2002). In 2000, the pesticides and/or metabolites most frequently found in the groundwater of "Beira Litoral"

Table 9.1. Sales of plant protection products in Portugal in 2003 (kg of active ingredients)

Total	17,030,910
Fungicides	12,954,391
Inorganic	11,084,776
Based on dithiocarbamates	1,110,567
Based on benzimidazoles	27,513
Based on diazoles or triazoles	23,396
Others	708,139
Herbicides	2,381,549
Based on phenoxy-phytohormones	84,407
Based on triazines	361,558
Based on amides	11,572
Based on carbamates	114,711
Based on dinitroaniline derivatives	18,924
Based on derivatives of urea, of uracil and sulphonylurea	113,751
Others	1,676,626
Insecticides and acaricides	505,458
Based on pyrethroids	7032
Based on chlorinated hydrocarbons	46,814
Based on carbamates	35,283
Based on organophosphorus products	315,670
Biopesticides	1236
Others	99,423
Molluscicides	27,467
Plant growth regulators	2766
Physiological regulating products	2276
Antisprouting products	490
Rodenticides	29
Mineral oil	582,325
Soil sterilants	549,185
Others	27,740

Source: MADRP (2005).

and "Ribatejo e Oeste", important agricultural regions of Portugal, were atrazine, used in maize, and its metabolites desethylatrazine and desisopropylatrazine; simazine used in vineyards and orchards; alachlor and metolachlor used in maize and metribuzin used in tomato and potato (Batista et al., 2002). The detection of pesticides in groundwater was relatively frequent, although their occurrence at levels above $0.1\,\mu g\,l^{-1}$ was rare. The highest detected values were for metolachlor ($17.0\,\mu g\,l^{-1}$), molinate ($16.3\,\mu g\,l^{-1}$), atrazine ($11.2\,\mu g\,l^{-1}$), alachlor ($8.0\,\mu g\,l^{-1}$), 3,4-dichloroaniline ($3.8\,\mu g\,l^{-1}$), α-endosulfan ($1.4\,\mu g\,l^{-1}$), dimethoate ($1.2\,\mu g\,l^{-1}$) and lindane ($1.1\,\mu g\,l^{-1}$). Most peak pesticide levels were detected after a period of pesticide application and during the irrigation period, i.e., from March to May (Batista et al., 2002). Leaching losses of pesticides from soil occur when rainfall (or irrigation) exceeds evapotranspiration losses and the soil water content rises above field capacity. The susceptibility of a pesticide to leaching depends upon the interaction of the chemical properties of the pesticide with the chemical and physical properties of the soil (Merrington et al., 2002).

The solubility in water and partition coefficient (K_{ow}) of pesticides varies considerably and has an important influence on their environmental persistence (Munz and Bachmann, 1993). The partition coefficient (K_{ow}) is calculated has the ratio between the distribution of the pesticide between the liquid layers of 1-octanol and water. As the value of log K_{ow} increases a pesticide tends to become more hydrophobic and it is more likely to bind to organic molecules. Pesticides with partition coefficients values greater than 7 do not generally move far in the agro-ecosystem, but tend to bind strongly to the organic matter in soils and sediments (Merrington et al., 2002). However, log K_{ow} between 4 and 7 indicate that the chemical is lipophilic and can accumulate in fatty tissues, with potential bioaccumulation effects.

Atrazine and simazine are two herbicides commonly used to control broadleaf and grassy weeds in agricultural lands and on non-cropped industrial lands. They are both significantly water soluble, have relatively low-soil adsorption coefficients and are persistent in both soil and groundwater (Wauchope et al., 1992). Like other organic pesticides, the fate of atrazine and simazine in soil and groundwater systems is controlled to a great extent by its physicochemical interactions with soil and aquifer material. Capriel et al. (1985) reported that nine years after application of atrazine to a field soil, as much as 90% of the herbicide residue was associated with soil organic matter. Systematic studies of the impact of contact time on sorption, desorption and extractability of agrochemicals to surface soils remain scarce. Such studies have the potential to contribute significantly to a better understanding of post-application herbicide dynamics in agricultural fields (Lesan and Bhandari, 2004).

9.4. Phenolic compounds in two dams of Alentejo region (south Portugal)

Water quality monitoring in dams used for human water consumption, carried out by the Alentejo Regional Authorities of the Environment (south Portugal), revealed seasonal peaks of phenolic compounds above the water-quality legislation. Owing to their toxicity and persistence in the environment, phenolic compounds are listed as priority pollutants (Pocurull et al., 1996). Such compounds originate from industrial applications or from the degradation of agricultural pesticides (Daví and Gnudi, 1999).

To identify the main phenolic compounds present in water and soil leachates, and to determine the sources of the seasonal concentrations of phenolic compounds, two catchments in Alentejo were selected, with different land-use patterns: Roxo dam, mainly dry land agricultural area, and Santa-Clara dam, mainly surrounded by shrubs and trees. Surface water and soil samples were collected in several points in the catchments basins of Santa-Clara and Roxo (Fig. 9.1) (Barrico, 2004).

The main phenolic compound detected was 2,4-dinitrophenol (2,4-DNP), both in stream water and soil leachates, with higher concentrations in Roxo catchments (Fig. 9.2). Roxo catchments presents a larger agricultural area than Santa Clara, and it is likely that the origin of the

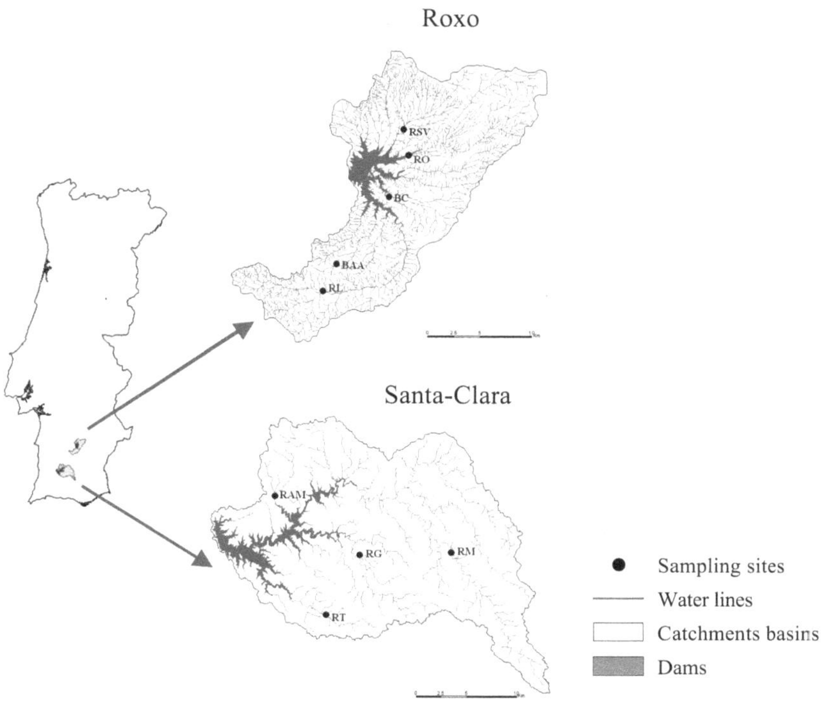

Figure 9.1. Location of the sampling sites in the catchments basins of Santa-Clara and Roxo (Alentejo, south Portugal). Santa Clara: Ribeira das Águas Muitas (RAM), Ribeira de Guilherme (RG), Rio Mira (RM), Rio Torto (RT); Roxo: Barranco da Água Azeda (BAA), Barranco dos Castelhanos (BC), Ribeira dos Louriçais (RL), Ribeira do Outeiro (RO), Ribeira de Santa Vitória (RSV).

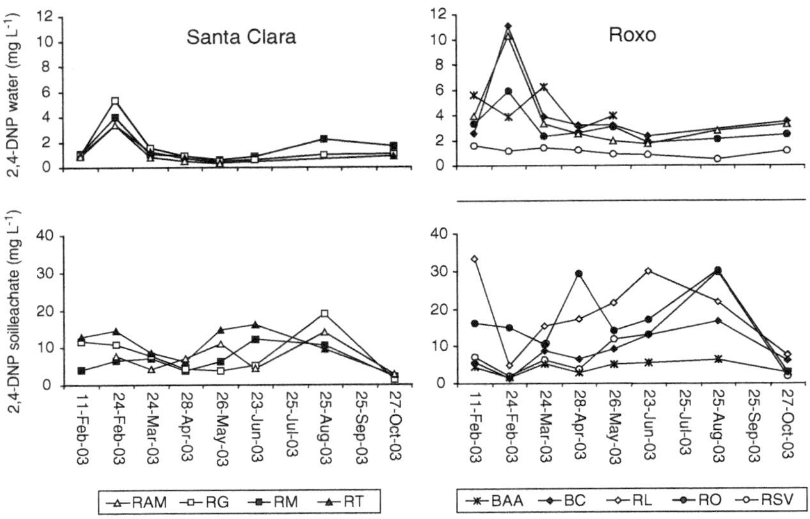

Figure 9.2. Seasonal concentrations of 2,4-dinitrophenol (2,4-DNP) in the water and soil leachate of Santa-Clara and Roxo catchments basins (for legend explanation, see caption of Fig. 9.1).

2,4-DNP is associated with the use of pesticides. According to the National Authorities for Crop Protection of the Portuguese Agricultural Ministry, approximately 20,000 kg fungicides with the active ingredient Dinocape (containing dinitrophenolic substances) were sold in Portugal during 2002.

A peak of 2,4-DNP concentration in stream water, both in Roxo and Santa-Clara catchments, was detected in February, when farmers plough their fields and apply pesticides before planting new crops (Fig. 9.2). The 2,4-DNP peak was probably caused by a precipitation event shortly after application of pesticides, increasing the loss of pesticides from agricultural lands into the adjacent streams.

The concentrations of 2,4-DNP in soil leachates were lower from February to April, and showed a general increase from April to August, somehow opposite to the seasonal trend of 2,4-DNP concentrations in water (Fig. 9.2). The lowest concentrations of 2,4-DNP from February to April could be explained by the precipitation events that "cleaned" the soil, as reflected by the increasing concentrations of 2,4-DNP in surface water (Fig. 9.2). The general increase from April to August could be attributed to: (1) fungicide application during the plant growing season, (2) the absence of precipitation, (3) the decrease on the rate of 2,4-DNP degradation due to lower activity of micro organisms in dry soils (contracted soils) and (4) facility to lixiviate the 2,4-DNP from contracted soils because its pore space does not contain water.

Although there were seasonal fluctuations of 2,4-DNP in soil leachates, it persisted in the soil during the whole sampling period (Fig. 9.2). The 2,4-DNP behaviour is highly pH-dependent due to its weak acid character ($pK_a = 4.09$) (Shea et al., 1983). Adsorption of 2,4-DNP in the soil increases with a decrease in pH when the concentration of non-ionized 2,4-DNP is higher (Shea et al., 1983; O'Connor et al., 1990). All soil samples showed acidic pH values and the leaching process is expected to be difficult, thus more 2,4-DNP remains in soil. Additionally, the microbial degradation (biodegradation), the most important process for the degradation/transformation of 2,4-DNP is not favoured by acidic pH values (Blasco et al., 1999).

The occurrence of 2,4-DNP, besides being related to agricultural activity, also depends on soil type. In Santa-Clara catchments, the highest concentrations of 2,4-DNP in soil leachates were observed in Rio Mira (RM) and Ribeiro do Guilherme (RG) (Fig. 9.2). These sub-basins also show the highest areas of agriculture. In Roxo catchments, Barranco dos Castelhanos (BC) and Ribeira dos Louriçais (RL) showed higher concentrations of 2,4-DNP compared to Ribeira do Outeiro (RO) (Fig. 9.2). These sub-basins showed similar percentages of agricultural area. The observed differences can be explained by the fact that the sub-basin of RO has soils with high-clay content (pellic vertisols) and, as 2,4-DNP is easily adsorbed to clay particles, its run-off from the soil to adjacent streams is probably lower (Shea et al., 1983; O'Connor et al., 1990). Ribeira de Santa Vitória (RSV) showed the lowest concentrations of 2,4-DNP in surface water (Fig. 9.2). Assuming that the agricultural practice is similar among the sub-basins, a likely explanation for the low concentrations of 2,4-DNP in the stream water of RSV, besides the fact that the soils have high-clay content (adsorbing 2,4-DNP), this stream has a high abundance of vegetation, mainly *Scirpus* spp. and *Typha* spp., that may play an important role in the removal of the 2,4-DNP from the water. Plant roots can produce peroxidases that are capable of oxidizing phenolic compounds to free radicals. The oxidation products can form water-insoluble oligomers, with no toxic effect (Agostini et al., 2003). Phytoremediation can constitute an important method to remove specific organic pollutants from contaminated waters.

9.5. Good agricultural practice for the protection of water resources

The production-oriented objectives of agricultural policy in the last 40 years have brought significant costs to the natural environment.

Diffuse pollution can be considered as the most relevant threat for the quality of surface and groundwater, as the majority of point source pollution has been intercepted in Europe and put under control (Lacorte et al., 2001). Diffuse pollution may result from the use of chemicals that are, after application, spread from the point of application to the environment by means of run-off, dispersion and diffusion. Poorly managed agriculture can lead to pollution of surface and groundwater by nitrates/nutrients and pesticides. The political agenda of most developed countries is therefore now shifting towards the integration of agricultural and environmental policy (Merrington et al., 2002). Best agricultural practices for water protection should be promoted and an appropriate network to monitor trends in the quality of groundwater concerning diffuse pollution (Paralta and Ribeiro, 2001). Current forestry research has demonstrated that planting or maintaining buffer strips of land near streams will trap excess agricultural chemicals in addition to providing timber and wildlife habitat (Task Force on Building a Science Roadmap for Agriculture, 2001).

The presence of pesticides residues in groundwater call for a need to improve plant protection and irrigation practices. It may be necessary to select carefully the pesticides to be used, in each agricultural area, according to its hydro-geological vulnerability (Batista et al., 2002). The adoption of less-polluting farming practices can only be made with the full cooperation of the agriculture community and largely depends on demonstrating both the environmental and economic benefits of such practices to land managers (Vandermeer et al., 2005).

REFERENCES

Abler, D., Shortle, J., Carmichael, J., Horan, R., 2002. Climate change, agriculture, and water quality in the Chesapeake bay region. Climatic Change 55, 339–359.

Agostini, E., Coniglio, M.S., Milrad, S.R., Tigier, H.A., 2003. Phytoremediation of 2,4-dichlorophenol by *Brassica napus* hairy root cultures. Biotechnol. Appl. Biochem. 37, 139–144.

Barrico, M.L., 2004. Natureza e origem de compostos fenólicos nas bacias hidrográficas das albufeiras de Santa Clara e do Roxo, Alentejo [Ms.C. Thesis]. Coimbra, University of Coimbra, Portugal, p. 105.

Batista, S., Silva, E., Galhardo, S., Viana, P., Cerejeira, M.J., 2002. Evaluation of pesticide contamination of ground water in two agricultural areas of Portugal. Int. J. Environ. Anal. Chem. 82(8–9), 601–609.

Blasco, R., Moore, E., Wray, V., Pieper, D., Timmis, K., Castillo, F., 1999. 3-Nitroadipate, a metabolic intermediate for mineralization of 2,4-dinitrophenol by a new strain of *Rhodococcus* species. J. Bacteriol. 181, 149–152.

Capriel, P., Haisch, A., Khan, S.U., 1985. Distribution and nature of bound (non-extractable) residues of atrazine in a mineral soil nine years after the herbicide application. J. Agr. Food Chem. 33, 567–569.

Davì, M.L., Gnudi, F., 1999. Phenolic compounds in surface water. Water Res. 33, 3213–3219.

Freire, D., Parkhurst, S., 2002. Where is Portuguese agriculture headed? An analysis of the common agricultural policy. Working Paper PRI-5, Institute of European Studies, University of California, Berkeley, USA.

Giordano, A., Bonfils, P., Roquero, C., Yassoglou, N., Sequeira, E., Peter, D., Briggs, D., Redda, A., Kormoss, I., Monsay, H., Maes, J., Chisci, G., Gabriels, D., Mancini, F., Noirfalise, A., Culleton, N., Eerkens, C., Reineger, P., Nychas, A., Briggs, D., Cornaert, M.-R., et al., 1992. CORINE soil erosion risk and important land resources in the southern regions of the European Community. An assessment to evaluate and map the distribution of land quality and soil erosion risk. Ed. Commission of the European Communities.

INAG, 1997. Definição, Caracterização e Cartografia dos Sistemas Aquíferos de Portugal Continental. Instituto Nacional da Água, Lisboa, Portugal, p. 236.

INAG, 2001. Plano Nacional da Água. Instituto Nacional da Água, Lisboa, Portugal. Available at www.inag.pt/inag2004/port/a_intervencao/planeamento/pna/pna_indice.html

Lacorte, S., Viana, P., Guillamon, M., Tauler, R., Vinhas, T., Barceló, D., 2001. Main findings and conclusions of the implementation of Directive 76/464/CEE concerning the monitoring of organic pollutants in surface waters (Portugal, April 1999–May 2000). J. Environ. Monitor. 3, 475–482.

Lesan, H.M., Bhandari, A., 2004. Contact-time-dependent atrazine residue formation in surface soils. Water Res. 38, 4435–4445.

Ma, L., Ahuja, L.R., Ascough II, J.C., Shaffer, M.J., Rojas, K.W., Malone, R.W., Cameira, M.R., 2001. Integrating system modelling with field research in agriculture: Applications of the Root Zone Water Quality Model (RZWQM). In: Sparks, D. (Ed.), Advances in Agronomy. Academic Press, USA, Vol. 71, pp. 233–292.

MADRP, 2005. Vendas de produtos fitofarmacêuticos em Portugal em 2003. Ministério da Agricultura, do Desenvolvimento Rural e das Pescas, Direcção-Geral de Protecção das Culturas, Direcção de Serviços de Produtos Fitofarmacêuticos, Lisboa, Portugal, p. 17. Available at www.dgpc.min-agricultura.pt/docs_pdf/fitofarmaceuticos/Relatorios/Vendas_2003.pdf

MAOT, 2000. Poluição provocada por nitratos de origem agrícola. Ministério do Ambiente e do Ordenamento do Território, Ministério da Agricultura, do Desenvolvimento Rural e das Pescas, Lisboa, Portugal, p. 25.

Merrington, G., Winder, L., Parkinson, R., Redman, M., 2002. Agricultural Pollution. Environmental Problems and Practical Solutions. London and New York, Spon Press, 243 p.

Munz, C., Bachmann, A., 1993. Documentation of an environmentally-sound soil remediation. In: Arendt, F., Annokkee, G.J., Bosman, R., van den Brink, W.J. (Eds.), Contaminated Soil '93. Kluwer Academic Publishers, Netherlands, pp. 1119–1126.

Ngoye, E., Machiwa, J.F., 2004. The influence of land-use patterns in the Ruvu river QJ;watershed on water quality in the river system. Phys. Chem. Earth 29, 1161–1166.

O'Connor, G.A., Lujan, J.R., Jin, Y., 1990. Adsorption, degradation, and plant availability of 2,4-dinitrophenol in sludge-amended calcareous soils. J. Environ. Qual. 19, 587–593.

Paralta, E., Ribeiro, L., 2001. Stochastic modelling and probability risk maps of nitrate pollution in the vicinities of Beja (Alentejo, south Portugal). 3rd International Conference on Future Groundwater Resources at Risk. Lisbon, Portugal, pp. 251–261.

Pimentel, D., 1995. Amounts of pesticides reaching target pests: Environmental impacts and ethics. J. Agr. Environ. Ethic. 8, 17–29.

Pocurull, E., Marcé, R.M., Borrull, F., 1996. Determination of phenolic compounds in natural waters by liquid chromatography with ultraviolet and electrochemical detection after on-line enrichment. J. Chromatogr. A 738, 1–9.

Shea, P.J., Weber, J.B., Overcash, M.R., 1983. Biological activities of 2,4-dinitrophenol in plant–soil systems. Res. Rev. 87, 1–41.

Task Force on Building a Science Roadmap for Agriculture, 2001. National Association of State Universities and Land-Grant Colleges, Experiment station Committee on Organization and Policy, "A Science roadmap for the Future". Available at www.nasulgc.org/comm_food.htm

Vandermeer, J., Perfecto, I., Green, R.E., Cornell, S.J., Scharlemann, J.P.W., Balmford, A., 2005. The future of farming and conservation. Science 308, 1257–1258.

Vitousek, P.M., Mooney, H.A., Lubchenco, J., Melillo, J.M., 1997. Human domination of Earth's ecosystems. Science 277, 494–499.

Wauchope, R.D., Buttler, T.M., Hornsby, A.G., Augustijn-Beckers, P.W.M., Burt, J.P., 1992. SCS/ARS/CES Pesticide properties database for environmental decision making. Rev. Environ. Contam. Toxicol. 123, 1–157.

Developments in Environmental Science, Volume 5
D. Sarkar, R. Datta and R. Hannigan (Editors)

ISSN: 1474-8177/DOI:10.1016/S1474-8177(07)05010-3

Chapter 10

Provenance and geochemistry of sediments in arsenic-affected areas of gangetic West Bengal, India

Sahadeb De, Chinmoy Chakrabarti, Gargi Chatterjee and S. Banerjee

Abstract

In parts of 80 blocks covering eight districts of gangetic West Bengal, arsenic has been found to occur in groundwater above the permissible limit. The area is a part of the Ganga–Brahmaputra delta having a near surface succession of Quaternary sediments of varying thickness. Two broad types of aquifers have been delineated—the single aquifer system to the north of Chakdah–Bongoan tract and the multiple aquifer system to the south. Six lithostratigraphic drillings were carried out—two in the single aquifer system and four in the multiple aquifer system. Five of these six drillings were located in areas with high concentration of arsenic in groundwater while the sixth one was carried out in a non-arsenious zone. Heavy mineral suites of these sediments were studied in detail. Six heavy mineral suites, viz. A, B, C, D, E and F, were identified. A mixed provenance of the Himalayas (dominantly igneous) and the Bihar Plateau regions (dominantly metamorphic) has been ascertained for the sediments corresponding to Suite A. The sediments corresponding to Suite B to E indicate they come from the Bihar Plateau region while those corresponding to Suite F indicate they come from the Himalayas. XRD analysis revealed that illite was the dominant clay minerals in the clay–silty clay partings in the area of the multiple aquifer type. No arsenic-bearing mineral phase were noted through XRD either in clays or in sands. Strong positive correlation coefficients exist among extractable arsenic, iron, manganese, lead and copper in the sediments of arsenious zones while zinc is aberrant.

10.1. Introduction

Arsenic toxicity in West Bengal, India, was first reported in 1978 in Maldah district and the first case of arsenic poisoning was reported in

1983. Since then several organizations worked on the genesis of arsenic in groundwater or on arsenic removal or on alternate source for drinking water to the vast population residing in those areas. Within this time (around 25 years) millions of people were affected in eight districts of West Bengal causing thousands of deaths. Till now, there is no concrete evidence of the source of arsenic in groundwater in these areas. There are mainly two schools of thought existing on the genesis of arsenic: (1) natural (geological) and (2) anthropogenic, caused by the application of arsenical pesticides in agricultural fields. Whatever the cause, the arsenic toxicity in the groundwater of West Bengal has been the "greatest arsenic calamity in the world" (Mandal et al., 1996). It is now considered as a natural disaster. Its unpredicted and unprecedented spatial growth has left millions of people in agony and uncertainty who thrive on groundwater for their drinking water needs.

Along a linear NNE–SSW tract extending from Kaliachak in the Maldah district through Raninagar block in the Murshidabad district and parts of Nadia and North 24-Parganas up to Baruipur block in South 24-Parganas, the groundwater was found in places to contain arsenic above the then permissible level of $0.05\,mg\,l^{-1}$ (WHO, 1993). Later, during 1996, arsenic was also found to contaminate the groundwater in parts of Balagarh block of the Hugli district (Mandal et al., 1996) and in Howrah city (Sikdar, 1996). This phenomenon first came to the notice of the Government of West Bengal around the year 1978, and sporadic studies were carried out from time to time by several agencies, i.e., School of Tropical Medicine, Calcutta; All India Institute of Hygiene & Public Health, Calcutta; Central Ground Water Board, Calcutta; Center for Study of Man and Environment (CSME), Calcutta; and Public Health Engineering Department of Government of West Bengal. During 1988–1991, a comprehensive investigation to inquire into the nature, degree and cause of arsenic pollution was carried out under the auspices of a Steering Committee set up by the Government of West Bengal, with participation of the above-noted organizations, plus the State Water Investigation Directorate (SWID), the Government of West Bengal, and with financial support from the National Drinking Water Mission, Government of India. The study attributed a geologic origin to the source of arsenic in the groundwater of West Bengal (Saha, 1991).

The present study was carried out during 1995–1998 with the primary objectives of (1) determining the source of the sediments in arsenic-affected areas of gangetic West Bengal through laboratory study of undisturbed core samples; (2) determining the mechanism of leaching of arsenic from sediment; and (3) working out the mechanisms of desorption release of a large quantity of arsenic in water-soluble form from sediment sources and to find out the chemical speciation of the arsenic present in sediments and groundwater.

10.2. Regional setting of the study area

The area under study lies in the alluvial terrain of the gangetic West Bengal. It is a part of the Ganga–Brahmaputra delta having a near surface succession of Quaternary sediments of varying thickness. The arsenious belt lies entirely within the Upper Delta Plain (UDP), a macro form-process region, characterized by a series of meander belts formed by rivers. The sediments of the UDP rests directly over the extensive clay beds of the Tertiary Period in the subsurface at a depth of about 150 m with a pronounced unconformity, and there is no evidence for the preservation of older Quaternary deposits at depth. It is suspected that all of them were eroded during the preceding falling stage of sea level when deeply entrenched valleys were formed (CSME, 1998). The sub-surface geology of the area is completely blanketed by these Quaternary sediments, which form the principal repository of the groundwater in the study area. The lithological characteristics of these formations are fairly well understood from the study of litho logs of the boreholes drilled in the area. However, the demarcation of the boundary between the Pleistocene and the Holocene periods in the vertical sequence is rather difficult.

Based on the results of the groundwater exploration in the eastern part of the Bhagirathi–Hugli basin, two broad types of aquifers—single and multiple aquifer types—can be delineated (CGWB, 1988). The single aquifer type, which is predominant in the northern part of the area, changes gradually to a multiple aquifer type south of the Chakdah–Bongaon tract (Fig. 10.1) in Nadia and 24-Parganas districts.

Groundwater occurs in a single aquifer and under water table conditions in the areas north of the Chakdah–Bongaon tract. The aquifer is mainly sands of varying grades and sizes, occasionally mixed with gravel/kankar. Sometimes in the levee zone as also in the back swamp, clay occurs at the top. This area represents one of the principal water-yielding zones in the state, where the yield of the individual tube well exceeds $150\,cm^3\,h^{-1}$.

The groundwater to the south of the Chakdah–Bongaon tract occurs in multiple aquifers and in a confined condition. A thin clay bed overlies the aquifers, the thickness of which increases appreciably further south, being as thick as 30–50 m to the south of Calcutta. The upper group of aquifers is separated from the lower group by a 20-m thick clay layer. Thickness of this clay layer shows a wide variation at different places depending upon

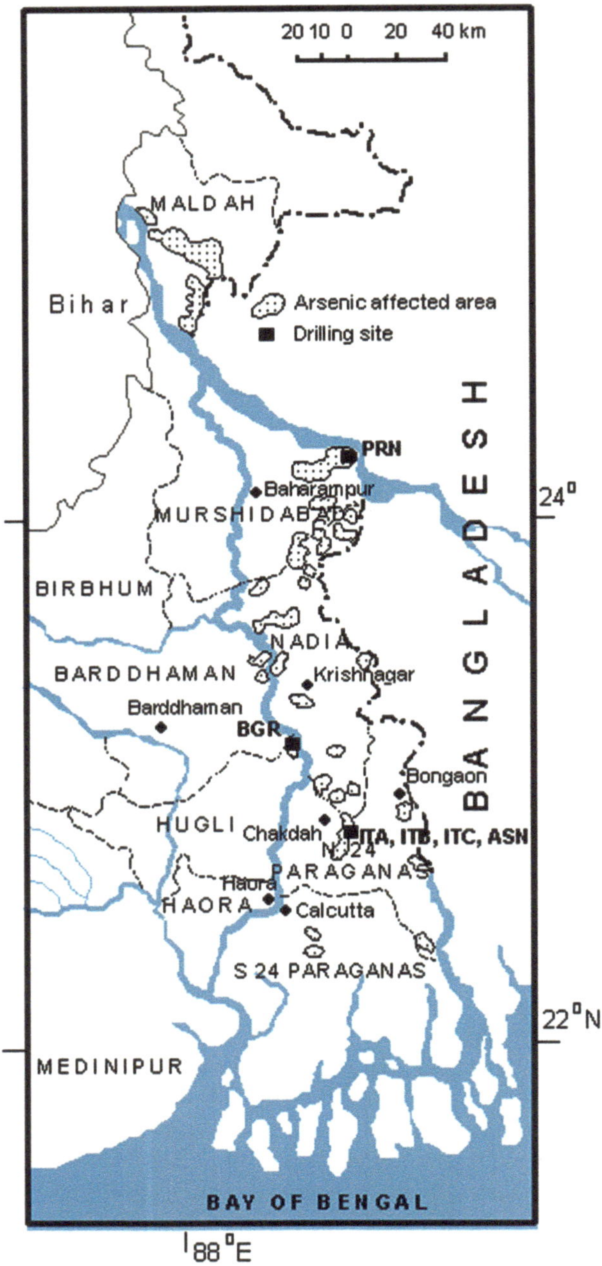

Figure 10.1. Areas affected by arsenic pollution in groundwater and location of drilling sites.

the depositional conditions. Impermeable clay layers of a lenticular nature again separate the individual aquifer in this multiple aquifer system. These clay layers control hydrochemical properties of the different groups of aquifers.

The upper group of aquifers, generally saline water bearing from the south of Calcutta to the southern most tip of the Bengal delta, consists of sand occasionally mixed with gravel and kankar. The lower group of aquifers, relatively fresh water bearing, is in general, fine grained, has greater thickness compared to the upper group and occurs between the depths of 170 and 350 m.

10.3. Materials and methods

10.3.1. Selection of study areas

The study was confined to three districts: Murshidabad, Hugli, and North 24-Parganas, located along a linear belt trending north to south (Fig. 10.1). The locations were selected on the premise that initial background information was available in order to facilitate the proposed study. Lithostratigraphic drilling sites were selected after a thorough analysis of arsenic concentration in groundwater from four villages spread over three districts in West Bengal. These four villages were (i) Itina, Habra block, North 24-Parganas district; (ii) Kachua, Habra block, North 24-Parganas district; (iii) Abdulpur, Balagarh block, Hugli district; and (iv) Paranpur, Raninagar II block, Murshidabad district (Fig. 10.1).

10.3.2. Background information on arsenic concentration in groundwater

A large number of groundwater samples (over 100) were analyzed for arsenic in the course of the present study. Some of the samples from Itina and Paranpur were also analyzed for temperature, pH, electrical conductivity (EC), phosphorous, manganese, lead, copper, zinc, iron and cadmium following the standard methods of APHA 1985.

At Itina and Kachua, Habra block, North 24-Parganas, groundwater occurs in a confined condition. There are two water-bearing horizons at Itina. Arsenic concentration in the water tapped from the upper one (light gray sand horizon, sampling depth: 14.0–16.7 m) ranged between <0.003 and $0.40\,mg\,l^{-1}$. On the other hand, the water tapped from the lower horizon (reddish-brown sand horizon, sampling depth: 43.0–45.0 m) yielded no detectable arsenic (detection level $= 0.003\,mg\,l^{-1}$). At Kachua,

Table 10.1. Concentration of arsenic in groundwater in and around drilling sites

Location		Depths of sampling (m)	*N*	Arsenic concentration ($mg\,l^{-1}$)			
Village	District			Min	Max	Mean	S.D.
Itina	N 24-Parganas	14.0–16.7	71	BDL	0.40	0.067	0.81
		43.0–45.0	8	BDL	BDL	BDL	0
Kachua	N 24-Parganas	26.0	9	0.16	0.40	N.D.	N.D.
Abdulpur	Hugli	9.6–87.6	50	BDL	0.60	0.077	0.137
Paranpur	Murshidabad	7.5–197	18	BDL	0.54	0.097	0.144

BDL, below detection level.

only one water-bearing horizon (light gray sand) was tapped. This horizon is similar to the upper one at Itina. Arsenic concentration in the water tapped from this horizon (sampling depth: 26.0 m) ranged between $0.16\,mg\,l^{-1}$ and $0.40\,mg\,l^{-1}$ (Table 10.1).

At Abdulpur and Paranpur, groundwater occurs in an unconfined condition. The State Water Investigation Directorate, Government of West Bengal, provided information regarding arsenic concentration in the groundwater of Abdulpur, Balagarh block. Here, the arsenic concentration (sampling depth: 9.6–87.6 m) ranged between BDL and $0.6\,mg\,l^{-1}$. At Paranpur, the concentration of arsenic (sampling depth: 7.5–197 m) in groundwater ranged between below detection level (BDL) and $0.54\,mg\,l^{-1}$ (Table 10.1).

10.3.3. Selection of drilling sites

Altogether, six lithostratigraphic drillings were carried out in the present study. Five of these six were located in the areas with high concentration of arsenic in groundwater: two at Itina village (Drilling codes: ITA and ITC), one at Kachua village (Drilling code: ASN) of Habra block, North 24-Parganas district; one at Abdulpur village (Drilling code: BGR) of Balagarh block, Hugli district; and one at Paranpur village (Drilling code: PRN) of Raninagar II block, Mursidabad district. The sixth drilling was carried out (Drilling code: ITB) in a non-arsenious zone at Itina, which was approximately 300 m away from the drilling site ITA, the zone of maximum arsenic concentration (Fig. 10.1).

Lithostratigraphic drillings were carried out up to a maximum depth of 50 m at six locations to collect undisturbed core samples. Two drilling sites (BGR and PRN) were located in the zone of the single aquifer system and four in the zone of the multiple aquifer system (ITA, ITB, ITC and ASN). Methods adopted in these lithostratigraphic drillings to recover undisturbed cores were 'cell and auger' in the zone of clay and silt and 'wash boring with bentonite slurry' in the sand zone.

10.3.4. Heavy minerals and clay minerals identification

Heavy mineral suites are immensely helpful in studies of the provenance and other related features of sediment (Folk, 1968). Six to fourteen sediment samples from each drilling were drawn for heavy minerals separation, ensuring that each litho-unit encountered in all the boreholes was represented in the samples. After preliminary disaggregation and washing of the sediment samples, heavy minerals were separated from them following the standard procedure of gravity separation using bromoform. The heavy minerals, thus separated, were identified under petrological microscope and X-ray diffraction (XRD) method. Clay minerals of some selected samples were also identified by XRD techniques.

10.3.5. Analytical procedure

Portions of the same samples used in heavy mineral identification were taken for geochemical analysis. To avoid contamination of sand with bentonite, the slurry was made thicker than normal, and the samples were taken out from the central part of the cores.

For estimation of the total arsenic in the sediments, 1 g of the sample and 5 g of a 3:1 mixture of quick lime and barium peroxide were put into a nickel crucible. The mixture was heated in a muffle furnace at 700°C for 1 h, was dissolved in HCl, and the volume was made up to 50 ml. From this liquid, arsenic was determined by the standard spectroscopic method.

For estimation of extractable arsenic in the sediments, 5 g of the sample were digested with 20 g of caustic soda for an hour. After cooling, the suspension was acidified with HCl, cooled, separated by filtering, and the volume was made up to 100 ml with distilled water. An aliquot of the solution was used for determination of arsenic by the standard spectroscopic method.

For estimation of other extractable selected heavy metals (Fe, Mn, Pb, Cu and Zn) in sediments, 5 g of the samples were digested with 25 ml of 1:1 nitric acid for about 24 h. The solute was separated by filtering, and its volume was made up to 50 ml by distilled water. The solution was used to determine the heavy metals with the help of the atomic absorption spectrophotometer (AAS).

10.3.6. Leaching experiments

Leaching experiments were done using both column test and cascade test. In the column test, the column used was fabricated from a high quality boro-silicate glass of 5 cm internal diameter and 20 cm height. 200 g of sediment was packed in the column. The contact liquid was made to flow upward at a rate of approximately $4.2\,l\,h^{-1}$ through the column. Leachates were collected in eight fractions and analyzed separately.

The cascade test involved 5 consecutive shake tests and the extraction of the corresponding fractions of leachates through filtration. The different fractions from the cascade were subjected to pH and conductivity measurements and then analyzed for arsenic.

10.3.7. Speciation of arsenic

The chemistry of arsenate is very similar to that of phosphate. Magnesium ammonium phosphate and magnesium ammonium arsenate are isomorphous and insoluble in dilute ammoniacal solutions. Precipitation of phosphate as magnesium ammonium phosphate by a magnesia mixture carries down all the arsenate present in the system as magnesium ammonium arsenate. Any trivalent arsenic or arsenite present in the system is not precipitated under this condition. This was the basis of the speciation experiment. The precipitated magnesium ammonium phosphate containing pentavalent arsenic was dissolved in dilute hydrochloric acid, and arsenic was determined by the standard method. The trivalent arsenic was determined from the filtrate of the magnesium ammonium phosphate.

10.3.8. Speciation of arsenic in groundwater

Twenty six groundwater samples collected from Itina, Habra block, North 24-Parganas district were subjected to precipitate as magnesium ammonium phosphate using a magnesia mixture in an ammoniacal medium. The pentavalent arsenic was precipitated with magnesium ammonium phosphate, and the trivalent arsenic remained in the filtrate. The arsenic content of both residue and filtrate was measured.

10.3.9. Speciation of arsenic in sediments

Valency of arsenic in sediments was determined by digesting the sediments with a strong NaOH solution, which brings the arsenic in solution without changing its valency. The filtrate was acidified and speciation was carried out using the same methodology given above.

For quality control, standard mixtures of known amounts of trivalent and pentavalent arsenic were subjected to analysis by the above-mentioned method. After a satisfactory result, the procedure was adopted for speciation of arsenic in groundwater and sediments.

10.4. Results and discussion

10.4.1. Lithological sequences

In the zone of single aquifer system, at Paranpur village, the aquifer encountered (Fig. 10.2) was composed of micaceous sands (fining upward sequence) of light brownish gray in color in the upper 14.5 m and of light gray to dark gray in color between 20.0 and 50.0 m. At Abdulpur, Balagarh, the same aquifer encountered at much shallower depth (1.36 m)

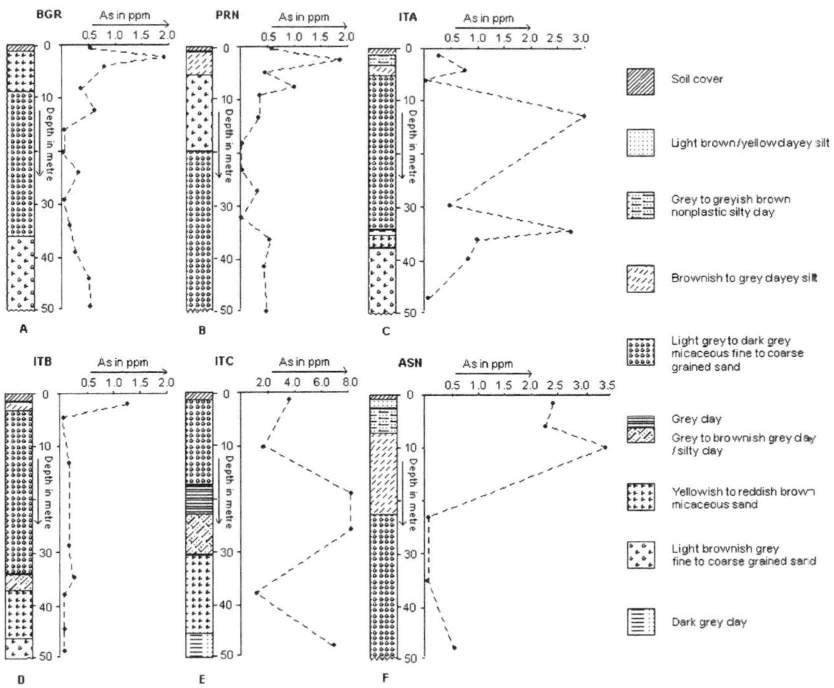

Figure 10.2. Depthwise distribution of arsenic in the sediments of six boreholes.

with yellowish to reddish brown color until 9.0 m, light gray to dark gray color between 9.0 and 36.0 m, and light brownish gray color there after (Fig. 10.2). The arsenic concentration in groundwater from this aquifer varied from <0.003 to $0.54\,mg\,l^{-1}$ at Paranpur and from <0.003 to $0.60\,mg\,l^{-1}$ at Abdulpur.

In the arsenious zone of multiple aquifer system, at Kachua village, Habra block, North 24-Parganas, only one aquifer was encountered, which was a micaceous light gray sand (fining upward sequence) occurring from 23.0 to 50.0 m below ground level (bgl). This aquifer is overlain by 15.7 m thick brownish to gray clayey silt, 4.7 m thick gray to grayish brown, non-plastic silty clay and 2.4 m thick light brown clayey silt, successively upwards (Fig. 10.2). The groundwater from this aquifer at Kachua contained arsenic as high as $0.4\,mg\,l^{-1}$. This aquifer was encountered at much shallower depths at Itina, 4 km east of Kachua, both in arsenious and non-arsenious zones. It occurs between 5.2 and 34.5 m in the arsenious zone and between 3.0 and 34.0 m in the non-arsenious zone. There is a second aquifer at Itina, which is reddish brown to light brownish gray in color and occurs between 35.5 and 50.0 m in the arsenious zone and between 36.9 and 50.0 m in the non-arsenious zone (Fig. 10.2). The groundwater from the upper aquifer at Itina also contained a variable amount of arsenic ranging from <0.003 to $0.40\,mg\,l^{-1}$, while that from the lower aquifer is free from arsenic. These two aquifers are separated by a 1-m thick, gray clay bed in the arsenious zone and a 2.9-m thick gray to brownish gray clay in the non-arsenious zone.

10.4.2. Environment of deposition

The fining upward sequence of the sediments indicates a fluvio-deltaic environment of deposition (Allen, 1965). Toward the close of sedimentation, bog and marshy conditions probably prevailed as indicated by the occurrence of peat in the upper horizon of the sediments. Occurrence of marsh or so-called salt lakes to the east of Calcutta also lends additional evidence for the bog environment. These lakes are nothing but remnants of extensive marshes, which were present in the geologic past.

The color of the sediments is also an important indicator of environment. Yellow or yellowish brown color generally indicates oxidizing conditions, and a dark gray to black color suggests sedimentation under euxinic conditions. From examination of borehole lithologs, it was observed that, in the depth span of 24–50 m, yellowish to brownish colored sands, very similar to those now occurring in the beds of the Damodar and Ajoy rivers originating from the Precambrian terrain of the Chhotanagpur plateau, were intersected. The occurrence of brown sand in this horizon suggests that during the relevant period of time, sediments were probably contributed to the rivers flowing from the west across the Precambrian terrain. The color of the sediments occurring below this horizon ranges from gray to light gray. These sediments were deposited under reducing conditions and were probably brought from the Himalayan domain during the late Quaternary period. The course of the rivers and their tributaries had undergone several shifts during the period of deltaic progradation, and the entire Quaternary sedimentary column of the Bengal Basin has evolved by vertical build-up and lateral coalescence of a number of deltaic formations.

10.4.3. Heavy mineral suites and the provenance of the sediments

In this alluvial zone of West Bengal, six heavy mineral suites were identified. The occurrences of heavy minerals in these six suites, in order of abundance, are depicted in Table 10.2.

The mineral grains in Suites A and F are, in general, rounded to sub-rounded, reflecting a fairly good degree of transport. The hornblende and biotite grains are mostly altered. The features of these assemblages indicate a distant provenance for the respective sediments. The mineral grains in Suites B, C, D, and E are, in general, sub-rounded to sub-angular, and the hornblende and biotite grains are mostly fresh, indicating a nearby provenance for the respective sediments.

On the basis of heavy mineral assemblages, a mixed low to medium grade metamorphic and basic igneous provenance is attributed to Suite A; a medium to high grade metamorphic provenance is attributed to Suite

Table 10.2. Heavy mineral suites in the study area

Suites	Heavy minerals in order of abundance
Suite A	Biotite (mostly altered)—chlorite—garnet—hornblende (mostly altered)—opaques (equidote)
Suite B	Hornblende (mostly fresh)—kyanite—garnet—sillimanite—opagues—tourmaline staurolite—biotite (mostly fresh)—chlorite
Suite C	Hornblende (mostly fresh)—kyanite—staurolite—garnet—sillimanite—opaques—tourmaline (rutile)—biotite (mostly fresh)—chlorite
Suite D	Hornblende (mostly fresh)—biotite (mostly fresh)—kyanite—sillimanite—garnet—opaques—rutile—tourmaline—chlorite
Suite E	Hornblende (mostly fresh)—kyanite—garnet—staurolite—tourmaline—epidote—opaques—rutile—chlorite—biotite (mostly fresh)
Suite F	Hornblende (mostly altered)—biotite (mostly altered)—garnet—chlorite—opaques—tourmaline (rutile) (zircon)

B to E, while a dominantly basic igneous provenance is attributed to Suite F (Tickell, 1965).

Suite A is analogous to Unit I of Saha (1991) and assemblage A of Ghosh and De (1995), indicating a mixed provenance of the corresponding sediments, partly the igneous terrains of the Himalayas and partly the Bihar Plateau region. Suite B is similar to assemblage B/I of Ghosh and De (1995), Suites C and D are similar to assemblage II of Ghosh and De (1995) and Suite E is similar to assemblage III of Ghosh and De (1995). These suites (B–E) is corresponding to Unit II of Saha (1991). These correlations clearly indicate the Rajmahal–Chhotonagpur plateau region to be the possible provenance for the corresponding sediments. Suite F is not identified by Ghosh and De (1995) while it corresponds to Unit III of Saha (1991), indicating the igneous terrains of the Himalayas to be the plausible provenance for the corresponding sediments.

The occurrences of these suites in the six boreholes in the area are given in Table 10.3. It is evident that the heavy mineral suites of arsenious and non-arsenious zones are similar. In the sediments of both the areas at Itina (ITA and ITB), Suites A, B, C and D are noted in equal extent, indicating that the sediments have been transported from the same provenance. But there is no reported arsenic mineralization in either of the provenances. The opaques in the sediments from both areas of Itina (ITA and ITB) were identified through XRD, but the arsenic-bearing mineral phase had not been recorded (Table 10.4). Therefore, the selective transport of arsenic from sulfide mineralization to the affected area had not occurred. It indicates that the arsenic contamination in groundwater may be a post-sedimentation phenomenon.

Table 10.3. Distribution of heavy mineral suites in six boreholes (up to 50 m)

Heavy mineral suites	PRN (m)	BGR (m)	ASN (m)	ITA (m)	ITB (m)	ITC (m)
A			0.20–23.0	1.12–5.2	1.32–5.2	
B	0.5–5.5	0.0–34.0	23.0–50.0	5.2–34.5	3.0–34.0	1.0–16.7
C				34.5–35.5	34.0–36.9	16.7–22.7
D	5.5–45.0	34.0–50.0		35.5–50.0	36.9–50.0	22.7–45.4
E						45.4–50.0
F	45.0–50.0					

PRN, at Paranpur village, Raninagar II block, Murshidabad district; BGR, at Abdulpur village, Balargarh block, Hugli district; ASN, at Kachua village, Habra block, North 24-Parganas district; ITA, at Itina village, Habra block, North 24-Parganas district; ITB, at Itina village, Habra block, North 24-Parganas district; ITC, at Itina village, Habra block, North 24-Parganas district.

Table 10.4. Results of X-ray diffraction analysis of clay–silty clay beds of Itina, North 24-Parganas

Zones	Lithology	Depth of sampling (m)	Major	Minor	Good amount	Considerable amount	Trace amount
Arsenious	Gray silty clay	1.59–1.6	Quartz	Feldspar	Illite, Chlorite		Kaolinite
	Gray clayey silt	4.27–4.28	Quartz	Illite		Chlorite, Feldspar	Calcite, Kaolinite
	Gray clay	34.90–34.91	Quartz	Feldspar	Illite	Calcite	Chlorite
Non-arsenious	Gray clayey silt	2.27–2.28	Quartz	Feldspar	Illite	Chlorite, calcite	Kaolinite
	Gray clay	35.17–35.18	Calcite	Feldspar	Illite	Chlorite, Feldspar	Kaolinite

10.4.4. Clay minerals

To identify the differences in the clays and silty clays of arsenious and non-arsenious zones, if any, five samples were selected from Itina only: three from the arsenious zone (ITA) and two from the non-arsenious zone (ITB). The difference lies in the type of major mineral. Quartz is dominant in all of the clay/silty clay horizons of the arsenious zone and in the upper clayey silt bed of the non-arsenious zone, while calcite is dominant in the clay at a depth of 35.17–35.18 m in the non-arsenious zone (Table 10.4). As far as clay minerals are concerned, illite is dominant in all the clay/silty clay beds of Itina, irrespective of the concentration of arsenic in groundwater. As such, there is no marked difference in the clay mineralogy of the arsenious and non-arsenious zones. In both the zones, no arsenic-bearing mineral phase had been detected.

10.4.5. Arsenic and some selected heavy metal concentrations in the sediments

Concentration of arsenic in the sediments varies from one arsenious zone to another. Generally, in the case of unconfined aquifers as found in Paranpur and Abdulpur, the concentration of arsenic was higher in the near surface horizon and then decreased to a depth of 30–32 m from where it again increased, although insignificantly, up to a depth of 50 m (Fig. 10.2).

In the case of confined aquifers as in Kachua and Itina of Habra block, North 24-Parganas district, the clay/silty clay horizons contained much higher concentrations of arsenic than in the sand, except for one arsenious zone in Itina (ITA), where the sand lying at a depth of approximately 13 m contained a maximum concentration of arsenic (Fig. 10.2). In all the cases except the one at Itina, the concentration of arsenic was higher in the argillaceous sediments than in the arenaceous ones. In the non-arsenious zone of Itina, the concentration of arsenic was insignificant except in the upper silty clay horizon (Fig. 10.2). Here the clay partings between the two water-bearing zones also contained a much lower amount of arsenic than compared to that of the arsenious zones. Iron (Fe), manganese (Mn), lead (Pb), copper (Cu), and zinc (Zn) were also estimated in these sediments (Tables 10.5–10.7).

It is important to note that a high, positive correlation coefficient exists among As, Fe, Mn, Pb, and Cu in the sediments of the arsenious zones of West Bengal, while Zn is aberrant (Table 10.8). In the non-arsenious zone at Itina, correlation coefficients are very low, except between arsenic and lead ($r = 0.8434$) and, to some extent, between arsenic and copper ($r = 0.4174$) (Table 10.9). The significant, high positive correlation of As

Table 10.5. Some selected heavy metal concentrations in the sediments at arsenious zones in the areas of the single aquifer type under confined condition

Lithology	*N*	As	Mn	Pb	Cu	Zn	Fe ($g\,kg^{-1}$)
		($mg\,kg^{-1}$)					
Grayish brown to bluish gray clayey silt	3	0.46–1.86 (0.97)	119.3–433.0 (255.0)	3.06–14.5 (7.4)	4.1–5.7 (5.1)	77.0–210.4 (159.2)	6.2–14.4 (9.3)
Brownish gray micaceous sand	8	<0.05–1.02 (0.41)	118.0–183.5 (140.2)	1.18–14.5 (5.37)	1.8–3.7 (2.95)	160.0–381.0 (235.8)	4.75–9.2 (6.20)
Dark bluish gray micaceous sand	12	<0.05–1.66 (0.39)	86.0–293.5 (170.36)	1.81–25.4 (6.77)	1.8–6.5 (4.10)	85.0–320.0 (221.7)	3.35–12.5 (7.22)

N = No. of samples. Figure in parentheses indicates mean value.

Table 10.6. Some selected heavy metal concentrations in the sediments at arsenious zones in the areas of the multiple aquifer type under confined condition

Lithology	N	As	Mn	Pb	Cu	Zn	Fe ($g\,kg^{-1}$)
		($mg\,kg^{-1}$)					
Light brown/yellow clayey silt	1	2.39	301.2	11.11	8.9	44.8	15.0
Gray silty clay	2	0.3–2.27 (1.28)	503.0–681.5 (592.2)	5.31–28.89 (17.1)	4.95–8.9 (6.92)	44.8–345.3 (1950)	11.5–30.0 (20.7)
Dark gray clayey silt	2	0.74–3.38 (2.06)	135.5–611.4 (373.4)	4.26–22.22 (13.24)	5.06–34.32 (19.7)	68.2–345.3 (206.7)	22.2–30.5 (26.3)
Dark gray micaceous sand	7	<0.05–3.2 (0.89)	125.9–720.5 (286.5)	2.22–17.51 (6.18)	2.81–20.19 (6.59)	22.5–290.5 (122.7)	9.7–28.3 (13.7)
Gray clay	3	2.8–8.26 (6.44)	440.6–997.0 (757.3)	14.25–37.78 (26.97)	15.76–25.0 (20.51)	77.1–462.2 (298.3)	27.3–34.0 (31.8)
Reddish brown sand	4	<0.05–1.26 (0.79)	185.0–420.5 (245.8)	2.12–6.12 (3.39)	2.12–8.15 (4.69)	60.4–205.6 (116.0)	7.1–18.1 (11.4)

N = No. of samples. Figure in parentheses indicates mean value.

Table 10.7. Some selected heavy metal concentrations in the sediments at non-arsenious zone in the areas of the multiple aquifer type under confined condition

Lithology	N	As	Mn	Pb	Cu	Zn	Fe ($g\,kg^{-1}$)
		($mg\,kg^{-1}$)					
Gray clay silty clay	1	1.22	188.5	5.81	5.15	180.3	12.5
Dark gray micaceous	3	<0.05–0.14 (0.11)	92.5–199.9 (137.0)	1.02–4.14 (2.75)	4.13–5.36 (4.89)	150.4–330.9 (238.7)	7.2–17.5 (13.5)
Gray clay	1	0.22	170.4	2.22	5.86	210.6	11.3
Reddish brown sand	3	<0.05–<0.05 (<0.05)	155.3–453.6 (333.1)	152–2.16 (1.87)	2.91–3.81 (3.22)	70.9–140.9 (77.1)	8.1–12.1 (9.8)

N = No. of samples. Figure in parentheses indicate mean.

Table 10.8. Correlation coefficent matrix ($n = 47$) for different metallic constituents in the sediments at arsenious zones of West Bengal

	As	Mn	Pb	Fe	Cu	Zn
As	1.0000					
Mn	0.7816*	1.0000				
Pb	0.7712*	0.6434*	1.0000			
Fe	0.8465*	0.8029*	0.6909*	1.0000		
Cu	0.7546*	0.6413*	0.8143*	0.8172*	1.0000	
Zn	0.1297	–0.0557	0.0098	0.0485	0.0309	1.0000

*Statistically significant at 99.9% confidence level.

Table 10.9. Correlation coefficient matrix ($n = 8$) for different metallic constituents in the sediments at non-arsenious zone in Itina

	As	Mn	Pb	Fe	Cu	Zn
As	1.0000					
Mn	–0.2033	1.0000				
Pb	0.8434*	–0.3783	1.0000			
Fe	0.1972	–0.3449	0.6036	1.0000		
Cu	0.4174	–0.8185*	0.5291	0.5393	1.0000	
Zn	0.1526	–0.8401*	0.5127	0.6527	0.8157*	1.0000

*Statistically significant at 99% confidence level.

with Fe, Mn, Pb, and Cu in the arsenious zones of gangetic West Bengal enables computation of a highly efficient regression equation directly from the raw data, by which arsenic can be predicted in terms of these heavy metal contents in the sediments.

$$[As] = 0.0022[Mn] + 0.1113[Pb] + 0.0916[Fe] - 0.0373[Cu] + 0.0022[Zn] - 1.6166.$$

Here, the number of samples (n) = 33 and the quality of fit of the equation is 84%, which is statistically very significant.

It seems from these strong, positive correlation coefficients that As, Fe, Mn, Cu, and Pb are cogenetic. In reality, the sources for As, Fe and Mn are not the same. Fe and Mn were co-precipitated along with As on the mineral grain surfaces, where they (Fe and Mn) both acted as scavengers. This phenomenon could be supported by the fact that coagulation of the Fe–Mn removal process can convert soluble As into insoluble reaction products, facilitating their subsequent removal from the water by filtration or sedimentation. The scanning electron microscope (SEM) study had also clearly indicated the presence of As as ultra-fine coatings along with Fe (Mn was not measured) on the mineral grains in the arsenious layers of the sediments at Kachua, Habra block and Halderbagan, Barasat block of North 24-Parganas district. Moreover, if there were a common source for As, Fe and Mn, the correlation between them would have been high in non-arsenious zones too—the lithologic characters being same in both (arsenious and non-arsenious) the areas.

The SEM study did not indicate any concentration of Cu (Pb was not measured) as coating on the arsenic-positive mineral grains in sediments from Kachua and Halderbagan, but Cu and Pb are positively correlated with As, both in the arsenious and non-arsenious zones (Tables 10.8 and 10.9). Being in the same geochemical classification of chalcophile (Mason and Moore, 1982), they may be cogenetic, occurring along with sulfide. If so, their concentrations in the sediments of the same geologic horizon within a short distance like 300 m would not have been as varied as in Itina. Moreover, Zn, being also a chalcophilic element, would have shown a high, positive correlation with As. But the correlation coefficient between As and Zn is very low ($r = 0.1297$) (Table 10.8). Hence, the source of As is not related to sulfide mineralization. The concentrations of As and other correlatable heavy metals are very high in the surficial regions of both arsenious and non-arsenious zones (Fig. 10.2), which indicates that the source of arsenic could be related to the surficial region of the sediments and has been active during the post-sedimentation period.

10.4.6. Speciation of arsenic and leaching experiments

Arsenic may be present in groundwater in three forms: (1) as organo-arsenic compound, (2) as inorganic trivalent arsenic, and/or (3) as inorganic pentavalent arsenic. Repeated attempts were made to collect the arsenic from the groundwater by solvent extraction. All of those attempts were abortive, indicating that the contaminant was purely inorganic in nature. Speciation of inorganic arsenic revealed that it was present in the groundwater of Itina mostly in the trivalent state (Fig. 10.3) while that in the sediments in pentavalent state (Table 10.10).

The leaching experiments were carried out under an oxidizing environment (distilled water containing small amounts of sodium bicarbonate), and arsenic did not leach out in any of the tests. Therefore, it can be concluded that arsenic did not leach out from the sediments under oxidizing conditions.

10.5. Conclusion

Important findings regarding arsenic occurrences in groundwater of gangetic West Bengal in the present study are as follows:

1. It occurs in places in the groundwater of rural sectors of the affected districts. Intensive agriculture is common land use practice of the area with groundwater as the major source of irrigation.

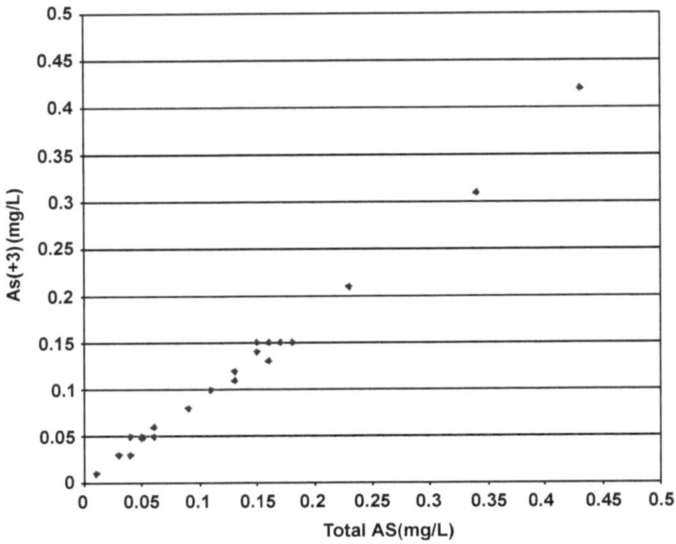

Figure 10.3. Scatter plot of As^{3+} versus total As in the groundwater samples of Itina. As^{5+} was below detection limit.

Table 10.10. Results of arsenic determination from the sediment samples from Itina, Habra block, North 24-Parganas

Sample No.	Total As	As^{3+}	As^{5+}
		$mg\,l^{-1}$	
ITA-5	0.3	0.07	0.27
ITA-12	0.74	0.08	0.69
ITA-SPT-29B	2.8	0.07	2.6
ITB-7	1.22	0.09	0.98
ITB-20	0.22	0.07	0.18

2. The lithologic succession of the area is mainly loose alluvial sands of late Quaternary age with intercalations of clay. Arsenic occurs mainly in the waters tapping from shallower aquifers (<100 m depth). At present, the occurrence of As in waters tapping from deeper aquifers (>100 m depth) is rare.
3. All waters tapping from similar depths of the same aquifer are not arsenious. Even within tens of meters, one tube well yields arsenious water with concentration as high as $0.4\,mg\,l^{-1}$ while the other yields of arsenic-free water.

4. Strong positive correlations exist between As (alkali digested) and extractable Cu, Pb, Fe, and Mn of the sediments.
5. Concentrations of As in the argillaceous sediments are higher than that of arenaceous sediments. It ranges between 2 and 8 ppm in the former while <0.05 and 3 ppm in the latter.
6. Heavy minerals of these sediments in order of abundance are hornblende—kyanite—garnet—staurolite—sillimanite—opague—biotite—chlorite—tourmaline—epidote indicating the Chhotonagpur plateau and the Himalayas to be the probable provenances.
7. XRD analysis reveals that the clays are dominantly of illite type, and there is no arsenic-bearing mineral phase either in clays or in sands.
8. Arsenic occurs in groundwater mostly in trivalent state while it occurs in sediments mostly in pentavalent state.

ACKNOWLEDGMENTS

We sincerely thank Late Dr. A. K. Saha for guiding us through the present work, which could not be initiated without him. We are thankful to Subhasish Ghosh, Suranjita Ghosh and Pradip Sikdar for their valuable inputs in various aspects of this study. We would also like to thank the authorities of Geological Survey of India, Calcutta for carrying out the XRD analysis of sediments, State Water Investigation Directorate, West Bengal for providing data on arsenic occurrences in groundwater at Balagarh block, Hugli district, Public Health Engineering Directorate, Haora Subdivision, West Bengal for funding the cost of drilling at Abdulpur, Science and Engineering Research Council, Department of Science and Technology, Govt. of India, for providing sediment samples from Kachua. Funding from Rajiv Gandhi National Drinking Water Mission, Ministry of Rural Areas and Employment, Government of India (Sanction no. 11046/24/95/TM II dated March 31, 1995) is also gratefully acknowledged.

REFERENCES

Allen, J.R.L., 1965. A review of origin and characteristics of recent alluvial sediments. Sedimentology, special volume.

American Public Health Association (APHA), 1985. Standard methods for the examination of water and waste water, sixteenth ed. APHA, AWWA, WPCF, pp. 1268.

Center for Study of Man and Environment (CSME), 1998. Geology and geochemistry of arsenic occurrences in groundwater of six districts of West Bengal, p. 47 (Unpublished report submitted to Rajiv Gandhi National Drinking Water Mission, Govt. of India.)

Central Ground Water Board (CGWB), 1988. Hydrogeological Atlas of West Bengal, CGWB Publication, Calcutta.

Folk, R.L., 1968. Petrology of Sedimentary Rocks. Hemphill, Austin, TX, pp. 170.

Ghosh, S., De, S., 1995. Sources of the arsenious sediments at Kachua and Itian, Habra Block, North 24-Parganas, West Bengal—A case study. Ind. J. Earth Sci. 22(4), 183–189.

Mandal, B.K., Roy Chowdhury, T., Samanta, G., Basu, G.K., Chowdhury, P.P., Chanda, C.R., Lodh, D., Karan, N.K., Dhar, R.K., Tamili, D.K., Das, D., Saha, K.C., Chakraborti, D., 1996. Arsenic in groundwater in seven districts of West Bengal, India—The biggest arsenic calamity in the world. Curr. Sci. 70(11), 976–986.

Mason, B., Moore, C.B., 1982. Principles of Geochemistry, fourth ed. Wiley, New York, p. 350.

Saha, A.K., 1991. Genesis of the arsenic in groundwater in parts of West Bengal, Annual Volume CSME, 1991, pp. 1–10.

Sikdar, P.K., 1996. Hydrogeology of the area in and around Calcutta and Howrah Municipal Corporation with special emphasis on the management of groundwater resources. Ph.D. Thesis, University of Calcutta, p. 163.

Tickell, F.G., 1965. The techniques of sedimentary mineralogy. Developments in Sedimentology, Elsevier, Amsterdam, Vol. 4, pp. 41–44, 132–153.

World Health Organization (WHO), 1993. Guidelines for Drinking Water Quality, second ed. Vol. 1, p. 188.

Developments in Environmental Science, Volume 5 229
D. Sarkar, R. Datta and R. Hannigan (Editors)

ISSN: 1474-8177/DOI:10.1016/S1474-8177(07)05011-5

Chapter 11

Rock–water interaction and its control on chemical composition of groundwater

L. Elango and R. Kannan

11.1. General

Groundwater chemistry is largely a function of mineral composition, the formation through which it flows due to rock–water interaction. Evaporation and concentration, dilution due to precipitation also can change the chemical composition of groundwater but rock–water interaction is the major process because solid phases (inorganic and organic matter) are the primary sources and sinks of dissolved constituents of groundwater. During groundwater movement along its path from recharge to discharge areas, a variety of chemical reactions with solid phases take place. These chemical reactions will vary spatially and temporally, depending on the chemical nature of the initial water, geological formations and residence time. The resulting concentrations of major ions of groundwater can be used to identify the intensity of rock–water interaction and chemical reactions. Elango et al. (2003) have inferred that carbonate weathering and dissolution, silicate weathering and ion exchange processes are responsible for groundwater chemistry in Lower Palar River Basin, India. Apodaca et al. (2002) have inferred that hydrogeochemical processes such as dissolution, ion exchange processes and the residence time along the flow path control the chemical composition of groundwater in the shallow alluvial aquifers of the Upper Colorado River, USA. Martinez and Bocanegra (2002) have identified that cation exchange processes and calcite equilibrium are the important hydrogeochemical processes that control groundwater composition in the Mar del Plata aquifer, Argentina. Sami (1992) demonstrated that leaching of surfacial salts, ion exchange processes and residential time cause hydrogeochemical variations of groundwater from a semiarid sedimentary basin, Eastern Cape, South Africa. Hence, knowledge on rock–water interaction that control groundwater chemical evolution will lead to improved understanding of hydrogeochemical characteristics of an aquifer. This would contribute in the effective management and utilization of groundwater resources.

11.2. Rock–water interaction

Generally different chemical processes occur during rock-water interaction, which include dissolution/precipitation, ion exchange processes, oxidation and reduction. Rock types are broadly classified as igneous, metamorphic and sedimentary. Minerals present in these rocks will completely or partially dissolve in water according to the resistance of chemical weathering (Table 11.1). Occurrence and intensity of these chemical reactions depend on the initial water's disequilibrium with rocks in contact. During rock–water interaction a variety of chemical processes occur such as weathering and dissolution, ion exchange processes, oxidation and reduction (Table 11.2). During these chemical processes the following changes may occur in the system

1. Concentration of ions in groundwater may increase or decrease
2. Mobility of the dissolved constituents may be affected
3. pH of groundwater may change

according to the chemical processes.

Table 11.1. Common rock types and its resistance to chemical weathering

Type of rock	Dominant minerals present in rock	Resistance to chemical weathering
Quartz-cemented sandstone	Quartz and k-feldspar	High
Calcite-cemented sandstone	Quartz, K-feldspar and clays	Low
Siltstone	Quartz, K-feldspar and clays	High
Shale	Quartz and clays	High
Limestone	Calcite	Low
Rock salt	Halite	Low
Rock gypsum	Gypsum	Low
Slate	Quartz, biotite and muscovite	High
Phyllite, schist	Quartz, K-feldspar, biotite and/or muscovite and amphibole	High
Gneiss	Quartz, K-feldspar, plagioclase, biotite and amphibole	High
Marble	Calcite	Moderate
Quartzite	Quartz	Very high
Granite	Quartz and K-feldspar	High
Basalt	Olivine, pyroxene	Moderate

Table 11.2. Common minerals and chemical processes during rock–water interaction

Mineral	Chemical formula	Relative resistance	Chemical weathering processes
Halite	NaCl	Very low	Dissolution
Gypsum	$CasO_4 \cdot 2H_2O$	Very low	Dissolution
Pyrite	FeS_2	Low	Dissolution and oxidation
Calcite	$CacO_3$	Low	Dissolution
Dolomite	$CaMg(CO_2)_2$	Low	Dissolution
Olivine	(Fe, Mg)SiO_4	Moderately low	Oxidation, hydrolysis
Pyroxene	Ca, Mg, Fe, Al silicates	Moderate	Oxidation, hydrolysis
Plagioclase	$CaAl_2Si_2O_8$; $NaAlSi_3O_8$	Moderate	Hydrolysis
Hornblende (Amphibole)	Na, Ca, Mg, Fe, Al silicates	Moderate	Oxidation, hydrolysis
Biotite	K, Fe, Mg, Al silicate	Moderate	Oxidation, hydrolysis
K-feldspar	$KalSi_3O_8$	Moderately high	Hydrolysis
Muscovite	K, Al silicate	High	Hydrolysis
Quartz	SiO_2	Very high	Dissolves very slowly
Clays	Al silicates	Very high	Hydrolysis

The chemical composition of groundwater is the imprints of the rock–water interaction and chemical processes. So, groundwater chemistry can be used to identify the rock–water interaction or chemical processes. Generally, presence of rock–water interaction was identified using TDS vs. Na/(Na + Ca) and TDS vs. Cl/(Cl + HCO_3) scatter diagrams as reported by Gibbs (1970) (Fig. 11.1). In these diagrams samples falling in the center of the curve are derived from rock–water interaction. Before studying the rock–water interaction, these diagrams will give a broad idea about the intensity of rock–water interaction. The samples falling along the rock–water interaction zone can be further interpreted with various techniques to identify the exact chemical reactions that control the chemical composition of groundwater.

11.2.1. Dissolution/precipitation of minerals

Mineral dissolution and precipitation involves transfer of mass to a phase or from a phase (Deutsch, 1997). Here, in this context mass transfer takes place between rock types and groundwater, which is called as dissolution and precipitation. Dissolution starts when groundwater tends to move toward equilibrium. This process continues until groundwater is saturated

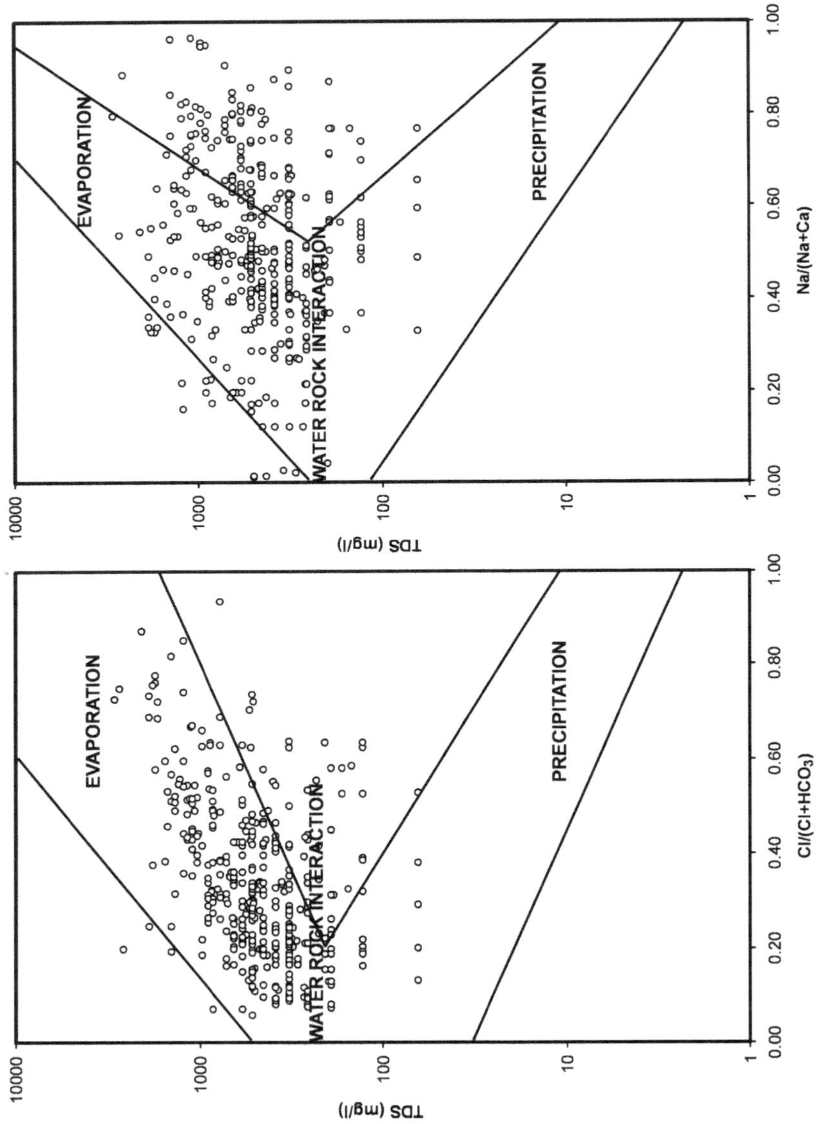

Figure 11.1. Rock–water interaction diagrams.

with those particular mineral components. At equilibrium concentration values, the mineral may continue to dissolve, that excess mineral components may combine in the solution to form precipitates of that particular mineral, and this reaction is called precipitation (Deutsch, 1997). Chemical equilibrium of groundwater for a particular mineral can be identified by calculating saturation indices. If the SI values are less than zero it will take more mineral into the solution, which is called dissolution

$$SI = \frac{\log IAP}{K_s}$$

where IAP is ion activity product, and K_s is solubility product of the mineral. (The use of SI is best limited to soluble simple salts like carbonate minerals.)

Carbonate weathering, silicate weathering, halite, gypsum dissolution and precipitation are common in an aquifer system. Rajmohan and Elango (2004) reported that the concentrations of the major ions increases mainly due to dissolution of minerals deposited during the preceding dry season in a part of Palar and Cheyyar River Basins. Similar kind of results was reported by Senthil Kumar (2004).

11.2.1.1. Carbonate weathering and dissolution

If calcium and magnesium are the dominant ions found in groundwater it reveals that dissolution of carbonate minerals is responsible for them. During infiltration or along the flow, groundwater may dissolve the $CaCO_3$, and $CaMg(CO_3)_2$ present in the rocks will increase the concentration of calcium ions in groundwater. Datta and Tyagi (1996) explained that in the (Ca + Mg) vs. (HCO_3 + SO_4) scatter diagram, the ionic concentrations falling above the equiline result from carbonate weathering, while those falling along the equiline are due to both carbonate weathering (Eqs. (1) and (2) and (3)) and silicate weathering (Eq. (4)). Such a (Ca + Mg) vs. (HCO_3 + SO_4) scatter diagram of the Lower Palar River Basin was given in Fig. 11.2 (Kannan, 2005), in this most of the sample points lie above the equiline with few along the equiline and below. This indicates that carbonate weathering is the main source for calcium ions in groundwater; whereas the few points below the equiline indicate silicate weathering. In this reaction, carbonic acid (CO_2 and water) and calcium carbonate in soil react to form bicarbonate and calcium ion as given below (Eqs. (1) and (2)):

$$CO_2 + H_2O \Rightarrow H_2CO_3 \quad \text{(Formation of carbonic acid)} \qquad (1)$$

$$CaCO_3 + H_2CO_3 \Rightarrow Ca^{2+} + 2HCO_3 \quad \text{(Calcite dissolution)} \qquad (2)$$

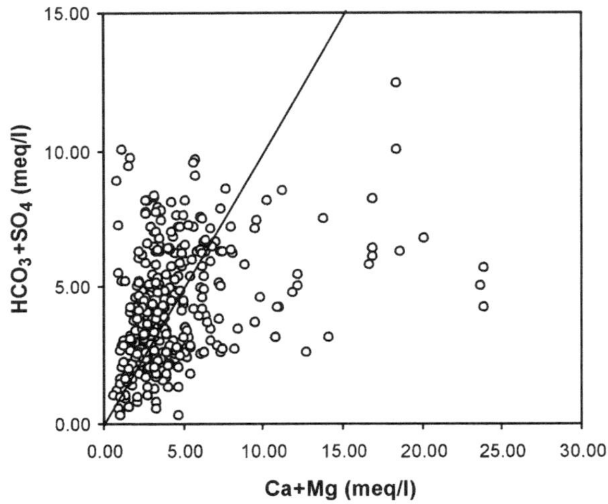

Figure 11.2. (Ca + Mg) vs. (HCO_3 + SO_4) scatter diagram of the lower Palar River Basin shows carbonate dissolution and silicate weathering.

Carbonate weathering by carbonic acid and water saturated with CO_2 is an intensive process. This water can easily dissolve the carbonate minerals available in its flow path. This process has increased the calcium, magnesium and bicarbonate ion contents in groundwater of the study area. Na vs. Ca scatter diagram (Fig. 11.3) shows that Ca/Na ion exchange also increases the concentration of Ca in groundwater. Magnesium may have come from the dissolution of magnesium calcite (Eq. (3)), gypsum and/or dolomite, as reported by Garrels (1976).

$$CaMg(CO_3)_2 + 2H_2CO_3 \Rightarrow Ca^{2+} + Mg^{2+} + 4HCO_3 \qquad (3)$$

(Magnesium calcite dissolution)

If carbonate weathering and silicate weathering takes place, bicarbonate is generally dominant among anions in groundwater. During infiltration of rainwater and irrigated water, the carbonate minerals available along the path will be dissolved depending on the availability of dissolved CO_2 and carbonic acid (Eq. (1)). During the recharge process HCO_3 and Ca will be released to groundwater. Similarly, silicate weathering also increases the concentration of HCO_3 in groundwater. The Na vs. HCO_3 (Fig. 11.4) scatter diagram shows that HCO_3 also comes from silicate weathering (Eq. (4)).

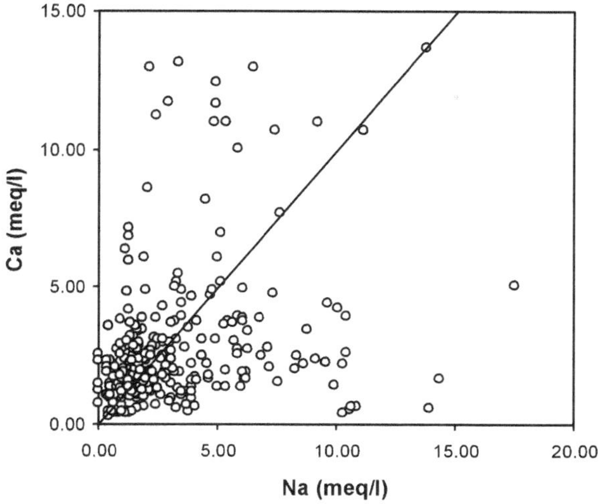

Figure 11.3. Na vs. Ca scatter diagram shows that Ca/Na ion exchange also increases the concentration of Ca in groundwater.

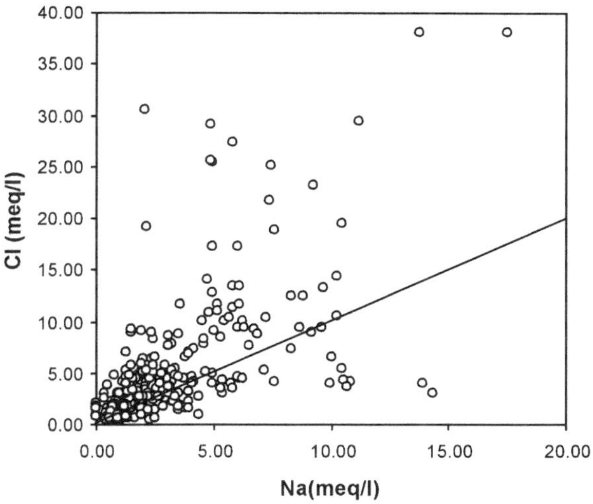

Figure 11.4. Na vs. HCO_3 diagram shows increased concentration of bicarbonate due to silicate weathering.

11.2.1.2. Silicate weathering processes

If bicarbonate and sulfate ions are dominant than calcium and magnesium that means silicate weathering is dominant (Elango et al., 2003). In Ca + Mg vs. $HCO_3 + SO_4$ diagram (Fig. 11.2) points falling along the bicarbonate side indicates the silicate weathering. Similarly, sodium is also high in groundwater. In general, 1:1 relationship between Na and Cl implies halite dissolution, whereas increased concentration of Na than Cl is typically interpreted as Na released from silicate weathering (Mayback, 1987). In the Lower Palar River Basin considerable amount of groundwater samples falling below the 1:1 line in the Na vs. Cl scatter diagram (Fig. 11.5) indicates silicate weathering. Samples falling all along the 1:1 comes from halite dissolution. But in the study area annual average rainfall is over 1000 mm, in this situation it is not possible for availability of free halite for dissolution in the soil zone, but the wind driven aerosol of sodium chloride from Bay of Bengal which is just 20 km from eastern side of the study area might have increased the concentration of Na and Cl in groundwater.

$$\underset{\text{Albite}}{2\,NaAlSi_3O_8} + 2H_2CO_3 + 9H_2O \Rightarrow \underset{\text{kaolinite}}{Al_2Si_2O_5(OH)_4} + 2Na^+ + 4H_4SiO_4 + 2HCO_3 \quad (4)$$

(Silicate weathering)

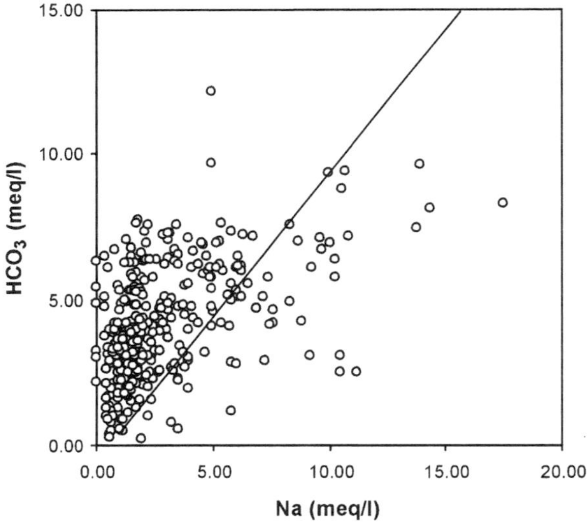

Figure 11.5. Na vs. Cl diagram shows that suggesting no halite dissolution it may be derived from silicate weathering.

The cation contribution to groundwater by silicate weathering can also be estimated by the (Na + K)/Total cations index (Stallard and Edmond, 1983; Sarin et al., 1989). The (Na + K) vs. TZ^+ (total cations) scatter diagram (Fig. 11.6) of the study area shows sample points falling both along and above the (Na + K) = 0.5 TZ^+ line. This suggests that the cations in groundwater might have been derived from silicate weathering. Datta and Tyagi (1996) observed that the contribution of cations may be derived from silicate weathering when (Na + K) = 0.5 TZ^+. In this case, however, the concentration of (Na + K) is about 50% of the total cations. The slightly lower concentration of (Na + K) is likely due to the Ca/Na exchange process as explained above, which might have reduced the amount of Na in groundwater. Potassium might have come from the illite present in the Lower Palar River Basin (Table 11.3).

11.2.2. Ion exchange processes

Ion exchange is a type of adsorption/desorption phenomenon. Clay minerals are the most common ion exchangers in soil and aquifer system. Cations present in the clay and groundwater are easily exchangeable than anion. pH of a solution also controls the exchange processes.

If huge amount of Na content is reduced from groundwater, it is attributed to ion exchange process. The Na vs. Ca (Fig. 11.3) scatter

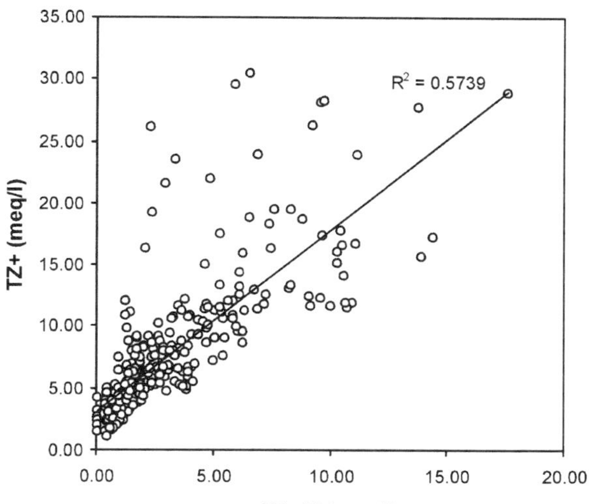

Figure 11.6. Na + K vs. Total cations diagram shows that 50 percent of Na and K comes from silicate weathering.

Table 11.3. Calculated cation exchange capacity of the clay present in the Lower Palar River Basin (after Elango et al., 2003)

Depth (m)	Sediment description	Clay %	Organic carbon %	CEC meq/100 g
1	Sand clay loam	28.4	0.12	20.30
1	Clay loam	35.9	0.30	26.18
1	Loam	23.4	3.00	26.88

diagram supports this process. Slightly elevated levels of calcium compared to sodium also confirm this, as reported by Lavitt et al. (1997). The adsorption of Ca ion and/or the release of Na to groundwater can reflect the same. Similarly, the calculated cation exchange capacity (CEC (meq/100 g) = 0.7 (%clay) + 3.5 (%OC)) value of selected well samples of the study area ranges from 20–27 meq/100 g (Table 11.1 of Chapter 5), which suggests that mixed type of clay is present in the study area. Various authors have reported CEC values for mixed clay minerals. Chapelle (1983) recorded 20–30 meq/100 g for the Aqua aquifer in Maryland, USA. Martinez and Bocanegra (2002) observed CEC values to vary from 20 to 40 meq/100 g for mixed clay sediments of The Mar del Plata aquifer. This type of mixed clay content is probably responsible for the ion exchange process. The following equations explain the ion exchange processes (Eqs. (5)–(8)).

$$2Na^{+} - X + Ca^{2+} \Rightarrow 2Na^{+} + Ca - X^{2} \quad (5)$$

$$2Na^{+} + Mg - X^{2} \Rightarrow 2Na - X + Mg^{2+} \quad (6)$$

$$2K^{+} - X + Ca^{2+} \Rightarrow 2K^{+} + Ca - X^{2}2K^{+} - X + Ca^{2+} \Rightarrow 2K^{+} + Ca - X^{2} \quad (7)$$

$$2K^{+} + Mg - X^{2} \Rightarrow 2K - X + Mg^{2+} \quad (8)$$

where X stands for matrix.

11.2.3. Oxidation/reduction processes

Lot of reactions in groundwater involves the transfer of electrons between dissolved and solid constituents. Electron loss results in oxidation and electron gain results in reduction. The most important oxidizing agents in groundwater are dissolved oxygen, oxo-anions such as nitrate and sulfate and water itself. Reducing agents include a wide variety of organic compounds, inorganic sulfides such as pyrite andiron silicates. At least one of the following groups will be available in groundwater. Sulfate and iron are common constituents in soils and groundwater that participates in the oxidation and reduction reaction. These reactions readily reflect changes in surrounding Eh/pH conditions. In groundwater systems iron occurs in one of two oxidation states: reduced soluble divalent ferrous iron (Fe^{2+}) or oxidized insoluble trivalent ferric iron (Fe^{3+}) (Eq. (9)). Similarly, the sulfate ions present in water may get reduced due to reduction process (Eq. (10)). If the concentrations of sulfate and iron in groundwater are very low and Eh of the groundwater samples are also in reduced condition then the ions might have reduced due to reduction. However, in some areas the lack of source or least occurrence of source of sulfate and iron might be the reason for the low concentration. (Dissolution of gypsum as in Eq. (10), which can supply sulfate to the groundwater.)

$$2Fe^{2+} + 3H_2O \Leftrightarrow Fe_2^{3+}O_3 + 6H^+ + 2e^- \tag{9}$$

$\rightarrow$ Oxidation

$\leftarrow$ Reduction

$$H_2O + CaSO_4 \cdot 2H_2O \Rightarrow Ca^{2+} + SO_4^{2-} + 3H_2O \tag{10}$$

Very low SO_4/Cl ratios (low concentration of SO_4) (Fig. 11.7) suggest that sulfate is being depleted, possibly by sulfate reduction (Lavitt et al., 1997). Earlier, Datta and Tyagi (1996) had observed that groundwater

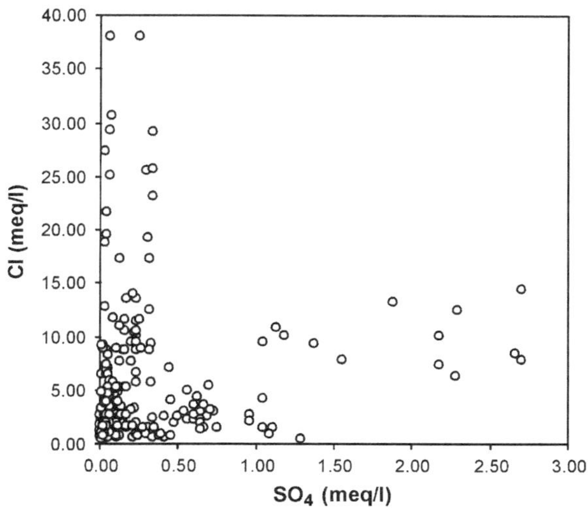

Figure 11.7. Cl vs. SO_4 scatter diagram shows that reduced amount of SO_4 is due to reduction process or lack of sulfate source.

with high Cl and low SO_4 probably indicates reduction. Thus, the low sulfate concentration in the groundwater of this area may be due to sulfate reduction and perhaps lack of natural sources in the area.

11.2.4. Hydrolysis

Hydrolysis is a chemical process in which a molecule is cleaved into two parts by the addition of a molecule of water. This is distinct from a hydration reaction, in which water molecules are added to a substance, but no cleavage occurs. Due to hydrolysis, rock and soil materials are decomposed into different products. Malov (2004) had reported that the principal role in the removal of Ca from rocks was probably played by the hydrolysis of calcium aluminosilicates that are in nonequilibrium state with all types of groundwaters present in Vendian sandy–clayey rocks aquifer. Zverev (1982) had reported that the most characteristic process is the hydrolysis of sodium aluminosilicates as follows Eqs. (11)–(13).

$$7NaAlSi_3O_8 + 6H^+ + 20H_2O^+ \Rightarrow 3Na_{0.33}Al_{2.33}Si_{3.67}O_{10}(OH)_2 + 10H_4SiO_4 + 6Na^+ \quad (11)$$

$$2NaAlSi_3O_8 + 2H^+ + 9H_2O \Rightarrow Al_2Si_2O_5(OH)_4 + 4H_4SiO_4 + 2Na^+ \quad (12)$$

$$NaAlSi_3O_8 + H^+ + 7H_2O \Rightarrow Al(OH)_3 + 3H_4SiO_4 + Na^+ \quad (13)$$

11.3. Rock–water interaction and hydrochemical patterns

Groundwater is classified into different types according to the percentage of chemical constituents present in it. This difference in groundwater rose mainly from different kinds of rock–water interaction. Generally, Ca–HCO_3, Ca–Mg–HCO_3, Ca–Cl, Na–HCO_3, Na–Cl, Ca–SO_4 and Na–SO_4 are the important groundwater types found throughout the world. A piper triplot (Piper, 1944) helps us to classify groundwater simply as in Fig. 11.8. Dissolution of calcite, dolomite, gypsum and halite will give rise to Ca–HCO_3, Ca–Mg–HCO_3, Ca–SO_4 and Na–Cl type of groundwaters, respectively. Huge amount of Ca and HCO_3 ions from calcite is responsible for Ca–HCO_3; Ca, Mg and HCO_3 ions that come from dolomite is responsible for Ca–Mg–HCO_3; Ca and SO_4 ions that come from gypsum is responsible for Ca–SO_4; Na and Cl that come from halite is responsible for Na–Cl. Further due to cation exchange process Na–HCO_3, Ca–Cl and Na–SO_4 groundwater types may form according

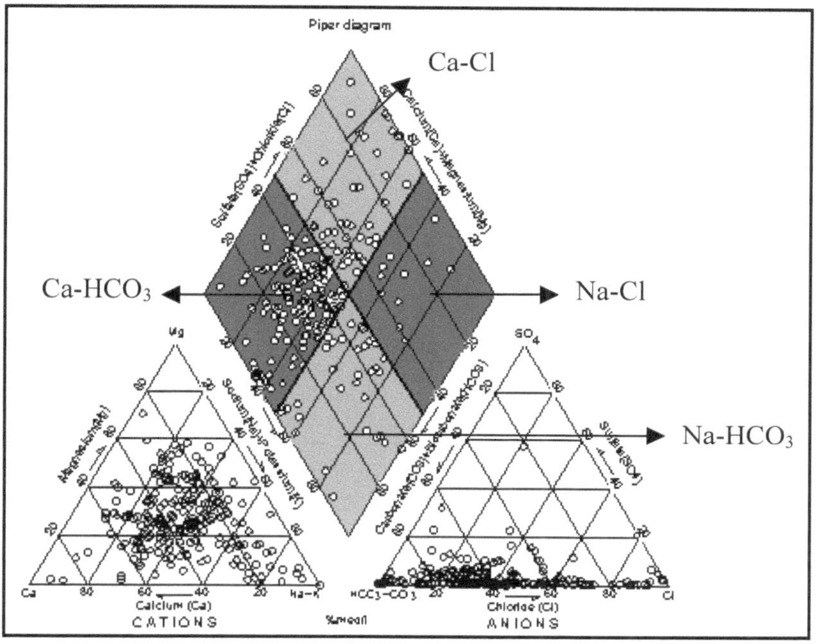

Figure 11.8. Piper triplot indicates different groundwater types.

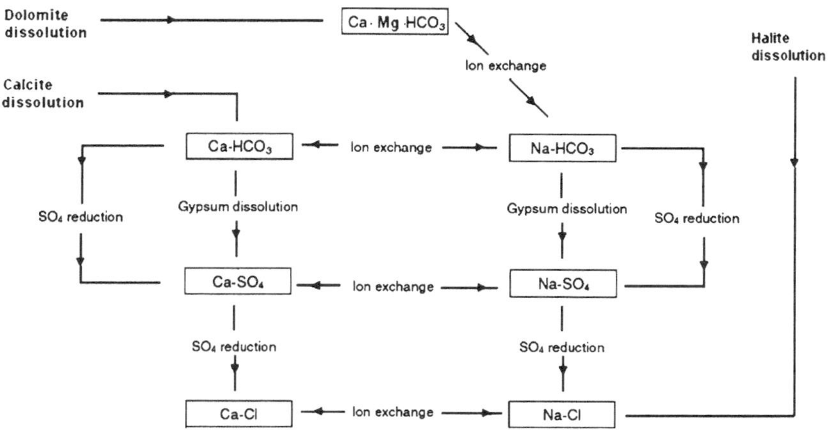

Figure 11.9. Rock–water interaction and resultant groundwater types.

to the cation exchange capacity of the matrix; this is a vice versa reaction. Sulfate reduction may also happen and it will give rise to Na–Cl, Ca–Cl, Ca–HCO_3 and Na–HCO_3 types of groundwater (Fig. 11.9). Thus, various chemical processes of rock–water interaction lead to groundwater chemical composition.

REFERENCES

Apodaca, L.E., Jeffrey, B.B., Michelle, C.S., 2002. Water quality in shallow alluvial aquifers, Upper Colorado River basin, Colorado, 1997. J. Am. Wat. Res. Ass. 38(1), 133–143.

Chapelle, F.H., 1983. Groundwater geochemistry and calcite cementation of the Aquia aquifer in Southern Maryland. Water Res. Res. 19(2), 545–558.

Datta, P.S., Tyagi, S.K., 1996. Major ion chemistry of groundwater in Delhi area: Chemical weathering processes and groundwater flow regime. J. Geol. Soc. India 47, 179–188.

Deutsch, W.J., 1997. Groundwater Geochemistry: Fundamentals and Applications to Contamination. Lewis Publishers, New York.

Elango, L., Kannan, R., Senthil Kumar, M., 2003. Major ion chemistry and identification of hydrogeochemical processes of groundwater in a part of Kancheepuram district, Tamil Nadu, India. Environ. Geosci. 10(4), 157–166.

Garrels, R.M., 1976. A survey of low temperature water mineral relations. In: Interpretation of Environmental Isotope and Hydrogeochemical Data in Groundwater Hydrology. International Atomic Energy Agency, Vienna, pp. 65–84.

Gibbs, R.J., 1970. Mechanisms controlling world water chemistry. Science 170, 795–840.

Kannan, R., 2005. Identification of aquifer vulnerability to pollution with special emphasis to unsaturated zone. Unpublished Ph.D. Thesis, Anna University, Chennai, India, p. 141.

Lavitt, N., Acworth, R.I., Jankowski, J., 1997. Vertical hydrogeochemical zonation in a coastal section of the Botany Sands aquifer, Sydney, Australia. Hydrogeol. J. 5, 64–74.

Malov, A.I., 2004. Water–rock interaction in vendian sandy–clayey rocks of the Mezen Syneclise. Lithol. Min. Resour. 39(4), 345–356. Translated from Litologiya i Poleznye Iskopaemye, No. 4, 2004, pp. 401–413.

Martinez, D.E., Bocanegra, E.M., 2002. Hydrogeochemistry and cation exchange processes in the coastal aquifer of Mar Del Plata, Argentina. Hydrogeol. J. 10, 393–408.

Mayback, M., 1987. Global chemical weathering of surficial rocks estimated from river dissolved loads. Am. J. Sci. 287, 401–428.

Piper, A.M., 1944. A graphical interpretation of water – analysis. Trans. Am. Geophys. Union 25, 914–928.

Rajmohan, N., Elango, L., 2004. Identification and evolution of hydrogeochemical processes in the groundwater environment in a part of the Palar and Cheyyar River Basins, Southern India. Environ. Geol. 46(1), 47–61.

Sami, K., 1992. Recharge mechanisms and geochemical processes in a semi-arid sedimentary basin, Eastern Cape, South Africa. J. Hydrol. 139, 27–48.

Sarin, M.M., Krishnaswamy, S., Dilli, K., Somayajulu, B.L.K., Moore, W.S., 1989. Major ion chemistry of the Ganga-Brahmaputra river system: Weathering processes and fluxes to the Bay of Bengal. Geochim. Cosmochim. Acta 53, 997–1009.

Senthil Kumar, M., 2004. Hydrogeological modelling and hydrogeochemical characterisation of an unconfined aquifer system of the Lower Palar River basin, Southern India. Unpublished Ph.D. Thesis, Anna University, Chennai, India, p. 166.

Stallard, R.F., Edmond, J.M., 1983. Geochemistry of the Amazon, the influence of geology and weathering environment on the dissolved load. J. Geophys. Res. 88, 9671–9688.

Zverev, V.P., 1982. Rol' podzemnykh vod v migratsii khimicheskikh elementov (Role of Groundwaters in the Migration of Chemical Elements), Moscow, Nedra.

Section III: Lithosphere–Hydrosphere Interactions: Applications of Geochemical Principles

Developments in Environmental Science, Volume 5
D. Sarkar, R. Datta and R. Hannigan (Editors)

ISSN: 1474-8177/DOI:10.1016/S1474-8177(07)05012-7

Chapter 12

Association of dissolved organic carbon with stream discharge and dissolved metals concentrations in black shale-draining streams

George M. Ogendi, Robyn E. Hannigan and Jerry L. Farris

Abstract

Watershed characteristics influence the concentrations of dissolved organic carbon (DOC) that affects the levels, fate, partitioning, bioavailability, and toxicity of dissolved metals in aquatic ecosystems. Despite increasing interest in black shales research, there are limited data relating influence of bedrock type upon DOC and dissolved metals in streams whose watersheds are dominated by black shales. Four streams in North-Central Arkansas, USA, were selected for study, three of which (Trace Creek, Begley Creek, and Cove Creek) drain black shales-dominated watersheds. A fourth nearby stream (Mill Creek) that flows over a geologically distinct limestone bedrock, was selected as a reference site to allow comparisons between these watersheds and the assessment of the role of bedrock lithology on DOC and metal concentrations. Spatial and temporal variations in DOC and dissolved metals among stream sites were determined. Streams draining black shales had significantly higher conductivity, and lower pH than the stream draining limestone. Dissolved metal concentrations at all sites were highest in June 2003 and lowest in September 2003, which coincided with a stormflow and a baseflow, respectively. The concentrations of DOC and metals (e.g. Cd, Cu, Hg, Ni, Pb, and Zn) in water collected from Trace, Begley, and Cove creeks were significantly higher than those of Mill Creek. Metals in streams draining black shales whose concentrations approached or exceeded US EPA's surface water criteria included Cr, Cu, Hg, Pb, and Se. A linear regression analysis of DOC with discharge as the predictor variable showed that variations in DOC concentrations were strongly explained by stream discharge with r^2-values ranged from 0.85 to 0.95, whereas DOC was inversely related to conductivity ($r^2 = 0.47$–0.89). The study also revealed that bedrock type was a significant predictor of the observed spatial variations in DOC. Streams draining black shale watersheds transported significantly higher ($p < 0.05$) amounts of DOC than those draining limestone, thereby emphasizing the important role of geology in determining the amount of DOC in streams. Owing to the strong positive correlations between DOC and dissolved metals, any attempt to quantify metals and predict their effect on the aquatic biota in these types of streams should consider DOC levels in addition to other physical and chemical variables.

12.1. Introduction

Watershed characteristics (e.g. hydrologic events, type of vegetation and geology) and anthropogenic activities in catchments influence the concentration of dissolved metals and dissolved carbon (inorganic and organic) in streams (Taylor et al., 1990; Shafer et al., 1997; Shiller, 1997; Sherrell and Ross, 1999; Meyer et al., 2004). Sherrell and Ross (1999) observed that both temporal and spatial variations in dissolved metal concentrations were related to hydrologic flow paths during high discharge. Whereas Shiller and Boyle (1987) and Taylor et al. (1990) have shown that dissolved metals in large rivers can be derived from natural rock weathering in the catchment, other studies have additionally demonstrated that discharge can be a strong predictor of dissolved metals with higher metal concentrations occurring during high discharge (Rember and Trefry, 2004).

Dissolved organic carbon (DOC) can comprise over 95% of total organic carbon (Bishop and Pettersson, 1996) and also constitutes between 40 and 50% of dissolved organic matter (DOM) in forested headwater streams (Allan, 1995; Kaplan and Newbold, 1993). It is a significant energy source for stream biota including macroinvertebrates that commonly form the energy base for food chains in aquatic ecosystems (Elder, 1988; Allan, 1995). DOC also plays an important role in the transport of metals within aquatic ecosystems thereby altering dissolved metal concentrations and, eventually, bioavailability to aquatic organisms. The ability of DOC to form soluble complexes with many metal ions is critical in the analysis of the distribution, accumulation, and effects of metals on biota in aquatic ecosystems. The role of DOC in metal complexation as a primary mechanism in the transport of metals into aquatic ecosystems has been proposed by a number of researchers (e.g. Allen and Hansen, 1996; Davies et al., 1998; Kim et al., 1999; Mounier et al., 1999). Interactions of DOC with metals lead to formation of DOC–metal complexes (Spitzy and Leenheer, 1988; Shafer et al., 1997) that affect the partitioning, fate, and bioavailability of metals (Apte and Batley, 1995; Gardner and Gunn, 1989). DOC–metal complexation renders the metals less bioavailable and less toxic to the aquatic organisms (Allen and Hansen, 1996).

Variations in levels of DOC in streams have also been linked to variations in precipitation in catchments with the highest DOC concentrations coinciding with maximum discharge in temperate and tropical streams (*this study*, Cauwet and Meybeck, 1987; Moore, 1989; Droppo and Jaskot, 1994; Hurley et al., 1996; Shafer et al., 1997; Rember and Trefry, 2004). Degens et al. (1988) also linked changes in stream DOC to climate; Thurman (1985) estimated the range of DOC concentrations for cool temperate and warm temperate climate streams to be between 2–8 and 3–15 $mg\,l^{-1}$, respectively. Differences in DOC in streams may also be attributed to geologic differences in the watersheds (Sherrell and Ross, 1999; Meyer et al., 2004). For instance, black shales have been shown to be organic carbon- and metal-enriched (Coveney and Martin, 1983; Kim and Thornton, 1993; Chon et al., 1996; Petsch et al., 2000; Ogendi et al., 2004b). These studies found that concentrations of As, Cd, Cr, Cu, Mn, Ni, Pb, V, and Zn in surface waters exceeded surface water criteria for metals in the United States, thus a threat to aquatic life. Throughout the United States, fractured black shales host shallow ground and surface waters with the water and soil in these regions often containing elevated metal concentrations in excess of US EPA maximum contaminant levels (Blette and Newton, 1996; Chon et al., 1996; US EPA, 2002b). The potential importance of metal release from black shales impacts areas of central and western Arkansas, eastern Missouri, Oklahoma, and Kansas as well as in western and central New York and northern Montana. Black shales also occur throughout central and northern Appalachians (Ayotte et al., 1999; Foley et al., 2002) and the southeastern United States (Seal et al., 2000). Despite the known impact of trace metals to aquatic biota and humans (e.g. Mize and Deacon, 2002; Jagoe et al., 2002; Riley et al., 2003; Moore, 2004), quantification of DOC and its relation to dissolved metal concentrations in streams is poorly understood. Increased knowledge of the relation between DOC on the spatial and temporal metal concentrations in headwater streams is critical to the linkages between chemical weathering of metal rich bedrock and metal bioavailability and toxicity.

Whereas a number of studies have shown that black shales are non-point sources for metals whose concentrations are sufficient to elicit toxicity (Kim and Thornton, 1993; Chon et al., 1996), there are limited data to the best of our knowledge, that show the role of DOC in metal transport and complexation in streams draining watersheds dominated by black shales. Increased metal concentrations in surface waters and stream sediments have been attributed to chemical weathering of the shales (Kim and Thornton, 1993; Chon et al., 1996; Hannigan, 1997; Loukola-Ruskeeniemi et al., 1998; Ogendi et al., 2004a). Increased metal concentrations significantly reduce water and sediment quality (Taylor et al., 1990) that may lead to fish kills (e.g. Pasava et al., 1995), reduced survival and growth of midge larvae (Ogendi et al., 2004b), and decreased taxa richness of other benthic macroinvertebrates (Mize and Deacon, 2002). In addition, adverse effects on the reproduction and recruitment of two endangered fish species (razorback sucker, *Xyrauchen texanus* and pike minnow, *Ptychocheilus lucius*) have also been attributed to elevated metal concentrations in surface water and sediments (Hamilton and Waddell, 1994; Deacon and Stephens, 1998). In this study, four bedrock headwater streams in two geologically distinct, but nearby catchments, were selected to assess the impact of DOC, discharge, and pH on spatial and temporal variations in dissolved metal concentrations.

12.2. Materials and methods

12.2.1. Study sites description

The study area consists of four headwater bedrock streams that are tributaries of the Little Red River, Arkansas, USA. The first three streams: Trace Creek, Begley Creek, and Cove Creek (Fig. 12.1) drain an extensive late Mississippian period (~350 mya) Fayetteville Shale outcrop of north-central Arkansas. The Fayetteville Shale is considered to rest conformably on the Batesville Sandstone (Hudson et al., 2001). The Fayetteville Shale consists of black shale that contains medium- to light-gray fetid septarian concretions as large as 0.6 m in diameter (Hudson et al., 2001; Murthy et al., 2004). In this region, the Fayetteville Shale embeds dark-gray, fine-grained limestones, and ranges in thickness from 3 to 120 m (AGC, 2005; Fig. 12.1). The lower part of the Fayetteville Shale outcrops along stream gullies, and is in intimate contact with surface and ground waters throughout the length of Trace, Begley, and Cove creeks (Ogendi et al., 2004a,b). The fourth stream, the Mill Creek drains the Mississippian Pitkin Limestone of north-central Arkansas that rests conformably on the Fayetteville Shale. The Pitkin Limestone is medium- to dark-gray, and its texture varies from micritic to coarse-grained. The Pitkin Limestone generally forms a prominent cliff, and its beds contain abundant crinoids, brachiopods, and corals (Hudson et al., 2001). Its thickness ranges from 0 to 18 m.

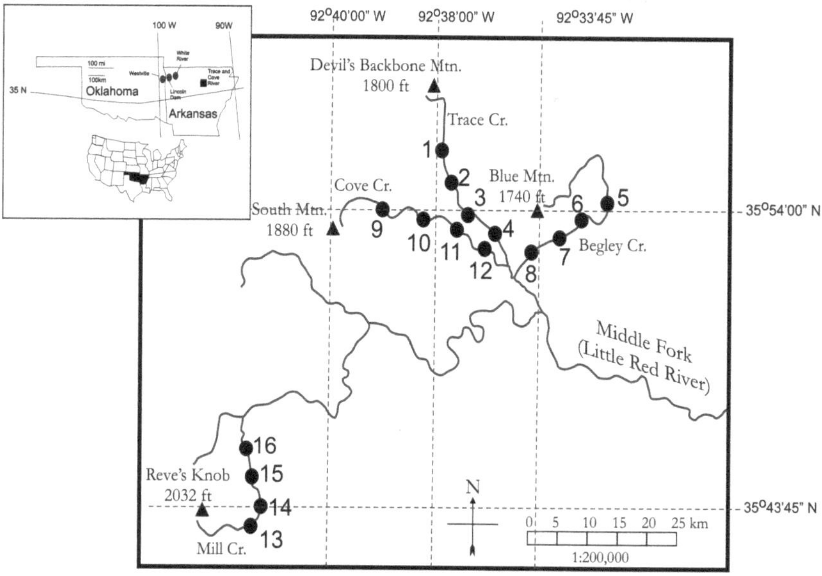

Figure 12.1. Location of sampling sites (numbered-solid circles) on streams draining black shales (i.e. Trace, Begley, and Cove creeks), and on a limestone-draining stream (Mill Creek). The streams are part of the Little Red River watershed (HUC 11010014) in north-central Arkansas, USA (shown on the top-left corner).

The mineralogy of the Fayetteville shale, based on X-ray diffraction (XRD) (Murthy, 2003) showed a dominance of quartz, micas, and feldspars dominate the bulk mineral composition with illite, kaolinite, and smectite the most abundant clay minerals. There is no significant spatial or temporal variation in the mineralogy of the shales exposed in the study area. The limestone mineralogy at Mill Creek, typical of intercalated limestones, is dominated by calcite with some illite. In the case of the organic carbon content, the shales contain 5–8 wt.% and the limestone contains no more than 1% (Murthy, 2003). Carbonate content in the shales is less than 10 wt.%. No vitrinite reflectance data are available and so thermal diagentic history is not well known. Based on organic matter content and petrography of the units the shales appear, on first order, to be mature to post-mature (Hannigan and Basu, 1997).

The Devil's Backbone (92° 38′ 00″ W, 35° 54′ 00′ N), Blue Mountain (92° 33′ 45″ W, 35° 51′ 35″ N), South Mountain (92° 40′ 33″ W, 35° 48′ 35″ N), and Reves Knob (92° 45′ 00″ W, 35° 43′ 45″ N) constitute the catchments for Trace, Begley, Cove, and Mill creeks, respectively. The Fayetteville shale bedrock streams were selected because they drain black shales that are potential non-point sources of metals (Ogendi et al., 2004b). The nearby Mill Creek, draining a geologically distinct limestone bedrock, was selected as a reference site to allow comparisons between these watersheds and for the assessment of the role of bedrock lithology on DOC and dissolved metals. The watersheds were identical in terms of principal land cover and hydrology, and consisted of over 90% forest riparian habitat with negligible anthropogenic disturbances. The natural vegetation along the study streams is primarily oak-hickory, oak-hickory-pine forests, and locally interspersed with cedar glades and short-leaf pines growing on steep slopes. The study area is characterized by a temperate climate with mild winters (summer maximum temperature 27.1°C and a winter minimum 4.8°C) with an annual average rainfall of 1370 mm.

12.2.2. Trace metal analysis

Water samples were collected from four randomly selected sites per stream (Fig. 12.1) using trace-metal clean procedures (Shelton and Capel, 1994; Shafer et al., 1997; APHA, 1998). Six selective sample collections representing the broad range of seasonal variations in DOC, dissolved metals, discharge, and other water quality parameters were carried out between June 2003 and May 2004. All equipment used for sample collection, storage, and analysis of trace metals were pre-cleaned using high-purity nitric acid (GFS Chemicals Inc.) and thoroughly rinsed with Milli-Q water. Such cleaning and storage procedures ensure that there are no detectable metal contaminants in the sampling equipment (Shafer et al., 1997). The samples were collected in polypropylene bottles and filtered immediately through 0.45 μm Gelman in-line filters and acidified with ultra-pure HNO_3 to $pH<2$ and stored at 4°C prior to trace metal analyses. Other water quality parameters known to affect dissolved metals and DOC were measured (i.e. dissolved oxygen, pH, and conductivity). DOC and trace metals in the filtrate (0.45 μm) are here operationally defined as "dissolved". We centered our attention on the dissolved fraction as this fraction is more likely to have measurable biological effects on aquatic organisms (Di Toro et al., 2000). In addition, the dissolved metals are also similar to the exposure conditions used in toxicity tests (US EPA, 2002a), allowing for comparisons between standard toxicity tests and field community surveys (Ogendi et al., 2004a,b). This study is part of a broader investigation on the Little Red River watershed designed to determine the impact of trace metals on the stream water and sediment quality, and the resident macroinvertebrate communities. Metal concentrations were determined by the dynamic reaction cell inductively coupled plasma mass spectrometry (DRC-ICP-MS; Elan 9000 PerkinElmer) following EPA 200.8 methodology. In brief, 15 ml of water sample was transferred into an autosampler vial into which an internal standard containing 40 $\mu g\,l^{-1}$ Li-6, Ge, In, and Bi was added. A standard calibration curve for all the analytes was established on standards prepared in a linear range from 1 ppb to 100 ppb. National Institute of Standards and Technology reference material (NIST 1640) and procedural blanks were analyzed for trace metals. The dissolved metal concentrations were within 3% of the reported values for the standards. Finally, the relative standard deviations (RSD) for sample dissolved metal concentrations were <5% for all analytes.

12.2.3. Determination of dissolved organic carbon

Water samples were collected in 1-liter glass bottles (I-Chem Nalge Company) equipped with TFE-backed septa. These bottles were washed scrupulously and rinsed thoroughly with water before drying. The Gelman in-line filters were flushed with sample water before collection and the same was done for the sample containers that were rinsed with an aliquot of sample before collection. Water samples for DOC analysis were taken from the same stations that trace metal samples were collected. They were also filtered through 0.45 μm (Gelman in-line filters) and stored in the dark at 4°C. The water samples were analyzed for DOC within 48 h of collection.

DOC analyses were accomplished by UV-persulfate analysis (EPA 415.2) using a Tekmar-Dohrman Phoenix 8000™ TOC analyzer. Briefly, the inorganic carbon (IC) is removed from the sample by acidification with phosphoric acid and sparging using nitrogen (99.999 pure nitrogen). Organic carbon (OC) is oxidized to CO_2 by sodium persulfate in the presence of ultraviolet (UV) light. The CO_2 produced is then purged from the sample, dried, and carried in a stream of nitrogen to a nondispersive infrared (NDIR) detector that was specifically tuned to the absorptive wavelength of CO_2 (~4.5 μm). This instrument can detect very low amounts (~2 ppb) of carbon while not sacrificing the ability to analyze widely varying amounts (Phoenix 8000™ User Manual, 1998). The instrument was calibrated every ten samples and the persulfate solution was prepared fresh each day of analysis. For every ten samples, procedural blanks, a quality control check standard, and a check standard duplicate were also analyzed to ensure quality control. The RSD for replicate standards and samples ranged between 2% and 3 % whereas the r^2-values for the calibration curves ranged between 0.9998 to 0.9999 for these analyses.

12.2.4. *Determination of pH, conductivity, and discharge*

Stream water pH and discharge were determined at the sites of sample collection. The stream pH was measured using a field pH meter and a glass electrode that was standardized immediately before each reading. Conductivity in microsiemens per centimeter ($\mu S\,cm^{-1}$) was measured using a YSI conductivity meter. The instrument was calibrated before use in the field. The values of conductivity are reported as conductivity of the water samples at 25°C. Discharge measured as $m^3\,s^{-1}$ was determined manually with the use of a flow meter, wading rod, and tape measure. In this study, a stream cross-section was divided into a number of smaller sections of equal width and uniform bed conditions. Velocity was measured at 60% (from the surface) of the depth of each unit and then multiplied by the unit area to give the unit discharge. Total discharge was then calculated as the sum of the discharges of the smaller sections.

12.2.5. *Statistical analyses*

The data were tested for normality and homogeneity of variance using Kolmogorov-Smirnov Normality Test ($p \leq 0.05$) and Levene's Test for equal variances ($p \leq 0.05$), respectively (MINITAB® Statistical Software for Windows ver. 14). Using data that satisfied the assumptions normality we compared the DOC and metal concentrations amongst the four study sites using analysis of variance (ANOVA) to test for differences among sites and sampling occasions ($\alpha = 0.05$). Linear regression analyses ($y = mx + b$) were carried out on the data that yielded regression slopes, coefficient of determination values (r^2-values), and p-values. In these analyses, except where defined, the criterion for significance was a p-value of ≤ 0.05.

12.3. Results

Both spatial and temporal variations in DOC, discharge, pH, and conductivity were observed among the study sites (Table 12.1). Significant variations in DOC concentrations among sites ($F_{(7,\ 179)} = 49.3$; $p < 0.001$) and sampling occasions ($F_{(5,\ 179)} = 1410.5$; $p < 0.001$) were observed with the highest and lowest values recorded in Trace Creek and Mill Creek, respectively (Fig. 12.2a). The average DOC values in Cove Creek were between 15% and 35% less than those observed in Trace Creek and Begley Creek. Despite its relatively lower DOC levels compared to Trace and Begley creeks, DOC levels at Cove Creek were 1.5 times higher than those of Mill Creek. On average, the highest and lowest DOC

Table 12.1. Means (± SD, in *italics*) of discharge ($m^3 s^{-1}$), pH, dissolved organic carbon ($mg\,l^{-1}$) and conductivity ($\mu S\,cm^{-1}$) measured at the study sites in north-central Arkansas (USA). TRA, TRB (Trace Creek A and B), BGA, BGB (Begley Creek A and B), CVA, CVB (Cove Creek A and B), MLA and MLB (Mill Creek A and B)

	TRA	TRB	BGA	BGB	CVA	CVB	MLA	MLB
Discharge								
6/16/03	0.95	0.95	0.80	0.83	0.77	0.78	1.02	1.05
	0.12	*0.07*	*0.03*	*0.03*	*0.02*	*0.03*	*0.06*	*0.10*
8/11/03	0.33	0.34	0.33	0.33	0.30	0.32	0.49	0.54
	0.02	*0.02*	*0.02*	*0.04*	*0.03*	*0.03*	*0.04*	*0.02*
9/17/03	0.07	0.07	0.06	0.06	0.04	0.05	0.11	0.13
	0.01	*0.01*	*0.02*	*0.01*	*0.01*	*0.01*	*0.02*	*0.01*
11/27/03	0.34	0.33	0.30	0.34	0.29	0.31	0.44	0.45
	0.01	*0.01*	*0.01*	*0.03*	*0.04*	*0.02*	*0.03*	*0.02*
1/9/04	0.19	0.19	0.21	0.19	0.18	0.15	0.30	0.31
	0.02	*0.02*	*0.02*	*0.02*	*0.04*	*0.02*	*0.02*	*0.02*
5/23/04	0.15	0.14	0.12	0.13	0.12	0.10	0.22	0.20
	0.01	*0.02*	*0.02*	*0.02*	*0.02*	*0.02*	*0.02*	*0.02*
pH								
6/16/03	4.83	5.10	5.80	5.60	6.68	6.78	7.60	7.60
	0.13	*0.08*	*0.08*	*0.08*	*0.05*	*0.05*	*0.08*	*0.08*
8/11/03	5.78	5.85	6.13	6.13	7.10	6.95	7.45	7.55
	0.05	*0.06*	*0.10*	*0.10*	*0.08*	*0.06*	*0.06*	*0.06*
9/17/03	5.95	5.93	6.33	6.38	7.35	7.23	7.70	7.70
	0.06	*0.10*	*0.05*	*0.05*	*0.06*	*0.05*	*0.08*	*0.05*
11/27/03	5.50	5.55	5.83	5.80	6.83	6.88	7.53	7.55
	0.08	*0.06*	*0.05*	*0.08*	*0.05*	*0.05*	*0.05*	*0.06*
1/9/04	5.65	5.63	5.95	5.98	6.83	6.85	7.50	7.58
	0.06	*0.05*	*0.06*	*0.05*	*0.05*	*0.06*	*0.08*	*0.05*
5/23/04	5.63	5.68	6.05	6.05	6.88	6.85	7.45	7.58
	0.05	*0.05*	*0.06*	*0.06*	*0.05*	*0.06*	*0.06*	0.05
DOC ($mg\,l^{-1}$)								
6/16/03	13.53	12.62	16.05	15.87	10.44	10.37	6.60	6.65
	0.15	*0.64*	*0.26*	*0.23*	*0.46*	*0.60*	*0.21*	*0.15*
8/11/03	10.68	10.90	10.64	9.96	7.57	7.07	5.81	5.56
	0.33	*0.42*	*0.43*	*0.35*	*0.35*	*0.14*	*0.17*	*0.15*
9/17/03	7.67	9.27	6.64	7.03	6.49	6.42	3.96	3.87
	0.11	*0.35*	*0.15*	*0.08*	*0.05*	*0.05*	*0.23*	*0.24*
11/27/03	10.89	10.56	9.77	9.93	7.48	7.23	5.50	5.24
	0.27	*0.28*	*0.43*	*0.31*	*0.28*	*0.19*	*0.59*	*0.03*
1/9/04	8.14	8.18	6.72	7.09	6.85	6.75	4.88	4.72
	0.20	*0.19*	*0.10*	*0.27*	*0.08*	*0.07*	*0.19*	*0.16*
5/23/04	8.31	7.79	9.06	8.42	7.03	7.11	4.52	4.56
	0.37	*0.16*	*0.22*	*0.44*	*0.08*	*0.15*	*0.24*	*0.19*
Conductivity								
6/16/03	165.00	173.75	165.50	169.50	161.25	156.00	55.75	58.75
	9.97	*5.62*	*8.39*	*4.65*	*9.22*	*2.94*	*2.63*	*2.22*

Table 12.1. (Continued)

	TRA	TRB	BGA	BGB	CVA	CVB	MLA	MLB
8/11/03	212.00	208.50	205.75	197.50	195.50	196.25	72.00	72.25
	4.97	*4.20*	*6.95*	*2.08*	*5.57*	*7.80*	*2.94*	*4.99*
9/17/03	365.50	362.25	368.50	365.50	333.00	334.00	170.00	165.25
	5.00	*3.50*	*3.51*	*7.55*	*5.94*	*6.78*	*3.56*	*8.46*
11/27/03	216.50	225.75	212.00	211.00	203.25	205.50	86.00	89.50
	4.43	*1.71*	*4.90*	*7.02*	*4.27*	*1.29*	*5.10*	3.51
1/9/04	248.25	241.75	237.75	233.00	245.50	249.50	130.50	130.50
	12.40	*14.20*	*4.20*	*4.51*	*4.80*	*5.07*	*2.38*	*7.14*
5/23/04	239.50	237.25	260.50	258.50	224.50	221.50	110.50	113.75
	3.10	*4.43*	*5.38*	*3.65*	*2.08*	*3.11*	*4.04*	*7.55*

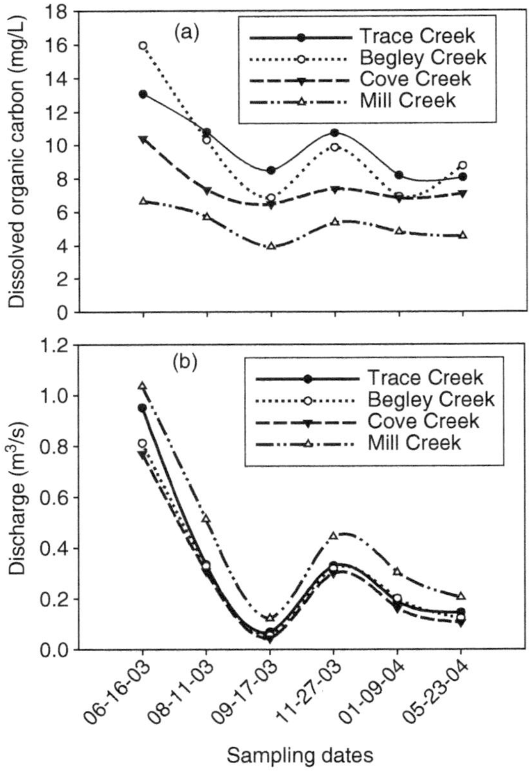

Figure 12.2. Temporal variations in (a) dissolved organic carbon and (b) stream discharge for Trace, Begley, Cove, and Mill creeks. Each point on the scatter plot is a mean of four sample measurements per sampling site during each sampling occasion.

concentrations were measured in June and September 2003, respectively (Table 12.1). The DOC concentrations for June 2003 were approximately twice as much as those measured in September 2003 for sampling sites. Discharge at Mill Creek was consistently greater than that observed at Trace, Begley, and Cove creeks (Fig. 12.2b). The highest and lowest flows occurred in June and September, 2003, respectively. However, the reverse was true of pH and conductivity with minimum and maximum measurements made in June and September, respectively (Table 12.1). A linear regression analysis of DOC with discharge as the predictor variable showed that variations in DOC concentrations were adequately explained by stream discharge with r^2-values ranging from 0.85 to 0.95 (Fig. 12.3). In addition, significant ANOVAs ($F_{(1,10)} = 43$, $p < 0.001$, Trace Creek; $F_{(1,10)} = 115$, $p < 0.001$, Begley Creek; $F_{(1,10)} = 144$, $p < 0.001$, Cove Creek; $F_{(1,10)} = 109$, $p < 0.001$, Mill Creek) were obtained for the corresponding DOC-discharge measurements. The concentrations of DOC were inversely related to specific conductivity with high r^2-values ranging from 0.47 to 0.89 (Fig. 12.4).

Dissolved metal concentrations at all sites were highest in June 2003 and lowest in September 2003 samples (Table 12.2). Except for manganese, dissolved metals in black shale-draining water samples were

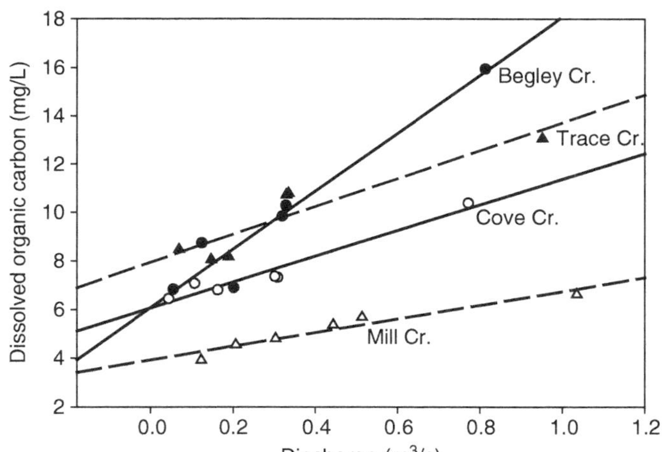

Figure 12.3. Linear regression plots of DOC vs. discharge for Trace Creek (solid triangle; $y = 7.9 + 5.8x$; $r^2 = 0.85$), Begley Creek (solid circle; $y = 6.1 + 12x$; $r^2 = 0.93$), Cove Creek (open circle; $y = 6.1 + 5.3x$; $r^2 = 0.95$), and Mill Creek (open triangle; $y = 3.9 + 2.8x$; $r^2 = 0.93$). Each point on the scatter plot is a mean of four sample measurements per sampling site per sampling occasion.

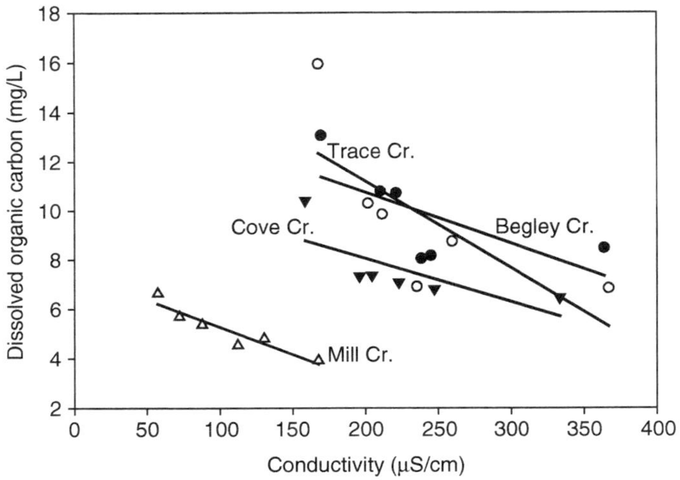

Figure 12.4. Linear regression plots of DOC vs. conductivity Trace Creek (solid circle; $y = 14.8–0.02x$; $r^2 = 0.47$), Begley Creek (open circle; $y = 12.3–0.03x$; $r^2 = 0.53$), Cove Creek (solid triangle; $y = 11.5–0.02x$; $r^2 = 0.54$), and Mill Creek (open triangle; $y = 7.5–0.02x$; $r^2 = 0.89$). Each point on the scatter plot is a mean from four sample measurements per sampling site per sampling occasion.

significantly higher ($p<0.05$) than those in the stream-draining limestone (Fig. 12.5). The concentrations of both Fe and Mn in all streams were consistently higher than those of other elements in all the sampling occasions. The mean dissolved V and Co were two to three times higher in Trace Creek than in Mill Creek, whereas dissolved Cr in both Trace Creek and Begley Creek was 2–36 times higher than Mill Creek. Dissolved mercury was significantly higher ($p<0.001$) in samples drawn from streams draining black shales than those collected from the limestone-draining stream (Table 12.2). Dissolved Ni and Pb were between 2 and 42 times higher in Trace, Begley, and Cove creeks than in the Mill Creek samples. In contrast, the concentrations of Zn and Sn were relatively higher in Mill Creek than the streams draining black shales (Table 12.2). Except for the November 2003 samples, dissolved Zn was between 10% and 30% higher in Mill Creek samples than those of Trace, Begley, and Cove creeks. Significant differences ($p<0.05$) were observed in dissolved Mn between the study streams (Fig. 12.5) with the highest concentrations occurring in Mill Creek. The concentration of Cd in June 2003 in Trace Creek was significantly higher ($p<0.05$) than that observed in the other streams (Table 12.2). Significantly higher ($p<0.05$) dissolved Cu was noted in Begley Creek than that of Trace, Cove, and Mill creeks. The

Table 12.2. Dissolved concentrations ($\mu g\,l^{-1}$) of selected elements in stream water as determined by inductively coupled plasma mass spectrometry. TRA, TRB (Trace Creek A and B), BGA, BGB (Begley Creek A and B), CVA, CVB (Cove Creek A and B), MLA and MLB (Mill Creek A and B). *CCC* is criteria continuous concentration—four-day average not to be exceeded more than once every three years on average, and *CMC* is criteria maximum concentration—one-hour average not to be exceeded more than once every three years on average (US EPA, 2002b). nd: not defined

	TRA	TRB	BGA	BGB	CVA	CVB	MLA	MLB	CCC	CMC
6/16/03										
V	5.04	5.34	3.53	3.12	1.21	0.60	1.11	0.50	nd	nd
Cr	14.52	15.93	10.99	8.16	5.14	1.31	0.60	0.40	11.00	16.00
Mn	14.92	20.16	7.06	9.58	14.52	13.41	31.75	56.85	nd	nd
Fe	336.47	338.47	258.15	259.05	184.24	170.13	159.13	96.32	nd	nd
Co	0.30	0.30	0.20	0.20	0.20	0.20	0.10	0.10	nd	nd
Ni	3.02	3.73	3.43	3.83	1.41	1.81	1.51	1.51	52.00	470.00
Cu	7.56	10.48	18.95	6.65	3.83	4.23	4.54	4.13	9.00	13.00
Zn	16.93	16.33	13.61	16.13	7.86	18.35	24.60	12.00	120.00	120.00
As	59.57	45.56	34.37	36.69	47.68	90.22	38.71	53.32	150.00	340.00
Se	3.63	5.95	9.48	8.67	5.24	9.17	3.12	2.32	5.00	nd
Cd	0.30	0.30	0.20	0.20	0.10	0.10	0.10	0.10	0.25	2.00
Hg	1.21	8.37	5.04	1.51	2.02	11.39	0.14	0.03	0.77	1.40
Pb	28.36	44.79	11.72	11.02	4.71	4.21	6.01	7.82	2.50	65.00
8/11/03										
V	1.04	1.11	0.75	0.65	0.27	0.16	0.26	0.12		
Cr	1.60	1.77	1.20	0.88	0.53	0.11	0.08	0.06		
Mn	2.76	3.99	0.94	1.52	2.68	2.41	6.69	12.58		
Fe	232.00	209.51	159.79	160.31	114.01	105.27	98.46	59.62		
Co	0.47	0.49	0.36	0.39	0.28	0.26	0.21	0.19		
Ni	1.79	1.99	1.67	1.87	0.77	0.91	0.85	0.85		
Cu	1.53	2.15	4.09	1.37	0.79	0.89	0.94	0.89		
Zn	9.22	8.45	7.51	8.29	4.68	8.02	11.33	5.38		
As	2.61	2.00	1.50	1.61	2.09	3.95	1.69	2.33		
Se	2.84	4.69	7.41	6.78	4.09	7.19	2.48	1.79		
Cd	0.06	0.06	0.09	0.09	0.03	0.04	0.06	0.06		
Hg	0.26	1.91	1.20	0.31	0.48	2.67	0.03	0.06		
Pb	17.01	0.37	0.34	0.15	0.12	0.14	0.22	0.24		
9/17/03										
V	1.11	1.22	0.78	0.67	0.33	0.11	0.33	0.11		
Cr	1.33	1.33	1.00	0.78	0.67	0.56	0.67	0.67		
Mn	9.77	13.43	4.55	6.22	9.55	8.77	21.20	38.41		
Fe	209.24	210.55	160.60	161.10	114.57	105.83	98.99	59.90		
Co	0.56	0.56	0.44	0.44	0.33	0.33	0.22	0.22		
Ni	2.66	3.22	2.89	3.33	1.22	1.55	1.33	1.33		
Cu	1.78	2.44	4.44	1.55	0.89	1.00	1.11	1.00		
Zn	8.77	7.99	7.10	7.88	4.44	7.66	10.77	5.11		
As	2.00	1.55	1.11	1.22	1.55	3.00	1.33	1.78		
Se	2.55	4.22	6.66	6.11	3.66	6.44	2.22	1.55		

Table 12.2. (*Continued*)

	TRA	TRB	BGA	BGB	CVA	CVB	MLA	MLB	CCC	CMC
Cd	0.11	0.11	0.11	0.11	0.00	0.00	0.11	0.11		
Hg	0.33	2.00	1.33	0.33	0.56	2.89	0.03	0.07		
Pb	12.93	0.30	0.30	0.10	0.10	0.10	0.20	0.20		
11/27/03										
V	3.72	3.96	2.64	2.30	0.91	0.47	0.85	0.35		
Cr	11.62	12.76	8.82	6.58	4.14	1.07	0.46	0.36		
Mn	14.81	20.05	7.06	9.51	14.44	13.33	31.52	56.54		
Fe	311.77	313.68	239.24	240.03	170.70	157.61	147.41	89.27		
Co	0.27	0.29	0.21	0.23	0.16	0.15	0.12	0.11		
Ni	2.77	3.39	3.06	3.44	1.25	1.68	1.37	1.40		
Cu	6.48	8.97	16.32	5.74	3.27	3.68	3.93	3.55		
Zn	16.82	16.20	13.57	16.08	7.77	18.23	24.43	11.97		
As	47.28	36.22	27.28	29.18	37.84	71.62	30.71	42.34		
Se	3.28	5.42	8.56	7.82	4.73	8.30	2.86	2.06		
Cd	0.25	0.28	0.26	0.24	0.08	0.09	0.08	0.07		
Hg	1.02	6.82	4.13	1.20	1.66	9.37	<0.01	<0.01		
Pb	31.82	50.30	13.22	12.39	5.30	4.75	6.76	8.83		
1/9/04										
V	1.98	2.11	1.43	1.26	0.55	0.33	0.52	0.27		
Cr	2.92	3.20	2.22	1.66	1.05	0.28	0.13	0.11		
Mn	3.64	4.83	1.87	2.43	3.55	3.30	7.45	13.16		
Fe	415.67	418.22	318.96	320.02	227.58	210.13	196.54	119.02		
Co	0.28	0.30	0.22	0.23	0.17	0.16	0.13	0.12		
Ni	1.86	2.27	2.05	2.31	0.84	1.13	0.92	0.94		
Cu	1.45	2.05	3.90	1.31	0.75	0.85	0.90	0.84		
Zn	8.94	8.19	7.28	8.03	4.54	7.77	10.99	5.22		
As	2.21	1.70	1.28	1.37	1.77	3.35	1.44	1.98		
Se	3.19	5.26	8.31	7.60	4.59	8.06	2.78	2.01		
Cd	0.08	0.09	0.09	0.08	0.03	0.03	0.03	0.02		
Hg	0.35	2.37	1.43	0.42	0.58	3.26	0.04	<0.01		
Pb	16.68	0.36	0.33	0.14	0.12	0.14	0.22	0.24		
5/23/04										
V	1.56	1.66	1.13	0.99	0.43	0.26	0.41	0.21		
Cr	2.23	2.45	1.69	1.27	0.80	0.22	0.10	0.08		
Mn	3.04	4.04	1.56	2.03	2.97	2.76	6.23	11.00		
Fe	326.29	328.30	250.38	251.21	178.65	164.95	154.28	93.43		
Co	0.22	0.23	0.16	0.18	0.13	0.12	0.10	0.09		
Ni	1.36	1.66	1.49	1.68	0.61	0.82	0.67	0.69		
Cu	1.10	1.54	2.94	0.99	0.57	0.64	0.68	0.64		
Zn	5.69	5.48	4.59	5.44	2.63	6.16	8.26	4.05		
Se	2.30	3.79	5.99	5.48	3.31	5.81	2.00	1.45		
Cd	0.07	0.07	0.05	0.06	0.02	0.02	0.01	0.02		
Hg	0.28	2.05	1.28	0.33	0.52	2.85	0.03	0.01		
Pb	11.50	0.25	0.23	0.10	0.08	0.10	0.15	0.17		

nd, not defined.

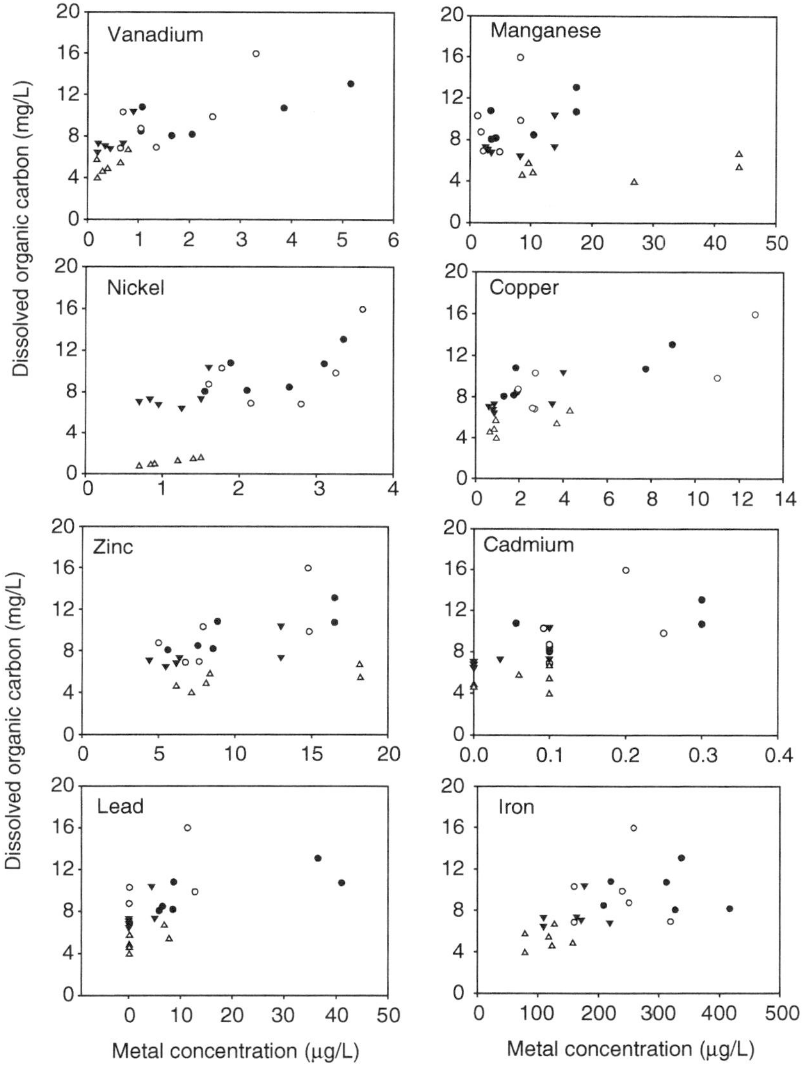

Figure 12.5. Plots of some selected elements vs. dissolved organic carbon for Trace Creek (solid circles), Begley Creek (open circle), Cove Creek (solid triangle), and Mill Creek (open triangle). Each point on the scatter plot is a mean from four sample measurements per sampling site per sampling occasion.

concentration of Ni in Begley Creek was consistently higher than those of Trace, Cove, and Mill creeks. Dissolved metals exhibited weak to strong positive relationships with DOC. The relationships were moderate to strong for V, Cu, Zn, and Pb as shown by the corresponding r^2-values (Table 12.3). Except for Mill Creek, significant temporal variations ($p<0.05$) in water pH were noted. DOC was inversely related to pH with weak to fairly strong relationships (Fig. 12.6).

12.4. Discussion

The pH and conductivity differences between the streams would be attributed to the geologic differences in their watersheds. The low pH values in Trace and Begley creeks may be explained by the dominance of organic-rich clay soils and unweathered materials derived from black shales in their watersheds. Low pH values have been encountered in streams whose watersheds are dominated by the organic-rich black shales (Loukola-Ruskeeniemi, 1994; Jambor et al., 2000; Tuttle et al., 2001; Foley et al., 2002; Hammarstrom et al., 2003). Overall, these studies noted that the observed low pH in streams draining shales was related to weathering and oxidation of the sulfide minerals in these rocks. Hammarstrom et al. (2003) observed that physical and chemical weathering of metal-enriched rocks not only contributes acidity into the streams but also metals, sediments, and secondary minerals which may sequester metals. The high conductivity values in the black shale-draining streams would be due to the presence of easily ionized materials in the watersheds. The notable conductivity increases during baseflow were attributed to groundwater, which derives its high conductivity signature from the black shale bedrock through which it flows. Whereas our conductivity values for streams flowing through blacks shales are 3 to 4 times higher, our pH values are comparable to those measured in Beech Flats Creek by Hammarstrom et al., 2003. In general bedrock type can significantly influence water quality as was shown by Flum and Nodvin (1995) at the Great Smoky Mountains National Park.

Time-series results (Fig. 12.2, 12.5; Table 12.2) reveal marked spatial and temporal variations in both dissolved organic carbon and dissolved metals. The DOC and metal concentrations in Trace, Begley, and Cove creeks, whose watersheds are dominated by black shales were significantly higher than those of the limestone-draining Mill Creek (Table 12.2). The highest and lowest V, Cu, Mn, Cr, Zn, and Pb concentrations coincided with the highest and lowest DOC values which in turn corresponded with the highest and lowest discharge, respectively. On average, the three

Table 12.3. R^2-square values of linear regression plots of dissolved metals vs. DOC; dissolved metals vs. pH; and dissolved metals and metalloids vs. discharge in Trace Creek, Begley Creek, Cove Creek, and Mill Creek in north-central Arkansas (USA)

	Metal	Trace Creek	Begley Creek	Cove Creek	Mill Creek	Metal	Trace Creek	Begley Creek	Cove Creek	Mill Creek
DOC										
	V	0.58	0.63	0.65	0.50	Zn	0.69	0.49	0.50	0.53
	Cr	0.48	0.52	0.75	0.24	As	0.49	0.68	0.84	0.45
	Mn	0.45	0.31	0.34	0.21	Se	0.71	0.58	0.67	0.52
	Co	0.10	0.15	0.10	0.15	Cd	0.48	0.31	0.53	0.11
	Ni	0.47	0.30	0.35	0.89	Pb	0.59	0.46	0.43	0.43
	Cu	0.70	0.61	0.60	0.53					
pH										
	V	0.84	0.71	0.61	0.63	Zn	0.63	0.59	0.65	0.28
	Cr	0.77	0.68	0.35	0.33	As	0.71	0.66	0.67	0.33
	Mn	0.33	0.34	0.23	0.61	Se	0.76	0.66	0.66	0.51
	Co	0.27	0.13	0.30	0.25	Cd	0.57	0.33	0.51	0.38
	Ni	0.67	0.38	0.61	0.51	Pb	0.64	0.49	0.38	0.29
	Cu	0.66	0.65	0.51	0.39					
Discharge										
	V	0.71	0.70	0.65	0.58	Zn	0.59	0.67	0.61	0.56
	Cr	0.67	0.69	0.64	0.14	As	0.66	0.69	0.69	0.63
	Mn	0.33	0.35	0.36	0.31	Se	0.66	0.59	0.64	0.33
	Co	0.12	0.10	0.09	0.34	Cd	0.33	0.56	0.65	0.17
	Ni	0.48	0.41	0.38	0.45	Pb	0.49	0.48	0.50	0.45
	Cu	0.65	0.65	0.66	0.61					

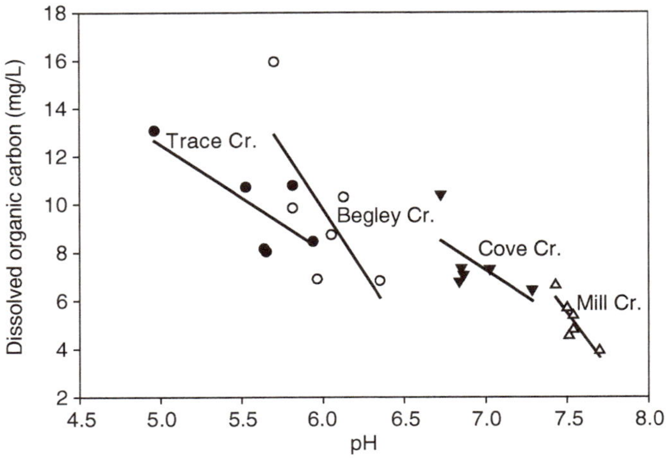

Figure 12.6. Linear regression plots of dissolved organic carbon vs. pH for Trace Creek (solid circle; $y = 34.6–4.4x$; $r^2 = 0.56$), Begley Creek (open circle; $y = 72.1–10.4x$; $r^2 = 0.51$), Cove Creek (solid triangle; $y = 38.4–4.4x$; $r^2 = 0.38$), and Mill Creek (open triangle; $y = 73.9–9.1x$; $r^2 = 0.73$). Each point on the scatter plot is a mean from four sample measurements per sampling site per sampling occasion.

streams draining through black shales had significantly higher dissolved metal levels than that draining limestone. Elevated metal concentrations have been encountered in soils and sediments derived from black shales as well as in streams whose watersheds are dominated by black shales (Coveney and Martin, 1983; Kim and Thornton, 1993; Chon et al., 1996; Loukola-Ruskeeniemi et al., 1998; Hammarstrom et al., 2003; Ogendi et al., 2004a, b). The elevated metal concentrations in Trace, Begley, and Cove creeks would be linked to weathering of the metal-enriched black shales over which these streams flow. Foley et al. (2002) attributed the high concentrations of As, Cu, and Mn in shale-draining streams in the Great Smoky Mountains National Park to near-surface reactions of sulfide minerals and water. Similar observations were made by Seal et al. (2000), and Hammarstrom et al. (2003). Short term perturbations in water quality due to acidity and elevated metal concentrations in creeks were also observed (Hammarstrom and Smith, 2002). Other studies that have documented elevated metal contents in surface waters draining pyritic-rich rocks include Bacon and Maas (1979) and Trumpf et al. (1979). In their study, Trumpf et al. (1979) found out that mass mortalities of native brook trout over an eight-km distance was due to metal toxicity notably aluminum. Ground waters in areas underlain by black shales may also be enriched in dissolved metals. A study by Ayotte et al. (1999) of shale bedrock wells in the Northern Appalachians of Maine and New Hampshire found that groundwaters contained elevated concentrations of dissolved metalloids (e.g. As) that exceeded the US EPA guidelines for drinking well water. The concentrations of Cr, Cu, Hg, Pb, and Se in streams draining black shales either approached or exceeded US EPA's metals criteria for surface waters (US EPA, 2002b). The elevated metals and metalloids in Trace, Begley, and Cove creeks during baseflow may be attributed to the groundwater inputs that contain these elements derived from soils and rocks through which the water flows. Ogendi et al. (2004b) measured between 4 and 10 ppm, and between 0.1 and 0.8 ppm of Hg in bedrock and sediments collected from streams draining black shales in the study area. Weathering of metal-enriched shales may thus lead to increased loading of toxic metals in surface waters. Enrichments of Hg up to 7.5 ppm in Proterozoic black shales have been encountered (Loukola-Ruskeeniemi, 1990). Chon et al. (1996) also encountered enrichments of AS, Mo, and Se in soils developed from black shales. The high concentrations of Fe in the black shale-draining streams are an inherent signature of the pyrite-rich black shales in the watersheds. Oxides of Mn play a significant role in metal sorption. Fe may alter the concentrations and partitioning of Co, Cu, Ni, and Zn, thus modifying metal bioavailability and toxicity to aquatic organisms (Jenne, 1968; Warren and Zimmerman, 1994; Erickson et al., 1996). Under aerobic conditions oxides of Fe and Mn can efficiently act as complexing ligands for metals (Hare et al., 1994; Erickson et al., 1996; Di Toro et al., 2001).

The study demonstrated that bedrock type and the amount of precipitation in the watersheds were strong predictors of the amount of DOC and dissolved metals in surface waters. On average, DOC concentrations in Trace Creek and Begley Creek were two times higher than those of Mill Creek for all the sampling occasions. Whereas the DOC-discharge correlations were very strong, the DOC-conductivity relationships were moderate to strong in the study streams. DOC levels in the streams draining black shales (with mean DOC ranging from 7 to 16 $mg\,l^{-1}$) are significantly higher ($p \leq 0.05$) than those of typical warm temperate streams (mean = 7; range = 3–15 $mg\,l^{-1}$) (Mulholland and Kuenzler, 1979; Thurman, 1985; Meybeck, 1993; Shafer et al., 1997). The higher DOC concentrations in our study sites may partly be explained by the presence of the organic-rich black shales in the watersheds. Black shales are rich in organic carbon (Petsch et al., 2000) and therefore may explain the significantly higher DOC in Trace, Begley, and Cove creek surface waters. In contrast, the DOC levels in Mill Creek are within the range of estimates for warm temperate streams (Thurman, 1985). Bedrock differences in the catchments appear to account for the spatial DOC differences because the vegetation types and principal land cover are the same for these watersheds. Temporal DOC variations are strongly correlated with changes in stream discharge (e.g. this study; Cauwet and Meybeck, 1987; Moore, 1989). The DOC data from this study were strongly positively correlated with the observed stream discharge with low- and high-DOC concentrations being encountered during baseflow and stormflow conditions, respectively. During storm events, DOC is likely accumulated by waters in contact with the organic matter-rich shale rock materials and soils, and then transported to the adjacent streams (Hendershot et al., 1992; Peters et al., 1995). The inverse relation between DOC and conductivity suggest that conductivity may indirectly influence metal transport in these streams (Fig. 12.4). The observed decline in conductivity with increased discharge and DOC levels suggests that DOC–metal complexes may be flushed from the watershed and banks into streams augmenting baseflow contents. Similarly, the strong positive linear relationships between discharge and DOC imply that significantly more dissolved metals are transported during peak discharge than during baseflow. Our findings are in agreement with similar studies by Depetris and Paolini (1991), Cauwet and Sidorov (1996), and Boyer et al. (1997) that also showed positive linear relations between discharge and the concentrations of DOC, with maximum DOC levels coinciding with peak discharge.

Metal–DOC relationships ranged from weak (Co, Cd, and Fe) to strong (Cu, Ni, V, and Pb) in the streams draining black shales. This tendency for dissolved metals to increase with increasing DOC concentration in the streams draining black shales supports the supposition that DOC is a major transport vector for dissolved metals. In Trace and Begley creeks, relatively high-DOC levels ranging from 8 to 16 $mg\,l^{-1}$ in relatively low pH environments (pH range: 4.8–5.8) were noted. Higher concentrations of DOC and dissolved metals in streams draining black shale than in streams draining limestone are most likely related to differences in lithology with lithology similarly influencing pH. These findings suggest that black shales in this region may be undergoing more intense chemical weathering as evidenced by higher concentrations of DOC and acid-generating potential, whereas the limestone buffers (e.g. Walker and Webber, 1979).

12.5. Conclusions

Our study found that the difference in lithology was a significant predictor of the observed spatial variations in DOC concentrations in headwater bedrock streams. Streams draining black shale watersheds transport significantly higher amounts of DOC than those draining over limestones thereby emphasizing the important role of geology in determining the amount of DOC in streams. Second, all streams exhibited a DOC maxima and minima that coincided with a stormflow and a baseflow, respectively. The watershed soils and/or rock surfaces are significant contributors of DOC to streams particularly during stormflow conditions. Third, our results indicate that the total variation in DOC in streams can be explained by discharge. It is also evident from this study that black shales are enriched in metals, which are transported in significantly large amounts during stormflow events as compared to the baseflow conditions. Oxidative weathering of the Mississippian Fayetteville Shale plays a crucial role in the release of metals that are subsequently transported into surface and groundwaters. Findings from numerous studies strongly suggest that DOC plays a significant role in altering the potential toxicity of metals to aquatic organisms (Diamond et al., 1997; Di Toro et al., 2001). The bioavailable fraction of a metal may be decreased significantly through complexation by DOC resulting in a reduction in the overall impact of metals on aquatic organisms (Newman and Jagoe, 1994; Erickson et al., 1996; Di Toro et al., 2001; Heijerick et al., 2003). The bioavailable fraction of the metal has been shown to correspond to the observed metal toxicity to aquatic organisms (Di Toro et al., 2001; De Schamphelaere and Janssen, 2004). Owing to the strong positive correlations between DOC and dissolved metals, we envisage low-metal impact upon resident aquatic communities in the streams draining black shales in this region. Thus, any attempt to quantify metals and predict their effect on aquatic biota in streams draining black shales should consider DOC levels in addition to other physical and chemical variables.

ACKNOWLEDGMENTS

This research was supported by a National Science Foundation grant (EAR 0233661—Hannigan and Farris). We are grateful to the personnel of the Water-Rock-Life Laboratory at Arkansas State University for their assistance in the stream water sample collection and processing including metal and dissolved organic carbon analyses. We appreciate the technical support we received from Dr. Ben Rougeau, Dr. Yingtao Chai, and Mr. David Clark while conducting the trace metal and DOC analyses on the DRC-ICP-MS and the Dorhman TOC analyzer. Finally, we wish to sincerely thank our two reviewers, Drs. Robert Seal (USGS, National Center, Reston, VA) and William Brumbaugh (USGS, Columbia

Environmental Research Center, MO) for their insightful comments and constructive critiques.

REFERENCES

Allan, J.D., 1995. Stream ecology: Structure and function of running waters. Chapman and Hall, New York.

Allen, H.E., Hansen, D.J., 1996. The importance of trace metal speciation to water quality criteria. Water Environ. Res. 68, 42–54.

American Public Health Association, American Water Works Association, Water pollution control Federation, 1998. Standard Methods for the Examination of Water and Wastewater, Twentieth ed. Washington, DC.

Apte, S.C., Batley, G.E., 1995. Trace metal speciation of labile chemical species: Other methods. In: Tessier, A., Turner, D.R. (Eds.), Metal Speciation and Bioavailability in Aquatic Systems. Wiley, Chichester.

Arkansas Geological Commission, 2005. Available at http://www.state.ar.us/agc/stratigr.html Accessed September 27, 2005.

Ayotte, J.D., Nielsen, M.G., Robinson, G.R., Jr., Moore, R.B., 1999. Relation of arsenic, iron, and manganese in ground water to aquifer type, bedrock lithogeochemistry, and land use in the New England Coastal Basins. U.S. Geological Survey Water Resources Investigations Report 99-4162.

Bacon, J.R., Maas, R.P., 1979. Contamination of Great Smoky Mountains trout streams by exposed Anakeesta formation. J. Environ. Qual. 8, 538–543.

Bishop, K., Pettersson, C., 1996. Organic carbon in the boreal spring flood from adjacent subcatchments. Environ. Int. 22, 535–540.

Blette, V.L., Newton, R.M., 1996. Application of integrated lake-watershed acidification study model to watershed liming at Woods Lake, New York. Biogeochemistry 32, 363–383.

Boyer, E.W., Hornberger, G.M., Bencala, K.E., McKnight, D.M., 1997. Response characteristics of DOC flushing in an alpine catchment. Hydrol. Process. 11, 1635–1647.

Cauwet, G., Meybeck, M., 1987. Seasonal fluctuations of carbon levels in a temperate river: The Loire (France). In: Degens, E.T., Kempe, S., Weibin, G. (Eds.), Transport of Carbon and Minerals in Major World Rivers. Pt. 4. Mitt.. Geol.-Paläont. Inst. Univ., Hamburg, SCOPE/UNEP Sonderbd.

Cauwet, G., Sidorov, I., 1996. The biogeochemistry of the Lena River: Organic carbon and nutrients distribution. Mar. Chem. 53, 211–227.

Chon, H.T., Cho, C.H., Kim, K.W., Moon, H.S., 1996. The occurrence and dispersion of potentially toxic elements in areas covered with black shales and slates in Korea. Appl. Geochem. 11, 69–76.

Coveney, R.M. Jr., Martin, S.P., 1983. Molybdenum and other heavy metals of the Mecca Quarry and Logan Quarry shales. Econ. Geol. 78, 132–149.

Davies, C.M., Apte, S.C., Johnstone, A.L., 1998. A bacterial bioassay for the assessment of copper bioavailability in freshwaters. Environ. Toxicol. Water Qual. 13, 263–271.

Deacon, J.R., Stephens, V.C., 1998. Trace metals in streambed sediment and fish liver at selected sites in the upper Colorado River Basin, Colorado, 1995–1996. U.S.G.S. Water-Resources Investigations Report 98-4124, p. 19.

Degens, E.T., S. Kempe and A.S. Naidu, (Eds.), 1988, Transport of Carbon and Minerals in Major World Rivers, Lakes and Estuaries, Pt. 5. Mitt.. Geol.-Paläont. Inst. Univ., Hamburg, SCOPE/UNEP Sonderbd.

Depetris, P.J., Paolini, J.E., 1991. Biogeochemical aspects of South American rivers: The Parana and Orinoco. In: Degens, E.T., Kempe, S., Richey, J.E. (Eds.), Biogeochemistry of Major World Rivers, Scope Report 42. Wiley, Chichester, UK.

De Schamphelaere, K.A.C., Janssen, C.R., 2004. Effects of dissolved organic carbon concentration and source, pH, and water hardness on chronic toxicity of copper to *Daphnia magna*. Environ. Toxicol. Chem. 23, 1115–1122.

Diamond, J.M., Koplish, D.E., McMahon, J. III, Rost, R., 1997. Evaluation of water-effect ratio procedure for metals in a riverine system. Environ. Toxicol. Chem. 16, 509–520.

Di Toro, D.M., Allen, H.E., Bergman, H.L., Meyer, J.S., Paquin, P., Santore, R.C., 2001. Biotic ligand model of the acute toxicity of metals. 1. Technical basis. Environ. Toxicol. Chem. 20, 2383–2396.

Di Toro, D.M., Allen, H.E., Bergman, H.L., Meyer, J.S., Santore, R.C., Paquin, P., 2000. The Biotic Ligand Model: A Computational Approach for Assessing the Ecological Effect of Metals in Aquatic Systems. The Intl. Copper Assoc., New York.

Droppo, I.G., Jaskot, C., 1994. The impact of river transport characteristics on contaminant sampling error and design. Environ. Sci. Technol. 29, 161–170.

Elder, J.F., 1988. Metal Biogeochemistry in Surface-Water Systems—A Review of Principles and Concepts. U.S. Geological Survey Circular 1013.

Erickson, R.J., Benoit, D.A., Mattson, V.R., Nelson, H.P., Leonard, E.N., 1996. The effects of water chemistry on the toxicity of copper to fathead minnows. Environ. Toxicol. Chem. 15, 181–193.

Flum, T., Nodvin, S.C., 1995. Factors affecting streamwater chemistries in the Great Smoky Mountains, USA. Water Air Soil Pollut. 85, 1707–1712.

Foley, N.K., Shultz, A.P., Southworth, C.S., Ayuso, R.A., Robinson, G.R., Seal II, R.R., 2002. Geochemical, mineralogical, and environmental aspects of metamorphosed metalliferous black shales of the central Appalachians, with comparisons to metalliferous shales of central Maine. U.S. Geological Survey Open-file Report 01-406.

Gardner, M.J., Gunn, A.M., 1989. The effect of natural ligands on trace metal partitioning. Chemosphere 19, 1251–1259.

Hammarstrom, J.M., Smith, K.S., 2002. Geochemical and mineralogic characterization of solids and their effects on waters in metal-mining environments. In: Seal II, R.R., Foley, N.K. (Eds.), Progress on Geoenvironmental Models for Selected Mineral Deposit Types: U.S Geological Survey Open-File Report 02-0195, pp. 8–54.

Hammarstrom, J.M., Seal, R.R. II, Meier, A.L., Jackson, J.C., 2003. Weathering of sulfidic shale and copper mine waste: Secondary minerals and metal cycling in Great Smoky Mountains National Park, Tennessee, and North Carolina, USA. Environ. Geol. 45, 35–57.

Hamilton, S.J., Waddell, B., 1994. Selenium in eggs and milt of razorback sucker (*Xyrauchen texanus*) in the middle Green River. Arch. Environ. Contam. Toxicol. 27, 195–201.

Hannigan, R.E., 1997. Trace and major elements in sedimentary and igneous processes: REE geochemistry of black shales and MORB and major element chemical variation in plume-generated basalts. PhD. Dissertation, University of Rochester.

Hannigan, R.E., Basu, A.E., 1997. Late diagenetic trace element remobilization in organic-rich black shales of the Taconic foreland basin of Quebec, Ontario, and New York. In: Shieber, J., Zimmerle, W., Sethi, P. (Eds.), Mudstones and Shales: Recent Progress in Shale Research. Schwizerbart'sche Verlagsbuchhandlung, Berlin, Germany, pp. 209–234.

Hare, L., Carignan, R., Huerta-Diaz, M.A., 1994. An experimental study of metal toxicity and accumulation by benthic invertebrates: Implications for acid volatile sulfide (AVS) mode. Limnol. Oceanogr. 39, 1653–1668.

Heijerick, D.G., Janssen, C.R., De Coen, W.M., 2003. The combined effects of hardness, pH, and dissolved organic carbon on the toxicity of Zn to *D. magna*: Development of a surface response model. Arch. Environ. Contam. Toxicol. 44, 210–217.

Hendershot, W.H., Savoie, S., Courschesne, F., 1992. Simulation of stream-water chemistry with soil solution and ground water flow contributions. J. Hydrol. 136, 237–252.

Hudson, M.R., Kyle, E.M., Pezzutti, D., 2001. Geologic map of the Jasper Quadrangle, Newton and Boone Counties, Arkansas. Miscellaneous field studies map MF-2356. U.S. Geological Survey, Denver, CO.

Hurley, J.G., Shafer, M.M., Cowell, S.E., Overdier, J.T., Hughes, P.E., Armstrong, D.E., 1996. Trace metal assessment of Lake Michigan tributaries using low-level techniques. Environ. Sci. Technol. 30, 2093–2098.

Jagoe, C.H., Bryan, L.A. Jr., Brant, H.A., Murphy, T.M., Brisbin, L.I., 2002. Mercury in bald eagle nestlings from South-Carolina, USA. J. Wildlife Dis. 38, 706–712.

Jambor, J.L., Nordstrom, D.K., Alpers, C.N., 2000. Metal-sulfates salts from sulfide mineral oxidation. In: Alpers, C.N., Jambor, J.L., Nordstrom, D.K. (Eds.), Sulfate Minerals—Crystallography, Geochemistry, and Environmental Significance. Reviews in Mineralogy and Geochemistry, Mineralogical Society of America, Washington, DC, USA, Vol. 40, pp. 303–350.

Jenne, E.A., 1968. Controls on Mn, Fe, Co, Ni, Cu, and Zn concentrations in soils and water: The significance of hydrous Mn and Fe oxides. Adv. Chem. Ser. 73, 337–387.

Kaplan, L.A., Newbold, J.D., 1993. Biogeochemistry of dissolved organic carbon entering streams. In: Ford, T.E. (Ed.), Aquatic Microbiology: An Ecological Approach. Blackwell Scientific Publishers, Boston, USA.

Kim, K.W., Thornton, I., 1993. Influence of uraniferous black shales on cadmium, molybdenum and selenium in soils and crop plants in the Deog-Pyong area of Korea. Environ. Geochem. Health 15, 119–133.

Kim, S.D., Ma, H., Allen, H.E., Cha, D.K., 1999. Influence of dissolved organic matter on the toxicity of copper to *Ceriodaphnia dubia*: Effect of complexation kinetics. Environ. Toxicol. Chem. 11, 2433–2437.

Loukola-Ruskeeniemi, K., 1990. Metalliferous black shales—A probable source of mercury in pike in lake Kolmisoppi Sotkamo, Finland. Bull. Geol. Soc. Finland 62, 167–175.

Loukola-Ruskeeniemi, K., 1994. Environmental geochemistry of Proterozoic black shales in eastern Finland: A pilot study. In: Bornhorst, T.J., McDowell, S.D. (Eds.), Institute on Lake Superior Geology. Proceedings of 40th annual meeting, Michigan Technological University, Houghton, MI.

Loukola-Ruskeeniemi, K., Utela, A., Tenhola, M., Paukola, T., 1998. Environmental impact of metalliferous black shales at Talvivaara in Finland, with indication of lake acidification 9000 years ago. J. Geochem. Explor. 64, 395–407.

Meybeck, M., 1993. C, N, P, and S in rivers: From sources to global inputs. In: Wollast, R., Machenzie, F.T., Chou, L. (Eds.), Interaction of C, N, P, and S Biogeochemical Cycles on Global Change. Springer-Verlag, pp. 163–193.

Meyer, J.L., Eggert, S.L, Wallace, J.B., Webster, J.R., 2004. Headwater streams as sources of dissolved organic carbon for downstream ecosystems. Presented at the NABS Annual meeting, Vancouver, BC, 2004 in Carbon and Biofilm Dynamics.

Mize, S.V., Deacon, J.R., 2002. Relations of benthic macroinvertebrates to concentrations of trace metals in water, stream sediments, and transplanted bryophytes and stream habitat conditions in non-mining and mining areas of the Upper Colorado River Basin, Colorado, 1995–1998. U.S. Geological Survey: Water Resources Investigations Report 02-4139.

Moore, M.R., 2004. A commentary on the impacts of metals and metalloids in the environment upon the metabolism of drugs and chemicals. Toxicol. Lett. 148, 153–158.
Moore, T.R., 1989. Dynamics of dissolved organic carbon in forested and disturbed catchments, Westland, New Zealand 1. Maimai. Water Resour. Res. 25, 1321–1330.
Mounier, S., Braucher, R., Benaim, J.Y., 1999. Differentiation of organic matter's properties of the Rio Negro basin by cross-flow ultrafiltration and UV spectro-fluorescence. Water Res. 33, 2363–2373.
Mulholland, P.J., Kuenzler, E.J., 1979. Organic carbon from upland and forested wetland watersheds. Limnol. Oceanogr. 24, 960–966.
Murthy, R., 2003. Trace and major element fractionation during diagenesis of phosphate nodules. Arkansas State University M.S. Thesis.
Murthy, R., Kidder, D., Mapes, R., Hannigan, R., 2004. Rare-earth chemistry of Mississippian-age phosphate nodules in the Fayetteville Shale of Oklahoma and Arkansas. Environ. Geosci. 11(2), 99–111.
Newman, M.C., Jagoe, C.H., 1994. Ligands and bioavailability of metals in aquatic environments. In: Hamelink, J.L., Landrum, P.F., Bergman, H.L., Benson, W.H. (Eds.), Bioavailability: Physical, Chemical and Biological Interactions. SETAC special publications series, Lewis Boca Raton FL, USA, pp. 39–62.
Ogendi, G.M., Farris, J.L., Hannigan, R.E., 2004a. Black shale trace metal concentrations and toxicity: Preliminary findings. In: Wanty, R.B., Seal II, R.R. (Eds.), Proceedings of the 11th International Symposium on Water Rock Interaction, Vol. 2. Taylor and Francis Group plc, London, UK, pp. 1359–1362.
Ogendi, G.M., Hannigan, R.E., Farris, J.L., Smith, D., 2004b. The impact of black shale weathering on sediment quality. J. Ark. Acad. Sci. 58, 84–90.
Pasava, J., Maclnnis, I., Graves, M., Zentilli, M., 1995. Black slates—A source of acid drainage at the Halifax International Airport, Nova Scotia, Canada. In: Pasava, J., Kribek, B., Zak, K. (Eds.), Mineral Deposits: From Their Origin to Their Environmental Impacts. Proceedings of the Third Biennial SGA Meetings, 28–31 August, Prague, Balkema, Rotterdam.
Peters, D.L., Buttle, J.M., Taylor, C.H., Lazerte, B.D., 1995. Runoff production in a forested shallow soil, Canadian shield basin. Water Resour. Res. 31, 1291–1304.
Petsch, S.T., Berner, R.A., Eglington, T.I., 2000. A field study of the chemical weathering of ancient sedimentary organic matter. Org. Geochem. 31, 475–487.
Phoenix 8000TM User Manual, 1998. Rev. C. Tekmar Dohrmann, Mason, OH.
Rember, R.D., Trefry, J.H., 2004. Increased concentrations of trace metals and organic carbon during snowmelt in rivers of the Alaskan Arctic. Geochem. Cosmochim. Acta 68, 477–489.
Riley, M.R., Boesewetter, D.E., Kim, A.M., Sirvent, F.P., 2003. Effects of metals Cu, Fe, Ni, V, and Zn on rat lung epithelial cells. Toxicology 190, 171–184.
Seal II, R.R., Wanty, R.B., Foley, N.K., 2000. Introduction to geoenvironmental models of mineral deposits. In Geoenvironmental Analysis of Ore Deposits, Chapter 1. Short Course at the Fifth International Conference on Acid Mine Drainage, May 21, Denver, CO, p. 8.
Shafer, M.M., Overdier, J.T., Hurley, J.P., Armstrong, D., Webb, D., 1997. The influence of dissolved organic carbon, suspended particulates, and hydrology on the concentration, partitioning and variability of trace metals in two contrasting Wisconsin watersheds (U.S.A.). Chem. Geol. 136, 71–97.
Shelton, L.R., Capel, P.D., 1994. Guidelines for collecting and processing samples from streambed sediment for analysis of trace elements and organic contaminants for the National Water Quality Assessment Program. USGS Open-File Report 94-458.

Sherrell, R.M., Ross, J.M., 1999. Temporal variability of trace metals in New Jersey Pinelands streams: Relationships of discharge and pH. Geochem. Cosmochim. Acta 19/20, 3321–3336.

Shiller, A.M., 1997. Dissolved trace elements in the Mississippi River: Seasonal, interannual and decadal variability. Geochem. Cosmochim. Acta 61, 4321–4330.

Shiller, A.M., Boyle, E.A., 1987. Variability of dissolved trace metals in the Mississippi River. Geochem. Cosmochim. Acta 51, 3273–3277.

Spitzy, A., Leenheer, J., 1988. Dissolved Organic in Rivers. SCOPE 42—Biogeochemistry of Major World Rivers. M. S. Swaminathan Research Foundation, Chennai, India.

Taylor, H.E., Garbarino, J.R., Brinton, T.I., 1990. The occurrence and distribution of trace metals in the Mississippi River and its tributaries. Sci. Total Environ. 97/98, 369–384.

Thurman, E.M., 1985. Organic Geochemistry of Natural Waters. Nijihoff/Junk, Dordrect.

Trumpf, W.F., Morgan, E.L., Herman, R., 1979. Man induced acid drainage impact on benthic macroinevertebrate communities in Great Smoky Mountans National Park. In: Lin, R.I. (Ed.), Conference on Scientific Research in the National Parks, Vol. l. National Park Service. Trans. Proc., Department of Interior, Washington, DC, USA, Vol. 5, pp. 643–647.

Tuttle, M.L.W., Goldhaber, M.B., Breit, G.N., 2001. Mobility of metals from weathered Black Shale: The role of salt efflorescences. Geol. Soc. Am. Abstracts with Programs 33, 191.

U.S. Environmental Protection Agency, 2002a. Short-term Methods for Estimating the Chronic Toxicity of Effluents and Receiving Waters to Freshwater Organisms, fourth ed., October 2002. U.S. Environmental Protection Agency Office of Water (4303T), Washington, DC.

U.S. Environmental Protection Agency, 2002b. National Recommended Water Quality Criteria. U.S. Environmental Protection Agency, Office of Water, EPA-822-R-02-047, Washington, DC.

Walker, D.A., Webber, P.J., 1979. Relationships of soil acidity and air temperature to the wind and vegetation at Prudhoe Bay, Alaska. Arctic 32, 224–236.

Warren, L.A., Zimmerman, A.P., 1994. Suspended particulate oxides and organic matter interactions in trace metal sorption reactions in a small urban river. Biogeochemistry 24, 21–43.

Developments in Environmental Science, Volume 5
D. Sarkar, R. Datta and R. Hannigan (Editors)

ISSN: 1474-8177/DOI:10.1016/S1474-8177(07)05013-9

Chapter 13

Mineral control of minor, trace and rare earth elements during black shale weathering at near-neutral pH

P.A. Abanda and R.E. Hannigan

Abstract

Mineral dissolution and alteration commonly influence the concentration of chemical species in natural water systems and consequently on water quality. This research looks at the role of mineral-water interactions during black shale weathering in influencing the chemical composition of streams and groundwaters with emphasis on minor and trace elements and the rare earth elements (REE). We performed batch weathering experiments under standard laboratory conditions standard laboratory conditions (standard ambient temperature (~25°C) and pressure (~100 kPa)) using three black shale samples from the Utica shale magnafacies of Quebec, Ontario and New York using buffered weathering solutions at pH 5 to determine the mineralogical controls of dissolved metal concentrations. R-mode factor analysis was used to describe the relationship among different elements (Ti, V, Cr, Ni, Zn, As, Sr, Ba, Th, U and the REE) in the weathering fluids. Five factors were extracted that accounted for 95% of the variance within the data. We interpret these five factors as representing the breakdown products of four major mineral phases (phosphates, carbonates, silicates and sulfides) in the black shale as well as the breakdown of organic matter. Our results also suggest that during weathering of black shales, the REE are fractionated leading to enrichment of the MREE (Sm-Dy) and HREE (Ho-Yb) relative to the LREE (La-Nd). We also observed that the average concentration of minor and trace elements and the REE in solution was highest during the first 24 h of reaction and decreased slowly with time. We interpret the changes in dissolved metal concentrations as the result of rapid breakdown or dissolution of easily weathered mineral phases, partial oxidation of organic matter and to the desorption of metals that are loosely bound to exchange sites on clay minerals. The decrease in metal concentration with time may be related to processes such as precipitation of secondary phases and sorption as well as dissolution of easily weathered mineral phases.

13.1. Introduction

Variations in stream water composition in many areas have been attributed to different weathering rates of the surrounding geological materials (Horton et al., 1999; Pretti and Stewart, 2002). Natural weathering of black shale may lead to increased abundances of dissolved minor and trace elements in solution, some of which at sufficient concentrations may impair water quality (Isaacs et al., 1999; Coveney, 2004). The minerals that weather to release these elements (phosphates, carbonates, sulfides and silicates) including organic matter are associated with the host rock and may weather at varying rates. Most laboratory studies of chemical weathering have looked at the weathering of pure mineral phases (Stumm, 1990; Lasaga et al., 1994; Zhang and Bloom, 1999; White and Brantley, 2003). Laboratory studies of chemical weathering of whole rocks to study how minerals weather in concert and the contribution of individual minerals to the concentration of chemical species in natural systems are few. Most experimental work to simulate chemical weathering of rocks in the field has resulted in mineral weathering rates in the laboratory being several orders of magnitude higher than those experienced in the field (Malmstrom et al., 2000; White and Brantley, 2003). Several hypotheses have been advanced to explain the observed differences between mineral weathering rates in the laboratory and in the field. Despite the availability of data on the weathering rates of pure minerals in the laboratory and extension to field weathering rates, it remains unclear how minerals weather in concert in whole rocks especially in fine-grained sedimentary rocks rich in organic matter. Because the weathering of a multimineralic rock such as black shale involves a complex interplay of processes such as oxidation, dissolution, precipitation and hydrolysis, it is important for any conclusive study to involve detail mineralogical studies of the whole rocks, mass balance calculations to determine the percentage of individual mineral contribution to the total elemental concentration in solution and consequently to use the data in modeling field weathering of black shale and release of trace metals to neighboring systems.

Field studies of black shale weathering have shown that chemical breakdown of these rocks may lead to elevated concentrations of trace metals and increased acidity in associated aquatic systems (Piper and Isaacs, 1995; Lee et al., 1998; Peucker-Ehrenbrink and Hannigan, 2000; Dalai et al., 2001; Coveney, 2001; Tuttle et al., 2001; Tuttle and Breit.,

2004). Owing to a combination of reduced minerals like pyrite, oxidizable organic matter and easily weathered minerals like calcite and the presence of high levels of metals in many black shales, these rocks upon exposure to surface conditions (oxygen- and moisture-rich conditions), may break-down easily releasing high levels of associated trace metals. Additionally, because black shales are made up mostly of fine-grained clay-size particles, the minerals provide a large surface area to chemical attack.

Primary as well as secondary minerals in black shale differ in their degree of resistance to weathering. Mineral weatherability has been studied by Loughnan (1969) and Carrol (1970). Berner and Berner (1996) and Drever (1997) have compiled tables of mineral weatherability based on their resistance to weathering. Bearing in mind the mineralogy of our samples (section "Methods"), the weatherability of minerals will increase in the order of clay minerals < biotite < Ca plagioclase < calcite < pyrite. However, it is important to note that this order depends on other factors including particle grain size or surface area of minerals exposed as well as the presence or absence of crystal defects in minerals. The black shale samples used in this study are made up mostly of clay-sized particles, so the problem of grain size in determining mineral dissolution rates is not critical. However in organic rich sediments, the breakdown or oxidation of organic matter may play an important role in determining the rate of primary mineral dissolution and release of trace metals (Stillings et al., 1996; Drever and Stillings, 1997; Zhang and Bloom, 1999). Zhang and Bloom (1999) suggested that organic anions tend to increase the rate of silicate weathering because these anions are capable of forming complexes with framework metals thereby enhancing both the solubility of the metals and the dissolution of minerals.

This chapter reports the results of a two-week weathering experiment to determine the mineral control of minor, trace and rare earth elements (REE) during black shale weathering using buffered weathering solutions at pH 5. This study involves a controlled laboratory leaching experiments of three black shale samples from the Utica Shale Magnafacies (USM) of Quebec, Ontario and New York. Specifically, we looked at the mineral control of minor and trace elements including the REE during weathering of a fine-grained multimineralic rock (black shale) at pH 5. We investigated the behavior of the following elements including Ti, V, Cr, Ni, Zn, As, Sr, Ba, Th, U and the REE (La-Lu) during an experimental period lasting 13 days (312 h). Here, we look at how different minerals in black shale breakdown together and which minerals exert greater control on the chemistry of the resulting weathering solutions. Our objective was to study elemental association of most easily weathered minerals in black shales of different grades of thermal maturity. We used R-mode statistical analysis to describe the relationship among Ti, V, Cr, Ni, Zn, As, Sr, Ba, Th, U and the REE in order to determine the dominant mineral control of these elements in solution. The results presented in this chapter are in no way conclusive about how minerals weather in concert and the contribution of individual host mineral to the elemental content in solution during black shale weathering. However, this study provides important baseline data critical in understanding the contribution of easily weathered minerals in black shale to the composition of chemical species in natural systems associated with black shale. The study sheds some light on the major control (mineral) on the release of minor and trace elements during chemical breakdown of black shale at pH 5.

13.2. Geological setting

In this study, we focused our attention on the release of Ti, V, Cr, Ni, Zn, As, Sr, Ba, Th, U and the REE during weathering of time correlative Middle Ordovician Utica Shale magnafacies (USM) of Quebec, Ontario and New York (Hannigan and Basu, 1998). These units were deposited in the Taconic Foreland Basin and have been constrained in time by graptolite biostratigraphy and bentonite chronostratigraphy. There is evidence that these units share a common provenance (Lehmann et al., 1995). The magnafacies thermal grade varies geographically; Quebec (thermally immature: T_{max} 20–50°C), Ontario (thermally mature: T_{max} 50–140°C) and New York (thermally post-mature: T_{max}200°C) Hannigan and Basu (1998). We selected rocks of different grade of thermal maturity not only to show how minerals in black shale weather in concert but also to investigate the effect of thermal maturation on the mobility of minor and trace elements, especially the REE during chemical breakdown of these rocks.

13.3. Methods

13.3.1. Sample selection and preparation

Samples used in this study were obtained from drill cores. Samples were carefully selected to include samples with similar amounts of organic carbon, mineral carbon and total sulfur. The core samples were broken into smaller fragments with a hammer. We selected rock fragments that ranged between 1 and 2 mm in diameter and weighed about 2.5 g of each sample into a 100 ml PTFE bottle. We submerged the rock fragments in deionized water (18.3 M Ohm) and subjected them to ultrasound for about 20 min to remove fine surface particles. This exercise was repeated five times using fresh deionized water. It is worth mentioning that before the end of this process the rock fragments were reduced further into even smaller size fractions because of the fragile nature of these fine-grained sediments. Because of the brittle nature of the samples and the minute sizes of mineral grains, it was difficult to estimate the surface area exposed to weathering. Isolation of the $<2\,\mu m$ clay-size fraction was done according to the method of Poppe et al. (2001). A fraction of the sample was pulverized ($<63\,\mu m$) and used to isolate the clay fraction. We used acetic acid to leach carbonate minerals associated with our samples and sodium pyrophosphate was used to put the clay fraction into suspension. The supernatant solution containing the $<2\,\mu m$ clay fraction was siphoned off and centrifuged at high speed (~3000 rpm) for 10 min to obtain the clay fraction. This fraction was mounted on glass slides and air dried before carrying out X-ray diffraction analysis.

13.3.2. Dissolution experiments

Batch weathering experiments have been applied by other workers in the study of mineral dissolution (Holdren and Berner, 1979; Stidson et al., 1998; Zhang and Bloom, 1999). These authors used initially reacted samples to overcome the effects of rapid initial dissolution, which is usually an artifact of sample pre-treatment (surface strain, ultrafine particles and high-energy crystal defects). Our samples were not pre-treated because of their fragile nature and concerns with easily weathered minerals like carbonate and sulfides. Because our study was concerned about the contribution of easily weathered minerals to the composition of chemical species in solution, we think pre-treatment would have significantly altered the composition of sulfides and carbonate minerals. However, to curb the effects of ultrafine particles on dissolution rates, our samples were treated as outlined in section "Sample selection and preparation" (washed several times in deionized water in an ultrasound bath). We prepared buffer solutions pH 5 using trace metal grade reagents. To each pre-weighed 2.5 g washed samples, we added 25 ml of buffered solution (pH 5) in a 100 ml PTFE bottle. After adding 25 ml of the buffered weathering solution to each sample, the bottles were capped and placed on a Lab-Line® orbital shaker at low speed through the whole length of the experiment (13 days). The dissolution experiment was carried out under standard laboratory conditions (standard ambient temperature (~25°C) and pressure (~100 kPa)). At various times during the experimental period (24, 48, 96, 144, 168, 264, 288 and 312 h) we extracted the weathering solutions for chemical analysis. Each time we obtained a sample, the entire weathering solution was emptied into a clean 100 ml PTFE bottle and the solid sample was washed using deionized water before adding fresh buffer solution (25 ml). By replacing the solution each time we collected a sample, our intention was to minimize oversaturation of the weathering solution and the precipitation of secondary mineral phases. The extracted solution was filtered through 0.45 μm millipore microfilter and stored in a refrigerator at 4°C for chemical analysis.

13.3.3. Analytical methods

Minor, trace and REE concentrations were measured by dynamic reaction cell inductively coupled plasma mass spectrometry (DRCII ICP-MS; PerkinElmer) following EPA 200.8 method (USEPA, 1991). We prepared standards in a linear range from 1 to 100 $mg\,l^{-1}$. Each sample and standard contained 40 $ug\,l^{-1}$ ^{115}In and ^{209}Bi internal standards. USGS SDO-1 (Ohio Devonian shale sample) (Kane et al., 1990) was digested and prepared in similar fashion and analyzed alongside samples and blanks to ensure accuracy of measurements. Elemental concentrations were within 5–7% of the reported values of these standards. Relative standard deviations for sample concentrations are better than 5% for all analytes.

Clay analyses were carried out at Arkansas State University using Rigaku X-ray diffractometer operated at 40 kV and 20 mA with CuKα radiation following the method of Moore and Reynolds (1989) on isolated clay size (<2 μm) fraction mounted on glass slides. Major elements analyses was carried out at the SGS analytical laboratories in Ontario, Canada using X-ray fluorescence spectroscopy (method XRF100) with lowest detection limits of 0.1% for all majors and loss on ignition. Total carbon and total sulfur analysis was carried out using the Leco analyzer. Inorganic or mineral carbon was done by coulometry. Total organic carbon was estimated as the difference between total carbon and inorganic carbon.

13.3.4. Statistical approach

We subjected the geochemical data to statistical analyses (R-mode factor analysis) to study the controls of soluble elemental composition in the weathering fluids (Reeder et al., 1972). Factor analysis is a multivariate technique designed to analyze the interrelationships within a set of variables. Factors are constructed such that the overall complexity of data is reduced based on the interdependencies of the variables considered. The statistical package Minitab (Minitab for Windows 95/98/2000 and NT

Upgrade Version 13.30 by Minitab Inc.) was used for the analysis. Factor analyses present us with a set of factors that are used to explain the variance in the data collected. The number of factors considered was small enough to allow reasonable interpretation of our data. Based on the number of factors given, we used these factors to explain the source of all elements considered. Elemental association gives us information about possible controls of metal release during breakdown of the major easily weathered minerals. Loadings represent how each factor is associated with a particular variable and have numerical values that range between -1 and $+1$. Factors were adjusted using a varimax rotation. Varimax rotation is preferred because it ensures that one variable has high loading only on one factor and low or no loading on other factors (Reeder et al., 1972). Interpretation of the factors is based on rotated factors and rotated factor loadings.

13.4. Results

13.4.1. Mineralogy of samples

Earlier mineralogical studies by Hannigan and Basu (1998) revealed that the Utica shale samples are made up mainly of clay, 69%; quartz, 14%; calcite, 13%; K-feldspar; plagioclase–feldspar and biotite all constituting about 1% and diagenetic pyrite making up <0.5% in most of the samples. Clay mineralogical analysis revealed that the major clay minerals in the Utica shales include illite, smectite, mixed layer illite/smectite and minor amounts of chlorite and kaolinite (Hannigan and Basu, 1998). Mineralogical analysis of the samples used in this study support earlier results obtained by Hannigan and Basu (1998). The major clay minerals in our samples include illite, mixed layer illite/smectite, quartz and minor amounts of kaolinite (Fig. 13.1). Average organic carbon content in these shales is about 2 wt%. The organic matter is made up mostly of marine planktonic material since these rocks were deposited some 420 Ma ago before the proliferation of vascular land plants. The samples selected for this study contain similar amounts of organic carbon, total sulfur and inorganic carbon. Average organic carbon content in the samples was 5.2 wt%, total sulfur averaged 1.20 wt% and mineral carbon averaged 4.20 wt%. Average major oxides composition of the samples was $SiO_2 = 63.97$ wt%, $Al_2O_3 = 12.66$ wt%, $CaO = 3.42$ wt%, $MgO = 3.03$ wt%, $Na_2O = 1.34$ wt%, $K_2O = 2.28$ wt%, $Fe_2O_3 = 4.69$ wt%, $MnO = 0.09$ wt%, $TiO_2 = 0.51$ wt%, $P_2O_5 = 0.17$ wt%, $Cr_2O_3 = 0.03$ wt% and $LOI = 11.70$ wt%. The chemical index of

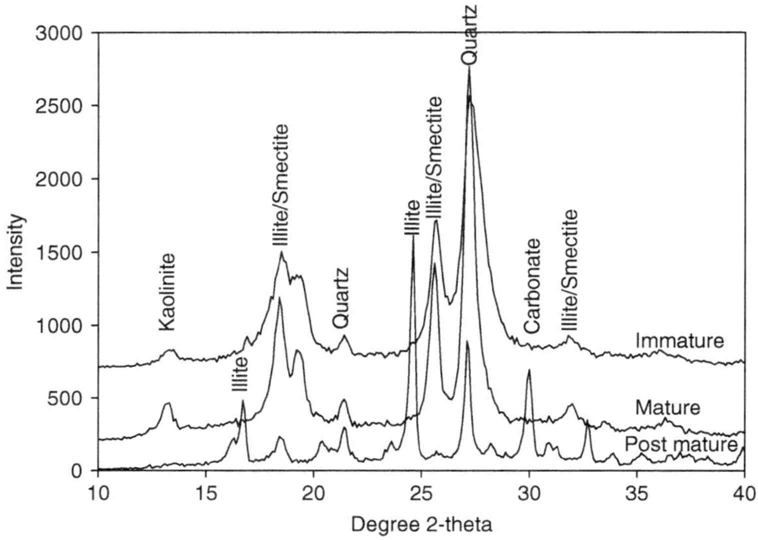

Figure 13.1. X-ray diffraction patterns of isolated clay-size fraction (air-dried) of thermally mature, thermally immature and thermally post-mature samples.

alteration (CIA) and chemical index of weathering (CIW) were computed from major element data as follows (Nesbitt and Young, 1982):

$$CIA = (Al_2O_3/(Al_2O_3 + Na_2O + K_2O + CaO))*100\% \text{ and}$$
$$CIW = (Al_2O_3/(Al_2O_3 + Na_2O + CaO))*100\%$$

The average CIA and CIW values were 64.26 and 72.67, respectively.

13.4.2. Rare earth elements (REE)

REE concentrations in weathering solutions are presented in Table 13.1. The concentrations of REE in the weathering fluids are low but significantly higher than the concentration of REE in natural waters. In ocean waters, the concentration of REE is < 10 ppt and about 200 ppt in river waters (Elderfield and Greaves, 1982; Taylor and McLennan, 1988). The sum of REE in the weathering fluids ranges from 2 ppb to 2 ppm. In the thermally immature sample, the highest REE concentration was observed after 24 h of reaction time. This concentration decreased to about 21 ppb after 288 h of reaction and increases to about 1 ppm at 312 h. In the thermally mature sample, the highest concentration of REE was observed during 144 h of reaction. In this sample, after 24 h of reaction the sum of

Table 13.1. Minor, trace and REE concentration in samples collected over 312-h reaction period

Analytes	TI 24	TI 48	TI 96	TI 144	TI 168	TI 264	TI 288	TM 24	TM 48	TM 96	TM 144	TM 168	TM 264	TM 288	TM 312	TPM 24	TPM 48	TPM 96	TPM 144	TPM 168	TPM 264
Sr	60.189	12.368	4.341	1.497	0.711	0.575	0.567	16.289	13.725	9.374	7.116	2.022	1.598	0.850	0.569	24.047	21.970	20.082	27.786	26.321	20.768
Ba	37.398	15.520	8.407	3.343	1.594	1.033	0.987	0.853	1.610	1.217	0.944	0.432	0.295	0.203	0.455	3.914	3.519	3.154	4.412	3.329	2.336
Ti	0.949	0.768	0.440	0.215	0.204	0.700	0.670	0.830	1.000	0.427	0.786	0.392	0.277	0.320	0.279	0.686	0.696	0.530	0.780	0.987	0.648
V	0.066	0.059	0.045	0.025	0.030	0.025	0.089	0.026	0.031	0.019	0.035	0.039	0.037	0.043	0.043	0.016	0.020	0.008	0.022	0.125	0.011
Cr	0.240	0.265	0.238	0.177	0.200	0.358	0.479	0.201	0.268	0.112	0.237	0.170	0.142	0.182	0.233	0.194	0.268	0.202	0.354	0.288	0.239
Ni	0.748	0.769	0.535	0.267	0.341	1.139	1.450	0.847	1.253	0.327	0.934	0.376	0.284	0.311	0.511	0.593	0.928	0.472	0.981	0.555	0.688
Zn	0.778	0.637	0.437	0.261	0.235	0.257	0.460	0.108	0.166	0.022	0.135	0.078	0.045	0.066	0.147	0.194	0.202	0.087	0.196	0.271	0.111
As	3.188	4.718	3.341	1.688	2.639	6.845	14.283	4.130	7.375	0.157	4.371	1.855	1.358	3.549	5.258	2.385	5.936	0.964	5.433	4.195	1.348
La	0.168	0.100	0.074	0.037	0.027	0.013	0.076	0.255	0.087	0.176	0.415	0.184	0.116	0.074	0.141	0.003	0.004	0.005	0.012	0.056	0.006
Ce	0.404	0.266	0.205	0.132	0.096	0.015	0.230	0.478	0.115	0.330	0.781	0.406	0.282	0.185	0.503	0.006	0.007	0.012	0.017	0.185	0.013
Pr	0.060	0.034	0.029	0.015	0.012	0.007	0.035	0.065	0.016	0.042	0.098	0.054	0.040	0.029	0.087	0.001	0.002	0.001	0.005	0.026	0.002
Nd	0.296	0.168	0.137	0.076	0.059	0.034	0.193	0.262	0.068	0.167	0.389	0.221	0.177	0.134	0.487	0.004	0.012	0.006	0.025	0.119	0.010
Sm	0.124	0.065	0.059	0.027	0.027	0.026	0.094	0.073	0.028	0.034	0.096	0.050	0.035	0.038	0.169	nd	0.011	nd	0.016	1.044	0.003
Eu	0.041	0.020	0.019	0.007	0.008	0.008	0.025	0.018	0.008	0.008	0.023	0.011	0.007	0.009	0.039	0.001	0.004	nd	0.005	0.011	0.001
Gd	0.253	0.118	0.101	0.045	0.039	0.032	0.106	0.131	0.043	0.074	0.174	0.089	0.061	0.054	0.190	0.008	0.016	0.008	0.023	0.058	0.011
Tb	0.024	0.012	0.012	0.005	0.006	0.004	0.014	0.014	0.005	0.007	0.019	0.009	0.006	0.007	0.022	nd	0.002	nd	0.002	0.007	nd
Dy	0.132	0.066	0.064	0.028	0.026	0.017	0.067	0.083	0.026	0.048	0.118	0.059	0.037	0.033	0.106	0.001	0.007	nd	0.010	0.035	0.002
Ho	0.026	0.012	0.013	0.005	0.006	0.004	0.013	0.017	0.005	0.009	0.023	0.011	0.006	0.007	0.016	nd	0.002	nd	0.002	0.006	nd
Er	0.070	0.034	0.035	0.013	0.014	0.015	0.040	0.048	0.017	0.025	0.065	0.031	0.015	0.017	0.036	0.001	0.006	nd	0.008	0.016	0.001
Tm	0.011	0.005	0.007	0.001	0.004	0.003	0.008	0.008	0.003	0.003	0.009	0.004	0.001	0.004	0.004	nd	0.002	nd	0.002	0.002	nd
Yb	0.059	0.029	0.030	0.009	0.013	0.017	0.038	0.047	0.016	0.020	0.056	0.027	0.010	0.014	0.022	nd	0.008	nd	0.010	0.013	0.001
Lu	0.018	0.007	0.012	nd	0.006	0.016	0.032	0.013	0.011	nd	0.014	0.007	nd	0.006	0.010	nd	0.007	nd	0.010	0.005	nd
Th	0.087	0.046	0.055	0.021	0.034	0.001	0.023	0.014	0.004	0.012	0.025	0.024	0.005	0.021	0.029	nd	nd	nd	nd	0.017	nd
U	0.067	0.032	0.022	0.008	0.009	0.039	0.019	0.039	0.051	0.021	0.033	0.011	0.010	0.009	0.016	0.033	0.039	0.030	0.044	0.013	0.027

TI, Thermally immature; TM, thermally mature; TPM, thermally post-mature and numbers represent time in hours; nd, below detection.

REE was 1.5 ppm. This amount decreased to 45 ppb after 48 h and then increased again to 2 ppm between 96 and 144 h and maintained an almost constant value up to the end of the experimental period. In the thermally post-mature sample, the sum of REE is much lower compared to the thermally immature and mature samples. After 24 h of reaction, the sum of REE in solution was only 2 ppb. This concentration increased to 58 ppb during168 h of reaction and decreased again to lower concentrations.

North American shale composite (NASC) normalized REE distribution patterns in weathering fluids are presented in Figs. 13.2–13.4. In the thermally immature samples, we observed a pattern that is significantly depleted in the LREE (La-Nd) and enriched in the MREE and HREE (Sm-Dy; Ho-Lu) with La/Sm ratios <1 and La/Yb ratio $<<1$. A similar pattern of LREE depletion, enrichment of the MREE and HREE was also observed in the weathering solutions of the thermally mature sample. In the thermally post-mature sample, we observed a slight depletion of the LREE, enrichment of the MREE and almost flat HREE pattern in the weathering solutions.

We calculated Ce/Ce* ratios of all weathering fluids collected over the entire experimental period (312 h). Ce/Ce* represents a cerium anomaly and was calculated using NASC-normalized values of Ce and adjacent

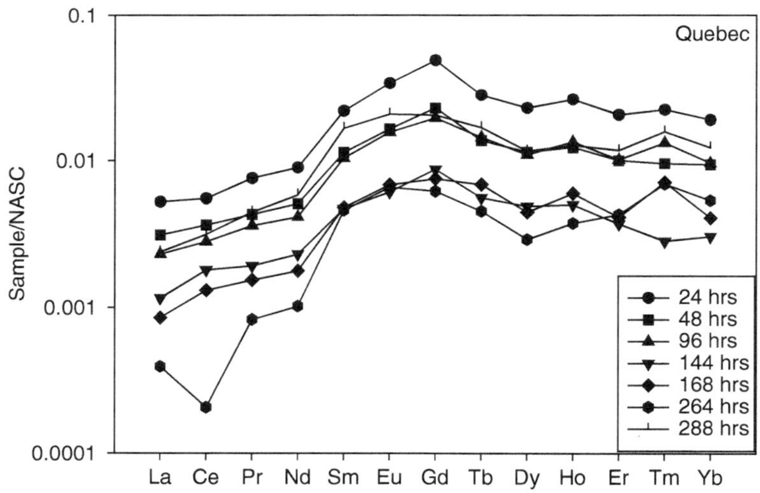

Figure 13.2. Rare earth elements pattern in weathering fluids of thermally immature sample collected over 288-h reaction period. Concentration of REE in sample was normalized to the concentration of REE in North American shale composite (NASC) (Gromet et al., 1984).

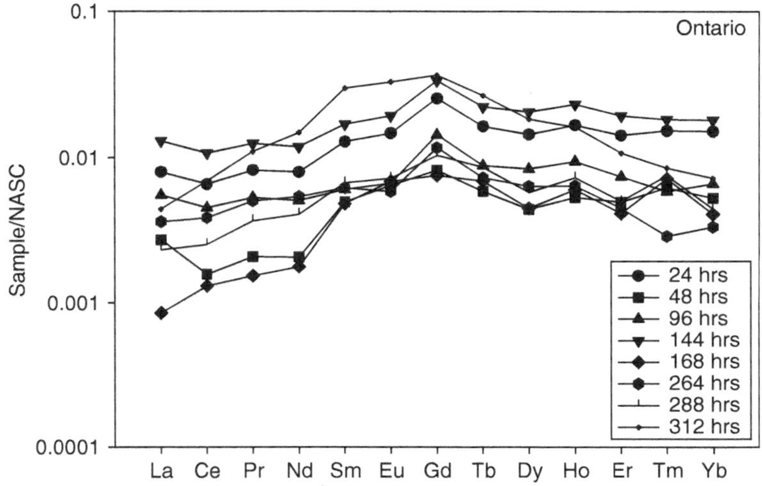

Figure 13.3. Rare earth elements pattern in weathering fluids of thermally mature sample collected over 312-h reaction period. Concentration of REE in sample was normalized to the concentration of REE in North American shale composite (NASC) (Gromet et al., 1984).

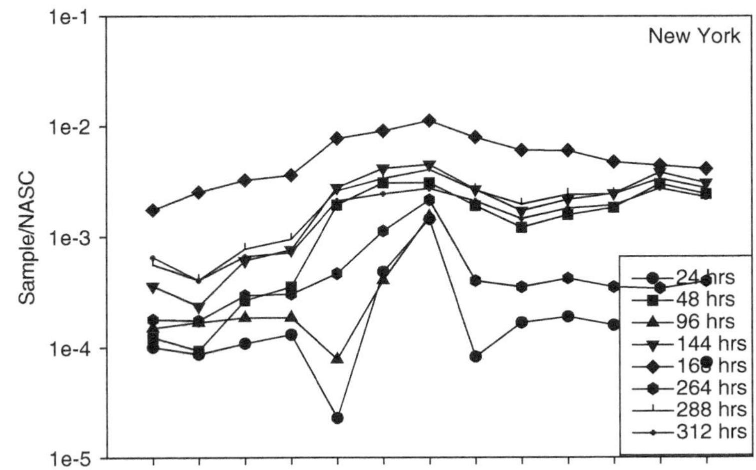

Figure 13.4. Rare earth elements pattern in weathering fluids of thermally post-mature sample collected over 288-h reaction period. Concentration of REE in sample was normalized to the concentration of REE in North American shale composite (NASC) (Gromet et al., 1984).

REE (La and Pr) (Taylor and McLennan, 1985). A majority of weathering fluids from the thermally immature shale sample show a slightly negative Ce anomaly (Ce/Ce* > 1). However, after 264 h of reaction time, a positive Ce anomaly of 1.29 was observed in the thermally immature sample. All weathering fluids from the mature sample collected over time result in a negative Ce anomaly with very little change over the entire experimental period. In the thermally post-mature sample, we observed variable Ce/Ce* with time as the experiment proceeded. Negative values were common during the beginning of the reaction up to 96 h of reaction time and slight positive values thereafter with punctuations of negative values.

13.4.3. Minor and trace elements

The concentrations of Ti, V, Cr, Ni, Zn, As, Sr, Ba, Th and U for all weathering fluids collected over time (24–312 h) are shown in Table 13.1. In the immature sample, the concentrations of several elements in solution tend to decrease with time with the highest concentrations occurring during the first 24 h of reaction. The rate of decrease in elemental concentration with time decreased steadily up to about 168 h of reaction time when the rate remains almost constant. However, certain elements, including Ti, Cr, Ni, As and U tend to increase in concentration after 168 h. In the thermally mature sample, a similar trend of decreasing elemental concentration with time was observed. With the exception of As, most trace elements in these samples maintained almost constant concentration after 168 h. Element release pattern in the post-mature sample was slightly different from that in the immature and mature samples. In the post-mature sample, the elemental concentration in the solutions collected over time stayed almost constant during the entire experimental period.

North American shale composite (NASC) normalized distribution pattern of Ti, V, Cr, Ni, Zn, As, Sr, Ba, Th and U for all samples are shown in Fig. 13.5–13.7. The following elements (Ca, Ti, V, Cr, Ni, Zn, As, Sr, Ba, Th and U) are all depleted in the weathering fluids relative to NASC. The degree of depletion is variable for different elements. Ti shows the highest degree of depletion in the immature and mature samples while Th shows the highest degree of depletion in the post-mature samples. Ti and V are almost equally depleted in the post-mature sample relative to NASC whereas in the mature and immature samples Ti is more depleted than V relative to NASC. However, the degree of depletion of Ti relative to V is higher in the immature sample compared to the mature sample. The concentration of As is close to that of NASC in all samples.

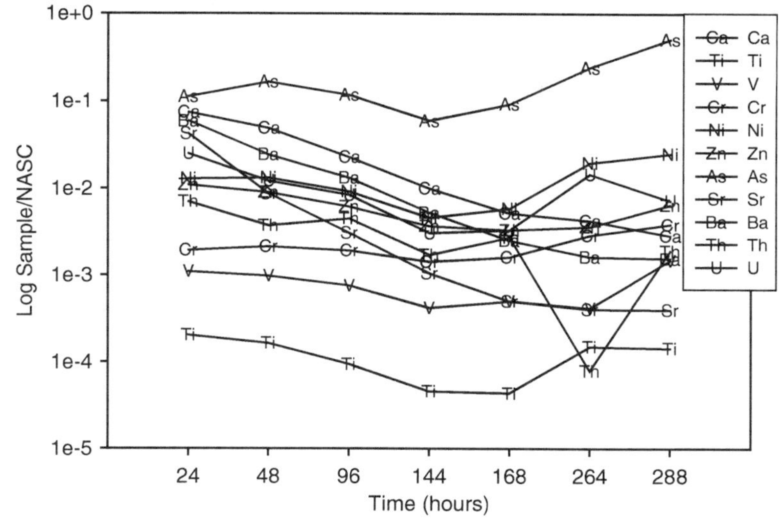

Figure 13.5. NASC normalized concentrations of minor and trace elements as a function of time in weathering fluids of thermally immature sample.

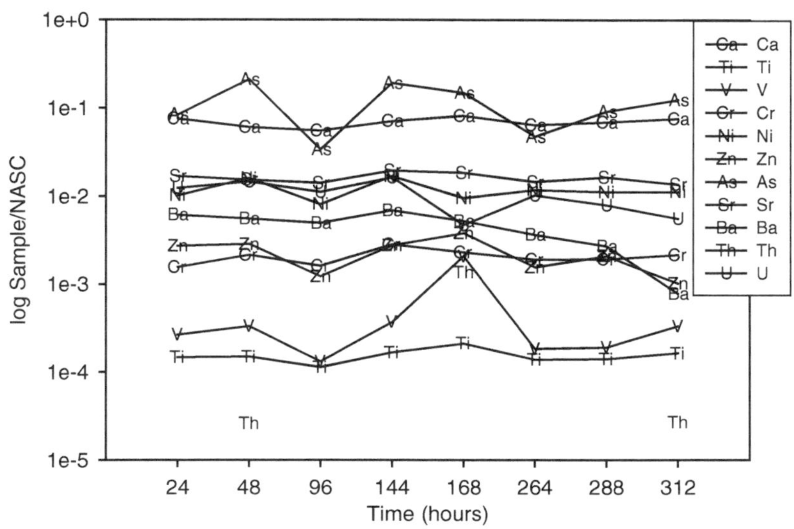

Figure 13.6. NASC normalized concentration of minor and trace elements as a function of time in weathering fluids of thermally mature sample.

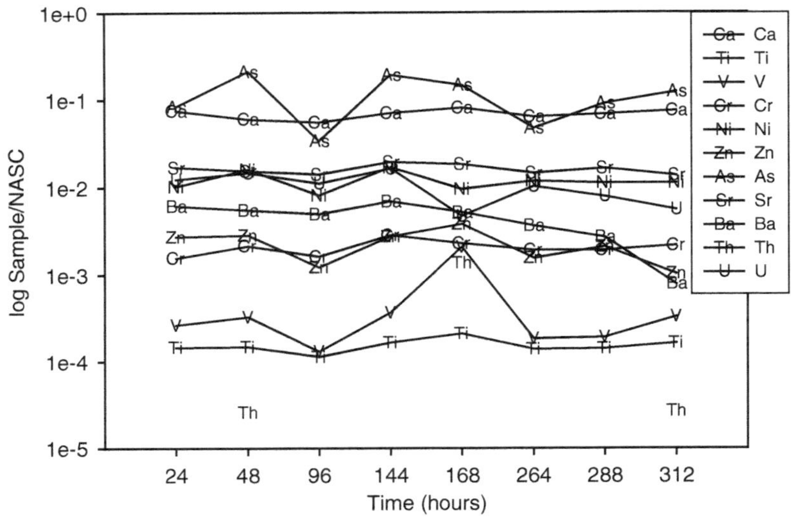

Figure 13.7. NASC normalized concentration of minor and trace elements as a function of time in weathering fluids of thermally post-mature sample.

We observed a general trend of decrease in elemental concentration with time for most of the elements analyzed. Since Ti is considered relatively immobile during chemical weathering (Ohmoto, 1996; Rye and Holland, 1998; White et al., 2001), we normalized the concentration of V, Cr, Ni, Zn, As, Sr, Ba, Th, U, La, Gd and Yb to the concentration of Ti in all weathering solutions and plotted these ratios on spider plots as shown in Fig. 13.8. When normalized to Ti, the ratio of all elements in the thermally immature sample excluding As, Ni and Cr decreased with time with a sharp decrease after 288 h. The ratio of As, Ni and Cr to Ti on the other hand increased with time over the entire experimental period. In the thermally mature samples, a similar trend of decreasing elemental concentration with time was observed with a sharp decrease during 48 h for the elements Th, U, La and Yb. In the thermally post-mature samples, the ratio of most elements to Ti remained almost constant over the entire reaction period.

13.5. Factor loadings

R-mode factor analysis was used to describe the relationship amongst dissolved constituents in the weathering fluids collected over a period of

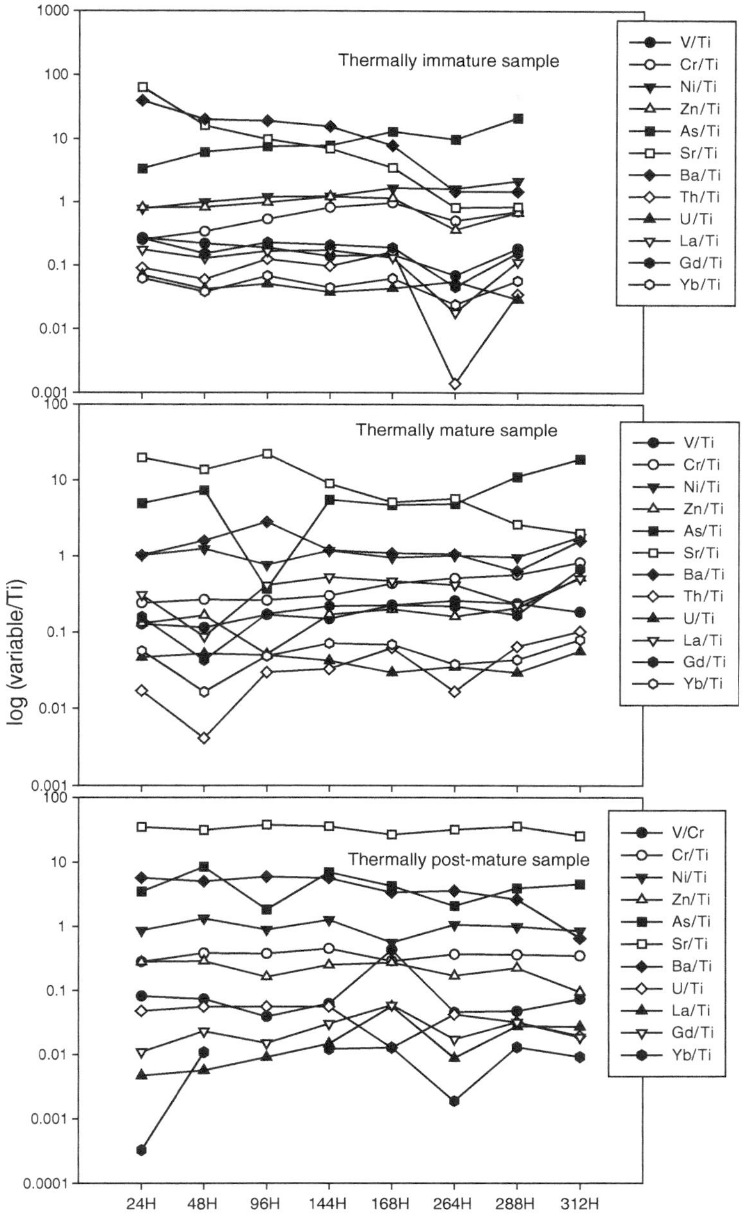

Figure 13.8. Minor and trace elements and selected REE concentrations in thermally immature, thermally mature and thermally post-mature samples normalized to concentration of Ti.

Table 13.2. Factor loadings and communalities of variables in time dependent weathering solutions (Extraction method: principal components; rotation: varimax)

Variable	Factor 1	Factor 2	Factor 3	Factor 4	Factor 5	Communality
Ti	0.048	−0.145	−0.464	0.846	0.140	0.947
V	0.106	−0.251	−0.254	0.088	0.907	0.968
Cr	−0.090	−0.113	−0.939	−0.005	0.245	0.963
Ni	0.082	−0.005	−0.929	0.325	−0.138	0.994
Zn	0.062	−0.883	−0.319	0.055	0.174	0.919
As	0.033	−0.021	−0.972	−0.067	0.137	0.969
Sr	0.035	0.620	0.060	0.711	0.114	0.907
Ba	0.148	−0.935	0.060	0.286	0.011	0.983
La	0.890	0.192	0.160	0.287	−0.118	0.952
Ce	0.970	0.079	0.173	0.095	0.028	0.986
Sm	0.831	−0.338	−0.180	−0.171	0.170	0.895
Eu	0.757	−0.544	−0.199	−0.100	0.142	0.938
Gd	0.854	−0.499	−0.011	0.070	0.057	0.987
Yb	0.810	−0.352	−0.195	0.270	−0.061	0.894
Th	0.398	−0.853	0.135	−0.071	0.133	0.927
U	0.185	−0.483	−0.362	0.618	−0.409	0.948
Percent of variance explained by factor						
	27.4	23.1	20	17.2	7.3	
Cumulative percent variance						
	27.4	50.5	70.5	87.7	95	

13 days of reaction time. Rotated factor loadings and communalities including the percent variance explained by each factor are shown in Table 13.2. The analysis resulted in five factors that we used in explaining the probable sources of Ti, V, Cr, Ni, Zn, As, Sr, Ba, Th, U and the REE (La, Ce, Sm, Eu, Gd and Yb) in the weathering solutions. These five factors explained about 95% of variance in our data. Factor 1 was the dominant factor and explained about 27.4% of the variance; factor 2 accounted for 23.1% of variance; factors 3 accounted for 20.0% of the variance while factors 4 and 5 accounted for 17.2% and 7.3% of the variance in our data, respectively. The remainder 5% could be attributable to noise.

Factor 1 (Fig. 13.9) had high loadings for La, Ce, Sm, Eu, Gd and Yb, intermediate loadings for Th and low loadings for Cr, Ba and U and almost zero loadings for Ti, V, Ni, Zn, As and Sr. Factor 2 (Fig. 13.10) had high loadings for V, Zn, Ba, Eu and Th, intermediate loadings for Sr, Sm, Gd and Yb, low loadings for Cr, Ni, La and U and almost zero loading for Ti, As and Ce. Factor 3 (Fig. 13.11) had high loadings for Cr, Ni and As, intermediate loadings for Ti, V and Zn, low loadings for Sr, Ba, La, Ce, Sm, Eu, Yb, Th and U and almost zero loading for Gd.

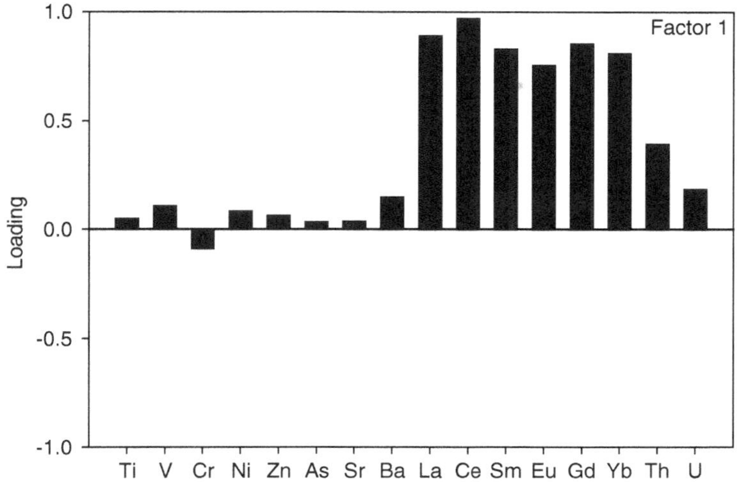

Figure 13.9. Bar plot of rotated factor loadings (factor 1). This factor shows high loadings for the REE and intermediate to low loadings for Th and U.

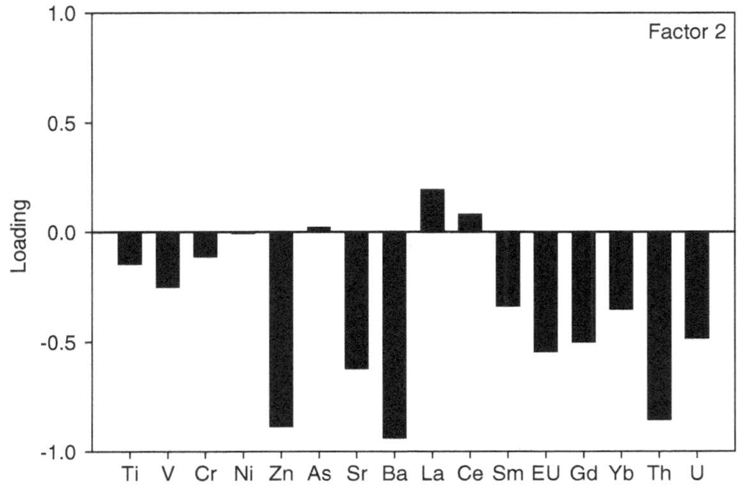

Figure 13.10. Bar plot of rotated factor loadings (factor 2). Factor 3 shows high loadings for Zn, Sr, Ba and Th and intermediate to low loadings for Sm, Eu, Gd, Yb and U.

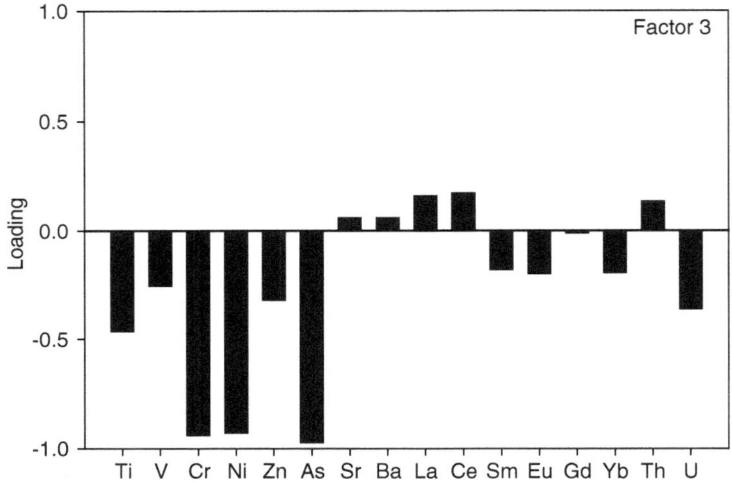

Figure 13.11. Bar plot of rotated factor loadings (factor 3). This factor shows high loadings for Cr, Ni and Zn and intermediate to low loading for Ti, V, Zn and U.

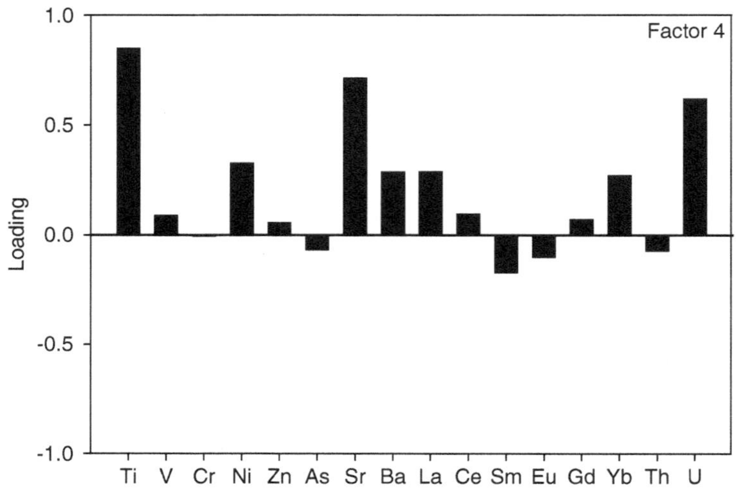

Figure 13.12. Bar plot of rotated factor loadings (factor 4). This factor shows high loadings for Ti, Sr and U and intermediate to low loading for Ni, Ba, La and Yb.

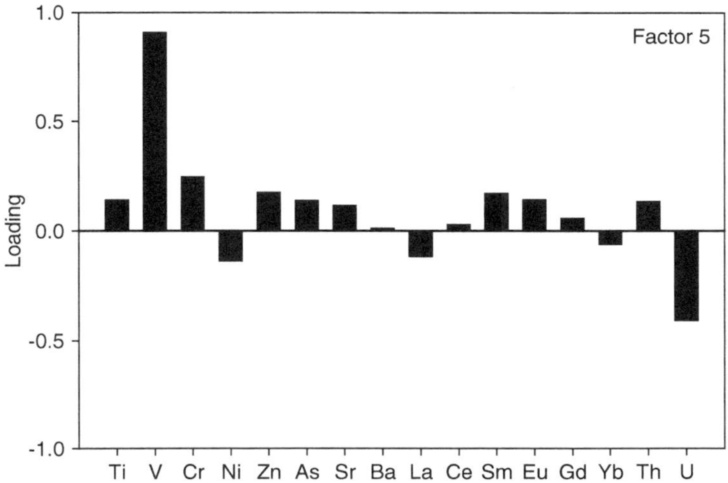

Figure 13.13. Bar plot of rotated factor loadings (factor 5). This factor shows high loading for V and intermediate for U.

Factor 4 (Fig. 13.12) had high loadings for Ti, Sr, Ba and U, intermediate loadings for Ni and Yb, low loadings for La, Sm and Th and almost zero loading for V, Cr, As, Ce and Eu. Factor V was essentially a Vanadium–Uranium factor, which had high loadings for V and low loadings for U and almost zero loading for all other variables (Fig. 13.13).

A second R-mode factor analysis was carried out using elements with known association to particular mineral phases in black shale. We selected La, Gd, Sr, Ni and V to represent elements associated with the clay, phosphate, carbonate and sulfide mineral phases, and organic matter, respectively (Patterson et al., 1986; Holly, 1990). Sorted rotated factor loadings and communalities for this analysis are presented in Table 13.2. This analysis resulted in four factors that explained over 96% of the variance in our data. In this analysis, we considered all variables with a factor score less than 0.3 as insignificant and gave those factors a zero value. Factor 1 had high loadings for La and Gd with zero loadings for V, Sr and Ni. Factor 2 had high loading for V, low loading for Gd and zero loading for Ni, La and Sr. Factor 3 had high loading for Sr and zero loading for V, Ni, La and Gd while factor 4 had high loading for Ni and zero loading for La, Gd, V, Sr and Ni.

To study whether these factors were related to each other, we performed a correlation analysis using the rotated factor scores (Fig. 13.14). The *r*-values were low in a majority of cases. The highest correlation

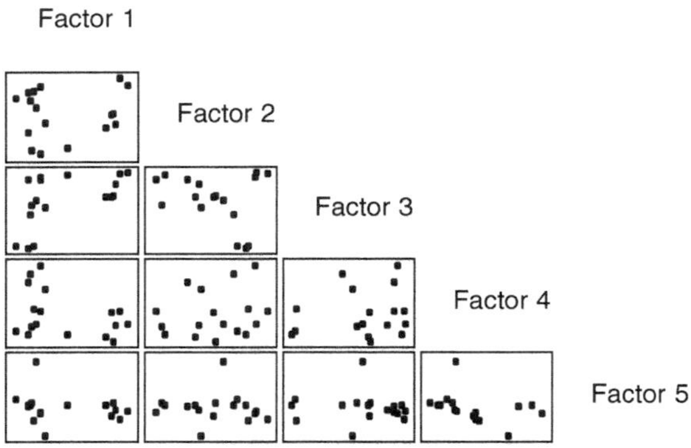

r-value	Factor 1	Factor 2	Factor 3	Factor 4	Factor 5
Factor 1	1				
Factor 2	0.13	1			
Factor 3	0.54	-0.27	1		
Factor 4	-0.36	0.19	0.16	1	
Factor 5	-0.20	-0.05	-0.06	-0.35	1

Figure 13.14. Correlation matrix plots of rotated factor scores. The table shows *r*-values for correlations in the figure.

($r = 0.54$) was between factors 1 and 3. The absence of a significant correlation between these factors suggests that they were independent of each other.

13.6. Discussion

13.6.1. Interpretation of factor loadings

Statistical analysis (R-mode factor analysis) has been applied in many geochemical studies to determine the source of different elements or the association among dissolved constituents in aquatic media (Oloronfemi, 1984; Caggianelli et al., 1992; Vreča et al., 2001). The work of Reeder et al. (1972) provides a classic example in the use of R-mode factor analysis to explain the chemistry of natural waters resulting from the chemical weathering of bedrock. We applied R-mode statistical analysis to show the association between different elements in solution in order to identify their source. We also compared the patterns of REE in the weathering fluids to reported patterns of REE in major and easily weathered minerals associated with black shale (Taylor and McLennan, 1985; Kidder and

Table 13.3. Correlation matrix for minor and trace elements including selected REE

	Sr	Ba	Ti	V	Cr	Ni	Zn	As	La	Gd	Yb	Th	U
Sr	1												
Ba	**0.71**	1											
Ti	**0.65**	0.35	1										
V	0.08	0.28	0.27	1									
Cr	0.07	0.04	0.49	0.39	1								
Ni	0.15	0.06	**0.69**	0.16	**0.83**	1							
Zn	0.40	**0.83**	0.33	**0.52**	0.41	0.31	1						
As	−0.14	−0.06	0.37	0.44	**0.88**	**0.84**	0.35	1					
La	−0.10	0.08	0.10	0.16	−0.24	0.04	−0.02	−0.02	1				
Gd	0.17	**0.57**	0.11	0.43	0.00	0.09	**0.52**	0.16	**0.73**	1			
Yb	0.16	0.49	0.27	0.36	0.12	0.29	0.48	0.26	**0.80**	**0.88**	1		
Th	0.47	**0.83**	0.04	0.34	−0.03	−0.12	**0.79**	−0.11	0.21	**0.74**	**0.62**	1	
U	**0.67**	**0.60**	**0.72**	−0.10	0.28	0.61	0.44	0.22	0.13	0.29	0.40	0.32	1

Note: La, Gd and Yb_{LA}, Gd and Yb represent the light, middle and heavy REE, respectively.

Eddy-Dilek, 1994; Hannigan and Sholkovitz, 2001; Abanda and Hannigan, 2004). Description and interpretation of factor composition is discussed on the basis of breakdown of mineral phases associated with black shales. We interpreted factor 1 as representing the breakdown products of a REE-Th-U-rich mineral phase because of the high-loading scores of La, Ce, Sm, Eu, Gd and Yb, intermediate loadings of Th and low loading of U. Taylor and McLennan (1989) suggested that Th shows similar geochemical behavior to the REE while Burt (1989) suggested that the actinide U and Th commonly substitute for the REE in nature. The positive correlation between Th, U and the REE (Table 13.3) in our samples suggests control of these variables by a common factor. We submit that a REE-Th-U-rich mineral phase or a REE-rich mineral phase in which U and Th substitute for the REE may account for the observed high loadings of the REE and as a result the similar behavior of these elements during black shale weathering. Generally in sedimentary rocks, the REE occur in essential as well as accessory mineral constituents (Burt, 1989). REE distribution patterns in the weathering solutions result in a pattern that is enriched in the MREE and HREE and depleted in the LREE. Earlier studies have shown that the enrichment of MREE in black shales or in streams draining black shale catchments is related to the presence of phosphate minerals or to the weathering of phosphate minerals (Kidder and Eddy-Dilek, 1994; Hannigan and Sholkovitz, 2001). Since phosphate is one of the most easily weatherable minerals in black shales we think that factor 1 is controlled by the weathering of a REE–phosphate mineral phase. Possible minerals include monazite, xenotime and apatite. High loadings of U associated with this factor may be due to weathering of a uranium-rich phase associated with phosphate. Uranium in the +4 and +6 states are the most common oxidation states in natural water systems. Uranium in the +4 oxidation state is essentially insoluble under oxidizing mildly acidic to alkaline conditions under which the leaching experiments were carried out. Conversely, U^{+6} ion is potentially much more mobile, because of the greater solubilities of most U^{+6} minerals (e.g., Osmond and Ivanovich, 1992). The above minerals have not been reported in the Utica shale samples, but it is likely that minor amounts of these minerals may exist that can account for the observed patterns of REE and U in our samples. It is also possible that desorption of REE bound to exchange sites on clay minerals may account for the observed high loadings of REE.

Factor 2 was interpreted as representing the weathering products of detrital clay mineral fraction. This factor had high loadings for Ba, Zn and Th that may be related to the breakdown of detrital material or source material from which the samples were derived. This may include weathering of clay minerals and plagioclase-feldspar, which may have contributed to high loadings of Eu thereby resulting in the observed positive Eu anomalies. We interpreted factor 3 as representing the weathering products (oxidation) of sulfide minerals as well as breakdown of detrital clay minerals because of the high loadings of Cr, Ni and As that are associated with sulfides (Patterson et al., 1986). This factor also had intermediate loading of Zn and Ti, elements that are usually associated with the detrital fraction. Since factor 1 was interpreted as the product of weathering of a REE-Th-U-rich phase and factor 3 as the product of breakdown of sulfide mineral phase as well as breakdown of detrital material, this may account for the slight positive correlation ($r = 0.52$) between factors 1 and 3 (Fig. 13.14). Factor 4 was interpreted as representing the breakdown products of carbonate and clay minerals due to the high loadings of Ti, Sr, Ba and U and intermediate loadings for Ni and Yb. Factor 5 was interpreted as representing the products of partial breakdown of organic matter because of the high loadings of Donald et al. (2000) showed that more vanadium is hosted by neomorphic illite than by organic matter in Pennsylvanian black shale from Indiana. It is possible that the high loadings of V associated with factor 5 may be the result of dissolution of illitic clays or the breakdown of organic matter associated with clay mineral surfaces (Kennedy et al., 2002). The leaching trends of V observed in this study, hint that V is more easily leached from the immature samples than the mature sample. In the post-mature sample, there is an abrupt increase in the concentration of V after 144 h of reaction but the concentration reduces significantly after 264 h. We submit that in the immature samples

V is associated more with organic matter while in the mature and post-mature samples V tends to associate with illite.

13.6.2. Identifying sources of trace element (mineral contribution) using R-mode factor analysis

R-mode factor analysis has been used in many geochemical studies to identify sources of metals (Reeder et al., 1972; Pattan et al., 1995). We used R-mode factor analysis to identify major controls of minor, trace and REE in solution from black shale weathering. Our results suggest that four major mineral phases including organic matter could be use to explain 95% of the observed elemental concentration in solution. A major REE mineral phase results in the release of high concentration of the REE in solution. Th and U are associated with this mineral phase, which suggests similar chemical behavior of these elements with the REE (Weimer et al., 1980). Weimer et al. (1980) have suggested the use of the REE as natural analogues for the actinides because of their similar chemical behavior. Sulfides, carbonates and detrital clay minerals also contribute substantially to the total dissolved load in the weathering fluids. Factor 5 that we ascribe to the weathering product of organic matter accounts for a small but reasonable portion of trace elements considering that the organic matter fraction in these shales is not readily oxidazable organic matter. Some research has shown that refractory organic matter occurs in soil profiles formed from weathering of Paleozoic black shale. It has also been reported that because of the refractory nature, this type of organic matter constitutes a sizable portion of total suspended solids in streams draining Paleozoic black shales basement (Petsch et al., 2000).

13.6.3. REE fractionation patterns in weathering solutions

REE fractionation during weathering of parent material has been reported by Nesbitt (1979). The patterns of REE in minerals associated with black shales have been reported in a number of studies (Taylor and McLennan, 1985; Kidder and Eddy-Dilek, 1994; Hannigan and Sholkovitz, 2001). Taylor and McLennan, (1985) in their review of the upper continental crust and its composition, reported the patterns of REE in common rock forming minerals as well as in common accessory minerals in sedimentary rocks. In black shales trace elements and the REE are associated with a number of phases including ion exchange sites on clay minerals, adsorbed on ferric oxides and aluminum oxides, adsorbed to organic matter, co-precipitated as carbonate and bound to crystalline mineral phases. Accessory (heavy minerals) associated with black shale also carry a significant amount of trace elements and the REE especially the HREE (Taylor and McLennan, 1985). As Taylor and McLennan (1985) suggested, an understanding of REE chemistry during black shale weathering will allow for greater understanding of the weathering process in these rocks. Proper understanding of trace element and REE mobility during weathering of black shale would allow for elimination of misinterpretation of sedimentary REE patterns and consequently interpretation of source material.

With increasing reaction time accompanied by decreasing concentration of REE in the weathering solutions we observed a decrease in the $(La/Sm)_N$ ratio from about 0.25 to 0.10 in the thermally immature sample. The thermally mature and thermally post-mature samples also show similar decrease in $(La/Sm)_N$ ratio with decreasing REE concentration and increasing reaction time. However, the average $(La/Sm)_N$ ratio is higher in the thermally mature sample compared to the thermally immature and thermally post-mature samples (thermally mature > thermally post-mature > thermally immature). This result suggests that the enrichment of the MREE is higher in the thermally mature sample relative to the thermally post-mature and thermally immature samples. Weathering of the Utica shales therefore results in the fractionation of REE leading to enrichment of MREE in solution as shown in Figs. 13.2–13.4 and as reported by other workers (Braun et al., 1998; Zhang et al., 1998; Leleyter et al., 1999; Aubert et al., 2001; Hannigan and Sholkovitz, 2001; Gnandi and Tobschall, 2003; Ji et al., 2004). It is likely that this fractionation is the result of different weathering rates of host mineral phases in rock and not related to source signature since the pattern of REE in the whole rock is different. In the whole rock the pattern of REE shows enrichment in the LREE and depleted in the MREE and HREE (Abanda, 2005). Earlier research has shown that the MREE are associated with phosphate mineral phase in black shale (Hannigan and Sholkovitz, 2001; Kidder et al., 2003). The breakdown of this phase during weathering may lead to preferential release of the MREE relative to the LREE as shown by enrichment of the MREE in our samples ($La/Sm < 1$).

We calculated the partition coefficients between the solid and liquid phases for all elements analyzed in this study using the concentration of the elements in the weathering solutions and their corresponding concentrations in the whole rock samples. Ideally, partition coefficients are computed for solutions in equilibrium with pure mineral phases. In our case, we cannot infer equilibrium and also multiple mineral phases were involved. In general, log K_d for most elements was highest at the beginning of the reaction but decreased slowly over time. For the following elements As, Ba, Sr, Zn, Ni, Cr, U, V, La, Gd and Yb, log K_d ranged between −0.7 and −3.5. log K_d for these elements follows the order As > Ba > Sr > ZnYbNi > GdLa > Cr > U > V. The observed high log K_d for As may be related to rapid breakdown of sulfide minerals. Ba and Sr are relatively more mobile during chemical weathering (Picouet et al., 2002; Singh et al., 2005), which may account for the high log K_d values of these elements. On the other hand, the REE and Cr are considered relatively immobile during chemical weathering (Taylor and McLennan, 1985) thereby lending support to the low log K_d values for these elements. Our results suggest that the Gd (MREE) and Yb (HREE) are relatively more mobile than La (LREE) during weathering of the Utica shales leading to higher log K_d for Gd and Yb (−2.3 and −2.0, respectively) relative to La (−2.6). log K_d for Sr decreased significantly from −1.1 after 24 h to −3.2 after 288 h of reaction. This result corroborates the association of Sr with the carbonate phase because we expect rapid dissolution of carbonate mineral at the beginning of the experiment. We suggested that V is associated with the organic fraction and it is not surprising that the log K_d for V is exceptionally low (~−3.5). Petsch et al. (2000) reported that refractory organic matter in Palaeozoic black shale is not very susceptible to chemical weathering and may therefore hold onto trace elements associated with this fraction.

13.7. Conclusions

The study of chemical weathering of black shales is important because of the abundance of these rocks in the sedimentary record and their role in controlling the composition of chemical species in associated natural water systems and consequently on water quality. Although much experimental work has been done on the weathering of pure minerals as well as weathering of rocks in which the modal mineralogy can easily be estimated, little experimental work has been done on studying the weathering of a fine-grained multimineralic rock rich in organic matter. This research reports the results of the application of multivariate statistical analysis (R-mode factor analysis) in assessing geochemical data from experimental study of black shale weathering thus allowing the identification of different mineral phases contributions to the composition of chemical species (minor trace and REE) in solution. R-mode factor analysis suggests that four major mineral phases (phosphates, carbonates, silicates and sulfides) including organic matter are important in controlling trace elements and the REE in solution from the weathering of black shale. Laboratory weathering of black shale results in elevated concentrations of minor and trace elements including the REE in weathering solutions relative to the concentrations of these elements in natural water systems. Extension of results from experimental studies of black shale weathering to model natural weathering of black shale in the field will necessitate detail mineralogical study of the rocks and mass balance calculations to determine the contribution of individual mineral phases and organic matter to the total elemental composition of chemical species in solution. We also showed that weathering of the black shale leads to significant fractionation of the REE with the MREE and the HREE being more mobile than the LREE. Slight differences occur between the weathering patterns of rocks of different thermal grades, which can be ascribed to diagenetic modifications. We observed that in the thermally post-mature sample the LREE are slightly depleted in the weathering solutions relative to the thermally mature and thermally immature samples.

ACKNOWLEDGMENTS

The authors would like to thank the Environmental Science PhD Program at Arkansas State University for providing stipend support to the graduate student Peter Azah Abanda. We would also like to thank Dr Raymond Coveney Jr and Dr Steve Lev for their reviews and insightful comments that helped improved the manuscript.

REFERENCES

Abanda, P.A., 2005. Mobility of trace elements during diagenesis and weathering of black shales. Unpublished PhD Dissertation Arkansas State University, p. 150.

Abanda, P. A., Hannigan, R. E., 2004. Laboratory study of chemical weathering of Middle Ordovician black shales. In: Wanty, R. B., Seal, R. R. II (Eds.), Proceedings of the 11th international symposium on water–rock interaction. Taylor and Francis, Lisse, The Netherlands, pp. 793–796.

Aubert, D., Stille, P., Probst, A., 2001. REE fractionation during granite weathering and removal by waters and suspended loads: Sr and Nd isotopic evidence. Geochim. Cosmochim. Acta 65, 387–406.

Berner, E., Berner, R., 1996. Global Environment: Water, Air, and Geochemical Cycles. Prentice Hall, New Jersey, p. 376.

Braun, J.-J., Viers, J., Dupré, B., Polve, M., Ndam, J., Muller, J.-P., 1998. Solid/liquid REE fractionation in the lateritic system of Goyoum, East Cameroon: The implication for the present dynamics of the soil covers of the humid tropical regions. Geochim. Cosmochim. Acta 62, 273–299.

Burt, D.M. 1989. Compositional and phase relations among rare-earth elements minerals. In: Lipin, B.R., McKay, G.A. (Eds.), Rev. Mineral. The Mineralogical Society of America, New York, Vol. 21, pp. 259–308.

Caggianelli, A., Fiore, S., Mongelli, G., Salvemini, A., 1992. REE distribution in the clay fraction of pelites from the southern Apennines, Italy. Chem. Geol. 99, 253–263.

Carroll, D., 1970. Rock Weathering. Plenum Press, New York, p. 203.

Coveney, R.M. Jr., 2001. Metal pollution from late Palaeozoic black shales of the Midwestern United States. GSA Abstracts with Program 33(6), A-187.

Coveney, R.M. Jr., 2004. An assessment of potential health hazard of Pennsylvanian black shales. GSA North-Central Section, Abstract with Programs 35(2), 6–7.

Dalai, T.K., Singh, S.K., Triverdi, J.R., Krishnaswami, S., 2001. Dissolved rhenium in the Yamuna river system and the Ganga in the Himalaya: Role of black shale weathering on the budgets of Re, Os, and U in rivers and CO_2 in the atmosphere. Geochim. Cosmochim. Acta, 66, 29–43.

Donald, R.P., Coveney, R.M. Jr., Gengmei, Z., 2000. Authigenic illite and organic matter: The principal host of vanadium in the Mecca Quarry shale at Velpen, Indiana. Clays and clay minerals 48, 311–316.

Drever, J.I., 1997. The Geochemistry of Natural Waters, Third ed.. Upper Saddle River, New Jersey, Prentice Hall, p.437.

Drever, J.I., Stillings, L.L., 1997. The role of organic acids in mineral weathering. Colloids and Surfaces A: Physicochemical and Engineering Aspects 120, 167–181.

Elderfield, H., Greaves, M.J., 1982. The rare earth elements in seawater. Nature 296, 214–219.

Gnandi, K., Tobschall, H.J., 2003. Distribution patterns of rare-earth elements and uranium in tertiary sedimentary phosphorites of Hahotoé–Kpogamé, Togo. J. Afr. Earth Sci. 37, 1–10.

Gromet, L.P., Dymek, R.F., Haskin, L.A., Korotev, R.L., 1984. The "North American Shale Composite": Its compilation, major and trace element characteristics. Geochem. Cosmochim. Acta 37, 419–438.

Hannigan, R.E., Basu, A., 1998. Late diagenetic trace element remobilization in organic-rich black shales of the Taconic Foreland Basin of Quebec, Ontario and New York Shales and mudstones II, pp. 209–233.

Hannigan, R.E., Sholkovitz, E.R., 2001. The development of middle rare earth element enrichments in freshwaters: Weathering of phosphate minerals. Chem. Geol. 175, 495–508.

Holdren, G.R. Jr., Berner, R.A., 1979. Mechanism of feldspar weathering: I. Experimental studies. Geochim. Cosmochim. Acta 50, 1161–1171.

Holly, L.O., 1990. Black shales: An economic geologist's perspective. In: Stivers, J. (Ed.), Eastern oil shale symposium; Institute for mining and mineral research. University of Kentucky, pp. 207–217.

Horton, T., Chamberlain, C., Fantle, M., Blum, J., 1999. Chemical weathering and lithologic controls of water chemistry in a high-elevation river system: Clark's Fork of the Yellowstone River, Wyoming and Montana. Water Resour. Res. 35, 1643–1656.

Isaacs, C.M., Keller, M.A., Bird, K.J., 1999. Potentially hazardous trace elements in the monterey formation, California. Am. Assoc. Petr. Geol. Bull. 83(4), 691.

Ji, H., Wang, S., Ouyang, Z., Zhang, S., Sun, C., Liu, X., Zhou, D., 2004. The mobility of rare earth elements during weathering. Chem. Geol. 203, 29–50.

Kane, J.S., Arbogast, B.F., Leventhal, J.S., 1990. Characterization of Devonian Ohio Shale SDO-1 as a USGS geochemical reference sample. Geostandard. Newslett. 14, 169–196.

Kennedy, M.J., Peaver, D.R., Hill, R.J., 2002. Mineral surface control of organic carbon in black shale. Science 295, 657–660.

Kidder, D.L., Eddy-Dilek, C., 1994. Rare-earth element variation in phosphate nodules from Midcontinent Pennsylvanian cyclothems. J. Sediment. Res. A 64, 584–592.

Kidder, D.L., Krishnaswamy, R., Mapes, R.H., 2003. Elemental mobility in phosphatic shales during concretion growth and implications for provenance analysis. Chem. Geol. 198, 335–353.

Lasaga, A.C., Soler, J.M., Ganor, J., Burch, T.E., Nagy, K.L., 1994. Chemical weathering rate laws and global geochemical cycles. Geochim. Cosmochim. Acta 58, 2361–2386.

Lee, J.-S., Chon, H.-T., Kim, J.-S., Kim, K.-W., Moon, H.-S., 1998. Enrichment of potentially toxic elements in areas underlain by black shale and slates in Korea. Environ. Geochem. Health 20, 135–147.

Lehmann, D., Brett, C.E., Cole, R., Baird, G., 1995. Distal sedimentation in a peripheral basin. Ordovician black shales associated with flysch of the western Taconic Foreland, New York State and Ontario. Geol. Soc. Am. Bull. 107, 708–724.

Leleyter, L., Probst, J-L., Depetris, P., Haida, S., Mortatti, J., Rouault, R., Samuel, J., 1999. REE distribution pattern in river sediments: Partitioning into residual and labile fractions. Comptes Rendus de l'Académie des Sciences—Series IIA—Earth Planet. Sci. 329, 45–52.

Loughnan, F.C., 1969. Chemical Weathering of Silicate Minerals. American Elsevier, New York, p. 169.

Malmstrom, M.E., Destouni, G., Banwart, S.A., Stromberg, B.H.E., 2000. Resolving the scale-dependence of mineral weathering rates. Environ. Sci. Technol. 34, 1375–1378.

McLennan, S.M., 1989. Rare earth elements in sedimentary rocks: Influence of provenance and sedimentary processes. Mineralogical society of America. Rev. Mineral. 21, 169–200.

Moore, D.M., Reynolds, R.C. Jr., 1989. X-ray Diffraction and the Identification and Analysis of Clay Minerals. Oxford University Press, Oxford, UK.

Nesbitt, H.W., 1979. Mobility and fractionation of rare earthy elements during weathering of a granodiorite. Nature 279, 206–210.

Nesbitt, H.W., Young, G.M., 1982. Early Proterozoic climates and plate motions inferred from major element chemistry of lutites. Nature 229, 715–717.

Ohmoto, H., 1996. Evidence in pre-2.2 Ga paleosols for the early evolution of atmospheric oxygen and terrestrial biota. Geology 24, 1135–1138.

Olorunfemi, B.N., 1984. Factors influencing the geochemistry of lateritic soils in the greenschist-amphibolite meta-tectonic belt, southwestern Nigeria. Chem. Geol. 42, 261–275.

Osmond, J.K., Ivanovich, M., 1992. Uranium-series mobilization and surface hydrology. In: Ivanovich, M., Harmon, R.S. (Eds.), Uranium Series Disequilibrium: Applications to Earth, Marine, and Environmental Sciences. Second ed. Oxford University Press, pp. 259–289.

Pattan, J.N., Rao, C.M., Higgs, N.C., Colley, S., Parthiban, G., 1995. Distribution of major, trace and rare-earth elements in surface sediments of the Wharton Basin, Indian Ocean. Chem. Geol. 121, 201–215.

Patterson, J.H., Ramsden, A.R., Dale, L.S., Fardy, J.J., 1986. Geochemistry and mineralogical residence of trace elements in oil shales from Julia Creek, Queensland, Australia. Chem. Geol. 55, 1–16.

Petsch, S.T., Berner, R.A., Eglinton, T.I., 2000. A field study of the chemical weathering of ancient sedimentary organic matter. Org. Geochem. 31(5), 475–487.

Peucker-Ehrenbrink, B., Hannigan, R.E., 2000. Effects of black shale weathering on the mobility of rhenium and platinum group elements. Geology 28, 475–478.

Picouet, C., Dupré, B., Orange, D., Valladon, M., 2002. Major and trace element geochemistry in the upper Niger River (Mali): Physical and chemical weathering rates and CO_2 consumption. Chem. Geol. 185, 93–124.

Piper, D.Z., Isaacs, C.M., 1995. Geochemistry of minor elements in the monterey formation, California. Seawater chemistry of deposition, U.S. Geological Professional Paper 1566, p. 41.

Poppe, L.J., Paskevich, V.F., Hathaway, J.C., Blackwood, D.S., 2001. A laboratory manual for X-ray powder diffraction. U.S. Geological Survey Open-File Report 01-041.

Pretti, V., Stewart, B., 2002. Solute sources and chemical weathering in the Owens Lake watershed, eastern California. Water Resour. Res. 38(8). Available at http://earthref.org/cgi-bin/er.cgi?s = er-s0-db-index.cgi?database_name = err&fl = D

Reeder, S.W., Hitchon, B., Levinson, A.A., 1972. Hydrogeochemistry of the surface waters of the Mackenzie River drainage basin, Canada: I. Factors controlling inorganic composition. Geochim. Cosmochim. Acta 36, 825–865.

Rye, R., Holland, H.D., 1998. Paleosols and the evolution of atmospheric oxygen. A critical review. Am. J. Sci. 298, 621–672.

Singh, M., Sharma, M., Tobschall, H.J., 2005. Weathering of the Ganga alluvial plain, northern India: implications from fluvial geochemistry of the Gomati River. Appl. Geochem. 20, 1–21.

Stidson, R., Hamilton-Taylor, J., Tipping, E., 1998. Laboratory studies of the chemical weathering of rock from the English Lake District. In: Arehart, G.B., Hulston, J.R. (Eds.), Proceedings of the 9th International Symposium on Water-Rock Interaction. Balkemia, The Netherlands, pp. 395–398.

Stillings, L.L., Drever, J.I., Brantley, S.R., Sun, Y., Oxburg, R., 1996. Rates of feldspar dissolution at pH 3–7 with 0–8 mM oxalic acid. Chem. Geol. 132, 79–89.

Stumm, W., 1990. Aquatic Chemical Kinetics, Reaction Rates of Processes in Natural Waters. Wiley, p. 544.

Taylor, S.R., McLennan, S.M., 1985. The Continental Crust: Its Composition and Evolution. Blackwell, Oxford, p. 312.

Taylor, S.R., McLennan, S.M., 1988. The significance of the rare earths in geochemistry and cosmochemistry. In: Gschneidner, K.A. Jr., Eyring, L. (Eds.), Handbook on the Physics and Chemistry of Rare Earth. Amsterdam (Elsevier), Vol. 11, pp. 485–578.

Tuttle, M.L.W., Breit, G.N., 2004. Metal Mobility, transport and fate during weathering of Devonian metaliferous black shale. In: Wanty, R.B., Seal, R.R. II (Eds.), Proceedings of the 11th International Symposium on Water–Rock Interaction. Taylor and Francis, Lisse, The Netherlands, pp. 883–886.

Tuttle, M.L.W., Goldhaber, M.B., Breit, G.N., 2001. Mobility of metals from weathered black shale: The role of salt efflorescence. GSA Abstracts with program 33(6), .

USEPA., 1991. EPA Method 200.8 (rev 4.4). Determination of trace elements in waters and wastes by inductively coupled plasma mass spectrometry. Research Triangle Park, NC.

Vreča, P., Pirc, S., Sajn, R., 2001. Natural and anthropogenic influences on geochemistry of soils in terrains of barren and mineralized carbonate rocks in the Pb–Zn mining district of Mežica, Slovenia. J. Geochem. Explor. 74, 99–108.

Weimer, W.C., Laul, J.C., Kutt, J.C., 1980. Prediction of the ultimate biological availability of transuranium elements in the environment. In: Baker, R.A. (Ed.), Contaminants and Sediments. Ann Arbor Science Publisher, Ann Arbor, pp. 465–484.

White, A.F., Brantley, S.L., 2003. The effect of time on the weathering of silicate minerals: Why do weathering rates differ in the laboratory and field?. Chem. Geol. 202, 479–506.

White, A.F., Bullen, T.D., Schultz, M.S., Blum, A.E., Huntington, T.G., Peters, N.E., 2001. Differential rates of feldspar weathering in granitic regoliths. Geochim. Cosmochim. Acta 65, 847–869.

Zhang, H., Bloom, P.H., 1999. Dissolution kinetics of hornblende in organic acid solutions. Soil Sci. Soc. Am. J. 63, 815–822.

Zhang, L., Min Sun, Wang, S., Yu, X., 1998. The composition of shales from the Ordos basin. China: Effects of source weathering and diagenesis. Sediment. Geol., 116, 129–141.

Developments in Environmental Science, Volume 5
D. Sarkar, R. Datta and R. Hannigan (Editors)

ISSN: 1474-8177/DOI:10.1016/S1474-8177(07)05014-0

Chapter 14

Hydrogeology of uranium-bearing groundwater in forest catchments in the humid temperate climate: A case study in the Kanamaru area, Yamagata, Japan

Yoji Seki, Kazuki Naito, Atsushi Kamei, Koichi Okuzawa, Naoto Takeno and Yoshio Watanabe

Abstract

Two small catchments containing first-order streams were studied to elucidate the principal features of hydrogeochemistry and hydrogeology of uranium-bearing groundwater. The study area (Kanamaru) is located in the northeastern part of Japan. It is in the foot of a mountainous region, with altitude of 150–310 m. Kanamaru has an average annual precipitation of 3000 mm and average temperature of 12°C. Cretaceous granite and overlying Miocene (16 Ma) sedimentary rocks are distributed around the drilling site. Twelve boreholes were drilled to get information of geological structure, lithology, rock chemistry, water chemistry, and hydrogeologic features in the area.

Groundwater flow matches surface topology and flows southward from the northern end of the small ridge to the southern end of the Kuzuresawa stream. Stepwise decrease of the hydraulic head with increasing depth is one of the most important hydrogeological features seen at this study site. The depth of the zone with the largest drop of hydraulic head matches the paleo-weathered zone of the Cretaceous granite where most of the cracks are cemented with hematite. Chemical change in groundwater with increasing depth from the water table to 50 m in depth is characterized by increase of pH, EC, Ca^{++} and HCO_3^-, and decrease of DO, Eh, SO_4^{--} and solute composition. Groundwater obtained from the shallower part of the sedimentary rocks has the major ion and a stable isotopic composition similar to those of the stream water. This suggests an important contribution of groundwater in the shallow part of the sedimentary rocks to the source of the stream water. Uranium concentrations of the water discharging from a nearby spring reach up to 6 ppb. Groundwater collected from boreholes at the depth of the uranium anomalous layer contains around 0.3 ppb U. If the removal rate is the same as the present rate, remaining time which is the time needed to remove all the uranium in the bedrock dissolved away is calculated to be 10 million years to 100 million years.

14.1. Introduction

The Kanamaru area, located in northeastern region Honshu, a main island of Japan, is a typical mountainous region with humid temperate climate, and has occurrence of uranium mineral in the Tertiary sedimentary rocks (Kadota, 1962). This study is significant because the behavior of uranium-bearing groundwater can be useful for such fields as ore-genesis, exploration study for uranium deposits, and to better understand groundwater migration scenarios in the high-level radioactive wastes disposal programs. This study will clarify the hydrogeological behavior of uranium-bearing groundwater in mountainous area under humid temperate climate. The full potential of this study has not been estimated yet, but the outlines of the hydrogeology of the area can be explained using obtained data sets at this moment. So we will present the principal structure of the hydrogeology and hydrogeochemistry of the Kanamaru area to show how the uranium-bearing groundwater acts in the slightly high relief topography in an eastern Asian island arc.

14.2. Topography, climate and geology

Altitude of the Kanamaru area ranges 150–310 m above sea level (asl), including a site for collecting precipitation samples (Fig. 14.1; Seki et al., 2004). The drilling survey area ("drilling site"), where drilling was intensively carried out, occupies about 50 × 200 m which intersects the general topographical trend and its altitude ranges 270–310 m asl (Watanabe et al., 2003).

Annual precipitation near the drilling site is about 3000 mm and an average temperature of about 12°C. Snow accumulation of up to 3 m covers the site from December to April in a typical year. Maximum monthly precipitation is in the winter snow season, followed by the rainy season in June and July. The monthly flow rate of nearby rivers reach maximum during snow melt season of April in a typical year. The occasional exception happens with heavy rainfalls in July and typhoons, which usually occur in summer and fall season.

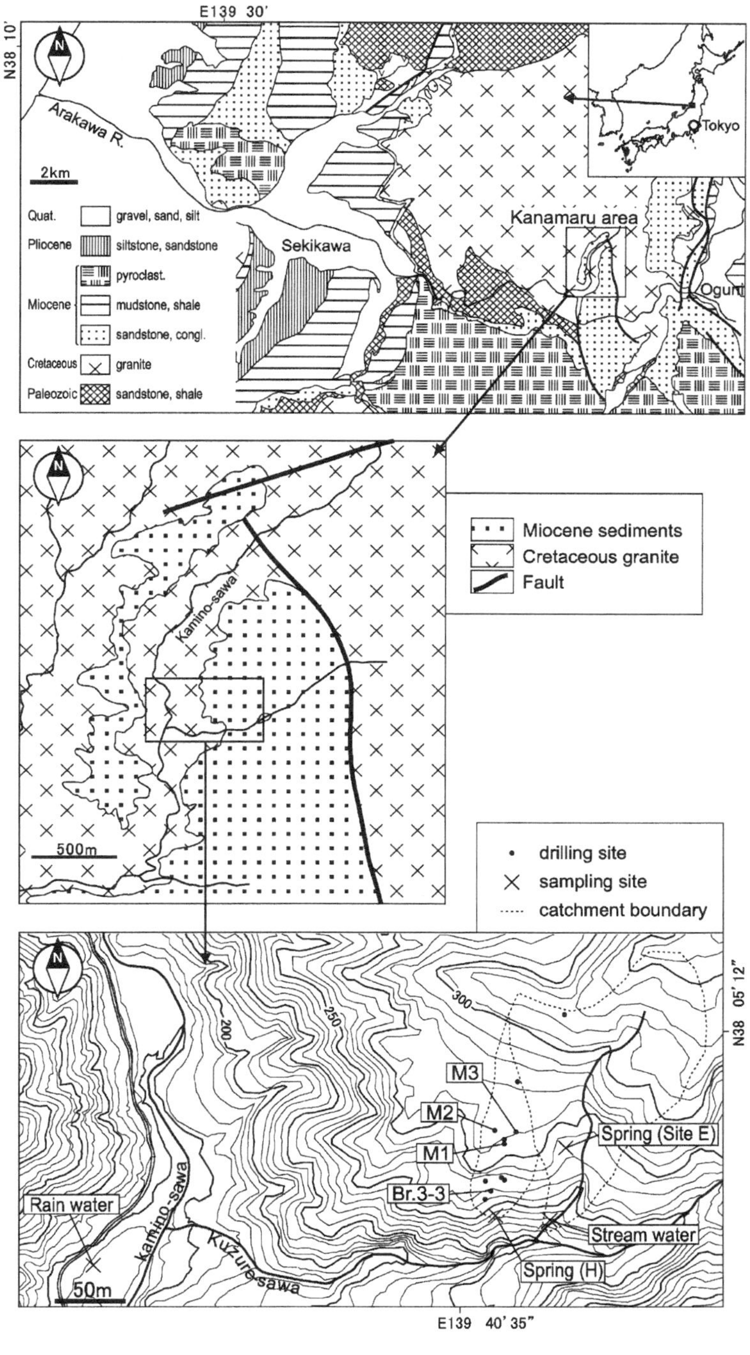

Figure 14.1. Regional geology (upper), local geology (middle) and geomorphology, locations of drilling and sampling sites (lower).

Topography around the drilling site has a general trend of westward decreasing altitude from the eastern upland of about 700 m asl to the western valley of the Kaminosawa which has an altitude of about 150 m asl.

Altitude difference is about 600 m to about 2 km from the upland to Kaminosawa stream. Topography adjacent to the drilling site is characterized by the northern gentle ridge running east to west, the first-order streams flow southward in the eastern and western edge. The Kuzuresawa stream is located in the southern edge of the area. The total area of these small catchments is about 0.1 km^2 (Fig. 14.1).

Cretaceous granitic rocks and overlying Miocene (16 Ma) sedimentary rocks are distributed around the drilling site. Granitic rocks are of porphyrytic biotite granite. Sedimentary rocks are coarse arkosic sandstone and arkosic sandstone interspersed with granitic pebbles, and intercalating muddy thin layers in several of the horizons. Uranium anomaly up to 100 ppm can be found in muddy thin intercalated layers in the Miocene arkosic sandstone as a result of domestic nuclear fuel exploration projects (Kadota, 1962; Hamachi and Igarashi, 1969).

14.3. Methods

Geochemical and isotropical analysis of spring water, stream water, and rainwater were carried out for geological logging, groundwater survey and hydrological survey. In total 12 boreholes were drilled during 2002–2004. Full core samples were recovered from all boreholes and were subjected to geological observation. Bottom hole depths of most of the boreholes reach 30–50 m, penetrating the unconformity between the Miocene sedimentary rocks and the Cretaceous granite (Fig. 14.2). MP system (Westbay Instruments Inc.; Black et al., 1986) was installed in the borehole of Br.3-3 in the southern part of the drilling site. The system divides 11 sections from the surface to the bottom by multiple packers to collect in-situ groundwater samples and measure pore water pressures (Fig. 14.3; Okuzawa et al., 2005).

Chemical parameters such as pH, electric conductivity (EC), dissolved oxygen (DO) and redox-condition (Eh) of the groundwater samples collected from the MP system were measured under low oxygen conditions by using nitrogen gas purge. Double-packer systems targeting a particular depth were installed in six boreholes, in which only the groundwater of interest depth can flow into PVC casings. PVC strainers for the complete depths were installed in most of the other boreholes. Collecting groundwater samples from the Br.3-3 was carried out about every two months, and collection of the other boreholes was monthly.

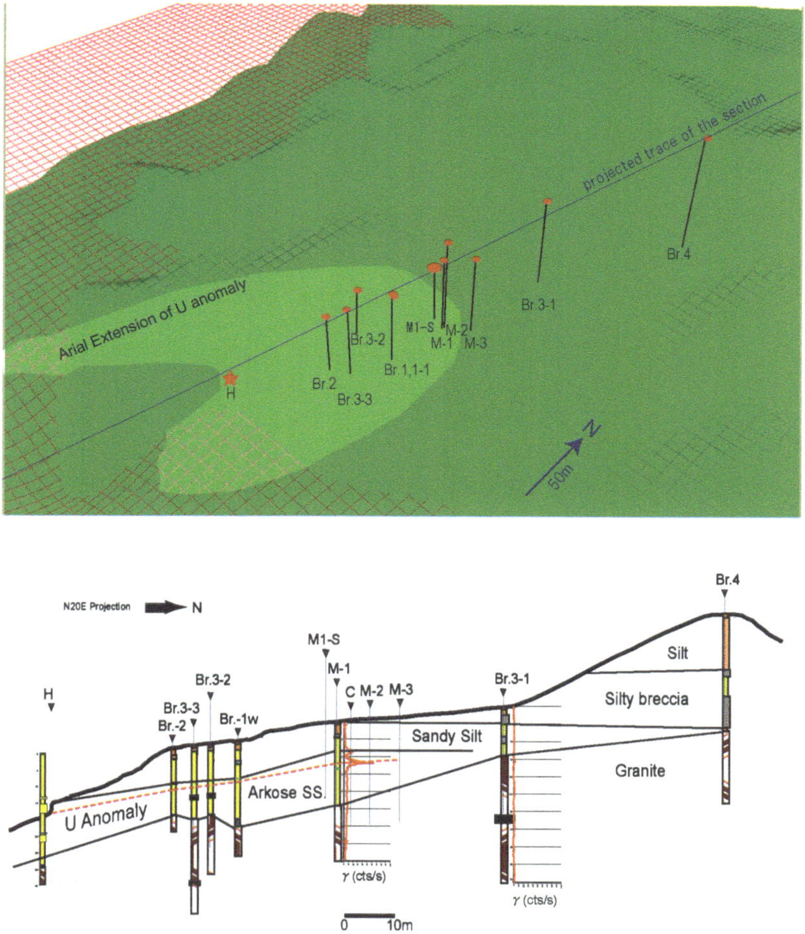

Figure 14.2. Distribution of the boreholes and underlying uranium anomalous layer (upper) and crosscut showing underground geology.

A spring found in a small valley head near the eastern edge of the drilling site ("Spring (E)") and another spring discharging at the outcrop of the high uranium muddy layer in the shoulder of the southern edge of the drilling site ("Spring (H)") was collected for analyses. Water samples from spring (site E) were collected at the immediate downstream of the spring. Those from spring (H) were collected from porous cup sampler with a one meter polyvinyl chloride (PVC) rod by suction method using a handy vacuum pump.

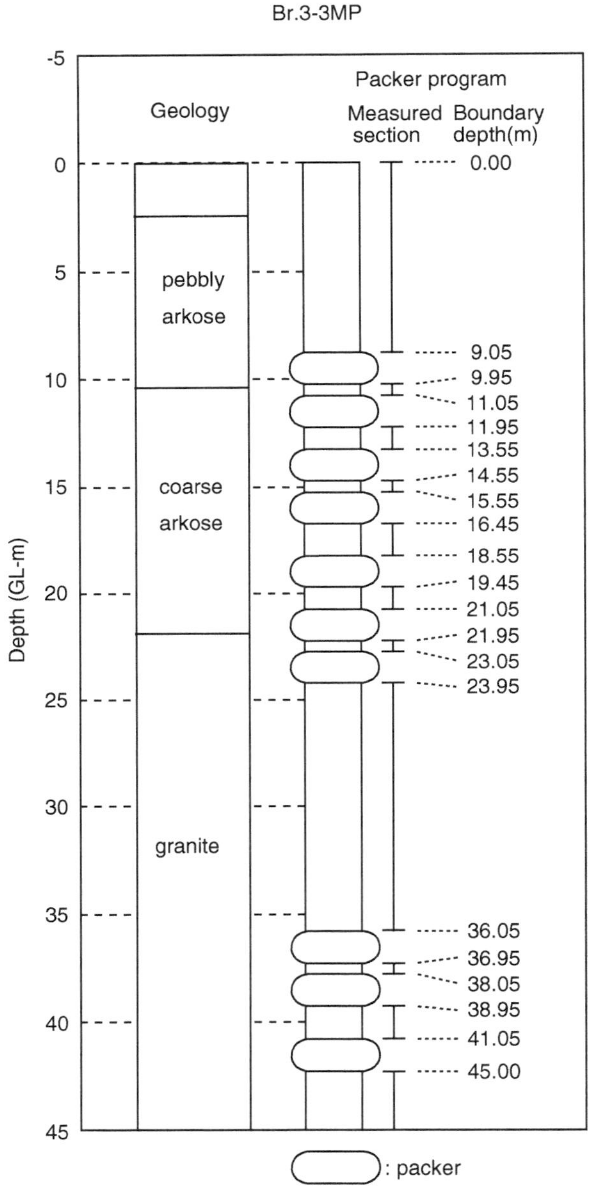

Figure 14.3. Outline of the MP system installed in the borehole Br.3-3.

Stream water samples were collected at the mouth of the first-order drainage located near the southeastern corner of the drilling site.

Rainwater samples were collected along the Kaminosawa stream about 300 m west of the Spring (H), using a funnel trap.

Time interval for sample collection of spring water, stream water, and rainwater are one month. Parameters such as temperature, pH, EC, DO and Eh were measured in the field. Water samples were collected to pre-washed polyethylene bottles and were filtered using 0.45 μm membrane filters. Major cations and minor elements were analyzed by ICP-MS. Major anions other than HCO_3^- were analyzed by ion chromatography. Bicarbonate ion was quantified by H_2SO_4 titration method.

Hydraulic conductivities were measured by permeability tests performed in the Br.3-3 shortly after completion of drilling. Permeability tests were carried out for several depths using a double-packered system, which forms a 2 m section hydrologically isolated from the other parts of the borehole. Hydraulic conductivity was calculated based on an attenuation curve of hydraulic pressure of each section when the groundwater was discharged at slow rate by a plunger pumping system.

Pore water pressures were measured by the MP system for the Br.3-3, and were also measured by water level observation for wells M1, M2 and M3, all of which are equipped with a double-packered system for a certain depth section of interest.

14.4. Results

14.4.1. Distribution of uranium anomaly in bedrocks

Cores with high uranium-bearing layer were found in most boreholes. Uranium anomaly is found in a dark gray muddy layer intercalated in the Tertiary arkosic sandstone. The muddy layer is distributed at about 10 m deep in the most drill holes, which gently slopes southward, and is concordant with the general structure of the Tertiary sedimentary rocks. Average thickness of the muddy layer is about 10–30 cm with a plane size of about 150 × 100 m (Watanabe et al., 2003; Fig. 14.2). Analyzed uranium concentration from the drilled cores ranges 50–150 ppm.

14.4.2. Water chemistry

Chemical parameters of various waters in the Kanamaru area can be summarized as follows (Fig. 14.4; based on Seki et al., 2004; Naito et al., 2004). The rainwater and the groundwater of the shallower part of

Tertiary sedimentary rocks are acidic with a pH of 4 to 5. Spring water and stream water are slightly acidic with a pH of 5 to 6. The groundwater of the deeper part of the Tertiary sedimentary rocks is near neutral pH of 6 to 7. The groundwater of the Cretaceous granite is of neutral pH, around 7. The rainwater, spring water and stream water have EC of $1–5\,\mathrm{mS\,m^{-1}}$, $5–15\,\mathrm{mS\,m^{-1}}$, and $5–10\,\mathrm{mS\,m^{-1}}$, respectively. The groundwater

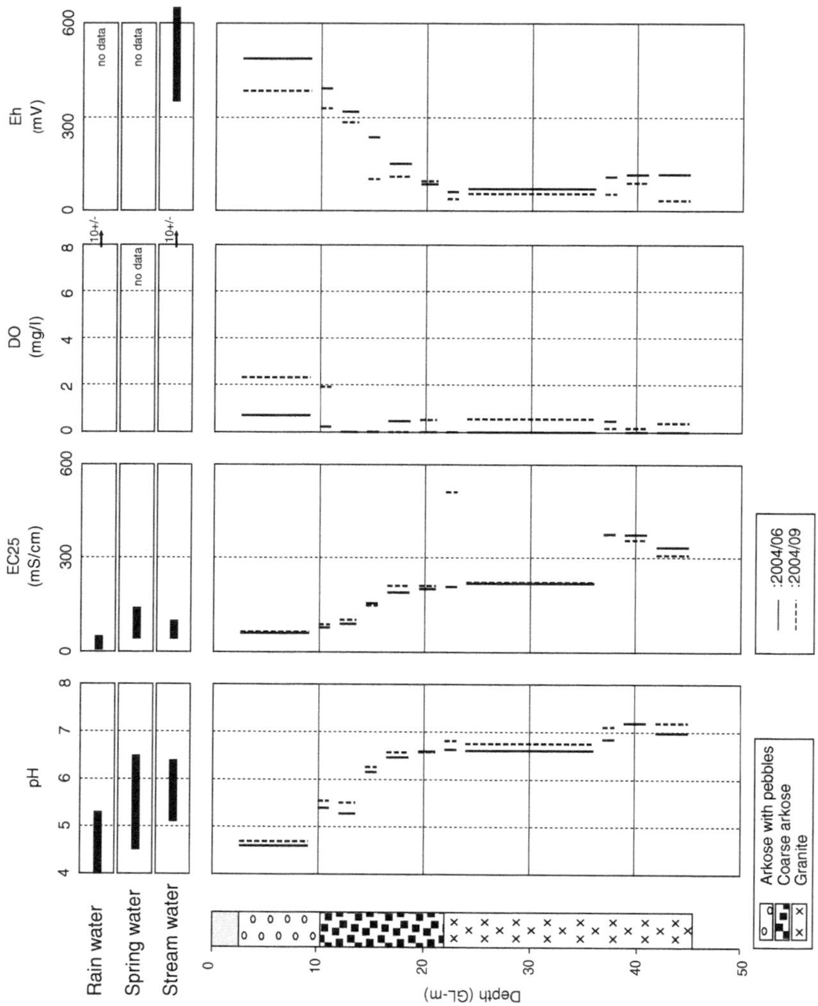

Figure 14.4. Chemical parameters of rainwater, stream water, spring water and groundwater.

of the shallower and the deeper part of the Tertiary sedimentary rocks, and those of the Cretaceous granite have EC of 5–10 mS m^{-1}, 10–20 mS m^{-1}, and 20–40 mS m^{-1}, respectively. The groundwater in the shallower part of the Tertiary sedimentary rocks has higher concentration of DO similar to surface water. The groundwater below the horizon of the uranium anomaly has DO close to zero. The groundwater with zero DO has an Eh of near 0 mV such as seen in the Cretaceous granite.

Rainwater has the lowest total dissolved solids (TDS). TDS increases, in ascending order from rainwater, to stream water, to spring water and then groundwater in the sedimentary rocks, to the groundwater of the granite. Major ions in the rainwater are Na^+, Cl^- and SO_4^{2-}. Major ions in the spring water and the stream water are Na^+ and Cl^-. Major ion composition of the groundwater changes from Na^+ and Cl^- in the shallower sedimentary rocks to Na^+, Ca^{2+} and HCO_3^- in the deeper granite (Fig. 14.5 and Table 14.1).

Concentrations of minor elements in the Kanamaru waters span six orders (Fig. 14.6). Many minor elements behave similar to TDS, that is, the lowest concentration in rainwater to the highest concentration in the Cretaceous granite. Exception is Al, which is lowest in concentration in groundwater. Al, Ti, Fe, Co, Cu, Pb, Th and U are at their highest concentration in Spring H. Cu, Pb, Zn, Sb and As have a relatively higher concentrations in rainwater (Fig. 14.7).

14.4.3. Stable isotopes

Deuterium and Oxygen-18 isotope composition of the rainwater from the Kanamaru area is plotted principally parallel to the "global meteoric waterline (Craig, 1961)". In northeast Japan the *d*-value (d = delta $D - 8 \times \text{delta}^{18}O$) of the meteoric line is larger than 20 in winter and smaller than 10 in summer (Waseda and Nakai, 1983). The Kanamaru rainwater seems to have *d*-values between 10 and 20, which is expected considering that no sample was taken in winter season. Isotopic compositions of the spring water and the stream water plotted in the limited area near deltaD is −9 per mill and delta^{18}O is −55 per mill (Fig. 14.8).

Isotopic composition plot of the groundwater collected from the borehole Br.3-3 using MP system shows two groups (Fig. 14.8). One is the groundwater from the Tertiary sedimentary rocks, which has deltaD of −9.3+/−0.1 per mill and delta^{18}O of −55+/−1 per mill. The other is the groundwater from Cretaceous granite, which has much lighter isotopic values. The isotopic composition of the groundwater from the Tertiary sedimentary rocks is almost the same as that of the stream water and the spring water.

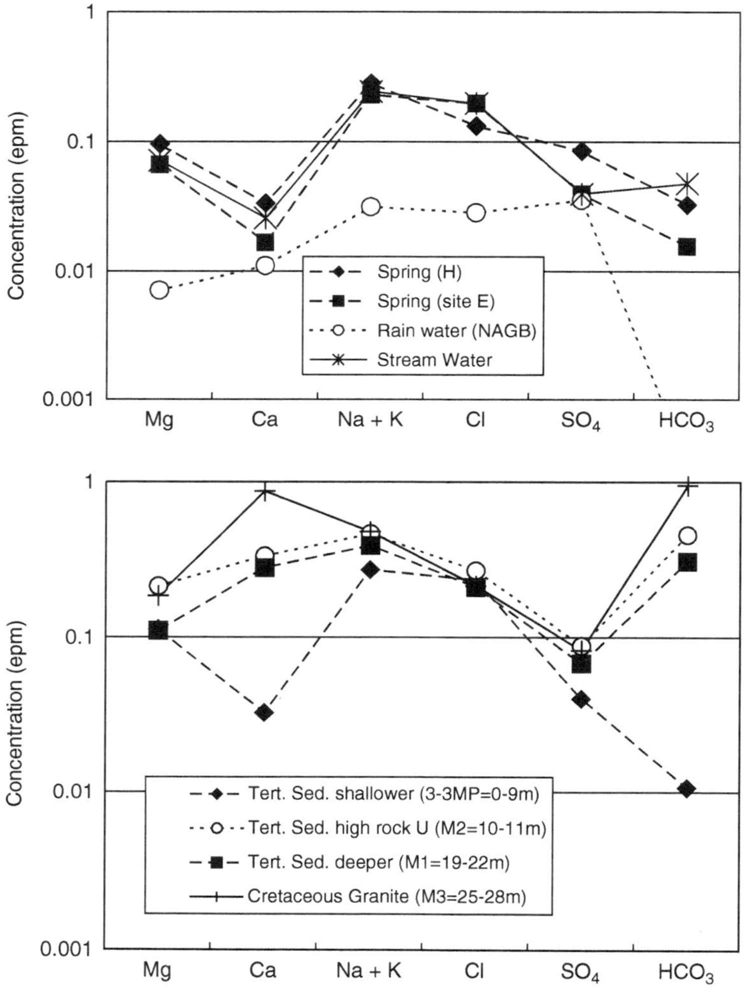

Figure 14.5. Composition of the major ions. A: Rainwater, stream water and spring water. B: Groundwater.

14.4.4. Hydraulic conductivity and hydraulic head

Pore pressure measurements by MP system in the borehole Br.3-3 show that the distribution of the hydraulic head of this area decreases stepwise downwards. A remarkable hydraulic gap is observed at around 23 m deep

Table 14.1. Chemical and isotopic compositions of the selected water samples from the Kanamaru area

Sample	Parameters				Major ions									Isotopes	
	pH	EC25	DO	Eh	HCO_3^-	F^-	Cl^-	NO_3^-	SO_4^{2-}	Na^+	K^+	Ca^{2+}	Mg^{2+}	δD	^{18}O
		$\mu S\,cm^{-1}$	$mg\,l^{-1}$	MV	ppm	$mg\,l^{-1}$	$mg\,l^{-1}$	$mg\,l^{-1}$	$mg\,l^{-1}$	$mg\,l^{-1}$	$mg\,l^{-1}$	$mg\,l^{-1}$	$mg\,l^{-1}$	‰	‰
Rain water	4.96	9	na	na	na	nd	0.30	0.43	0.26	0.21	0.07	0.11	0.03	−66	−9.6
Spring (site E)	4.67	59	6.48	394	0.0	nd	7.92	0.29	1.81	4.24	0.16	0.28	0.68	−54	−9.1
Spring (H)	5	48	11.43	388	0.7	nd	6.79	nd	4.08	4.24	0.87	0.29	0.68	−62	−9.6
Stream water	5.1	54	8.93	379	2.1	nd	6.61	0.46	2.07	4.74	0.32	0.44	0.74	−55.0	−9.0
GW from Tert.sed. (shallower)	4.8	63	1.66	363	2.1	nd	8.49	0.11	2.70	5.09	0.35	0.84	1.13	−56.5	−9.0
GW from Tert. sed. (high U)	6.4	147	0.28	511	27.9	0.03	9.52	nd	4.25	8.73	3.36	6.70	2.59	nd	nd
GW from Tert. sed. (deeper)	6.8	130	na	na	18.8	nd	7.41	1.13	3.27	7.78	2.01	5.60	1.34	nd	nd
GW from Cret. gran.	6.6	189	0.00	na	57.8	nd	7.56	nd	4.00	8.19	4.85	17.40	2.24	nd	nd

Sample	Minor elements															
	Al	Si	Ti	Mn	Fe	Co	Cu	Pb	Zn	As	Br	Rb	Sr	Ba	Th	U
	$\mu g\,l^{-1}$	$\mu g\,l^{-1}$	$\mu g\,l^{-1}$	$\mu g\,l^{-1}$	$\mu g\,l^{-1}$	$\mu g\,l^{-1}$	$\mu g\,l^{-1}$	$\mu g\,l^{-1}$	$\mu g\,l^{-1}$	$\mu g\,l^{-1}$	$\mu g\,l^{-1}$	$\mu g\,l^{-1}$	$\mu g\,l^{-1}$	$\mu g\,l^{-1}$	$\mu g\,l^{-1}$	$\mu g\,l^{-1}$
Rain water	5	nd	0.0	2.8	nd	0.019	0.7	0.40	218	0.13	nd	0.140	0.4	1.9	nd	nd
Spring (site E)	27	2050	0.2	4.0	nd	0.267	nd	0.15	2	0.07	45	0.571	4.9	7.8	nd	0.012
Spring (H)	201	2350	0.6	9.0	28	0.698	66.4	4.53	173	0.16	41	2.826	3.8	10.8	0.160	6.607
Stream water	17	2960	0.7	1.1	nd	0.029	nd	0.03	1	0.06	45	0.979	5.2	6.8	0.003	0.004
GW from Tert.sed. (shallower)	11	2550	0.6	22.2	nd	0.432	0.8	0.70	54	0.08	56	1.084	5.9	9.2	nd	0.077
GW from Tert. sed. (high U)	3	9400	0.7	98.0	nd	0.675	2.9	2.29	>250	0.33	41	2.380	26.3	47.8	nd	0.266
GW from Tert. sed. (deeper)	7	4700	0.4	25.2	nd	0.185	3.0	0.44	>250	0.24	34	1.870	28.0	9.2	0.001	0.064
GW from Cret. gran.	6	7400	0.7	118.0	60	0.446	5.6	3.54	>250	0.35	38	3.430	153.0	45.8	0.002	0.087

na, not analyzed; nd,: not detected.

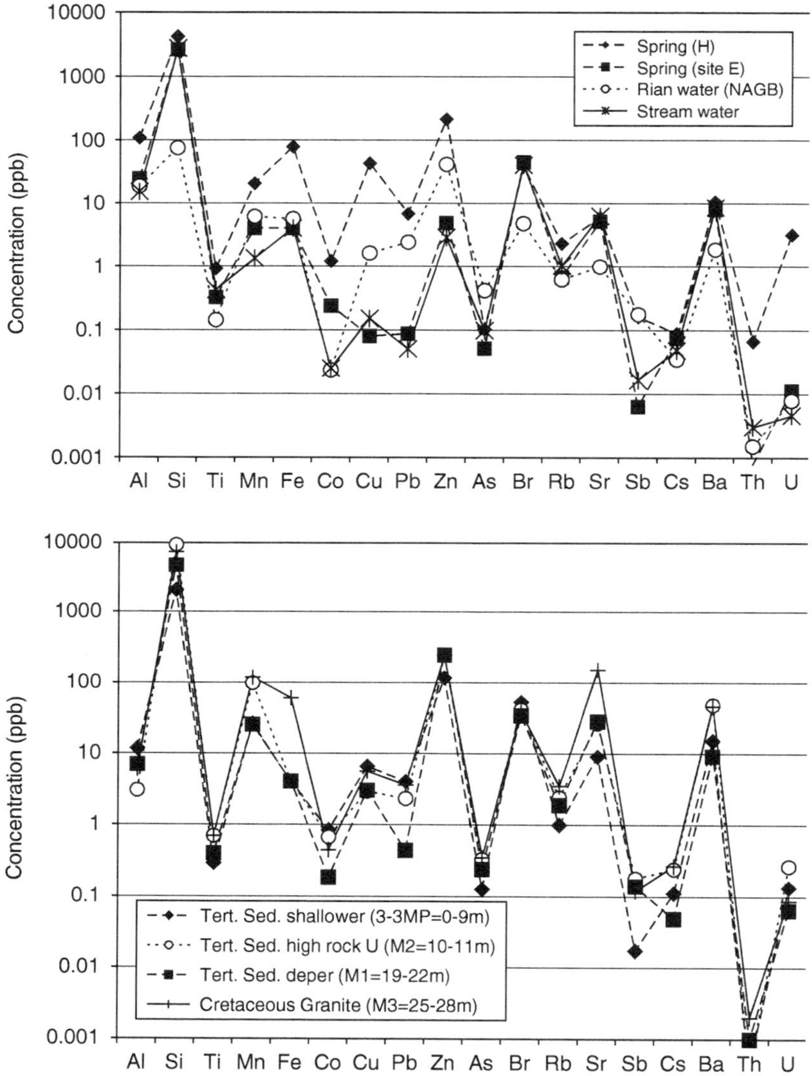

Figure 14.6. Composition of the minor elements. A: Rainwater, stream water and spring water. B: Groundwater.

in Br.3-3, where hydraulic head decreases by 89 m. Above that depth the hydraulic head of the sedimentary rocks ranges Gl-2 to GL-8 m. Below that depth the hydraulic head of the granite is GL-15 m showing no change toward depth (Fig. 14.9).

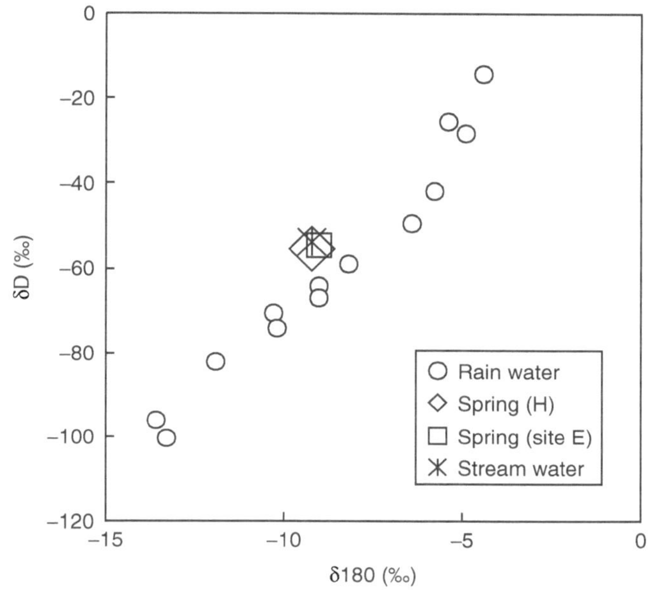

Figure 14.7. Stable isotopic composition of rainwater, stream water and spring water.

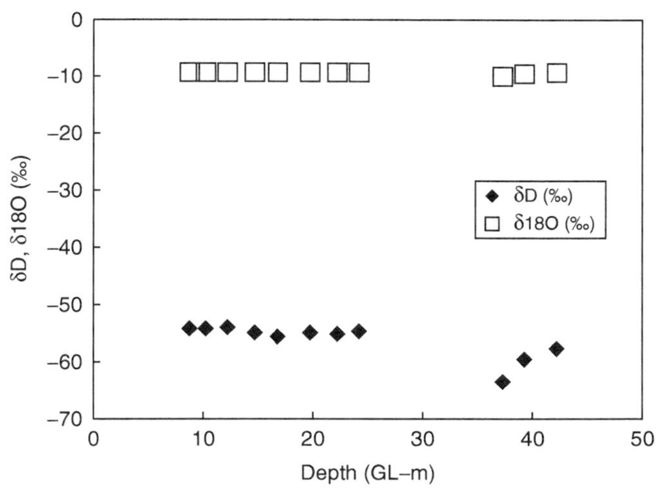

Figure 14.8. Stable isotopic composition of groundwater measured at the borehole Br.3-3 using the MP system.

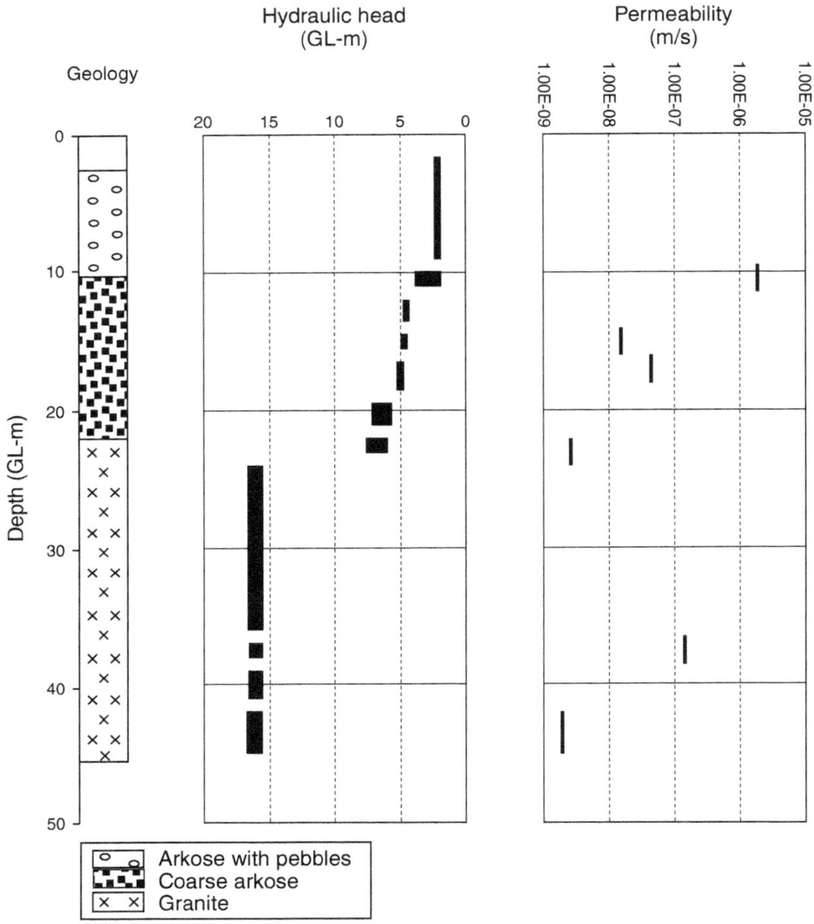

Figure 14.9. Hydraulic head and hydraulic conductivity measured at the borehole Br.3-3.

Hydraulic head of the deeper part of the sedimentary rocks measured at the borehole M1, that of the high uranium horizon in sedimentary rocks at the M2, and that of the granite at the M3 show that the hydraulic head is higher in the northern part of the drilling site and lower in the southern part toward the Kuzuresawa. Distribution of the hydraulic heads in the granite shows little change both horizontally and vertically.

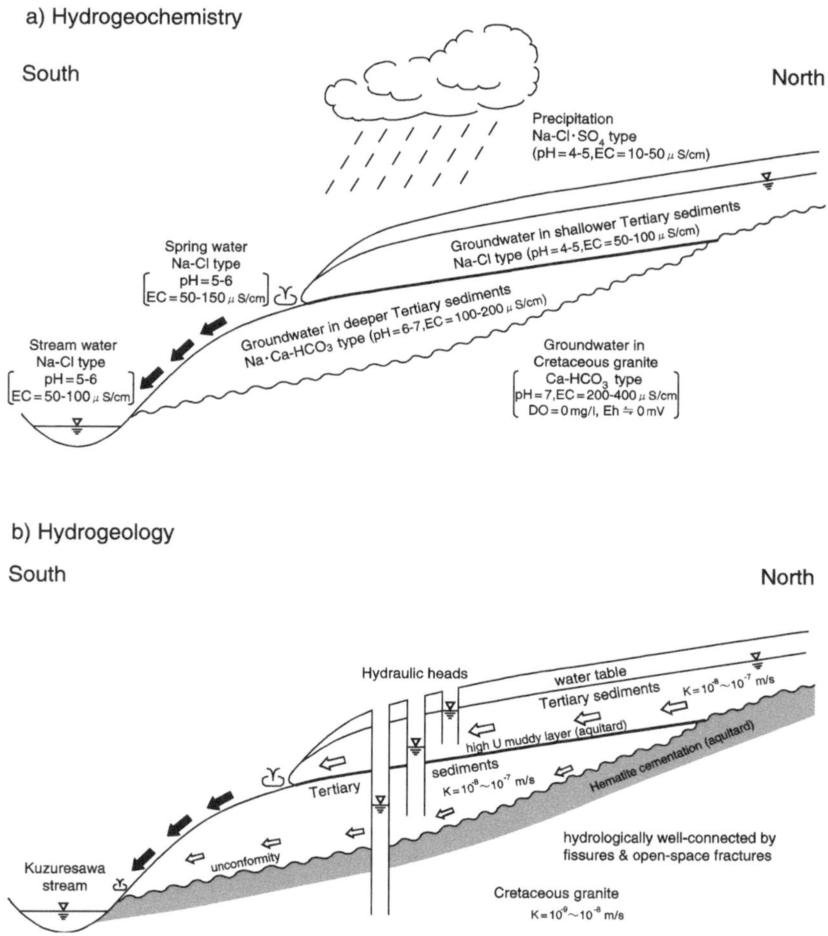

Figure 14.10. Sketch models illustrating hydrogeochemistry (upper) and hydrogeology (lower) for the Kanamaru area.

Hydraulic conductivities measured by the permeability test in the borehole Br.3-3 are 10^{-6} to 10^{-8} m s^{-1} for the sedimentary rocks and 10^{-7} to 10^{-9} m s^{-1} for the granite. For both geologic bodies the more permeable rock section has open-space fracture(s) which can act as an effective groundwater path (Seki et al., 2005). Rocks without open-space fractures show 10^{-7} to 10^{-8} for the sedimentary rocks and 10^{-8} to 10^{-9} for the granite (Fig. 14.9).

14.5. Discussion

14.5.1. Hydrogeological structure and groundwater flow

Principal movement of groundwater in the Kanamaru area is thought to be concordant with its topography, flowing southward from the northern end of the small ridge to the southern end of the Kuzuresawa (Fig. 14.10).

Stepwise decrease of the hydraulic head with increasing depth is one of the most important hydrogeological features. The depth zone of the largest hydraulic head drop agrees to the weathered zone of the Cretaceous granite where most cracks are perfectly cemented with hematite (Seki et al., 2005). The granite in the Kanamaru area usually has many hair cracks and larger joints. But this type of thorough cementation is usually only observed near the top of the granite, the first few meters. Therefore this hematite cementation is thought to make hydraulic conductivity of the zone much lower compared to the other zones of the granite, and that causes the largest gap of the hydraulic head in the zone. Relatively small gaps of the hydraulic head are observed in the Tertiary sedimentary rocks. These smaller gaps are thought to be made by intercalated thin muddy layers which have lower permeability compared to the surrounding arkosic sandstone.

Thus the hydraulic head gaps are attributed to the low permeable zone at the top of the granite and the thin muddy layers in the sedimentary rocks. The sedimentary structure and the shape of the top plane of the granite are concordant to the topography in this area. In addition, distribution of the observed hydraulic heads is also concordant to the topography. Therefore, it is rational to think that the principal groundwater flows in the sedimentary rocks are concordant with the topography approaching southward from the northern ridge to the southern Kuzuresawa.

Hydraulic head in the granite is unchanging both vertically and horizontally and that makes very small hydraulic gradient in the granite. Judging from the drill cores and borehole images of the granite a large number of hair cracks and joints develop in the granite and some of them have higher permeability (Seki et al., 2005). Hydraulic conductivity of granite with little cracks is as small as $10^{-9}\,m\,s^{-1}$. The facts mentioned above indicate that the principal pathways in the granite are open-space fractures and those pathways are well connected to maintain the hydraulic head of the large rock mass uniform. If so, once interference of the hydraulic pressure occurs somewhere in the rock mass, rapid groundwater flow in the pathways along fractures may take place to change even the hydraulic pressure in the whole rock mass.

The groundwater obtained from the shallower part of the sedimentary rocks has the major ion and the stable isotopic composition similar to those of the stream water. This suggests that groundwater in the shallow part of the sedimentary rocks makes an important contribution to the source of the stream water.

14.5.2. Geochemical evolution of groundwater

Chemical change in groundwater with increasing depth from the water table to 50 m deep in the Kanamaru area is described as an increase of pH and EC, decrease of DO and Eh, increase of Ca^{2+} and HCO_3^-, and decrease of SO_4^{2-} (Fig. 14.10).

Principal water–rock interaction in the Kanamaru area is thought to be that of meteoric water and some of the granite-composing minerals such as quartz, plagioclase, K-feldspar, biotite and muscovite. Compositional change of groundwater with increasing depth mentioned above indicates plagioclase dissolution by CO_2 gas, which was described in the shallower part of the granite in the Tono area (Iwatsuki and Yoshida, 1999). Undersaturation of albite calculated for most of the borehole waters using geochemical code (PHREEQE; Parkhurst et al., 1980) also supports that idea. Decreasing CO_2 partial pressure with increasing depth calculated from pH and HCO_3^- concentration (Stumm and Morgan, 1996) also supports the hypothesis of plagioclase dissolution by CO_2.

Decreasing Eh with increasing depth is thought to be the result of a change in dissolved sulfur species. H_2S concentrations of 12 ppm were obtained for most groundwater samples collected from granite zone. Calculated Eh values based on the measured concentrations of H_2S and SO_4^{2-} are similar to those directly measured by Eh probes. Such Eh decrease in relatively shallower level is probably related to the existence of pyrite and/or organic carbon in the thin muddy layers in the sedimentary rocks.

14.5.3. Uranium behavior

Uranium concentrations of the water discharging from Spring H reach up to 6 ppb. On the other hand, uranium concentrations of groundwater collected from boreholes (e.g., M2) at the depth of the uranium anomalous layer are around 0.3 ppb. There is 1 order of difference of uranium concentrations between the two waters collected from the same geological unit of the uranium anomalous layer. There are some possible reasons for this difference. The first is the differences in the redox condition. Because Spring H faces open air it must have near-atmospheric redox condition.

Meanwhile the borehole waters have much lower DO. Near-atmospheric oxidizing condition of Spring H can promote the dissolution of uranium from high uranium bedrock because uranium is much more mobile under oxidizing condition (Nash et al., 1981). The second reason may relate to secondary mineralization. If secondary uranium minerals exist near Spring H and they form a secondary enrichment zone, uranium concentration of the adjacent bedrocks may become higher than in original rocks. The oscillation of water level could also explain uranium concentrations. Because spring H is located on the steep slope beneath the shoulder of the topography, the groundwater is easy to discharge which makes the water table lean toward the Kuzeresawa. But in an event of heavy precipitation the water table goes up. Difference of water table level between dry and wet condition must be much larger than in areas adjacent to the boreholes. This can cause large changes of redox conditions compared to the areas of the boreholes and may lead to accelerate the dissolution of uranium.

No uranium mineral of oversaturation or equilibrium for the various levels of the groundwater in the Kanamaru area is found (Naito et al., 2004) based on calculation using geochemical code (PHREEQE; Parkhurst et al., 1980). This means that any uranium minerals in the area face possibility to be dissolved.

14.5.4. Uranium in future

It is possible to calculate remaining time to remove all the uranium in the bedrock by dissolving. This calculation is based on the total area of the high uranium bedrock layer, average uranium concentration of the bedrock layer, average uranium concentration in the groundwater adjacent to the layer, and flow direction and flow speed of the groundwater adjacent to the layer.

Total uranium mass is expressed as

$$tU = V \times d \times rU$$

where tU is total uranium mass, V is the total volume of the high uranium bedrock layer, d is the average density of the layer, and rU is the average concentration of uranium in the layer.

Although accurate numbers of these factors were not obtained yet, it can be calculated as follows:

$$tU = 150\ \text{m (NS)} \times 100\ \text{(EW) m} \times 0.2\text{m (thickness)}$$
$$\times 2000\ \text{kg m}^{-3} \times 100\ \text{mg kg}^{-1} = 600\ \text{kg}$$

Groundwater mass flowing in unit time is expressed as

$$Q = k \times A \times l$$

where Q is the groundwater mass in unit time, k the hydraulic conductivity, A the cross-sectional area for groundwater flow, and I the hydraulic gradient.

So again rough calculation can be made as follows:

$$Q = 10^{-7}\ \mathrm{m\ s^{-1}}\ (\text{or } 10^{-8}\ \mathrm{m\ s^{-1}}) \times 100\ \mathrm{m}\ (\text{EW}) \times 0.2\ \mathrm{m}\ (\text{thickness}) \times 0.2\ (\text{tangent of approximate average slope}) = 12,600\ \mathrm{kg}\ (\text{or } 1260\ \mathrm{kg})$$

Uranium mass move out in unit time is expressed as

$$oU = Q \times wU$$

where oU is the uranium mass removed in unit time, and wU the average concentration of uranium in groundwater.

Then it can be calculated as follows:

$$oU = 12,600\ \mathrm{kg}\ (\text{or } 1260\ \mathrm{kg}) \times 5\ \mu\mathrm{g\ kg^{-1}} = 6.3\ \mathrm{mg}\ (\text{or } 63\ \mathrm{mg})$$

This means 6.3–63 mg is removed from high uranium bedrock layer in the Kanamaru area annually. Therefore, if the removal rate is the same as the present rate, remaining time is calculated as follows:

$$600\ \mathrm{kg}/6.3\ \mathrm{mg}\ (\text{or } 63\ \mathrm{mg})/\text{year} = 10\ \text{million years (or 100 million years)}$$

In this long period of time some geological processes or events such as progression of erosion may occur which could accelerate to remove uranium in the area. So the remaining time obtained from the above calculation means maximum continuing time of existence of the uranium in the Kanamaru area.

14.6. Conclusions

(1) Groundwater from water table (GL-2 to 5 m) to 50 m deep shows stepwise decrease of hydraulic head with increasing depth. Low permeable muddy layers intercalated in the arkosic sandstone in the Tertiary sedimentary rocks or a low permeable zone of hematite cementation along the top of the Cretaceous granite are responsible for the hydraulic head gaps. Water chemistry is also divided by the hydraulic head gaps.

(2) Hydraulic gradient in the sedimentary rocks is concordant to the surface topography. Groundwater in the sedimentary rocks principally flows southward from the northern end of the small ridge to the southern end of the Kuzuresawa.

(3) Chemical change in groundwater with increasing depth from the water table to 50 m deep is characterized by increase of pH and EC, decrease of DO and Eh, increase of Ca^{2+} and HCO_3^-, and decrease of SO_4^{2-}.

(4) Compositional change of groundwater with increasing depth can be explained by plagioclase dissolution by CO_2 gas.

(5) Decreasing Eh with increasing depth is thought to result from change of dissolved sulfur species. Eh decrease in relatively shallower level is probably related to the existence of pyrite and/or organic carbon in the thin muddy layers in the sedimentary rocks.

(6) The groundwater obtained from the shallower part of the sedimentary rocks has the major ion and the stable isotopic composition similar to those of the stream water. This suggests an important contribution of the groundwater in the shallow part of the sedimentary rocks to the source of the stream water.

(7) Uranium concentrations of the water discharging from Spring H reach up to 6 ppb. Meanwhile that of groundwater collected from boreholes at the depth of the uranium anomalous layer are around 0.3 ppb.

(8) If the removal rate is the same as the present rate, remaining time needed to remove all the uranium in the bedrock by dissolving it away is calculated to be 10 million years to 100 million years.

REFERENCES

Black, W.H., Smith, H.R., Patton, F.D., 1986. Multiple level groundwater monitoring with the MP system. Proc. NWWA-AGU Conference on surface and borehole geophysical methods and groundwater instrumentation, pp. 41–61.

Craig, H., 1961. Isotopic variations in meteoric waters. Science 133, 1702–1703.

Hamachi, T., Igarashi, T., 1969. Uraniferous deposits of the Oguni-Kanamaru district, Niigata and Yamagata prefectures. Geological Survey of Japan, Report no. 222, 581–593.

Iwatsuki, T., Yoshida, H., 1999. Groundwater chemistry and fracture mineralogy in the basement granitic rock in the Tono uranium mine area, Gifu Prefecture, Japan—Groundwater composition, Eh evolution analysis by fracture filling minerals. Geochem. J. 33, 19–32.

Kadota, O., 1962. Uranium deposits in the Oguni-Kanamaru district on the border of Niigata and Yamagata prefectures. Min. Geol. 12, 199–210.

Naito, K., Kamei, A., Tsukamoto, H., Watanabe, Y., Seki, Y., 2004. The behavior of the groundwater containing uranium in wells: An example of Kanamaru district, Yamagata. Abstracts of the 111th Annual Meeting, Geological Society of Japan, p. 284.

Nash, J.T., Granger, H.C., Adams, S.S., 1981. Geology and concepts of genesis of important types of uranium deposits. Econ. Geol. 75th Anniversary, 63–116.

Okuzawa, K., Seki, Y., Naito, K., Kamei, A., Watanabe, Y., 2005. Propagation processes of artificial disturbance in in-situ groundwater samples: An example of Kanamaru area

Yamagata, Japan. Abstracts of Japan Earth and Planetary Science Joint Meeting. G018–P012.

Parkhurst, D.L., Thorstenson, D.C., Plummer, L.N., 1980. PHREEQE-A computer program for geochemical calculations. U.S. Geological Survey Water-Resources Investigations, pp. 80–96.

Seki, Y., Hishida, S., Konishi, H., Naito, K., Watanabe, Y., 2005. An application of heat pulse flowmeter logging: Identification of water passage and detection of seepage flow in low permeability rock mass. Eng. Geol. 46, 190–197.

Seki, Y., Naito, K., Kamei, A., Tsukamoto, H., Watanabe, Y., 2004. Behavior of uranium bearing groundwater in Neogene sedimentary rocks. Observation at the Kanamaru district, Yamagata, Japan. Abstracts of GSA annual meeting.

Stumm, W., Morgan, J.J., 1996. Aquatic Chemistry. Wiley, New York.

Waseda, A., Nakai, N., 1983. Isotopic compositions of meteoric and surface waters in Central and Northeast Japan. Chikyukagaku (Geochemistry) 17, 83–91.

Watanabe, Y., Seki, Y., Naito, K., Suzuki, M., Kamei, A., Nakajima, T., 2003. Uranium anomaly in the shallow Neogene sediments in Kanamaru Yamagata: A natural analogue study of the radionuclide transport. Abstracts of the 110th Annual Meeting, Geological Society of Japan, p. 302.

Section IV:
Geochemistry in Soils Research

Developments in Environmental Science, Volume 5
D. Sarkar, R. Datta and R. Hannigan (Editors)

ISSN: 1474-8177/DOI:10.1016/S1474-8177(07)05015-2

Chapter 15

Effects of incubation time and arsenic load on arsenic bioaccessibility in three Florida soils amended with sodium arsenate

Rupali Datta, Konstantinos C. Makris and Dibyendu Sarkar

Abstract

The potential for human exposure to arsenic (As), a group A carcinogen, has increased tremendously due to the encroachment of suburban areas into former agricultural lands, where arsenical pesticides were used extensively prior to the 1990s. Soil ingestion is the no. 1 route of As exposure, attributable to incidental hand-to-mouth activities by children in playgrounds or house yards having As-contaminated soil. Previous studies have shown that As bioaccessibility in soils is mostly a function of their physicochemical properties. We selected three Florida soils covering a wide range in chemical properties, such as, Fe/Al hydroxides, organic matter, and Ca/Mg content, which are most likely to influence As bioavailability. The soils were amended with sodium arsenate pesticide at loads ranging from 45 to 450 mg As per kg, and subsequently incubated up to 12 months. The overall objectives of this study were to evaluate the effects of incubation time and As load on soil speciation, hence bioaccessibility of As. Results showed a reduction in the water-soluble, and plant-available (water + NH_4Cl-extractable) As fractions of the three soils after four months of incubation, which remained unchanged up to 12 months. This reduction with time was accompanied by a concurrent increase in the NaOH-extractable As fraction, suggesting As sorption by amorphous Fe/Al hydroxides could decrease As bioaccessibility. The effect was most pronounced for the Pahokee Muck soil, which had the greatest amount of amorphous Fe/Al hydroxides (2000 mg kg^{-1}) of all soils. Arsenic associated with the water- and plant-available As fractions is most bioaccessible as indicated by the significant ($\alpha = 0.05$) correlation between the water-soluble, plant-available, and the in-vitro bioaccessible As fraction in the soils. The NaOH-extractable As fraction was negatively correlated with the in-vitro bioaccessible As fraction, suggesting that the presence of Fe/Al hydroxides could decrease As lability in soils.

15.1. Introduction

Arsenic is a human health concern because it can contribute to skin, bladder, and other cancers (National Research Council-NRC, 1999). Arsenic can also be toxic to plants; an average toxicity threshold of $40\,\mathrm{mg\,kg^{-1}}$ soil was established for several crop plants (Sheppard, 1992). Inorganic arsenicals are classified as the number one toxin in the U.S. Environmental Protection Agency (USEPA) list of prioritized pollutants. Arsenic can be found in surface and subsurface water bodies, soils and in many foods; the risk of exposure being the greatest in drinking water (Brown and Ross, 2002). Recommendations by the NRC to lower the current maximum contaminant level (MCL) allowed for As in drinking water from 50 to $10\,\mathrm{\mu g\,l^{-1}}$ have been accepted by the USEPA.

While As occurs naturally in the geosphere, there are several anthropogenic sources of As in the environment. Naturally occurring As can be found in soils as the result of the weathering of primary and secondary As-bearing minerals. Sixty percent of the anthropogenic As input comes from only two sources: Cu-smelting and coal combustion (Matschullat, 2000). Other anthropogenic sources of As include herbicide, pesticide, and rodenticide use, as well as waste incineration, mining, batteries, semiconductor industries, steel/glass production, and pressurized wood production (Matschullat, 2000). All of the above listed As sources have significantly increased As levels in soils (total soil As range 7–970 $\mathrm{mg\,kg^{-1}}$, Matschullat, 2000).

Arsenic exists in soils mostly in the +5 or +3 oxidation states. Arsenate (As(V)) is the oxidized form and occurs in well-aerated soils, whereas in chemically -reduced soil environments, arsenite (As(III)) is the prevalent As form. Although arsenite is more toxic than arsenate, arsenate can also have deleterious effects on animals, plants, and microorganisms. Sodium arsenate (SA) was extensively used as a pesticide in the past in agricultural land, elevating soil As concentrations above background levels. The risk of human contact with soil–As has greatly increased in the last two decades as residential areas expanded into former agricultural land. Elevated As concentrations were found in soils used for residential developments located on former apple orchards that had been amended with arsenical pesticides for years (Murphy and Aucott, 1998).

Soil ingestion is the no. 1 route of exposure to As from incidental hand-to-mouth activity of children playing in playgrounds or back yards with As-contaminated soils (Cohen et al., 1998). Bioavailability of As depends upon soil speciation of As, which in turn is a function of the physicochemical properties of the soil. However, current practices utilize the conservative estimate of 100% As bioavailability in all soil types, ignoring differences in physicochemical properties between soil types. Several in-vivo animal studies have shown that bioavailability of As in soils is significantly lower. Since calculated health risk is a direct function of the input value for chemical dose, the assumption of using an input value of 100% bioavailability for exposure to As-enriched soils potentially overestimates the actual risk, thereby elevating the expenses associated with potential site-cleanup. Hence, there is an increasing need for more studies on the bioavailability of As in contaminated soils of varying physicochemical properties.

In-vivo bioavailability studies using animal models are expensive and time-consuming. The elevated cost associated with performing in-vivo bioavailability studies has encouraged the use of in-vitro methods to estimate bioavailable- or bioaccessible-As (Davis et al., 1992; Sheppard et al., 1995; Rodriguez et al., 1999). We chose to utilize the recently developed term *bioaccessibility* to better define the ambiguous use of *bioavailability* that has been used for both in-vivo biological and in-vitro chemical extraction assays. According to Semple et al. (2004), the *bio-accessible* compound is available to cross an organism's cellular membrane from the environment, if the organism has access to the chemical. However, the chemical may be either physically removed from the organism or only bioavailable after a period of time (Semple et al., 2004). A *bioavailable* compound is defined as the compound which is freely available to cross an organism's cellular membrane from the medium the organism inhabits at a given time (Semple et al., 2004).

Most of the in-vitro methods to estimate bioaccessibility developed for As have dealt with severely contaminated waste disposal sites, but were not applied to arsenical pesticide-treated soils, which usually contain marginally higher than background As concentrations. A few recent studies have utilized soils contaminated with lower levels of As ($\sim 45\,mg\,kg^{-1}$) to evaluate the effect of different soil types on the magnitude of bioaccessible As, where dimethylarsinic acid (DMA) (Sarkar et al., 2004) and sodium arsenite (Datta and Sarkar, 2004) were used as the As pesticide source. Sarkar et al. (2004) emphasized the influence of soil type on the geochemical fractionation and bioaccessibility of As; soils with high concentrations of amorphous Fe/Al hydroxides retained more As–DMA, rendering As less bioaccessible.

The current study aims at addressing the effect of soil physicochemical properties on As fractionation and bioaccessibility in three soils incubated with sodium arsenate. We hypothesized that poorly As sorbing soils (low in amorphous Fe/Al hydroxides) would exhibit high degree of As bioaccessibility, whereas soils with higher amorphous Fe/Al content would show increased As retention. The main objectives of this study were: (i) to evaluate the effects of incubation time and As load on As geochemical fractionation and bioaccessibility in three FL soils and (ii) to determine the degree of correlation between various soil–As forms, as obtained by sequential extraction and different measures of As bioaccessibility obtained by in-vitro tests.

15.2. Materials and methods

Soil sampling and characterization. Three surface (0–12 cm depth) FL soils were collected for this study: soil samples from the Immokalee series were collected from Southwest Florida Research and Education Center, Immokalee, Florida; Millhopper soil samples were collected from the University of Florida campus at Gainesville, Florida, and the Pahokee Muck soil was collected from Everglades Research and Education Center at Belle Glade, Florida. The soils were selected based on their presumed As retention capacities. Immokalee soil is a sand (93%) with minimum As retention capacity, Millhopper soil is a sandy loam with relatively high concentration of Fe/Al hydroxides, thus, higher As sorption capacity than Immokalee soil. The third soil used was a Pahokee Muck soil series, which is expected to show the highest affinity for As, since it is characterized by large amounts of Fe/Al and Ca/Mg contents, as well as, increased organic matter content.

Soils were air-dried, passed through a 2-mm sieve, and analyzed for pH, electrical conductivity, particle size, and water content using standard protocols. Organic matter content was determined using the loss-on-ignition method (Klute, 1996). Exchangeable cations were extracted in 1 M ammonium acetate (pH 7.0) and cation exchange capacity was determined with the ammonium acetate method (Rhoades, 1982). Plant-available Ca, Mg, and P were determined using the Mehlich III soil test (Mehlich, 1984). Oxalate-extractable Fe and Al concentrations of the soils were determined using Tamm's reagent (Klute, 1996). Total-recoverable Ca, Mg, Fe, Al, P, and As concentrations were measured in acid digests according to USEPA method 3050B (US EPA, 2000).

Phosphorus was colorimetrically measured with a UV/vis spectrophotometer, using the molybdate–ascorbic acid method (Watanabe and

Olsen, 1965). Soluble Fe, Al, Ca, and Mg were analyzed using flame atomic absorption spectrometry (FAAS). Total soluble As was analyzed using graphite furnace atomic absorption spectrometry (GFAAS).

Soil amendments. Two hundred grams of the Immokalee, Millhopper, and Pahokee soils were amended with low ($45\,mg\,kg^{-1}$), medium ($225\,mg\,kg^{-1}$), and high ($450\,mg\,kg^{-1}$) As loads as sodium arsenate (SA). These As loads represent average As concentrations in soils treated with arsenical pesticides ($45\,mg\,kg^{-1}$), or loads that may be encountered in soils with a prolonged history of arsenical pesticide applications ($450\,mg\,kg^{-1}$ load) (Chisholm et al., 1955). Water content was maintained at 70% of the water holding capacity of the soils. Samples were thoroughly mixed and were stored in tightly sealed bags at room temperature. Soils were aerated regularly and constant water content (70% of the water holding capacity) was maintained biweekly. Extractable As was quantified in soil sub-samples taken at time zero (immediately after sodium arsenate application), 4 and 12 months of incubation, using sequential extractions and in-vitro procedures as described below.

Sequential As extractions. Sequential extraction of As was performed using the method of Chunguo and Zihui (1988) with a few modifications (Datta and Sarkar, 2004) for the following operationally defined As forms: (1) Water-soluble phase, (2) Exchangeable phase, (3) Fe- and Al-bound phase, (4) Ca- and Mg-bound phase, (5) Organic matter and sulfide-bound phase, and (6) residual phase. The sum of water- and NH_4Cl-extractable As pools is designated as plant-available pool, based on earlier phosphorus (P) fractionation experiments (Hedley et al., 1982). Extracts were filtered and analyzed for total soluble As using GFAAS.

In-vitro test. Bioaccessible As was determined following the method of Rodriguez et al. (1999) who developed the "physiologically based extraction test" (PBET) to predict metal bioavailability in human gastric juices (stomach phase).

Statistical analyses. Data were analyzed for the main and interaction effects of the incubation time and As load treatments in a 3×3 completely randomized design using the Design-Expert software (Design-Expert, 2001). Potential outliers were identified and were eliminated from statistical analyses. Pearson correlation coefficients were computed using the Statistix 7 software package. Data are reported as the means of triplicate measurements $\pm$ one standard deviation.

15.3. Results and discussion

Soil properties. The three surface FL soils were acidic, with electrical conductivities and cation-exchange capacities in the order of

Table 15.1. General chemical properties of the three FL soils

		Immokalee	Millhopper	Pahokee
pH		6.0	6.4	5.9
EC ($\mu S\,cm^{-1}$)		59	145	558
CEC ($cmol\,kg^{-1}$)		777	2356	18,908
Soil organic matter ($g\,kg^{-1}$)		8.4	43.8	854
Total-recoverable As ($mg\,kg^{-1}$)		0.8	1.8	1.9
P ($mg\,kg^{-1}$)	Mehlich 3	4.0	134	36.0
	Total	210	4900	6800
Ca + Mg ($mg\,kg^{-1}$)	Mehlich 3	270	900	10,000
	Total	1200	3200	41,000
Fe + Al ($mg\,kg^{-1}$)	Oxalate	66	704	2000
	Total	212	4800	6000

EC, Electrical conductivity; CEC, cation-exchange capacity.

Immokalee < Millhopper < Pahokee (Table 15.1). Soil organic matter content was least for the Immokalee (0.84%), higher in Millhopper (4.3%), and extremely high (85%) for the Pahokee Muck soil. Total As concentrations of the three soils were < 2 mg kg^{-1}, and within the range of background soil As concentrations (0.1–40 mg kg^{-1} soil) (Kabata-Pendias and Pendias, 1992). Total-recoverable P, Ca + Mg, and Fe + Al concentrations increased in the Immokalee < Millhopper < Pahokee order (Table 15.1). Oxalate-extractable Fe + Al concentrations for all three soils were < 40% of the total concentrations, suggesting the presence of Fe/Al hydroxides and potential for oxyanion (As or P) retention.

Sequential As extraction. Soil sub-samples taken at 0, 4, and 12 months were fractionated into different operationally defined As pools. Fractionation data for the unamended (no As added) soils showed As concentrations in the order of < 0.5 mg kg^{-1} in each fraction, and the sum of all As fractions was ~2 mg kg^{-1} for all three soils, being consistent with total-recoverable As concentrations measured independently via acid digestion (data not shown). Right after pesticide amendment to the soils (time zero), most of the added As was found in the water-soluble phase, representing > 60% of the total As for all As loads of the Immokalee and Pahokee soils (Fig. 15.1). Immokalee soil had negligible As retention capacity due to the absence of As-adsorbents such as Fe/Al oxides, or As-precipitators, such as Ca and Mg. The amount of water-soluble As was high in Pahokee Muck soil despite the high oxalate-extractable Fe/Al and Ca + Mg concentrations (Table 15.1). The large amount of organic matter (85%) in this soil probably covered the mineral surfaces and retarded As sorption by stabilizing As into solution. The relatively high amorphous Fe/Al content (700 mg kg^{-1}) of the Millhopper soil could

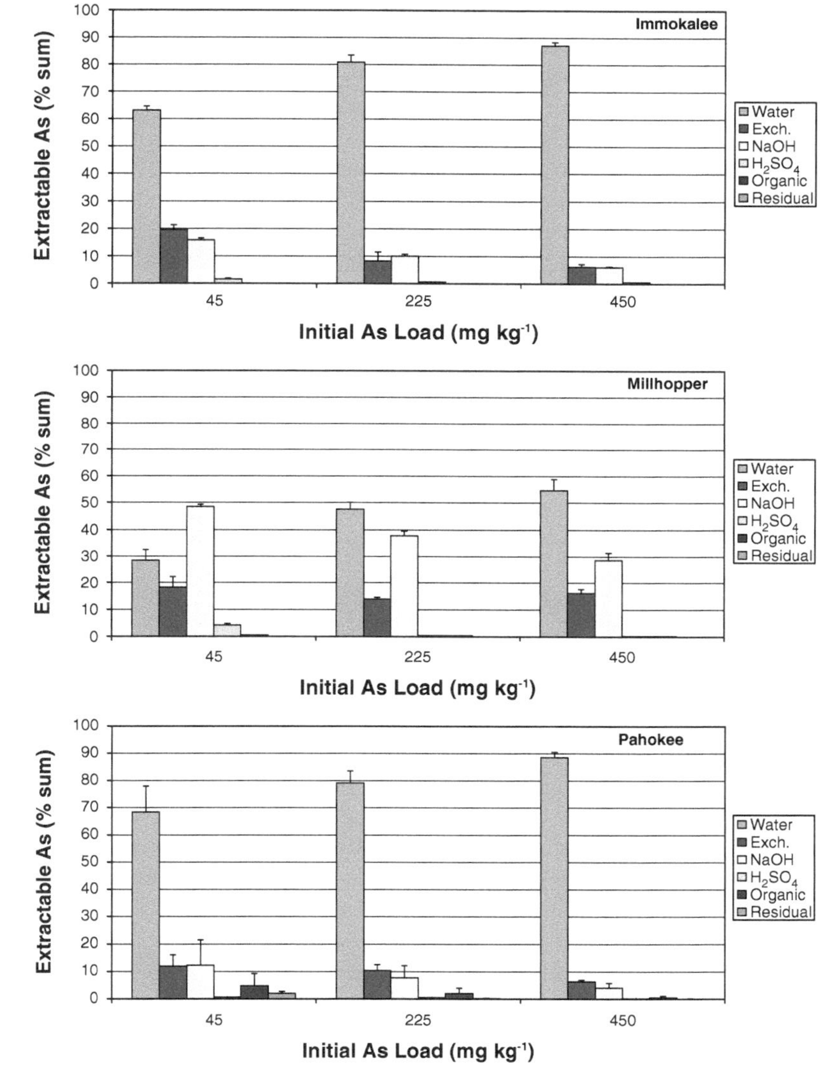

Figure 15.1. Sequential As fractionation data as a function of As load for the three soils at time zero. Data are the average of three replicates $\pm$ one standard deviation.

serve as a sink for As and resulted in low amounts of As in the water-soluble phase at time zero (30–50%, all As loads) (Fig. 15.1). Increasing As loads from 45 to 450 mg kg^{-1} resulted in greater amount of As extracted in the water-soluble phase for all soils, suggesting greater risk of As mobility (Fig. 15.1). Concurrently, As in the exchangeable and NaOH-extractable phase decreased with increasing As load.

Sub-samples were removed after 4- and 12-month incubation, and subjected to the sequential As fractionation scheme (Fig. 15.2). Water-soluble As in the Immokalee and Pahokee soils decreased after 4- and 12-month incubation, paralleling increases in the NaOH-extractable As fraction. Results for the Millhopper soil were less clear, but within the trends observed with the other two soils (Fig. 15.2). Reductions in water-soluble As fractions of the soils with incubation time suggest that soluble As was allowed to react and sorb onto Fe/Al- or Ca/Mg-bearing soil phases. Analysis of variance showed that there was a significant ($\alpha = 0.05$) interaction between incubation time and As load, rendering impossible to draw conclusions from the significant ($\alpha = 0.05$) main effects of time and As load as shown in Figs. 15.1 and 15.2.

The sum of water-soluble and exchangeable-As is frequently cited as the plant-available As fraction. Incubation time and As load showed an interactive effect on the plant-available As fraction of the Immokalee soil; plant-available As concentrations decreased with incubation time only at the lowest As load (45 mg kg^{-1}) (Fig. 15.3). At 225 and 450 mg kg^{-1}, plant-available As remained unchanged with time, suggesting that the minimum As sorption capacity of the Immokalee soil was unable to retain As added at high loads (225 and 450 mg As per kg). No significant interaction between As load and incubation time was observed for the plant-available fraction of the Pahokee soil (Fig. 15.3); plant-available As concentrations significantly ($\alpha = 0.05$) increased with As load at all incubation times (Fig. 15.3). Plant-available As of the Pahokee soil was significantly ($\alpha = 0.05$) reduced within four months of incubation from ~83% to 53%, and remained unchanged thereafter (Fig. 15.3). Data for the Millhopper soil were similar and are not shown. Incubation experiments with the Immokalee and Millhopper soils amended with dimethylarsinic acid(V) (DMA) showed a similar decrease in the water-soluble As fraction with incubation time (Sarkar et al., 2004). Despite the obvious decrease in plant-available As concentrations with incubation time, a significant amount of As (~50% of the total As added) remains in solution, and it is accessible to plants and microorganisms, posing a serious health risk for living organisms.

The observed decreases in plant-available As concentrations with incubation time could be explained on the basis of a concurrent increase in the NaOH-extractable As fraction with incubation time (Fig. 15.4). In the

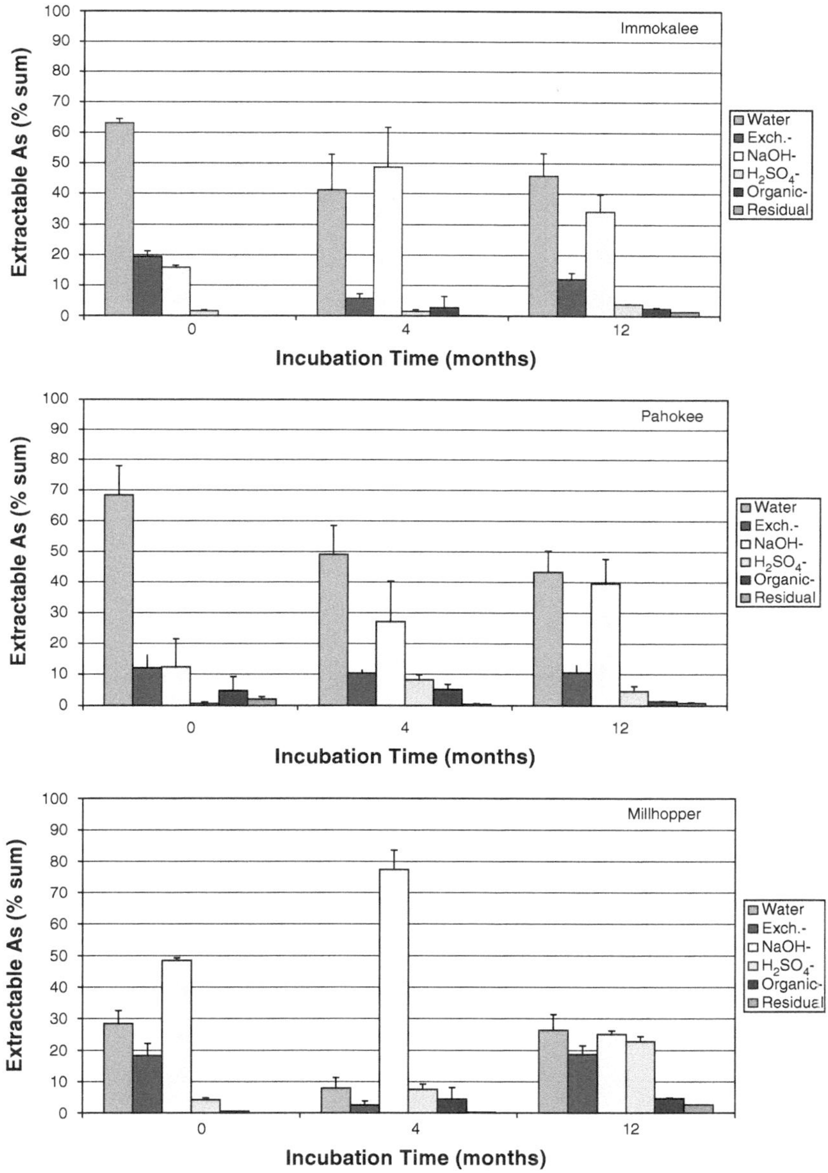

Figure 15.2. Effect of incubation time on the different As fractions for the three soils amended with 45 mg As per kg. Data are the average of three replicates ± one standard deviation.

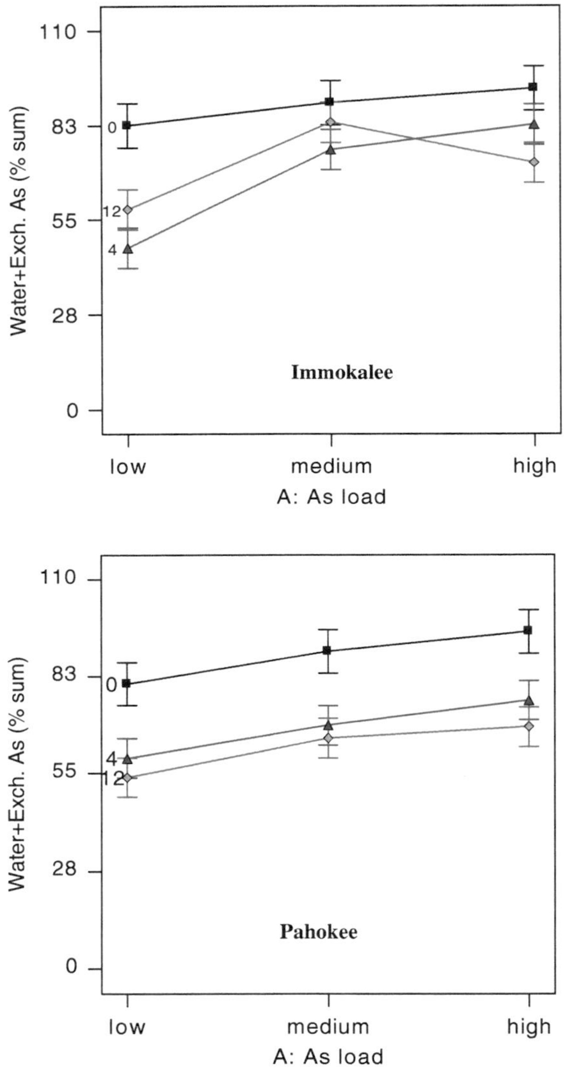

Figure 15.3. Effects of sodium arsenate application rate (low = 45, medium = 225, and high = 450 mg kg^{-1}) and incubation time (0, 4, and 12 months after amendment application) on the plant-available As (water + NH_4Cl – extractable As) fractions of the Immokalee and Pahokee soils (expressed as percent of the sum of all As fractions). The error bars represent the least significant difference (LSD) at the 95% confidence level. Data are the mean of three replicates. Data for the Millhopper were similar (not shown).

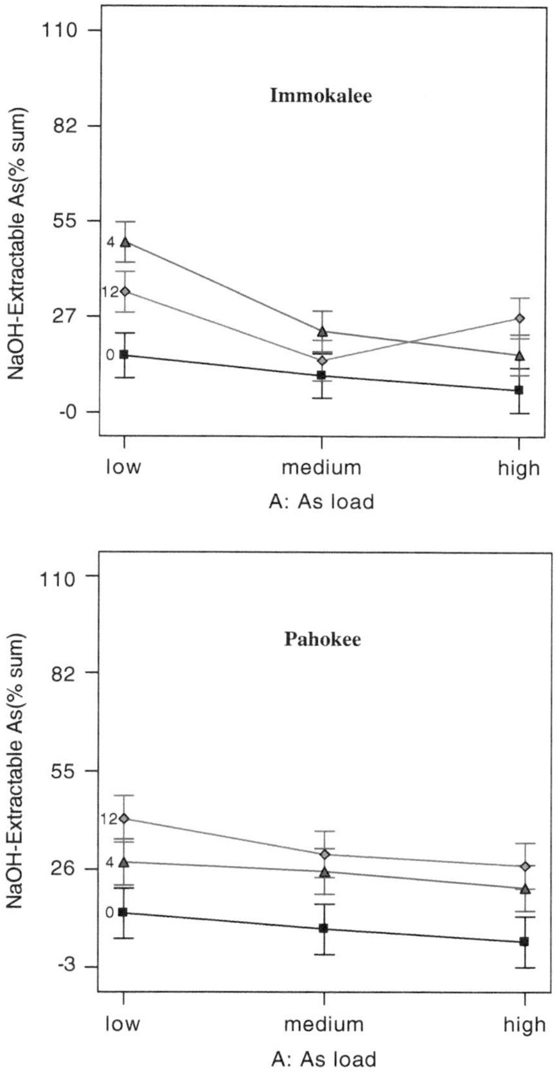

Figure 15.4. Effects of sodium arsenate application rate (low = 45, medium = 225, and high = 450 mg kg^{-1}) and incubation time (0, 4, and 12 months after amendment application) on the NaOH-extractable As fractions of the Immokalee and Pahokee soils (expressed as percent of the sum of all As fractions). The error bars represent the least significant difference (LSD) at the 95% confidence level. Data are the mean of three replicates. Data for the Millhopper were similar (not shown).

case of Immokalee soil, there was a significant ($\alpha = 0.05$) increase in NaOH-extractable As concentration with time, but only in the 45 mg kg^{-1} As load, consistent with the observed temporal decreases in plant-available As concentrations (Fig. 15.4). A positive incubation time effect on the NaOH-extractable As concentration in the Pahokee soil was observed for all As loads, which is also consistent with the temporal trends of the plant-available As (Fig. 15.4). The Millhopper soil showed similar trends (data not shown). Regardless of the initial As load (by mass), the relatively high amorphous Fe/Al content of the Pahokee soil (2000 mg kg^{-1}) was responsible for decreasing labile As concentrations, rendering sorbed As less bioaccessible. Arsenic adsorption is positively correlated with the Fe and Al hydroxide content of soils (Goldberg, 2002). Similar to the NaOH-extractable As, the H_2SO_4-extractable concentrations increased with time for all three soils, suggesting that As was transferred toward less-labile pools, such as Ca/Mg compounds (data not shown).

The two mineral soils (Immokalee and Millhopper) showed no significant effect of incubation time or As load on the organically bound As fraction (data not shown). Interestingly, the soil with the highest organic matter content (Pahokee Muck) also showed no effect of incubation time, or As load on the magnitude of the organic-bound As fraction, indicating that the major As sorbents in the three FL soils, are mostly inorganic (Fe/Al and Ca/Mg pools).

In-vitro bioaccessibility of As. Sub-samples of the three FL soils were subjected to the in-vitro stomach phase to simulate soil–As particle bioaccessibility when present in human gastric juices. For the stomach phase of the Immokalee soil, there was a significant ($\alpha = 0.05$) increase in bioaccessible As concentrations with increasing As load, and a concurrent decrease in As bioaccessibility with incubation time (Fig. 15.5). Despite the significance of the decrease in bioaccessible As with time, remaining levels (~55% of total As added) were still high, being consistent with the large water-soluble and plant-available As concentrations that remained even after 12 months of incubation (Figs. 15.1–15.3). Similar results were obtained with the Pahokee Muck and Millhopper soils; there was a significant reduction in bioaccessible As of the stomach phase with time (Fig. 15.5), but large concentrations of bioaccessible As remained, which could pose a serious threat to human health. It is noteworthy that the remaining bioaccessible As levels were approximately equal for all three soils (50–60%). These data mandate the need for extensive research to develop novel and cost-effective methods to lower As bioaccessibility and, thus, bioavailability in As-contaminated soils.

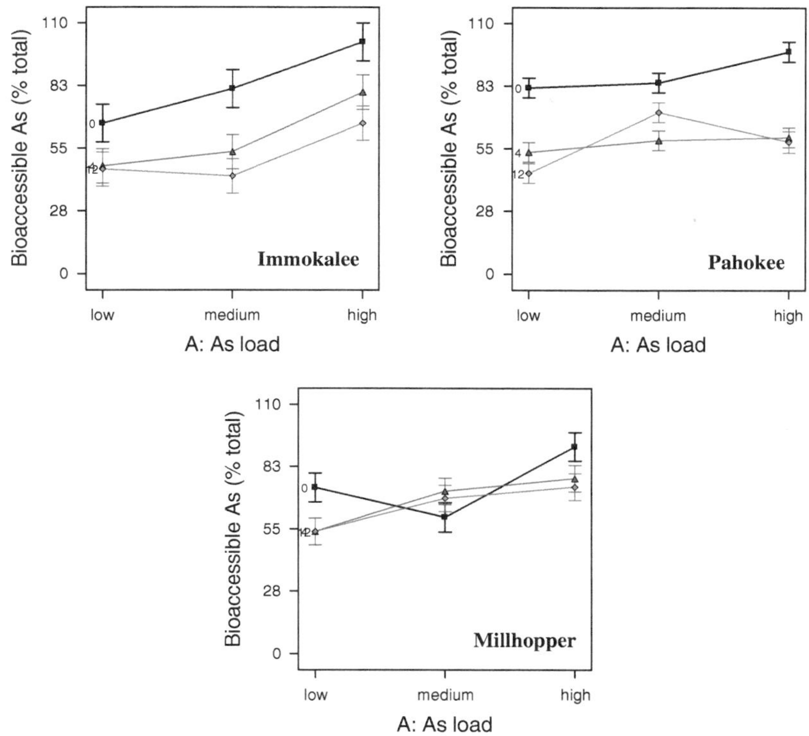

Figure 15.5. Effects of sodium arsenate application rate (low = 45, medium = 225, and high = 450 mg kg^{-1}) and incubation time (0, 4, and 12 months after amendment application) on the bioaccessible As fractions as measured by the in-vitro gastric phase of the three soils (expressed as percent of the sum of all As fractions). The error bars represent the least significant difference (LSD) at the 95% confidence level. Data are the mean of three replicates.

Correlations. We attempted to linearly correlate the different As fractions of the sequential fractionation scheme with the in-vitro As bioavailable fraction, i.e. the stomach-As phase, utilizing all incubation times and As load data. For the Immokalee soil, the only As fraction that was significantly ($\alpha = 0.05$) correlated with the in-vitro stomach method was the water-soluble ($r^2 = 0.56$) and the (water + NH_4Cl)-extractable As fractions ($r^2 = 0.53$) (Table 15.2). The NaOH-extractable As fraction of the Immokalee soil was negatively correlated with the stomach method ($r^2 = 0.54$), suggesting resistance of the Fe/Al-bound As to dissolution, when present in the acidic content of the in-vitro gastric test.

Table 15.2. Pearson correlation coefficients for different measures of As bioaccessibility in the soils

Extractable As	Water	NH_4Cl^-	Plant-available	$NaOH^-$	$H_2SO_4^-$	$H_2O_2^-$	HNO_3^-	Stomach phase
Immokalee								
NH_4Cl^-	−0.26	1.00						
Plant-available[@]	0.94**	0.09	1.00					
$NaOH^-$	−0.93**	0.10	−0.99**	1.00				
$H_2SO_4^-$	−0.61*	0.31	−0.52*	0.43	1.00			
$H_2O_2^-$	−0.52*	−0.18	−0.60*	0.49	0.42	1.00		
HNO_3^-	0.05	0.18	0.40	0.38	0.86**	0.46*	1.00	
Stomach phase	0.56*	−0.15	0.53*	−0.55*	−0.17	−0.22	0.09	1.00
Pahokee Muck								
NH_4Cl^-	−0.39	1.00						
Plant-Available[@]	0.97**	−0.17	1.00					
$NaOH^-$	−0.95**	0.20	−0.96**	1.00				
$H_2SO_4^-$	−0.57*	0.03	−0.61*	0.43	1.00			
$H_2O_2^-$	−0.11	−0.05	−0.13	0.08	0.28	1.00		
HNO_3^-	0.05	0.04	0.04	0.02	0.12	0.15	1.00	
Stomach phase	0.77**	−0.11	0.79**	−0.72**	−0.71**	−0.10	0.08	1.00
Millhopper								
NH_4Cl^-	0.91**	1.00						
Plant-Available[@]	0.99**	0.95**	1.00					
$NaOH^-$	0.56*	0.51*	0.56*	1.00				
$H_2SO_4^-$	0.45	0.68*	0.52*	0.43	1.00			
$H_2O_2^-$	0.11	0.11	0.13	0.24	0.0.36	1.00		
HNO_3^-	0.11	0.23	0.15	−0.22	0.59*	0.16	1.00	
Stomach phase	0.70**	−0.61*	0.70**	0.47*	0.29		0.09	1.00

*Significant at the $\alpha = 0.05$ probability.
**Significant at the $\alpha = 0.01$ probability.

For the Pahokee soil, positive linear correlations were observed between the water-soluble ($r^2 = 0.77$) and (water + NH_4Cl)-extractable ($r^2 = 0.79$) As fractions with the bioaccessible-As of the stomach phase (Table 15.2). Similar positive correlations were observed for different batches of the Immokalee and Millhoper soils amended with DMA (Sarkar et al., 2004). Similar to Immokalee, a negative linear correlation was observed for the NaOH-extractable As fraction with bioavailable As in the gastric phase of the Pahokee and Millhopper soils. The H_2SO_4-extractable As fraction was also correlated with bioaccessible As obtained from the in-vitro test in the Pahokee Muck soil, consistent with the results of Sarkar et al. (2004). It seems that Fe/Al, and to a lesser extent, Ca/Mg solid phases control As bioaccessibility in Pahokee Muck soil. This may be the result of the greater amorphous Fe/Al content of the Pahokee soil (oxalate-extractable Fe + Al = 2000 mg kg^{-1}) when compared to 704 and 66 mg kg^{-1} in the Millhopper and Immokalee soils, respectively.

15.4. Conclusions

This study demonstrated that soil physicochemical properties impact the relative As distribution between labile and non-labile soil phases, and thus, influence the magnitude of bioaccessible As, as estimated by the in-vitro test. Soil solid phases such as amorphous Fe/Al hydroxides as well as Ca/Mg compounds (to a lesser extent) may be the predominant sorbents for As in soils. Incubating the soils with different As loads up to 12 months showed that there was a significant reduction in the water-soluble and the plant-available As fractions with time, but a significant bioaccessible pool of As remained in the soils. Observed decreases in the plant-available As fraction with incubation time were paralleled by significant temporal increases in the NaOH-extractable As fraction of all three soils, which represents the amorphous Fe/Al-bound As fraction. Another As fraction that increased with time was the H_2SO_4-extractable As fraction, which represents As bound to Ca/Mg compounds. This effect was seen in the Millhopper and the Pahokee soils that had large amounts of total Ca/Mg concentrations, but not in the Immokalee soil. There was no effect of time or As load on the organic-bound As fraction, suggesting that mineral phases control As availability in the soils studied.

Results from the in-vitro test confirmed the trends observed in the sequential fractionation experiments. Bioaccessible As measured in the stomach phase decreased with incubation time at all As loads for the three soils, but not to levels that would minimize the risk associated with exposure to soil–As particles. Arsenic associated with the water- and plant-available As fractions is most bioaccessible as indicated by the significant ($\alpha = 0.05$) correlation between the water-soluble, plant-available As concentrations, and the in-vitro bioaccessible As fraction in the soils. The NaOH-extractable As fraction was negatively correlated with the in-vitro bioaccessible As fraction, suggesting that the presence of Fe/Al hydroxides could decrease As lability in soils.

The results obtained in this study provide an useful estimation of As bioavailability in soils varying in physicochemical properties. Despite the decrease in bioaccessible As concentrations with time for all soils, a large amount of bioaccessible As remained in the soils even after 12-month incubation, which could potentially pose a serious health hazard for humans. This information can be used for As risk assessment studies in FL soils. However, since the data were generated in a static incubation study, caution should be exercised in extrapolating the data to the field scale. A greenhouse study is currently being conducted to evaluating As bioaccessibility using the same soils to validate the observed trends in the incubation experiments.

ACKNOWLEDGMENTS

We acknowledge the USEPA-STAR Program for funding this study. We would like to thank Drs. E. Hanlon and S. Daroub, Mr. Schwandes and Mr. Taylor of the University of Florida for providing the soils used in this study.

REFERENCES

Brown, K.G., Ross, G.L., 2002. Arsenic, drinking water, and health: A position paper of the American Council on Science and Health. Regul. Toxicol. Pharm. 36, 162–174.

Chisholm, D., MacPhee, W., MacEachern, C.R., 1955. Effect of repeated applications of pesticides to soil. Can. J. Agr. Sci. 35, 433–439.

Chunguo, C., Zihui, L., 1988. Speciation of arsenic in water, suspended solids and sediment of Xiangjiang river. China Sci. Total Environ. 77, 69–82.

Cohen, J.T., Beck, B.D., Bowers, T.S., Bornschein, R.L., Calabrese, E.J., 1998. An arsenic exposure model: Probabilistic validation using empirical data. Hum. Ecol. Risk Assess. 4, 341–377.

Datta, R., Sarkar, D., 2004. Arsenic geochemistry in three soils contaminated with sodium arsenite pesticide: An incubation study. Environ. Geosci. 11, 87–97.

Davis, A., Ruby, M.V., Bergstrom, P.D., 1992. Bioavailability of arsenic and lead in soils from the Butte, Montana, mining district. Environ. Sci. Technol. 26, 461–468.

Design-Expert, 2001. Software for design of experiments, version 6.0.5, Stat-Ease Inc. Minneapolis, MN.

Goldberg, S., 2002. Competitive adsorption of arsenate and arsenite on oxides and clay minerals. Soil Sci. Soc. Am. J. 66, 413–421.

Hedley, M.J., Stewart, J.E.B., Chauhan, B.S., 1982. Changes in phosphorus fractions in soils under intensive plant growth. Soil Sci. Soc. Am. J. 46, 970–976.

Kabata-Pendias, A., Pendias, H., 1992. Trace Elements in Soils and Plants, second ed.. CRC Press, Boca Raton, FL.

Klute, A., 1996. Methods of Soil Analysis: Part 1: Physical and Mineralogical Methods. SSSA Publications, Madison, WI.

Matschullat, J., 2000. Arsenic in the geosphere—A review. Sci. Total Environ. 249, 297–312.

Mehlich, A., 1984. Mehlich No 3 soil test extractant: A modification of Mehlich No 2 extractant. Commun. Soil Sci. Plant Anal. 15, 1409–1416.

Murphy, E.A., Aucott, M., 1998. An assessment of the amounts of arsenical pesticide used historically in a geographic area. Sci. Total Environ. 218, 89–101.

National Research Council, 1999. Arsenic in Drinking Water. National Academy Press, Washington, DC, p. 273.

Rhoades, J.D., 1982. Methods of Soil Analysis. Part 2. In: Agron. Monogr 9. ASA and SSSA, Madison, WI, pp. 149–158.

Rodriguez, R.R., Basta, N.T., Casteel, S.W., Pace, L.W., 1999. An in-vitro gastrointestinal method to estimate bioavailable arsenic in contaminated soils and solid media. Environ. Sci. Technol. 33, 642–649.

Sarkar, D., Datta, R., Sharma, S., 2004. Fate and bioavailability of arsenic in organo-arsenical pesticide-applied soils. Part-I: Incubation study. Chemosphere 60, 188–195.

Semple, K.T., Doick, K.J., Jones, K.C., Burauel, P., Craven, A., Harms, H., 2004. Defining bioavailability and bioaccessibility of contaminated soil and sediment is complicated. Environ. Sci. Technol. 38, 228A–231A.

Sheppard, S.C., 1992. Summary of phytotoxic levels of soil arsenic. Water Air Soil Pollut. 64, 539–550.

Sheppard, S.C., Evenden, W.G., Schwartz, W.J., 1995. Ingested soil: Bioavailability of sorbed lead, cadmium, cesium, iodine, and mercury. J. Environ. Qual. 24, 498–505.

US EPA, 2000. Test methods for evaluating solid waste, Physical/Chemical Methods, USEPA-65 FR 70678, Draft Update IVB SW-846. U.S. Governmental Printing Office, Washington, DC.

Watanabe, F.S., Olsen, S.R., 1965. Test of an ascorbic acid method for determining phosphorus in water and $NaHCO_3$ extracts from soil. Soil Sci. Soc. Am. Proc. 29, 677–678.

Developments in Environmental Science, Volume 5
D. Sarkar, R. Datta and R. Hannigan (Editors)

ISSN: 1474-8177/DOI:10.1016/S1474-8177(07)05016-4

Chapter 16

A greenhouse study on soil-arsenic forms and their bioaccessibility in two chemically variant Florida soils amended with sodium arsenate pesticide: Preliminary results

Shahida Quazi, Dibyendu Sarkar, Rupali Datta and Saurabh Sharma

Abstract

Long-term application of arsenical pesticides in agricultural lands has resulted in high levels of arsenic (As) in certain soils. Conversion of former agricultural lands to residential areas has increased human contact with soil–As. Soil ingestion from incidental hand-to-mouth activity by children is now a very important issue in assessing human health risk associated with exposure to arsenical pesticide-applied former agricultural soils. Human health risk from exposure to soil–As is restricted only to those fractions of As in the soil that are available to the human gastrointestinal system. This study followed up on a static incubation experiment aimed at addressing the issue of soil variability on As bioaccessibility as a function of soil chemical properties, but in a greenhouse column system accounting for dynamic interactions between soils, water, plants, and pesticides. Two chemically variant soil types were chosen based on their potential differences with respect to As reactivity. The soils were amended with sodium arsenate pesticide at two high rates. Rice (*Oryza sativa*) was used as the test crop. A sequential extraction scheme was employed to identify the geochemical forms of As in soils (soluble, exchangeable, organic, Fe/Al-bound, Ca/Mg-bound, residual) immediately after spiking and after six months of equilibration. Concentrations of these As forms were correlated with the in-vitro fractions of As to identify those As species that are most likely to be bioaccessible in the human gastrointestinal system. Results from this study verified those obtained from the static incubation experiment, and demonstrated that As bioaccessibility is a function of soil speciation of As and that soil–As forms are a function of soil chemical properties.

16.1. Introduction

Arsenic is one of the most abundant elements on the earth (ranked as the 20th most abundant element) and occurs in many chemical forms. Varying amounts of As are naturally present in many soils, with concentrations ranging from 0.2 to 40 $\mu g\, g^{-1}$ (Walsh et al., 1977). However, elevated levels of As can be present in the environment as a result of mineral weathering and dissolution, geothermal activity, and numerous anthropogenic activities, including but not limited to mining and smelting activities, pesticide use, and coal combustion (Manning et al., 1998). As a result, it is the most common contaminant at hazardous waste sites in the United States and has been classified by the U.S. Environmental Protection Agency (EPA) as a group A human carcinogen (Southworth, 1995).

During the 1970s, agricultural products within the U.S. accounted for approximately 81% of the total As used (Adriano, 2001). By the 1980s, agricultural application accounted for only 46% of As use in the U.S. (Adriano, 2001). The average As contents in agricultural fields that received As-containing pesticides and defoliants range from 5 to 2553 $mg\, kg^{-1}$ (Walsh and Keeney, 1975). Although the use of arsenical pesticides has reduced, long-term application of such pesticides in agricultural lands has resulted in high levels of As residues in certain soils (Murphy and Aucott, 1998). Moreover, leaching of As from chromated copper arsenate (CCA)-treated wood may also elevate soil arsenic levels (Cao and Ma, 2004). Arsenic concentrations as high as 350 $mg\, kg^{-1}$ have been observed in soils near CCA-treated decks in Connecticut (Stillwell and Gorny, 1997). Murphy and Aucott (1998) reported As-contaminated soils in residential developments located on former apple orchards.

During the last two decades, rapid encroachment of suburban development on lands previously used for agriculture in fast metropolitan areas has tremendously increased the potential for human contact with soil–As (Datta and Sarkar, 2004). Moreover, since CCA-treated wood has been used extensively as fences and utility poles in urban neighborhood, As exposure via consumption of home-grown vegetables and fruits may constitute a potential health risk (Schoof et al., 1999). Soil ingestion from incidental hand-to-mouth activity by children is now a very important issue in assessing human health risk associated with exposure to arsenical pesticide-applied agricultural soils, which are now being used for residential purposes (Sarkar and Datta, 2004).

A critical parameter that allows for realistic health risk assessment in arsenic-contaminated soils is an estimate of "bioaccessible" As. Bioaccessibility refers to the extent of absorption of a chemical into the bloodstream from the gastrointestinal tract, lungs, or skin (Halmes and

Roberts, 1997). Studies investigating As ingested by humans suggest that close to 100% of soluble inorganic As is absorbed by the gastrointestinal tract (US EPA, 1992). In contrast to As in drinking water (soluble As), As in soils generally exists as mineral forms or as soil–As complexes that will be incompletely solubilized during transit through the gastrointestinal tract. Research indicates that As must be dissolved in order to be absorbed (Freeman et al., 1995); therefore, As in soil will be less well absorbed than As in drinking water (Ruby et al., 1996). This is because the majority of the soil forms of As are geochemically stable and/or insoluble in human gastric/intestinal juices and, hence, are not likely to be available for systemic absorption.

Adriano (2001) summarized several soil properties that are most likely to influence soil availability of As, namely, pH, texture (clay content), amorphous Fe–Al oxides, organic matter content, sulfur content, phosphorus concentration, and soil redox conditions. Incubation studies on As bioaccessibility and speciation have shown that As speciation is a direct function of soil properties and that these geochemical forms of As again affect the bioaccessibility (Sarkar and Datta, 2004). However, since incubation studies are static equilibrium experiments that fail to provide the more realistic viewpoint of a dynamic system, either a field study or a greenhouse study would generate data encompassing more realistic scales and growing condition. However, given the hazardous nature of As and its potential toxic effects on human health, a field scale study to investigate As bioaccessibility is an unrealistic proposition. Thus, the reported greenhouse study aimed at addressing the issue of soil variability on As speciation and bioaccessibility in a greenhouse column study. The primary objective of this study was to evaluate the relationship between geochemical speciation of As and bioaccessibility involving dynamic interactions between soil, water, and plants in a temperature and humidity controlled greenhouse setting. Preliminary results obtained after six months of soil–plant–pesticide equilibration have been presented in this report.

16.2. Materials and methods

16.2.1. Soil sampling, preparation and characterization

Two soils were used for the greenhouse study. Surface soil from the Immokalee series (the State soil of Florida) was collected from the Southwest Florida Research and Education Center, Immokalee, Florida, and the Millhopper series soil was collected from the University of

Florida campus at Gainesville, Florida. The soil samples were air-dried and passed through a 2-mm sieve and characterized for various soil properties. Soil pH, electrical conductivity, particle size, water content, and cation exchange capacity were measured using standard protocols (Sparks, 1996). Organic matter was measured using the loss-on-ignition method (Sparks, 1996). Plant-available Ca, Mg, and P were extracted by Mehlich III solution (Mehlich, 1984). Oxalate-extractable Fe and Al was obtained using Tamm's reagent (Sparks, 1996). Total P was extracted using the ignition method (Sparks, 1996). Total recoverable Ca, Mg, Fe, Al, P, and As was obtained by soil digestion according to USEPA method 3050B (USEPA, 1996). Phosphorus was measured colorimetrically by an UV/Visible light spectrophotometer using the molybdate–ascorbic acid method (Sparks, 1996). Ca, Mg, and Al were analyzed using the flame atomic absorption spectrometry (FAAS) and As was analyzed by graphite furnace atomic absorption spectrometry (GFAAS).

16.2.2. Greenhouse study

16.2.2.1. Soil amendments and plant growth

Both Immokalee and Millhopper soils were spiked with sodium arsenate at two rates 675 and 1500 mg As/kg soil. These two rates were selected to represent the higher and lower end of the superfund soil–As concentrations. PVC columns (13″ tall × 6″ id) were used in this greenhouse study. The bottom 7″ of the column was filled with white sand that had no arsenic retention capacity. Pesticide amended soil was packed in the top 6″ of the PVC column. Each column was provided with a reservoir compartment to hold the excess leachate and a hole was fitted with nalgene tubing to collect the leachate. The columns were arranged in a randomized block design and were rotated periodically to account for variances in temperature and sunlight within the greenhouse. Rice was used as the test crop because of their reported tolerance to As-enriched irrigation water in Bangladesh and India. The number of PVC columns with treatment were 12 (2 soils × 2 rates × 3 replicates) plus 6 controls (2 soils × 3 replicates). The first leaching was induced after two weeks of pesticide application. Rice was harvested in those columns where there was germination of rice seedlings. The second leaching took place after rice was harvested in late August. Leachate water was analyzed for soluble As using the GFAAS. Soils were collected from the surface layer immediately after spiking (0 time) and after six months of soil–pesticide equilibration and assessed for soil–As forms and in-vitro bioaccessible As.

16.2.3. Sequential extraction procedure

Sequential extraction scheme developed by Chungao and Zihui (1988) was employed to identify the various operationally defined forms of As, as follows:

1. *Water-soluble phase.* One gram of soil was extracted at room temperature for 30 min with 50 ml of deionized water under continuous shaking. The samples were centrifuged, filtered, and analyzed for soluble As.
2. *Exchangeable phase.* The residual soil from the water-soluble fraction step was shaken at room temperature for 30 min with 50 ml of 1 M NH_4Cl. The mixture was centrifuged, and the supernatant was analyzed for exchangeable As.
3. *Fe- and Al-bound phase.* The soil residue from step 2 was shaken at room temperature for 17 h with 50 ml of 0.1 M NaOH. The residue after centrifugation was washed twice using 25 ml of saturated NaCl solution. The supernatant from these washes were pooled and analyzed to determine the As fraction bound to Fe/Al oxides.
4. Ca- *and Mg-bound phase.* The residual soil was extracted using 50 ml of 0.25 M H_2SO_4 by shaking for 1 h at room temperature. The residue after centrifugation was washed twice using 25 ml of saturated NaCl solution. The supernatant from these washes were pooled and analyzed to determine the As fraction bound to Ca/Mg compounds.
5. *Organic matter and sulfide-bound phase.* The residual soil was digested using 3 ml of 30% H_2O_2 (adjusted to pH 2.0 using HNO_3) at 85°C for 3 h. To prevent adsorption of extracted As to the oxidized soil, the samples were cooled and 5 ml of 3.2 M NH_4OAc in diluted HNO_3 was added. The samples were diluted to 20 ml with deionized water, followed by shaking for 30 min at room temperature, and then analyzed to determine the soil organic matter and sulfide-bound As.
6. *Residual phase.* The remaining soil was extracted using 25 ml of concentrated HNO_3 at 105°C, until approximately 5 ml of solution remained. The samples were diluted to 25 ml with deionized water, centrifuged, and the supernatants were analyzed for As-bound to the silicate framework.

All the above separations were done by centrifugation at 3000 × g for 30 min. The supernatants were filtered and analyzed for As using GFAAS.

16.2.4. In-vitro procedures

The fraction of bioaccessible As was estimated following the method of Rodriguez et al. (1999) with certain modifications made by Sarkar and Datta (2003). Reactions were carried out in 250 ml beakers in a 37°C water bath to stimulate body temperature. Anaerobic conditions were maintained by passing argon gas through the solutions. The pH of the solutions was continuously monitored. The in-vitro gastrointestinal method was conducted in two sequential phases, a low pH gastric phase, followed by a higher pH intestinal phase.

1. *Stomach phase.* The gastric-phase solution consisted of 0.15 M NaCl and 1% porcine pepsin (Sigma Chemical Co., St. Louis, Missouri). One gram of soil sample was added to 150 ml of gastric solution, and the pH of the solution was adjusted to 1.8 using 1 N HCl. The solution was incubated for 1 h, at the end of which 10 ml of the solution was collected, centrifuged at 3000 × g for 30 min and analyzed by GFAAS.
2. *Absorbed-intestinal phase.* At the end of the stomach phase, 10 ml of the gastric solution was added to replace the solution removed for analysis The pH of the solution was adjusted to 7.0 using a saturated solution of $NaHCO_3$, followed by the addition of 525 mg of porcine bile extract and 52.5 mg of porcine pancreatin (Sigma Chemical Co., St. Louis, Missouri). To simulate absorption through the intestinal lining a 40-cm^2 filter paper strip coated with ferric oxide was used (Sarkar and O'Connor, 2001). The ferric oxide strip was placed in a square bag (sides 6.5 cm) made of nylon membrane filter of 8 μm pore size. The bag was tied with a string and suspended in the reaction vessel. The solution was incubated for 1 h, at the end of which 10 ml of the solution was collected, centrifuged at 5000 rpm for 30 min and analyzed by GFAAS. As adsorbed by the ferric oxide strip was desorbed by shaking it vigorously in 80 ml of 1 N HNO_3 for 1 h.
3. *Preparation of ferric oxide strips.* Ferric oxide strips were prepared according to Sarkar and O'Connor (2001). In brief, Whatman no. 8 filter papers were immersed in a 10-g/100-ml ferric chloride solution for 1 h. The filter papers were air-dried, followed by immersion in a 2.7 M NH_4OH solution for 1 min for deposition of ferric oxide. The iron oxide-coated papers were air-dried and used to mimic As adsorption in the in-vitro absorbed-intestinal phase.

All analyses were carried out in triplicates and the results are shown as mean values. Replicates had to fall within 95–105% to be considered acceptable. Recoveries of 90–110% of spikes and external standards were considered acceptable.

16.3. Results and discussion

16.3.1. Soil properties

The soils for this greenhouse study were selected based on their potential differences with respect to As reactivity. It has been reported that soil properties, such as clay content, pH, CEC, total C, P, Ca and Mg contents, and oxalate-extractable Fe and Al contents influence soil speciation of As and hence, bioavailability (Adriano, 2001). Soil characterization data for Immokalee and Millhopper soils are shown in Table 16.1. The primary difference between these two soils was in their Fe/Al, Ca/Mg, and P contents. While the Immokalee soil is an acid sand with low Fe/Al, Ca/Mg, and P contents; Millhopper is an acid sandy loam with high extractable Fe/AL oxides, as well as high Ca/Mg and P content. Being sandy and lacking positively charged adsorptive surfaces (e.g. amorphous Fe/Al oxides), the Immokalee soil is likely to have minimal As retention capacity (Pierce and Moore, 1980; Oscarson et al., 1981) and hence, potential of increased As bioaccessibility. Millhopper soils, on the other hand, have high oxalate extractable Fe and Al (i.e., high amorphous Fe–Al oxide contents) and is likely to have higher As retention capacity, and hence, potential of decreased As bioaccessibility. Based on the incubation study results (Sarkar and Datta, 2004), Immokalee was used as a control soil to study the effect of soil properties such as Fe/Al, Ca/Mg, and P content in Millhopper in a dynamic greenhouse column setting. Both soils were slightly acidic with pH ranging from 6.0 to 6.4.

Table 16.1. Selective chemical properties of soils (Immokalee and Millhopper)

Properties		Immokalee	Millhopper
pH		6.0	6.4
EC ($\mu S\,cm^{-1}$)		59	145
CEC ($Cmol\,kg^{-1}$)		777	2356
Sand (%)		99.7	96.8
SOM (%)		0.84	4.38
As ($mg\,kg^{-1}$)		15.0	16.5
P ($mg\,kg^{-1}$)	Mehlich 3	4.0	134
	Total	208	4875
Ca + Mg ($mg\,kg^{-1}$)	Mehlich 3	266	886
	Total	1178	3155
Fe + Al ($mg\,kg^{-1}$)	Oxalate	66	704
	Total	212	4745

EC, Electrical conductivity; CEC, cation exchange capacity; SOM, soil organic matter.
Note: Data represent mean of three replicates.

Moreover, since Millhopper soils have high P content, retained As could potentially desorb from the pesticide-applied soil. Phosphorus has similar properties like As, and is likely to compete with As for sorption sites in soils, resulting in desorption and, hence, increased bioaccessibility (Woolson et al., 1973).

16.3.2. Total arsenic in soil, water, and plants after six months of soil equilibration

Figure 16.1 shows the distribution of As in soil and leachate water in the Immokalee and Millhopper soils amended with sodium arsenate at two rates 675 mg kg^{-1} and 1500 mg kg^{-1}. In case of Immokalee, for both pesticide application rates, the majority of As was present in the leachate. Very small amount of As was present in the surface soil. This is expected for Immokalee since it is a sandy soil with low extractable Fe/Al and, hence, very low As retention capacity. Therefore, the majority of As moved downwards to the bottom of the column, and became a part of the leachate water (Fig. 16.1). In soil, As is mainly associated with Fe oxides and hydroxides (Fassbender, 1974; Akins and Lewis, 1976; Hale et al., 1997). It has been found that the sorption of dissolved As on soil is controlled by Fe oxides (Fordham and Norrish, 1983; Elkhatib et al., 1984) and increases with Fe- and Al-oxide contents (Jacobs et al., 1970). Since Millhopper has high Fe/Al content, it has high As retention capacity compared to that of Immokalee. As a result, a significant amount of As was retained in the surface soil, to a similar degree in that of the leachate water (Fig. 16.1). Arsenic uptake by the rice plants was minimal in Immokalee soil (data off-scale in Fig. 16.1), and there was no growth of rice in the Millhopper soil. A review of phytotoxic levels of soil As found that inorganic As is five times more toxic in sands than in clay soils (Sheppard, 1992). This agrees with observations that plant uptake of As is usually relatively low on clays and slits and high on sands and sandy loams (Woolson, 1973; O'Neill, 1995). Because of the high concentration of As in the surface soil, the seedlings did not germinate in the Millhopper soil. However, in Immokalee, the majority of As leached downwards; hence, there was plant growth for both pesticide application rates. However, the amount of As taken up by the rice plants was negligible compared to the amount of As presents in the soil and the leachate.

16.3.3. Geochemical forms of arsenic in soils

Soil speciation is the process of identification and quantification of the various "operationally defined" species, forms or phases of metals and

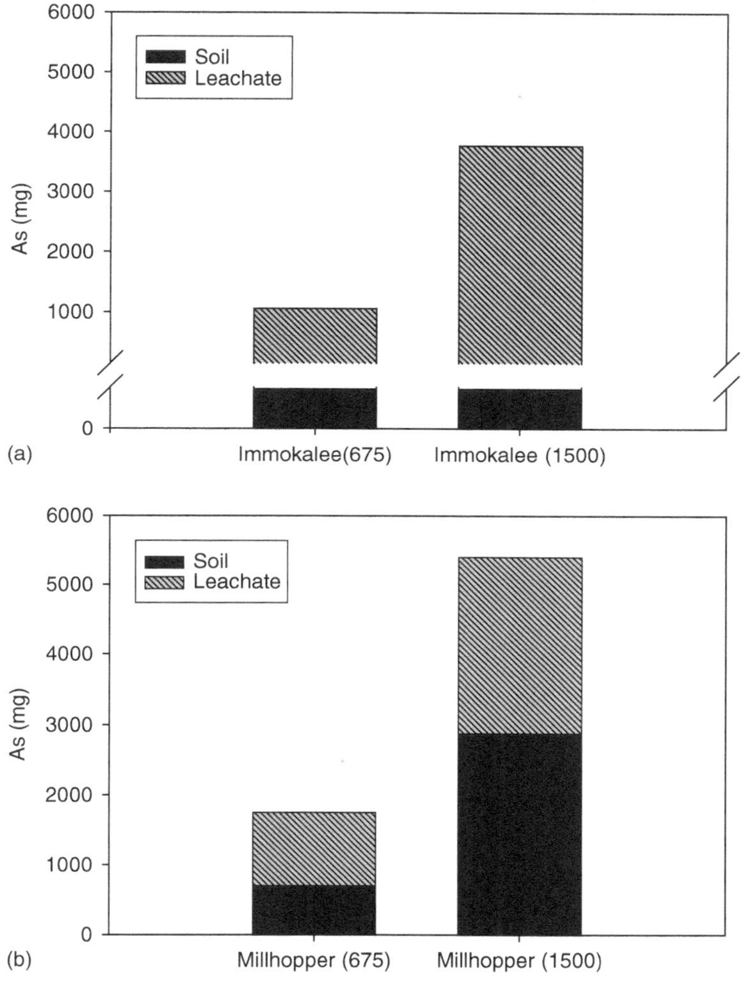

Figure 16.1. Mass balance for As in (a) Immokalee and (b) Millhopper soils. Plant As concentration in the Immokalee soil was minimal and there was no plant growth. So they are not shown here.

metalloids occurring in the soil (Pickering, 1981). Metals may occur in soils and sediments in various fractions, chemical species or forms: adsorptive-exchangeable, carbonate-bound, oxide-bound, organic matter-bound, and detrital or crystal lattice metals (Salomon and Forstner, 1980). These geochemical forms of heavy metals in soils affect their solubility, the risk of ground water pollution, and bioavailability (Harrison,

1981; Xian, 1989). It has been increasingly accepted that certain geochemical forms of arsenic are not bioavailable (Anawar et al., 2004).

Figure 16.2 shows the geochemical speciation of As for both soils amended with 675 mg kg^{-1} of As. In case of Immokalee soil, about 94%

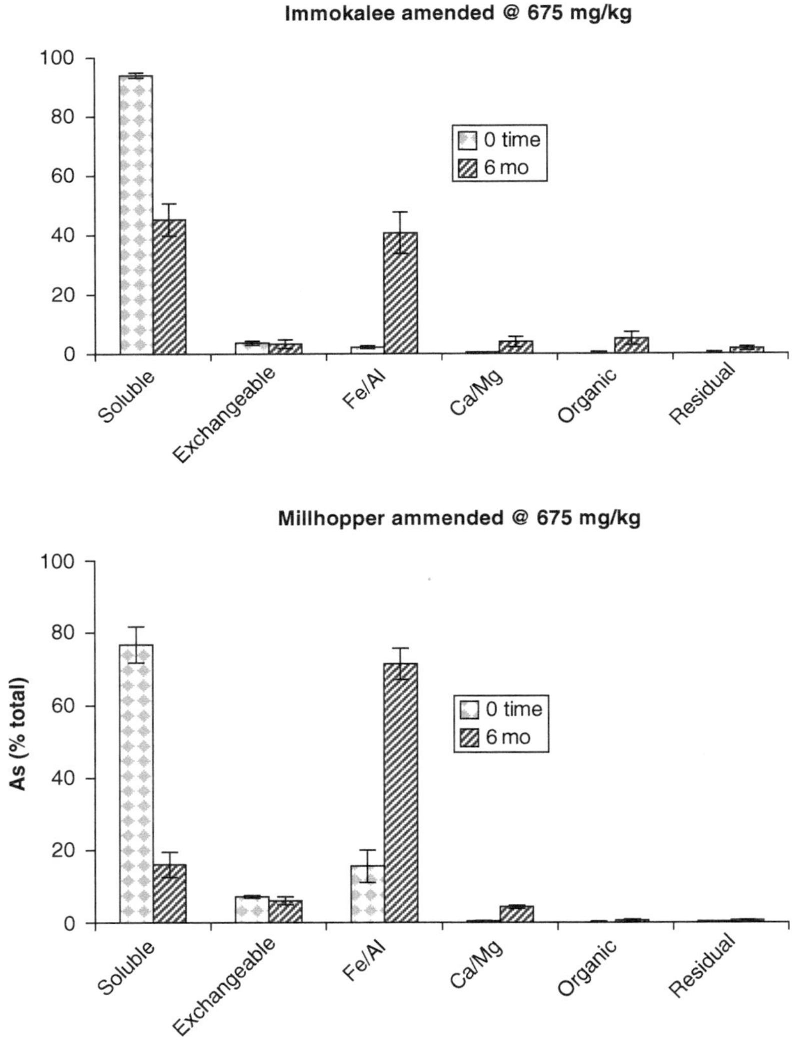

Figure 16.2. Geochemical speciation of As in pesticide-applied soil (675 mg kg^{-1} As) at time 0 and after 6 months of equilibration.

of the total applied As was extracted in the soluble fraction immediately after spiking the soil with pesticides (0 time). Since soluble fraction of As is considered both bioaccessible and phytoavailable (Sarkar and Datta, 2004), about 94% soluble fraction of the total As would have been available for plant uptake immediately after pesticide application. Owing to the phytotoxic effect of As seeds were not germinated. The phytotoxicity of As residues is influenced more by its chemical form than by amount. According to Woolson et al. (1971), soils with water-soluble As generally were more phytotoxic than those with no detectable water-soluble As. However, the soluble fraction of As decreased to 45% after six months of soil equilibration with pesticides. Moreover, after six months of soil equilibration, the Fe/Al-bound fraction increased from 2% to 41%. This indicated the strong tendency of the As oxyanions to be adsorbed by amorphous Fe/Al oxides, despite a low concentration of Fe/Al oxides in Immokalee soil natively. Very little As was extracted in the remaining fractions; cumulatively, the exchangeable, Ca/Mg-bound, organic bound, sulfide-bound, and residual As accounted for less than 10% of the total As in the surface soil at both times (Fig. 16.2). For the Millhopper soil, about 76% of the total applied As was extracted in the soluble form immediately after spiking; approximately 7% was extracted in the exchangeable fraction, and the Fe/Al-bound fraction contributed to about 16% (Fig. 16.2). After six months, the amount of As-bound to the soluble fraction decreased to 16%. However, at the same time, the amount of As retained by the Fe/Al fraction increased to 71%. Thus, after six months of soil–pesticide equilibration, the Fe/Al oxides proved to be the primary phase controlling As mobility in Millhopper soil, which had abundant oxalate-extractable Fe/Al natively. Similar results were obtained in both Immokalee and Millhopper contaminated with 1500 mg kg^{-1} of As (data not shown).

16.3.4. Bioaccessible arsenic in soils

Human health risk from direct exposure to soil arsenic via hand-to-mouth action should ideally be restricted only to those fractions of As in the soil that are bioaccessible, i.e., available to the human gastrointestinal system (Datta and Sarkar, 2004). Arsenic exists in many geochemical and soil mineralogical forms, many of which are geochemically stable and/or insoluble in human gastric/intestinal juices and, hence, are not likely to be available for systemic absorption (Sarkar and Datta, 2004). According to Rodriguez (1998), As adsorbed hysterically to Fe/Al oxide is considered to be unavailable for plant uptake, and is also not likely to be bioavailable to human gastrointestinal system. Because calculated health risk is a direct function of the input value for chemical dose (Ng, 1999), using total soil arsenic concentrations to quantify daily chemical intake for risk characterization typically results in carcinogenic risk results greater than the lowest possible amount (i.e., 10^{-6}) for soils in naturally occurring background settings (Rodriguez et al., 2003). Thus, bioaccessible As should be used instead of total soil As (as is the current practice) to determine the risk associated with exposure to contaminated soil (Sarkar et al., 2004).

Bioaccessibility experiments were conducted to estimate the amount of arsenic that would be available to the human stomach as well as to the intestine. In the stomach phase (for 675 mg kg^{-1} As application rate), approximately 80% of the total applied As was bioaccessible at 0 time in the Immokalee soil, while in Millhopper 85% of As was bioaccessible. However, after six months of soil–pesticide equilibration, bioaccessibility decreased to 51% and 62% in Immokalee and Millhopper soils, respectively (Fig. 16.3). In case of Immokalee soil (having less Fe/Al content and thus less As retention capacity) most of As was leached out after the first leaching. So the bioaccessibility for Immokalee soil was less compared to Millhopper soils which had more As owing to its high Fe/Al content. However, there was wide variability in experimental data represented by the large error bars. In the absorbed intestinal phase, in both soils, more than 85% of the applied As was bioaccessible at time 0; this decreased to 53% in Immokalee and 37% in Millhopper after six months of equilibration (Fig. 16.3). Since Millhopper soil has higher Fe/Al content (i.e., higher As retention capacity), a significant portion of the applied As was expected to be unavailable to the human gastrointestinal system. However, the results indicated that the virtually irreversible Fe/Al-bound As re-solubilized in the strong chemical environment that exists in the human gastrointestinal system, and hence, became bioaccessible. Similar trends were observed in soils contaminated with 1500 mg kg^{-1} of As (data not shown).

16.3.5. Correlation between geochemical speciation and bioaccessibility of soil arsenic

Figure 16.4a,b shows the correlation between bioaccessible As (stomach phase) and As extracted by the different steps of the sequential extraction method. In case of Immokalee (Fig. 16.4a,b), at 0 time, the correlation between the bioaccessible and the soluble fraction was 68% and with the Fe/Al-bound forms, it was 62%. However, with an increase in equilibration time (six months), the correlation between bioaccessible and Fe/Al-bound As forms increased to 91% and the correlation between the

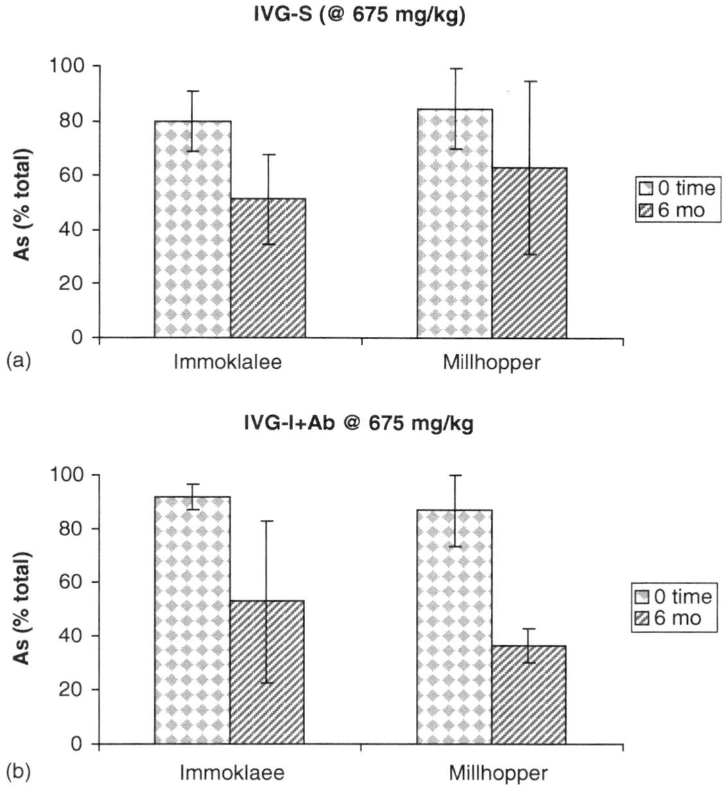

Figure 16.3. Arsenic bioavailability in pesticide-applied soils (675 mg kg^{-1}) using two in-vitro methods at time 0 and after 6 months equilibration.

bioaccessible and soluble As fraction decreased to 35%. Increase in correlation of bioaccessible As with the Fe/Al-bound As fraction was rather unexpected, because, as mentioned in the earlier paragraph, Fe/Al oxides are reported to adsorb As hysterically, thereby decreasing potentially bioaccessible As (Chen et al., 1999). Apparently, As-bound to Fe/Al oxides is not stable in the highly acidic (pH 1.8) environment of the stomach, resulting in re-dissolution of a fraction of adsorbed As, thereby accounting for the high correlation (91%) between bioaccessible As and that adsorbed by the Fe/Al phases in the soil. According to Rodriguez et al. (2003), the amount of arsenic dissolved in the stomach (i.e., potentially bioavailable) is comprised of As fractions between the surficially complexed (desorbable) As and As inside the mineral matrix associated with iron oxides.

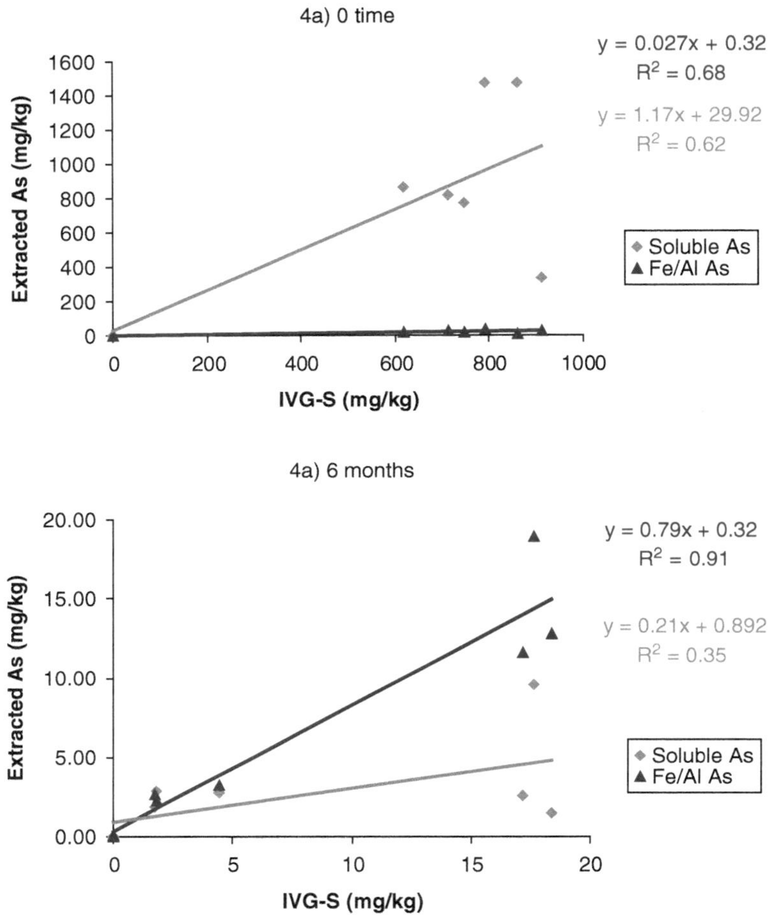

Figure 16.4. Relationship between bioavailable arsenic estimated using the in-vitro gastrointestinal stomach (IVG-S) method and As geochemical forms at time 0 and after 6 months equilibration in (a) Immokalee and (b) Millhopper.

For Millhopper soil (Fig. 16.4c,d) at 0 time, bioaccessible As strongly correlated with both soluble (86%) and Fe/Al (95%) bound As fractions. However, after six months of soil–pesticide equilibration the correlation between bioaccessible and Fe/Al-bound As decreased to 63%. This result is similar with an earlier incubation study on Millhopper soil contaminated with sodium arsenite pesticide (Datta and Sarkar, 2004). At 0 time when there is no soil–pesticide equilibration, the metastable As–Fe/Al phase that formed re-dissolved in the highly acidic environment of the

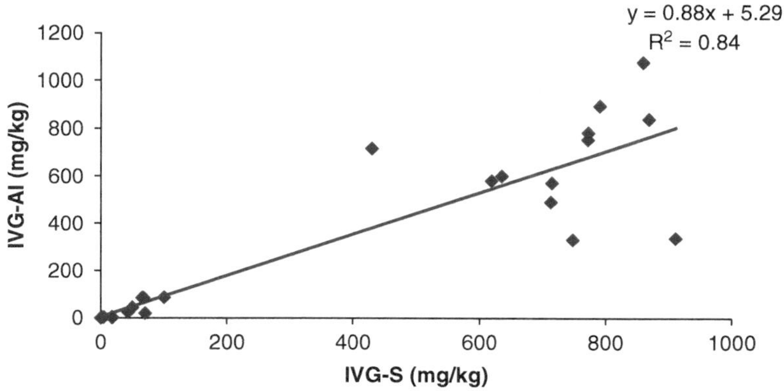

Figure 16.5. Correlation between As extracted by the two in-vitro methods. Data have been pooled for two soils contaminated with sodium arsenate at three rates for 0 and 6 months ($n = 36$). In-vitro gastrointestinal absorbed-intestinal (IVG-AI).

human stomach (pH 1.8). However, with time this phase stabilized and a fraction of As adsorbed onto Fe/Al oxides exhibited the classic hysteric character (Datta and Sarkar, 2004). Since hysterically adsorbed As is not bioaccessible, the correlation between the bioaccessible fraction and the Fe/Al-bound As decreased.

Similar results were obtained when As obtained from various steps of sequential extraction was correlated with bioaccessible As obtained by the IVG absorbed-intestinal method (data not shown). Figure 16.5 shows the correlation between As extracted by the two in-vitro bioaccessibility methods. A significant ($p \leqslant 0.01$) correlation coefficient (0.84) was obtained, indicating that both methods were extracting As from similar soil–As pools.

16.4. Conclusions

The reported greenhouse study demonstrated, similar to the earlier incubation experiment that soil properties have a marked impact on geochemical speciation of As, with the majority of As extracted in the soluble form at 0 time in both soils. Immokalee, being an acid sand with minimal oxalate-extractable Fe/Al had very low As retention capacity and thus, a very small amount of As was present in the surface soil after six months of soil–pesticide equilibration. The majority of the pesticide-applied As moved downwards to become a part of the leachate. As a result, some plant growth was observed in Immokalee soil. However, in case of

Millhopper, As was found in equivalent amounts in the surface soil and in the leachate. This resulted in no plant growth in Millhopper soil. After six months of soil–pesticide equilibration, the majority of As extracted in the Millhopper soil was in the Fe/Al-bound form. This indicated that As adsorbed onto the amorphous Fe/Al oxides, which resulted in decreased As bioaccessibility. However, the fact that not all As adsorbed onto the Fe/Al oxides were unavailable to the human gastrointestinal system indicated that part of the pre-adsorbed As re-dissolved in the harsh chemical environment of the simulated human stomach/intestinal juices. The reported study demonstrated that As bioaccessibility is a function of geochemical speciation of As, which is a function of soil chemical properties.

ACKNOWLEDGMENT

We acknowledge the USEPA-STAR Program for funding this study and the Environmental Geochemistry Laboratory of UTSA for providing the analytical facilities.

REFERENCES

Adriano, D.C., 2001. Trace Elements in Terrestrial Environments: Biogeochemistry, Bioaccessibility, and the Risk of Metals, second ed. Springs, New York.

Akins, MB., Lewis, RJ., 1976. Chemical distribution and gaseous evolution of arsenic-4 added to soils as DSMA-74As. Soil Sci. Soc. Am. 40, 655–658.

Anawar, H.M., Akai, J., Sakugawa, H., 2004. Mobilization of arsenic from subsurface sediments by effect of bicarbonate ions in groundwater. Chemosphere 54, 753–762.

Cao, X., Ma, LQ., 2004. Effects of compost and phosphate on plant arsenic accumulation from soils near pressure-treated wood. Environ. Pollut. 132, 435–442.

Chen, M., Ma, LQ., Harris, W.G., 1999. Baseline concentrations of 15 trace elements in Florida surface soils. J. Environ. Qual. 28, 1173–1181.

Chungao, C., Zihui, L., 1988. Speciation of arsenic in water, suspended solids and sediment of Xiangjiang river. China: Sci. Total Environ. 77, 69–82.

Datta, R., Sarkar, D., 2004. Arsenic geochemistry in three soils contaminated with sodium arsenite pesticide: An incubation study. Environ. Geosci. 11, 53–63.

Elkhatib, E.A., Bennett, O.l., Wright, RJ., 1984. Arsenite sorption and desorption in soils. Soil Soc. Am. 48, 1025–1030.

Fassbender, H.W., 1974. Content, forms and fixation in forest soils of arsenate in comparison with phosphate. Z Pflanzenernahr Dung Bodenkd 137, 188–203.

Fordham, A.W., Norrish, K., 1983. The nature of soil particles particularly those reacting with arsenate in a series of chemically treated samples. Aust. J. Soil Res. 21, 455–477.

Freeman, G.B., Schoof, R.A., Ruby, M.V., Davis, A.O., Dill, J., Liao, S.C., lapin, C.A., Bergstorm, P.D., 1995. Bioaccessibility of arsenic in soil and house dust impacted by smelter activities following oral administration in Cynomologus monkeys. Fundam. Appl. Toxicol. 28, 215–222.

Hale, JR., Foos, A., Zubrow, J.S., Cook, J., 1997. Better characteriztion of arsenic and chromium in soils: A field-scale example. J. Soil Contam. 6, 371–389.

Halmes, N.C.H., Roberts, S.M., 1997. Arsenic bioaccessibility: A review of the literature. Technical Report 97-02. Center for Environmental and Human toxicology. University of Florida Lindau, L.1977. Emissions of arsenic in Sweden and their reduction. Environ. Health Perspect. 19, 25–29.

Harrison, R.M., 1981. Chemical association of Pb, Cd, Cu and Zn in street dusts and roadside soils. Environ. Sci. Technol. 15, 1378–1383.

Jacobs, L.W., Seyers, J.K., Keeney, D.R., 1970. Arsenic sorption by soils. Soil Sci. Soc. Am. 34, 750–754.

Manning, B.A., Fendorf, S.E., Goldberg, S., 1998. Surface structures and stability of As (III) on goethite: Spectroscopic evidence for inner-sphere complexes. Environ. Sci. Tecnol. 32, 2383–2388.

Mehlich, A., 1984. Mehlich No 3 soil test extractant: A modification of Mehlich No 2 extractant. Commun. Soil Sci. Plant 15, 1409–1416.

Murphy, E.A., Aucott, M., 1998. An assessment of the amounts of arsenical pesticides used historically in a geographically area. Sci. Total Environ. 218, 89–101.

Ng, J.C., 1999. Speciation, bioavailability and toxicology of arsenic in the environment. PhD Dissertation, University of Australia, Queensland, Australia.

O'Neill, P., 1995. Arsenic. In: Alloway, B.J. (Ed.), Heavy Metals in Soils. Blackie Academic & Professional, London.

Oscarson, D.W., Huang, P.M., Defosse, C., Herbillon, A., 1981. Oxidative power of Mn (IV) and Fe (III) oxides with respect to As (III) in terrestrial and aquatic environments. Nature 291, 50–51.

Pickering, W.F., 1981. Selective chemical extraction of soil components and bound metal species. Crit. Rev. Anal. Chem. 12, 233–266.

Pierce, M.L., Moore, C.B., 1980. Adsorption of arsenite on amorphous iron hydroxide from dilute aqueous solution. Environ. Sci. Technol. 14, 214–216.

Rodriguez, R.R., 1998. Bioaccessibility and biomethylation of arsenic in contaminated soils and solid wastes. PhD Dissertation, Oklahoma State University, Stillwater, OK.

Rodriguez, R.R., Basta, N.T., Casteel, S.W., Page, L.W., 1999. An in-vitro gastrointestinal method to estimate bioaccessible soils and solid media. Environ. Sci. Technol. 33, 642–649.

Rodriguez, R.R., Basta, N.T., Casteel, S.W., Armstrong, F.P., Ward, D.C., 2003. Chemical extraction methods to assess bioavailable arsenic in soils and solid media. J. Environ. Qual. 32, 876–884.

Ruby, M.V., Davis, A., Schoof, R., Eberle, S., Sellstone, C.M., 1996. Estimation of lead and arsenic bioavailabilty using a physiologically based extraction test. Environ. Sci. Technol. 30, 422–430.

Salomon, W., Forstner, U., 1980. Trace metal analysis on polluted sediments. Part 2. Evaluation of environmental impact. Environ. Technol. Lett. 1, 506–517.

Sarkar, D., O'Connor, G.A., 2001. Using the Pi soil, test to estimate available P in biosolids-amended soils. Commun. Soil Sci. Plant 32(13–14), 2049–2063.

Sarkar, D., Datta, R., 2003. A modified in-vitro method to assess bioaccessible arsenic in pesticide-applied soils. Environ. Pollut. 126, 363–366.

Sarkar, D., Datta, R., 2004. Arsenic fate and bioaccessibility in two soils contaminated with sodium arsenate pesticide: An incubation study. Bull. Environ. Contam. Toxicol. 72, 240–247.

Sarkar, D., Parra-Noonan, M., Datta, R., 2004. Distribution of arsenic in chemically variant dipping vat site soils. Bull. Environ. Contam. Toxicol. 73, 838–845.

Schoof, R.A., Yost, L.J., Eickhoff, J., Crecelius, E.A., Cragin, D.W., Meacher, D.W., Menzel, D.M., 1999. A market basket survey of inorganic arsenic in food. Food and Chemical. Toxicology 37, 839–846.

Sheppard, Sc., 1992. Summary of phytotoxic levels of soil arsenic. Water Air Soil Pollut. 64, 539–550.

Southworth, R.M., 1995. Part 503 land application pollutant limit for arsenic. U.S. Environment Protection Agency, Washington, DC.

Sparks, D.L., (Ed.), 1996, Methods of Soil Analysis: Part 3, Chemical Methods. Soil Science Society of America Publications, Madison WI.

Stillwell, D.E., Gorny, K.D., 1997. Contamination of soil with copper, chromium, and arsenic under decks built from pressure treated wood. Bull. Environ. Contam. Toxicol. 58, 22–29.

US EPA, 1992. Relative absorption factors for risk assessment. U.S. Environment Protection Agency, Exposure Assessment Group, Washington, DC.

USEPA, 1996. Test methods for evaluating solid waste. SW 846, third ed. Office of solid waste and emergency response, Washington, DC.

Walsh, L., Keeney, D., 1975. Behavior and phytotoxicity of inorganic arsenicals in soils. In: Woolson, E.A. (Ed.), Arsenical Pesticides, ACS Symposium Series 7. American Chemical Society, Washington, DC.

Walsh, L.M., Sumner, M.E., Keeney, D.R., 1977. Occurrence and distribution of arsenic in soils and plants. Environ. Health Perspect. 19, 67–71.

Woolson, E.A., Axley, J.H., Kearney, P.C., 1973. The chemistry and phytotoxicity of arsenic in soils: II. Effects of time and phosphorous. Soil Sci. Soc. Am. 37, 254–259.

Woolson, E.A., Axley, J.H., Kearney, P.C., 1971. The chemistry and phytotoxicity of arsenic in soils. I. Contaminated field soils. Soil Sci. Soc. Am. Pro. 35, 938–943.

Woolson, E.A., 1973. Arsenic phytotoxicity and uptake in six vegetable crops. Weed Sci. 21, 524–527.

Xian, X., 1989. Effect of chemical forms of cadmium, zinc, and lead in polluted soils on their uptake by cabbage plants. Plant Soil 113, 257–264.

Developments in Environmental Science, Volume 5
D. Sarkar, R. Datta and R. Hannigan (Editors)

ISSN: 1474-8177/DOI:10.1016/S1474-8177(07)05017-6

Chapter 17

Dissolution chemistry of inorganic selenium in alkaline mine soils

Shankar Sharma and George F. Vance

Abstract

Selenium (Se) is a natural geological constituent in many mining areas of the western United States. While Se is an essential micronutrient for humans and animals, its low toxicity threshold poses health concerns. Oxidative exposure of Se in excavated mine soils leads to increased concentrations of selenite ($Se^{IV}O_3^{2-}$) and selenate ($Se^{VI}O_4^{2-}$) species, which can potentially enter the food chain. To ascertain the geochemical processes contributing to Se bioavailability, we examined Se dissolution, speciation, fractionation, adsorption, and solid phases in coal and uranium mine soils from Wyoming. The soils were alkaline in nature, and majority (86%) of the labile Se remained in the exchange phase. Despite 49% of total Se levels identified in siliceous minerals, the exchangeable Se exceeded the critical toxicity limit in >60% samples. Se^{VI} and Se^{IV} were the dominant species in aqueous and exchangeable fractions, respectively. Se^{VI} was the major Se species that dissolved under alkaline condition. Se^{IV} was primarily related to ligand exchange or complexation processes. Mine soils adsorbed up to 99% of Se^{IV} and 95% of Se^{VI}, and the relative sorption was on an average six-fold higher for Se^{IV}. Occurrence of oxides and oxyhydroxides of Ca, Mn, and Fe in soil resulted in greater Se retention, which was mostly augmented in presence of Fe-minerals. The soil solutions were undersaturated with respect to Se solid phases, thereby suggesting that Se precipitation would less likely prevail. $CaSeO_3 \cdot 2H_2O$ and $CaSeO_4$ were the only Se^{IV} and Se^{VI} solids in the vicinity of the saturation equilibrium line. Fe_2O_3, $CuFe_2O_4$, $ZnFe_2O_4$, $CuFeO_2$, FeOOH, Fe_3O_4, $CaFe_2O_4$, $MgFe_2O_4$, $Al(OH)_3$, Al_2O_3, CuO, and $Cu_4(SO_4)(OH)_6$ were the primary saturated mineral phases, and could potentially bind Se. These results are germane to Se chemistry of earth materials under pedochemical weathering.

17.1. Introduction

Selenium (Se) occurs indigenously within geological materials associated with coal and uranium (U) mining areas of the western United States (Valkovic, 1983; Boon, 1989). While Se is an essential micronutrient for humans and animals, its low toxicity thresholds (as related to 0.5 and 5 mg Se/kg soil and vegetation, respectively) pose potential concern for land use (Rosenfeld and Beath, 1964; Lakin, 1972; Maas, 1998; Vance, 2000). This is critical as animal grazing is the primary post-reclamation land use practice in many mining areas of the western United States. Revegetation of mine backfill materials high in Se may result in plants that contain elevated level of Se. Seleniferous mine land environments and areas impacted by mine drainage waters may, therefore, be problematic to wildlife and domestic grazing animals that consume vegetation containing greater than 5 mg kg^{-1} Se for extended periods of time (NRC, 1976; Vance, 2000). During mining operations, excavated soils and subsurface materials are exposed to oxidative conditions, thereby elevating the potential levels of bioavailable selenite ($Se^{IV}O_3^{2-}$) and selenate ($Se^{VI}O_4^{2-}$) species, which ultimately could enter the food chain via grazing by wildlife and domestic animals. Se related mortality in waterfowls, shorebirds, migratory and aquatic birds have been well documented (Ohlendorf, 1989).

The bioavailability of Se is dependent on the presence of different soil Se species including selenide (Se^{2-}), elemental Se (Se^0), selenite ($Se^{IV}O_3^{2-}$) and selenate ($Se^{IV}O_4^{2-}$); both $Se^{IV}O_3^{2-}$ and $Se^{VI}O_4^{2-}$ are the predominant Se species in oxidized and alkaline environments (Elrashidi et al., 1987; Jayaweera and Biggar, 1996). Details on the natural occurrence and geochemistry of Se in various seleniferous environments other than mines have been reviewed by McNeal and Balistrieri (1989) and Mayland et al. (1989). There are numerous reports on Se geochemistry in diverse terrestrial systems such as alluvial soils (Neal et al., 1987), biogenic particles and sediments (Cutter, 1985), oil shale (Spackman et al., 1990), goethite (Balistrieri and Chao, 1987), kaolinite (Bar-Yosef and Meek, 1987), montmorillonite (Goldberg and Glaubig, 1988), groundwaters (Vance et al., 1998), and agricultural and forest soils (Pezzarossa et al., 1999). Siddique et al. (2005) studied agricultural drainage waters that were amended with rice straw and reported that Se^{VI} could be reduced by bacteria to Se^{IV} and Se^0. Vegetative and microbial remediation of Se from agricultural drainage water was discussed within several recently published studies (Lin et al., 2002; Gao et al., 2003; Bañuelos et al., 2004).

Limited information exists on the geochemical behavior and bioavailability of Se in alkaline mine soils. Blaylock et al. (1995) assessed

$Se^{IV}O_3^{2-}$ sorption by coal mine soils using various equilibrium models. Their study found that mine soils could sorb >90% of the added Se, and that the initial mass (IM) model was a better predictor of Se sorption than Freundlich and Langmuir isotherms. In a field study that assessed plant and soil Se at a reclaimed U mine site, Sharmasarkar and Vance (2002a) observed that grass, forbs, and shrub type vegetation accumulated high level (10–1000 mg kg^{-1}) of Se. Lovett (2005) evaluated the chemistry and mitigation of Se in coal-mine drainage waters and observed this system had "a far more substantial biochemistry than traditional mine drainage contaminants" thereby facilitating the release of Se back into solution. For Se in coal-mine drainage waters from Canada and Alaska, Sobolewski (2005) suggested that in-situ treatments and phyto techniques were the most feasible remediation approaches. Cummings (2000) suggested the possible use of microbial reduction of Se for the reclamation of mine spoil leachates.

Several areas within the western United States have overburden geological materials that contain Se. Mine soils within these areas are potential sites for environmental Se problems if not managed correctly because Se toxicity can occur when Se is present even at low levels. To determine the prospective of Se transfer through the food chain and possible development of biological (e.g., plant and microbial) remediation techniques, it is important to understand the chemistry related to Se bioavailability. Therefore, the objectives of this study were to assess geochemical processes causing Se dissolution and speciation chemistry of solution Se in alkaline mine soils. This information could be utilized for developing mine land reclamation strategies.

17.2. Materials and methods

17.2.1. Soil sampling

Soil samples were collected from coal and U mine sites located within the Powder River Basin of eastern Wyoming (Fig. 17.1). The taxonomy of the native soils within the contiguous areas of the mine sites were comprised of Ustic Torrifluvents, Ustic Haplargids, Ustic Torriorthents, Ustic Calciargids, Mesic Haplargids, Ustic Torrpsamments, Ustic Paleargids, Ustic Haplocambids, Typic Haplocambids, and Typic Torriorthent. In Wyoming, high levels of Se occur naturally within several regions known as super toxic (Rosenfeld and Beath, 1964).

We were interested in specific mine sites that contained high Se indicator plants within their surrounding native areas. The sites consisted of

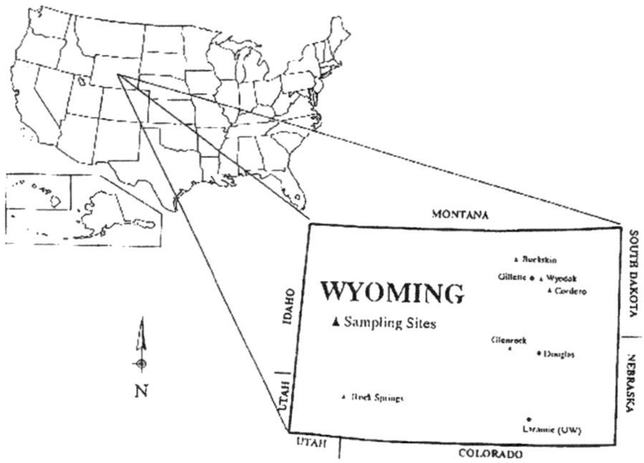

Figure 17.1. Mine soil-sampling areas in Wyoming.

Table 17.1. Physicochemical properties of mine soils

pH	EC (dS m^{-1})	Sand (%)	Clay (%)	Organic carbon (%)	Sulfate (mg kg^{-1})	Total Se (mg kg^{-1})	Fe (g kg^{-1})	Mn (g kg^{-1})	Ca (g kg^{-1})
7.3–8.5	0.3–10	3–84	8–59	0.1–7	48–2940	0.24–26	1.3–4.1	0.05–0.19	0.9–15

overburden materials collected from both coal and U mine pits, which were deeper than the depths of our soil-sampling cores. Vegetation at the sites was primarily reseeded plant species; however, some native species were also evident and resulted from dispersion of soils that contained native plant seeds. The native areas surrounding the sites were slightly undulating with a predominance of bare soil, scattered grasses, forbs, and shrubs. No trees existed in the dry, harsh landscape associated with our study sites. Soil moisture was greatest in sampling site located in depressions, particularly during late spring and early summer due to accumulation of snowmelt and rainfall. However, during the time of sampling (e.g., June and July) there were no noticeable differences in the moisture status of the different sampling sites.

Sixty soil samples (0–200 cm depth) were collected from the different mine sites. The samples were of contrasting nature to ensure diversity within the mine environment. Soils had alkaline pH levels, moderate to high-salt contents, low organic carbon levels, and a predominance of sand particle-sized materials (Table 17.1). These soils were enriched in kaolinite, smectite, illite, Fe, Mn, Ca, and high levels of sulfate. The soil Se concentrations were generally above the toxic level.

17.2.2. Analytical techniques

Soils were air-dried, finely ground (<0.25 mm), and stored at room temperature in polyethene bags following established processing and analytical protocols for mine soil materials (Spackman et al., 1994). Inorganic Se (Se_i) was analyzed after extraction with deionized water (DW), 0.1% $CaCl_2$ background electrolyte in hot water (HW), ammonium bicarbonate–diethylenetriamine pentaacetic acid at pH 7.6 (AB), 0.01 M NaOH at pH 11.6 (OH), and 1 M KH_2PO_4 at pH 4.7 (PO). These dissolution agents cover a wide array of pH, and some of these extractants have been reported to be significant predictors of soil–plant Se correlations under field conditions (Pasch and Vance, 1995). In a study of Se isotopes in agricultural drainage sediments, Herbel et al. (2002) also utilized NaOH extraction to assess soluble Se. Total Se (Se_T) was determined in mixed acid (15.8 *M* HNO_3 + 11.7 *M* $HClO_4$ + 28.9 *M* HF) digests. Net inorganic Se ($Se^{IV}O_3^{2-} + Se^{VI}O_4^{2-}$) was determined with atomic absorption spectrophotometry—hydride generation (AAS-HG) after an HCl reduction step. Solution $Se^{IV}O_3^{2-}$ concentrations were measured prior to the reduction process, and then subtracted from the net Se values to obtain $Se^{VI}O_4^{2-}$ concentrations. Extraction protocol was based on correspondence of soil Se with plant availability (Orso et al., 1993; Spackman et al., 1994; Sharmasarkar and Vance, 1995).

The fractional partitioning technique of Chao and Sanzolone (1989) was followed by dissolving soil Se sequentially using 0.25 *M* KCl, 1 *M* KH_2PO_4, 4 *M* HCl, $KClO_3$ + 12 *M* HCl, and mixed acid (15.8 *M* HNO_3 + 11.7 *M* $HClO_4$ + 28.9 *M* HF). There was occasional modification based on the standard methodologies for Se in mine soils (Spackman et al., 1994). For example, 1 *M* KH_2PO_4 (pH 4.7) was used instead of 0.1 *M* KH_2PO_4, as Se extracted by the former has been reported to correspond more closely to plant-available Se (Orso et al., 1993). Each fraction was then analyzed for Se by AAS-HG.

Soil organic carbon and particle size distributions were analyzed by the Walkley-Black method (Nelson and Sommers, 1982) and the pipette method (Gee and Bauder, 1986), respectively. The pH, electrical conductivity (EC) ,and soluble sulfate (SO_4^{2-}) were determined in saturated paste extracts. Crystalline Fe, Mn, and Ca were determined using citrate–dithionite buffer (Holmgren, 1967) method.

Se sorption experiments were conducted using a range of 0–2.5 ppm Se. Aqueous extracts of selected samples containing high Se levels were analyzed for dissolved ions by inductively coupled plasma mass spectroscopy (ICP-MS) and ion chromatography (IC) for estimating solution and solid phases using a geological equilibration model PHREEQC

(Parkhurst and Appelo, 1999). All experiments were conducted in triplicate at room temperature ($23 \pm 1°C$), and statistical evaluations were conducted using Minitab (1992).

17.3. Results and discussion

17.3.1. Dissolution and speciation

Total dissolved inorganic Se in different extracts and the relative distribution of bioavailable selenate (Se^{VI}) and selenite (Se^{IV}) in these fractions are shown in Fig. 17.2. Concentrations of Se in the various extracts followed the order: PO > OH > AB > HWDW. On a relative scale, PO and OH dissolved up to 56 and 30% of Se_i, respectively, as compared to only 3–7% by water and AB. More than 50% of the total soluble Se was in inorganic form. On an absolute scale, DW, HW, AB, OH, and PO extracted 67, 50, 57, 68, and 52% of Se_t, respectively, with $r = 0.80$–0.99 ($p < 0.01$). About 70% of DW and OH extractable Se fractions were inorganic in nature. Selenium dissolved by PO and OH in > 60% of our samples exceeded the Wyoming Department of Environmental Quality critical limit of 0.3–0.8 $mg\,kg^{-1}$ for mine soils (Vance et al., 1998). The Se^{VI} concentrations were greater in OH, DW, and HW solutions as compared to PO and AB solutions. The OH-DW-HW fractions contained equiproportional (nearly 30% each) fraction of Se^{VI}, albeit with an occurrence of a very low level (< 5%) of Se^{IV}. Conversely, AB and PO extracts together contained > 90% Se^{IV} species. Thus, Se^{VI} may be inferred as the major species to be released under aqueous or alkaline dissolution conditions, whereas, Se^{IV} would be primarily solvated by ligand exchange or complexation phenomena during pedochemical processes.

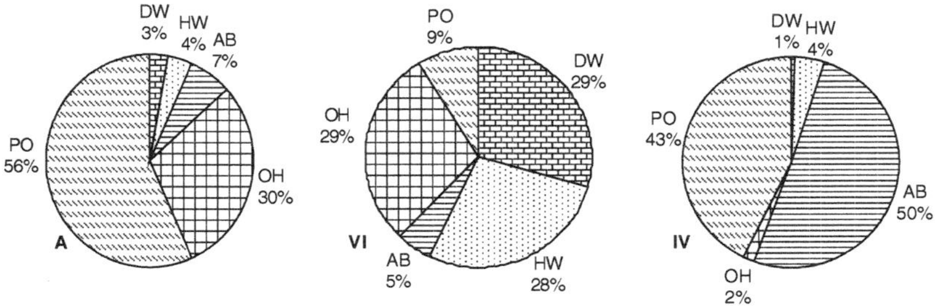

Figure 17.2. Selenium dissolution and speciation in mine soils: (A) dissolved Se as % total Se; (VI) bioavailable selenate (Se^{VI}); and (IV) bioavailable selenite (Se^{IV}).

A correlation matrix indicated dissolved Se_i and anionic Se (IV and VI) species in various extracts were linearly related within the $p < 0.01$ probability level. Selenium concentrations in PO, AB, and HW extracts resulted in better correlations (r0.90) with Se in the OH extracts. The r-values corresponding to Se^{VI} were always higher in HW than DW solutions. Even greater r-values were observed for Se^{IV} in AB solutions when correlated with SE^{VI} in OH and Se^{IV} in PO solutions. Correlation coefficients for DW-extractable Se were lower than other solvents, even though they were significant at the $p < 0.01$ level. These correlations are statistical validations of thermodynamic reactions among the Se species, for example, oxidation of Se^{IV} to Se^{VI} (Elrashidi et al., 1989). Such correlations also validate the importance of various Se extraction methods; for example, a linear correlation ($r = 0.98$, $p < 0.01$) between Se fractions in DI and HW extracts suggested that HW extractable inorganic Se could be readily dissolved by soil moisture present in pore spaces. Thus, statistical correlations may be used as indicators of plausible geochemical processes.

Our findings for Se bioavailability in mine soils were consistent with Se reported in other ecosystem studies. Martens and Suarez (1998) reported that Se^{IV} and Se^{VI} species were dominant in phosphate and aqueous extracts, respectively. Conceptually, the Se^{VI} species extracted by distilled-deionized and/or hot water is thought to be loosely bound to the soil particles and hence brought into solution by solubilization or low thermal energy processes. In contrast to the inner-sphere complexation of Se^{IV}, bonding of Se^{VI} is considered to occur mainly through diffuse-ion swarm association and outer-sphere complexation (Sposito, 1989). The relatively lower strength of Se^{VI} stems from its electrostatic nature of bonding as compared with the more stable nature of combined ionic and covalent bonds involving Se^{IV}. Subsequently, the stabilization energy (heat of bond formation) for Se^{VI} bonding is larger than that for Se^{VI}. The hydroxyl ions (OH^-) in the hot water extract can substitute for nonspecifically adsorbed $Se^{VI}O_4^{2-}$ that is weakly bound in outer-sphere complexes. This is also consistent with thermodynamic predictions regarding $Se^{VI}O_4^{2-}$ solubility (Elrashidi et al., 1987, 1989). Phosphate can solubilize specifically adsorbed Se^{IV} species by means of anionic competition and/or ligand exchange mechanisms (Parfitt, 1978). The Se^{IV} species are generally present as inner-sphere complexes associated with clay minerals such as kaolinite and montmorillonite (Bar-Yosef and Meek, 1987). Extraction with AB-DTPA, a strong chelating ligand, can result in metal cation complexes that release Se compounds, thereby solubilizing Se^{IV} species. However, the predominance of a particular species does not exclude the occurrence of other species in solution in relatively lower concentrations. Thus, $Se^{VI}O_4^{2-}$ was also detected in KH_2PO_4 and AB-DTPA extracts, as well as $Se^{VI}O_4^{2-}$ was determined in the other extracts.

17.3.2. Sequential fractionation

Results of sequential fractionation of Se are illustrated in Fig. 17.3. Fractions included Se components that were aqueous (AQ; 0.25 *M* KCl), exchangeable (EX; 1 *M* KH_2PO_4), associated with amorphous and carbonate materials (AC; 4 *M* HCl), sulfide and humic bound (SH; $KClO_3$ + 12 *M* HCl), and structured into siliceous minerals (SI; 15.8 *M* HNO_3 + 11.7 *M* $HClO_4$ + 28.9*M* HF). Aqueous and exchangeable fractions are available for leaching and plant uptake, whereas the acid extractable fraction may be conditionally available. The last two fractions are weathered with time and pedochemical changes.

Our samples indicated Se distributions in fractions followed the order AQ < EX < AC < SH < SI. There was a greater abundance of sulfide-humic bound and siliceous Se. The sum of all the five fractions (0.24–1.45 mg kg^{-1}) was very close to the total soil Se (0.26–1.55 mg kg^{-1}). We were able to recover > 95% of the total soil Se by fractional partitioning technique ($r = 0.99$ at $p < 0.001$). Se^{VI} and Se^{IV} were the dominant species in AQ and EX, respectively. Fraction AQ would primarily correspond to the nonspecifically adsorbed Se formed through outer-sphere complexation. Since this interaction is a weak-bond force, chloride (from 0.25 *M* KCl) can easily provide the dissolution energy for SeO_4^{2-} and bring it into solution by means of anion exchange and mass action mechanisms. Selenium in fraction EX would be specifically adsorbed on exchange sites of oxides (e.g., goethite (Balistrieri and Chao, 1987)) and clay minerals (kaolinite and montmorillonite (Bar-Yosef and Meek, 1987)). This interaction is a strong-field innersphere complex. Phosphate (1 *M* KH_2PO_4) would solubilize SeO_3^{2-} by means of anionic competition or ligand exchange mechanisms (Parfitt, 1978).

Fraction AC would correspond to Se associated with amorphous materials (Rajan and Watkinson, 1976) and carbonates (Goldberg and Glaubig, 1988), which would be solubilized by HCl (4 *M*) through hydrolysis and dissolution mechanisms. The Se in this fraction would be subjected to sudden chemical changes in the soil environments; for example, decrease in redox potential or pH (during acid leaching process) might enhance the solubilization of Se in fraction AC. Thus, Se in fraction AC would be only conditionally available. Both Se^{IV} and Se^{VI} were evenly distributed in this fraction.

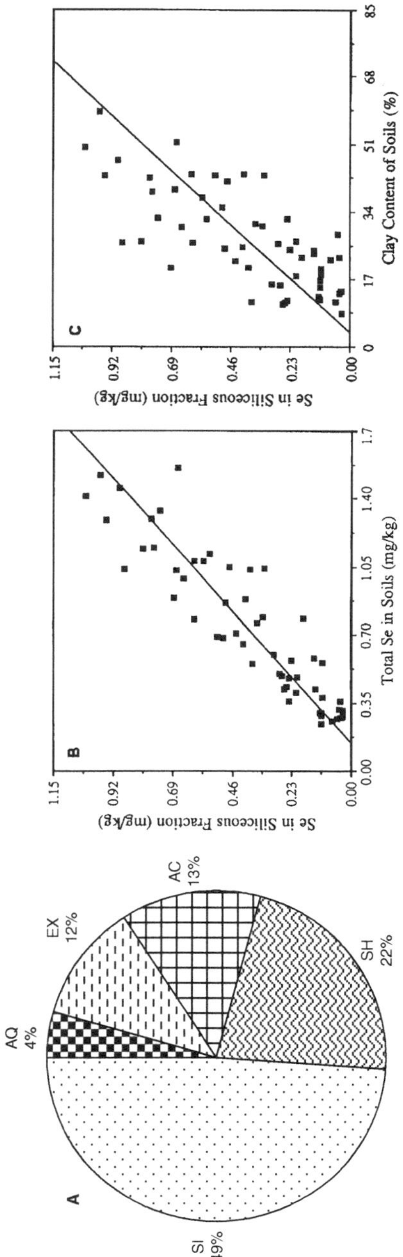

Figure 17.3. Fractional partitioning of mine soil Se: (A) % distribution; and correlations of siliceous Se with (B) total Se and (C) clay content.

Selenium could also be released from its association with sulfides (Coleman and Delevaux, 1957; Hawley and Nichol, 1959) and organic matter (Rosenfeld and Beath, 1964) during oxidative weathering (fraction SH). Associations of Se with resistant sulfide minerals (e.g., pyrites) could be due to the comparable ionic radii of Se^{-II} and S^{-II} ($r_{Se-2} = 0.191$ nm, and $r_{S-2} = 0.184$ nm; Weast, 1985). Combination of $KClO_3$ and 12 *M* HCl acts as a strong oxidant that can dissolve such materials, which otherwise may not be easily attacked by 4 *M* HCl.

Fraction SI corresponds to highly resistant Se in layer lattices of siliceous minerals. The SI component contained the largest fraction of Se, which increased linearly by 68% with proportional increases in total soil Se content ($r = 0.91$ at $p < 0.001$, Fig. 17.3b). The siliceous Se was also directly correlated with clay content of soils ($r = 0.74$ at $p < 0.001$, Fig. 17.3c), thereby suggesting the presence of Se in silicate minerals. Though there have been no mineralogical studies done to identify Se in the clay lattice, Chao and Sanzolone (1989) observed that Se "compounds may occur in accessory minerals imbedded within the silicate matrices". The Se in SI fraction increased linearly with total Se ($r = 0.91$ at $p < 0.001$) and clay content ($r = 0.74$ at $p < 0.001$) of the soils. This suggested possible presence of Se within the silicate minerals.

Total soil Se increased with increasing clay content ($r = 0.81$ at $p < 0.001$), while the reverse was observed with sand content ($r = -0.69$ at $p < 0.001$). This suggested a probable association of Se with clay (Bar-Yosef and Meek, 1987; Mayland et al., 1989). Our finding conformed to Chao and Sanzolone (1989) who studied altogether different types of soils (soils from Hawaii and certain reference sample materials from U.S. Geological survey, namely, GXR 5).

Selenium speciation data (e.g., Se^{IV} and Se^{VI}) for SH and SI fractions were not conclusive. This could be due to any drastic changes in the oxidation states of the Se species caused by the strong oxidative reagents. Similar changes in oxidation status during sequential extraction of Se were also reported by Neal et al. (1987). Selenium in these two fractions could be transformed into simpler available forms through weathering and pedochemical changes.

17.3.3. Adsorption

High adsorption coefficients (K10 kg/10 g) suggested the mine soils could retain significant amounts of both Se^{IV} and Se^{VI} species, and that Se^{IV} adsorption was on an average six-fold higher than that for Se^{VI} (Fig. 17.4). Overall, the soils adsorbed 40–99% of the added Se^{IV} and

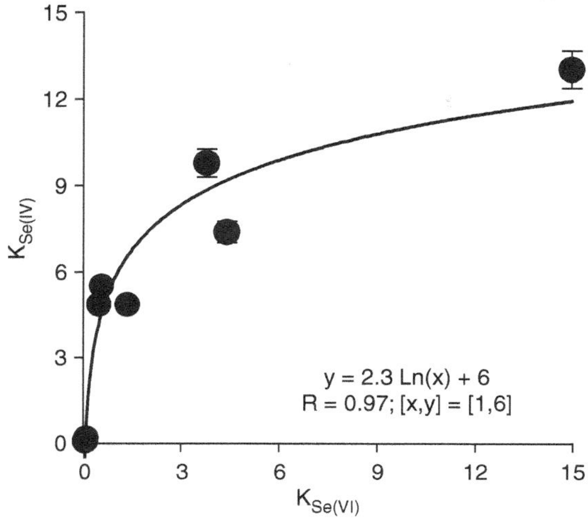

Figure 17.4. Relative adsorption of Se^{IV} and Se^{VI} by mine soils.

30–95% of added Se^{VI}, and the relative sorption was always higher for Se^{IV}. The large magnitude of sorbate to sorbent ratio suggested a possible multilayer sorption mechanism. This was verified by applying various sorption isotherm models to the data. The poorest fit ($r = 0.01$–0.56) was observed with the Langmuir monolayer model. In using the Langmuir model, the energy of adsorbed molecules is assumed to be uniform, which is possible only for an intrinsically homogeneous surface. Based on this assumption, intra- and inter-molecular forces of attraction or repulsion should be zero, which is unlikely in a heterogeneous mine soil system. Hence, it is not unexpected that >90% of our data did not conform to the Langmuir model.

In contrast with the Langmuir model, multilayer Temkin and Freundlich models fit the Se^{VI} and Se^{IV} data for 60–100% of the mine soils at varying degree of correlations within the significance level of $p < 0.01$. The best conformation was noticed for the Initial Mass model, which was fitted to both Se species for all of our samples with high correlation ($r = 0.99$–1.00). We already reported various model fittings elsewhere (Sharmasarkar and Vance, 2002b), and therefore, presented here only a relevant synopsis. The continued sorptive loading suggested that at low Se concentrations sorption sites on mine soil particles were probably never saturated. This was consistent with our observation of undersaturated soil conditions as described below.

In a study with SeO_3^{2-}, Blaylock et al. (1995) concluded that the IM isotherm was able to describe Se sorption behavior better than Freundlich and Langmuir isotherms. However, it should be realized that the model fitting would be ecosystem specific. For example, Singh et al. (1981) reported conformity of Se sorption to the Langmuir isotherm in Indian agricultural soils. In a study with oil shales, Spackman et al. (1990) concluded that both Langmuir and Freundlich isotherms described SeO_3^{2-} adsorption, with data fitting best with the Freundlich model. Freundlich fitting of Se sorption was reported by Fio et al. (1991), who observed that $Se^{VI}O_4^{2-}$ was not retained by alluvial soils, whereas $Se^{IV}O_3^{2-}$ was rapidly adsorbed and conformed to the Freundlich model under alkaline soil conditions. Greater sorption capacity of Se^{IV} than Se^{VI} was also reported in other types of soils by Balistrieri and Chao (1987). Application of the Temkin model to Se sorption is a new approach; this isotherm was used by Sanyal et al. (1993) to evaluate phosphate sorption by acid soils.

In contrast to the inner-sphere complexation of $Se^{IV}O_3^{2-}$, sorption of $Se^{VI}O_4^{2-}$ is considered to occur mainly through diffuse-ion swarm association and outer-sphere complexation (Sposito, 1989). Relatively lower sorption strength of the Se^{VI} species is due to its electrostatic nature of bonding as compared to the more stable coupling of ionic and covalent bonds for Se^{IV} species. The stabilization energy of bond formation for Se^{IV} sorption is subsequently higher than that for Se^{VI} sorption.

Adsorption parameters were multivariate functions of chemical and physical properties of the mine soils. For example, amounts of both Se species in PO were directly related to Freundlich *K* values. The correlation was more significant for Se^{IV}, which could be attributed to Se^{IV} being the predominant Se species in PO, as we have noted earlier. Both phosphate and SeO_3^{2-} adsorb principally as inner-sphere complexes via ligand exchange mechanism (Sposito, 1989), and so, their mutual relationship is not unusual. Similar correlations were also observed between Se_t and *K* values for both Se species. These relationships are important because Se in PO has been reported to be an indicator of plant Se uptake in mine environment (Sharmasarkar and Vance, 1995).

Presence of oxides or oxyhydroxides of Ca, Mn and Fe also would result in greater Se retention in the mine soils. This was evident from direct correlations of these dissolved metals with Se sorption parameter such as Freundlich-*n*, which was also a function of electrical conductivity, clay content, and specific surface area. Also, complexes of these metals with Se would control Se solid phases and their dissolution under pedochemical changes. Se sorption increased with increasing metal concentrations, with the effect of metals being in the order of Fe > MnCa.

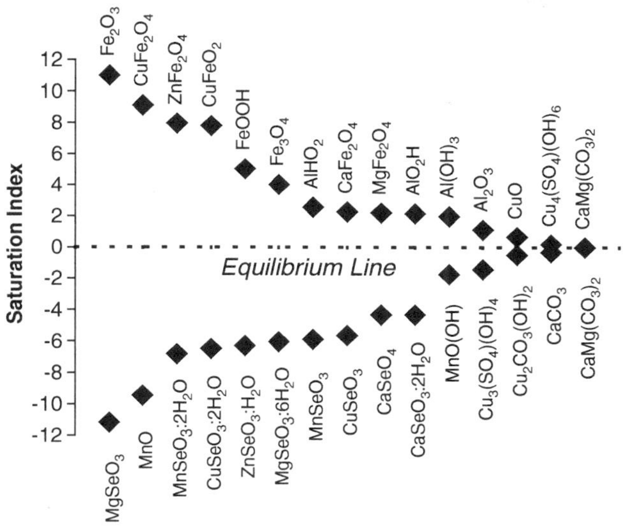

Figure 17.5. Saturation equilibrium of mineral solid phases in mine soils.

The functional relationships with soil properties could be used as predictors of Se loads and bioavailability in mine environments.

17.3.4. Solid phases

Although dissolved Se levels surpassed the Se contaminant threshold, mineral saturation index values in these mine soils indicated they were undersaturated with respect to various Se solid phases (Fig. 17.5). This observation suggests that Se availability in these mine soils is less likely to be controlled by a precipitation mechanism; rather, desorption—adsorption equilibrium would plausibly control solution Se concentrations. Solution phase computation by the geochemical modeling predicted the predominance of $[>SeO_3]^0$ and $[>SeO_4]^0$ over corresponding metal complexes of Se^{IV} and Se^{VI}. Detailed review of the data indicated that $CaSeO_3 \cdot 2H_2O$ and $CaSeO_4$ were the closest Se^{IV} and Se^{VI} solids to the saturation equilibrium line. These solid phases, therefore, have the future potential for partially controlling the solution Se chemistry if the saturation equilibria are reached through long-term pedochemical weathering.

The supersaturated solid phases in our mine soils included Hematite (Fe_2O_3), Ferrite–Cu ($CuFe_2O_4$), Ferrite–Zn ($ZnFe_2O_4$), Delafossite ($CuFeO_2$), Goethite (FeOOH), Magnetite (Fe_3O_4), Diaspore ($AlHO_2$),

Ferrite–Ca ($CaFe_2O_4$), Ferrite–Mg ($MgFe_2O_4$), Boehmite (AlO_2H), Gibbsite ($Al(OH)_3$), Corundum (Al_2O_3), Tenorite (CuO), and Brochantite ($Cu_4(SO_4)(OH)_6$), with Dolomite ($CaMg(CO_3)_2$) on the equilibrium line. These Fe, Cu, Zn, Al, Ca, and Mg-based minerals would play important roles in solution phase chemistry and subsequently impact the bond strength of inner- or outer- sphere complexes of Se species during its sorption or desorption processes.

17.4. Conclusion—Physicochemical and environmental significance

Our data indicate Se^{IV} was the dominant Se species in AB and PO extracts, whereas Se^{VI} was the major Se species in the DW, HW, and OH extracts. Conceptually, the Se^{VI} extracted by DI and HW extracts are loosely bound to the soil particles and hence brought into solution by mass action mechanisms or low-thermal energy. The hydroxyl ions (OH^-) in the NaOH extract can substitute for non-specifically adsorbed selenate that is weakly bound in outer-sphere complexes. This is consistent with thermodynamic predictions on Se solubility (Elrashidi et al., 1987). Phosphate can solubilize specifically adsorbed $Se^{IV}O_3^{2-}$ by means of anionic competition or ligand exchange mechanisms (Parfitt, 1978); SeO_3^{2-} is present as an inner-sphere complex associated with clay minerals such as kaolinite and montmorillonite (Bar-Yosef and Meek, 1987). The AB extracts, being a strong chelating ligand, can form complexes with metal cations associated with Se compounds, thereby solubilize Se oxyanions. It should be realized that the predominance of a particular species does not exclude the occurrence of other species in solution in relatively lower concentrations. Thus, both Se^{IV} and Se^{VI} species were detected in all the extracts. From a thermodynamic stand-point, SeO_3^{2-} remains stable within a pE + pH range of 7.5–14.5, and above this value $Se^{VI}O_4^{2-}$ predominates in soil solutions (Elrashidi et al., 1989); thus, occurrence of these two Se species is probable in coal mine environments where mining activity exposes soils to atmospheric oxidative conditions.

The plausibility of multilayer sorption suggested that at low Se concentrations the sorption sites were probably never saturated. Subsequently, it was not unusual to observe >90% sorption for both Se species. Differences in the sorption mechanisms for $Se^{IV}O_3^{2-}$ and $Se^{VI}O_4^{2-}$ can be attributed to the structural difference of the two Se species, both of which are formed from the same cationic and anionic counterparts, i.e., Se and O with respective electronic configurations of $[Ar]3d^{10}4s^24p^4$ and $[He]2s^22p^4$. Applying the concepts of the valence shell electron pair repulsion theory and hybridization, structures of SeO_3^{2-} and SeO_4^{2-} can be concluded to be distorted pyramidal (Se is sp^2) and tetrahedral (Se is sp^3), respectively. There are three-bond pairs (BP) and one lone pair (LP) of electrons in the valence shell of SeO_3^{2-}, whereas, SeO_4^{2-} valence shell contains four-bond pairs only with no lone pairs at all. The distortions in structure are caused by the BP–BP and BP–LP repulsions. In SeO_3^{2-}, the hybridized Se possesses 33.3% *s*-character as compared to 25% *s* in SeO_4^{2-}. However, the *p*-character of Se is much higher in SeO_4^{2-} (75%) in comparison to SeO_3^{2-} (66.7%).

Owing to undersaturated soil conditions with respect to various Se solid phases, albeit with high levels of phosphate-extractable Se surpassing the Wyoming Department of Environmental Quality critical limits, sorption–desorption processes appeared to be the more plausible mechanism governing mine Se immobilization chemistry. The mine soils were rich in siliceous Se fraction. The aqueous Se was dependent on the exchangeable fraction, which was again directly related to the acid extractable Se. In aqueous and exchange phases, Se^{VI} and Se^{IV} were the respective major species, whereas, both Se species were distributed almost evenly in acid extractable fraction. Aqueous Se^{VI} and exchangeable Se^{IV} were also functions of exchangeable Se^{IV} and acid extractable Se^{IV}, respectively.

The results of this study also indicated that Se must be present in the structure of clay minerals. Statistical analysis indicated that there were several significant functional correlations between different partitioning and speciation parameters. These relationships indicate a possibility of the geochemical processes that may occur with time. Solution-phase Se extraction and subsequent speciation provide a means of evaluating the distribution of different Se components in various soluble forms. This information is also important for examining the mutual relationships among Se forms and species, which can be useful in understanding the solubility and toxicity behavior of Se in coal mine environments. For future research, we recommend mineralogical studies that investigate Se in soil minerals.

REFERENCES

Balistrieri, L., Chao, T., 1987. Selenium adsorption by goethite. Soil Sci. Soc. Am. J. 51, 1145–1151.

Bañuelos, G.S., Sharmasarkar, S., Pasakdee, S., 2004. Utilization of biosolid as a fertilizer for Canola. Compost. Sci. Util. 12, 61–68.

Bar-Yosef, B., Meek, D., 1987. Selenium sorption by kaolinite and montmorillonite. Soil Sci. 144, 11–19.

Blaylock, M.J., Tawfic, T., Vance, G., 1995. Modeling selenite sorption in reclaimed coal mine soil materials. Soil Sci. 159, 43–48.

Boon, D.Y., 1989. Potential selenium problems in Great Plains soils. In: Jacobs, L.W. (Ed.), Selenium in Agriculture and Environment, SSSA special publication no. 23, Madison, WI, pp. 107–121.

Chao, T.T., Sanzolone, R., 1989. Fractionation of soil selenium by sequential partial dissolution. Soil Sci. Soc. Am. J. 53, 385–392.

Coleman, R.G., Delevaux, M., 1957. Occurrence of selenium in sulfides from some sedimentary rocks of the western United States. Econ. Geol. 52, 499–529.

Cummings, D.E., 2000. Microbial reduction of iron and selenium in mine spoils. Thesis, University of Idaho. TD 899 .M5 C8 2000. p. 216.

Cutter, G.A., 1985. Determination of selenium speciation in biogenic particles and sediments. Anal. Chem. 57, 2951–2955.

Elrashidi, M., Adriano, D., Lindsay, W., 1989. Solubility, speciation, and transformations of selenium in soils. In: L.W. Jacobs (Ed.), Selenium in Agriculture and the Environment. SSSA special publication no. 23, Madison, WI, pp. 51–63.

Elrashidi, M., Adriano, D., Workman, S., Lindsay, W., 1987. Chemical equilibria of selenium in soils: A theoretical development. Soil Sci. 144, 141–152.

Fio, J.L., Fujii, R., Deverel, S., 1991. Selenium mobility and distribution in irrigated and nonirrigated alluvial soils. Soil Sci. Soc. Am. J. 55, 1313–1320.

Gao, S., Tanji, K., Lin, Z., Terry, N., Peters, D., 2003. Selenium removal and mass balance in a bonstructed flow-through wetland system. J. Environ. Qual. 32, 1557–1570.

Gee, G.W. Bauder, J.W., 1986. Particle-size analysis. In: Klute, A. (Ed.), Methods of Soil Analysis, Pt. 1. American Society of Agronomy, WI, pp. 383–411.

Goldberg, S., Glaubig, R., 1988. Anion sorption on a calceous, montmorillonitic soil-selenium. Soil Sci. Soc. Am. J. 52, 954–958.

Hawley, J.E., Nichol, I., 1959. Selenium in some Canadian sulfides. Econ. Geol. 54, 608–628.

Herbel, M.J., Johnson, T., Tanji, K., Gao, S., Bullen, T., 2002. Selenium stable isotope ratios in California agricultural drainage water management systems. J. Environ. Qual. 31, 1146–1156.

Holmgren, G.G.S., 1967. A rapid citrate-dithionite extractable iron procedure. Soil Sci. Soc. Am. Proc. 31, 210–211.

Jayaweera, G.R., Biggar, J.W., 1996. Role of redox potential in chemical transformation of selenium in soils. Soil Sci. Soc. Am. J. 60, 1056–1063.

Lakin, H.W., 1972. Selenium accumulation in soils and its absorption by plants and animals. In: Cannon, H.L., Hopps, H.C. (Ed.), Geochemical Environment in Relation to Health and Disease. Special Paper No. 140, Geol. Soc. Am., Boulder, CO, pp. 27–34.

Lin, Z.Q., de Souza, M., Pickering, I.J., Terry, N., 2002. Evaluation of the macroalga, muskgrass, for the phytoremediation of selenium-contaminated agricultural drainage water by microcosms. J. Environ. Qual. 31, 2104–2110.

Lovett, R.J. 2005. Chemistry of Selenium in Mine Drainage and During Treatment. http://www.wvu.edu/~agexten/landrec/2005TFPPRS/Lovett.pdf

Maas, J., 1998. Selenium metabolism in grazing ruminants: Deficiency, supplementation, and environmental implications. In: Frankenberger, W.T. Jr., Endberg, R.A. (Eds.), Environmental Chemistry of Selenium. Marcel Dekker, Inc., New York, NY, pp. 113–128.

Martens, D.A., Suarez, D.L., 1998. Sequential extraction of selenium oxidation states. In: Frankenberger, W.T. Jr., Endberg, R.A. (Eds.), Environmental Chemistry of Selenium. Marcel Dekker, Inc., New York, NY, pp. 61–79.

Mayland, H.F., James, L., Panter, K., Sonderegger, J., 1989. Selenium in seleniferous environments. In: Jacobs, L.W. (Ed.), Selenium in Agriculture and the Environment. SSSA special publication no. 23, Madison, WI, pp. 15–50.

McNeal, M.J., Balistrieri, L., 1989. Geochemistry and occurrence of selenium: An overview. In: Jacobs, L.W. (Ed.), Selenium in agriculture and the environment. SSSA special publication no. 23, Madison, WI, pp. 1–13.

MINITAB statistical software standard version, 1992. Minitab Inc., Penn State College, PA.

National Research Council (NRC). 1976. Selenium. National Academy of Science, Washington, DC, p. 203.

Neal, R.H., Sposito, G., Holtzclaw, K., Traina, S., 1987. Selenite adsorption on alluvial soils: I. Soil composition and pH effects. Soil Sci. Soc. Am. J. 51, 1161–1165.

Nelson, D., Sommers, L., 1982. Total carbon, organic carbon, and organic matter. In: Klute, A. (Ed.), Methods of Soil Analysis, Pt. 2. American Society of Agronomy, WI, pp. 539–579.

Ohlendorf, H.M., 1989. Bioaccumulation and effects of selenium in wildlife. In: Jacobs, L.M. (Ed.), Selenium in Agriculture and the Environment. SSSA special publication no. 23, pp. 133–177.

Orso, W., Blaylock, M., Vance, G., 1993. Predicting selenium bioavailability with phosphate extractions of soils and mine backfill materials. Agronomy Abstracts. ASA-CSSA-SSSA Annual Meetings, p. 43.

Parfitt, R.L., 1978. Anion adsorption by soils and soil materials. Adv. Agron. 30, 1–50.

Parkhurst, D.L., Appelo, C.A.J., 1999. User's guide to PHREEQC (Version 2)—A computer program for speciation, batch-reaction, one-dimensional transport, and inverse geochemical calculations: U.S. Geological Survey Water-Resources Investigations Report 99-4259, p. 310.

Pasch, R.N., Vance, G., 1995. Soil selenium: Laboratory comparisons and soil/backfill variability. In: Schuman, G.E., Vance, G.F. (Eds.), Decades Later: A Time for Reassessment. American Society for Surface Mining and Reclamation, Princeton, WV, Vol. I, pp. 258–269.

Pezzarossa, B., Piccotino, D., Petruzzelli, G., 1999. Sorption and desorption of selenium in different soils of the Mediterranean area. Commun. Soil Sci. Plant Anal. 30, 2669–2679.

Rajan, S.S.S., Watkinson, J., 1976. Adsorption of selenite and phosphate on an allophane clay. Soil Sci. Soc. Am. J. 40, 51–54.

Rosenfeld, I., Beath, O.A., 1964. Selenium—Geobotany, Biochemistry, Toxicity, and Nutrition. Academic Press, New York.

Sanyal, S.K., De Datta, S.K., Chan, P.Y., 1993. Phosphate sorption-desorption behavior of some acidic soils of south and southeast Asia. Soil Sci. Soc. Am. J. 57, 937–945.

Sharmasarkar, S., Vance, G., 1995. Characterization and correlation of soil and plant selenium in some range and coal mine environments of Wyoming. Commun. Soil Sci. Plant 26, 2577–2591.

Sharmasarkar, S., Vance, G., 2002a. Soil and plant selenium at a reclaimed uranium mine. J. Environ. Qual. 31, 1516–1521.

Sharmasarkar, S., Vance, G., 2002b. Selenite–selenate sorption in surface coal mine environment. Advances in Environmental Research 7, 87–95.

Siddique, T., Okeke, B., Zhang, Y., Arshad, M., Han, S., Frankenberger, W. Jr., 2005. Bacterial diversity in selenium reduction of agricultural drainage water amended with rice straw. J. Environ. Qual. 34, 217–226.

Singh, M., Singh, N., Relan, P.S., 1981. Adsorption and desorption of selenate and selenite on different soils. Soil Sci. 132, 134–141.

Sobolewski, A., 2005. Treatment Options for Selenium at Northern Mines. Proceedings, Northern Latitudes Mining Reclamation Workshop, Dawson City, Canada, p. 17.

Spackman L., Vicklund, L., Vance, G., Carroll, P., Steward, D., Luther, J., 1994. Standard operating procedures for the sampling and analysis of selenium in soil and overburden/

spoil material. Research Publication MP-82, College of Agriculture, University of Wyoming, Laramie, WY, p. 13.
Spackman, L.K., Hartman, K., Harbour, J., Essington, M., 1990. Adsorption of oxyanions by spent western oil shale: II. Selenite Enviro. Geol. Water Sci. 15, 93–99.
Sposito, G., 1989. The Chemistry of Soils. Oxford University Press, Inc., New York.
Valkovic, F., 1983. Trace Elements in Coal. CRC Press Inc., Boca Raton, FL.
Vance, G.F., 2000. Problems associated with selenium leaching from waste shale. In: Daniels, W.L., Richardson, S.G. (Eds.), A New Era for Land Reclamation. American Society for Surface Mining and Reclamation, Lexington, KY, Vol. 17, pp. 71–82.
Vance, G.F., See, R., Reddy, K., 1998. Selenite sorption by coal-mine backfill materials in the presence of organic solutes. In: Frankenberger, W.T. Jr., Endberg, R.A. (Eds.), Environmental Chemistry of Selenium. Marcel Dekker, Inc., New York, NY, pp. 259–280.
Weast, R.C., (Ed.), 1985. Handbook of Chemistry and Physics, sixty-sixth ed. The Chemical Rubber Co., Cleveland, OH.

Developments in Environmental Science, Volume 5
D. Sarkar, R. Datta and R. Hannigan (Editors)

ISSN: 1474-8177/DOI:10.1016/S1474-8177(07)05018-8

Chapter 18

Factors affecting spatial patterns of vadose-zone nitrate in south-central Kansas

Margaret A. Townsend and Richard O. Sleezer

Abstract

Vadose-zone nitrate concentrations vary spatially and temporally as a function of multiple factors including soil properties, land use, and climate. Field studies in south-central Kansas used soil core and lysimeter techniques to study factors affecting temporal and spatial variations in vadose-zone nitrate. Statistical analysis of data from 34 irrigated and 31 dryland sites indicated that vadose-zone nitrate concentrations at the end of the growing season were most strongly influenced by: (1) the nitrogen fertilizer-application rate, (2) the amount of nitrate present below the root zone before planting, and (3) the amount of nitrate in the root zone prior to planting. Soil-water lysimeter data indicated that nitrate was moving through most soils but not leaching out of all soils. Analysis of within-field variability of plant-available nitrogen (PAN = nitrate and ammonium) at two of the 65 fields mentioned above (1 dryland and 1 irrigated) indicated that a higher percentage of the PAN was nitrate in the dryland farmed soils. Irrigated soils had more total PAN but a greater percent of the PAN in these soils was ammonium. Nitrate was apparently preferentially used and/or leached from these soils by irrigation water. Spatial patterns of PAN were strongly influenced by topography and internal stratigraphy with more PAN occurring at lower landscape positions with less permeable stratigraphy at both sites.

18.1. Introduction

Much research has been conducted to determine soil characteristics that affect the movement or retention of nitrogen particularly in the upper root zone. Examples of these properties include the hydraulic conductivity, bulk density, pH, and cation exchange capacity as well as the underlying stratigraphy and the surface topography (Burrough, 1991; Hall and Olson, 1991; Lammers and Johnson, 1991; Townsend et al., 1997). In addition, studies in Canada, South Dakota, and California illustrate the impact that landforms have on soil moisture and other soil properties influencing the field-scale variability of soil nitrogen and ultimately on crop yield and yield predictions (Pennock, 1998; Franzen and Kitchen, 1999; Moulin et al., 2002; Onsoy et al., 2005). It is also known that farming practices such as nitrogen fertilizer application rates and irrigation versus dryland cropping practices can affect retention and leaching rates for nitrate (Townsend et al., 1995).

18.1.1. Spatial and temporal variability of vadose-zone nitrate

Spatial representation of soil variability is important in order to evaluate the soil properties that affect the vertical movement of water and solutes, including nitrogen, through the soil and underlying vadose zone. Several important properties include the underlying stratigraphy, surface topography, and land use, as well as measurements of nitrate, ammonium, and total nitrogen (Burrough, 1991; Hall and Olson, 1991; Lammers and Johnson, 1991; Townsend et al., 1997).

Spatial variability in vadose-zone nitrate can result from variations in land use (e.g., types of crops, tillage practices, fertilization-application rates, types of fertilizer applied, and irrigated versus dryland farming techniques), which can cause between-field variations in nitrogen loading and nitrate movement/retention. Soil variability with respect to organic-matter content, permeability, bulk density, pH, stratigraphy, and other properties causes within-field variations in initial nitrogen content (before fertilization), nitrate retention, nitrate transformations, plant uptake, and ultimately nitrate leaching through the unsaturated zone to underlying aquifers. This type of variability can be difficult to quantify at different spatial scales due to the inherently large variability in soil and underlying unconsolidated sediments that are alluvial in origin (Wilding and Drees, 1973; Weerts and Bierkens, 1993; Townsend et al., 1996). All the listed factors can impact the analysis of nitrate data when trying to determine soil-nitrogen loading rates, nitrate movement, and sources of nitrate contamination as measured in soils and groundwater samples from wells. Determination of the spatial variability of soil stratigraphy, topography, and land use at a field scale and how this spatial variability affects the movement and retention of potential contaminants such as nitrate is necessary for vadose-zone studies as well as aquifer impact studies.

Temporal variability can be evaluated on the basis of anthropogenic and environmental factors. Temporal variability of nitrate is related to crops grown and activities related to farming type (irrigated or dryland), and water application (Bruckler et al., 1997). Land uses, including such factors as the timing and quantity of fertilization, or irrigation and crop growth stage strongly influence the amount of total nitrogen and nitrate in the soil that is available for plant uptake or leaching at any given point in time. Environmental factors such as precipitation events and soil temperature also impact temporal variability by providing water for nitrate leaching and by influencing the rates of microbial nitrification/denitrification processes. Most studies have evaluated these properties with respect to potential groundwater contamination (Whittemore et al., 1987; Townsend and Sleezer, 1995; Igbal et al., 1997; Townsend et al., 1997) and not to the impact on spatial variability of soil properties in the vadose zone.

18.1.2. Description of study area

This paper summarizes results from two studies that were designed to evaluate factors influencing the temporal and spatial variability of vadose-zone nitrate. The study areas are in Harvey County in south-central Kansas, USA (Fig. 18.1). Land uses in the area that have the potential to impact groundwater nitrate concentrations include irrigated and dryland farming, dairies, and cattle feedlots. Harvey County is an area underlain by a significant aquifer, locally referred to as the Equus Beds (Macfarlane, 2000), that is part of the larger High Plains aquifer system as defined by the U.S. Geological Survey (2005). The Equus Beds aquifer is a thick and productive system that provides water for domestic, agricultural, and industrial use for approximately 21 percent of the population of Kansas (Sleezer, 2001), and it has been a major water supply system for the area for more than 70 years (Williams and Lohman, 1949; Stramel, 1967). It is, therefore, an important water resource that is potentially at risk for contamination by nitrate because it is overlain in places by permeable soils, relatively thin unsaturated zones (<10 m), and numerous sources for nitrate pollution.

18.1.3. Description of study objectives

The first study was designed to evaluate the relationship between farming type (irrigated versus dryland) and fertilization differences between

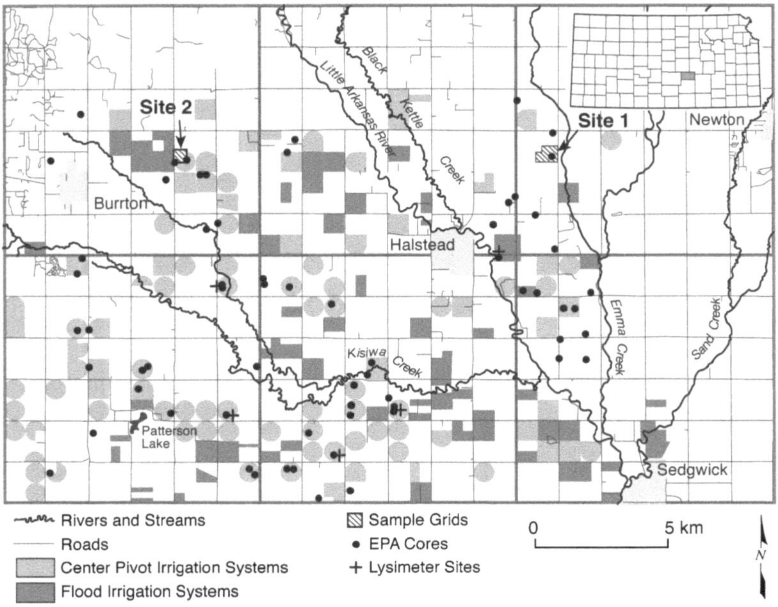

Figure 18.1. Location of soil-coring sites, soil-water lysimeter sites, land use by irrigation or dryland-farming practice, and two field sites (site 1 and site 2) for investigation of effects of topography and soil stratigraphy on nitrate movement in the vadose zone in south-central Kansas.

farming types on the occurrence of vadose-zone nitrate and observed groundwater nitrate values. A variety of temporal and land-use factors were statistically evaluated in this study to determine their influence on soil-nitrate retention and movement (Townsend and Sleezer, 1995). These factors included crop type, cropping history, irrigated versus dryland farming practices, and fertilizer-application rates. The study was funded by the U.S. EPA (1992–1994; Townsend and Sleezer, 1995) and will be referred to as the EPA study. It was a two-year soil-nitrogen monitoring study. Sampling by cores was performed at fixed locations in 65 fields (34 irrigated and 31 dryland) prior to planting in the spring and after harvest in the fall for two years. Methods used for soil sampling and analyses are presented in Appendix 1.

The second study evaluated field-scale spatial variability of nitrate using soil cores, soil-characterization pits, electromagnetic-induction measurements, and measurements of root-zone plant-available nitrogen (PAN, $NO_3 + NH_4$) at two study sites used previously in the EPA study. Sampling methods used in this study are presented in Appendix 1. This study was funded by the Kansas Water Resources Research Institute (KWRRI; USGS funds, 1994–1995; Townsend et al., 1997) and will be called the KWRRI study.

During the course of the EPA study two general observations were made about soils in the study area: (1) descriptions of soils sampled often did not match with soils described in the Harvey County soil survey (Hoffman and Dowd, 1974); and (2) most soil descriptions indicated similar surface textures (e.g., loamy sand or sandy loam), but subsurface horizons in many cases had strongly contrasting and highly variable textures (e.g., loam, clay loam, silty clay loam, or silty clay). Both of the chosen sites for the KWRRI study had soils with sandy surface textures and rolling sand-dune topography, but cores from the previous EPA study indicated that subsurface textures could be expected to vary. Soils at the two sites were not mapped as the same units in the Harvey County soil survey (Hoffman and Dowd, 1974) but, due to the observations from the EPA study, the soils were considered to be relatively similar. Site 1 was farmed using dryland methods and site 2 was irrigated with a center-pivot system. While the EPA study focused on temporal variations in vadose-zone nitrate at single core or lysimeter locations within fields, the KWRRI study focused on characterizing within-field spatial variability in vadose-zone nitrate and comparing soil-nitrate variability between fields farmed with different techniques (dryland versus irrigated).

18.2. Results

18.2.1. Temporal and land-use affects on nitrate movement—EPA study

Figure 18.1 shows the spatial patterns of irrigated cropland (center-pivot versus flood), the location of soil-coring sites at irrigated and dryland farms, and soil-water lysimeter sites that were evaluated in the EPA study. The two sites used for the evaluation of the field-scale spatial variability of soil nitrate in the KWRRI study are also indicated. The majority of the irrigated sites sampled in this study were center-pivot irrigated sites.

At the sites sampled in the EPA study, dryland crops included milo (grain sorghum, 80%) and wheat (15%) while irrigated crops included corn (45%), soybeans (32%), and milo (20%). Median fertilizer-application rates by farming method (dryland versus irrigated) and crop grown are shown in Fig. 18.2 (Appendix 2, Table A1). This figure illustrates the much higher nitrogen fertilizer-application rates that are used with irrigated farming practices. Differences in fertilizer-application rates were evaluated

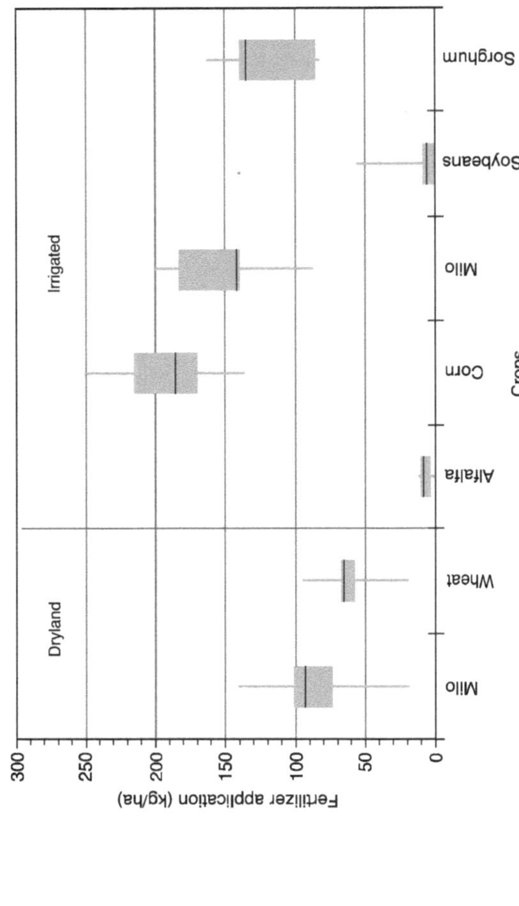

Figure 18.2. Median fertilizer-application rates (shown by black lines in box plot) by farm type and crop grown in study area. Note that irrigation fertilizer rates are higher than dryland.

using the non-parametric Kruskal–Wallis test (Appendix 1; Appendix 2; Table A1). The Kruskal–Wallis value of $\rho = 0.0013$ indicates that differences in fertilizer-application rates between dryland and irrigated farming practices were statistically significant ($\alpha = 0.1$; Townsend and Sleezer, 1995).

The higher nitrogen fertilizer-application rates result in increased nitrogen loading on these soils. Soil-nitrate concentrations were therefore expected to be higher in soils in irrigated fields. The combination of higher soil-nitrate concentrations and the addition of irrigation water to these fields were expected to represent a greater risk for nitrate leaching through the vadose zone into the underlying aquifer.

Because most soil nutrient testing methods focus on evaluating the root zone (upper 1.2 m or shallower) (Fjell et al., 1994), soil-nitrate values were evaluated in two sections: upper root zone (0–1.2 m) and below the root zone (1.2–3 m). Figures 18.3A and B show the separated soil-nitrate values for both years and land uses. In all cases median soil-nitrate concentrations were higher in irrigated soils regardless of sampled interval (Appendix 2, Table A2, significance level $\alpha = 0.1$, range of ρ values from 0.0002 to 0.089). This can largely be explained by higher nitrogen fertilizer-application rates applied to irrigated fields with higher crop-yield goals. The values for 1993 were lower than those for 1992 due to the intense rainfall and flooding that occurred during the growing season (Townsend and Sleezer, 1995). In most cases the range in soil-nitrate concentrations was greater in irrigated fields as compared to dryland fields reflecting the variability in nitrate concentrations associated with plant uptake, soil properties, leaching, and fertilizer application differences.

18.2.1.1. Soil-water lysimeter study

Six lysimeter sites were chosen as a subset of the original EPA irrigated core study sites for a more time-intensive study of soil nitrate (Fig. 18.1). Each site was an irrigated field in which corn or sorghum was planted and nitrogen fertilizer-application rates were between 196 and 246 kg ha^{-1}. Soils varied from very sandy throughout the upper 3 m to soils that were silt loam at the surface with silty clay subsoil and coarse sand at 3-m depth. The soil-water lysimeters were installed at 0.6-, 1.2-, 1.8-, and 3-m depth at each site without regard for differences in stratigraphy (Appendix 1). Soil cores and lysimeter samples were obtained and analyzed at four different times during the growing season (May, June, July, and August).

The fertilizer-application information provided by the farmers indicated either a spring, pre-plant application and an application later in the

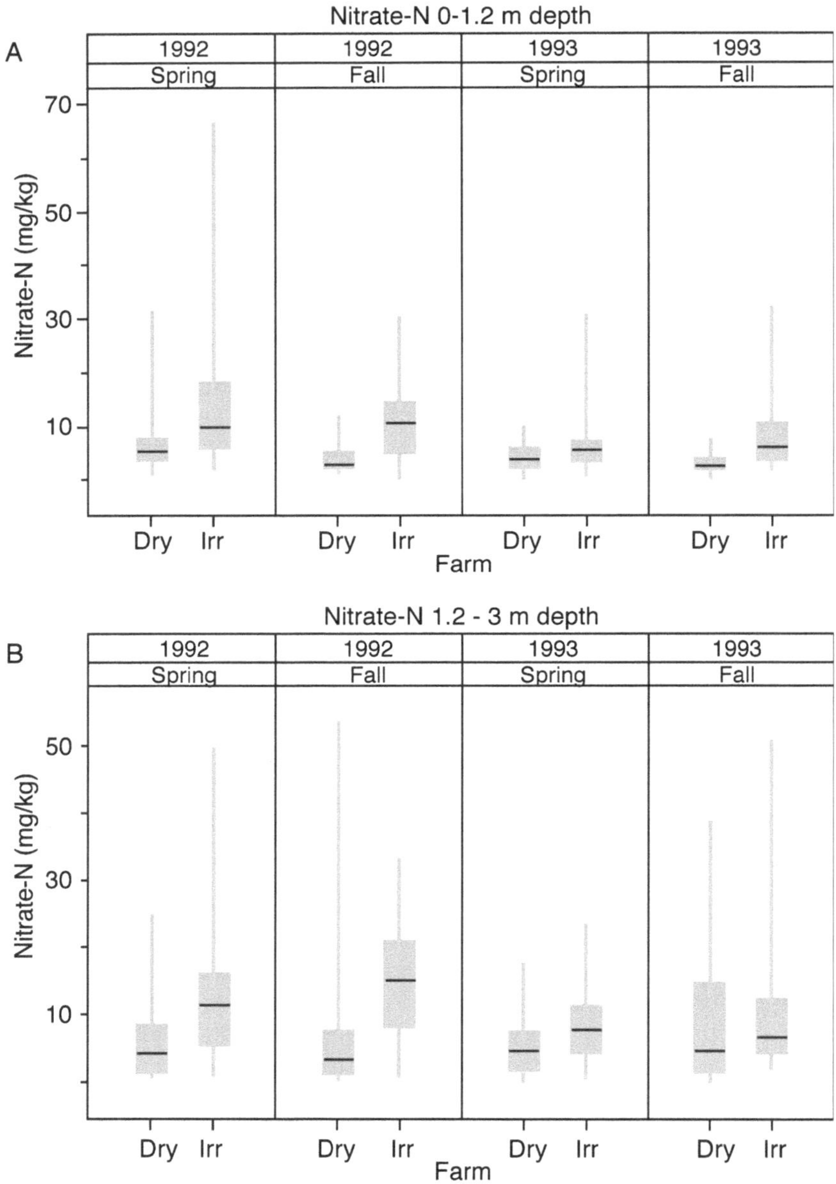

Figure 18.3. A and B Comparison of range and median nitrate-N concentration for dryland and irrigated site soil cores. Median values shown by lines in boxes. Irrigated sites show higher soil-nitrate concentrations than dryland sites for all time periods except fall 1993.

season or just a spring pre-plant application. In either case there was not a continuous input of nitrogen into these systems. The soils at these six sites had a variety of surface textures and subsurface horizons. Surface-soil textures ranged from coarse (loamy-fine sand) to fine (silty-clay loam). At four of the sites the lysimeters installed at 3 m were located within a horizon described as sand or coarse sand with finer-textured horizons directly above them. Soil water perched in or directly above the finer-textured horizons, and no sample was recovered from the lysimeters installed at a 3-m depth in coarse sand. The permeability contrast between the overlying fine-textured layers and the underlying coarse sand (Fetter, 1980) may have prevented water and nitrate movement downward to the 3-m depth. Also, the large pores in the unsaturated coarse sand probably allowed the lysimeters to lose vacuum relatively rapidly. This may have caused the lysimeters to miss any flow through these coarse sandy layers if flow occurred too long after vacuum was applied to the lysimeters (Townsend and Sleezer, 1995).

Table 18.1 contains the soil-water nitrate-concentration data obtained from the six lysimeter sites. The highest concentrations ($>200\,mg\,l^{-1}$) were obtained at site I-14. Soils at this site were classified as a Punkin series (Fine, mixed, superactive, mesic Leptic Vertic Natrustolls). These soils were salty with electrical-conductivity measurements of soil water

Table 18.1. Soil-water nitrate concentrations from lysimeters installed at six sites in Harvey county, Kansas summer of 1992

Month	Depth (m)	I-14 ($mg\,l^{-1}$)	I-15 ($mg\,l^{-1}$)	I-18 ($mg\,l^{-1}$)	I-29 ($mg\,l^{-1}$)	I-30 ($mg\,l^{-1}$)	I-32 ($mg\,l^{-1}$)
May	0.6	221.2	99.1	39.5	32.7	26.9	N.D.
	1.2	76.1	89.2	40.2	33.6	15.3	19.9
	1.8	46.7	66.8	62.3	98.9	38.8	N.D.
	3.0	N.D.	N.D.	N.D.	31.6	52.8	N.D.
June	0.6	230.2	101.8	47.9	58.7	50.1	64.1
	1.2	79.2	97.5	36.3	42.0	21.9	19.4
	1.8	53.3	76.7	55.8	86.9	22.4	22.2
	3.0	N.D.	N.D.	N.D.	37.7	60.1	N.D.
July	0.6	224.2	129.1	37.0	61.6	75.4	9.9
	1.2	84.4	104.3	41.8	78.3	31.2	30.0
	1.8	58.2	85.8	47.0	51.7	8.8	20.3
	3.0	N.D.	N.D.	N.D.	36.3	67.0	N.D.
August	0.6	N.D.	164.8	8.1	84.4	67.3	0.8
	1.2	94.1	111.1	47.9	131.2	37.2	36.1
	1.8	61.4	90.1	40.4	62.5	10.4	41.8
	3.0	N.D.	N.D.	N.D.	32.1	76.7	N.D.

N.D., not determined.

extracted in lysimeters as high as 20,000 μmhos cm^{-1}. Nitrate concentrations in soil water remained relatively constant throughout the growing season at this site probably related to evapoconcentration of solutes in the soil profile, lower plant uptake due to salty conditions, and low denitrification rates due to high pH (Townsend and Sleezer, 1995).

The lowest nitrate concentrations in soil water were measured at I-32, which was the only flood-irrigated site and had silty-clay loam textures throughout its profile (Elandco—fine-silty, mixed, superactive, thermic Cumulic Haplustoll). The highest soil-water nitrate concentrations measured at sites I-15 (unnamed series—fine, mixed, superactive, thermic Udertic Haplustalf) and I-29 (Pratt—sandy, mixed, thermic Psammentic Haplustalf) were sampled toward the end of the growing season and at shallow depths (0.6 and 1.2 m, respectively). Clearly defined pulses of nitrate movement were not readily observable in the lysimeter data, and nitrate concentrations remained relatively constant in some soils (e.g., I-14, I-15, I-29 in Table 18.1).

The lysimeter data show that high nitrate concentrations occur in irrigated soils with higher fertilizer-application rates (Fig. 18.2). The data also indicate that if the soil is not uniformly sandy with depth, other factors such as organic-matter content, silt and clay content, and perching layers in the vadose-zone impacted the movement of nitrate through the profile.

18.2.3. Factors affecting spatial variability of plant-available nitrogen—KWRRI study

The second study in Harvey County (KWRRI) evaluated the field-scale spatial variability of plant-available nitrogen (PAN = $NO_3 + NH_4$) and determined the effects of topography, soil stratigraphy, and farming method (irrigated versus dryland) on the spatial variability of PAN. Its focus was on the upper part of the root zone (upper 75 cm of soil).

18.2.3.1. Stratigraphic effects

Both sites had soils with an eolian sand component in the near surface. The soil map units delineated in the Harvey County soil survey at the two sites were representative of soils found in approximately 30 percent of the study area. The sites were also previously sampled during the EPA study (Townsend and Sleezer, 1995). The locations of the two sites are shown in Fig. 18.1 (site 1 and site 2). Soils at each site were sampled by 19 soil cores and one soil characterization pit for a total of 20 sample points at each site (Appendix 1). The sample points were selected based on topographic and electromagnetic surveys conducted at regularly spaced grid-sampling points across each study site (Figs. 18.4 and 18.6) to represent the field-scale variability in topographic position and electromagnetic ground conductivity (Townsend et al., 1997).

Site 1 is dryland farmed. Figure 18.4 shows the soil map units as delineated by the Harvey County soil survey (Hoffman and Dowd, 1974), the grid points where elevation and electromagnetic-conductivity

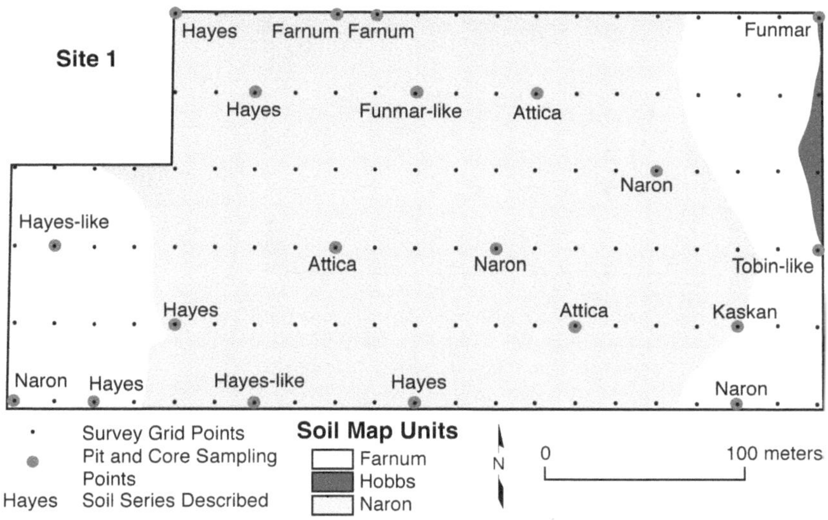

Figure 18.4. Soils map of site 1 based on USDA soil map (Hoffman and Dowd, 1974) and soil cores collected at the field site. Note variability of soil textures across the site.

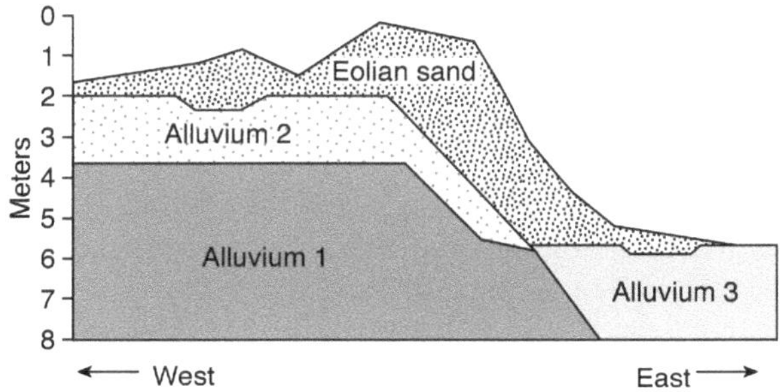

Figure 18.5. Simplified conceptual model of stratigraphy at site 1. Note the two different terrace levels (top of alluvium 2 and 3). Textural characteristics for the three alluvial strata are not uniform.

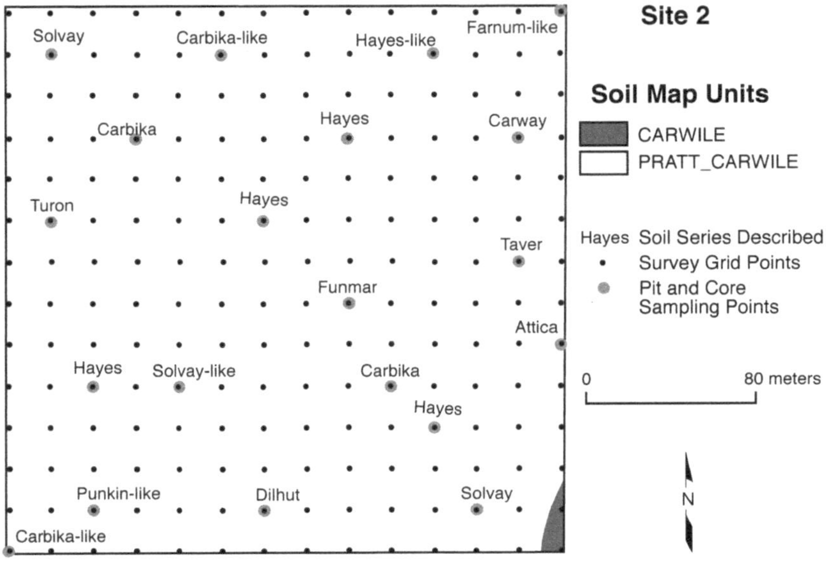

Figure 18.6. Soils map of site 2 based on USDA soil map (Hoffman and Dowd, 1974) and soil cores collected at the field site. Note variability of soil textures across the site. Soil textures are more similar to each other at this site than at site 1.

measurements were performed, and the locations where the soils were sampled by pit or core.

Deep cores (>4 m) described at site 1 indicated that four different stratigraphic units were observable in the vadose zone at site 1. Surface-soil horizons are formed in eolian sand at all sampled locations across the site, but in the subsurface three different alluvial units, two of which occur at apparent terrace breaks at the site were observed. Figure 18.5 shows a conceptual cross section of the soil stratigraphy at site 1. At least two distinct terrace surfaces exist under the eolian sand (Fig. 18.5). The boundary between the eolian and alluvial sediments is therefore not level.

The textural characteristics of the alluvium are not uniform at site 1. Alluvium 1 consists of horizons with silty clay loam, silty clay, or clay textures, and generally contains more than 35 percent clay. Alluvium 2 is much coarser textured than alluvium 1. It consists of horizons with sandy clay loam or heavy sandy loam textures, and clay contents range from about 15 to 32 percent. Alluvium 3 consists of stratified sediments with a wide range of textures (silt loam, loam, sandy loam, loamy sand, sandy clay loam, and clay loam). Clay contents range from 5 to >30 percent. Water at this site tended to perch above the less-permeable alluvium and move laterally toward the lower landscape positions carrying with it nitrate-rich vadose-zone waters.

Site 2 is irrigated. Figure 18.6 shows the soil map units as delineated by the Harvey County soil survey (Hoffman and Dowd, 1974), the grid points where elevation and electromagnetic-conductivity measurements were performed, and the locations where the soils were sampled by pit or core (Appendix 1).

The soil stratigraphy at site 2 is much simpler. Most of the soils at site 2 have formed in only two parent material strata (Fig. 18.7). The upper layer consists of eolian sand, which ranges in thickness from approximately 15 to 225 cm. Most of this material has a texture of sandy loam, fine sandy loam, or loamy fine sand. The clay content is less than 20 percent in all horizons, and the depth-weighted average clay content ranges from 7.4 to 14.5 percent. The lower layer consists of fine-textured alluvium whose upper surface appears to be nearly level. Topography at site 2 is therefore, principally a function of the thickness of the eolian sand. Textures within the alluvium are most commonly silty clay loam, silty clay, clay, or clay loam. Clay content within this layer ranges from 28 to 60 percent. Maximum clay contents are most commonly between 35 and 50 percent.

The boundary between these horizons is most commonly described as abrupt or clear. In quantitative terms, this means that the transition zone from eolian sand to clayey alluvium occurs in less than 5 cm. Wetness and redoximorphic features at the base of the eolian sand indicate that infiltrating water tends to perch for extended periods of time above the alluvium. In most cases the alluvium represents an aquitard or aquiclude for infiltrating water. This has potential to impact the movement of nitrate through the vadose zone at this site.

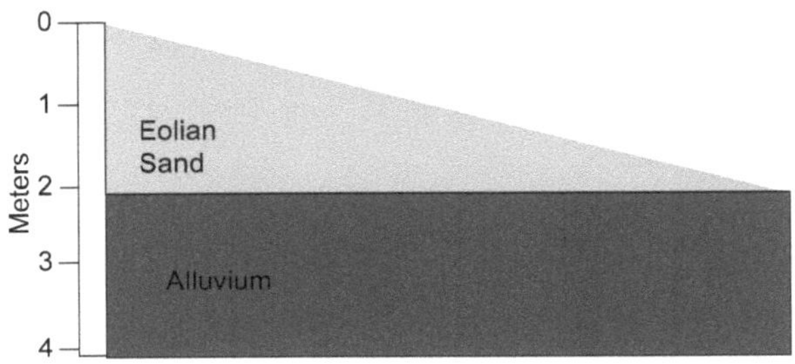

Figure 18.7. Simplified conceptual model of stratigraphy at site 2. The surface of the clay layer is relatively flat. Topographic relief is a function of the thickness of the eolian sand.

18.2.3.2. Spatial variation of plant-available nitrogen

Soil tests are routinely used to evaluate residual plant-available nitrogen in the soil between growing seasons to determine the quantity of fertilizer needed for the next crop to be planted (Fjell et al., 1994). In most soils, nitrate is the most common form of plant-available nitrogen in the soil profile, although in the Harvey County area measurable quantities of ammonium were also present. Plant-available nitrogen includes both nitrate and ammonium forms of nitrogen.

Table 18.2 contains summary statistics for nitrate-N, ammonium-N, and Total Plant Available Nitrogen (TPAN) for two depth intervals (0–30 and 30–75 cm). Median nitrate-N concentrations in the upper 30 cm are significantly different at the two sites. Nitrogen values were calculated in kg ha^{-1} in order to compare the results with fertilizer-application rates.

What is surprising is that they are generally lower at the irrigated site (site 2) where fertilizer-application rates are considerably higher. However, median ammonium-N values in the upper 30 cm are higher at the irrigated site. The net effect of these disparities is that there is no statistically significant difference ($\rho = 0.7359$) in the amount of Total Plant Available Nitrogen in the upper 30 cm of soil between the two sites even though the nitrogen fertilizer-application rate is considerably higher at the irrigated site. A similar pattern of observations occurs in the deeper depth interval sampled (30–75 cm). There are statistically significant differences in nitrate-N and ammonium-N measured at the two sites with

Table 18.2. Plant-available nitrogen and farm type

Number of samples, depth interval (cm) and nitrogen sample type	Mean (kg ha^{-1})		Median (kg ha^{-1})		Kruskal–Wallis ρ value
	Dryland (Site 1)	Irrigated (Site 2)	Dryland (Site 1)	Irrigated (Site 2)	
Number of samples	19	20	19	20	
0–30					
Nitrate-N	14.6	11.0	14.0	9.9	0.0022
Ammonium-N	12.7	17.9	10.4	18.4	0.0492
TPAN	27.3	28.9	26.6	28.5	0.7359
30–75					
Nitrate-N	35.2	26.4	30.5	24.6	0.0018
Ammonium-N	21.0	45.3	16.9	37.6	0.0003
TPAN	56.2	71.7	54.4	63.9	0.1032

TPAN, total plant available nitrogen.

Note: One sample from dryland site not used in statistical analyses due to unusually high value. See Fig. 18.9 for outlier values.

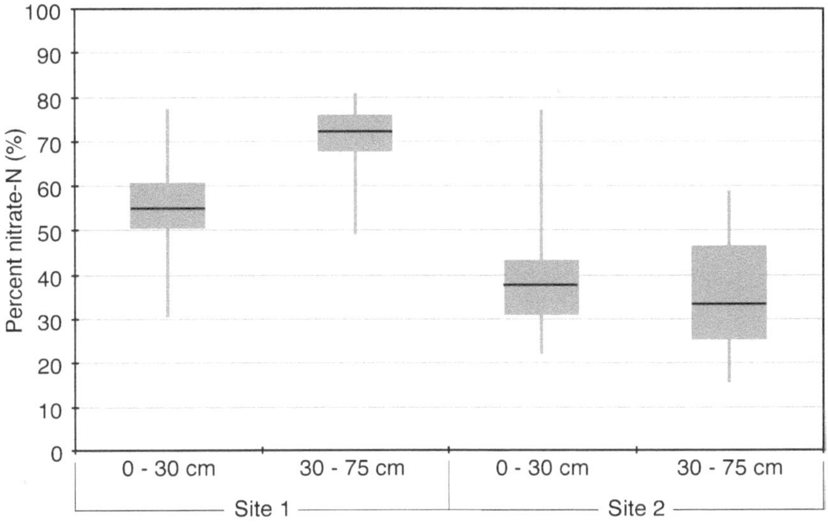

Figure 18.8. Percent of plant-available nitrogen that is nitrate-N in the upper and lower root zone. Lines in boxes indicate median values for each site.

higher concentrations of nitrate-N at the dryland farm (site 1) and higher concentrations of ammonium-N observed at the irrigated site (site 2).

Figure 18.8 shows the percentage of plant-available nitrogen that is nitrate available for movement in the upper and lower root zone at each site. The dryland site (site 1) shows a higher percentage of nitrate available for movement. However, the lack of additional water means that this nitrogen may be stored for a longer period of time within the root zone instead of moving. At the irrigated site (site 2), the median percentage of plant-available nitrogen that is nitrate is similar for both the upper and lower root zone suggesting that the additional water causes a homogenizing effect resulting in more uniform nitrate concentration throughout the vadose zone in irrigated soil.

Figure 18.9 shows the quantity of soil nitrogen in kilograms of plant-available nitrogen per hectare that are stored in the upper 75 cm of the soil profiles at sites 1 (dryland) and 2 (irrigated). The median quantities in the upper root zone are similar at both sites with a slightly higher median value measured at site 2. At both sites lower root-zone plant-available nitrogen values are higher than those calculated for the upper root zone, but this is partly attributable to the greater soil thickness (45 versus 30 cm) used in this calculation. More plant-available nitrogen is available at greater depth at the irrigated site (site 2) as compared with the dryland

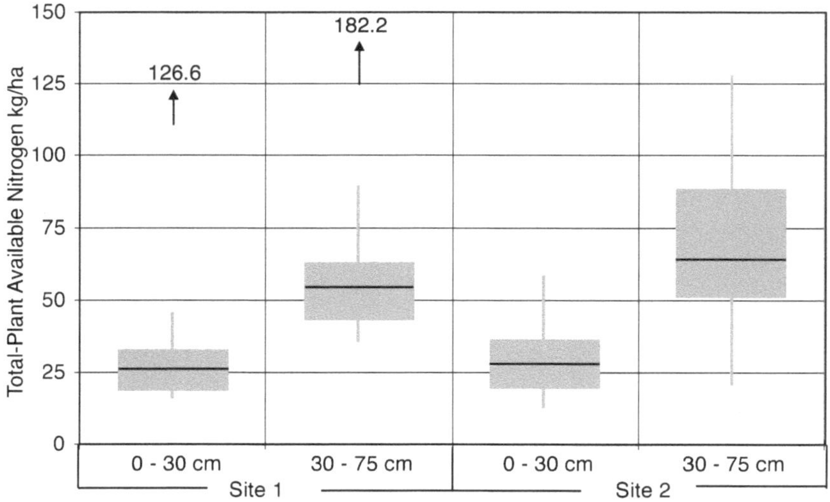

Figure 18.9. Quantity of plant-available nitrogen potentially stored in soil based on concentrations measured at 19 points (dryland, site 1) and 20 points (irrigated, site 2). Note similarity of stored nitrogen in upper 30 cm at both sites. Differences occur at depth most likely due to increased water and fertilizer application at the irrigated sites.

site (site 1) probably due to increased leaching of the higher fertilizer application at the site.

18.2.3.3. Impact of topographic position and stratigraphy on nitrogen variability

In addition to the vertical variability of plant-available nitrogen in the vadose zone, the KWRRI study looked at the impact of topography on the occurrence of plant-available nitrogen in a given field. Soil scientists have recognized for some time that topography has a strong influence on soil properties including soil nitrogen (Aandahl, 1949) and that analysis of terrain variables has the potential to increase the accuracy of soil-attribute prediction (Moore et al., 1993). Both sites had observable topographic features that are either simple or complex internally.

Site 1 has eolian sand overlying three different alluvial strata (Fig. 18.5). The topography at this site is a result of dune-sand deposits overlying fluvial terraces. Some of the topographic relief at site 1 is therefore a function of the thickness of the dune sand, and some is a function of the terrace treads and risers that are beneath it. At site 2 the internal stratigraphy is much simpler. The eolian sand is deposited on what appears to be a relatively flat alluvial surface. Topography is therefore a function of the thickness of the dune sand sitting on the alluvium (Fig. 18.7). At the time of sampling the landscape position (summit, shoulder, backslope, footslope, toeslope) was described and recorded for each core and pit location. The desire was to determine if there were observable relationships between plant-available nitrogen within the root zone and landscape position.

Figure 18.10 shows the relationship between plant-available nitrogen and topographic position for the combined data from both sites. In general, more plant-available nitrogen is in soils at lower topographic

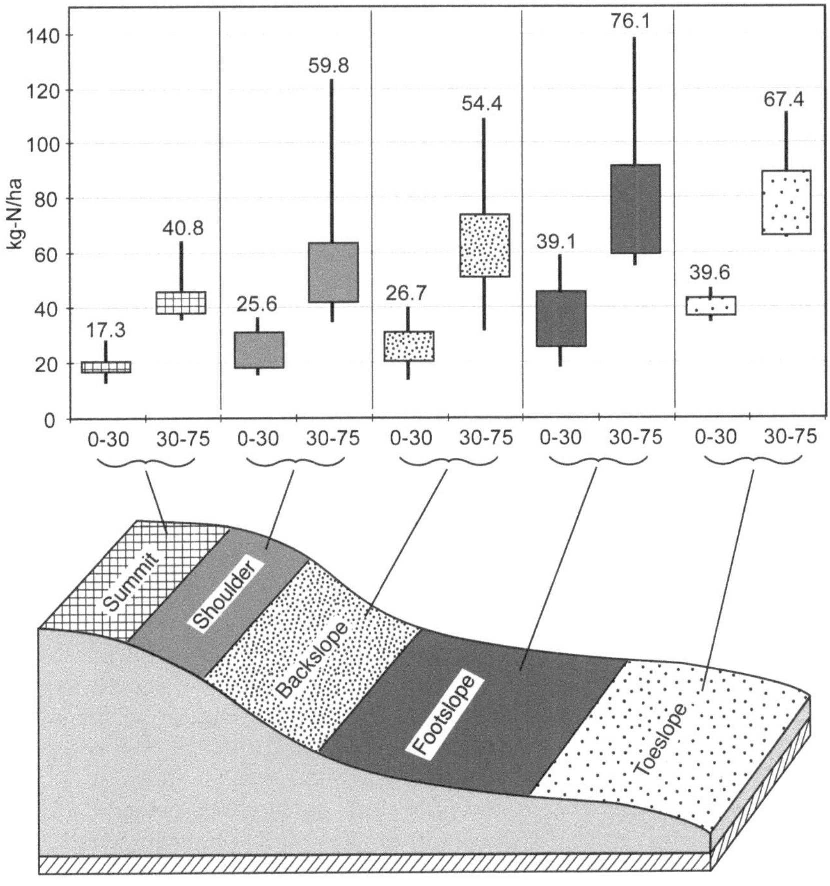

Figure 18.10. Relationship of increasing plant-available nitrogen in the vadose zone with decreasing topographic position in a field. Note that the highest concentrations occur in the foot-slope and toe-slope positions (modified from Hall and Olson, 1991).

positions (backslope, footslope, toeslope) and less plant-available nitrogen in soils at higher slope positions (summit and shoulder). Concentrations of plant-available nitrogen are also different between the upper and lower root zones at higher and lower topographic positions (Kruskal–Wallis $\rho = 0.0099$ upper root zone (TPAN 0–30 cm); $\rho = 0.0325$ lower root zone (TPAN 30–75); Appendix 2, Table A3). It would appear that there is a tendency for higher plant-available nitrogen concentration at lower topographic positions. These findings suggest that topographic position needs to be considered when sampling for soil nitrogen, when monitoring soil-nitrate concentrations, and when attempting to model nitrate retention versus movement in the vadose zone.

18.3. Discussion

The EPA study was originally designed to monitor changes in soil-nitrate concentrations and to determine what factors most strongly influence changes in vadose-zone soil nitrate content that occur during the growing season. Sampling by cores was performed at fixed locations in 65 fields (34 irrigated and 31 dryland) prior to planting in the spring and after harvest in the fall for two years. Not all sites were sampled each time due to weather, field conditions, and changes in crop grown. At six of the irrigated sites, soil-water samplers (lysimeters) were also installed to monitor the concentrations of nitrate in soil water throughout one growing season. Of particular concern in this project was determining how much nitrate was below the root zone after harvest. Soil-nitrate data indicated that the three factors most strongly influencing the amount of nitrate below the root zone after harvest were (1) the nitrogen fertilizer-application rate, (2) the amount of nitrate already below the root zone before planting, and (3) the amount of nitrate in the root zone prior to planting. A vadose zone with high nitrate concentration prior to planting to which additional nitrogen fertilizer is added at a high application rate is likely to result in a high concentration of nitrate below the root zone at the end of the growing season. If sufficient precipitation or irrigation water is available, this nitrate has the potential to leach downward to underlying aquifers and to contaminate groundwater.

A number of unexpected complications and findings occurred in the EPA project. It is easy to make the assumption that all nitrate moves, but this is not always the case. At site I-14 for example, soil-water nitrate concentrations were very high and did not change significantly throughout the growing season. In these salty soils, nitrate concentrates with evaporation just like any other salt. Its concentration is apparently not significantly reduced by denitrification or plant uptake, but it also is not removed by leaching due primarily to the low permeability of the soil. The net effect is that nitrate can build up in some soils without immediate adverse effects on groundwater quality.

The core sampling for the EPA project also served to inspire a number of questions about the soils and vadose-zone stratigraphy in the study area. Early in the project many discrepancies between the soils mapped in the Harvey County soil survey (Hoffman and Dowd, 1974) and the soils observed in the cores became apparent. The extreme variability in vadose-zone properties both vertically within a single core description (e.g., dune sand, over alluvial clay, over coarse sandy alluvium) and spatially within each field (sandy to clayey subsoil transitions over short distances) was considerable. In many cases cores described in close proximity (< 10 m) to each other were drastically different (Townsend et al., 1997).

In the KWRRI project intensive core sampling was performed at two fields after harvest. Farming at site 1 was dryland and site 2 was irrigated. Both sites had what appeared to be sand-dune topography and though they were mapped as different soil series in the Harvey County soil survey (Hoffman and Dowd, 1974), they were deemed similar enough based on cores described during the EPA project to allow for useful comparisons. Looks were deceiving and the internal vadose-zone stratigraphies at the sites were very different. The range in soil-nitrate concentrations at these two sites was similar to the range in soil-nitrate values found in the EPA cores taken at 65 fields across the study area. This finding indicates that within-field variability in soil nitrate is quite large, and that single cores probably do not represent whole-field soil-nitrate conditions very well.

Soil-nitrate concentrations were more uniform with depth in the vadose zone at site 2, which was irrigated as compared with site 1, which was farmed with dryland techniques. It is likely that this is a result of irrigation providing more water for deeper movement of nitrate combined with the higher nitrogen fertilizer-applications providing more nitrate for movement and increased soil moisture producing greater crop uptake of nitrogen. The fact that the soil stratigraphy within the root zone was less complex at site 2 also may have contributed to the uniformity of nitrate concentrations with depth at individual core locations.

Another important question within the KWRRI study was to determine how much of the plant-available nitrogen in the root zone is nitrate and how much is ammonium. Differences between the two fields may be important to point out. At site 1, greater than 50 percent of the plant-available nitrogen was nitrate, and nitrate concentrations generally increased with depth. At site 2 greater than 50 percent of the plant-available nitrogen was ammonium and the relative percentages of nitrate and ammonium did not appear to change with depth. Nitrate appeared to be retained more readily in the dryland-farmed soils even though the total amount of plant-available nitrogen at depth was greater at the irrigated site. This is most likely related to lack of sufficient water to move solutes through the vadose zone. Retention of nitrogen in the vadose zone of dryland farms is similar to that noted in rangeland areas that do not have nitrogen inputs (McMahon et al., 2003).

Finally, topography seems to be an important factor influencing spatial patterns of plant-available nitrogen in these soils. Soils in the upper landscape positions (summit and shoulder) tend to have less plant-available nitrogen than do the soils at lower landscape positions (backslope, footslope, and toeslope). This may be due to a variety of factors, but two seem to be most likely: (1) the underlying alluvial materials are less permeable and therefore soils at lower landscape positions are not leached as rapidly as the sandy soils near crests of landforms; and (2) water perches above the less-permeable alluvium and moves laterally toward the lower landscape positions carrying with it nitrate-rich vadose-zone waters. The net result is that lower landscape positions have higher plant-available nitrogen concentrations.

18.4. Conclusions

The results from the two studies in Harvey County have broad implications for vadose-zone studies. The exclusive use of soil surveys for soil-characterization information may result in erroneous decisions concerning vadose-zone stratigraphy at a given site. Site characterization of soil stratigraphy using many cores, characterization pits, and/or geophysical instruments is strongly suggested if the data are needed for nitrate flux calculations and/or modeling.

Overland- and subsurface-flow processes are strongly impacted by topography and stratigraphy. Careful analysis of landforms may provide a means to better understand spatial patterns of plant-available nitrogen and the processes that cause the spatial patterns to exist. Field analysis of landforms should also help to design field-sampling methods that better account for spatial variability of soil properties potentially affecting vadose-zone nitrate. If these features are not characterized on the surface by site evaluation, topographic-map analysis, and aerial-photo studies, then site evaluation of nitrate movement in the vadose zone may not be correct.

Analyses of soil cores for nitrate, ammonium, total organic nitrogen, organic matter, bulk density, moisture content, and permeability are necessary for estimation of nitrate flux in the vadose zone. Work from these two studies strongly indicates how the lack of information results in the inability to accurately characterize spatial and temporal patterns of soil nitrogen that are essential for modeling nitrate fluxes. In addition, information from farmers concerning fertilizer applications, crop histories, irrigation rates, and precipitation in study areas are also vital pieces of information necessary to evaluate field- and regional-scale patterns of vadose-zone nitrate occurrence and flux.

The data obtained from the two studies discussed in this paper show that movement of nitrate is impacted by many factors, anthropogenic as well as natural. The studies illustrate the impacts of farming practices on nitrate retention and movement in the vadose zone as well as the impacts of topographic features and stratigraphic information. These studies illustrate the need for more detailed information in order to do more complex evaluation of nitrate movement and retention in the vadose zone.

ACKNOWLEDGMENTS

The authors wish to acknowledge the assistance of the NRCS personnel with the electromagnetic-conductivity measurements and soil coring and descriptions for the KWRRI study, and Groundwater Management District #2 personnel and the Harvey County Soil Conservation committee for assistance in obtaining landowners permission for access to their fields in the EPA-funded study. Also, we wish to thank the personnel at the Kansas State University soils laboratory for their help with the soil-chemistry analyses and the USGS for assistance in collecting cores. We also wish to acknowledge the Kansas Geological Survey for field support and use of equipment.

REFERENCES

Aandahl, A.R., 1949. The characterization of slope positions and their influence on the total nitrogen content of a few virgin soils of Western Iowa. Soil Sci. Soc. Am. Proc. 13, 449–554.

Alpkem Corporation, 1986a. RFA Methodology No. A303-S021, Ammonia Nitrogen. Clackamas, OR 97015.

Alpkem Corporation, 1986b. RFA Methodology No. A303-S170, Nitrate/Nitrite Nitrogen. Clackamas, OR 97015.

Bruckler, L., de Cockborne, A.M., Renault, P., Claudot, B., 1997. Spatial and temporal variability of nitrate in irrigated salad crops. Irrigation Sci. 17, 53–61.

Burrough, P.A., 1991. Sampling designs for quantifying map unit composition. In: Mausbach, M.J., Wilding, L.P. (Eds.), Spatial variabilities of soils and landforms. Soil Sci. Soc. Am. Special Publication 28, 89–126, Madison, WI.

Davis, J.C., 1986. Statistics and Data Analysis in Geology second ed. Wiley, New York, p. 646.

Fetter, C.W. Jr., 1980. Applied Hydrogeology: Columbus. Charles E. Merrill Publishing, Columbus, OH, 488 pp.

Fjell, D.L., Vanderlip, Roozeboom, R.L., Kraig, L., 1994. Corn Production Handbook: Kansas State University Agricultural Extension Agency C560, pp. 12–15, http://www.oznet.ksu.edu/library/crpsl2/c560.pdf (verified October 2005).

Franzen, D.W., Kitchen, N. R., 1999. Developing management zones to target nitrogen applications. In: Site specific management guidelines: Potash and Phosphate Institute. Norcross, GA, http://www.ppi-far.org/ssmg (verified April 2005).

Gelderman, R.H., Beegle, D., 1998. Recommended chemical soil test procedures for the north central region: North Central Regional Publication No. 221 (Revised). University of Missouri Agricultural Experiment Station, Columbia, MO, pp. 18–19.

Hall, G.F., Olson, C.G., 1991. Predicting variability of soils from landscape models. In: Mausbach, M.J., Wilding, L.P. (Eds.), Spatial variabilities of soils and landforms. Soil Sci. Soc. Am. Special Publication 28, 9–24.

Hoffman, B.R., Dowd, L.W., 1974. Soil survey of Harvey County, Kansas: U.S. Department of Agriculture, National Resource Conservation Service, p. 55.

Igbal, M.Z., Kroethe, N.C., Spalding, R.F., 1997. Nitrogen isotope indicators of seasonal source variability to groundwater. Environ. Geol. 32, 210–218.

Keeney, D.R., Nelson, D.W., 1982. Nitrogen—Inorganic Forms. In: Page, A.L., Miller, R.H., Keeney, D.R. (Eds.), Methods of Soil Analysis, Part 2, Chemical and Microbiological Properties (second ed.). American Society of Agronomy, Agronomy Series 9 (Part 2), Madison, WI, pp. 648–649.

Lammers, D.A., Johnson, M.G., 1991. Soil mapping concepts for environmental assessment. In: Mausbach, M.J., Wilding, L.P. (Eds.), Spatial variabilities of soils and landforms. Soil Sci. Soc. Am. Special Publication 28, 149–160.

Macfarlane, P.A., 2000. Revisions to the nomenclature for Kansas aquifers: Kansas Geological Survey Current Research in Geologic Sciences, Bulletin 244, part 2, 14 p, http://www.kgs.ku.edu/Current/2000/macfarlane/macfarlane.pdf

McMahon, P.B., Dennehy, K.F., Michel, R.L., Sophocleous, M.A., Ellett, K.M., Hurlbut, D.B., 2003. Water movement through thick unsaturated zones overlying the central High Plains aquifer, southwestern Kansas, 2000–2001. U. S. Geological Survey Water-Resources Investigations Report 03–4171,p. 32.

Moore, I.D., Gessler, P.E., Nielsen, G.A., Peterson, G.A., 1993. Soil attribute prediction using terrain analysis. Soil Sci. Soc. Am. J. 57, 443–452.

Moulin, A., Derksen, D., McLaren, D., Grant, C., 2002. Spatial variability of soil fertility and identification of management zones on hummocky terrain. University of Manitoba, 2nd Annual Manitoba Agronomists conference, http://www.umanitoba.ca/afs/agronomists_conf/2002/pdf/P4.pdf (verified April 2005).

Onsoy, Y.S., Harter, T., Ginn, T.R., Horwath, W.R., 2005. Spatial variability and transport of nitrate in a deep alluvial vadose zone. Vadose Zone J. 4, 41–54.

Pennock, D., 1998. Field scale variability and soil productivity. Saskatchewan Soil Conservation Association, http://ssca.usask.ca/conference/1998proceedings/PENNCOK.html (verified April 2005).

Sleezer, R.O., 2001. Groundwater vulnerability to nitrate pollution from non-point sources in Harvey County, Kansas [PhD. Thesis]. Lawrence, University of Kansas, p. 320.

S-Plus 6 for Windows Users Guide, 2001. S-Plus 6 for Windows Users Guide. Insightful Corporation, Seattle, WA.

Stramel, G.J., 1967. Progress report on the ground-water hydrology of the Equus-Beds area, Kansas—1966. Kansas Geological Survey Bulletin 187, part 2, p. 27.

Townsend, M.A., Sleezer, R.O., 1995. Pollution prevention demonstration project, Harvey County, Kansas, final report. Kansas Geological Survey Open-file Report 95–74, p. 130.

Townsend, M.A., Sophocleous, M.A., Sleezer, R.O., 1996. Effects of soil variability on nitrate transport. Kansas Water Resources Research Institute Contribution No. 319, Kansas State University, p. 57.

Townsend, M.A., Sophocleous, M.A., Sleezer, R.O., 1997. Field-scale variability of soil properties influencing retention of plant-available nitrogen in central Kansas. Kansas Water Resources Research Institute, Contribution No. 326, p. 65.

U.S. Geological Survey, 2005. National Water-Quality Assessment (NAWQA) Program, High Plains Regional Ground Water (HPGW) Study. U.S. Geological Survey, http://co.water.usgs.gov/nawqa/hpgw/HPGW_home.html (verified May 2005).

Weerts, H.J.T., Bierkens, M.F.P., 1993. Geostatistical analysis of overbank deposits of anastomosing and meandering fluvial systems, Rhine-Meuse Delta, The Netherlands.. Sediment. Geol. 85, 221–232.

Whittemore, D.O., Merchant, J.W., Whistler, J., McElwee, C.D., Woods, J.J., 1987. Ground water protection planning using the ERDAS geographic information system, automation of DRASTIC and time-related capture zones. In: Proceedings, National Water Well Association FOCUS Conference on Midwestern Ground Water Issues. Dublin, OH, National Water Well Association, pp. 359–374.

Wilding, L.P., Drees, L.R., 1973. Spatial variability and pedology. In: Wilding, L.P., Smeck, N.E., Hall, G.F. (Eds.), Pedogenesis and Soil Taxonomy: Developments in Soil Science 11A. Elsevier, Amsterdam, pp. 83–116.

Williams, C.C., Lohman, S.W., 1949. Geology and ground-water resources of a part of south-central Kansas, with special reference to the Wichita municipal water supply. Kansas Geol. Surv. Bull. 79, 455.

Appendix 1. Methods

Sampling methods for the EPA study consisted of soil sampling with a 5-cm-diameter Shelby tube sampler and an auger rig. In the EPA study soils were sampled in 0.6-m intervals to depths of 3-m twice a year for two years to monitor soil nitrate in the upper 3-m of the unsaturated zone at 32 irrigated and 28 dryland fields. In addition soil-water samplers (lysimeters) were installed at six of the irrigated fields to monitor the movement of nitrate through the unsaturated zone. Methods of lysimeter installation and collection and analysis of soil-water and groundwater samples were described by Townsend and Sleezer (1995).

Sampling for the KWRRI study involved use of a trailer-mounted Giddings Soil probe and 5–7.5-cm-diameter soil cores. Samples were collected by soil horizon to depths of 3 m. Each soil profile was described using standard National Resource Conservation Service (NRCS) techniques and correlated to the soil series that was deemed the closest match to its properties. Of the 40 soil profiles described, only two of the sampled points matched the soil described in the existing soil survey.

In both studies sub-samples from soil cores were placed in plastic bags, labeled, and stored on ice until delivered to the Soil Testing Laboratory at Kansas State University where they were refrigerated until analyzed. Both inorganic nitrogen forms, NH_4^+ and NO_3^-, were extracted with 20 ml of 1 M KCl, using 2 g of prepared soil (Keeney and Nelson, 1982; Gelderman and Beegle, 1998). Cadmium reduction and colormetric procedures were used for nitrate and an indophenol colorimetric reaction was used for ammonium. Both analyses were run in separate channels in a flow analyzer to measure these ions simultaneously (Alpkem Corporation, 1986a, b).

Statistical analysis

Statistical analyses were done using non-parametric methods available in S-Plus 6 for Windows Users Guide (2001). The Kolmogorov test for normality was performed and all data were found to have a non-normal distribution. The non-parametric Kruskal–Wallis test for comparison of independent samples was used. The non-parametric Wilcoxon rank sum test was used for hypothesis testing of whether values of one type of sample were greater or less than another. A level of significance of $\alpha = 0.10$ was used for all tests because of the high level of uncertainty involved when dealing with spatially and temporally variable data (Davis, 1986). For both the Kruskal–Wallis and Wilcoxon test if the reported ρ value is less than $\alpha = 0.10$ the results are considered significant. The S-Plus for Windows v.6 was also used for graphing.

Appendix 2. Statistical test results

Table A1. Fertilizer application by farming type (dryland vs. irrigated)

Crop	Number of samples	Maximum	3rd Quartile	Mean	Median	1st Quartile	Minimum
Dryland							
Milo	46	141.1	100.8	89.3	91.8	72.8	20.2
Wheat	46	94.1	67.2	62.2	67.2	57.4	20.2
Irrigated							
Alfalfa	56	11.2	10.1	6.9	10.1	2.5	137.8
Corn	56	249.8	215.0	193.6	185.4	169.1	88.5
Milo	56	199.4	183.7	150.3	142.2	140.0	0.0
Soybeans	56	56.0	9.0	7.5	7.8	0.0	84.0

Note: Hypothesis: H_0: Fertilizer application rate independent of farming method; H_1: Amount of fertilizer used in irrigated farming is different from dryland farming; Kruskal–Wallis $\rho = 0.0013$ ($\alpha = 0.10$).

Table A2. Median nitrate-N values by year, season, and farm type

Season and sample interval (m)	Median nitrate-N (mg kg^{-1})		Wilcoxon rank sum test ρ value	Number of samples	
	Dryland	Irrigated		Dryland	Irrigated
1992					
Spring 0.0–1.2	5.9	9.9	0.0040	27	30
Spring 1.2–3.0	4.5	11.5	0.0011	27	30
Spring 0.0–3.0	10.1	20.2	0.0002	27	30
Fall 0.0–1.2	2.9	11.3	0.0002	17	24
Fall 1.2–3.0	3.5	15.4	0.0011	17	24
Fall 0.0–3.0	8.7	27.1	0.0002	17	24
1993					
Spring 0.0–1.2	4.2	6.1	0.0340	26	40
Spring 1.2–3.0	4.6	8.1	0.0113	26	40
Spring 0.0–3.0	8.6	13.5	0.0081	26	40
Fall 0.0–1.2	2.9	6.3	0.0002	22	17
Fall 1.2–3.0	4.7	6.7	0.0892	22	17
Fall 0.0–3.0	7.9	13.2	0.0097	22	17

Note: Hypothesis tested: H_0: Irrigated values = dryland values for each sample interval. H_1: Dryland values < irrigated values for each sample interval. Wilcoxon rank sum test used (S-plus 7.0, 2005).

Table A3. Nitrogen values for topographic position and depth

Topographic hillslope position	Nitrate-N 0–30 cm		Ammonium-N 0–30 cm		TPAN 0–30 cm		Nitrate-N 0–75 cm		Ammonium-N 0–75 cm		TPAN 30–75 cm	
	Mean kg ha^{-1}	Median kg ha^{-1}	Mean kg ha^{-1}	Median kg ha^{-1}	Mean kg ha^{-1}	Median kg ha^{-1}	Mean kg ha^{-1}	Median kg ha^{-1}	Mean kg ha^{-1}	Median kg ha^{-1}	Mean kg ha^{-1}	Median kg ha^{-1}
Summit	10.7	11.7	8.6	7.9	18.7	17.3	25.1	26.0	18.9	15.8	44.0	40.8
Shoulder	10.9	10.5	14.0	12.4	24.9	25.6	27.6	27.1	33.5	20.8	61.1	59.8
Backslope	11.9	11.0	14.8	12.8	26.7	26.7	29.2	26.9	31.7	25.5	61.0	54.4
Footslope	15.5	17.8	21.4	23.5	36.9	39.1	36.8	34.4	44.7	34.6	76.1	76.1
Toeslope	19.2	16.5	20.8	21.8	40.0	39.6	41.4	37.3	39.6	38.7	80.9	67.4
Kruskal–Wallis ρ value	0.0735		0.0474		0.0099		0.0653		0.1744		0.0325	

Note: Hypothesis: H_0 = No difference in nitrogen content between topographic positions sampled; H_1 = Differences between sampled sites related to topographic position sampled.

Developments in Environmental Science, Volume 5
D. Sarkar, R. Datta and R. Hannigan (Editors)

ISSN: 1474-8177/DOI:10.1016/S1474-8177(07)05019-X

Chapter 19

Using GIS to display complex soil salinity patterns in an inland salt marsh

Matthew Grunstra and O.W. Van Auken

Abstract

Inland salt marshes are found near salt lakes and desert springs in areas of high evaporation and low precipitation. Spatial and seasonal fluctuations of the soil salinity levels are assumed to be associated with local climate. However, topography complicates the general soil salinity patterns attributed to climate. To better demonstrate and display these spatial and temporal relationships, soil salinity data were gathered over a ten-month period and then examined with a geographic information system software package. A variety of surface contour plots can be generated depending on the software used, the interpolation method, and the input variables chosen. To demonstrate some of this variation, surface plots were interpolated using the inverse distance weighted method in ArcView 3.3 and the ordinary kriging method in ArcGIS 8. Various methods and settings should be examined to create the clearest interpolated surface to present the collected data. Using the same data, ordinary kriging generated a gradual, smooth surface while the overall quality of the inverse distance weighted surface was irregular. The ordinary kriging method produced mean error and root-mean-square error statistics closer to zero indicating it would generally yield a better estimation of the soil salinity. The surfaces generated showed some seasonal fluctuation in soil salinity but spatial changes were more distinct. The low-elevation areas in the center of the salt marsh had low levels of soil salinity while the marsh edges at slightly higher elevations had increased levels of soil salinity. The spatial patterns of soil salinity appear to depend on seasonality, specifically rainfall and plant activity and the amount of water present in the salt marsh. However, local surface anomalies can mask these patterns. Through the use of global information systems (GIS) and interpolation some of the masking effects can be reduced revealing the underlying patterns.

19.1. Introduction

Geographical information systems (GIS) and global positioning systems (GPS) provide new ways to investigate and display areas both spatially and temporally. The surface contour plot is a tool readily available with the emergence of cheaper GIS software. A surface contour plot allows a limited point dataset to be expanded to display estimated values at any point within a study area, thereby possibly reducing the amount and expense of sampling.

There are numerous interpolation methods available for the creation of these surface contours (Lam, 1983; Isaaks and Srivastava, 1989; Burrough and McDonnell, 1998). Each of these individual interpolation methods allow the user to modify characteristics and input variables that can yield large variations in the output surface. As of yet there is no universally accepted systematic approach to method selection. This chapter will demonstrate different output surfaces created using two of the more common interpolation techniques, Inverse Distance Weighted (IDW) and Ordinary Kriging (Franzen and Peck, 1995; Weisz et al., 1995). It will also demonstrate how these surfaces can be applied to field research by examining the spatial and temporal fluctuations of soil salinity levels in an inland salt marsh.

The IDW method is a local deterministic interpolation technique (ESRI, 1999; Ormsby and Alvi, 1999; Johnston et al., 2001). A local technique calculates the estimated values for the entire study area based on small spatial areas or neighborhoods. The IDW method is an exact interpolator which predicts a value at a sample location that is equal to the measured value in the dataset (ESRI, 1999; Johnston et al., 2001). The calculations used with this procedure are based on the principle that each data point has a local influence that is reduced with distance (ESRI, 1999; Ormsby and Alvi, 1999; Johnston et al., 2001). The IDW method produces accurate surface interpolation as long as a regularly distributed sampling pattern is used in data collection (ESRI, 1999, 2000; Ormsby and Alvi, 1999; Johnston et al., 2001). Low concentration or an uneven distribution in the sampling points often produces sharp peaks or troughs in the output surface (Ormsby and Alvi, 1999; ESRI, 2000; Johnston et al., 2001).

The ArcGIS 8 software allows the use of the geostatistical interpolation method of ordinary kriging to create surface contour plots. Kriging has theoretical advantages over the IDW method because it is a geostatistical technique which is based on the principle that direction or distance between points reflects a spatial correlation that can explain the variation of the surface (Isaaks and Srivastava, 1989; Burrough and McDonnell,

1998; ESRI, 2000; Johnston et al., 2001; ESRI, 2003). Ordinary kriging is based on the principle that direction or distance between points reflects a spatial correlation that can explain the variation of the surface (ESRI, 2000; Johnston et al., 2001).

The Geostatistical Analyst extension in the ArcGIS 8.0 software also allows cross-validation of the prediction errors, which serves as a diagnostic tool to evaluate the surface plot generated (Johnston et al., 2001). Cross-validation uses all of the dataset to estimate the model and then it removes each data location, one point at a time, and predicts the associated data value (Johnston et al., 2001). The prediction error statistics include a mean error, root-mean-square error, average standard error, and the root-mean-square standardized error (Johnston et al., 2001).

Many studies have compared the performance of these two methods using cross-validation techniques and generation of error statistics (Warrick et al., 1988; Laslett et al., 1987; Wollenhaupt et al., 1994; Gotway et al., 1996; Kravchenko and Bullock, 1999; Brodsky et al., 2001; Mueller et al. 2001; Jones et al., 2003; Kravchenko, 2003; Mueller et al., 2004). Several studies found kriging to produce lower root-mean-square errors than the IDW method for their data (Lefohn et al., 1987; Kravchenko and Bullock, 1999; Bekele et al., 2003). Others authors have indicated IDW as a more suitable method with the IDW method generating a lower root-mean-square error than the kriging method when investigating soil properties in western Australia (Robinson and Metternicht, 2005). Some studies used other generated error statistics such as mean error, mean absolute error, and error standard deviation (Phillips et al., 1997) or log mean squared error (Zimmerman et al., 1999) to evaluate performance between the different methods. Most often sampling location, spatial correlation and grid scale were found to influence the optimal method (Bucher and Vckovski, 1995; Gotway et al., 1996; Brodsky et al., 2001; Dille et al., 2003; Jones et al., 2003; Kravchenko, 2003; Mueller et al., 2004). The varied results in each of these studies indicate careful consideration should be employed when determining the interpolation method and settings to be used with a given dataset.

Many studies of soil salinity, vegetation and salt marshes have incorporated GIS into their analysis or presentation of their data. GIS interpolations have been used to determine the spatial dynamics of soil salinity in arid and in semi-arid conditions (Jordan et al., 2004) as well as to determine the temporal and spatial variability of soil salinity in a coastal saline field (Shi et al., 2005) and a cotton field irrigated with low-quality water (Cetin and Kirda, 2003). Recent and subfossil vegetation data was imported into a GIS software package and used to investigate and determine long-term salt marsh vegetation dynamics on Frisian barrier islands off the coast of Germany (Freund et al., 2003). GIS combined with aerial photography has indicated the broad distribution of the vegetational patterns in a Georgia tidal salt marsh were due to salinity (Higinbotham et al., 2004).

Saline marshes are of two general types, marine and inland, and occur in many different geographical areas worldwide (Odum, 1988). Marine marshes are found along large bodies of saltwater, and are usually considered coastal salt marshes (Chapman, 1974). Inland salt marshes typically occur in desert areas usually without a large body of water nearby. Inland marshes develop within high-evaporation basins, usually next to inland saline lakes, and lowlands associated with desert springs (Odum, 1988). This is due to the high rate of evaporation and low level of precipitation in these desert ecosystems. The level of salts and the kind of salts found in the soil vary in both time and location (Borchert, 1971; Ungar, 1974). This is due to the amount of precipitation, timing of the precipitation in the system and the underlying rock type (Brune, 1981).

The Diamond Y Spring is located on a 6.1 km^2 nature preserve owned by the Nature Conservancy of Texas located approximately 16 km north of Fort Stockton, Texas (Fig. 19.1). The Diamond Y Spring is the last major spring still flowing in Pecos County, Texas (Veni, 1991). The Diamond Y Spring Preserve protects six federally endangered or threatened species. Included are the Puzzle Sunflower (*Helianthus paradoxus*), Leon Springs Pupfish (*Cyprinondon bovinus*), Pecos Gambusia (*Gambusia nobilus*), Diamond Y Spring snail (*Tryonia adamantia*), Gonzalez Spring Snail (*Tryonia stocktonensis*), and Pecos Assiminea snail (*Assiminea pecos*) (McDonald, 1999; Bush and Van Auken, 2004; Van Auken et al., 2007).

The types of soil, water chemistry, as well as, the plant communities present are all indicators that the low elevation area at the Diamond Y Spring Preserve is a salt marsh (Chapman, 1974; TPWD, 2003). Soils in the study area are composed mainly of three associations: the Balmorhea, Orla, and Lozier (Rives, 1980; Grunstra, 2002) (Fig. 19.1). These soils are found in specific areas of the Diamond Y Spring Preserve due to elevation and soil water. Balmorhea soils tend to be deep (>152 cm) moderately saline and poorly drained often occupying lower elevation areas in and around spring fed marshes where they are usually anaerobic due to continuous saturation (Fig. 19.1) (Rives, 1980; Lavelle and Spain, 2001; Bush, 2002; Grunstra, 2002). Balmorhea soils are Mollisols and are classified as fine-silty, thermic Cumulic Haplaquolls (Schoeneberger et al., 1998). These soils are usually gray silt loam containing fine concretions of calcium carbonate (Rives, 1980; Lavelle and Spain, 2001). Orla soils are Aridisols and are classified as fine-loamy, gypsic, thermic Typic Gypsiorthids (Schoeneberger et al., 1998). Orla soils tend to be located in

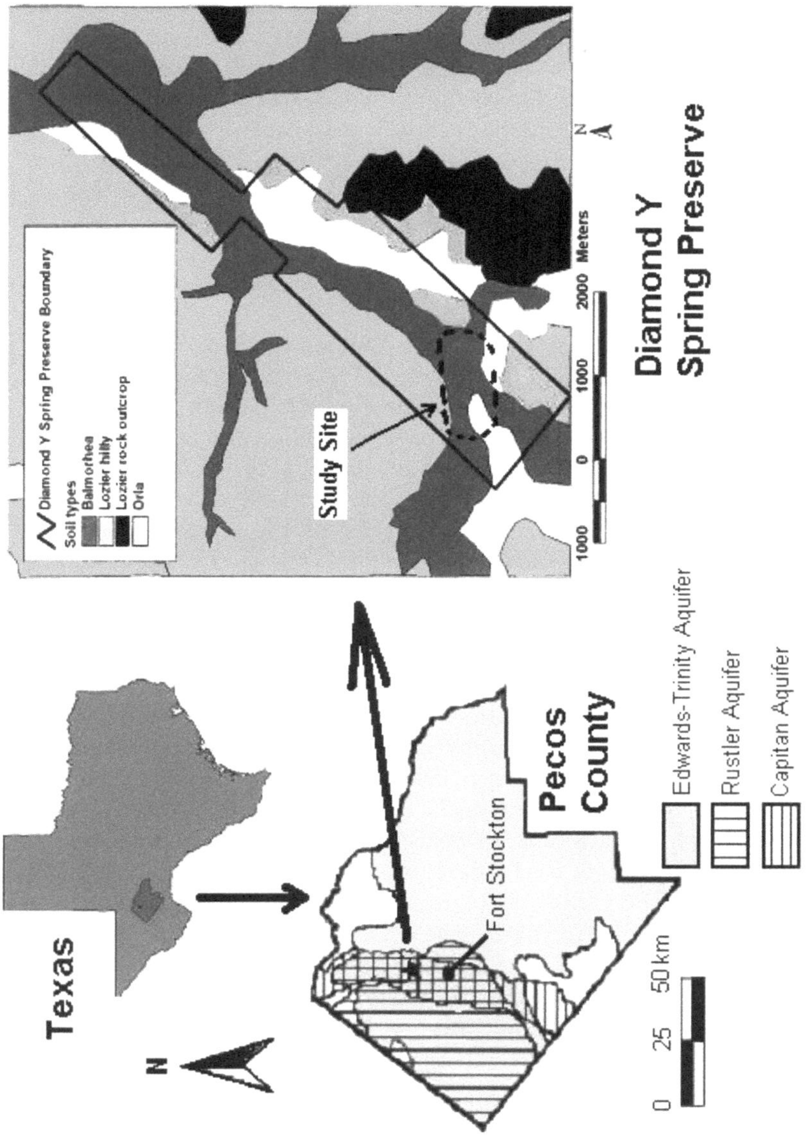

Figure 19.1. Pecos County is located in west Texas, USA. The major aquifers underlying the Fort Stockton area and Pecos County, Texas include the Edwards-Trinity, Rustler, and Capitan Reef. Soil types present in the area of the Diamond Y Spring Preserve include Balmorhea, Orla, and Lozier association soils (Rives, 1980).

the low to medium elevations of the study area, usually higher in elevation than the Balmorhea soils and above the salt marsh (Fig. 19.1) (Grunstra, 2002). Gypsum and loam comprises 40–80% of the composition of these soils giving them a moderately alkaline pH, moderate permeability, and a low water holding capacity (Rives, 1980; Schoeneberger et al., 1998; Lavelle and Spain, 2001). Lozier soils occupy the higher elevations surrounding the salt marsh, above the Orla soils, lying on the limestone outcrops (Fig. 19.1) (Grunstra, 2002). Lozier series soils are Aridsols and are classified as loamy-skeletal, carbonatic, thermic Lithic Calciorthids (Schoeneberger et al., 1998). Lozier soil is well drained, shallow, moderately alkaline and characterized as gravely and stoney, light brownish gray in color with high limestone content (Rives, 1980; Lavelle and Spain, 2001).

There have been a number of studies done on the area hydrology but nothing specific to the Diamond Y Spring Preserve (Boghici, 1997). The general physiographic setting of the Diamond Y Spring Preserve is a semi-arid karst terrain with an average annual precipitation of 33 cm year^{-1} and an evaporative rate of 204 cm year^{-1} (Fig. 19.2) (Larkin and Bomar, 1983). The aquifers underlying the area are: the Capitan Reef Complex, Rustler, and Edwards-Trinity (Fig. 19.1).

The Capitan Limestone formation lies approximately 1200 m below the Diamond Y Spring Preserve and generally contains highly mineralized calcium–magnesium bicarbonate water with dissolved-solids concentrations that range from 850 to 2500 mg l^{-1} (Veni, 1991; USGS, 2002). The Rustler aquifer lies in the carbonates and evaporites of the Rustler Formation at a depth of approximately 220 m under the Diamond Y Spring Preserve (Veni, 1991) and uniformly yields water high in calcium and low in bicarbonate (Small and Ozuna, 1993; Boghici, 1997). Total dissolved solids range from 1500 to 80,000 mg l^{-1} (Veni, 1991). The Edwards-Trinity aquifer is found beneath most of Pecos County and covers areas underlain by both the Capitan Reef Complex and Rustler aquifers (Fig. 19.1). The depth of the Edwards-Trinity aquifer under the Diamond Y Spring Preserve is approximately 85 m. The Edwards-Trinity aquifer water is dominated by sulfate and chloride with high concentrations of sodium, calcium, and magnesium cations (Veni, 1991; USGS, 2002). In Pecos county, Edwards-Trinity aquifer water can be fresh to brackish containing less than 1500 mg l^{-1} total dissolved solids or with total dissolved solids up to 80,000 mg l^{-1} due to mixing with Rustler water (Veni, 1991; Boghici, 1997).

The aquifer that feeds the Diamond Y Spring is a point of slight contention. In 1991, Veni suggested it was the Edwards-Trinity aquifer. However, in 1997, Boghici suggested that the Rustler aquifer feeds the

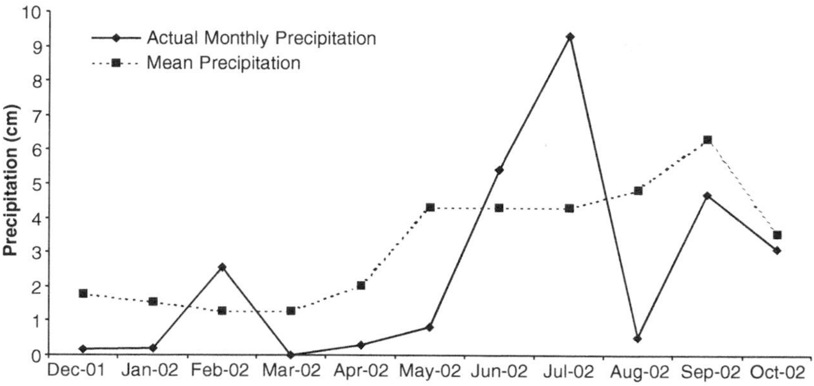

Figure 19.2. Total monthly precipitation data from December 2001 to October 2002. The values are plotted for the center of the months. The solid line is the actual monthly precipitation reported from Fort Stockton, Texas. The mean precipitation data are based on 1231 months from 1859 to 1987 (NCDC, 2002).

preserve. In either case, the water has been found to be dominated by a sodium–chloride–sulfate chemistry with moderate amounts of calcium and magnesium (Veni, 1991; USGS, 2002). The Diamond Y Spring is located approximately 1000 m southwest of the current study site flowing into the drainage area where it joins with Leon Creek flowing from the west creating the salt marsh. The water drains eastward mostly through subsurface flow out of the marsh approximately 25 km to the Pecos River. The surface flow is ephemeral due to the low rainfall and high-evaporative rate (Van Auken and Bush, 1998; Grunstra, 2002; Hart, 2002).

The plant communities located within the salt marsh are in fairly distinct locations across the marsh landscape (Grunstra, 2002; Grunstra and Van Auken, 2007; Van Auken et al., 2007). This distribution may be due to the different species specific water requirements, salt tolerance, or ability to out-compete rivals in the differing salinity levels of the soil (Chapman, 1974; Niering and Warren, 1980; Bertness, 1991; Van Auken and Bush, 1998; Bush, 2002; Grunstra and Van Auken, 2007; Van Auken et al., 2007). Surrounding the marsh at a slightly higher elevation are *Prosopis glandulosa* woodlands (Van Auken and Bush, 1998). Slightly lower within the marsh are three distinct grassland communities. The *Sporobolus airoides* grassland is slightly higher and drier than the *Distichlis spicata* community that is usually located in areas that are flooded for at least part of the year. In the deepest soils of the central areas of the marsh in soils that are usually flooded year round is the *Schoenoplectus Americana* community.

To gain a greater understanding of the distribution of the soil salinity throughout the marsh and its possible relationship with the salt marsh vegetation, soil salinity data were gathered. The point data were then placed into a GIS software package in which a surface contour plot was generated. The surface contour plot interpolated values to fill in the study area and provide a representative overview of the soil salinity levels throughout the salt marsh. Different geographic information system software, interpolation methods, and input variables will generate visually different surface contour plots and should be investigated for the creation of the most useful interpolated surface of the data.

19.2. Methods

The borders of the study area were designated on the northern, western, and eastern edges by fencing that allows controlled grazing on the marsh. A sharp change in elevation out of the lowland and into the slightly higher elevation of the surrounding limestone outcropping set the southern limit of the study area. A grid pattern of observation points was chosen to cover the study area based on an existing pattern of observation wells located in the southwestern corner of the marsh (Bush, 2002). The grid pattern used in the present study was expanded to a larger scale in order to increase the size of the study area and encompass a larger area of the salt marsh. The grid pattern consisted of 7 transects with a total of 87 observation points (Fig. 19.3).

A GeoExplorer III GPS receiver unit along with a Beacon-on-a-Belt both manufactured by Trimble were used to map the study area and the 87 observation points. For the observation points, a mean of ten measurements per position were collected which helps to increase the precision of the location data collected by increasing the size of the sample set. The position mode was set to collect 3D data using a minimum of four satellites. The majority of the GPS measurements for each location were collected using six to eight satellites while only three locations used five satellites. The elevation mask was set for twenty degrees to reduce the amount of ionosphere noise, which is above the recommended minimum setting of fifteen degrees (Trimble, 1999). The signal-to-noise ratio mask was set at ten to ensure acceptance of only strong satellite signals and reduce noise distortion. The Position Dilution of Precision mask was set at five to ensure the use of an even geometric distribution of satellites (Trimble, 2001).

Trimble's Beacon-on-a-Belt allowed the data to be real time differentially corrected. Differential correction calculates the errors associated

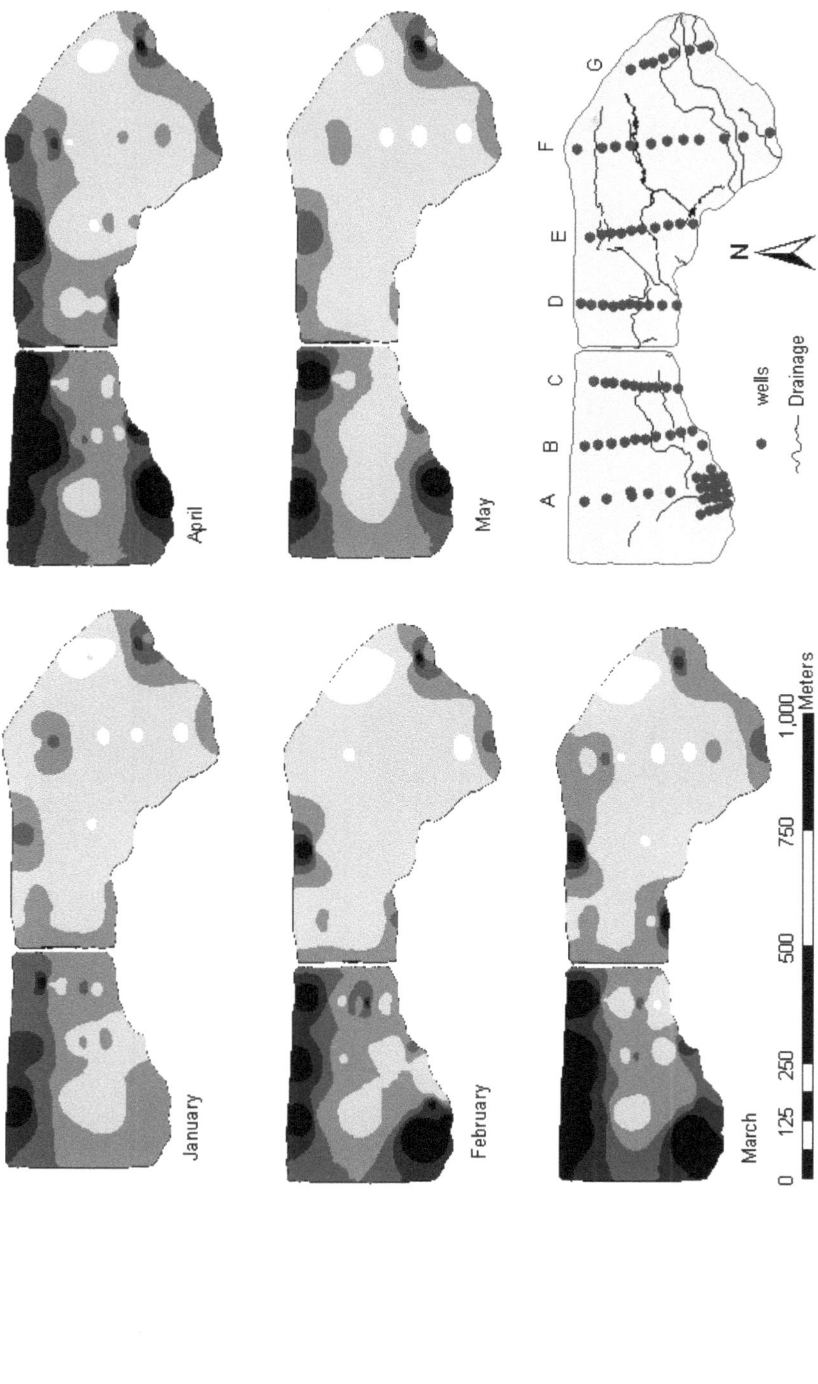

Figure 19.3. Interpolated soil salinity for ten months of 2002 created using the IDW method. The locations of the soil collection sites and transects (A-G) are also presented.

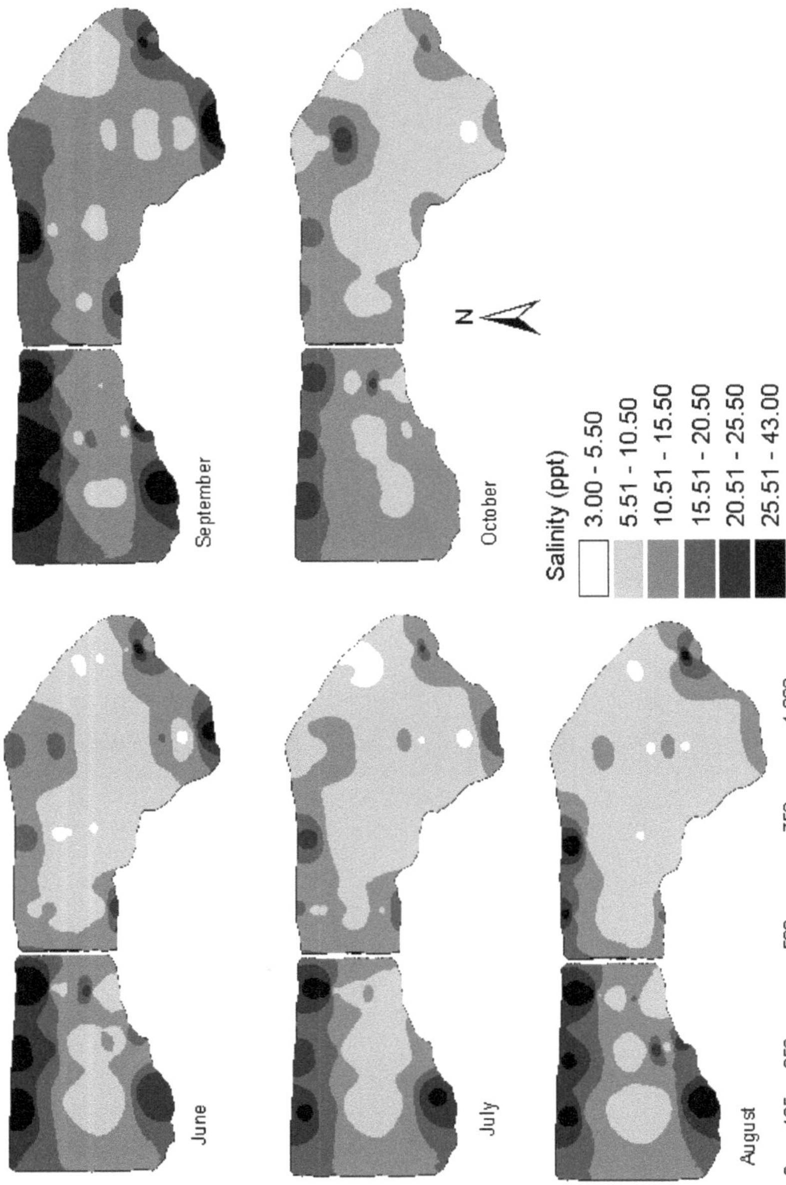

Figure 19.3. (*Continued*)

with a known point and applies them to an unknown point's location (Hurn, 1989; Hurn, 1993; Trimble, 1999; Trimble, 2001). Real-time differential correction involves receiving radio signals to correct the data at the time of collection. It works by receiving GPS differential correction information from local Minimum Shift Keying beacons run by the United States government and transmitting that information to the GeoExplorer III (Trimble, 2001). The beacon used in this study was located in Odessa, Texas, approximately 100 km from the study area.

Using these settings minimized error and allowed attainment of sub-meter horizontal precision on 65% of the 87 observation site locations at a 95% percent level of confidence as calculated by the pathfinder software. The highest calculated error of the horizontal precision estimate was 2.7 m with a mean error of 1.2 m based on the 87 observation site locations.

All of the GPS data collected in the field was imported from the Geo-Explorer III into GPS Pathfinder Office software. The data files were then exported as ArcView shapefiles from the GPS Pathfinder Office Software. ArcView 3.3 and ArcGIS 8 were then used to create maps and surfaces that were used to present and/or analyze the collected measurements (ESRI, 1999; Ormsby and Alvi, 1999).

Soil samples from each of the 87 observation sites in the study area were collected on a monthly basis beginning in January 2002 and continuing until October 2002 for 870 total observations. A trowel was used to gather approximately 300 g of surface soil from each observation site. Surface litter was removed and the sample was collected from the top 1 cm of soil (TAES, 1983). When a sampling site was under water, a sample of water was collected. The soil was gathered approximately 2 m from each observation point in a cross pattern. The samples were placed in plastic bags and sealed to prevent evaporation. The sample bags were taken to the University of Texas at San Antonio greenhouse where the soil was analyzed. Soil salinity was measured by making a 1:1 paste (soil:de-ionized water, V:V) (Westerman, 1990; Rowell, 1994). The paste was stirred thoroughly and allowed to sit for 30 min (TAES, 1983). After the 30-min period, the paste was stirred again and measured with a salinity probe. Salinity was measured in parts per thousand. These measurements were then entered into the attribute table of the GIS shapefiles created in ArcView 3.3 and ArcGIS 8 from the collected GPS field sampling site locations.

The soil salinity point measurements were converted to a raster grid to produce a continuous surface or contour plot across the study area. The software creates a raster grid by first dividing the total area into a matrix of equally sized cells (ESRI, 1999; Ormsby and Alvi, 1999; Johnston et al., 2001). Each cell is then assigned a value depending on the method of interpolation used for estimation. In this interpolation, the area was divided into 250 rows and 657 columns for grid cells of approximately 2 m by 2 m. In ArcView 3.3, the Spatial Analyst extension was used to do this interpolation by the IDW method. In ArcGIS 8, the Geostatistical Analyst extension was used to create the surface plots by the kriging method.

Both the IDW and ordinary kriging surface plots were created for the soil salinity measurements for each location and for each of the ten months of the study. Several iterations were performed for each interpolation method using various initial settings of neighborhood size of analysis, lag spacing, and power of magnification. This was performed in a systematic fashion in order to produce the most logical and representative interpolated surface for each method. The settings used for the IDW surface displayed in this study were a neighborhood size of 15 with a power magnification of 3. The settings that produced the best cross-validation error statistics for the ordinary kriging method were for an elliptical search neighborhood with no offset that is divided into four sectors including five points per sector. Both methods allowed for the display and investigation of spatial and temporal trends and patterns of soil salinity levels in the salt marsh. The outputs from the two separate interpolation methods were compared visually as well as by the error statistics produced through cross-validation. The ArcGIS 8.0 software was used to perform cross-validation and generate the mean prediction error and the root-mean-square prediction error.

The monthly soil salinity was also examined for possible significant differences between months using SAS and GraphPad's Prism software packages (Milliken and Johnston, 1992; SAS, 1995, 1999). Bartlett's test was used to test for equal variances (Milliken and Johnston, 1992). In SAS, the Shapiro-Wilk's statistic was used to test for normality (Milliken and Johnston, 1992). The soil salinity data were found in all cases to be non-normal, so transformations were performed to meet the normality assumption. The following transformations were examined in the SAS software; log, reciprocal, square root, exponential, and square. None of the transformations yielded normalized data so nonparametric tests were used. GraphPad's Prism software was used to run the Friedman nonparametric one-way analysis of variance test to account for the repeated measures inherent in the datasets. Dunn's multiple comparison test was used to compare monthly differences.

19.3. Results

The ArcView 3.3 soil surface salinity contour plots created using the IDW method show a range of salinity levels from a low of 3 ppt to a high of

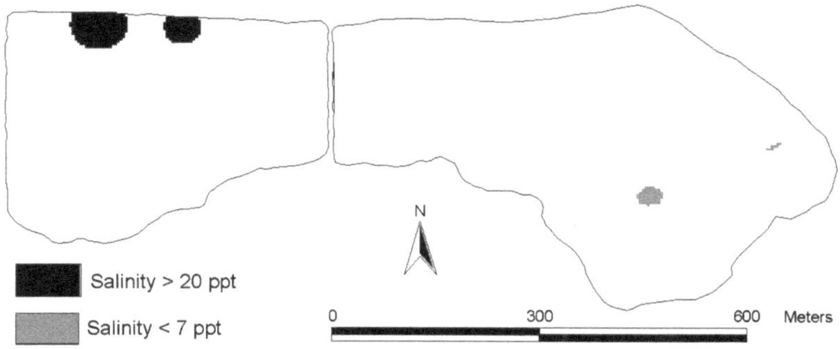

Figure 19.4. Locations of areas where the soil salinity stayed in the identified limit for the entire ten-month study period. The solid black areas are the locations where the salinity level stayed continually high with values always greater than 20 ppt. The light gray colors are consistently low-salinity levels never getting above 7 ppt.

43 ppt (Fig. 19.3). The solid black areas represent the highest salinity levels while lower levels are shown with various patterns. High salinity levels were found along the borders or edges of the study area especially in the northwest with lower salinity levels found toward the center or lower elevations of the marsh associated with the drainage (Fig. 19.3). The surface soils in the eastern half of the study area consistently showed large areas with soil salinity levels in the 3–10 ppt salinity range (Fig. 19.3). There was little variation in the soil salinity levels in these locations through the ten-month study period.

A few locations in the study area had consistently high or low soil salinity (Fig. 19.4). Even though the levels of surface soil salinity maintained a fairly stable pattern on a month-per-month basis, there were months when the levels fluctuated greatly leaving only small areas with constant high or low values for every month. There were two small areas located in the southeastern region of the study area that consistently had soil surface salinity values less than 7 ppt. Surface soil salinity levels greater than 20 ppt were found in locations along the edge of the northwestern part of the study area (Fig. 19.4).

The spatial variation of soil salinity across the study area was examined for each of the seven transects. The seven transects were plotted on separate bar graphs depicting the ten-month soil salinity mean for each location (Fig. 19.5). The nonparametric Kruskal-Wallis test showed significant differences for each transect across the marsh ($P<0.05$). The trend in the north to south direction across the study area can be most seen with locations at the ends of the transects having significantly higher soil salinity values than areas located in the central region of the study.

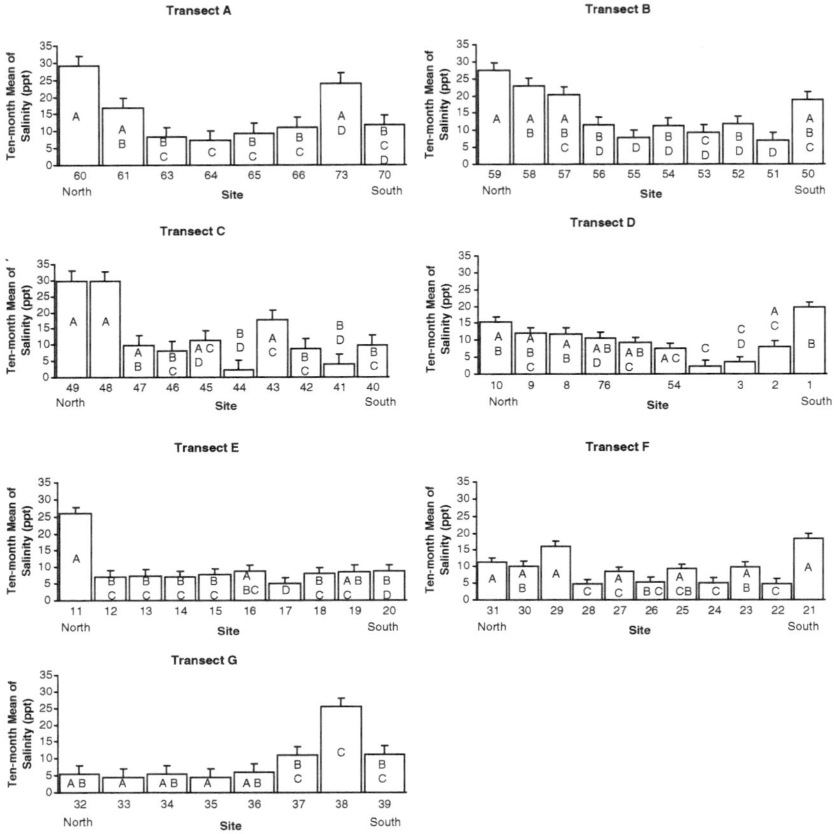

Figure 19.5. Ten-month mean salinity values for the soil-sampling sites located on the seven transect lines across the study area. The bars are the standard errors. The nonparametric Kruskal-Wallis test demonstrated significant differences for each transect ($P<0.05$). Means with the same letter are not significantly different (Dunn's multiple comparison test, $P>0.05$).

These statistically higher soil salinity locations coincided with the high-salinity values displayed in the GIS surface contours (Fig. 19.3).

The monthly means for the salinity data were also plotted for each of the ten months of the study with error bars indicating the 95% confidence interval (Fig. 19.6). The nonparametric Kruskal-Wallis test demonstrated significant differences for each month ($P<0.05$). From January to March, there is only a slight variation in the salinity mean values with 11.10, 11.62, and 12.86 ppt, respectively (Fig. 19.6). April and September were the months with the highest means with September's mean of

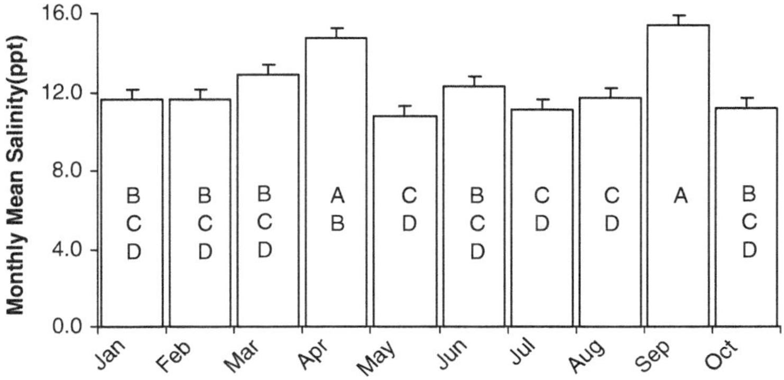

Figure 19.6. Monthly means for the soil salinity data. The mean salinity is presented for each of the ten months in the study period. The nonparametric Kruskal-Wallis test demonstrated significant differences for each transect ($P < 0.05$). The error bar shows the standard error. Means with the same letter are not significantly different between the using Dunn's multiple comparison test ($P > 0.05$).

15.43 ppt being slightly higher than April's 14.72 ppt (Fig. 19.6). May had the lowest mean surface soil salinity with 10.65 ppt. In September, salinity was found to be significantly different from all of the other months except April (Fig. 19.6). For surface soil salinity, April values were significantly different than those for May, July, and August (Fig. 19.6).

The surface contours created using the ordinary kriging method were smoother or less ragged than those created by the IDW procedure (comparing Figs. 19.3 and 19.7). In addition, they also show a large portion of the study area with soil salinity in the range from 5 to 10 ppt (Fig. 19.7). The majority of the higher levels of soil salinity in the range of 25–43 ppt were found along the northwestern border of the study area with occasional occurrences near the southwestern border. February and October show most of the study area with lower levels of salinity with very little area in the higher ranges (Fig. 19.7). April and September show larger areas with higher levels of salinity covering more of the study area than the rest of the time period (Fig. 19.7). The trend shows higher levels always near the borders and lower levels toward the middle of the study area.

Through visual examination both general differences and similarities of the surface plots created using the IDW and ordinary kriging methods could be observed (comparing Figs. 19.3 and 19.7). In the IDW surface plots, the sampling points and transects can more readily be observed (Fig. 3). The ordinary kriging method depicts more gradual and smoother transitions between soil salinity values (Fig. 19.7). The IDW depicts a

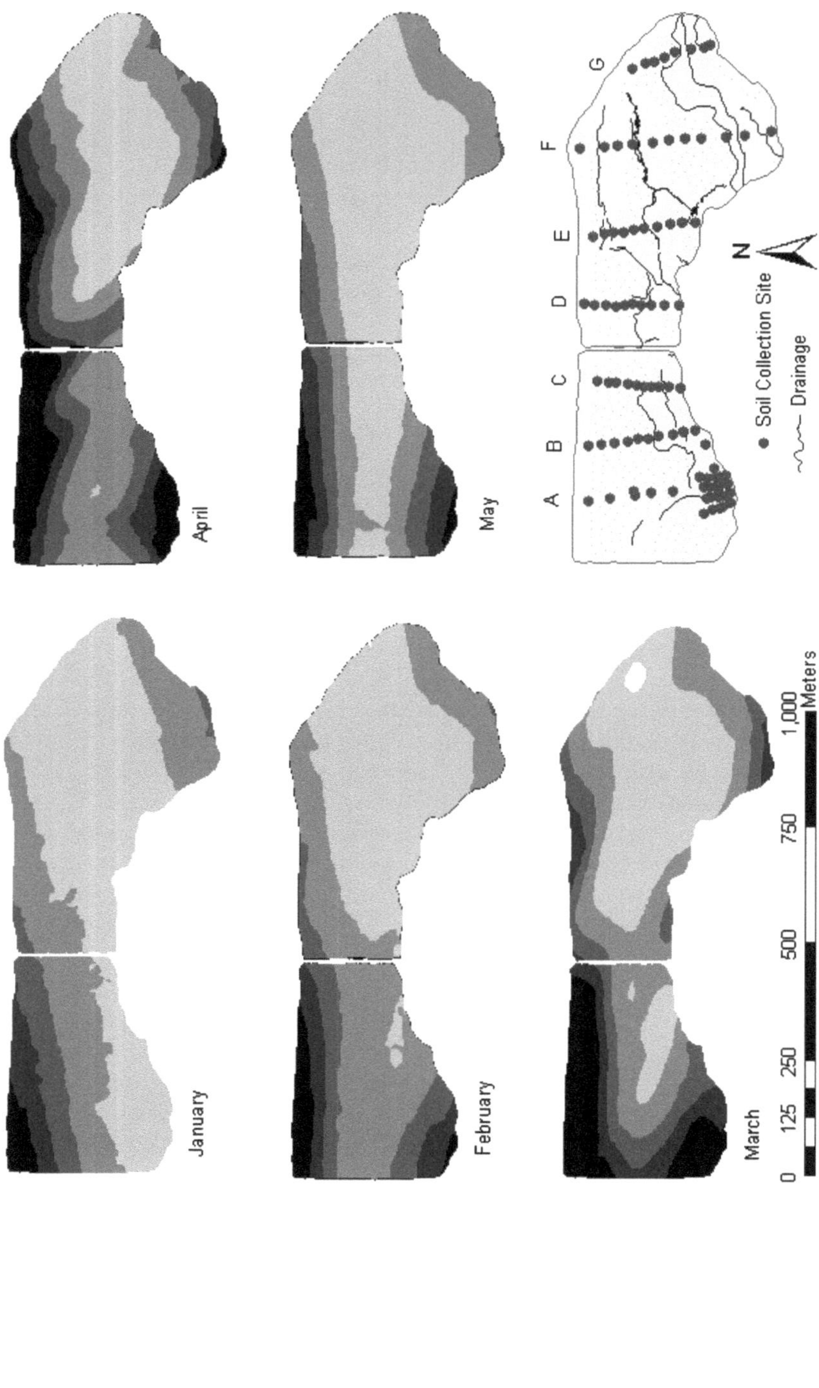

Figure 19.7. Interpolated soil salinity values per ten months of 2002 created using the Ordinary Kriging method in ArcGIS 8.0 software.

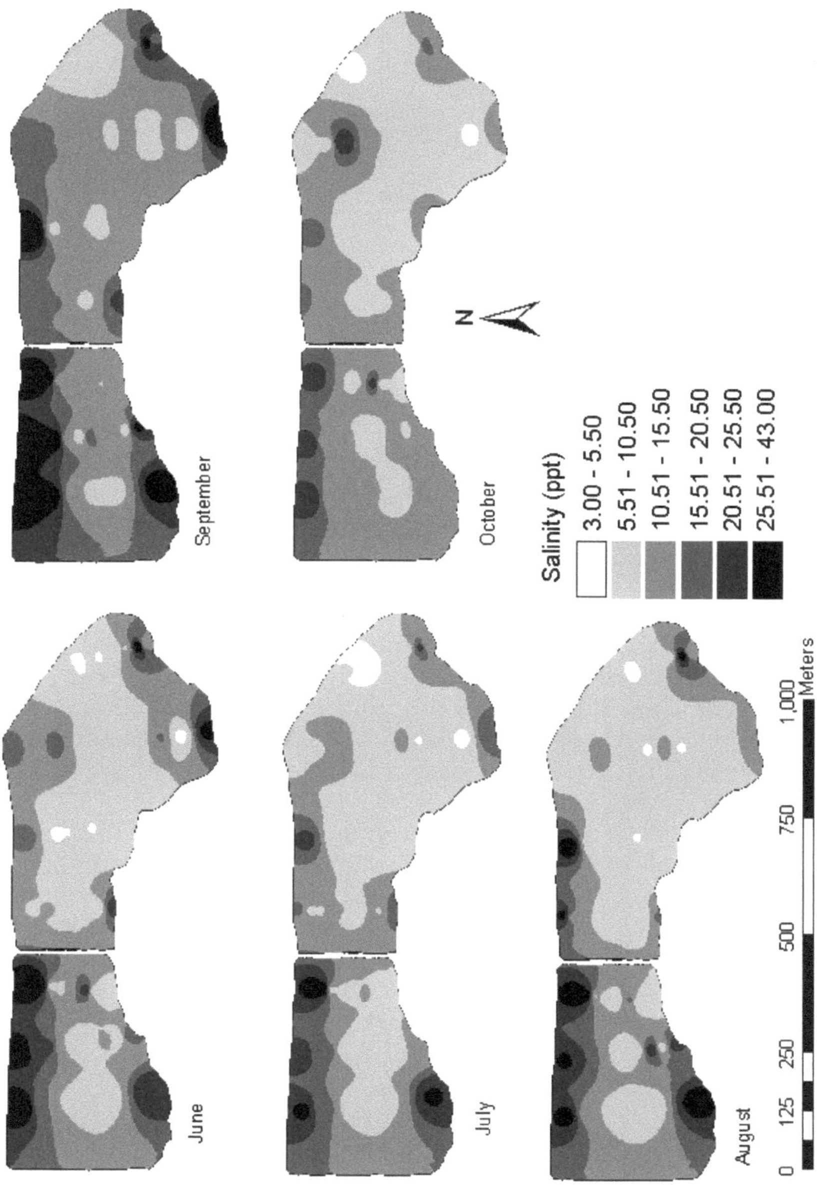

Figure 19.7. (*Continued*)

Table 19.1. Mean and root-mean-square error statistics generated for the interpolated surfaces created using the IDW and ordinary kriging methods. The ordinary kriging method consistently produced values closer to zero indicating it would generally yield a better estimation of the soil salinity

Month	Mean		Root-mean-square	
	IDW	Kriging	IDW	Kriging
January	−0.15	0.05	5.48	5.10
February	−0.69	−0.05	7.31	6.97
March	−1.13	−0.06	7.67	7.30
April	−1.26	−0.29	10.1	8.57
May	−1.02	−0.13	7.63	6.38
June	−1.06	−0.18	8.27	6.59
July	−0.89	−0.36	6.17	5.20
August	−0.87	0.01	7.74	6.07
September	−1.33	−0.27	7.65	6.76
October	−0.46	−0.05	5.97	4.97

larger surface area of the salt marsh covered by the lower values of 3–5 ppt while the ordinary method estimates those same areas to have slightly higher values between 5 and 10 ppt (Figs. 19.3 and 19.7). On the northern border, the IDW method shows localized concentrations of high-soil salinity around the sampling locations while the ordinary kriging method shows a banding pattern in those same areas. The IDW method tends to show less area covered by higher salinity and more area covered with lower salinity while the ordinary kriging method tends to show the reverse. Although both methods of visual representation show similar trends, the ordinary kriging surface contour plots are much smoother (Figs. 19.3 and 19.7).

The IDW and the ordinary kriging methods were evaluated by comparing the overall mean prediction error and root-mean-square prediction error for their surface contour plots (Table 19.1). The ordinary kriging method consistently produced values closer to zero indicating it would generally yield a better estimation of the soil salinity. Both methods depict the northern and southern borders of the salt marsh consistently, with higher levels of soil salinity while the center of the marsh had lower levels.

19.4. Discussion

The surface contour plots available in GIS software provide many new ways to investigate and display diverse results. They facilitate the ability to produce surface contour plots that can estimate and display values across a large surface area from a limited point dataset (Isaaks and Srivastava, 1989; Burrough and McDonnell, 1998). This allows investigators to quickly identify spatial and temporal patterns and trends which they can then examine more critically. These surface plots also allow one the ability to quickly visualize and identify possible interactions and influences that different factors may have in their study area.

The GIS user has numerous interpolation techniques from which to choose depending on the software package selected (Isaaks and Srivastava, 1989; Burrough and McDonnell, 1998; ESRI, 1999; Johnston et al., 2001). Furthermore, variables such as the neighborhood size of analysis, lag spacing, and other settings can be changed within a given interpolation method. The surface contour plots created by these methods may show similar characteristics and trends but will often produce visually dissimilar surface contours with considerable differences at specific locations (Bucher and Vckovski, 1995; Gotway et al., 1996; Brodsky et al., 2001; Dille et al., 2003; Jones et al., 2003; Kravchenko, 2003; Mueller et al., 2004). These dissimilarities are inherent to the mathematical procedures used to create the final surface contour such as the mathematical equations, calculations, and estimations used by those methods (Isaaks and Srivastava, 1989). Some methods are deterministic exact interpolators predicting values at sample locations that are equal to the measured value in the dataset utilized (ESRI, 1999; Johnston et al., 2001). While other methods are geostatistical in their calculations using mathematical models combined with statistical methods to create the final surface contour plots (Johnston et al., 2001). Deterministic interpolators seem to be more sensitive to the distribution and data point value variance entered into the model producing surfaces that are more irregular. While geostatistical interpolation seem to be less sensitive to the distribution and large variances of the point data and consequently produce smoother surface plots.

In this study, both the IDW and the ordinary kriging methods were used to provide examples of different visual outputs created using the same data. The kriging method produced a smooth, more regular interpolated surface, whereas the IDW method produced a surface that was more strongly influenced by local measurements or the values of the specific soil samples in this case (comparing Figs. 19.3 and 19.7). Most likely this was caused by the irregularly spaced pattern of the observation sites that were considerably closer in the west–east direction than the north–south direction. The IDW method produces a fairly exact surface interpolation as long as a regularly distributed sampling pattern is employed (ESRI, 1999; Ormsby and Alvi, 1999; ESRI, 2000; Johnston et al., 2001).

High point value variance or uneven distribution in the sampling patterns often produces sharp peaks or troughs in the output surface (Ormsby and Alvi, 1999; ESRI, 2000; Johnston et al., 2001). The point is that care must be taken with different methods and different settings when creating an interpolated surface in order to avoid interjecting various biases (Gotway et al., 1996; Dille et al., 2003; Jones et al., 2003; Kravchenko, 2003; Mueller et al., 2004). The results can be useful both visually and in predicting values for variables (in this case salinity) between sample points.

When deciding upon the appropriate interpolation method to use to investigate or display data, one must critically evaluate various methods and program settings in order to obtain the best visual representation of logical values between the sample points (Bucher and Vckovski, 1995; Gotway et al., 1996; Brodsky et al., 2001; Dille et al., 2003; Jones et al., 2003; Kravchenko, 2003; Mueller et al., 2004). In the current study, the IDW results are ragged with little smoothing between sampling locations because the IDW method uses exact interpolation with contours formed on the specific measurements entered into the program. Consequently, when the plots are examined it is easy to see where the measurements were actually made. For example, when one examines Fig. 19.3, the locations of the transects (and many soil collection sites) are obvious because of the sharp local differences in the places on the plots and the consequent lack of smoothing. In the kriging method, the northern and southern borders of the salt marsh consistently show the highest levels of soil salinity. These same areas coincide with elevations that are higher than the center of the salt marsh. The higher salinity values at the northern and southern borders of the salt marsh are attributed to a shallow soil or deeper water table that allows the soil to dry and therefore increase the soil surface salt concentration. More of the salts would be washed out of the surface soils in an area with the water table closer to the surface (Neill, 1993; Ala et al., 1995). The areas with high soil salinity can be seen to grow larger as the water table gets deeper and dry areas of the marsh increase in size (see Grunstra, 2002). The same high surface soil salinity areas then recede when the water table rises and the salts are flushed out of the soil by the presence of surface water.

Surface soil salinity in the Diamond Y Spring salt marsh was previously found to be at its lowest level in early spring and increasing during the summer months (Schmidt, 1986; Van Auken and Bush, 1993; Grunstra and Van Auken, 2007; Van Auken and Bush, 1995; Bush, 2006). This fluctuation in surface soil salinity was thought to occur in unison with the cyclic pattern of the water table. A higher water table would allow for more of the salts to be washed out of the surface soils while a lower water table would allow the soil to dry and therefore increase the surface soil salt concentration. A study on the effects of soil salinity on the production of a cool season grass found that spring flooding decreased the soil salinity (Neill, 1993). In another study, high levels of soil water allowed salts to be distributed throughout the soil profile while low levels of soil water caused salt accumulation in the upper soil layers due to the high evaporative rates (Ala et al., 1995).

The expected annual cycle of salinity in the Diamond Y Spring salt marsh was not as apparent as expected due to variation in the annual rainfall pattern (Fig. 19.2) (NCDC, 2002). January of the study year received no precipitation when it usually receives approximately 2 cm while June and July received greater amounts of precipitation than normal (Fig. 19.2). The month of August shows very little precipitation (0.4 cm) compared to the mean precipitation normally expected during that month (5 cm) (Fig. 19.2). Thus, the mean levels for the soil salinity did not show the expected pronounced temporal cyclic pattern. The monthly mean soil salinity fluctuations were observed but they were not as large as expected (Figs. 19.3, 19.6 and 19.7).

The spatial and temporal patterns of abiotic factors and their interactions have been found to play an important role in the distribution of the salt marsh vegetation in many areas and plant communities. Temporal and spatial distributions of soil moisture, pH, and ionic composition were significant in determining plant community locations in a Mediterranean salt marsh (Rogel et al., 2001). Bush (2002, 2006) found that surface salinity had a negative effect on all growth parameters and aboveground dry mass of *Helianthus paradoxus* at the Diamond Y Spring Preserve. In addition, these effects were time dependent. Spatial and temporal fluctuations in three halophyte species in upper coastal salt marsh communities were influenced by saline stress and soil nutrient level (Omer, 2004). Temporal change in soil salt levels was found to determine plant community locations along the shoreline of a desert basin lake (Toft and Elliot-Fisk, 2002). Soil salinity and moisture were also found to effect the spatial and temporal variation in plant germination and establishment in upper tidal marshes of three southern California wetlands (Noe and Zedler, 2001). Plant zonation was related to spatial and temporal variations in soil salinity in southeastern Spain (Ortiz et al., 1995). Vegetation distribution was also determined by soil salinity in spring fed salt marshes in western Utah (Bolen, 1964) and around the Great Salt Lake (Flowers, 1934).

Through the use and application of GIS, greater knowledge of the spatial and seasonal fluctuations of the soil salinity levels in salt marshes can be obtained. GIS interpolations have been used to determine the spatial dynamics of soil salinity in arid and semi-arid conditions (Jordan et al., 2004) as well as to determine the temporal and spatial variability of soil salinity in a coastal saline field (Shi et al., 2005) and a cotton field irrigated with low-quality water (Cetin and Kirda, 2003). In this study, the surface contours created for the Diamond Y Spring salt marsh have identified seasonal fluctuations and spatial distribution in the soil salinity across the salt marsh. The varying soil salinity levels may indicate zonation and probable locations of salt marsh vegetation although this is most likely coupled with the interaction of water level at different points in time during the growing season.

REFERENCES

Ala, F., Ismael, S., Ahmad, R., Shasheen, R., 1995. Effects of salinity and waterlogging on physiological processes and ionic regulation in Atriplex amnícola. Pakistan J. Bot. 27, 283–295.

Bekele, A., Downer, R.G., Wolcott, M.C., Hudnall, W.H., Moore, S.H., 2003. Comparative evaluation of spatial prediction methods in a field experiment for mapping soil potassium. Soil Sci. 168, 15–28.

Bertness, M.D., 1991. Interspecific interactions among high marsh perennials in a New England salt marsh. Ecology 72, 138–148.

Boghici, R., 1997. Hydrogeological Investigations at Diamond Y Springs and Surrounding Area, Pecos County, Texas. M.A. Thesis. University of Texas at Austin, Austin, TX.

Bolen, E.G., 1964. Plant ecology of spring-fed marshes in western Utah. Ecol. Monogr. 34, 143–166.

Borchert, J.R., 1971. The dust bowl in the 1970s. Ann. Assoc. Am. Geogr. 61, 1–22.

Brodsky, L., Vanek, V., Szakova, J., Stipek, K., 2001. Spatial heterogeneity of soil properties. Rost. Vyroba 47, 529–535.

Brune, G.M., 1981. Springs of Texas. Branch-Smith, Inc., Ft. Worth, TX.

Bucher, F., Vckovski, A., 1995. Improving the selection of appropriate spatial interpolation methods. In: Frank, A.U., Werner, K. (Eds.), The Proceedings of the International Conference COSIT '95. Spatial Information Theory: A Theoretical Basis for GIS, Semmering, Austria.

Burrough, P.A., McDonnell, R., 1998. Principles of Geographical Information Systems. Oxford University Press, Oxford, United Kingdom.

Bush, J., 2002. The effects of soil moisture, soil oxygen, and soil salinity on the growth of *Helianthus paradoxus*. Doctoral Dissertation. Department of Environmental Science, The University of Texas at El Paso, El Paso, TX.

Bush, J., 2006. The role of soil moisture, salinity, and oxygen on the growth of *Helianthus paradoxus* (Asteraceae) in an inland salt marsh of west Texas. J. Arid Environ. 64, 22–36.

Bush, J., Van Auken, O.W., 2004. Relative competitive ability of *Helianthus paradoxus* and it progenitors, *H. annus* and *H. petiolaris* (Asteraceae), in varying soil salinities. Int. J. Plant Sci. 165, 303–310.

Cetin, M., Kirda, C., 2003. Spatial and temporal changes of soil salinity in a cotton field irrigated with low-quality water. J. Hydrol. 272, 238–249.

Chapman, V.J., 1974. Salt Marshes and Salt Deserts of the World. Interscience Publishers, Inc., New York, NY.
Dille, J.A., Milner, M., Groeteke, J.J., Mortensen, D.A., Williams, M.M., 2003. How good is your map? A comparison of spatial interpolators. Weed Sci. 51, 44–55.
ESRI, 1999. Getting to Know ArcView GIS. Environmental Systems Research Institute, Redlands, CA.
ESRI, 2000. ArcView Spatial Analyst, An ESRI White Paper. Environmental Systems Research Institute, Redlands, CA.
ESRI, 2003. ArcGIS 8.0 Geostatistical Analyst Extension. Environmental Systems Research Institute, Redlands, CA.
Flowers, S., 1934. Vegetation of the Great Salt lake region. Bot. Gaz. 95, 353–418.
Franzen, D., Peck, A., 1995. Field soil sampling density for variable rate fertilization. J. Prod. Agric. 8, 568–574.
Freund, H., Petersen, J., Pott, R., 2003. Investigations on recent and subfossil salt marsh vegetation of the East Frisian barrier islands in the southern North Sea (Germany). Phytocoenologia 33, 349–375.
Gotway, C.A., Ferguson, R.B., Hergert, G.W., Peterson, T.A., 1996. Comparison of kriging and inverse-distance methods for mapping soil parameters. Soil Sci. Soc. Am. J. 60, 1237–1247.
Grunstra, M.B., 2002. Spatial and temporal hydrogeological description and ecological community comparison of the Diamond Y Spring Preserve. M.S. Thesis. Department of Earth and Environmental Science, University of Texas at San Antonio, San Antonio, TX.
Grunstra, M.B., Van Auken, O.W., 2007. Factors that influence the distribution and cover of *Helianthus paradoxus* in a west Texas salt marsh. Phytologia 89, 24–42.
Hart, D., 2002. Description and comparison of the upland plant communities of the Diamond Y Spring Preserve. M.S. Thesis. Department of Biology, University of Texas at San Antonio, San Antonio, TX.
Higinbotham, C.B., Alber, M., Chalmers, G., 2004. Analysis of tidal marsh vegetation pattern in two Georgia estuaries using aerial photography and GIS. Estuaries 27, 670–683.
Hurn, J., 1993. Differential GPS Explained. Trimble Navigation Limited, Sunnyvale, CA.
Hurn, J., 1989. GPS A Guide to the Next Utility. Trimble Navigation Limited, Sunnyvale, CA.
Isaaks, E.H., Srivastava, R.H., 1989. Appl. Geostatistics. Oxford University Press, New York, NY.
Johnston, K., Ver Hoef, J.M., Krivoruchko, K., Lucas, N., 2001. Using ArcGIS Geostatistical Analyst. ESRI Inc., Redlands, CA.
Jones, N.L., Davis, R.J., Sabbah, W., 2003. A comparison of three-dimensional interpolation techniques for plume characterization. Ground Water 41, 411–419.
Jordan, M.M., Navarro-Pedreno, J., Garcia-Sanchez, E., Mateu, J., Juan, P., 2004. Spatial dynamics under arid and semi-arid conditions: Geological and environmental implications. Environ. Geol. 45, 448–456.
Kravchenko, A., 2003. Influence of spatial structure of accuracy of interpolation methods. Soil Sci. Soc. Am. J. 67, 1564–1571.
Kravchenko, A., Bullock, D.G., 1999. A comparative study of interpolation techniques for mapping soil properties. Agronomics J. 91, 393–400.
Lam, N., 1983. Spatial interpolation methods: A review. Am. Cartographer 10, 129–149.
Larkin, T.J., Bomar, G.W., 1983. Climatic Atlas of Texas. Texas Department of Water Resources, LP-192, Austin, TX.

Laslett, G.M., McBratney, A.B., Pahl, P.J., Hutchinson, M.F., 1987. Comparison of several spatial prediction methods for soil pH. J. Soil Sci. 38, 325–341.
Lavelle, P., Spain, A.V., 2001. Soil Ecology. Kluwer Academic Publishers, Dordrecht, The Netherlands.
Lefohn, A.S., Knudsen, H.P., Logan, J.L., Simpson, J., Bhumralkar, C., 1987. An evaluation of the kriging method to predict 7-h seasonal mean ozone concentrations for estimating crop losses. J. Air Pollut. Control Assoc. 37, 595–602.
McDonald, C., 1999. Endangered and threatened wildlife and plants; determining the status for the plant *Helianthus paradoxus* (Pecos sunflower). Fed. Register 64, 56582–56590.
Milliken, G.A., Johnston, D.E., 1992. Analysis of Messy Data. Chapman & Hall/CRC, Boca Raton, FL.
Mueller, T.G., Pierce, F.J., Schabenberger, O., Warncke, D.D., 2001. Map quality for site specific fertility management. Soil Sci. Soc. Am. J. 65, 1547–1558.
Mueller, T.G., Pusuluri, N.B., Mathius, K.K., Cornelius, P.L., 2004. Map quality for ordinary kriging and inverse distance weight interpolation. Soil Sci. Soc. Am. J. 68, 2042–2047.
National Climatic Data Center, 2002. National Oceanic and Atmospheric Administration. U.S. Department of Commerce, Washington, DC.
Neill, C., 1993. Seasonal flooding, soil salinity and primary production in northern prairie marshes. Oecologia 95, 499–505.
Niering, W.A., Warren, R.S., 1980. Vegetational patterns and processes in New England salt marshes. BioScience 30, 301–307.
Noe, G.B., Zedler, J.B., 2001. Spatio-temporal variation of salt marsh seedling establishment in relation to the abiotic and biotic environment. J. Veg. Sci. 12, 61–74.
Odum, W.E., 1988. Comparative ecology of tidal freshwater and salt marshes. Annu. Rev. Ecol. Syst. 19, 147–176.
Omer, L., 2004. Small-scale resource heterogeneity among halophytic plant species in an upper salt marsh community. Aquat. Bot. 78, 337–448.
Ormsby, T., Alvi, J., 1999. Extending Arcview GIS. Environmental Systems Research Institute, Redlands, CA.
Ortiz, R., Rogel, J.A., Alcaraz, F., 1995. Soil-vegetation in 2 coastal salt marshes in southeastern Spain. Arid Soil Res. Rehab. 9, 481–493.
Phillips, D.L., Lee, E.H., Herstrom, A.A., Hogsett, W.E., Tingley, D.T., 1997. Use of auxiliary data for spatial interpolation of ozone exposure on southeastern forests. Environmetrics 8, 43–61.
Rives, J.L., 1980. Soil Survey of Pecos County, Texas. United States Department of Agriculture Soil Conservation Service, Washington, DC.
Robinson, T.P., Metternicht, G., 2005. Comparing the performance of techniques to improve the quality of yield maps. Agr. Syst. 85, 19–41.
Rogel, J.A., Silla, R.O., Ariza, F.A., 2001. Edaphic characterization and soil ionic composition influencing plant zonation in a semiarid Mediterranean salt marsh. Geoderma 99, 81–98.
Rowell, D., 1994. Soil Science: Methods and Applications. Longman Group UK Limited, Essex, England.
SAS Institute, 1999. The Analyst Application, first ed. SAS Institute Inc., Cary, NC.
SAS Institute., 1995. Logistic Regression Examples Using the SAS System, Version 6, first ed. SAS Institute Inc., Cary, NC.
Schmidt, R.H.J., 1986. Chihuahuan Climate. In: Barlow, J.C., Powell, A.M., Timmermann, B.N. (Eds.), Invited papers from the Second Symposium on the Resources of the Chihuahuan Desert Region. Chihuahuan Desert Research Institute, Fort Davis, TX, pp. 40–63.

Schoeneberger, P.J., Wysocki, D.A., Benham, E.C., Broderson, W.D., 1998. Field book for describing and sampling soils. Natural Resources Conservation Services, USDA, National Soil Survey Center, Lincoln, NE.

Shi, Z., Li, Y., Wang, R., Makeshine, F., 2005. Assessment of temporal and spatial variability of soil salinity in a coastal saline field. Environ. Geol. 48, 171–178.

Small, T.A., Ozuna, G.B., 1993. Ground-water conditions in Pecos County, Texas. USGS Water-Resources Investigations, Report 92-4190.

T.A.E.S., 1983. Soil Analysis Texas Agriculture Extension Service. College Station, TX.

T.P.W.D., 2003. Wetland Ecology. Texas Parks and Wildlife Department, Austin, TX.

Toft, C., Elliot-Fisk, D., 2002. Patterns of vegetation along a spatiotemporal gradient on shoreline strands of a basin desert lake. Plant Ecol. 158, 21–39.

Trimble, 1999. GeoExplorer 3, Operations Manual. Trimble Navigation Limited, Sunnyvale, CA.

Trimble, 2001. GPS Mapping for GIS with GeoExplorer 3, Classroom Manual. Trimble Navigation Limited, Sunnyvale, CA.

Ungar, I.A., 1974. Inland halophytes of the United States. In: Reinhold, R.J., Queen, W.H. (Eds.), Ecology of Halophytes. Academic Press, New York, NY, pp. 235–306.

USGS., 2002. U.S. Geological Survey. Ground Water Atlas of the United States. Oklahoma, TX. HA 730-E.

Van Auken, O.W., Bush, J.K., 1993. Annual Report. The Nature Conservancy of Texas. San Antonio, TX.

Van Auken, O.W., Bush, J.K., 1995. Annual Report. The Nature Conservancy of Texas. San Antonio, TX.

Van Auken, O.W., Bush, J.K., 1998. Spatial relationships of *Helianthus paradoxus* (Compositae) and associated salt marsh plants. Southwest. Nat. 43, 313–320.

Van Auken, O.W., Grunstra, M.B., Brown, S.C., 2007. Composition and structure of a west Texas salt marsh. Madroño. (In press).

Veni, G., 1991. Delineation and preliminary hydrogeologic investigations of the Diamond Y Spring, Pecos County, Texas Report to the Nature Conservancy of Texas.

Warrick, A.W., Zhang, R., El-Harris, M.K., Myers, D.E., 1988. Direct comparisons between kriging and other interpolators. In: Wierenga, P.J., Bachelet, D. (Eds.), Validation of Flow and Transport Models for the Unsaturated Zone. Proceedings. New Mexico State University, Las Cruces, NM.

Weisz, R., Fleischer, S., Smilowitz, Z., 1995. Map generation in high-value horticultural integrated pest management: Appropriate interpolation methods for site-specific pest management of Colorado potato beetle (Coleoptera: Chrysomelidae). J. Econ. Entomol. 88, 1650–1657.

Westerman, R.L., 1990. Soil Testing and Plant Analysis, third ed.. Soil Science Society of America, Inc., Madison, WI.

Wollenhaupt, N.C., Wolkowski, R.P., Clayton, M.K., 1994. Mapping soil test phosphorus and potassium for variable-rate fertilizer application. J. Prod. Agric. 7, 441–448.

Zimmerman, D., Pavlik, C., Ruggles, A., Armstrong, M.P., 1999. An experimental comparison of ordinary and universal kriging and inverse distance weighting. Math. Geol. 31, 375–390.

Section V: Environmental Biogeochemistry—Concepts and Case Studies

Developments in Environmental Science, Volume 5
D. Sarkar, R. Datta and R. Hannigan (Editors)

ISSN: 1474-8177/DOI:10.1016/S1474-8177(07)05020-6

Chapter 20

Understanding spatial variability and its application to biogeochemistry analysis

Sabine Grunwald, Rosanna L. Rivero and K. Ramesh Reddy

Abstract

Transforming the conceptual ideas of biogeochemical cycling into spatially explicit context has been hampered by ecosystem complexity, multiple nested levels of interrelated physical, biological and chemical processes, and the lack of sufficient quantitative data. The mosaic of ecosystem structures and functions distributed across a landscape represent the combined effects and interactions of a variety of biotic and abiotic factors. As a result, existing landscape patterns implicitly contain information about the processes that generated these patterns. Deductive science has generated extensive knowledge of how individual parts of aquatic ecosystem's function derived from site-specific studies. Yet, understanding of how the parts interact as a whole requires a holistic perspective that considers the spatial variability, distribution and interaction of all components of a landscape. The predicament entailed by the complexity of aquatic ecosystems requires a synergistic approach that integrates knowledge from different disciplines including biogeochemistry, geography, statistics/geostatistics, ecology, hydrology and others. Interdisciplinary collaboration will be key to reconcile deductive and inductive science and allow us to understand the linkages between biogeochemical properties and processes at landscape scale. In this chapter we provide an overview of a variety of geostatistical and hybrid methods that can be used to characterize the spatial variability, distribution and uncertainty of biogeochemical properties. A case study demonstrates the application of these methods to predict soil total phosphorus across a wetland in the Greater Everglades.

20.1. Introduction

Conceptually, scale represents the *window of perception*, the filter, or the measuring tool through which a landscape may be viewed or perceived (Levin, 1992). Thus, changing the scale changes the properties that we observe—the patterns of reality, which has implications for understanding the dynamics of any environmental system. Since parameters and processes important at one scale are frequently not important or predictive at another, it is essential to gain a better understanding of the spatial variability and distribution of ecosystem properties. Turner et al. (1989) pointed out that ecological problems often require the upscaling of fine-scale measurements for the analysis of coarse-scale phenomena. A holistic view of landscapes is required to understand ecosystem processes at micro, meso and macro spatial scales.

Biogeochemistry is defined as the scientific study of the interactions among the biological, geological and chemical systems of Earth, including the cycling of matter and energy through them. Numerous conceptual models have been developed that describe the cycling of matter and material in aquatic ecosystems (Mitsch and Gosselink, 2000). The carbon (C), nitrogen (N), phosphorus (P) cycles play a critical role in the functioning of wetland ecosystems, their structure, resilience and sensitivity to natural and human-induced disturbances.

In this chapter we discuss the importance of spatially explicit mapping of biogeochemical properties that applies to aquatic and terrestrial ecosystems in the context of spatial autocorrelation and covariation of these properties. Understanding the spatial variability and distribution of biogeochemical properties is a necessity for holistic assessment of environmental quality at landscape-scale.

20.2. Concepts and methods

The mosaic of system structures and functions distributed across a landscape represent the combined effects and interactions of a variety of biotic and abiotic factors. As a result, existing landscape patterns implicitly contain information about the processes that generated these patterns (Holling and Gunderson, 2002). Spatial and temporal ecosystem attributes are neither uniform nor scale independent. Likewise, biogeochemical properties and processes vary gradually and continuously in space and time. Discrete boundaries that cause abrupt shifts from high to low values of physico-chemical and biological properties are limited to crisp physical boundaries (e.g., hydrologic boundaries, land use) or anthropogenic disturbances. To restrict observations to few sites (e.g., along a transect) limits the ability to capture the biogeochemical signatures of a landscape.

Some biogeochemical properties are labor-intensive and costly to derive (e.g., phosphorus fractionation schemes, alkaline phosphatase activity) whereas other measurements are standard routine (e.g., bulk density, total phosphorus—TP). Hence, in the past, numerous studies focused on sparse spatially distributed sampling or experimental lab/field studies to understand biogeochemical cycling (Reddy et al., 1998; White and Reddy, 2000; Fisher and Reddy, 2001; Craft and Chiang, 2002; Karathanasis et al., 2003; Morris et al., 2004) resulting in a tremendous amount of knowledge. The strength of such site-specific studies is to describe feedback processes in response to a localized change in biogeochemical conditions. For example, changing plant communities provide feedback on soil biogeochemical properties and microbial communities that respond to changing environmental conditions (e.g., organic matter, hydroperiod, temperature, light, etc.). Numerous site-specific studies have ignored the *spatial* interrelationships, i.e., the *spatial covariation*, among different biogeochemical properties and assume independence of observations. Deductive science has generated extensive knowledge of how individual parts of aquatic ecosystems function (compare Eq. (1)). Yet, understanding of how the parts interact as a whole requires a holistic perspective that considers the spatial interaction of all components of a landscape. Transforming the conceptual ideas of biogeochemical properties and processes into spatially explicit context has been hampered by ecosystem complexity, multiple nested levels of interrelated physical, biological and chemical processes, and the lack of sufficient quantitative data.

$$y(x_i) = f\{z_j(x_i)\} \tag{1}$$

where $y(x_i)$ is the response variable observed at location x_i and $z_j(x_i)$ the biogeochemical properties ($j = 1, 2, 3, \ldots, n$) observed at location x_i.

20.2.1. The concept of spatial autocorrelation

The deductive approach is rooted in the concept developed by Fisher (1925), who pointed out the importance of random selection ensuring that estimates are unbiased. This assumption of independence is a prerequisite for many statistical tests including regression analysis, analysis of variance, *t*-tests and others (Berthouex and Brown, 2002). Classical statistical methods have been used extensively to document significant differences in biogeochemical properties/behavior using controlled experiments (e.g., mesocosms), plots, or blocks. Such analyses are based on the assumption that the investigated properties show no spatial autocorrelation. *Spatial autocorrelation* is a term referring to the degree of relationship that exists between two or more spatial variables, such that when one changes, the other(s) also change (Webster and Oliver, 2001). This change can either be in the same direction, which is a positive autocorrelation, or in the opposite direction, which is a negative autocorrelation (Burrough, 1986; Isaaks and Srivastava, 1989). Popular measures of spatial autocorrelation are Moran's I (Moran, 1950) and Geary's C (Geary, 1954) coefficients. Positive values of Moran I and value smaller than 1 for Geary's C coefficient correspond to positive autocorrelation. True spatial independence of properties in aquatic and terrestrial landscapes does not really exist due to the interconnectedness of biogeochemical processes of N, C and P and other nutrients and transport processes that move water, material and energy through the ecosystem. Rather, the sampling density and spacing between observations determine if the underlying spatial autocorrelation can be captured by a model or not. Studies that limit observations to sparse sampling locations far apart from each other ignore spatial autocorrelation and focus on the identification of significant differences between sites with contrasting biogeochemical conditions. However, numerous studies documented that spatial autocorrelation of biogeochemical soil properties is inherent at landscape scale [$>1\,km^2$] (Newman et al., 1997; DeBusk et al., 2001; Grunwald et al., 2004, 2006; Bruland and Richardson, 2005). To identify spatial autocorrelation of a specific biogeochemical property, observations need to be collected with a spatially distributed design throughout an ecosystem that accounts for short-, medium- and long-range variability. Constraining biogeochemical observations to few point locations (sites) cannot possibly explain the underlying spatial autocorrelation of properties as well as interrelationships between different biogeochemical properties (e.g., TP and labile P), which is called the spatial covariation. The predicament entailed by the complexity of aquatic and terrestrial ecosystems requires a synergistic approach integrating knowledge from different disciplines including biogeochemistry, geography, statistics/geostatistics, ecology, hydrology and others. Interdisciplinary collaboration will be key to reconcile deductive and inductive science to understand the linkages between properties and processes at landscape scale. Each soil biogeochemical property exhibits specific behavior of spatial autocorrelation that is related to environmental factors such as nutrient cycling, hydrology, topography, anthropogenic-induced nutrient inputs, vegetation types, etc. and spatial scale. For example, TP showed long-range correlation lengths across the Water Conservation Area-2A (WCA-2A), a wetland in the northern Everglades, in 1990 and 1998 with spatial autocorrelations of 6500 and

7549 m, respectively (Grunwald et al., 2004). Similar long-range spatial autocorrelation for TP was found in the Blue Cypress Marsh Conservation Area, a wetland in eastern Florida, with 7240 m (Grunwald et al., 2006).

20.2.2. The concept of regionalized variable theory

We can formalize the concept of spatial variation outlined above using the theory of regionalized variables (Webster and Oliver, 2001). Since the environment and its component attributes result from many interactive physical, chemical and biological processes that are non-linear and/or chaotic, the outcome is so complex that the variation appears to be random. If we adopt a stochastic view then at each point in geographic space there is not just one value for an attribute but a whole set of values. Consider a region R that comprises an infinite number of points x_i, $i = 1, 2, \ldots, \infty$. Whereas in the classical statistical approach the values of an observed biogeochemical attribute, z, at these points constitute the population, in the geostatistical approach this population is assumed to be just one realization of a random process or random function that could generate any number of such populations. Then, at each place x_0 the attribute is considered a random variable $Z(x_0)$. Thus, at a location x_0 a biogeochemical attribute is treated as a random variable with mean (μ), a variance ($\sigma 2$) and a cumulative distribution function (cdf). Note, that the actual value $z(x_0)$ is just one drawn at random from that distribution. The set of random variables, $Z(x_1)$, $Z(x_2)$, $\ldots Z(x_n)$ constitute a random function or a stochastic process (Webster and Oliver, 2001). Regionalized variable theory assumes that the spatial variation of any biogeochemical variable can be expressed as the sum of three major components (Eq. (2)) (Burrough and McDonnell, 1998): (i) a structural component having a constant mean or trend that is spatially dependent, (ii) a random, but spatially correlated component, known as the variation of the regionalized variable and (iii) a spatially uncorrelated random noise or residual term. A non-stationary trend extends over the whole study area and is therefore called a global model. In contrast, local trends, also called drift, describe localized variation (Webster and Oliver, 2001). [Note: a header "$\hat{Z}$" indicates that the variable is estimated]

$$\hat{Z}(x_i) = m(x_i) + \varepsilon'(x_i) + \varepsilon'' \qquad (2)$$

where $\hat{Z}(x_i)$ is the value of a random variable at locations x_i with $i = 1, 2, \ldots, n$; $m(x_i)$ the trend model—deterministic function describing the "structural" component of $\hat{Z}(x_i)$; $\varepsilon'(x_i)$ the stochastic, locally varying but spatially dependent component (the regionalized variable); ε'' the A residual, spatially independent noise term having zero mean and variance; and x_i the geographic position in 1, 2 or 3 dimensions.

Observations obtained close to each other are more likely to be similar than observations taken further apart from each other. This spatial correlation of $\varepsilon'(x_i)$ is described by the semivariance $\hat{\gamma}(h)$ (Eq. (3)). If γ is plotted as a function of the lag distance h that separates x_i and x_i+h, the semivariogram is obtained. One implicit assumption is that the mean, variance and covariance depend only on the separation distance h and not on the absolute position. This assumption constitutes second-order stationarity (Isaaks and Srivastava, 1989). $\hat{\gamma}(h)$ is estimated as

$$\hat{\gamma}(h) = \frac{1}{2N(h)} \sum_{i=1}^{N(h)} [z_u(x_i) - z_u(x_i + h)]^2 \tag{3}$$

where $\hat{\gamma}(h)$ is the semivariance at lag h; h the distance between data pairs $(z_u(x_i) - z_u(x_i+h))$ (or lag); N the total number of data pairs $(z_u(x_i) - z_u(x_i+h))$; and $z_u(x_i)$ the biogeochemical variable u.

Commonly, least square fitting of the experimental semivariogram model is used. However, visual inspection of the statistical fitting process is essential. Alternatively, interactive fitting can be used to model the semivariogram. Characteristic parameters that can be derived from semivariograms are the *nugget*, *sill* and *range*. The nugget describes the measurement error and fine-scale variability. Semivariograms of properties or processes that exhibit second-order stationarity reach upper bounds at which they remain after their initial increases. This upper bound, or maximum is known as the sill. The range describes the spatial autocorrelation and marks the limit of spatial dependence (Webster and Oliver, 2001). The univariate semivariance can be extended to consider two biogeochemical variables, z_u and z_v, to derive the cross-semivariances $\hat{\gamma}_{uv}(h)$.

$$\hat{\gamma}_{uv}(h) = \frac{1}{2N(h)} \sum_{i=1}^{N(h)} [z_u(x_i) - z_u(x_i + h)][z_v(x_i) - z_v(x_i + h)] \tag{4}$$

Regionalized variable theory and semivariogram analyses are described in detail by Goovaerts (1997), Chilès and Delfiner (1999), Webster and Oliver (2001, 2005). The semivariogram provides input for kriging, which is a weighted interpolation technique to create continuous prediction maps of biogeochemical properties. Major limitations of the univariate geostatistical technique of kriging are due to the assumption of stationarity, which is often not met by the field-sampled datasets and the large amount of data (>100 observations; recommended >150 observations) required to characterize the spatial autocorrelation of a property (Webster and Oliver, 2001).

20.2.3. Global models

Global models use all available observations to provide predictions for the whole area of interest, while local interpolators operate within a small zone around the point being interpolated to ensure that estimates are made only with data from locations in the immediate neighborhood (Burrough and McDonnell, 1998). Trend surfaces are the simplest global, geospatial model that requires fitting some form of polynomial equation through biogeochemical attribute values. These are least square methods that model the long-range spatial variation. Such models assume that the spatial coordinates are the independent variables and z (attribute of interest) is the dependent variable (Webster, 1994). As trend surfaces are simplified representations of reality, it is difficult to ascribe any physical meaning to complex, higher-order polynomials. Therefore, the main use of trend surface analysis is not as an interpolator, but as a way of removing broad features of the data prior to using some complex local interpolator. The concept is based on partitioning the variance between trend and the residuals from the trend. Compare Eq. (2) that partitions the variability of a property into (i) a global trend component, (ii) stochastic, locally varying but spatially dependent component and (iii) a residual noise term.

20.2.4. Local models

Local, deterministic interpolation methods focus on modeling short-range local variations. The interpolation involves: (i) defining a search neighborhood around the point to be interpolated, (ii) finding the observations within this neighborhood, (iii) choosing a mathematical function to represent the variation over this limited number of points or area, and (iv) evaluating it for the point on a regular grid. The procedure is repeated until all the points on a grid have been computed (Burrough and McDonnell, 1998). For example, commonly used local interpolation methods are inverse distance weighting (IDW), splines and kriging (Burrough and McDonnell, 1998). Splines (local fitting functions) estimate values using a mathematical function that minimizes overall surface curvature, resulting in a smooth surface that passes exactly through the observation points, while at the same time ensuring that the joins between one part of the curve (or surface) and another are continuous. In contrast to trend surfaces and weighted averages, splines retain small-scale features. A disadvantage of splines and IDW is that there are no direct estimates of the errors associated with these forms of interpolation (Burrough and McDonnell, 1998). Local interpolators in its simplest form are based weights that are computed from a linear function of distance between sets of data points and the point to be predicted (Eq. (5)).

$$\hat{Z}(x_0) = \sum_{i=1}^{n} \lambda_i * z(x_i) \quad \sum_{i=1}^{n} \lambda_i = 1 \tag{5}$$

where $\hat{Z}(x_0)$is the predicted attribute value at unsampled location x_0; λ_i the weights; $z(x_i)$ the observed attribute value at locations x_i; and n the number of observations.

In Eq. (5) $z(x_1)$, $z(x_2)$, … $z(x_n)$ are the measured values of the biogeochemical property z at locations x_1, x_2, …, x_n and λ_i are the weights that sum to 1 to assure unbiasedness. The expected error is $E[\hat{Z}(x_0) - Z(x_0)] = 0$ and the prediction variance is

$$\begin{aligned} \text{var}[\hat{Z}(x_0)] &= E[\{\hat{Z}(x_0) - Z(x_0)\}^2] \\ &= 2\sum_{i=1}^{N} \lambda_i \gamma(x_i, x_0) - \sum_{i=1}^{N}\sum_{j=1}^{N} \lambda_i \lambda_j \gamma(x_i; x_j) \end{aligned} \tag{6}$$

where $\gamma(x_i; x_0)$ is the semivariance of Z between the ith sampling point and the target point x_0 at an unsampled location and $\gamma(x_i; x_j)$ the semivariance of Z between sampling points x_i and x_j.

20.2.5. Kriging

Kriging is a generic term adopted by geostatisticians for a family of generalized least-squares regression algorithms. In the kriging system the goal is to minimize the kriging variance, where the weights sum to 0 (Eq. (7)).

$$\sum_{i=1}^{n} \lambda_i(x_i; x_j) - \psi(x_0) \text{ with } \sum_{i=1}^{n} \lambda_i = 0 \tag{7}$$

where $\psi(x_0)$ is the Lagrange multiplier.

All kriging estimators are but variants of the basic linear regression estimator $\hat{Z}(x_0)$ defined as

$$\hat{Z}(x_0) - m(x_0) = \sum_{i=1}^{n} \lambda_i(x_0)[Z(x_i) - m(x_i)] \tag{8}$$

where $\hat{Z}(x_0)$ is the linear regression estimator at unsampled location x_0 and $\lambda_i(x_0)$ the weights assigned to datum x_0 interpreted as a realization of the random variable $Z(x_i)$ and located within a given neighborhood $W(x)$ centered on x.

The weights are chosen to minimize the error variance $\sigma 2(x_0) = \text{var}\{\hat{Z}(x_0) - Z(x_0)\}$ under the constraint of unbiasedness of the estimator. These weights are obtained by solving simultaneously a system of linear equations which is known as the kriging system (Goovaerts, 1999).

The local neighborhood characteristics including (i) the number of observations ($z(x_i)$) within a neighborhood, (ii) major and minor semiaxis (size of the neighborhood) and (iii) the neighborhood shape (uniform or discretized into subunits) determine the strengths of smoothing. In general, a small neighborhood generates localized predictions based on few observations, which are prone to the influence of extreme values producing blocky/crisp looking maps. In contrast, large neighborhoods smooth over a larger area providing conservative predictions for areas with high observed values.

20.2.6. Kriging variants

Differences in kriging variants reside in the model considered for the trend $m(x)$ (Goovaerts, 1999):

(a) *Simple kriging* (*SK*) considers the mean $m(x)$ known and constant throughout the study area.
(b) *Ordinary kriging* (*OK*) accounts for local fluctuations of the mean by limiting the domain of stationarity of the mean to the local neighborhood ($W(x)$). The mean is considered constant but unknown.
(c) *Universal kriging* (*UK*), also known as kriging with a trend model, considers that the unknown local mean $m(x')$ smoothly varies within each local neighborhood, and the trend is modeled as a linear combination of functions $f_k(x)$ of the coordinates

$$m(x') = \sum_{k=0}^{K} a_k(x') f_k(x') \tag{9}$$

with $a_k(x')$ constant within each local neighborhood $W(x)$ but unknown.
(d) *Regression kriging* (*RK*): Alternatively, the trend function $m(x)$ can be modeled separately, where kriging is combined with regression or a regression variant (Odeh et al., 1994, 1995; Odeh and McBratney, 2000; Hengl et al., 2004). The deterministic component $m(x)$ is considered dependent on some exogenous factors such as climate, hydrology, topography, vegetation or other environmental factors that can be described via multivariate regression, Generalized Linear

Models, Classification and Regression Trees, Generalized Additive Models or other functions. Odeh et al. (1994, 1995) defines regression kriging where model *f*(.) is used to describe the relationship between predictors and environmental factors:

$$\hat{Z}(x_i) = f(Q, x_i) + \hat{R}(x_i) \tag{10}$$

where $f(Q, x_i)$ is the function describing the structural component of $\hat{Z}(x_i)$as a function of Q environmental variables at location x_i; $\hat{R}(x_i)$ the stochastic, locally varying but spatially dependent residual from $f(Q, x_i)$. In RK, the target biogeochemical property at an unvisited site is first predicted by *f*(.), followed by kriging of the residuals of the model.

(e) *Kriging with an external drift* (*KED*) models the trend $m(x)$ as a linear function of a smoothly varying secondary (external) variable $y(x)$ instead of a function of the spatial coordinates. Besides the difficult inference of the residual semivariogram, this method requires that the relation between the primary trend and secondary variable is linear and makes physical sense (Goovaerts, 1999). In KED the secondary exhaustive data are only used to inform on the shape of the trend of the primary variable. In contrast, cokriging exploits more fully the secondary information by directly incorporating the values of the secondary variable and measuring the degree of spatial association with the primary variable through the cross-semivariogram. Cokriging is the extension of ordinary kriging of a single variable to two or more variables. There must be some kind of coregionalization among the variables for it to be profitable. It is particularly useful if some property that is cheap to measure (e.g., remote sensing image) is spatially correlated with another property that is expensive to measure and/or labor intensive to collect and available at few sites (Webster and Oliver, 2001).

In many environmental applications a few hot spots (high values) coexist with many small values that vary continuously in space. Depending on whether large values are clustered or scattered in space, our physical interpretation of the biogeochemical processes controlling high and low values may change. The characterization of the spatial distribution of z-values above or below a given threshold value z_k requires prior coding of each observation as an indicator datum *ind* $(x_i;\ z_k)$ defined as (Goovaerts, 1999):

$$\begin{aligned} ind(x_i; z_k) &= 1 \quad \text{if } z(x_i) \leq z_k \\ &= 0 \quad \text{otherwise} \end{aligned} \tag{11}$$

Indicator semivariograms can be computed by substituting indicator data *ind* $(x_i;\ z_k)$ for z-data:

$$\hat{\gamma}(h, z_k) = \frac{1}{2N(h)} \sum_{i=1}^{N(h)} [ind(x_i; z_k) - ind(x_i + h; z_k)]^2 \qquad (12)$$

The indicator variogram value $\hat{\gamma}(h, z_k)$ measures how often two z-values separated by a vector h are on opposite sides of the threshold value z_k.

20.2.7. Spatial stochastic simulations

Least-square interpolation algorithms such as kriging tend to smooth out local details of the spatial variation of the attribute, with small values typically overestimated and large values underestimated (Isaaks and Srivastava, 1989). This is a serious shortcoming if large pollutant concentrations or biogeochemical property values are of interest.Kriging aims at local accuracy through minimizing a covariance-based error variance, while spatial stochastic simulation aims at reproducing spatial structure through a covariance model. Unlike kriging, spatial conditional stochastic simulations do not aim at minimizing a local error variance but focus on the reproduction of statistics such as the sample histogram or the semivariogram model in addition to honoring of data values. The output results, i.e., a set of alternative realizations, provide a visual and quantitative measure of the spatial uncertainty. A probability distribution (ccdf) for attributes at a particular location can be built for a set of multiple realizations of the joint distribution of attribute values in space (Goovaerts, 1997). Kriging produces one output map. In contrast, stochastic simulation generates multiple realizations of the spatial distribution of (biogeochemical) attribute values and it uses differences among simulated maps as a measure of uncertainty. Therefore, kriging is preferred for local estimation whereas simulation is increasingly preferred for assessment of spatial uncertainty and reproduction of global statistics, risk assessment, flow modeling and water quality simulation modeling (Goovaerts, 1997; Grunwald et al., 2004; Chilès and Allard, 2005). The latter situations require knowledge about the uncertainty of environmental attribute values at many locations simultaneously (multiple-point or spatial uncertainty).

Conditional Sequential Gaussian Simulation (CSGS) is a stochastic simulation method for the generation of partial realizations using normal random functions. The method uses the Gaussian model type and is ergodic, which means that simulations have a sample mean close to the theoretical mean and a sample covariance to the theoretical covariance

$C(h)$. This implies that all simulations are drawn from the realizations of a random function that is ergodic in the mean value and the covariance (second-order ergodicity) (Chilès and Delfiner, 1999). Conditioning is the operation that ensures that simulation values match values at sample points. Conditional Sequential Gaussian Simulation provides a measure of local uncertainty because each conditional cdf relates to a single spatial location x.

20.2.8. Summary

A flow chart that provides an overview of statistical, geostatistical and hybrid geospatial modeling techniques presented above is shown in Fig. 20.1. Depending on the spatial autocorrelation and the spatial covariation, different methods are suggested to generate the best possible predictions of biogeochemical properties within a given aquatic ecosystem. Many variations of these methods exist that can be further explored in Goovaerts (1997), Webster and Oliver (2001, 2005), McBratney et al. (2000, 2003) and Grunwald (2005).

20.3. Case study

In this case study we illustrate the concepts and methods outlined above to characterize spatial variability of TP within a subtropical wetland, WCA-2A, located in the Greater Everglades, Florida. The Everglades is a naturally P-limited aquatic ecosystem that has been impacted by nutrient inputs from agricultural land use over the last decades (Noe et al., 2001). In addition, this system has been manipulated hydrologically (canals, levees, hydrologic control structures) altering the historic uninterrupted sheet flow through the whole system (Porter and Porter, 2002). Water Conservation Area-2A is 418 km^2 in size consisting primarily of Histosols (soils with at least 12–18% organic C by weight), which developed during the past 5000 years (McCollum et al., 1976). Vegetative communities are dominated by sawgrass (*Cladium jamaciense Crantz*), cattail (*Typha spp.*), mixed sawgrass and cattail communities and few tree islands (Porter and Porter, 2002). The wetland is surrounded by canals and levees with two major surface water inflow points (S-7 and S-10 pump stations). Water moves through WCA-2A as sheet flow from the north-east to the outflow located south (DeBusk, 2001). Elevated soil P concentration in WCA-2A has been strongly linked to productivity and community structure of macrophytes (McCormick et al., 2002).

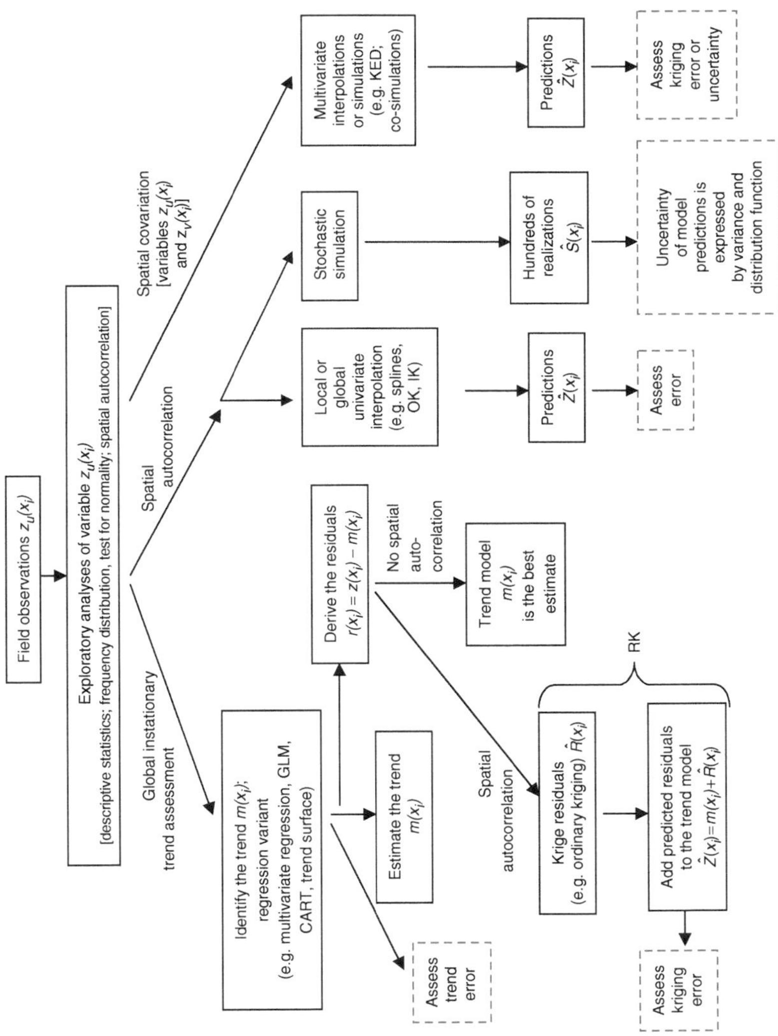

Figure 20.1. Decision flowchart for spatial modeling of biogeochemical soil properties.

In 2003, we collected soil samples at 0–10 cm depth at 111 sites based on a stratified random sampling design, spatially distributed throughout WCA-2A that were analyzed for TP. Strata were derived using historic ecological data layers such as the Normalized Difference Vegetation Index as a proxy for vegetative communities, as well as soil and hydrologic data. The sampling design was optimized to account for short-, medium- and long-range variability of attributes. A map that shows the distribution of soil observations is shown in Fig. 20.2. Total P was measured with a dry ashing procedure (Anderson, 1976) followed by determination with an automated colorimetric procedure (U.S. Environmental Protection Agency, 1993, Method 365.1). We complemented the soil dataset with an exhaustive ancillary dataset derived from Advanced Spaceborne Thermal Emission and Reflection Radiometer (ASTER) satellite image. The ASTER sensor covers a wide spectral range and high spatial and radiometric resolutions. The spectral region is covered by three telescopes, three Visible and Near Infrared Radiometer (VNIR) bands with a spatial resolution of 15 m, six Short Wave Infrared Radiometer (SWIR) bands with a spatial resolution of 30 m and five thermal Infrared Radiometer (TIR) bands with a spatial resolution of

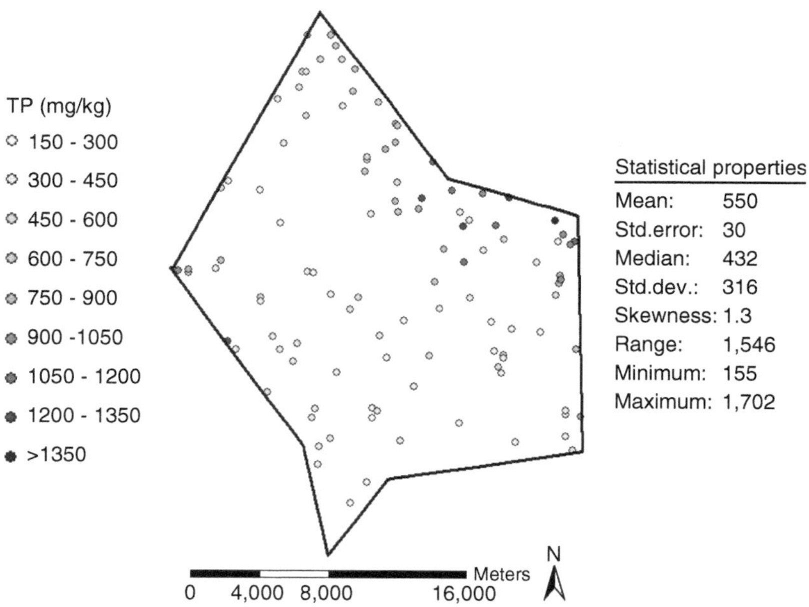

Figure 20.2. Observations of total phosphorus (TP) at 0–10 cm depth measured at 111 locations across WCA-2A.

90 m (Abrams et al., 2002). For the purpose of this study, we used ASTER bands 2 and 3 (red and near-infrared) from the VNIR spectral region in order to calculate the Normalized Difference Vegetation Index (NDVI) (Rouse et al., 1974), which is slightly different when compared to the NDVI derived from Landsat satellite images that use bands 3 and 4.

Descriptive statistics of TP are enclosed in Fig. 20.2 and indicate a positively skewed distribution with a skewness coefficient of 1.3, mean of 550, median of 432, minimum of 155 and maximum of 1702 mg kg^{-1}. To predict TP across the WCA-2A we used completely regularized splines with different neighborhood options (Fig. 20.3). The spline map in Fig. 20.3c shows the smoothing effect of using a large local neighborhood with 15 observations (minimum of 5) to predict TP at unsampled locations. In contrast, spline map 1 (Fig. 20.3a) shows blocky spatial patterns due to the small local neighborhood with few (6) observations (minimum of 2) that contributed to the predictions of TP at unsampled locations. The mean prediction error for was lowest for spline map 2 with 0.4688, followed by spline map 3 with −3.631 and spline map 1 with −4.598. The RMSE was similar for spline map 3 with 252 and spline map 2 with 253 and highest for spline map 1 with 293. Here, a heuristic approach was used to identify the interpolation method that showed the lowest prediction errors.

To explore the spatial autocorrelation of TP, h-scattergrams and semivariograms were generated (Fig. 20.4). Just as the scattergram is a plot of all pairs of values related to two different attributes measured at the same location, the h-scattergram is a plot of all pairs of measurements ($z(x_i)$, $z(x_i+h)$) on the same attribute z at locations separated by a given distance h in a particular direction. A perfect correlation would entail that all points in the h-scatter diagram lie on the line of equal values. As seen in Fig. 20.4, the spread of the points indicated the variability of TP values. The increasing inflation of the cloud in all h-scattergrams computed in four different directions reflected the increasing dissimilarity between observations farther apart. Overall, no distinct different patterns in the directional h-scatter diagram were found indicating isotropic spatial distributions. Isotropic means that the spatial patterns are the same/similar in different directions.

Because the TP data were non-Gaussian we transformed TP values with a log transformation to approximate a normal distribution. The omnidirectional experimental semivariogram of log TP is shown in Fig. 20.4b that was fitted with an exponential model with a nugget of 0.0185, sill 0.0512 and range of 7468 m. At the range the spatial autocorrelation becomes 0 indicating that this separation distance h marks the limit of spatial dependence. Observations taken further apart than the

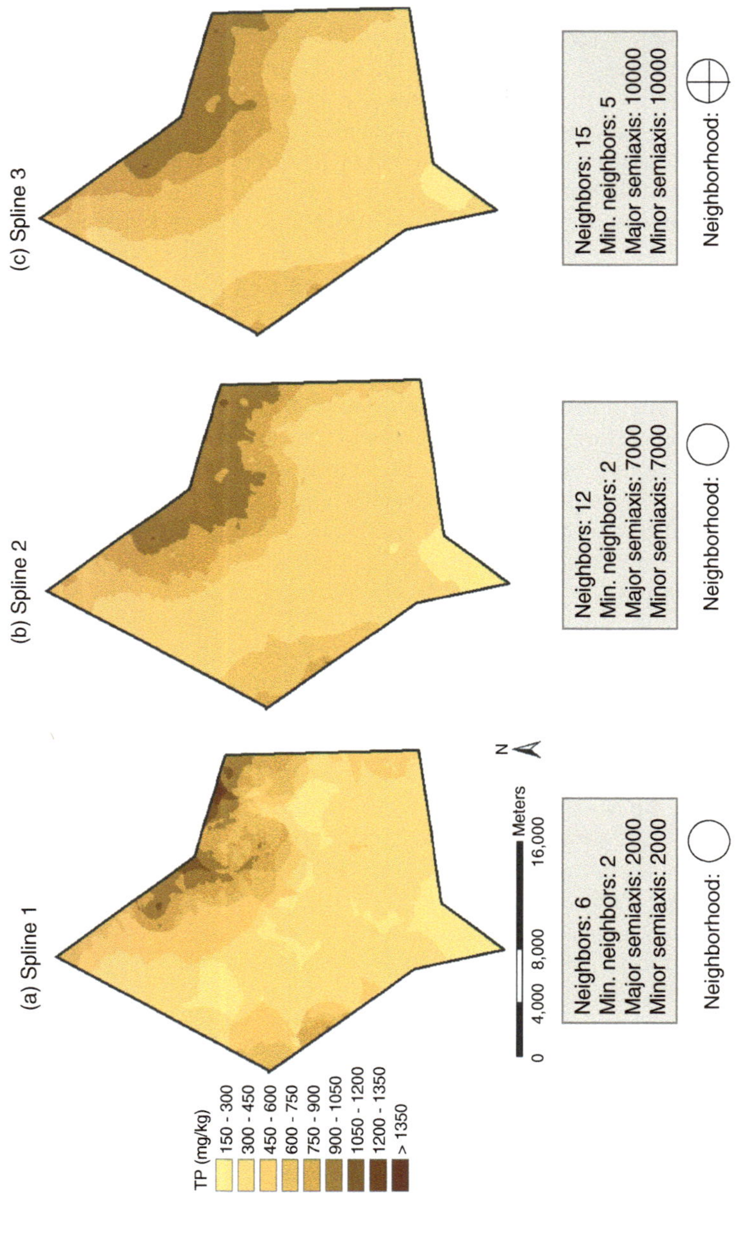

Figure 20.3. Predictions of TP derived from Completely Regularized Splines with different local neighborhood options.

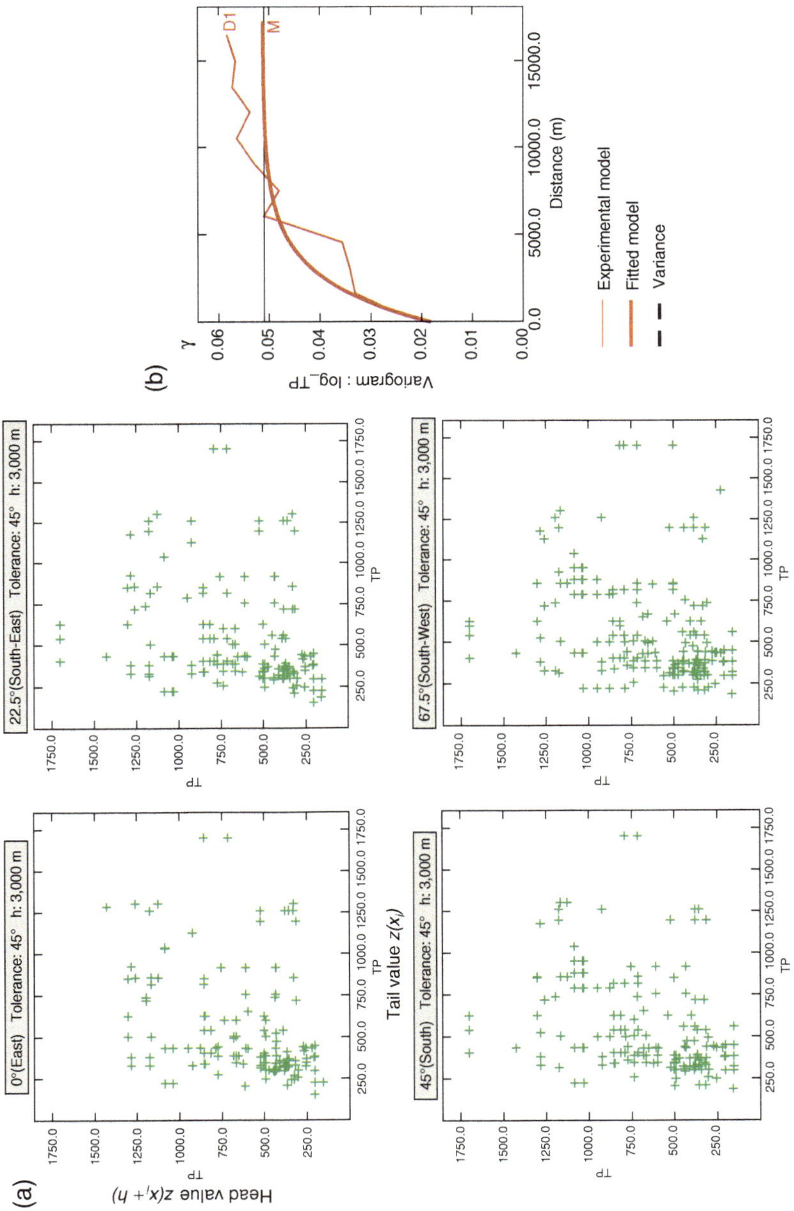

Figure 20.4. (a) h-scattergrams and (b) semivariogram of log TP.

range are spatially independent. Experimental semivariograms were computed using Eq. (3) in four directions (0, 45, 90 and 135°) to reveal anisotropy in the variation. Anisotropy describes directional spatial structures. Due to the limited amount of data pairs available for different directions it was difficult to identify any clear directional trends. Although there is a hydraulic gradient in WCA-2A that extends from the northeastern canal into the west-south direction we could not confirm these trends in the soil TP dataset. Therefore, we proceeded with the omnidirectional fitted semivariogram model. We used a search neighborhood of 7468 m; 1 angular sector; a minimum number of observations of 4 and an optimum number of observations of 10 to predict TP at unsampled locations using OK. The OK prediction map is shown in Fig. 20.5. Based on cross-validation the mean prediction error was −41.77 indicating a slight underestimation of true observations and a root mean square prediction error of 257 mg kg^{-1}. The coefficient of determination (R^2) between measured TP and estimated TP was 0.82 indicating good predictions. To highlight differences between TP

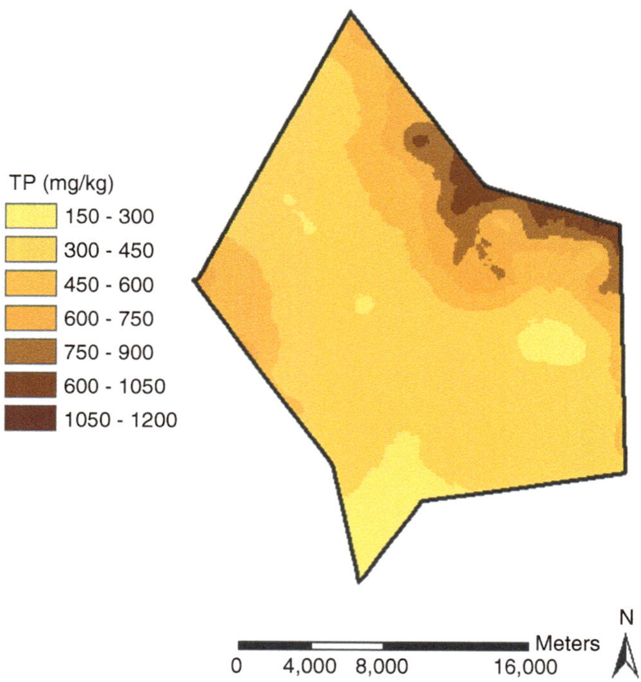

Figure 20.5. Predictions of TP based on Ordinary Kriging.

predictions derived from our spatially distributed observations and observations along a linear corridor (transect) we created a subset of observations extending along a previously identified TP gradient (Reddy et al., 1997; DeBusk et al., 2001). The corridor extended from the hydrologic boundary in the east into the interior of the marsh to the south-west boundary (Fig. 20.6a). Total phosphorus observations were predicted based on 10 observations along the linear corridor using OK. We derived a difference map by subtracting the TP predictions made along the linear corridor from the TP predictions derived from the spatially distributed design (Fig. 20.6b). The difference map shows deviations of more than −240 up to 180 mg kg^{-1} TP illustrating the large differences between both prediction maps. This example illustrates the limitations of sparse biogeochemical datasets based on only 10 observations along a corridor. Sparse datasets along corridors/transects have major limitations to map the underlying spatial variability. In the eastern part of WCA-2A, low TP values (328 and 380 mg kg^{-1}) coexisted with very high TP values (1259 mg kg^{-1}) in close proximity (Fig. 20.6b). In contrast, over large distances from the interior of the marsh to the south-west hydrologic boundary, TP values were relatively invariant at different lags ranging from 257 to 371 mg kg^{-1} TP.

Prediction maps are valuable to characterize the spatial distribution and variability of biogeochemical properties such as TP. However, to assess the impact of biogeochemical properties exceeding thresholds that stimulate net productivity and turnover rates is important to characterize the ecological integrity and structure of an ecosystem. Thus, we used IK to derive the probabilities of TP being above a respective cutoff value (z_k) of 450, 500, 550, 600, 650 and 700 mg kg^{-1} (Fig. 20.7). For reference purposes, historical background TP concentrations in WCA-2A have been estimated to be 500 mg kg^{-1} (DeBusk et al., 2001). Probability maps are well suited for decision makers demonstrating the outcome ranging from conservative assumptions (low threshold value) to liberal assumptions (high threshold value). Hot spots can be well identified using IK highlighting the extreme values that exceed the background concentrations.

Stochastic simulations emphasize the uncertainty associated with predictions of biogeochemical properties generating hundreds or thousands of realizations. We used CSGS to generate 100 realizations of TP (Fig. 20.8) using the following modeling options: dilation radius of 20 in the x and y direction; 15 maximum number of data nodes; and 5 maximum number of simulation nodes. The E-type map of the mean showed similar patterns of TP as seen on the OK prediction map. However, the TP patterns in Fig. 20.8 were much more speckled when compared to the

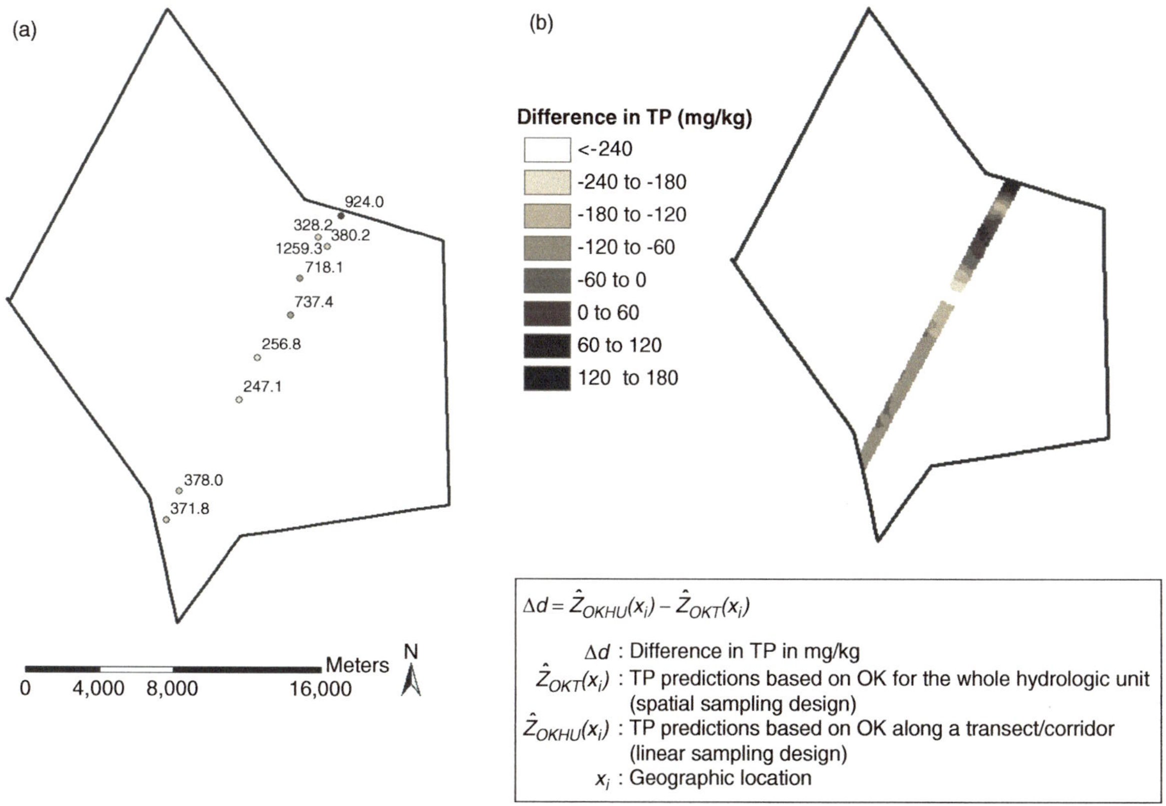

Figure 20.6. (a) Transect/corridor along TP gradient extending from the hydrologic boundary into the marsh interior. Map shows point TP observations in mg kg^{-1}. (b) Map shows the difference in TP predictions along the transect/corridor derived from a spatial and linear sampling design.

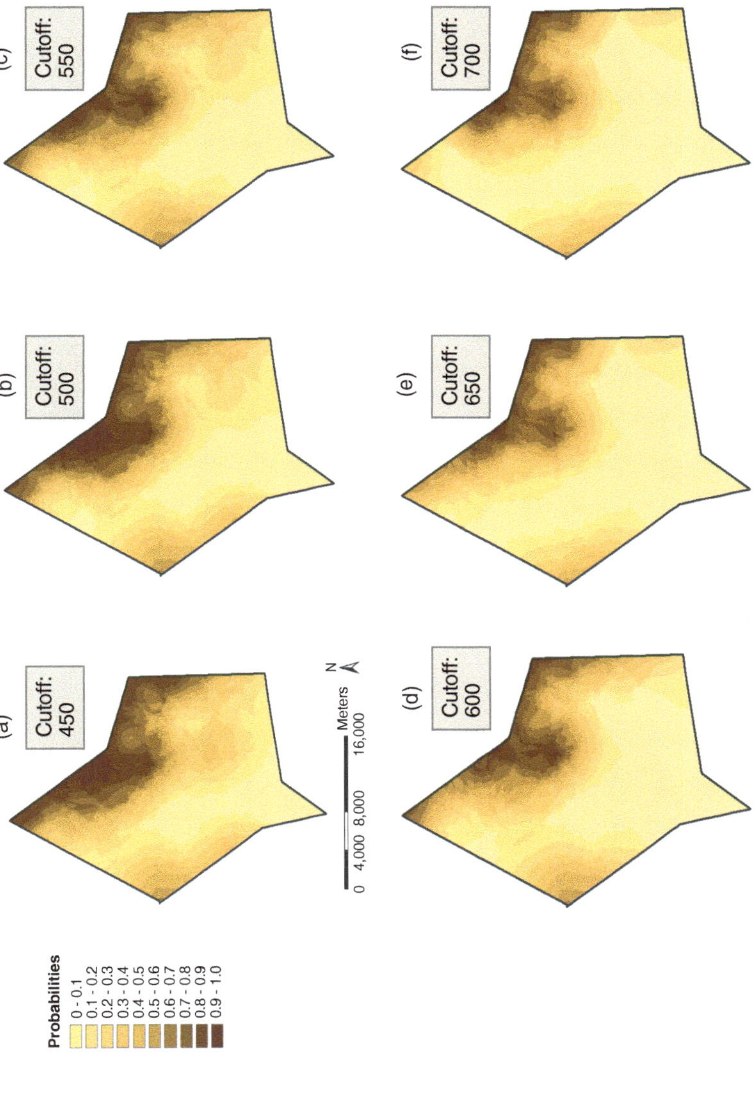

Figure 20.7. Probabilities of being above respective TP cutoff values derived from Indicator Kriging.

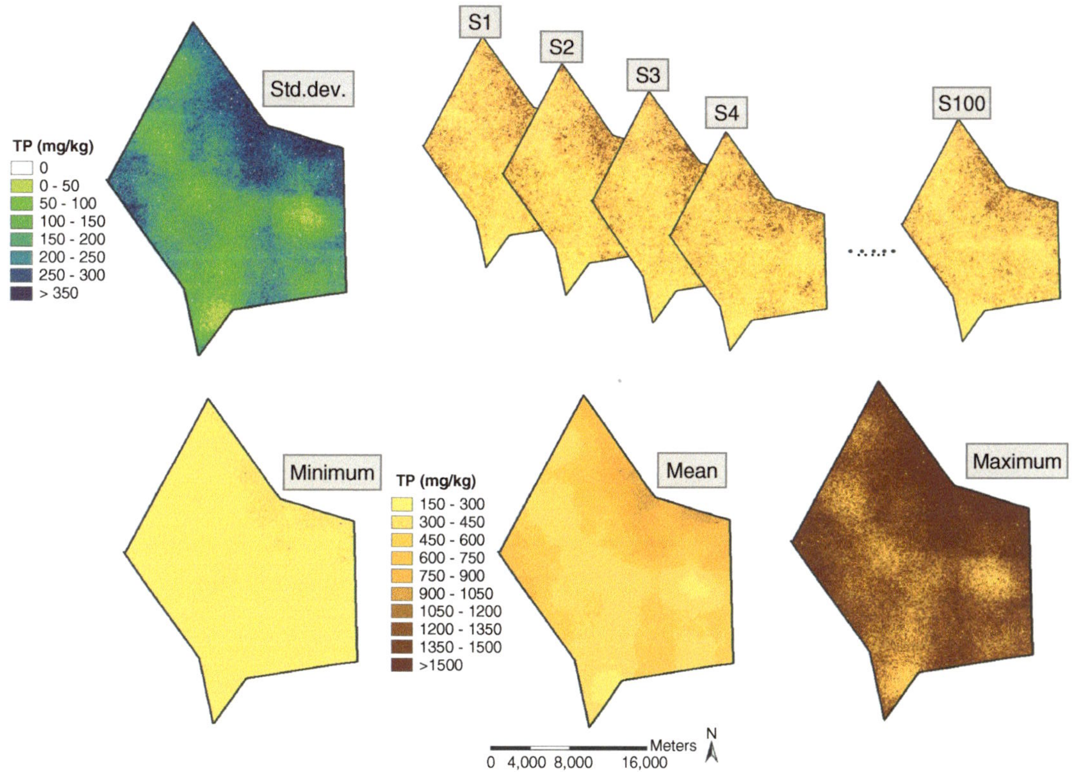

Figure 20.8. Minimum, mean, maximum and select realizations (S1, S2, S3, S4, S5, ..., S100) of TP generated with Conditional Sequential Gaussian Simulation. The uncertainty of predictions is expressed by the standard deviation (std. dev.)

smooth OK prediction map. The uncertainty of TP predictions based on CSGS was expressed using minimum and maximum realization maps that show the range of realizations and the standard deviation map. The largest uncertainty was found in the eastern part of WCA-2A coinciding with high TP predictions. The lowest uncertainty was found in the interior of the marsh extending south and eastwards mirroring the lowest observed and predicted TP.

To improve predictions of TP we analyzed the relationships between soil TP, spectral bands and the NDVI. The Spearman Correlation Coefficient (r) between TP and NDVI was high with 0.65. We used a stepwise multiple regression to quantify the relationship between soil and spectral data and derived the following algorithm:

$$\text{TP} = 401.78 + 1890.77 * \text{NDVI} \quad \text{with an } R^2 \text{ of } 0.37\ (p = 0.001)$$

This linear relationship was used as trend model to predict TP across WCA-2A using the exhaustive NDVI pixel dataset. We derived residuals by subtracting the trend model from the observations. The spatially autocorrelated residuals were then kriged and residual predictions added to the global trend model to generate the TP prediction map shown in Fig. 20.9. The RK procedure showed a R^2 of 0.93, which was higher than the one derived for OK (R^2 of 0.82). Overall, the TP prediction map derived from RK showed similar patterns as seen on the OK prediction map (Fig. 20.5) and the CSGS realization maps (Fig. 20.8). However, predictions in Fig. 20.9 highlighted landscape features such as vegetative patterns, slough/ridge systems, open water, tree islands, etc. mapped through the NDVI index that showed quantitative linkages to soil TP. Predictions derived from RK do not honor observed data values and have limitations to describe the uncertainty of predictions. This is the strength of stochastic simulations that are focused to characterize the uncertainty of predictions.

In Fig. 20.10 we contrast the different geospatial methods and TP observations in terms of their distribution functions. Predictions based on RK and realizations derived from CSGS showed the closest match to the distribution of TP observations. In particular, the high TP values above $1050\,\text{mg kg}^{-1}$ were well represented in the RK and CSGS maps.

20.4. Remarks

Considering the complexity of wetland ecosystems it is essential to employ spatially distributed sampling designs and geospatial prediction methods that have the ability to characterize the underlying spatial

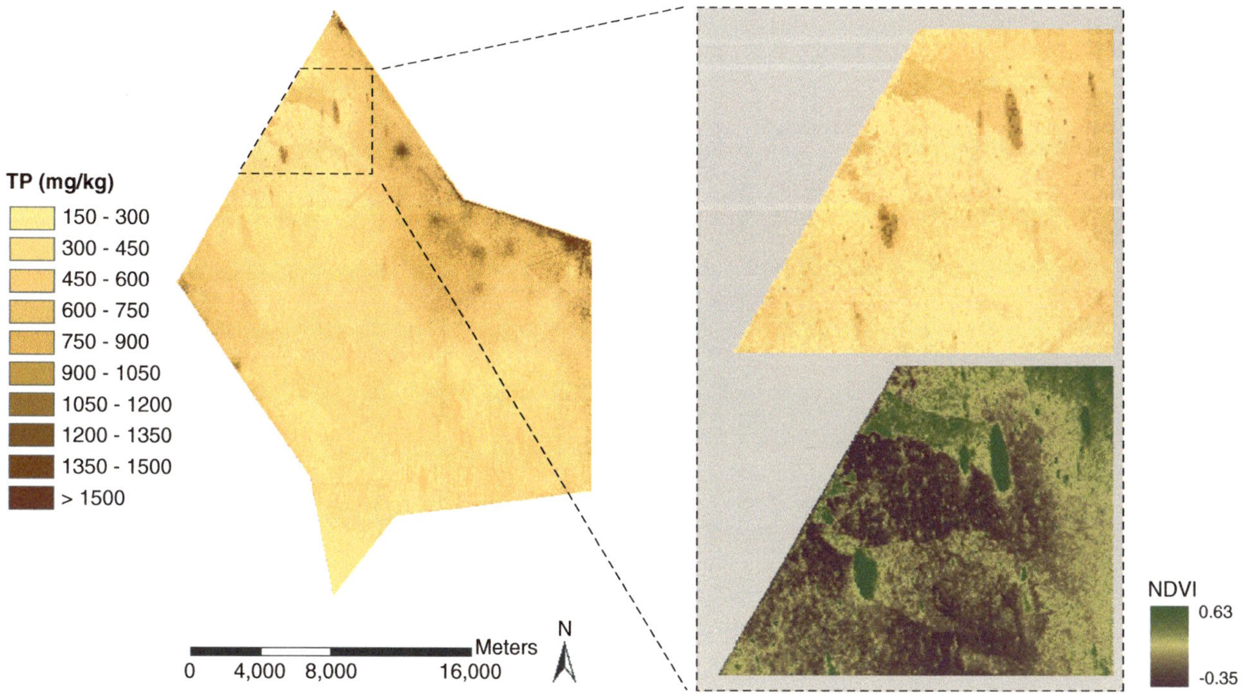

Figure 20.9. Left: Predictions of TP derived from Regression Kriging. Right: Close-up of TP predictions and NDVI map depicting landscape features such as tree islands and vegetation structures.

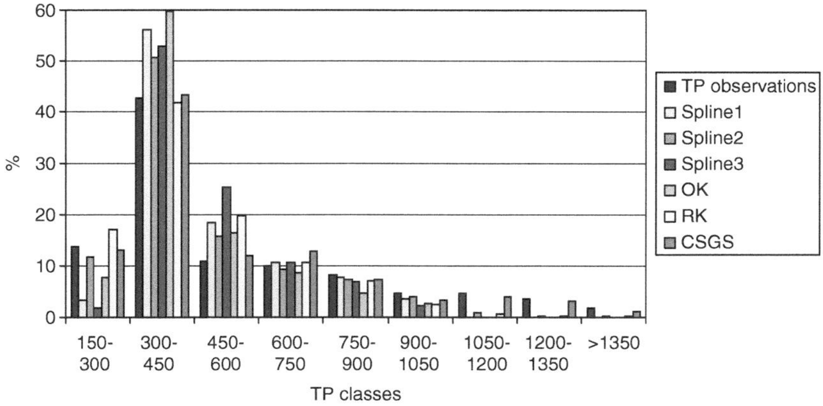

Figure 20.10. Distribution functions of TP observations, predictions and realizations generated with different interpolation methods.

distribution and variability of biogeochemical properties. The integration of sparse point observations of biogeochemical properties and dense ancillary environmental datasets has the potential to improve spatially explicit predictions of properties throughout aquatic (and terrestrial) ecosystems. For example, remote sensing data are cost-effective and provide exhaustive, high-resolution information at landscape-scale. In this chapter we focused on spatially explicit mapping of soil biogeochemical properties exemplified by TP. The case study demonstrated that OK, IK, RK and CSGS produced different maps using the same TP observations. All maps are valid in the sense that they describe the spatial patterns of TP in this specific wetland. Since different geostatistical methods aim at different goals (e.g., to minimize the variance, assess the uncertainty of model predictions, etc.) we have to use them cautiously. Multi-variate geostatistical and hybrid methods that incorporate ancillary environmental data to model biogeochemical properties have much potential to improve our understanding of aquatic and terrestrial ecosystems. To document the evolution of biogeochemical properties through time more monitoring programs are in need. Bridging the gaps between micro, meso and macro spatial scales will be the challenge of future research in aquatic and terrestrial biogeochemistry.

ACKNOWLEDGEMENTS

Funding was provided by the South Florida Water Management District and varied other federal sources. We would like to thank Todd Z.

Osborne (Soil and Water Science Dept, UF), Sue Newman (South Florida Water Management District) and Terry Jones (Aircoastal Helicopter) to assist with data collection; Yu Wang of the Wetland Biogeochemistry Laboratory for her assistance with the laboratory analyses. This research was supported by the Florida Agricultural Experiment Station and approved for publication as Journal Series No. R-10817.

REFERENCES

Abrams, M., Hook, S., Ramachandran, B., 2002. ASTER user handbook. Version 2. Pasadena, CA; Jet Propulsion Laboratory, Sioux Fall, SD, EROS Data Center.

Anderson, J.M., 1976. An ignition method for determination of total phosphorus in lake sediments. Water Res. 10, 329–331.

Berthouex, P.M., Brown, L.C., 2002. Statistics for Environmental Engineers. Lewis, New York, p. 489.

Burrough, P.A., 1986. Principles of geographical information systems for land resources assessment. Clarendon Press, UK, p.194.

Burrough, P.A., McDonnell, R.A., 1998. Principles of geographical information systems—spatial information systems and geostatistics. Oxford University Press, Oxford, UK, p. 333.

Bruland, G.L., Richardson, C.J., 2005. Spatial variability of soil properties in created, restored and paired natural wetlands. Soil Sci. Soc. Am. J. 69, 273–284.

Chilès, J.-P., Allard, D., 2005. Stochastic simulation of soil variations. In: Grunwald, S. (Ed.), Environmental Soil-Landscape Modeling—Geographic Information Technologies and Pedometrics. CRC Press, New York, pp. 289–323.

Chilès, J.-P., Delfiner, P., 1999. Geostatistics—Modeling Spatial Uncertainty. Wiley-Interscience, New York, p. 695.

Craft, C.B., Chiang, C., 2002. Forms and amounts of soil nitrogen and phosphorus across a longleaf pine-depressional wetland landscape. Soil Sci. Soc. Am. J. 66(5), 1713–1721.

DeBusk, W.F., Newman, S., Reddy, K.R., 2001. Spatial and temporal patterns of soil phosphorus in Everglades Water Conservation Area 2A. J. Environ. Qual. 30, 1438–1446.

Fisher, R.A., 1925. Statistical methods for research workers. Oliver and Boyd, Edinburgh, Scotland, p. 189.

Fisher, M.M., Reddy, K.R., 2001. Phosphorus flux from wetland soils affected by long-term nutrient loading. J. Environ. Qual. 30(1), 261–271.

Geary, R.C., 1954. The contiguity ratio and statistical mapping. Incorporated Statistician 5, 115–145.

Goovaerts, P., 1997. Geostatistics for natural resources evaluation. Oxford University Press, New York, p. 483.

Goovaerts, P., 1999. Geostatistics in soil science: State-of-the-art and perspectives. Geoderma 89, 1–45.

Grunwald, S., 2005. What do we really know about the space-time continuum of soil-landscapes?. In: Grunwald, S. (Ed.), Environmental Soil-Landscape Modeling—Geographic Information Technologies and Pedometrics. CRC Press, New York, pp. 3–37.

Grunwald, S., Corstanje, R., Weinrich, B.E., Reddy, K.R., 2006. Spatial patterns of labile forms of phosphorus in a subtropical wetland 10 years after a sustained nutrient impact. J. Environ. Qual. 35, 378–389.

Grunwald, S., Reddy, K.R., Newman, S., DeBusk, W.F., 2004. Spatial variability, distribution and uncertainty assessment of soil phosphorus in a south Florida wetland. Environmetrics 15, 811–825.

Hengl, T., Heuvelink, G.B.M., Stein, A., 2004. A generic framework for spatial prediction of soil variables based on regression-kriging. Geoderma 120, 75–93.

Holling, C.S., Gunderson, L.H., 2002. Resilience and adaptive cycles. In: Gunderson, L.H., Holling, C.S. (Eds.), Panarchy: Understanding Transformations in Human and Natural Systems. Island Press, Washington, D.C., p. 287.

Isaaks, E.H., Srivastava, R.M., 1989. An introduction to applied geostatistics. Oxford University Press, New York, p. 289.

Karathanasis, A.D., Thompson, Y.L., Barton, C.D., 2003. Long-term evaluations of seasonally saturated wetlands in western Kentucky. Soils Sci. Soc. Am. J. 67(2), 662–673.

Levin, S.A., 1992. The problem of pattern and scale in ecology. Ecology 73, 1943–1967.

McBratney, A.B., Odeh, I.O.A., Bishop, T.F.A., Dunbar, M.S., Shatar, T.M., 2000. An overview of pedometric techniques for use in soil survey. Geoderma 97, 293–327.

McBratney, A.B., Mendonça Santos, M.L., Minasny, B., 2003. On digital soil mapping. Geoderma 117, 3–52.

McCollum, S.H., Carlisle, V.W., Volk, B.G., 1976. Historical and current classification of organic soils in the Florida Everglades. Proc. Soil Crop Sci. Soc. FL 35, 173–177.

McCormick, P.V., Newman, S., Miao, S., Gawlik, D., Marley, D., Reddy, K.R., Fontaine, T., 2002. Effects of anthropogenic phosphorus inputs on the Everglades. In: Porter, J.W., Porter, K.G. (Eds.), The Everglades, Florida Bay, and Coral Reefs of the Florida Keys: An Ecosystem Sourcebook. CRC Press, Boca Raton p. 1000.

Mitsch, W.J., Gosselink, J.G., 2000. Wetlands. Wiley, New York, p. 920.

Moran, P.A.P., 1950. Notes on continuous stochastic phenomena. Biometrika 37, 17–23.

Morris, D.R., Glaz, B., Daroub, S.H., 2004. Organic matter oxidation potential determination in a periodically flooded Histosol under sugarcane. Soil Sci. Soc. Am. J. 68, 994–1001.

Newman, S., Reddu, K.R., DeBusk, W.D., Wang, Y., Shih, G., Fisher, M.M., 1997. Spatial distribution of soil nutrients in a northern Everglades marsh: Water Conservation Area 1. Soil Sci. Soc. Am. J. 61, 1275–1283.

Noe, G.B., Childers, D.L., Jones, R.D., 2001. Phosphorus biogeochemistry and the impact of phosphorus enrichment: Why is the Everglades so unique. Ecosystems 4, 603–624.

Odeh, I.O.A., McBratney, A.B., 2000. Using AVHRR imageries for spatial prediction of clay content in the lower Namoi Valley of Eastern Australia. Geoderma 97, 237–254.

Odeh, I.O.A., McBratney, A.B., Chittleborough, D.J., 1994. Spatial prediction of soil properties from landform attributes derived from a digital elevation model. Geoderma 63, 197–214.

Odeh, I.O.A., McBratney, A.B., Chittleborough, D.J., 1995. Further results on prediction of soil properties from terrain attributes: Heterotopic cokriging and regression-kriging. Geoderma 67, 215–226.

Porter, J.W. and K.G. Porter, (Eds.), 2002, The Everglades, Florida Bay and Coral Reefs of the Florida Keys—An Ecosystem Sourcebook. CRC Press, New York p. 1000.

Reddy, K.R., Wang, Y., DeBusk, W.F., Fisher, M.M., Newman, S., 1998. Forms of soil phosphorus in selected hydrologic units of Florida Everglades ecosystems. Soil Sci. Soc. Am. J. 62(4), 1134–1147.

Reddy, K.R., White, J.R., Wright, A., Chua, T., 1997. Influence of phosphorus loading on microbial processes in the soil and water column of wetlands. In: Reddy, K.R., O'Connor, G.A., Schelske, C.L. (Eds.), Phosphorus Biogeochemistry in Subtropical Ecosystems. Lewis, New York, p. 707.

Rouse, J.W., Haas, R.H., Schell, J.A., Deering, D.W., 1974. Monitoring vegetation systems in the Great Plains with ERTS. Proceedings, third Earth Resources Technology Satellite-1 Symposium, Greenbelt: NASA SP-351, pp. 3010–3017.

Turner, M.G., O'Neill, R.V., Gardner, R.H., Milne, B.T., 1989. Effects of changing spatial scale on the analysis of landscape pattern. Landscape Ecol. 3(3/4), 153–162.

U.S. Environmental Protection Agency, 1993. Methods for determination of inorganic substances in environmental samples. Environmental Monitoring Systems Lab, Cincinatti, OH, USA.

Webster, R., 1994. The development of pedometrics. Geoderma 62, 1–15.

Webster, R., Oliver, M.A., 2001. Geostatistics for environmental scientists. Wiley, New York, p. 271.

Webster, R., Oliver, M.A., 2005. Modeling spatial variation of soil as random functions. In: Grunwald, S. (Ed.), Environmental Soil-Landscape Modeling Geographic Information Technologies and Pedometrics. CRC Press, New York, pp. 241–289.

White, J.R., Reddy, K.R., 2000. Influence of phosphorus loading on organic nitrogen mineralization of Everglades soils. Soil Sci. Soc. Am. J. 64(4), 1525–1534.

Developments in Environmental Science, Volume 5
D. Sarkar, R. Datta and R. Hannigan (Editors)

ISSN: 1474-8177/DOI:10.1016/S1474-8177(07)05021-8

Chapter 21

Use of plants in biotechnology: Synthesis of metal nanoparticles by inactivated plant tissues, plant extracts, and living plants

J.G. Parsons, J.R. Peralta-Videa and J.L. Gardea-Torresdey

Abstract

The biological production of metal nanoparticles is becoming a very important field in chemistry, biology, and materials science. Metal nanoparticles have been produced chemically and physically for a long time; however, their biological production has only been investigated very recently. The biological reduction of metals by plant extracts has been known since the early 1900s; however, the reduction products were not studied. Furthermore, the synthesis of nanoparticles using plant materials, for the most part, has only recently been studied within the last three decades, while the production of nanoparticles using living plants has only been studied in the last half decade. The synthesis of metal nanoparticles using biological materials has been shown to produce nanoparticles of the same shapes and sizes as those produced through chemical or physical methods. For example alfalfa, wheat, and oat biomasses are able to reduce Au(III) to Au(0) forming face centered cubic (FCC), tetrahedral, decahedral, hexagonal, icosahedral multitwinned, irregular shape, and rod shape gold nanoparticles. Most of these nanoparticles had sizes ranging from 10 to 50 nm in diameter. Also it has been demonstrated that living alfalfa plants are capable of reducing Au(III) and Ag(I), producing gold and silver nanoparticles. The use of high resolution-TEM showed that the plants formed small silver nanoparticles with sizes averaging from 2 to 20 nm in diameter. These results have proven that biological materials provide an environmentally friendly or green chemistry method to produce invaluable materials because the biosynthesis eliminates the need to use harsh or toxic chemicals. The synthesis of metal nanoparticles using plant-derived materials is a simple yet effective method for the production of nanoparticles. Finally, the applications of nanoparticles are making the fields of nanoscience and nanotechnologies grow rapidly. Researchers have found uses for nanoparticles in medicine, catalysis, bio-sensors, chemical sensors, environmental remediation, semi-conductors, and photovoltaic devices, among others.

21.1. Introduction

Microscopic particles whose size is measured in nanometers (10^{-9} m) are called nanoparticles. These particles have very special properties which are used for the production and applications in technological devices. This new technology, known as nanotechnology, has applications in catalysis, chemical industry, electronics, and electro-optical devices, among others (Brust et al., 1998; Martin and Mitchell, 1998; Armendariz et al., 2004a,b). In the long term, nanotechnology intends to be able to manipulate individual atoms and molecules for applications in different fields of industry (Tanaka, 1999).

The existence of inorganic nanoparticles was recognized about 435 years ago when Paraselsus described the preparation of *aurum potabile*, which was found to contain very thinly divided gold (Caseri, 2000). Since then, several methods to produce metal nanoparticles have been described. Current techniques for manufacturing nanoparticles have been catalogued as wet and dry methods. Wet methods regularly entail the use of substances such as sodium borohydride, hydroxylamine tetrahydroxymethylphosphonium chloride, and polyvinylpyrrolydone (Westcott et al., 1998; Han et al., 1999; Esumi et al., 2001; Meltzer et al., 2001; Vorobyova et al., 2001; Armendariz et al., 2004a). Dry methods comprise lithography, ultraviolet irradiation, and aerosol technology (Tolles, 1996; Deppert et al., 1998; Zhou et al., 1999; Magnusson et al., 1999a,b; Kammler et al., 2001; Armendariz et al., 2004b).

So far, using wet and dry methods researchers have produced nanoparticles of metals including magnesium, silicon, titanium, vanadium, chromium, manganese, iron, cobalt, nickel, copper, zinc, niobium, barium, cesium, and tungsten (Bae and Oh, 2003; Bunge et al., 2003). Other researches have obtained high-purity metal nanoparticles of gold, silver, platinum, palladium, rhodium, ruthenium, copper, and/or indium (Yanagimoto et al., 2005). Additionally, efforts have been made to control the particle size. Goia and Matijevic (1999) found that by changing the pH of a tetrachloroaurate-isoascorbic acid solution it is possible to keep primary particles or to get much larger and uniform particles due to aggregation. On the other side, Magnusson et al. (1999a,b) produced size-selected gold nanoparticles $\leqslant 30$ nm by passing gold aerosol though a device which uses UV light to create free electrons and positively charged particles. Sau et al. (2001) used UV to irradiate gold particles ranging in size from 5 to 20 nm and they obtained larger particles ranging from 20 to 110 nm. Using ultrasonic field, varying solution temperature and the position of the reactor, Okitsu et al. (2001) were able to form gold nanoparticles of definite size. Other researchers have used size exclusion chromatography to separate gold nanoparticles (Wei and Liu, 1999; Wei et al., 1999). In addition, using surfactant-directed growth Busbee et al. (2003) obtained nanorods from spherical gold nanoparticles used as seeds. Recently, Gardea-Torresdey and coworkers reported the formation of metal nanoparticles by inactivated and living tissues of (*Medicago sativa*) alfalfa and (*Chilopsis linearis*) Desert willow (Gardea-Torresdey et al., 1999, 2002a,b,c,d, 2003, 2005).

21.2. Metal uptake by plants

The metal uptake by plants has been documented elsewhere (Blaylock and Huang, 2000; Reeves and Baker, 2000). This review will be mainly focused on the uptake of gold and silver by plants. According to Reeves and Baker (2000), the first report on metal accumulating plants date back to 1855. Other reports concerning the unusual high accumulation of certain metals in plant tissues appeared more than 50 years ago (Reeves and Baker, 2000; Blaylock and Huang, 2000). More recently, researchers have focused their attention on the uptake of precious metals and the possibility of using plants for phytomining. Gold has been one of the most studied metals. Starting in the late 1970s, Girling, Peterson, and coworkers published a series of papers describing the gold accumulation by plants (Girling and Peterson, 1978; Girling et al., 1979; Girling and Peterson, 1980). At the beginning of the 1980s appeared a review authored by Brooks (1982), where he reviewed the biological method of prospecting gold. In this review, the author stated the fact that gold can be found in vegetation as product of a mineralization process. The author summarized the number of plants species, origin, gold content, and type of soil where the plants were grown. The data reviewed by the author indicated that the gold content of dry vegetation is usually about 10% of that of the substrate. According to Erdman and Olson (1985), the wood ash of Douglas-fir from Idaho grown in an Au-bearing stockwork contained up to 14 mg kg^{-1} of Au. However, the use of plants for Au prospecting has problems such as the localization of Au in plant tissues and the variation of Au concentrations due to season. Later on, other researchers gave attention to the use of plants for phytomining and phytoremediation of valuable metals.

Researchers from New Zealand have dedicated time and effort to determine the application of plants for metal extraction, especially gold (Anderson et al., 1998; Lamb et al., 2001). The possibility of phytomining of metals such as nickel and thallium has also been studied (Anderson et al., 1999). More recently, Gardea-Torresdey and coworkers have studied the uptake of gold and silver by alfalfa and desert plants (Gardea-Torresdey et al., 2002a,b,c,d, 2003, 2005). Gomez (2002) cultivated alfalfa plants in precious metals-enriched solid media. He found that the plants cultivated for 14 days in medium containing platinum at $10\,mg\,l^{-1}$ accumulated more than 100 mg platinum per kg of shoot dry weight. However, the amounts of gold and silver accumulated were lower. For gold the plants exposed to $10\,mg\,l^{-1}$ concentrated 56 mg Au per kg of gold in the shoot dry tissues, but the plants exposed to $40\,mg\,l^{-1}$ of silver concentrated 11 mg Ag per kg of shoot dry tissue. Attempts have been made to increase the uptake of gold by plants adding naturally occurring chemicals such as thiocyanate or thiosulfate to a gold-rich soil on which plants are growing. It is hypothesized that plants may accumulate the complexed gold as a function of evapotranspiration; however, there is no knowledge about the form of gold distributed through plant tissues. Gardea-Torresdey and coworkers (2005) found that in desert willow (*Chilopsis linearis*), the addition of $10^{-4}\,mol\,l^{-1}$ NH_4SCN to 2.5×10^{-5} M Au solution increased the gold concentration by approximately 595%, 396%, and 467% in the roots, stems, and leaves, respectively, compared to the Au concentration obtained without the use of NH_4SCN.

21.3. Formation of metal nanoparticles by inactivated plant biomass

The use of biomass for metal removal from aqueous solutions has been amply documented in the literature. This technology, known as biosorption, emerged in the 1980s and obtained attention because it has shown to be very promising for the removal of contaminants from effluents in an environmentally friendly approach (Gardea-Torresdey et al., 1996a; Volesky, 2001). Biosorbents are generally low-priced because they are either naturally abundant or found as waste material from certain processes (Bailey et al., 1999). Common sources of biosorbents are bacteria, algae, fungi, plant-derived, and animal-derived. Inactivated biomass does not need maintenance and is not sensitive to high concentrations of pollutants. On the other hand, living organisms are frequently affected by high concentrations of contaminants (Gardea-Torresdey et al., 1996b).

Table 21.1. Plant biomaterials studied for the removal of metals from aqueous solutions

Pollutant	Biomaterial	Source
Heavy metals	Pine bark Alfalfa Oat Hops Orange peel Papaya Chickpea Saltbush	Al-Asheh and Duvnjak, 1997; Gardea-Torresdey et al., 2000a, 2002a,b; Ajmal et al., 2000; Saeed et al., 2005a,b; Sawalha et al., 2005
Precious metals	Alfalfa Hops	Gardea-Torresdey et al., 2000b, 2002c; Parsons et al., 2003; Herrera et al., 2003; Lopez et al., 2005
Lanthanides and actinides	Alfalfa	Gardea-Torresdey et al., 2004; Parsons et al., 2005

Several researchers have studied plant-derived materials to remove a wide variety of elements from aqueous solutions. Examples of biomaterials that have been successfully used in the removal of heavy metals are shown in Table 21.1. As shown in this table, plant-derived materials have been extensively studied for the removal toxic heavy metals. Less attention has been given to the use of plant biomass for recuperating precious metals and actinides and lanthanides elements. However, the earliest reports of silver and gold reduction by non-living plant tissues occurred in the 1920s and 1930s. In 1921, a researcher by the name of Molisch reported on the microchemistry of plants and the reduction of silver by chlorophyll (Molisch, 1921). Later, Iwase (1928) reported on the production of gold sols produced by the aqueous extract of fresh leaves. He noted that the color of the solution was not as vivid red as particles obtained by the Veimarn tartrate method for the reduction of gold salts. In the 1950s, the reduction of silver nitrate by plant cells was investigated (Nagai, 1951). The study was aimed at determining the chemical species in cells of leaf extracts responsible for the reduction of the silver nitrate. It was determined that ascorbic acid was responsible for silver nitrate in spinach, white clover, sorrel, gladiolus, *Vicia sativa*, broad bean, *Acuba japonica*, normal and elongated seedlings of barley, and the fruit of garden pepper. Nagai also noted that 3,4-dihydroxyphenylalanine in *V. faba* has a strong reducing capacity. Dalal (1979) investigated the reduction of silver ions on three plant materials and concluded that the ascorbic acid was responsible for the reduction. In addition, he concluded that silver nitrate was a specific reagent for localizing ascorbic acid in plant tissues. For the most part, with the exception of the gold sols production using plant extract, the aforementioned studies only report on the reduction of silver ions by plant materials as a tool to determine ascorbic acid. However, these studies are the first indications that plant materials have the ability to reduce metal ions to form particles either nanoparticles or microparticles.

Gardea-Torresdey et al. (1999) reported for the first time the production of gold nanoparticles by a plant biomass. These researchers reacted a 0.1 M Au(III) solution (from potassium tetrachloroaurate) with 10 mg of ground alfalfa biomass at various pH values. The decanted biomass was observed by using a high resolution JEOL-4000 Fx microscope and a JEOL 2010 microscope fitted with EDS analysis. The results demonstrated that the biomass was effective at binding and reducing the Au(III) to Au(0) forming nanoparticles of different forms and sizes. The particles found had tetrahedral, face centered cubic (FCC), hexagonal, irregular, decahedral, and icosahedral forms. The results also suggested that the most frequent shape for the smaller particles was the icosahedral, while for the largest the decahedral and tetrahedral were most abundant shapes. The authors concluded that the particles formed by alfalfa biomass were similar in structure to the particles obtained by chemical methods. As for the mechanisms involved in the nanoparticle formation by the biomass, the authors concluded that three moles of chloride ions are released per mole of gold adsorbed. In addition, they found that both amino and carboxyl functional groups on the alfalfa biomass are involved in the reduction process. The X-ray absorption near edge structure (XANES) and extended X-ray absorption fine structure (EXAFS) spectroscopy suggested that the Au(III) reduction goes through Au(I) to Au(0) (Gardea-Torresdey et al., 2002c).

The formation of gold nanoparticles by other biomasses has also been studied by Gardea-Torresdey's research group. Armendariz et al. (2004a,b) characterized the gold nanoparticles formed by wheat biomass exposed to a 0.3 mM potassium tetrachloroaurate solution at pH values of 2–6 for three and a half hours at room temperature. Through the observation of the reacted biomass with a JEOL-400 EX microscope, the researchers concluded that wheat biomass was able to reduce Au(III) to Au(0) forming FCC tetrahedral, decahedral, hexagonal, icosahedral multitwinned, irregular shape, and rod shape nanoparticles. This was the first report about the formation of rod shape nanoparticles by biomaterials. Most of the observed particles were 1030 nm in diameter. The stem biomass of oat (*Avena sativa*) was also studied as a reducing agent for Au(III). A 0.1 mM solution of Au(III) (from tetrachloroaurate) was reacted for 1 h with powdered oat biomass at pH values of 2–6. As in the case of wheat, oat biomass formed FCC tetrahedral, decahedral, hexagonal, icosahedral multitwinned, irregular shape, and rod shape nanoparticles. This is the second report about the formation of rod shape nanoparticles by biomaterials. Fig. 21.1 shows TEM micrographs to provide a visualization of some of the typical nanoparticles created using biomass (in this case hops biomass) in aqueous solution. Fig. 21.2A shows the TEM micrograph of a triangular (twinned) nanoparticle while Fig. 21.2B shows the structure of a rod-shaped nanoparticle, Fig. 21.2C shows an icosahedral nanoparticle, and Fig. 21.2D shows a decahedral nanoparticle. On the other hand, Table 21.2 shows that pH controls the size of the nanoparticle formed by the biomasses. At pH values of 3 and 4, oat and wheat biomasses formed nanoparticles whose diameter varied from 10 to 20 nanometers. In addition, pH controlled the shape of the

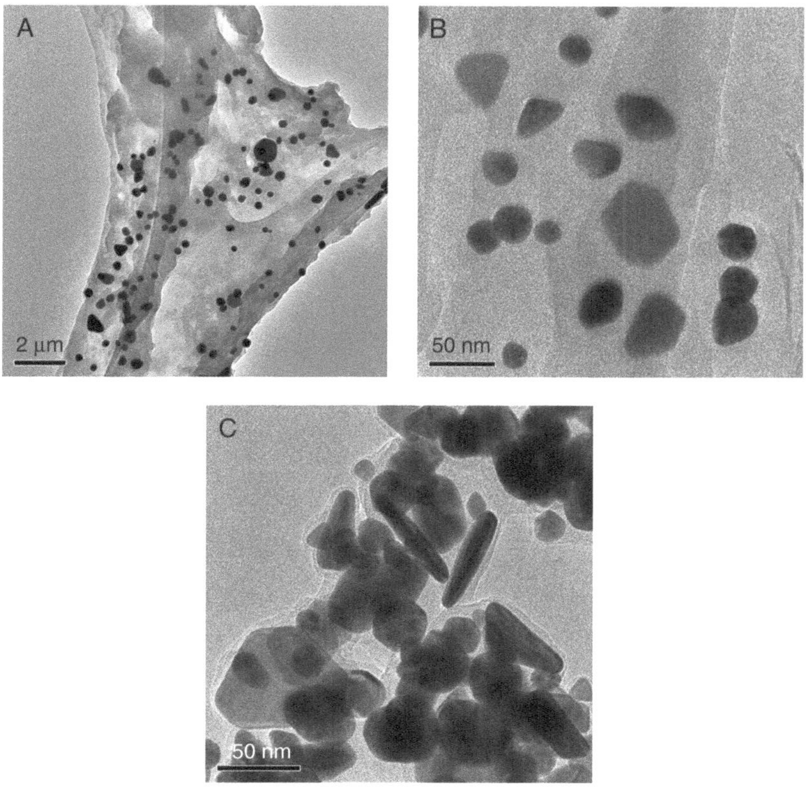

Figure 21.1. Low-resolution TEM of gold nanoparticles formed on hops biomass and in the aqueous reaction solution. A. Gold nanoparticles immobilized on the inactivated hops biomass, B. Magnified view of gold nanoparticles immobilized on the biomass showing a large hexagonal platelet, and C. Gold nanoparticles formed in solution during the reaction of hops biomass with tetrachloroaurate.

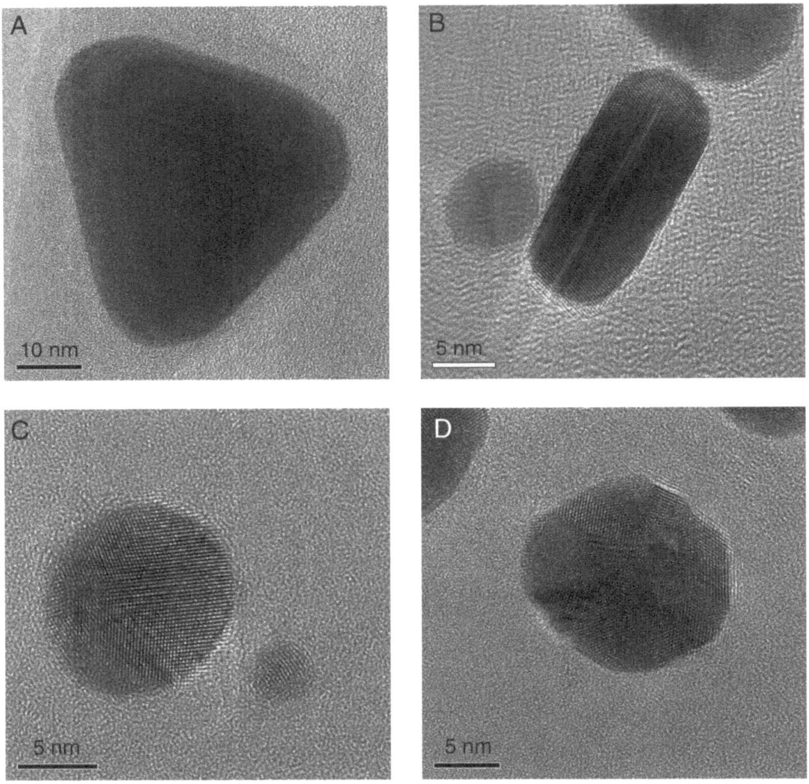

Figure 21.2. TEM micrographs of typical gold precipitated by hops biomass in aqueous solution at room temperature. A. A triangular nanoparticle (size bar indicates 10 nm), B. Rod-shaped gold nanoparticle (size bar indicates 5 nm), C. Icosahedral gold nanoparticle (size bar indicates 5 nm), and D. Decahedral shaped gold nanoparticle (size bar indicates 5 nm).

nanoparticles formed. In alfalfa, oat, and wheat, most of the nanoparticles formed at pH 2 had an irregular shape (Gardea-Torresdey et al., 1999; Armendariz et al., 2004a). However, pH has a major impact on the size rather than on the shape of the nanoparticle formed.

Lopez et al. (2005) studied the effect of chemical modification on the Au(III) reduction capacity of hops biomass. Using X-ray absorption spectroscopy (XAS) analysis, XANES (X-ray absorption near edge structure), and EXAFS (extended X-ray absorption fine structure), these researchers determined that Au(III) (from potassium tetrachloroaurate) was reduced to Au(0) at approximately 81%, 70%, and 83% by native,

Table 21.2. Gold nanoparticles formed by alfalfa, wheat, and oat biomasses reacted with Au(III) solution at various pH values

pH	Biomass/nanoparticle size		
	Alfalfa[a]	Wheat[b]	Oat[c]
2	15 nm (60%)	20–50 nm (75%)	20–40 nm (60%)
3	–	10–20 nm (72%)	10–20 nm (65%)
4	–	10–15 nm (65%)	10 nm (75%)
5	–	10–20 nm (70%)	10–30 nm (nd)
6	–	10–30 nm (90%)	–

[a]Gardea-Torresdey et al., 1999.
[b]Armendariz et al., 2004a.
[c]Armendariz et al., 2004b.

esterified, and hydrolyzed hop biomass, respectively. The average particle size for the native and esterified hops biomass showed particle radii of 17.3 Å and 9.2 Å, respectively.

Researchers are looking for new method for nanoparticle formation using plant extracts. Egorova and Revina (2000) obtained stable silver and copper nanoparticles with quercetin, a natural plant pigment, in the presence of air–oxygen in water by the reduction of silver and copper salts. These researchers obtained particle size ranging between 1 and 1.5 nm. Shankar et al. (2003a) reported that the leaf biomass of *Pelargonium graveolens* reacted with a 10^{-3} M $HAuCl_4$ solution formed gold nanoparticles of a variety of shapes including rods, flat sheet, and triangles. Shankar et al. (2005) have explored the use of lemon grass extract for the Au(III) reduction and the formation of gold nanoparticles. These researchers added different amounts of lemongrass extract to the Au solution. The results have shown that increasing the amount of the extract reduces the average size of the triangular and hexagonal particles. However, the ratio of the number of spherical particles to triangular/hexagonal particles increased as the amount of extract increased. According to the authors, the nanoparticles formed can be extended on a glass substrate in the form of films which absorb IR radiation. In addition, Shankar et al. (2004a) have shown that the particle size changes with time. After two minutes the particles had an average size of approximately 1 micron, at 90 min particles of a 0.1 micron started to appear, and at 360 min the distribution of the particle size was bimodal with 0.1 and 1.0 micron particles. It has also been shown by Ankamwar et al. (2005) that tamarind leaf extract is also capable of forming gold nanotriangles in aqueous solution. These authors showed that the reduction process was rapid resulting in thin flat single crystal gold nanotriangles ranging in thickness from 20 to 40 nm. The authors also tested the effects of different organic solvent vapors on the conductivity of the nanotriangles and suggested the application of these nanoparticles in future chemical sensors. In another study by Shankar et al. (2003b), silver nanoparticles were created using a crude extract from geranium leaves. The authors boiled cleaned and finely cut geranium leaves for 5 min in distilled water and added the extract to a 10^{-3} M solution of Ag(I) ions and produced silver nanoparticles ranging in size from 16 to 40 nm with an average size of 27 nm.

Ascencio et al. (2004) have shown the production of ytterbium nanoparticles using ground alfalfa biomass in aqueous solution. The ytterbium nanoparticles were produced by alfalfa biomass in solution at pH values ranging from 2 to 10. The analysis performed using TEM showed the presence of FCC nanoparticles. Another study by Ascencio et al. (2005) demonstrated that alfalfa biomass reacted with a samarium solution is able to produce samarium nanoparticles. In this study, the author reported that at pH 4 a larger number of small nanoparticles ranging from 1 to 3 nm were formed; however, at pH 6 the majority of the nanoparticles had sizes ranging from 2 to 8 nm.

The production of core-shell nanoparticles has also been performed using plant extracts. Shankar et al. (2004b) used an extract from Neem (*Azadirachta indica*) leaves and a mixture of tetrachloroaurate and silver nitrate and produced gold core-silver shell nanoparticles. In this study, the authors showed the production of monometallic silver and gold nanoparticles by the plant extracts. The particle sizes and shape for the gold nanoparticles produced using this plant extract were poly-dispersed both in size and shape; however, the silver nanoparticles were spherical in shape and ranged in size from 5 to 35 nm. The authors also noted that the particles were stable in solution up to four weeks after synthesis. The bimetallic core-shell nanoparticles had sizes ranging from 50 to 100 nm, but did not have a continuous shell of silver over the gold. Instead the silver layer is a compilation of silver nanoparticles. In another study, investigating the effects of plant extracts on metal ions, Mallikarjuna et al. (2005) showed the reduction of palladium and silver ions by coffee and tea extracts. They observed a surface plasmon peak at approximately 430 nm, using a UV-Vis instrument that indicates the production of silver nanoparticles. In addition, in the palladium reduction they observed a continuous absorption in the UV region, indicative of colloidal palladium. Further investigation using TEM showed that the silver nanoparticles were poly-dispersed and the palladium particles showed a layered structure. The structures obtained from this study indicate that the particles were capped with an organic molecule from the parent solution.

In another study performed with ground alfalfa, Ascencio et al. (2000) using TEM analysis showed the production of a new type of gold nanoparticle, a truncated icosahedron. The new particle found through the bioreduction of tetrachloroaurate on alfalfa biomass was thought to be formed through slow growth or equilibrium processes. Ascencio et al. (2003) have also prepared europium–gold nanoparticles. In this study, it was shown that the size and shape was dependent on the pH of the solution. The authors used TEM to perform the particle analysis and showed that the highest distribution of small nanoparticles occurred at pH 8; whereas the larger bimetallic nanoparticles were formed at 6. Furthermore, the authors showed that the structure of the Eu–Au nanoparticles were multitwinned with a gold core and a thin layer of Eu atoms over the gold core.

21.4. Formation of metal nanoparticles by living plants

The production of nanoparticles by living systems dates back to the early 1900s where it was shown that colloidal silver was deposited in organs of living organisms (Voigt, 1914). Around the same time it was shown that fungi could remove colloidal gold from solution through adsorption (Williams, 1918). It was shown in the early 1970s that bacteria could adsorb colloidal iron particles from solution (Macrae and Edwards, 1972). However, the actual formation of metal nanoparticles by both fungi and bacteria was not known until the 1990s (Karamushka, et al., 1990; Moiseenko et al., 1999). As cited above, the characterization of gold nanoparticles by plant extracts has only been performed in the last few years. However, the reduction of silver by plant roots has been noted in the literature since 1935. Gautheret (1935) reported the production of a green color when plant roots were treated with silver nitrate. The color indicates the reduction of silver and the production of nanoparticles; green-colored silver nanoparticles have been observed with the reduction of silver nitrate in the presence of ascorbic acid and light (Jiang et al., 2001).

Although the reduction of silver by living plants was previously reported, the products of the reduction were not characterized. The nanoparticle production by living plants and their characterization has only recently reported by Gardea-Torresdey and coworkers (2002a,b,c,d). Gardea-Torresdey et al. (2002d) showed the production of gold nanoparticles by alfalfa plants. In this study the authors showed the accumulation of Au(0) nanoparticles inside living alfalfa plants using XAS and TEM analytical techniques to characterize the gold particles. The XAS studies showed that the nanoparticles consisted of Au(0) with no oxidized gold present. To complement the XAS data, TEM images of the plants containing the gold nanoparticles were taken. The images showed that the particles were in the FCC configuration, and were multi-twinned and the lattice parameter calculated was 0.23, which is approximately the spacing between the 111 planes of Au(0). It was further shown in this study that the gold was present in an icosahedral structure, a low-energy configuration for Au(0), indicating that the reduction was performed slowly or under equilibrium conditions. The same researchers, Gardea-Torresdey et al. (2003), again using alfalfa, showed for the first time the production of Ag(0) nanostructures inside a living plant. In this study a combination of XAS and TEM was used to characterize the silver nanoparticles. The XANES spectroscopy showed that silver was present as Ag(0) in the plants. From the EXAFS of the root sample, Ag(0) in the plant had a coordination of approximately four Ag atoms. The low coordination found in the EXAFS is indicative of colloidal silver or silver nanoparticles because of the incomplete first shell coordination. An example of the EXAFS obtained from silver nanoparticles is represented in Fig. 21.3. Fig. 21.3A shows the reduced amplitude in the raw EXAFS of the nanoparticle compared to the bulk silver. Fig. 21.3B shows the Fourier-transformed EXAFS of the samples shown in Fig. 21.3A. It is noticeable that the Fourier-transformed EXAFS of the silver nanoparticle have less amplitude than the bulk silver. Bulk silver or large particles of silver have a first shell coordination of 12 nearest neighboring Ag atoms. Fig. 21.3C shows the XANES of the Ag(I) model compound ($AgNO_3$), Ag(0) metal foil, and Ag bound to the alfalfa biomass. From the energy position and the shape of the Ag bound to the alfalfa biomass it is clear that the silver present on the biomass is present as Ag(0). It was also shown using EXAFS that the estimated average particle size of the silver nanoparticles was about 9 Å. Furthermore, the dark-filed imaging TEM investigation showed that the nanoparticles or nanostructures were dispersed throughout the plant in small assemblies like nanowires. The use of high-resolution TEM showed small silver nanoparticles with sizes averaging from 2 to 20 nm in diameter.

Yates et al. (2005) attempted to synthesize pure metal nanoparticles of silver, gold, titanium, chromium, zinc, and cobalt using the metabolic pathways of plants. More specifically, they studied the possible reduction of the aforementioned metals using morning glory (*Ipomoea lacunosa*) and alfalfa (*Medicago sativa*) followed by deposition of the particles on stainless steel supports. The metals were prepared in a hydroponics growth solution and supplied to the plants to investigate the natural process of bio-precipitation or bio-mineralization.

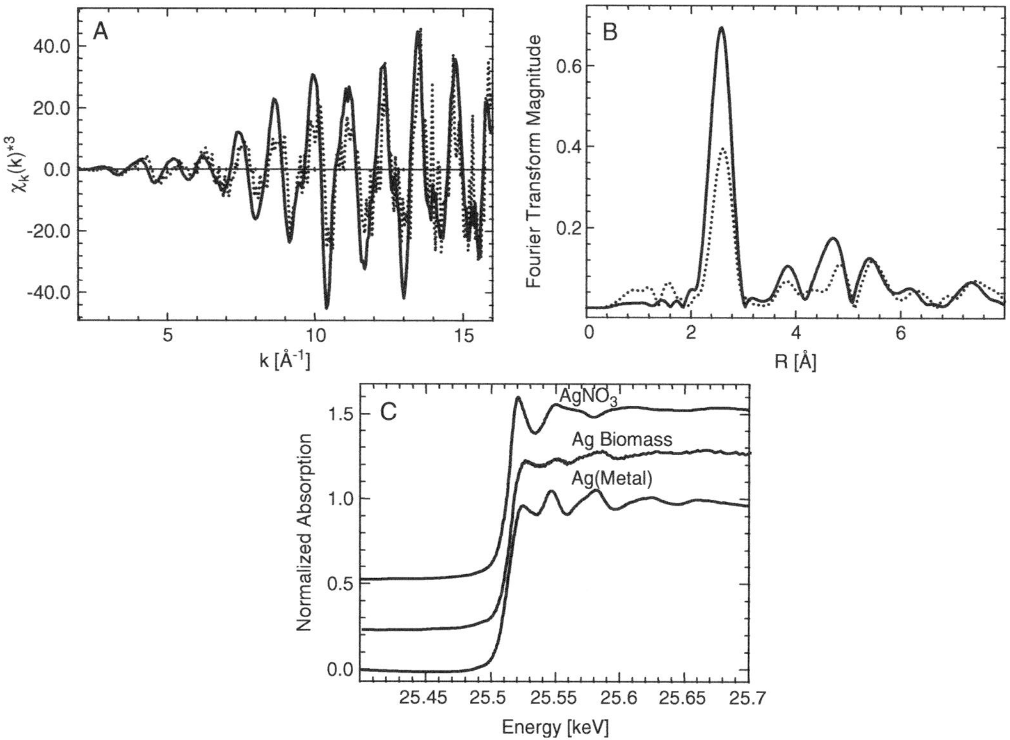

Figure 21.3. A. Raw EXAFS of silver nanoparticles formed on inactivated tissues of alfalfa biomass (dotted line) and the EXAFS of a metallic silver standard (solid line), B. Fourier-transformed EXAFS of silver nanoparticles formed on inactivated tissues of alfalfa biomass (dotted line) and the EXAFS of a metallic silver standard (solid line),and C. XANES spectra from 25.40 to 25.70 keV showing the change in the energy and the reduction from Ag(I) to Ag(0) on the biomass.

In another study, Gardea-Torresdey et al. (2005) investigated the effects of using thiocyanate as a complexing agent for the enhancement of gold uptake by desert willow (*Chilopsis linearis*). These researchers, using ICP-OES, showed that thiocyanate does increase the uptake of gold from a gold enriched media. In addition, through the use of XAS the authors showed that the Au in the plants was present as Au(0) forming Au nanoparticles.

21.5. Nanoparticle extraction

The synthesis of metal nanoparticles using chemical and physical methods is well documented in the literature, and the extraction of nanoparticles from inorganic materials is as well (Praharaj et al., 2004; Abed et al., 2005). Also, the phase transfer of nanoparticles from organic solvents of aqueous solvents and micelles has been shown in the literature (Zhang et al., 2001; Zhang et al., 2002; Wei et al., 2004). In the solvent extraction or phase transfer of nanoparticles, generally a micelle or a non-ionic liquid is used to transport the nanoparticles from the organic phase to a polar phase. Finally, the extraction of nanoparticles from micelles is performed through the dissolution of an antisolvent into solution that disrupts the micelles, thus releasing the nanoparticles into solution.

The synthesis of both monometallic and bimetallic nanoparticles by living and non-living biological materials has been reported in the literature as shown above. However, the extraction of biologically synthesized nanoparticles for further application is not well studied or well understood. In fact, there are only two references about the extraction of nanoparticles from either living or non-living plant materials (Yates et al., 2005; Armendariz, 2005). Yates et al. (2005) mentioned the solventless extraction of titanium nanoparticles and the subsequent deposition onto a stainless steel support. The deposition of the titanium nanoparticles was performed by grinding the biomass sample and preparing it in 2.0 ml of water and coating the stainless steel plate with the solution. Armendariz (2005) extracted gold nanoparticles from inactivated tissues of oat and wheat biomasses. Studies using different instrumental techniques were performed to extract the nanoparticles from the biological materials. Primarily, the extraction process involved the use of CTAB and citrate combined with sonication to transfer the nanoparticles from the biomass samples into aqueous solution. The UV-Vis spectroscopy showed that some of the gold was extracted from the plant material. In the extraction solutions using either CTAB or citrate the 540 nm plasmon resonance peak was present. However, it should be noted that the citrate extractions were more successful than the CTAB extractions. The combination of XAS and TEM data showed that a higher number of smaller nanoparticles were extracted from the biomass in the first extraction, and by the fourth extraction of the same biomass, larger nanoparticles were being extracted. The XAS analysis demonstrated that the effective radius of the nanoparticles left on the biomass was increasing after the first extraction. While the TEM analysis of the extraction solutions revealed the smaller radius of the nanoparticles extracted at first.

21.6. Applications of nanoparticles

Nanoparticles whether chemically, physically, or biologically synthesized have a number of different potential applications which make them an invaluable material. A broad overview of the different applications of metal nanoparticles includes the following: medical, bioanalysis, catalysis, environmental remediation, sensors, semi-conductors, among other applications.

Recently, Paciottie et al. (2005) showed the ability of gold nanoparticles to be used for tumor-targeted drug delivery systems. The authors showed the ability to attach a very powerful yet very toxic anticancer protein tumor necrosis factor. The study showed that the mice treated with this had a 10-fold increase in concentrations of the anticancer protein tumor necrosis factor as compared with mice treated with just the anticancer protein tumor necrosis factor. The attachment of this antitumor treatment to gold nanoparticles increased the efficiency of the factor. In another study by Hainfeld et al. (2004), 1.9 nm gold nanoparticles were injected into mice with mammary carcinomas to investigate the enhancement of radiotherapy. The authors found that 86% of the mice injected with the gold nanoparticles survived compared with 20% treated with the X-rays alone. The use of the gold nanoparticles in this study allowed for high Z-radioenhancement. Carpenter et al. (2005) showed that iron nanoparticles could be used for magnetodynamic therapy for cancer patients. A combination of the magnetic nanoparticles and 100 MHz generated localized heating of the nanoparticles facilitated cellular destruction or targeted drug delivery. In addition, Nie (2003) talked about nanotechnology applications with regard to nanoprobes for molecular imaging and nanoscaffolds for tissue engineering. This author discusses the applications of pure metal and metal oxide nanoparticles in imaging and sensing, luminescence for cancer cell identification and profiling, drug delivery, and tissue engineering properties of nanoparticles.

In medicine, nanoparticles can be used in bioanalysis and as biosensors. The bioanalysis can have a variety of applications. For example, nanoparticles can be used to induce signal transduction, as quantitation identifiers, in bioassays, and finally nanoparticles can be used for specific functions in biological systems (Penn et al., 2003). Maxwell et al. (2002) showed that colloidal gold could be used to create biosensors to identify specific DNA sequences and base mutations. It has also been shown that copper–gold bimetallic nanoparticles can be used as oligonucleotides labels for the electrochemical stripping detection of DNA hybridization (Cai et al., 2003).

Related to health and environmental issues it has been shown that nanoparticles can be used in the remediation of organic pollutants in the environment (Obare and Meyer, 2004). Iron nanoparticles have been proven to be effective in the dechlorination of polychlorinated biphenyls (Paciottie et al., 2005; Johnson and Lowry, 2003). It is also shown in the literature that bimetallic nanoparticles can be used for groundwater treatment (Elliott and Zhang, 2001). This study looked at the destruction of tricholorethene (TEC) and other chlorinated hydrocarbons using bimetallic nanoparticles. The results of this study showed that 96% of the TCE was eliminated within four weeks of injecting the nanoparticles.

The environmental applications of nanoparticles for the most part concern the catalysis of environmental reactions resulting in the destruction of a contaminant. However, from an industrial/chemical stand point nanoparticles are applicable in the catalysis of different reactions. For example, supported gold nanoparticles have been shown to be useful in hydrogenation and oxidation reactions (Schimpf et al., 2002). Another example of catalysis using nanoparticles is the three-way catalysis (Baker et al., 2001). In the three-way catalysis application, the authors showed that platinum, rhodium, and palladium nanoparticles supported on a cesium–zirconium mixed oxide was an excellent material for the three-way catalysis in automobiles.

Another application of nanoparticles is in sensors. For example, Krasteva et al. (2002) showed that thin film resistors composing of gold and different organic-based dendrimers could be used to detect different organic vapors. Ishibashi (2002) showed that noble metal nanoparticles with diameters ranging between 5 and 20 nm with a tertiary amine-protecting group could be used in paints. The nanoparticles in this range for silver and gold exhibit yellow and red colors, respectively, which could be used as unique colorants. Finally, Thomas and Kamat (2003) investigated chromophore-functionalized metal nanoparticles from the ground- and excited-state interactions of the nanoparticles and the chromophores. They concluded that these functionalized nanoparticles could be used in photovoltaics, for the light mediated binding of biologically important molecules, and fluorescence applications in fluorescence display devices.

21.7. Conclusions

The field of nanoscience is growing rapidly with regard to the synthesis, characterization, and applications of metallic nanoparticles. In conjunction with this rapid growth, the field of bio-nanotechnology is also growing rapidly in all areas. The newest subcategory of bio-nanotechnology is the use of plants and plant-derived materials for the synthesis of metallic nanoparticles as has been shown in this brief review. The use of plants and plant-derived materials for the synthesis of nanoparticles has shown to be effective in producing size and shape-controlled nanoparticles. In addition, the bio-synthesized nanoparticles have the same defects, structures and properties of nanoparticles that have been synthesized through chemical and physical methods. The only problem with using plant material and living plant-derived nanoparticles is the extraction of these nanoparticles for future applications. However, studies are being performed to solve the problems of extracting nanoparticles from these biologically derived materials for further application. As mentioned earlier, the potential uses for nanoparticles include medicine, catalysis, bio-sensors, chemical sensors, environmental remediation, semi-conductors, and photovoltaic devices, among others.

ACKNOWLEDGMENTS

The authors would like to acknowledge the National Institutes of Health (grant S06 GM8012-33) and the University of Texas at El Paso's Center for Environmental Resource Management (CERM) through funding from the Office of Exploratory Research of the U.S. Environmental Protection Agency (cooperative agreement CR-819849-01). We also thank the financial support from the Southwest Consortium for Environmental Research and Policy (SCERP) program, and the HBCU/MI, Environmental Technology Consortium that is funded by the Department of Energy. We also thank the Stanford Synchrotron Radiation Laboratory, a national user facility operated by Stanford University on behalf of the U.S. Department of Energy, Office of Basic Energy Sciences. The SSRL Structural Molecular Biology Program is supported by the Department of Energy, Office of Biological and Environmental Research, and by the National Institutes of Health, National Center for Research

Resources, Biomedical Technology Program. Dr. Gardea-Torresdey acknowledges the Dudley family for the Endowed Research Professorship in Chemistry and also funding from the NIEHS (grant R01ES11367-01).

REFERENCES

Abed, O., Vaskevich, A., Arad-Yellin, R., Shanzer, A., Rubinstein, I., 2005. Preparative manipulation of gold nanoparticles by reversible binding to a polymeric solid support: Chemistry. Eur. J. 11(9), 2836–2841.

Ajmal, M., Rao, R.A., Ahmad, R., Ahmad, J., 2000. Adsorption studies on citrus reticulata (fruit peel of orange): Removal and recovery of Ni(II) from electroplating wastewater. J. Hazard. Mater. 79(1–2), 117–131.

Al-Asheh, S., Duvnjak, Z., 1997. Sorption of cadmium and other heavy metals by pine bark. J. Hazard. Mater. 56(1–2), 35–51.

Anderson, C.N.W., Brooks, R.R., Chiarucci, A., LaCoste, C.J., Leblanc, M., Robinson, B.H., Simcock, R., Stewart, R.B., 1999. Phytomining for nickel, thallium and gold. J. Geochem. Explor. 67, 407–415.

Anderson, C.N.W., Brooks, R.R., Stewart, R.B., Simcock, R., 1998. Harvesting a crop of gold in plants. Nature 395, 553–554.

Ankamwar, B., Chaudhary, M., Sastry, M., 2005. Gold nanotriangles biologically synthesized using tamarind leaf extract and potential application in vapor sensing. Synthesis and Reactivity in Inorganic, Metal-Organic, and Nano-Metal Chemistry 35(1), 19–26.

Armendariz, V., 2005. Bio-reduction of gold(III) to gold(0) and nanoparticles formation by oat and wheat biomasses. The use of plants in nanobiotechnology. MS Thesis. El Paso, University of Texas, p. 107.

Armendariz, V., Herrera, I., Peralta-Videa, J.R., Jose-Yacaman, M., Troiani, H., Santiago, P., Gardea-Torresdey, J.L., 2004a. Size controlled gold nanoparticle formation by *Avena sativa* biomass: Use of plants in nanobiotechnology. J. Nanoparticle Res. 6, 377–382.

Armendariz, V., Jose-Yacaman, M., Duarte-Moller, A., Peralta-Videa, J.R., Troiani, H., Herrera, I., Gardea-Torresdey, J.L., 2004b. HRTEM characterization of gold nanoparticles produced by wheat biomass. Rev. Mex. Fís. 50(Suppl. 1), 7–11.

Ascencio, J.A., Mejia, Y., Liu, H.B., Angeles, C., Canizal, G., 2003. Bioreduction synthesis of Eu–Au nanoparticles. Langmuir 19(14), 5882–5886.

Ascencio, J.A., Perez, M., Jose-Yacaman, M., 2000. A truncated icosahedral structure observed in gold nanoparticles. Surf. Sci. 447(1–3), 73–80.

Ascencio, J.A., Rincon, A.C., Canizal, G., 2005. Synthesis and theoretical analysis of samarium nanoparticles: Perspectives in nuclear medicine. J. Phys. Chem. B 109(18), 8806–8812.

Ascencio, J.A., Rodriguez-Monroy, A.C., Liu, H.B., Canizal, G., 2004. Synthesis and structure determination of ytterbium nanoparticles. Chem. Lett. 33(8), 1056–1057.

Bae, S.J., Oh, S.I., 2003. Method for forming metal nanoparticles and method for forming conductive layer using the same. Repub. Korean Kongkae Taeho Kongbo. KR 2003083961 A 20031101 Patent written in Korean. Patent Application No. KR 2002-22386 20020424. Abstract.

Bailey, S.E., Olin, T.J., Bricka, R.M., Adrian, D.D., 1999. A review of potentially low-cost sorbents for heavy metals. Water Res. 33(11), 2469–2479.

Baker, R.T., Bernal, S., Calvino, J.J., Gatica, J.M., Cartes, C.L., Omil, J.A.P., 2001. Pt, Rh and Pd nano-particles supported on Ce0.68Zr0.32O2 mixed oxide for application in automotive three-way catalysts. Institute of Physics Conference Series, vol. 168 (Electron Microscopy and Analysis), pp. 417–420.

Blaylock, M.J., Huang, J.W., 2000. Phytoextraction of metals. In: Raskin, I., Ensley, B.D. (Eds.), Phytoextraction of toxic metals-using plants to clean the environment. Wiley, New York, NY, pp. 53–69.

Brooks, R.R., 1982. Biological methods of prospecting for gold. J. Geochem. Explor. 17, 109–122.

Brust, M., Betchell, D., Kiely, C.J., Schiffrin, D.J., 1998. Self assembled gold nanoparticles thin films with nonmetallic optical and electronic properties. Langmuir 14, 5425–5429.

Bunge, S.D., Boyle, T.J., Headley, T.J., 2003. Synthesis of coinage-metal nanoparticles from Mesityl precursors. Nano Lett. 3(7), 901–905.

Busbee, B.D., Obare, S.O., Murphy, C.J., 2003. An improved synthesis of high-aspect-ratio gold nanorods. Adv. Mater. 15(5), 414–416.

Cai, H., Zhu, N., Jiang, Y., He, P., Fang, Y., 2003. Cu–Au alloy nanoparticle as oligonucleotides labels for electrochemical stripping detection of DNA hybridization. Biosens. Bioelectron. 18(11), 1311–1319.

Carpenter, E.E., Michaels, M., Radwan, F., Morrison, S., 2005. Iron Nanoparticles for Magnetodynamic Therapy. American Chemical Society, 230th National Annual Meeting, Washington, DC, p. 169.

Caseri, W., 2000. Nanocomposites of polymers and metals or semiconductors: Historical background and optical properties. Macromol. Rapid Commun. 21, 705–722.

Dalal, S.L., 1979. Specificity of silver nitrate reduction by plant cells. Natl. Acad. Sci. Lett 2(4), 127–129.

Deppert, K., Nielsch, K., Magnusson, M.H., Kruis, F.E., Fissan, H., 1998. Feasibility study of nanoparticle synthesis from powders of compounds with incongruent sublimation behavior by the evaporation/condensation method. Nanostructur. Mater. 10(4). 565–573.

Egorova, E.M., Revina, A.A., 2000. Synthesis of metallic nanoparticles in reverse micelles in the presence of quercetin. Colloid. Surface A 168(1), 87–96.

Elliott, D.W., Zhang, W.-X., 2001. Field assessment of nanoscale bimetallic particles for groundwater treatment. Environ. Sci. Technol. 35(24), 4922–4926.

Erdman, J.A., Olson, J.C., 1985. The use of plants in prospecting for gold: A brief overview with a selected bibliography and topic index. J. Geochem. Explor. 24, 281–304.

Esumi, K., Kameo, A., Suzuki, A., Torigoe, K., Yoshimura, T., Koide, Y., Shosenji, H., 2001. Preparation of gold nanoparticles using 2-vinylpyridine telomers possessing multi-hydrocarbon chains as stabilizer. Colloid Surface A 176(2–3), 233–237.

Gardea-Torresdey, J., Hejazi, M., Tiemann, K., Parsons, J.G., Duarte-Gardea, M., Henning, J., 2002b. Use of hop (*Humulus lupulus*) agricultural by-products for the reduction of aqueous lead(II) environmental health hazards. J. Hazard. Mater. 91(1–3), 95–112.

Gardea-Torresdey, J.L., Cano-Aguilera, I., Webb, R., Tiemann, K.J., Gutierrez- Corona, F., 1996a. Copper adsorption by inactivated cells of *Mucor rouxii*: Effect of esterification of carboxyl groups. J. Hazard. Mater. 48(1–3), 171–180.

Gardea-Torresdey, J.L., Dokken, K., Tiemann, K.J., Parsons, J.G., Ramos, J., Pingitore, N.E., Gamez, G., 2002a. Infrared and X-ray absorption spectroscopic studies on the mechanism of chromium(III) binding to alfalfa biomass. Microchem. J. 71(2–3), 157–166.

Gardea-Torresdey, J.L., Gomez, E., Peralta-Videa, J.R., Parsons, J.G., Troiani, H., Jose-Yacaman, M., 2003. Alfalfa sprouts: A natural source for the synthesis of silver nanoparticles. Langmuir 19, 1357–1361.

Gardea-Torresdey, J.L., Parsons, J.G., Gomez, E., Peralta-Videa, J.R., Troiani, H.E., Santiago, P., Jose-Yacaman, M., 2002d. Formation of Au nanoparticles inside live alfalfa plants. Nano Lett. 2, 397–401.

Gardea-Torresdey, J.L., Rodriguez, E., Parsons, J.G., Peralta-Videa, J.R., Meitzner, G., Cruz-Jimenez, G., 2005. Use of ICP and XAS to determine the enhancement of gold phytoextraction by *Chilopsis linearis* using thiocyanate as a complexing agent. Anal. Bioanal. Chem. 382, 347–352.

Gardea-Torresdey, J.L., Tiemann, K.J., Armendariz, V., Bess-Oberto, L., Chianelli, R.R., Rios, J., Parsons, J.G., Gamez, G., 2000a. Characterization of Cr(VI) binding and reduction to Cr(III) by the agricultural byproducts of Avena monida (Oat) biomass. J. Hazard. Mater. 80(1–3), 175–188.

Gardea-Torresdey, J.L., Tiemann, K.J., Gamez, G., Dokken, K., Cano-Aguilera, I., Furenlid, L.R., Renner, M.W., 2000b. Reduction and accumulation of gold(III) by Medicago sativa alfalfa biomass: X-ray absorption spectroscopy, pH, and temperature dependence. Environ. Sci. Technol. 34(20), 4392–4396.

Gardea-Torresdey, J.L., Tiemann, K.J., Gamez, G., Dokken, K., Tehuacanero, S., Jose-Yacaman, M., 1999. Gold nanoparticles obtained by bioprecipitation from gold(III) solutuions. J. Nanoparticle Res. 1, 397–404.

Gardea-Torresdey, J.L., Tiemann, K.J., Gonzalez, J.H., Henning, J.A., Townsend, M.S., 1996b. Ability of silica-immobilized *Medicago sativa* (alfalfa) to remove copper ions from solution. J. Hazard. Mater. 48(1–3), 181–190.

Gardea-Torresdey, J.L., Tiemann, K.J., Parsons, J.G., Gamez, G., Herrera, I., Jose-Yacaman, M., 2002c. XAS investigations into the mechanism(s) of Au(III) binding and reduction by alfalfa biomass. Microchem. J. 71(2–3), 193–204.

Gardea-Torresdey, J.L., Tiemann, K.J., Peralta-Videa, J.R., Parsons, J.G., Delgado, M., 2004. Binding of erbium(III) and holmium(III) to native and chemically modified alfalfa biomass: A spectroscopic investigation. Microchem. J. 76(1–2), 65–76.

Gautheret, R.J., 1935. The formation of chlorophyll in roots and the reduction of silver salts by chloroplasts. Rev. Gen. Bot. 47, 401–421, 484–512.

Girling, C.A., Peterson, P.J., 1978. Uptake, transport and localization of gold in plants. Trace Substance. Environ. Health 12, 105–118.

Girling, C.A., Peterson, P.J., 1980. Gold in plants. Gold Bull. 13(4), 151–157.

Girling, C.A., Peterson, P.J., Warren, H.V., 1979. Plants as indicators of gold mineralization at Watson Bar, British Columbia, Canada. Econ. Geol. Bull. Soc. 74(4), 902–907.

Goia, D.V., Matijevic, E., 1999. Tailoring the particle size of monodispersed colloidal gold. Colloids Surf., A 146, 139–152.

Gomez, E., 2002. The nanoparticle formation and uptake of precious metals by living alfalfa plants. MS Thesis. El Paso, University of Texas, p. 81.

Hainfeld, J.F., Slatkin, D., Smilowitz, H.M., 2004. The use of gold nanoparticles to enhance radiotherapy in mice. Phys. Med. Biol. 49(18), N309–N315.

Han, M.Y., Quek, C.H., Huang, W., Chew, C.H., Gan, L.M., 1999. A simple and effective chemical route for the preparation of uniform nonaqueous gold colloids. Chem. Mater. 11(4), 1144–1147.

Herrera, I., Gardea-Torresdey, J.L., Tiemann, K.J., Peralta-Videa, J.R., Armendariz, V., Parsons, J.G., 2003. Binding of silver(I) ions by alfalfa biomass (*Medicago sativa*): Batch pH, time, temperature, and ionic strength studies. J. Hazard. Subst. Res. 4(1), 1–16.

Ishibashi, H., 2002. Preparation of highly concentrated pastes of gold and silver nanoparticles and their application to paint coloring. Techno-Cosmos 15, 2–7.

Iwase, E., 1928. The preparation of red-gold sols by using as reducing agents the extracts of fresh leaves and plants. Kolloid Zeitschrift 44, 42–43.

Jiang, Z.-L., Zhong, F.-X., Li, T.-S., Li, F., Liu, Q.-Y., Liang, H., 2001. Study on the resonance scattering spectra of green silver nanoparticles. Huaxue Xuebao 59(3), 438–441.

Johnson, K.M., Lowry, G.V., 2003. Iron mediated PCB destruction for in situ applications. American Chemical Society, 225th National Annual Meeting, New Orleans, LA, extended abstract, 43(1), pp. 91–95.

Kammler, H.K., Madler, L., Pratsinis, S.E., 2001. Flame synthesis of nanoparticles. Chem. Eng. Technol. 24, 583–596.

Karamushka, V.I., Ul'berg, Z.R., Gruzina, T.G., 1990. The role of membrane processes in Au(III) and Au(0) accumulation by bacteria. Ukr. Biokhim. Zh. 62(1), 76–82.

Krasteva, N., Besnard, I., Guse, B., Bauer, R.E., Muellen, K., Yasuda, A., Vossmeyer, T., 2002. Self-assembled gold nanoparticle/dendrimer composite films for vapor sensing applications. Nano Lett. 2(5), 551–555.

Lamb, A.E., Anderson, C.N.W., Haverkamp, R.G., 2001. The extraction of gold from plants and its application to phytomining. Chem. New Zeeland 65(2), 31–33.

Lopez, M.L., Parsons, J.G., Peralta Videa, J.R., Gardea-Torresdey, J.L., 2005. An XAS study of the binding and reduction of Au(III) by hop biomass. Microchem. J. 81(1), 50–56.

Macrae, I.C., Edwards, J.F., 1972. Adsorption of colloidal iron by bacteria. Appl. Microbiol. 24(5), 819–823.

Magnusson, M.H., Deppert, K., Malm, J.-O., Bovin, J.-O., Samuelson, L., 1999. Gold nanoparticles: Production, reshaping, and thermal charging. J. Nanoparticle Res. 1(2), 243–251.

Magnusson, M.H., Deppert, K., Malm, J.-O., Bovin, J.-O., Samuelson, L., 1999. Size-selected gold nanoparticles by aerosol technology. Nanostruct. Mater. 12, 45–48.

Mallikarjuna, N.N., Zhang, X., Wu, A., Kolla, H., Manohar, S., 2005. Green chemistry approach to silver and palladium nanoparticles using coffee/tea extracts. American Chemical Society, 229th National Meeting, San Diego, California, Abstract, p. 143.

Martin, C.R., Mitchell, D.T., 1998. Nanomaterials on analytical chemistry. Anal. Chem. 9, 322A–327A.

Maxwell, D.J., Taylor, J.R., Nie, S., 2002. Self-assembled nanoparticle probes for recognition and detection of biomolecules. J. Am. Chem. Soc. 124(32), 9606–9612.

Meltzer, S., Resch, R., Koel, B.E., Thompson, M.E., Madhukar, A., Requicha, A.A.G., Will, P., 2001. Fabrication of nanostructures by hydroxylamine seeding of gold nanoparticle templates. Langmuir 17(5), 1713–1718.

Moiseenko, V.G., Kuimova, N.G., Makeeva, T.B., Pavlova, L.M., 1999. Formation of biogenic gold particles by fungi. Dok. Akad. Nauk 364(4), 535–537.

Molisch, H., 1921. The microchemistry of plants. XVI. The reduction of silver by chlorophyll granules. Ber. Botan. Ges. 39, 136–139.

Nagai, S., 1951. The reduction of silver nitrate by plant cell. II. Nature and responsibility of substances which cause the reduction. J. Inst. Polytech., Osaka City University, Series D: Biology, 2, 1–8.

Nie, S., 2003. Biomedical nanotechnology: from nanoprobes in molecular imaging to nanoscaffolds in tissue engineering. American Chemical Society, 225th National Meeting, New Orleans, LA, Abstract, p. 194.

Obare, S.O., Meyer, G.J., 2004. Nanostructured materials for environmental remediation of organic contaminants in water. J. Environ. Sci. Health A 39(10), 2549–2582.

Okitsu, K., Yue, A., Tanabe, S., Matsumoto, H., Yobiko, Y., 2001. Formation of colloidal gold nanoparticles in an ultrasonic filed: Control of rate of gold(III) reduction and size of formed gold nanoparticles. Langmuir 17, 7717–7720.

Paciotti, G.F., Myer, L., Kingston, D.G.I., Ganesh, T., Tamarkin, L., 2005. Colloidal gold nanoparticles: A versatile platform for developing tumor targeted cancer therapies. Nanotechnology Conference and Trade Show, Anaheim, California, Abstract, pp. 7–10.

Parsons, J.G., Gardea-Torresdey, J.L., Tiemann, K.J., Gamez, G., 2003. Investigation of trace level binding of PtCl6 and PtCl4 to alfalfa biomass (Medicago sativa) using Zeeman graphite furnace atomic absorption spectrometry. Anal. Chim. Acta 478(1), 139–145.

Parsons, J.G., Peralta-Videa, J.R., Tiemann, K.J., Saupe, G.B., Gardea-Torresdey, J.L., 2005. Use of chemical modification and spectroscopic techniques to determine the binding and coordination of gadolinium(III) and neodymium(III) ions by alfalfa biomass. Talanta 67(1), 34–45.

Penn, S.G., He, L., Natan, M.J., 2003. Nanoparticles for bioanalysis. Curr. Opin. Chem. Biol. 7(5), 609–615.

Praharaj, S., Nath, S., Ghosh, S.K., Kundu, S., Pal, T., 2004. Immobilization and recovery of Au nanoparticles from anion exchange resin: Resin-bound nanoparticle matrix as a catalyst for the reduction of 4-nitrophenol. Langmuir 20(23), 9889–9892.

Reeves, R.D., Baker, A.J.M., 2000. Metal accumulating plants. In: Raskin, I., Ensley, B.D. (Eds.), Phytoextraction of toxic metals-using plants to clean the environment. Wiley, New York, NY, pp. 193–229.

Saeed, A., Akhtar, M.W., Iqbal, M., 2005b. Affinity relationship of heavy metal biosorption by the husk of Cicer arietinum (chickpea var. black gram) with their atomic weights and structural features. Fresen. Environ. Bull. 14(3), 219–223.

Saeed, A., Akhter, M.W., Iqbal, M., 2005a. Removal and recovery of heavy metals from aqueous solution using papaya wood as a new biosorbent. Sep. Purif. Technol. 45(1), 25–31.

Sau, T.K., Pal, A., Jana, N.R., Wang, Z.L., Pal, T., 2001. Size controlled synthesis of gold nanoparticles using photochemically prepared seed particles. J. Nanoparticle Res. 3(4), 257–261.

Sawalha, M.F., Gardea-Torresdey, J.L., Parsons, J.G., Saupe, G., Peralta-Videa, J.R., 2005. Determination of adsorption and speciation of chromium species by saltbush (Atriplex canescens) biomass using a combination of XAS and ICP-OES. Microchem. J. 81(1), 122–132.

Schimpf, S., Lucas, M., Mohr, C., Rodemerck, U., Bruckner, A., Radnik, J., Hofmeister, H., Claus, P., 2002. Supported gold nanoparticles: In-depth catalyst characterization and application in hydrogenation and oxidation reactions. Catal. Today 72(1–2), 63–78.

Shankar, S.S., Ahmad, A., Pasricha, R., Sastry, M., 2003a. Bio reduction of chloroaurate ions by geranium leaves and its endophytic fungus yields gold nanoparticles of different shapes. J. Mater. Chem. 13, 1822–1826.

Shankar, S.S., Ahmad, A., Sastry, M., 2003b. Geranium leaf assisted biosynthesis of silver nanoparticles. Biotechnol. Progr. 19(6), 1627–1631.

Shankar, S.S., Rai, A., Ahmad, A., Sastry, M., 2004b. Rapid synthesis of Au, Ag, and bimetallic Au core-Ag shell nanoparticles using Neem (*Azadirachta indica*) leaf broth. J. Colloid Interf. Sci. 275(2), 496–502.

Shankar, S.S., Rai, a., Ahmad, A., Sastry, M., 2005. Controlling the optical properties of lemon grass extract synthesized gold nanotriangles and potential application in infrared-absorbing optical coatings. Chem. Mater. 17, 566–572.

Shankar, S.S., Rai, A., Ankamwar, B., Singh, A., Ahmad, A., Sastry, M., 2004a. Biological synthesis of triangular gold nanoprisms. Nature Mater. 3(7), 482–488.

Tanaka, K., 1999. Nanotechnology towards the 21st century. Thin Solid Films 341, 120–125.

Thomas, K.G., Kamat, P.V., 2003. Chromophore-functionalized gold nanoparticles. Accounts Chem. Res. 36(12), 888–898.

Tolles, W.M., 1996. Nanoscience and nanotechnology in Europe. Nanotechnology 7(2), 59–105.

Voigt, S., 1914. Biological investigations of colloidal silver with a new method for the demonstration of very fine metal deposits in organs. Deut. Med. Wochenschr. 40, 483–484.

Volesky, B., 2001. Detoxification of metal-bearing effluents: Biosorption for the next century. Hydrometallurgy 59(2–3), 203–216B.

Vorobyova, S.A., Sobal, N.S., Lesnikovich, A.I., 2001. Colloidal gold, prepared by interphase reduction. Colloid Surface A 176(2–3), 273–277.

Wei, G.-T., Liu, F.-K., 1999. Separation of nanometer gold particles by size exclusion chromatography. J. Chromatogr. A 836, 253–260.

Wei, G.-T., Liu, F.-K., Wag, C.R.C., 1999. Shape separation of nanometer gold particles by size-exclusion chromatography. Anal. Chem. 71, 2085–2091.

Wei, G.-T., Yang, Z., Lee, C.-Y., Yang, H.-Y., Wang, C.R.C., 2004. Aqueous-organic phase transfer of gold nanoparticles and gold nanorods using an ionic liquid. J. Am. Chem. Soc. 126(16), 5036–5037.

Westcott, S.L., Oldenburg, S.J., Lee, T.R., Halas, N.J., 1998. Formation and adsorption of clusters of gold nanoparticles onto functionalized silica nanoparticle surfaces. Langmuir 14(19), 5396–5401.

Williams, M., 1918. Absorption of gold from colloidal solutions by fungi. Ann. Bot.. 32, 531–534.

Yanagimoto, H., Kuroda, K., Iwamoto, M., Egusa, E., Ikutake, N., Hayashi, S., 2005. Manufacture of metal nanoparticles and nanoparticle dispersion: Japan Kokai Tokkyo Koho, Patent Application JP 2005220435 A2 20050818 Patent written in Japanese. Abstract, p. 14.

Yates, B.J., Myre, E., Breetz, D., Dionysiou, D.D., 2005. Biotemplating of nanoparticles for environmental applications using phytomining techniques. American Chemical Society, 230 National Meeting, Washington, DC, Preprints of extended Abstracts, 45(2), pp. 740–742.

Zhang, J., Han, B., Liu, J., Zhang, X., He, J., Liu, Z., Jiang, T., Yang, G., 2002. Recovery of silver nanoparticles synthesized in AOT/C12E4 mixed reverse micelles by antisolvent CO_2: Chemistry. Eur. J. 8(17), 3879–3883.

Zhang, J., Han, B., Liu, J., Zhang, X., Liu, Z., He, J., 2001. A new method to recover the nanoparticles from reverse micelles: Recovery of ZnS nanoparticles synthesized in reverse micelles by compressed CO_2. Chem. Commun.(24), 2724–2725.

Zhou, Y., Wang, C.Y., Zhu, Y.R., Chen, Z.Y., 1999. A Novel ultraviolet irradiation technique for shape-controlled synthesis of gold nanoparticles at room temperature. Chem. Mater. 11(9), 2310–2312.

Developments in Environmental Science, Volume 5
D. Sarkar, R. Datta and R. Hannigan (Editors)

ISSN: 1474-8177/DOI:10.1016/S1474-8177(07)05022-X

Chapter 22

Phytorestoration of metal-contaminated industrial wasteland: A greenhouse feasibility study

Z.-Q. Lin, H. Hussein, Z.H. Ye and N. Terry

Abstract

Metal-contaminated industrial waste sites pose potential toxic impacts on the surrounding environment. Phytostabilization can be a cost-effective and sustainable technology for the restoration of industrial wasteland. This greenhouse study was conducted to demonstrate the feasibility of phytorestoration at a metal-contaminated industrial waste site near San Francisco Bay, California. The contaminated soil in the field was unable to support vegetation systems due to phytotoxicity of contaminants and poor soil quality. Concentrations of arsenic, mercury, lead, and selenium in the contaminated soil were 3030, 1830, 12600, and 2370 mg kg^{-1}, respectively, and soil pH ranged from 3.6 to 3.8. To improve soil quality for plant growth, the effects of four soil amendments (dolomitic limestone, mushroom compost, gypsum, and organic fertilizer) on soil acidity reduction, seed germination, and shoot biomass production were evaluated with eight plant species, including Fawn tall fescue (*Festuca arundinacea* Schreber), Perennial ryegrass (*Lolium perenne*), Harding grass (*Phalaris aquatia* var Perla), California brome (*Bromus carinatus*), Carolina poplar (*Populus canadensis* L.), Streaker redtop (*Agrostis gigantean* Roth), Reubens Canada blue grass (*Poa campressa* L.), and Indian mustard (*Brassica juncea* L.). The soil amended with dolomite at 8.8 g kg^{-1} and organic fertilizer at 12.5 g kg^{-1} effectively reduced soil acidity and ameliorated soil conditions for optimum plant growth. Among the tested species, Fawn tall fescue was the most tolerant species to the contaminated soil, and, along with the soil amendments, it can be suitable for phytorestoration at metal-contaminated industrial sites.

22.1. Introduction

High levels of metal contaminants in soils are potentially toxic to plants, animals, and humans (Kabata-Pendias and Pendias, 1992). Metal-contaminated industrial wasteland is often times devoid of vegetation and therefore, aesthetically displeasuring, and becomes hazardous to the surrounding environment. To manage such highly contaminated industrial waste sites, plants can be used to stabilize and inactivate soil contaminants in situ, a process termed phytostabilization (US EPA, 2000). Plants forming dense canopies and large root systems on contaminated sites will physically stabilize the contaminated soil against rain impact or leaching, as well as restrict off-site transport of contaminants through lateral wind erosion. Therefore, establishment of vegetation (i.e., phytorestoration) on industrial wasteland effectively reduces migration of pollutants from the contaminated site (Winterhalder, 1996).

Phytorestoration on industrial wasteland is a sustainable and cost-effective approach for pollution control. It may be achieved by the application of soil amendments and subsequent revegetation of the area with tolerant plant species (Adriano et al., 1997). Soil amendments significantly change the chemical speciation and fractionation of metals in the contaminated soil so as to reduce the chemical and biological availability of the contaminants present, as well as to improve the soil as a medium for plant growth. For example, application of ground limestone to acidic and toxic soils increases soil pH from 3.5 to about 6.5 resulting in immediate detoxification (Winterhalder, 1996). Toxic metals would be precipitated as carbonates or hydroxides. If high-soil acidity is corrected by using dolomitic limestone that contains both calcium and magnesium, the soil also benefits from the additional Ca and Mg supplied in dolomite, because high levels of Ca and Mg protect vegetation against trace element toxicity in acidic soils by reducing Al toxicity (Adriano, 2001).

Soil may also be ameliorated by other amendments. For instance, the application of fertilizers and organic materials increases the contents of bioavailable nutrients, e.g. N, P, and K, resulting in better plant growth. Higher levels of organic matter in soil increase adsorption and fixation of metal contaminants in soils. In general, toxic elements become less bioavailable through the physical and chemical processes of adsorption, ion exchange, and precipitation in soil. Mohamed et al. (2005) reported that the biomass yield of rice in As-contaminated soil treated with amorphous Fe/Al oxides and cationic polymers increased 5-fold more than in the control soil.

A suitable plant species for the phytorestoration of industrially contaminated soils should have the following general characteristics: (1) *high tolerance to a wide range of soil pH*. The selected plant species should be able to grow well with a wide range of soil pH change, if the succeeding lime treatment could not be made promptly or the neutralizing power of the limestone would deplete after lime application; (2) *high tolerance to low-soil fertility*. Industrially contaminated soils generally have low levels of soil fertility, especially deficiency in soil phosphorus (Winterhalder, 1996). After several years of plant growth, the nitrogen deficiency may also become another major limiting factor; (3) *high tolerance to high levels of contaminants in soil*. Industrial waste sites are usually co-contaminated with different metals at high concentrations (oftentimes $>1000\,mg\,kg^{-1}$). Therefore, the selected plant species needs to be genetically tolerant to such high levels of contaminants in soil; (4) *high-biomass production*. A greater biomass production will ensure broader plant coverage on ground and intensive root systems in soil. Grass species are highly recommended at the early stage of the phytorestoration because of their fast growth and high tolerance to droughts, insects, and diseases; and (5) *least field management*. A cost-effective phytorestoration technique should require less labor, machinery investment, and other site maintenance costs.

The goal of this research was to study the feasibility of revegetating a metal-contaminated industrial waste site. The specific objectives of this greenhouse study were to (1) determine the effective soil amendment and (2) identify the tolerant plant species that will facilitate phytorestoration of metal-contaminated industrial wasteland.

22.2. Materials and methods

22.2.1. Sampling in the field

The metal-contaminated industrial waste site was located near the San Francisco Bay, California. There was no vegetation growing in the contaminated soils. The soil profiles showed that the industrial chemical wastes were mixed with soil, and that the contaminated soil layer extended down to 150 cm in depth. Five randomly selected soil profiles were sampled at 0–20 and 20–40 cm depth, respectively. Soil samples were stored and transported in clean plastic bags.

22.2.2. Greenhouse experiments

The contaminated soil was brought back from the site to the UC Berkeley greenhouse for pot experiments. The soils were air-dried and large-size rocks were removed. The dry contaminated soil was thoroughly mixed with the selected amendment materials (see below). Each plastic pot (6″) contained 1.6 kg the experimental soils. There were three replicates of each treatment.

Soil amendment materials. It was hypothesized that low-soil pH or strong acidity of the contaminated soil, and consequently low levels of soil fertility (e.g., P), might be important factors responsible for the lack of vegetation on the contaminated site. Therefore, the selection of soil amendments was focused on three primary aspects: soil acidity, soil fertility, and soil structure. The soil amendment treatments included the addition of dolomitic limestone, organic fertilizer, mushroom compost, and gypsum. The general properties of the four selected soil amendment materials are as follows: (1) Dolomite: Commercially available GreenAll® (Suisun, CA) dolomite was used. It was derived from natural dolomite limestone, which contained 24.9% of Ca and 11.4% of Mg (or 49% of calcium carbonates and magnesium carbonate). (2) Organic fertilizer: The organic fertilizer used was derived from the bone meals, kelp, alfalfa, dried poultry waste, bat guano, and blood meal. It was a natural organic transplanting fertilizer, containing 2% water insoluble total nitrogen, 4% available phosphorous as P_2O_5, and 2% soluble potassium as K_2O. (3) Mushroom compost: The compost was moist, containing 69.4% of water, and with a pH of 6.8. (4) Gypsum: This was obtained commercially from GreenAll® (Suisun, CA). The gypsum (92% $CaSO_4 \cdot 2H_2O$) contained 21.0% Ca and 17% S.

Plant species. Eight plant species were selected, including Fawn tall fescue (*Festuca arundinacea* Schreber), Perennial ryegrass (*Lolium perenne*), Harding grass (*Phalaris aquatia* var Perla), California brome (*Bromus carinatus*), Carolina poplar (*Populus canadensis* L.), Streaker redtop (*Agrostis gigantean* Roth), Reubens Canada bluegrass (*Poa campressa* L.), and Indian mustard (*Brassica juncea* L.). Except Carolina poplar that was cultured from shoot-stakes, seeds of the other selected plant species were sowed into ~0.5 cm below soil surface. Pots were watered with tap water every two days.

Experiment 1. To determine the effectiveness and application rates of amendments, each of the four selected materials was tested with Fawn tall fescue, a plant species that is generally tolerant to poor soil conditions (Duble, 1999). The three application rates were 2.9, 4.8, and 6.7 $g\,kg^{-1}$ for mushroom compost; 3.8, 6.3, and 8.8 $g\,kg^{-1}$ for Gypsum; 3.8, 6.3, and 8.8 $g\,kg^{-1}$ for Dolomite; and 6.3, 9.4, and 12.5 $g\,kg^{-1}$ for organic fertilizer. The original contaminated soil was used as the control. Germination rates and shoot-biomass yields of Fawn tall fescue and soil pH were determined on Day-140 after seed germination.

Experiment 2. To determine the plant species that is most tolerant to the contaminated soil, a total of 150 seeds of each selected species, except for Indian mustard with 15 seeds and Carolina poplar by three shoot-stakes, were sowed into the contaminated soil amended with dolomite at a rate of $8.8\,g\,kg^{-1}$. The eight plant species were also grown in the clean soil (i.e., UCMix, growth medium containing peat moss, 36%, Colma sand, 64%, and 9 kg organic fertilizer m^{-3}, Ye et al., 2003) and the original contaminated soil as controls. The same observations (see Experiment 1) were conducted on Day-80 after seed germination. The tolerance index (%) of each tested species was calculated as the ratio of the total dry shoot biomass on dolomite-amended contaminated soil to the total dry shoot biomass on the clean soil (Lin and Huang, 1990).

Experiment 3. To determine the effectiveness of various combinations of the selected amendments for phytorestoration on the contaminated soil, four combinations of the selected amendments were examined for Fawn tall fescue growth and soil acidity neutralization. The application rate of each amendment was $6.7\,g\,kg^{-1}$ for mushroom compost, $8.8\,g\,kg^{-1}$ for Gypsum, $8.8\,g\,kg^{-1}$ for Dolomite, and $12.5\,g\,kg^{-1}$ for organic fertilizer. Measurements of soil pH, seed germination, and shoot biomass were conducted on Day-60 after seed germination. The contaminated soil without amendment treatments was used as the control.

22.2.3. Physio-chemical properties of soils

The soil sample (about 1 kg) was air-dried and passed through a 2-mm sieve. Measurements of soil physio-chemical properties were conducted by ANR Analytical Laboratory (University of California, Davis) according to the standard methods (Page et al., 1982), except for organic matter (OM) (Nelson and Sommers, 1982), particle size distribution (Gee and Bauder, 1979), SMP soil buffer pH (a measure of the soil lime requirement) (Shoemaker et al., 1961), and soil test P (Horneck et al., 1989; Diamond, 1995). Soil pH was measured in deionized distilled water (1:1 soil to water, w/w) using a Corning Checkmate Modular Testing System (Corning, NY).

22.2.4. Chemical analysis of metals

Soil samples were ground and passed though a 0.25-mm sieve prior to wet digestion with HNO_3, H_2O_2, and HCl for chemical analysis (Method 3050B, see US EPA, 1996). A total of 25 elements (Al, As, B, Ba, Ca, Cd,

Co, Cr, Cu, Fe, Hg, K, Mg, Mn, Mo, Na, Ni, Pb, S, Sb, Sc, Se, Sr, V, and Zn) were determined. Total element concentrations in digests were determined following the procedures described in Method 7742 (US EPA, 1994) by inductively coupled plasma emission spectroscopy, except for Se and Hg determined using atomic absorption spectrophotometry with a hydride generator. The standard reference material of SRM-2709 (the San Joaquin soil) (NIST, 1996) was analyzed as an external quality control.

22.2.5. Statistical analysis

Statistical analysis was performed using the statistical analysis system (SAS Institute, 1988a,b). The multiple comparisons of the means were performed using PROC GLM with *Tukey* option.

22.3. Results and discussion

22.3.1. Physical and chemical properties of soil

The physical and chemical properties of the contaminated soil are compiled in Table 22.1. The soil properties, particularly in the topsoil (0–20 cm depth), are important for evaluating the site for future phytorestoration. The contaminated soil was characterized by low-soil pH varying from 3.6 to 3.8 and low-clay content (from 14 to 16% dry weight). The total and extractable macronutrients in the contaminated soil, such as N, P, K, Ca, and Mg, were found generally within their range of variation in agricultural soils. The low levels of CEC and clay content indicated that the contaminated soil had a low pH-buffering capacity. When soil pH values lay below 4, most plant nutrients would become unavailable in soil, except Fe and Mn (Hewitt and Smith, 1974). The reduced availability of P would be an important limiting factor to plant growth, because it can be bounded strongly by Al- and Fe-oxides in the acidic soils. Additionally, the contaminated soil with low levels of clay content and CEC (i.e., 6 to 10 cmol (+) kg^{-1}) had a low-adsorption capacity to bind metal contaminants.

22.3.2. Soil metal concentrations

The chemical analysis suggested that the potentially phytotoxic metal contaminants in the soil (0–20 cm and 20–40 cm depths) at the

Table 22.1. Soil physio-chemical properties at the contaminated site. Values are means ± standard deviations ($n = 5$)

	Soil depth (cm)	
	0–20	20–40
Soil pH (water) (2:1)	3.8 ± 0.4	3.6 ± 0.2
Organic matter (OM) (%)	2.9 ± 0.2	4.1 ± 1.1
Salinity (EC) ($dS\,m^{-1}$)	5.42 ± 1.88	4.95 ± 0.71
Cation exchange capacity (cmol (+) kg^{-1})	6.2 ± 4.3	10 ± 11.3
Sand, silt, and clay (%)	33 ± 8; 53 ± 7; 14 ± 1	33 ± 4; 50 ± 5; 16 ± 1
Total $CaCO_3$ (%)	<0.2%	0.2%
Total C (%)	4.3 ± 1.2	0.2 ± 0.1
Total N (%)	0.17 ± 0.03	0.17 ± 0.06
Total P (%)	0.05 ± 0.01	0.07 ± 0.05
Extractable K (cmol (+) kg^{-1})	0.1 ± 0	0.15 ± 0.1
Extractable Ca (cmol (+) kg^{-1})	42.2 ± 24.6	41.0 ± 53.5
Extractable Mg (cmol (+) kg^{-1})	4.1 ± 1.7	5.2 ± 1.1
Soil test P (Bray-2) ($\mu g\,g^{-1}$)	13.7 ± 9.9	14.2 ± 5.9

contaminated site include Al, As, Pb, Hg, and Se (Table 22.2). Concentrations of these five selected elements were generally well above 1000 $\mu g\,g^{-1}$. For example, Al concentrations ranged from 23,600 to 38,900 $\mu g\,g^{-1}$, As from 1600 to 3000 $\mu g\,g^{-1}$, Pb from 5790 to 12,600 $\mu g\,g^{-1}$, Hg from 1700 to 1800 $\mu g\,g^{-1}$, and Se from 2000 to 2400 $\mu g\,g^{-1}$. In addition, concentrations of B, Cd, Cu, Mn, Ni, and Zn were also excessively high compared to normal agricultural soils (Angelone and Bini, 1992). The phytotoxicity of metal contaminants in the soil is greatly aggravated by a low-soil pH of <4. Although the contents of bioavailable elements were not determined in the contaminated soil, a low-soil pH likely increases the bioavailability of Al, Mn, Cu, and Ni, this in turn leads to uptake of toxic quantities of these metals in plants (Adriano et al., 1997).

Concentrations of metal contaminants generally increased with soil depth at the contaminated site (Table 22.2). In particular, Al, As, B, Fe, Pb, Sc, and V showed the most striking changes (2- to 70-fold) with soil depth (from 0–20 to 20–40 cm depth). For example, Pb concentrations varied from 5790 $\mu g\,g^{-1}$ (at 0–20 cm depth) to 12,600 $\mu g\,g^{-1}$ (at 20–40 cm). Nickel, V, and Cr concentrations, however, decreased with soil depth. The higher levels of contaminants at a deeper soil depth suggest that the downward movement of contaminants (e.g., As and Pb) may have occurred in the profile of this acidic and light texture soil.

High-soil acidity may cause direct injurious effects on plants by inhibiting root growth. It also indirectly affects bioavailability of P in soil, because of the active P-adsorption to Al- and Fe-oxides at low pH. Such

Table 22.2. Concentrations ($mg\,kg^{-1}$) of selected elements in soils collected from the contaminated site in February 1998. Values are means (AVG) and standard deviation (STD)

Element	0–20 cm		20–40 cm	
	AVG	STD ($n = 5$)	AVG	STD ($n = 5$)
Al	23,600	2930	38,900	15,700
As	1590	641	3030	2580
B	75	26	164	47
Ba	153	28	265	119
Ca	8150	3340	12,800	7620
Cd	13	3	18	7
Co	8	1	12	5
Cr	83	7	195	107
Cu	319	25	833	655
Fe	106,000	16,800	149,000	51,500
Hg	1580	570	1830	360
K	2610	233	7420	4220
Mg	3470	512	7850	4570
Mn	152	20	219	82
Mo	21	1	52	29
Na	1710	437	4750	4290
Ni	33	5	53	20
Pb	5790	2270	12,600	11,800
S	13,800	4460	20,400	7690
Sb	110	30	146	43
Sc	10	4	723	656
Se	2090	740	2370	740
Sr	55	14	100	49
V	60	2	170	99
Zn	707	151	1060	459

indirect acidic effects on soil fertility are especially critical at the study site due to the extremely high-soil concentrations of Al and Fe (Table 22.2). The low levels of cation exchange capacity (CEC) and low-clay contents in the contaminated soil could also reduce the water and mineral nutrient capacity of the soil, as well as provided fewer electrostatic adsorbing sites available for the reduction of toxic metals in the soil solution.

22.3.3. Selecting effective soil amendments on contaminated soils

The addition of different amendments to the contaminated soil has different effects on soil acidity, seed germination, and plant biomass production of Fawn tall fescue (Table 22.3). The best soil amendment for improving plant growth was organic fertilizer; an application rate of

Table 22.3. Biomass production of Fawn tall fescue in the contaminated soil treated with different soil amendments. The shoot biomass of Fawn tall fescue and soil pH were determined 140 days after seed germination. Values are means and standard deviations ($n = 3$)

Treatment	Rate ($g\,kg^{-1}$)	pH	Germination (%)	Biomass ($g\,pot^{-1}$)
Unamended soil	Control	3.7±0.3	1±1	ns
Mushroom compost	2.9	4.0±0.0	17±4	ns
	4.8	4.0±0.0	23±6	ns
	6.7	3.8±0.3	23±20	ns
Gypsum	3.8	3.8±0.3	13±12	ns
	6.3	3.8±0.3	15±4	ns
	8.8	3.8±0.3	13±4	ns
Dolomite	3.8	4.5±0.0	37±9	2.5±0.9
	6.3	4.5±0.0	49±12	6.8±2.7
	8.8	4.8±0.3	56±11	9.9±0.8
Organic fertilizer	6.3	4.2±0.6	31±14	11.9±11.9
	9.4	3.8±0.3	43±15	34.3±4.1
	12.5	4.3±0.3	50±16	45.5±1.6

ns, seedlings did not survive after seed germination.

12.5 $g\,kg^{-1}$ yielded the highest biomass of 45.5 g per pot. In addition, the only other effective soil amendment was dolomite, with a yield of 9.9 g per pot at the application rate of 8.8 $g\,kg^{-1}$. The biomass production was significantly ($P<0.05$) lower at the lowest application rate of dolomite (i.e., 3.8 $g\,kg^{-1}$), compared to the other two application rates. In contrast, the addition of gypsum or mushroom compost did not enable the tested species to grow on the amended soil, even though the treatments had similar germination rates compared with the other treatments of organic fertilizer or dolomite. The addition of dolomite limestone at a rate of 8.8 $g\,kg^{-1}$ significantly increased soil pH from 3.7 to 4.8. The other soil amendments did not change soil pH significantly. Seeds of all tested species were unable to germinate (or $<1\%$) on the original contaminated soil. Clearly, Fawn tall fescue survived only when the contaminated soil was amended with organic fertilizer or dolomite.

Since plant biomass production increased with increasing the rate of organic fertilizer in Table 22.3, the growth of Fawn tall fescue would have been increased more if the fertilizer had been applied at a higher rate. The addition of fertilizer improved soil fertility by increasing the supply of N, K, and P, as well as an increase of soil pH from 3.7 to 4.3. Although the addition of mushroom compost also provided organic materials for the improvement of soil physical property, it did not improve plant growth because it provided little bioavailable nutrients during the course of this experiment.

There was no significant difference between the 6.3 and 8.8 g kg^{-1} application rates of dolomite (*P*0.05). This indicates that biomass production would likely not increase significantly if the application rate of dolomite increased. It would appear that the important factors for revegetation of the contaminated soil were soil fertility and acidity. The addition of dolomite provided more Ca and Mg that are known to ameliorate the toxic effects of Al on roots. Possible roles of Ca and Mg ions in the recovery of metal toxicity by dolomitic limestone may be in the competitive exclusion of metal ions from the root-hair exchange complex, and by improving root membrane integrity (Hutchinson and Collins, 1978). Lime application increases soil capacity for metal binding and restricts the solubility of metals in soil. A similar effect can be expected following the application of organic fertilizer (Chaney et al., 1999). Additionally, toxic metals may also precipitate as carbonate or hydroxide compounds when soil pH increases.

22.3.4. Screening plant species for tolerance to the contaminated soil

Different plant species differ in their ability to grow in metal contaminated soils. Eight plant species were tested for germination rates and aboveground biomass production on the soil amended with 8.8 g kg^{-1} of dolomite. Dolomite was selected over organic fertilizer because the dolomite-treatment has less biomass production or greater growth stress than the organic fertilizer treatment (Table 22.3). Results show that the tested plant species were not able to grow on the contaminated soil without soil amendment. With the dolomite treatment, most of the tested plant species had very low-seed germination rates (i.e., $<20\%$), except for Fawn tall fescue and perennial ryegrass (Table 22.4). Indian mustard seeds were not able to germinate on the dolomite-amended contaminated soil. Of the eight plant species tested, Fawn tall fescue had the highest germination rate and produced the highest level of shoot biomass, followed by the perennial ryegrass.

The tolerance indexes of the eight plant species to the contaminated soil were shown in Fig. 22.1, indicating that the Fawn tall fescue was a good candidate for phytorestoration at the contaminated site. Tolerance is thought to be the constitutive property presenting in every cell, tissue, organ of the plant, and/or the whole plant (Ernst et al., 1992). Plant tolerance to metal contaminants can be demonstrated at different levels of integration. For phytorestoration of the contaminated sites, high levels of the aboveground biomass production are desired, because they would generate effective plant coverage of the contaminated soil. Therefore, our

Table 22.4. Tolerance of the tested plant species to the contaminated soil amended with 8.8 g dolomite per kg soil. The shoot biomass and soil pH were measured at Day-80 after seed germination. Values are means $\pm$ standard deviations ($n = 3$)

Species	pH	Germination (%)	Biomass (g pot^{-1})
Fawn tall fescue (*Festuca arundinacea* Schreber)	5.0±0.2	35±5	4.0±2.3
Perennial ryegrass (*Lolium perenne*)	4.9±0.1	27±11	1.2±0.9
Harding grass (*Phalaris aquatia var Perla*)	4.9±0.1	4±2	0.9±0.6
California brome (*Bromus carinatus*)	4.8±0.1	14±4	0.4±0.3
Carolina poplar (*Populus canadensis* L.)	4.9±0.1	N/A	0.3
Streaker redtop (*Agrostis gigantean* Roth)	4.9±0.1	1±1	0.1±0.1
Reubens Canada bluegrass (*Poa campressa* L.)	4.8±0.1	18±10	0.1±0.1
Indian mustard (*Brassica juncea* L.)	5.0±0.2	0	0

N/A, not applicable.

selection of high-tolerant plant species to the contaminated soil was primarily based on aboveground biomass production, and the tolerance index was defined as the ratio (%) of the total dry shoot biomass on the dolomite-treated contaminated soil to the total dry shoot biomass on the clean soil. Clearly, a dense vegetative groundcover can greatly reduce soil erosion and human exposure to metal contaminants from the contaminated site (US EPA, 2000).

22.3.5. Interactive effects of selected amendments on growth of Fawn Tall Fescue

Effects of different interactions between the dolomite and organic fertilizer, as well as with other two selected soil amendments (i.e., mushroom compost and Gypsum), on seed germination, soil pH, and shoot biomass production of Fawn tall fescue are compiled in Table 22.5. Different combinations of the four selected amendment materials showed different

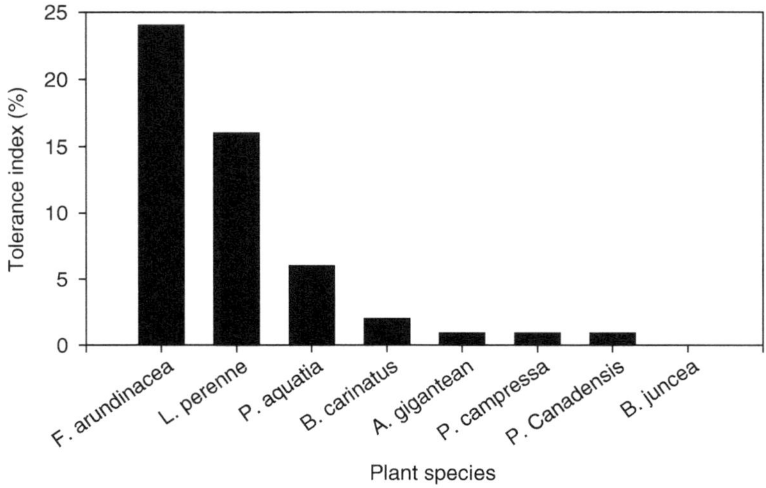

Figure 22.1. Tolerance index of the tested plant species to the contaminated soil amended with 8.8 g dolomite per kg soil. The dry plant biomass on the clean soil (i.e., UCMIX®, containing peat moss, 36%, Colma sand, 64%, and 9 kg organic fertilizer m^{-3}) is not shown.

Table 22.5. Interactive effects of different soil amendments on soil pH, seed germination, and shoot biomass production of Fawn tall fescue. The biomass and soil pH were determined 60 days after seed germination. The data are presented in means with the standard deviations ($n = 3$). The means with different letters in the same column are significantly different ($P < 0.05$)

Combination of amending materials	pH	Germination (%)	Biomass (g pot^{-1})
Control (the contaminated soil)	3 ± 0.2^{c}	0.5 ± 1.0	ns
Dolomite	4 ± 0.3^{ab}	57 ± 9	4.9 ± 0.5^{b}
Organic fertilizer	4.1 ± 0.2^{b}	44 ± 18	12.1 ± 2.3^{a}
Dolomite + organic fertilizer	5.3 ± 0.1^{a}	45 ± 18	10.1 ± 0.5^{a}
Dolomite + organic fertilizer + gypsum	4.8 ± 0.5^{ab}	81 ± 7	6.5 ± 1.3^{b}
Dolomite + organic fertilizer + compost	5.4 ± 0.1^{a}	59 ± 11	13.9 ± 1.6^{a}
Dolomite + organic fertilizer + gypsum + compost	5.3 ± 0.1^{a}	69 ± 9	11.1 ± 0.4^{a}

ns, seedlings did not survive after seed germination.

effects on the ability to support Fawn tall fescue growth. Data in Table 22.5 show that soil acidity was reduced significantly when the soil was treated with both dolomite and organic fertilizer. The biomass production data showed that the addition of organic fertilizer most effectively improved the soil quality for plant growth. Shoot biomass productions on the contaminated soils amended with dolomite or with dolomite + gypsum + organic fertilizer were significantly lower than on the soils treated with organic fertilizer, dolomite + organic fertilizer, or dolomite + organic fertilizer + mushroom compost. The addition of gypsum did not show any beneficial effects compared to the other three amendments, likely due to low-soil acidity and Al-toxicity to plant roots. The most cost-effective soil amendment, therefore, was the combination of the dolomite and organic fertilizer.

Fawn tall fescue is a variety of tall fescue, a cool-season bunchgrass that is native to Europe and North Africa. Tall fescue (*Festuca arundinacea* Schreb.) was introduced into the United States in the early 1800s, and Fawn tall fescue was first developed in Oregon in 1954 (Duble, 1999). Fawn tall fescue was found to be the best plant species among the eight plant species screened in our greenhouse experiments. It meets the criteria (see Introduction) for selecting suitable plant species for the phytorestoration of the contaminated site. Fawn tall fescue grows well in soil with a pH of 4.7–9.5, although forage production is best when soil pH is maintained between 5.5 and 8.5 (Duble, 1999). Furthermore, it can adapt to a wide range of soils and is tolerant of low-soil fertility. Fawn tall fescue has a massive root system that aids in erosion control, and in surviving drought and flood. It has greater forage and seed yield, compared to other fescue varieties. Although tolerant of low-fertility conditions, fescue requires moderate fertility levels for good production; it is very responsive to nitrogen fertilization (Duble, 1999). Proper watering is important to the survival of Fawn tall fescue, as it requires frequent watering during the summer months (Duble, 1999).

22.4. Conclusions

The limitation for the phytorestoration of the metal contaminated site can be overcome by increasing soil fertility and decreasing soil acidity through the applications of soil amendments. The application of dolomitic limestone at 8.8 g kg^{-1} or organic fertilizer at 12.5 g kg^{-1} in the topsoil layer was essential for plant growth on the contaminated soil. The combination of dolomite and organic fertilizer was suggested to be a cost-effective soil amendment for the phytorestoration of the contaminated site. Fawn tall fescue was identified to be the best plant species for the revegetation, due to its highest tolerance to contaminants in the contaminated soil, and a high aboveground biomass production. Further field trials are needed to validate the findings from this greenhouse study and to implement the phytorestoration technology under field conditions.

ACKNOWLEDGMENTS

The authors thank Dr. E.A. Elkhatib at Alexandria University, Egypt, and Dr. T.Q. Zhang at Agriculture and Agri-Food Canada for their constructive review comments, and Mr. Peter Cushing at the University of California at Berkeley for his assistance in conducting the greenhouse experiments.

REFERENCES

Adriano, D.C., 2001. Trace Elements in Terrestrial Environments: Biogeochemistry, Bioavailability, and Risks of Metals. Springer-Verlag, New York.

Adriano, D.C., Chlopecka, A., Kaplan, D.I., Clijsters, H., Vangronsveld, J., 1997. Soil contamination and remediation: Philosophy, science and technology. In: Prost, R. (Ed.), Contaminated Soil. INRA, Paris, pp. 465–504.

Angelone, M., Bini, C., 1992. Trace element concentrations in soils and plants of Western Europe. In: Adriano, D.C. (Ed.), Biogeochemistry of Trace Metals. Lewis Publishers, Ann Arbor, pp. 19–60.

Chaney, R.L., Ryan, J.A., Brown, S.L., 1999. Metals. In: Anderson, W.C., Loehr, R.C., Smith, B.P. (Eds.), Environmental Availability in Soils: Chlorinated Organics, Explosives, Metals. American Academy of Environmental Engineers, Annapolis, MD, pp. 111–154.

Diamond, D., 1995. Phosphorus in soil extracts. QuikChem Method 10-115-01-1-A. Lachat Instruments, Milwaukee, WI.

Duble, R.L., 1999. Tall fescue. Available at http://www.aggie-horticulture.tamu.edu/Plantansers/turf/publication/tallfesc.html

Ernst, W.H.O., Verkleij, J.A.C., Schat, H., 1992. Metal tolerance in plants. Acta Bot. Neerl. 41, 229–248.

Gee, G.W., Bauder, J.W., 1979. Particle size analysis by hydrometer: A simplified method for routine textural analysis and a sensitivity test of measurement parameters. Soil Sci. Soc. Am. J. 43, 1004–10007.

Hewitt, E.J., Smith, T.A., 1974. Plant Mineral Nutrition. John Willey & Sons, New York.

Horneck, D.A., Hart, J.M., Topper, K., Koespell, B., 1989. Methods of soil analysis used in the soil testing laboratory at Oregon State University. Agric. Exp. Stn. SM 89, 4.

Hutchinson, T.C., Collins, F.W., 1978. Effect of H^+ ion activity and Ca^{2+} on the toxicity of metals in the environment. Environ. Health Perspect. 25, 47–52.

Kabata-Pendias, A., Pendias, H., 1992. Trace elements in soils and plants, second ed.. CRC press, Boca Raton, p. 365.

Lin, Z.Q., Huang, H.-Y., 1989. Tolerance of woody-plants to mercury. Acta Ecolog. Sinica 9(4), 315–319.

Mohamed, H., Therapong, C., Andra, S., Datta, R., Sarkar, D., 2005. Phyto-availability of arsenic in pesticide-applied soils: Effect of chemical remediation. American Geochemistry Union. The 2005 Joint Assembly, May 23–27. New Orleans, Louisiana.

Nelson, D.W., Sommers, L.E., 1982. Total carbon, organic carbon, and organic matter. In: A.L. Page (Ed.), Methods of Soil Analysis: Part 2. Chemical and Microbiological Properties. Monograph Number 9, second ed. ASA, Madison, WI, pp. 539–579.

NIST (National Institute of Standards and Technology), 1996. Standard Reference Materials. NIST Special Publication 260.

Page, A.L., Miller, R.H., Keeney, D.R., (Eds.), 1982. Methods of Soil Analysis. Part 2—Chemical and Microbiological Properties. ASA Monograph No. 9.

SAS Institute., 1988a. SAS procedures guide. Version 6.03 Edition. Statistical Analysis Systems Institute, Inc., Cary, NC, p. 441.

SAS Institute., 1988b. SAS/STAT. User's Guide. Release 6.03 Edition. Statistical Analysis Systems Institute, Inc., Cary, NC, p. 1028.

Shoemaker, H.E., McLean, E.O., Pratt, P.F., 1961. Buffer methods for determining lime requirement of soils with appreciable amount of exchangeable aluminum. Soil Sci. Soc. Am. Proc. 25, 274–277.

US EPA., 1994. Method 7742. Selenium (atomic absorption, borohydride reduction). In: Methods for Chemical Analysis of Water and Wastes.

US EPA. 1996. Method 3050B. Acid digestion of sediments, sludge, and soils. p. 11. In: Methods for Chemical Analysis of Water and Wastes.

US EPA, 2000. Introduction to Phytoremediation. Office of Research and Development. U.S. Environment Protection Agency, Cincinnati, Ohio, EPA 600-R-99-107.

Winterhalder, K., 1996. Environmental degradation and rehabilitation of the landscape around Sudbury, a major mining and smelting area. Environ. Rev. 4, 185–224.

Ye, Z.H., Lin, Z.-Q., Whiting, S.N., de Souza, M.P., Terry, N., 2003. Possible use of constructed wetland to remove selenocyanate, arsenic, and boron from electric utility wastewater. Chemosphere 52, 1571–1579.

Developments in Environmental Science, Volume 5
D. Sarkar, R. Datta and R. Hannigan (Editors)

ISSN: 1474-8177/DOI:10.1016/S1474-8177(07)05023-1

Chapter 23

Linkages between diet and metal accumulation in crayfish

Matthew Horton and Robyn Hannigan

Abstract

Crayfish plays an important role in the environmental biogeochemical cycling of metals in terrestrial aquatic ecosystems by processing sediments, organic matter, algae, and phytoplankton, as well as consuming detritus, carrion, and various fish species.

Although their importance in biogeochemical cycling has been long established little is known of the fractionation of metals from food source to tissue. Understanding of elemental behavior at this level therefore provides critical insight into the role of sediment processors in the biogeochemical cycling of metals. Crayfish can potentially store metals in both soft and hard tissue. This has consequences for bioaccumulation. For example, if crayfish shed most accumulated metals through molting, and the exoskeleton is buried or otherwise left undisturbed, one can consider the exoskeleton a net sink for metals. To explore the partitioning of metals in crayfish tissue, crayfish were collected at the Spring River in Hardy, AR. Treatment groups were selected with each being fed a diet consisting of known concentrations of metals as well as two control groups which were fed standard, non-spiked diets. Metal concentrations in the gills, muscle, and exoskeleton over the duration of the experiment were analyzed by DRC-ICP-MS. In the case of tissues there is a consistent and predictable relation between diet and tissue chemistry. Certain metals such as Cd and Pb were found in the exoskeleton in amounts proportional to that of the diet. Concentrations in the exoskeleton increased over time as expected. Our data suggests that biomagnification of metals in crayfish tissue is directly related to environmental concentrations with crayfish acting as passive "processors" rather than "fractionators" of metals.

23.1. Introduction

The objectives of this study are to identify what metals are taken up by crayfish and determine if they are stored or excreted, determine if the metals are taken in through the gills or the diet, and identify if crayfish can be used to biomonitor metals. It is hypothesized that diet is a pathway for metal storage in the crayfish tissues and through a metal-spiked diet, crayfish will store and accumulate some metals.

The crayfish is an invertebrate, common to many rivers, streams, lakes, ponds, and floodplains throughout North America (Hobbs, 1989). Crayfish occupy a wide range of habitats because of ecological, behavioral, and physiological adaptations through nearly 300 million years of evolution (Holdich and Lowery, 1988; Hasiotis and Mitchell, 1993; Hasiotis, 1999). Most species of crayfish are able to survive equally well in both moving and standing water bodies provided there are suitable physical, biological, and chemical conditions. Some have the ability also to colonize aquatic systems with such extremes in water quality as varying pH and temperature (Holdich and Lowery, 1988). Crayfish have the ability to accumulate high concentrations of toxic metals without experiencing unfavorable effects (Geisy et al., 1980).

Some species of crayfish have the ability to regulate such physiologically important metals as Zn (Mirenda, 1986) and Cu (Anderson, 1978). Other metals like Cd and Pb, which are not physiologically important, have been shown to accumulate passively in crayfish in proportions similar to those in their environments (Anderson, 1978). Regardless of the mechanism of uptake, active or passive, crayfish play an important role in the distribution of metals in the food web of aquatic environments by feeding on algae and organic matter, processing sediment and serving as a major prey species for many kinds of fish. For this reason, if metals are accumulated in crayfish through diet or by intake through the gills, they could be accumulated subsequently in fish and other vertebrates higher up in the food chain or effectively immobilized by storage in the molt. By interacting with sediments, crayfish may also play an important role in the redistribution of metals. Periphyton and phytoplankton, upon which crayfish feed, are known to participate in exchange processes of metals. For example, high concentrations of lead, zinc, and copper were found in algae samples from a metal-polluted Missouri stream (Hasset et al., 1980). Periphyton and phytoplankton can therefore serve as a source of metals in crayfish (Hart, 1982).

Studies show that crayfish accumulate some metals such as Cd and Pb based on relative concentrations in the environment (Anderson, 1978) and also through increased exposure to low pH waters (Young and

Harvey, 1990; DiStefano and Neves, 1991). Crayfish from river sites with high Cd and Pb inputs were found to have high levels of the metals (2.22 ppm Cu and 27.39 ppm Pb per dry weight of crayfish). If crayfish muscle, gills, and exoskeleton are sinks for certain metals it may be possible to use their chemistry for bio-monitoring of those metals in the aquatic ecosystem.

23.2. Methods

The experimental conditions consisted of eight 10-gallon aquaria tanks with two of these tanks serving as controls and the other six as treatments. Six crayfish were placed in each control and treatment. Six pieces (7.62 cm each) of PVC pipe were placed in each tank to serve as hiding areas for the crayfish and to minimize carnivorous behavior.

The metal-enriched food was prepared using Wardley shrimp pellets. The pellets were powdered using a Zircon mill. 283.5 g (10 ounces) of the shrimp meal was used for each treatment, with the two control groups receiving an unaltered diet. Using the prepared stock solution (Table 23.1), aliquots, along with the shrimp meal, were added to 500 ml of milli-Q water (18.2 Mohm) to obtain the desired concentrations given in Table 23.1. All of the metal standards were metal salts. Upon mixing there is no assumption that these elemental metals were speciated to a more bioavailable form. Rather ingestion of dilute metals salts, inorganic species, and associated accumulation of the metals can be attributed to diet and metabolic transformation of non-bioavailable metals. After thorough homogenization, a 60 cc flushing syringe was used to re-pellet the mixture onto trays that were then heated at 250°C for 2 h to dry out the pellets. While the crayfish food (excluding control) was metal spiked in increasing concentrations, the sub-samples taken do not reflect these increments.

Water samples were collected every two months over the course of the study for a total of four samples per tank. Food samples were taken before and after the duration of the experiment and analyzed for metal

Table 23.1. Concentrations of metals in each treatment in $\mu g\, g^{-1}$. Standards for spiking of shrimp pellets were made from the metal salts as shown

Formula	Stock	T1	T2	T3	T4	T5	T6
$HgCl_2$	2715	271.5	27.15	2.715	0.2715	0.02715	0.002715
$Cd(CH_3CO_2)_2 \cdot 2H_2O$	26653.4	2665.34	266.534	26.6534	2.66534	0.266534	0.026653
$Pb(NO_3)_2$	33120	3312	331.2	33.12	3.312	0.3312	0.03312

content. The lack of homogeneity of the food samples could be a limitation to the experimental design because unless all the food is consumed, the crayfish do not receive the total intended dose of metals.

A single species of crayfish (*Cambarus hubbsi)*, collected from the Spring River at Hardy, Arkansas, were fed a metal-enriched diet for an eighth month period from early August 2003 to end of March 2004. Molts and carcasses of dead crayfish were collected 2–3 times a week and frozen for subsequent analysis. Although the hepatopancreas is an important organ for food digestion and waste processing, it was not sampled. In this study we were only concerned with metal storage in the soft tissues and the exoskeleton from the diet, not in the digestive and physiological organs.

Upon termination of the study, remaining crayfish were euthanized and frozen. Crayfish were dissected and the gill, muscle tissue, claw, and exoskeleton separated. All crayfish parts and molts were placed, separately, into 50 ml Teflon tubes. 1 ml of ultra-pure nitric acid and 1 ml of 35% hydrogen peroxide were added to each tube. Weights of crayfish parts varied from 10–1000 mg and the reported concentrations are weight corrected. Watch glasses were placed on the top of each tube and then placed on a digestion block at 80°C for about 3 h. Upon complete digestion, the watch glasses were removed and the samples were dried to a cake in the digestion block. After drying, the samples were brought into a solution by adding 0.5 ml of ultra-pure nitric acid along with 2 ml of an internal 1000 ppb In standard. Final internal standard concentration was 40 ppb. The samples were then brought to 50 ml volume in volumetric tubes using 18.2 Mohm water.

Concentrations of Cd, Hg, and Pb were measured by inductively coupled plasma mass spectrometry (PerkinElmer ELAN 9000) following EPA 6020. Concentrations were calculated based on least squares regression of five elemental calibration standards (0.1 ng g^{-1} to 1 μg g^{-1}, spiked with 40 ng g^{-1} Indium). Quality control was measured by two

Table 23.2. Average treatment concentrations (μg g^{-1}) for whole body total metal content

	Cd	Hg	Pb
T1	365.5953	2395.846	383.3205
T2	203.7375	531.6307	161.7169
T3	165.9837	360.6048	96.01894
T4	67.84018	15.04965	86.31192
T5	48.84657	437.3458	93.20399
T6	383.998	148.8616	266.9582
C1	85.54701	1.587691	42.50354

Table 23.3. Average treatment concentrations ($\mu g\,g^{-1}$) for exoskeleton

	Cd	Pb	Hg
T1	34.61539	154.8646	747.3635
T2	218.9685	239.4492	708.596
T3	42.21743	63.3588	46.80132
T4	134.9583	170.366	44.05
T5	106.5266	138.9873	1092.944
T6	250.4659	278.2298	1.772144
C1	69.15562	16.29173	1.764735

Table 23.4. Average treatment concentrations ($\mu g\,g^{-1}$) for tissue

	Cd	Pb	Hg
T1	606.5466	568.5739	3894.128
T2	106.7864	0.756923	
T3	44.35711	140.0057	607.2782
T5	0.574571	0.932901	0.273001
T6	542.5269	283.2979	259.3928
C1	97.84055	62.1624	1.469662

quality control standards run as unknowns ($5\,ng\,g^{-1}$, $150\,ng\,g^{-1}$) with precision better than 95%. Relation between weight-corrected concentration data was explored using ANOVA, parametric and non-parametric bivariate correlations, and linear regression (InStat 4.0). For the purpose of data analysis, the carapace and claws were grouped together as "exoskeleton", with the gills and muscle tissue grouped as "tissue". Mercury, cadmium, and lead are reported in $\mu g\,g^{-1}$ (ppb) (Tables 23.2, 23.3, 23.4).

23.3. Results

Figure 23.1 shows results of the treatment concentration versus whole body total Hg (sum of all tissues). Control and treatment groups are abbreviated with C and T respectively. It is important to note that because the crayfish did not consume all of the metal-enriched food, the relationships presented here can only be considered as loose generalizations or as initiations of a trend. As can be seen from this figure, there appears to be a relationship between food Hg concentration and Hg content in the whole crayfish although this relation is not statistically significant. This trend was also recorded by the soft tissue (Fig. 23.2) but not by the exoskeleton (Fig. 23.3). Overall, there was a moderate

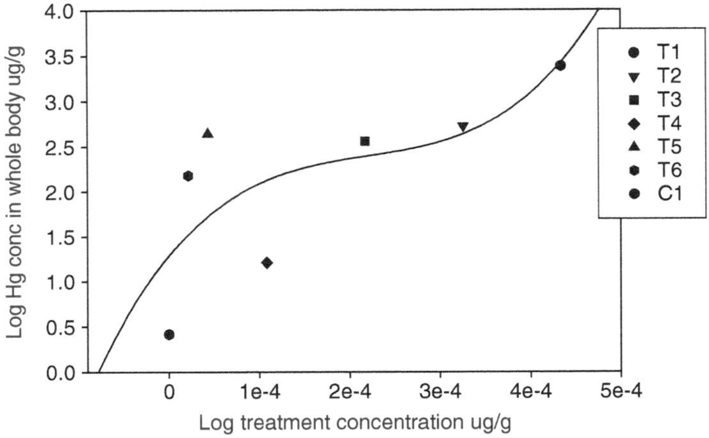

Figure 23.1. Mercury concentration ($\mu g\,g^{-1}$) of the whole crayfish compared to the concentration of individual treatments. Highest concentration treatment on far right with control group on the far left.

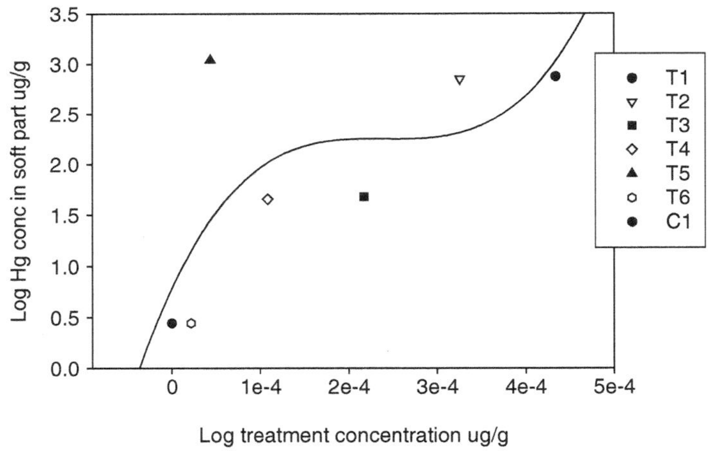

Figure 23.2. Mercury concentration of the crayfish tissue compared to individual treatment groups.

treatment effect however not statistically significant at an alpha of 0.05. Although neither regression analysis nor ANOVA substantiated the trend suggested by these results, further investigation of these trends may elucidate this relation more discretely. Figure 23.4 shows results of the treatment concentration versus whole body total Cd. As can be seen from

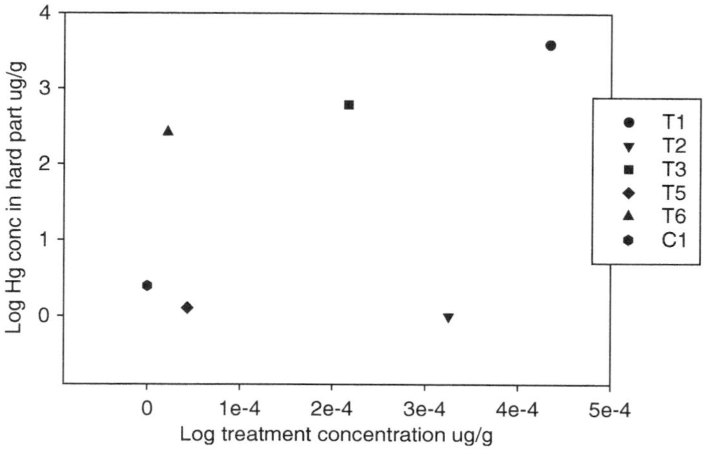

Figure 23.3. Mercury concentration of the crayfish exoskeleton compared to individual treatment groups. Treatment 4 failed to pass QA/QC (EPA 6020) and so no data are presented here.

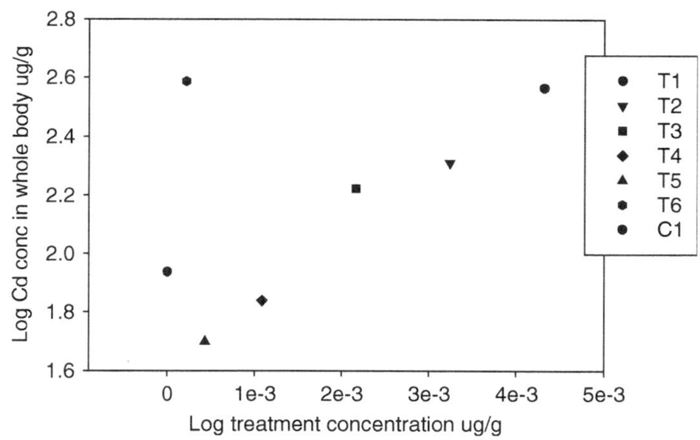

Figure 23.4. Cadmium concentration ($\mu g\, g^{-1}$) of the whole crayfish compared to the concentration of individual treatments.

this figure, there is a relationship between food Cd concentration and Cd content in the whole crayfish. This trend was not recorded by the soft tissue (Fig. 23.5) or the exoskeleton (Fig. 23.6). Data points for treatments 5 and 6 are off-scale compared to the higher concentrated treatments in all figures for Cd. This could be due to a threshold concentration

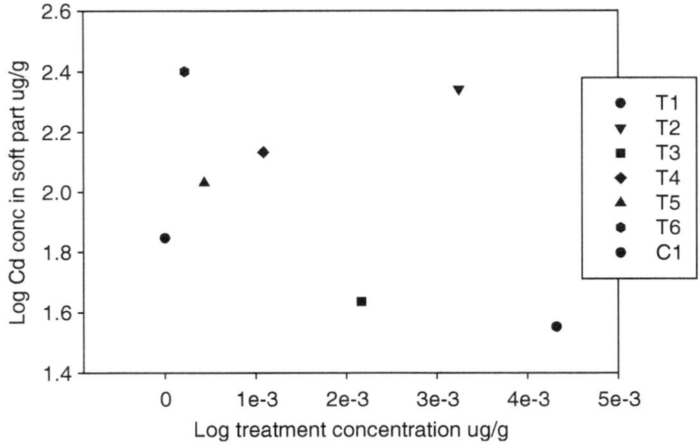

Figure 23.5. Cadmium concentration of the crayfish tissue compared to individual treatment groups.

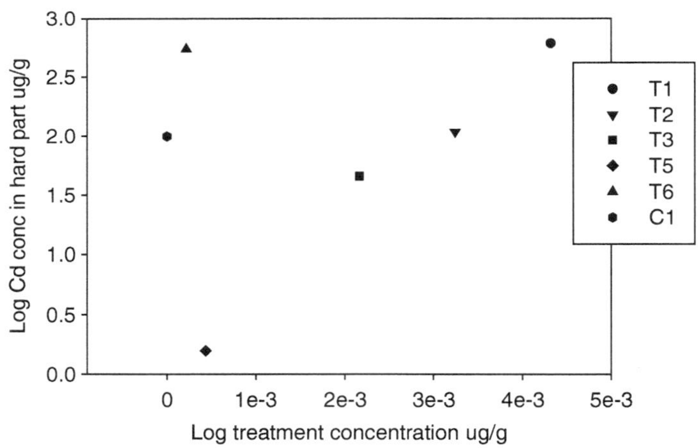

Figure 23.6. Cadmium concentration of the crayfish exoskeleton compared to individual treatment groups. Treatment 4 failed to pass QA/QC (EPA 6020) and so no data are presented here.

below which it is difficult to resolve interferences. Overall, there is only a significant treatment difference in the whole body total Cd treatments when compared to the control. No treatment effect was found for Lead. As can be seen in Fig. 23.7, there seems to be a treatment effect for the first three treatments but the data means are nearly identical.

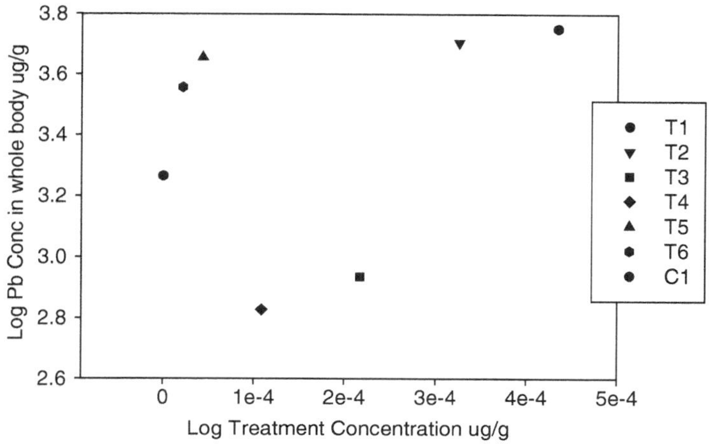

Figure 23.7. Lead concentration of the crayfish exoskeleton ($\mu g\ g^{-1}$) compared to individual treatment groups.

23.4. Discussion

Trace elements enter and leave a stream in the form of soluble, colloidal, and particulate matter. Crayfish interact and assist with the cycling of these forms of matter by residing in or near the water column where water is continually flushed over the gills, and by serving as detritivores, herbivores, carnivores, and scavengers. For example, crayfish can process several types of metal complexes (soluble, colloidal, and particulate matter) associated with the biota (phytoplankton, periphyton, and bacteria) upon which they feed. As carnivores and scavengers, crayfish are susceptible to metal uptake by predating on fish and carrion. Some freshwater fish, whose diet often includes crayfish, are known to accumulate methyl mercury and may be a sink for some other metals as well (Gilmour et al., 1991). As detritivores, crayfish assist in the breakdown of organic matter (both coarse particulate organic matter and fine particulate organic matter). Natural organic matter has also been shown to aid in the cycling of metals by complexing the metals and keeping them in solution. Dissolved and colloidal organometals can then be washed over the gills, crossing the gill–blood barrier being accumulated in crayfish tissue. It is deduced that because crayfish interact with a variety of metals in most river systems, some of these metals are being stored in the soft tissue where they are not lost during the molting process.

When mercury enters an aquatic system, it can be converted by sediment residing bacteria to MeHg, a more toxic form that is bioaccumulated up the food chain (Gilmour et al., 1991). It is well known that mercury, in the form of methyl mercury, is accumulated and stored in the tissues of many aquatic organisms (Campbell et al., 2003), and Hg has been found to accumulate in detoxifying organs (Wright and Welbourne, 1993; Mason et al., 2000). It is therefore logical to expect bioaccumulation of mercury in crayfish. The introduction of mercury into the food chain by crayfish can provide a potential pathway for exposure to humans due to crayfish being a major food source for many types of sport fish (Roel and Orth, 1993) which are consumed by humans. When the fish consume the MeHg-contaminated crayfish, the MeHg is transferred to the soft tissue in the fish where it is available to be accumulated by humans upon consumption. Understanding accumulation of mercury in crayfish from their diet is important because thousands of tons of mercury are released annually from power plants and incinerators, as well as from natural geological processes that make its way into the food web (Baker, 1998). This means that crayfish in many aquatic environments have readily accessible sources of Hg. Crayfish have been shown to be good monitors of heavy metal input into aquatic ecosystems (Vermeer, 1972), and results from this study indicate that crayfish may be good indicators of mercury inputs.

A treatment effect (increasing Hg in crayfish as Hg concentration increased in food) was observed for mercury for the whole crayfish and for the crayfish tissue. No treatment effect was found for the crayfish exoskeleton. There was a treatment effect for mercury storage in the soft tissue however. The mercury species used in this study was $HgCl_2$ which is a form that is 15–20% bioavailable.

Crayfish also accumulate Cd and Pb. Dickson et al. (1979) have shown that longer-ived species of crayfish accumulate more Cd and Pb in the tissue than shorter-lived species. Alikahn et al. (1990) found that the crayfish, *Cambarus bartoni*, from three lakes near a Sudbury smelter in Northwestern Ontario, Canada, had high concentrations of Cd in the hepatopancreas and gut of the crayfish. These concentrations varied with distance from the smelter with those closest to the site having the highest Cd concentration. Mwangi and Alikhan (1993) found high Cd concentrations in crayfish hepatopancreas after 3 days of lab exposure to Cadmium (12.5, 62.5, and 125 $\mu g\, l^{-1}$) but found no significant concentrations in the soft tissues when compared to the control.

Our results show an overall trend of increasing Cd in the whole crayfish with increasing food content. Unlike the whole crayfish, which revealed a treatment effect, the tissue and the crayfish exoskeleton did not reveal a treatment effect, indicating that there may be another tissue type or organ, such as the hepatopancreas, responsible for this trend.

No treatment effect for the tissue or exoskeleton was found for Pb in our data but other researchers have found significant cases of metal storage by crayfish. For example, adult crayfish exoskeleton and muscle tissue have been used to monitor concentrations of Pb and Cd in aquatic environments (Khan et al., 1995). In a study designed to monitor heavy metals in Tuskegee Lake (TL) and National Forest Creek (NFC), in Tuskegee, Alabama, Khan et al. (1995) found significant concentrations of Pb in the TL crayfish exoskeleton which was attributed to environmental contamination at the sample site. Naqvi et al. (1993) discovered that both Cd and Pb are readily accumulated in field-collected crayfish, *Procambarus clarkii*, although no significant biomagnification occurred from the main food source, alligator weed, to the crayfish. Anderson et al. (1997) found increasing Pb concentrations in the gills, muscle tissue, and exoskeleton of crayfish (*Procambarus clarkii*) after seven weeks of laboratory exposure to lead nitrate. Overall our results suggest that shedding of the exoskeleton is one possible mechanism by which crayfish void their system of metals. It is entirely possible that in addition to shedding that bodily waste may contain significant amounts of metals; however, we did not test the metal content of fecal material.

The results of this study show that Hg and Cd accumulated in the body of crayfish through the diet. Only a moderate correlation was established between metal accumulation and the spiked diet. This could be due to the fact that such key organs for metal storage as the hepatopancreas and gut were not sampled. In addition, it was not clear until the end of the study the degree to which homogenization of food would affect the results. Beyond grinding the food, no attempt was made to homogenize the food but in a controlled environment with only one food source, this is particularly important. While other researchers have found significant metal concentrations in crayfish in field conducted studies, where many forms of metal species were available, in the lab, where only one species of each metal was available, these results were not entirely duplicated.

ACKNOWLEDGMENTS

We thank Dr. Nate Bickford for helping MH initiate this project and Sarah Webb and David Clarke for all their help with sample preparation. We are especially grateful to the thorough reviews of the initial manuscript provided by Dr. Stephen Hasiotis and Dr. JoAnn Holloway. Funding for this research was provided by the Water-Rock-Life contract laboratory. Musical inspiration provided by Bob Dylan.

REFERENCES

Alikhan, M.A., Bagatto, G., Zia, S., 1990. The crayfish as a biological indicator of aquatic contamination by heavy metals. Water Res. 24(9), 1069–1076.

Anderson, M.B., Preslan, J.E., Jolibois, L., Bollinger, J.E., George, W.J., 1997. Bioaccumulation of lead nitrate in Red Swamp Crayfish (Procambarus clarkia). J. Hazard. Mater. 54, 15–29.

Baker, B., 1998. Mercury in fish: Agencies work to find common ground. Bioscience 48(11), 900.

Campbell, L.M., Osano, O., Hecky, D.G., 2003. Mercury in fish from three rift valley lakes (Turkana, Naivasha and Baringo), Kenya, East Africa. Environ. Pollut. 125, 281–286.

Dickson, G.W., Briese, L., Geisy, J.P., 1979. Tissue metal concentrations in two crayfish species co-habitating a Tennessee cave stream. Oecologia 44, 8–12.

DiStefano, R.J., Neves, R.J., 1991. Response of the crayfish *bartonii bartonii* to acid exposure in southern Appalachian streams. Can. J. Zool. 69, 1585–1591.

Geisy, J.P., Bowling, J.W., Kania, H.J., 1980. Cd and zinc accumulation and elimination by freshwater crayfish Cambarus robustus and Cambarus bartoni. Can. J. Zool. 63, 2313–2322.

Hart, B.T., 1982. Uptake of trace metals by sediments and suspended particulates: A review. Hydrobiologia 91, 299.

Hasiotis, S.T., 1999. Crayfish fossils and burrows from the upper triassic chinle formation, Canyonlands National Park, Utah. Paleontol. Res. 2, 83–90.

Hasiotis, S.T., Mitchell, C.E., 1993. A comparison of crayfish burrow morphologies: Triassic and Holocene paleo- and neoichnological evidence, and the identification of their burrowing signatures. Ichnos 2, 291–314.

Hasset, J.M., Jennett, J.C., Smith, J.E., 1980. Heavy metals accumulation by algae. In: Baker, R.A. (Ed.), Contaminants and Sediments, Vol. 2, pp. 409–424.

Hobbs, H.H., 1989. An illustrated checklist of American crayfishes (Decapoda: Astacidae, Cambaridae, and Parastacidae). Smithson. Contrib. Zool. 480, 236.

Holdich, D.M., Lowery, R.S., 1988. Freshwater Crayfish, Biology, Management, and Exploitation. Timber Press, Portland, OR.

Khan, A.T., Forester, D.M., Mielke, H.W., 1995. Heavy Metal Concentration in Two Populations of Crayfish. Vet. Human Toxicol. 37(5), 426–428.

Mason, R.P., Laporte, J., Andres, S., 2000. Factors controlling the bioaccumulation of mercury, methylmercury, arsenic, selenium, and cadmium by freshwater invertebrates and fish. Arch. Environ. Contam. Toxicol. 38(3), 283–297.

Mirenda, R.J., 1986. Acute toxicity and accumulation of zinc in the crayfish, *Orconectes virillis* (Hagen). Bull. Environ. Contam. Toxicol. 37, 387–394.

Mwangi, S.M., Alikhan, M.A., 1993. Cadmium and nickel uptake by tissues of Cambarus bartoni (Astacidae, Decapoda, Crustacea): Effects on copper and zinc stores. Water Res. 27(5), 921–927.

Naqvi, S.M., Howell, R.D., Sholas, M., 1993. Cadmium and lead residues in field-collected red swamp crayfish (Procambarus clarkia) and uptake by alligator weed, Alternanthera philoxiroides. J. Environ. Sci. Health, B 28(4), 473–485.

Roel, M.J., Orth, D.J., 1993. Trophic basis of production of stream-dwelling Smallmouth Bass, Rock Bass, and Flathead Catfish in relation to Invertebrate Bait Harvest. Trans. Am. Fisher. Soc. 122, 46–62.

Wright, D.A., Welbourn, P.M., 1993. Effects of mercury exposure on ionic regulation in the crayfish Orconectes propinquus. Environ. Pollut. 82(2), 139–142.

Developments in Environmental Science, Volume 5
D. Sarkar, R. Datta and R. Hannigan (Editors)

ISSN: 1474-8177/DOI:10.1016/S1474-8177(07)05024-3

Chapter 24

Relations among land cover, vegetation index, and nitrate concentrations in streams of the Enoree River Basin, piedmont region of South Carolina, USA

Suresh Muthukrishnan, Gregory P. Lewis and C. Brannon Andersen

Abstract

Globally, high nitrate concentrations and fluxes in rivers are correlated with human population density and can lead to eutrophication of estuaries and coastal oceans. Although elevated nitrate concentrations often are associated with agricultural land cover, urban land cover also can contribute substantially to elevated nitrate concentrations in streams and rivers. In the piedmont region of the southeastern United States, urban areas typically are located in the headwater areas of watersheds. Because headwaters account for the majority of stream channel length in a watershed, the effect of urbanization on the biogeochemical cycling of nitrogen is magnified. We examined the relations between stream nitrate concentrations, land cover, and vegetation density in watersheds of nineteen tributaries of the Enoree River in northwestern South Carolina, USA. Based on data from 134 sample localities, stream nitrate concentrations generally increased with increasing urban land cover and decreased with increasing forest cover and vegetation density (normalized density vegetation index). Although watersheds with the highest percent urban land cover typically had the highest nitrate concentrations, nitrate concentrations were most variable spatially in drainage basins with 5 to 20% urban land cover. The relations between land cover, vegetation density, and nitrate concentrations are complicated by variation in the intensity of urbanization and spatial location of urban and forested land within the drainage basin of each sample locality. Artificial ponds in urban areas appear to play an important role in lowering stream nitrate concentrations and contribute to the spatial variability of nitrate concentrations.

24.1. Introduction

Globally, nitrate concentrations in rivers correlate positively with human population density (Peierls et al., 1991). Furthermore, human activities have increased greatly the flux of nitrogen, especially nitrate–nitrogen, from terrestrial to aquatic ecosystems within the last two centuries (Meybeck, 1982; Meybeck and Helmer, 1989; Vitousek et al., 1997; Galloway and Cowling, 2002). Elevated nitrate concentrations and fluxes in river water threaten drinking water quality and contribute to eutrophication of estuaries and coastal oceans (Vitousek et al., 1997). Humans have increased nitrate loading to rivers in several ways. For example, both agricultural non-point sources and point-source inputs from wastewater treatment plants (WWTPs) contribute nitrates to rivers (Vitousek et al., 1997; Paul and Meyer, 2001). Atmospheric deposition of nitrogen to watersheds also appears to contribute to river nitrogen (Jaworski et al., 1997). Clearing forest vegetation for timber, agriculture, urban development, or other purposes reduces the uptake and storage of nitrogen in plant biomass, thereby allowing more nitrogen to enter streams and rivers (e.g., Likens et al., 1970).

Land use and land cover in watersheds often are effective predictors of water quality in rivers (e.g., Hunsaker and Levine, 1995) because they integrate the effects of various human activities. Previous studies have incorporated remote sensing data and geographical information systems (GIS) with field-based data to better understand the relationships between land use/land cover and biophysical processes. Satellite data and aerial photographs are commonly used to create land use/land cover, vegetation density, and biodiversity data (Stoms and Estes, 1993; Filoso et al., 2003). Satellite data also have been used widely in assessing the health and density of vegetation biomass using several indices, including the commonly used normalized density vegetation index (NDVI) (Tucker and Sellers, 1986; Bannari et al., 1995; Ji and Peters, 2003). GIS has been used commonly to organize and display data and to study spatial and temporal relations between variables such as water chemistry and land use and land cover within a watershed (Ballester et al., 2003).

Based on previous studies of land cover and land use, elevated stream nitrate concentrations are associated with both agriculture and urban land covers. For example, agricultural land use contributes significant amounts of nitrogen in large watersheds such as the Mississippi, the Seine, and the Changjiang (Moreau et al., 1998; Roy et al., 1999; Goolsby, 2000; Rabalais et al., 2002; Zhiliang et al., 2003), as well as in smaller watersheds such as the River Vilaine (Moreau et al., 1998). Agricultural sources contribute 20% or more of the dissolved inorganic nitrogen in rivers within these watersheds. In some cases, however, streams in urbanized watersheds have higher nitrate concentrations than streams in agricultural watersheds (e.g., Boyle et al., 1997; Douglas et al., 2002). In contrast, streams draining forested watersheds typically have very low nitrate concentrations (Clark et al., 2000).

In regions undergoing rapid urban development, several types of land cover may be found within a single watershed (e.g., urban residential, urban commercial, forested, agricultural). This variety of land covers may increase the variability of nitrate concentrations within the watershed. Thus, the distribution of land use and land cover in watersheds may be important in controlling the concentration of nitrogen in streams.

In this study, we examined the relations among stream nitrate concentrations, land cover and vegetation density in tributary watersheds of the 1893 km^2 Enoree River basin in the Piedmont Province of South Carolina, USA (Fig. 24.1). Water samples were collected from localities representing drainage areas from about 0.2 km^2 to 307 km^2. Remote sensing and GIS were used to develop land use and land cover data, to calculate NDVI, to delineate drainage areas for individual sample

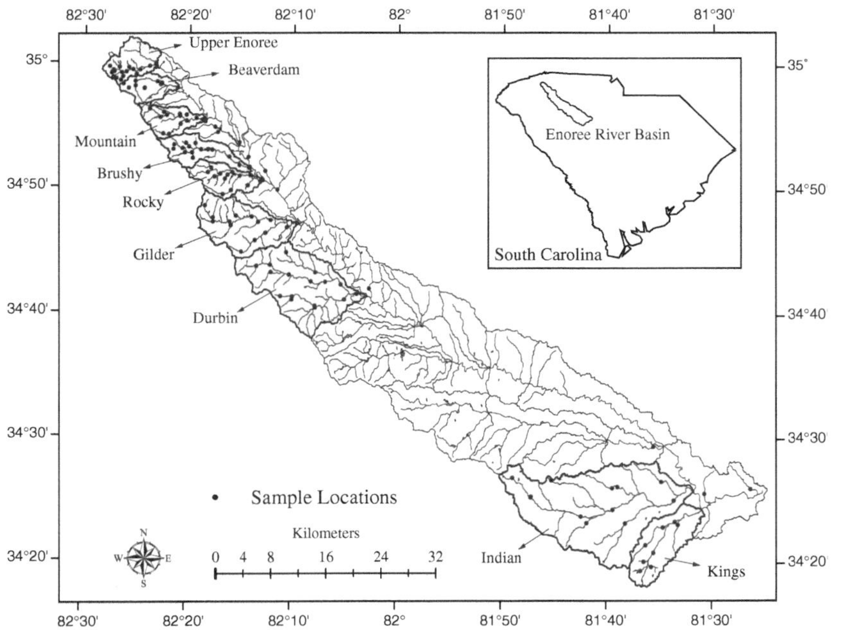

Figure 24.1. Location map of watersheds and sample localities within the Enoree River Basin.

localities, and to analyze the spatial distribution of urban land cover and vegetation within the drainage area of each sampling site.

24.2. Study area

24.2.1. Climate and hydrology

The Enoree River basin is located within the lower Broad River basin, which in turn comprises the northern half of the Santee River basin, one of the United States Geological Survey's National Water Quality Assessment Program watersheds. The Enoree River basin is classified as sixth-order (Strahler, 1952).

The climate of the region is subtropical, with daily high temperatures averaging 22°C and daily lows averaging 11°C. Rainfall averages 120 cm per year with a distinctly rainy winter and dry late summer and fall (Camp, 1960, 1975; Camp et al., 1960, 1975). During the period of the study, South Carolina experienced the most severe drought since the mid-1950s, and many rivers in the Piedmont Province, including the Enoree River, experienced record low discharges and few storm events during the summer months (Andersen et al., 2001; Andersen et al., 2004).

24.2.2. Geology, soils, and vegetation

The Enoree River basin is underlain by igneous and high-grade metamorphic rocks of the Inner Piedmont and Charlotte belts that are comprised mainly of silicate minerals (see review and references in Andersen et al., 2001). These minerals have low solubility, resulting in relatively dilute stream waters with conductivities often $< 100\,\mu S\,cm^{-1}$ (Andersen et al., 2001). The soils in the area are predominantly ultisols, with some alfisols in the southern third of the basin (Camp, 1960, 1975; Camp et al., 1960, 1975). Forests are typically second-growth and are composed mainly of pines (*Pinus* spp.) and/or hardwoods, especially oaks (*Quercus* spp.), hickories (*Carya* spp.), red maple (*Acer rubrum*), sweet gum (*Liquidambar styraciflua*), and tulip poplar (*Liriodendron tulipifera*). Pine plantations are especially common in the southern half of the basin.

24.3. Methods

24.3.1. Sample collection and analysis

Grab samples were collected four to seven times at each of 134 localities in 19 tributary watersheds during June through August 1999 or 2000 (Fig. 24.1, Table 24.1). Nine of the watersheds had multiple sample

Table 24.1. Nitrate and land cover data for sample drainages in the Enoree River basin

Locality	Year			Nitrate ($mg\,l^{-1}$)			Percent Land Cover			
		n	Area (km^2)	Mean	Minimum	Maximum	Urban	Forested	Grass	NDVI
Upper Enoree River										
UE01	1999	8	0.63	2.93	2.46	3.76	40.0	24.6	34.9	111
UE02	1999	10	0.71	3.23	2.51	4.10	38.1	24.7	37.3	112
UE03	1999	7	1.72	1.41	1.09	1.64	9.8	37.1	53.0	120
UE04	1999	9	0.74	2.72	1.72	8.95	38.3	25.2	36.4	112
UE05	1999	6	1.75	1.11	0.77	1.83	10.5	37.2	52.4	120
UE06	1999	8	7.38	1.71	1.13	2.33	9.6	45.2	44.6	124
UE07	1999	7	0.27	5.35	3.46	13.59	15.6	20.3	64.1	110
UE08	1999	5	0.58	0.75	0.00	1.04	3.8	32.0	64.4	123
UE09	1999	8	18.07	1.67	1.41	2.07	9.8	47.5	42.5	126
UE10	1999	7	10.05	1.27	0.93	1.53	13.5	42.1	44.0	123
UE11	1999	6	0.63	4.53	0.85	12.67	43.4	16.0	40.4	114
UE12	1999	6	0.18	2.05	1.65	2.31	0.0	40.2	59.8	129
UE13	1999	7	1.63	0.39	0.26	0.52	0.5	57.4	40.8	129
UE14	1999	8	34.55	1.53	1.13	1.76	6.1	51.1	42.6	128
UE15	1999	8	19.13	1.64	1.12	2.51	9.3	47.9	42.6	126
UE16	1999	5	1.77	2.63	2.35	2.96	34.1	25.1	41.0	118
UE17	1999	7	13.55	2.52	0.91	7.85	10.5	46.5	42.8	125
UE18	1999	5	0.20	1.76	1.34	2.89	36.7	14.2	50.0	118
UE19	1999	4	2.10	1.42	1.16	1.67	1.6	64.7	33.7	131
Beaverdam Creek										
BD01	1999	7	24.38	1.54	1.27	2.52	8.3	58.2	33.4	130
BD02	1999	7	15.05	0.44	0.15	1.73	11.3	51.6	36.9	127
BD03	1999	7	0.34	1.56	1.40	1.84	0.0	49.6	50.4	128
BD04	1999	7	1.55	0.84	0.32	2.07	4.0	47.2	48.4	125
BD05	1999	7	13.76	1.33	0.70	1.54	12.4	51.5	36.0	127
BD06	1999	7	0.72	1.26	1.09	1.70	5.5	61.8	32.3	131
BD07	1999	7	0.92	1.84	1.62	2.35	3.7	57.9	38.0	130
BD08	1999	7	7.08	2.08	1.75	2.66	21.2	54.0	24.5	126
BD09	1999	7	3.90	1.75	1.39	2.37	4.1	73.2	22.7	135
BD10	1999	7	0.44	1.13	0.44	1.32	8.1	64.0	28.0	129
BD11	1999	5	0.83	3.37	2.74	3.60	60.6	24.2	14.3	108
Mountain Creek										
MC01	2000	6	30.36	1.75	0.83	5.39	23.9	53.0	22.8	127
MC02	1999	7	29.56	1.45	1.17	2.24	24.3	53.6	22.2	127
MC03	1999	7	29.03	1.34	1.14	1.55	23.4	53.6	22.6	127
MC04	1999	7	24.54	1.13	0.01	2.04	20.1	55.8	24.0	129
MC05	1999	7	14.58	0.66	0.09	3.26	5.9	72.6	21.1	136
MC06	1999	7	7.95	1.33	1.03	1.67	0.0	96.2	4.9	143
MC07	1999	7	1.02	2.15	1.68	2.65	40.7	23.5	35.6	121
MC08	1999	7	8.41	0.72	0.43	1.30	42.8	30.0	27.2	118
MC10	1999	7	1.36	1.98	1.72	2.39	0.0	98.7	1.3	144
MC11	1999	7	3.77	1.64	1.42	2.13	34.1	38.9	23.7	123
MC12	1999	7	1.93	1.39	1.25	1.54	9.6	43.0	47.4	126
MC13	1999	7	0.89	0.57	0.44	0.66	0.0	67.0	33.1	136

Table 24.1. (*Continued*)

Locality	Year	n	Area (km^2)	Nitrate ($mg\,l^{-1}$)			Percent Land Cover			NDVI
				Mean	Minimum	Maximum	Urban	Forested	Grass	
MC14	1999	7	2.69	0.82	0.58	1.42	30.9	46.1	20.6	124
MC16	1999	7	0.68	1.41	1.07	2.21	30.8	27.6	41.1	122
MC17	1999	6	2.93	0.26	0.14	0.34	0.0	92.5	6.9	142
Brushy Creek										
BY01	2000	7	38.20	1.96	1.74	2.55	64.7	21.0	14.1	112
BY02	2000	7	36.11	2.05	1.66	2.52	64.9	20.5	14.3	113
BY03	2000	7	32.06	2.20	1.83	2.70	65.4	21.1	13.1	113
BY04	2000	7	24.99	2.46	2.14	3.04	68.3	20.8	10.7	113
BY05	2000	7	23.25	2.48	2.29	2.99	67.8	21.5	10.4	113
BY06	2000	7	21.39	2.54	2.30	2.98	68.4	21.5	9.8	113
BY07	2000	7	1.83	5.93	5.30	8.24	61.7	24.0	14.2	113
BY08	2000	7	13.73	2.88	2.65	3.25	67.8	22.8	9.1	112
BY09	2000	7	1.22	1.34	0.80	2.33	72.0	7.8	20.2	112
BY10	2000	7	1.80	3.32	2.96	3.92	76.0	19.8	4.2	114
BY11	2000	7	2.19	1.31	1.09	1.56	69.9	23.9	6.2	116
BY12	2000	7	6.12	2.13	1.68	2.73	64.4	23.5	11.8	111
BY13	2000	7	3.25	3.37	3.05	3.69	82.0	14.1	3.9	110
BY14	2000	7	0.64	5.73	4.95	7.83	84.4	9.6	5.9	108
BY15	2000	7	1.94	5.41	5.14	5.98	66.1	15.2	18.7	103
Rocky Creek										
RC01	1999	7	36.37	2.23	1.77	3.41	50.2	21.9	26.8	108
RC02	1999	7	30.78	2.47	1.79	3.54	52.4	20.9	25.6	107
RC03	1999	7	29.02	2.34	2.04	2.63	53.3	20.7	24.8	107
RC04	1999	7	18.34	2.23	1.73	2.80	56.4	20.2	22.2	109
RC05	1999	7	13.55	2.46	2.05	2.87	59.9	20.3	18.9	110
RC06	1999	6	3.45	5.11	2.05	6.06	68.8	15.2	16.0	116
RC07	1999	7	3.98	3.31	2.77	3.77	68.3	15.0	16.5	104
RC08	1999	7	2.93	1.27	0.60	2.69	47.9	18.3	32.9	112
RC09	1999	7	5.87	2.14	1.74	2.46	55.4	21.2	23.3	99
RC10	1999	7	3.75	2.03	1.67	2.63	59.1	21.7	19.0	97
RC11	1999	7	0.86	2.42	1.30	7.88	30.5	12.2	57.7	103
RC12	1999	7	1.21	0.78	0.34	1.36	34.2	12.2	51.2	106
RC13	1999	7	1.76	1.83	0.63	2.84	34.5	27.3	35.7	113
Gilder Creek										
GC01	1999	7	81.26	2.48	1.84	3.01	35.9	28.0	35.8	113
GC02	1999	7	4.89	1.63	1.13	2.48	6.4	62.1	28.4	126
GC03	1999	7	50.41	2.53	2.26	3.05	49.9	19.1	30.9	109
GC04	1999	7	17.14	1.68	1.44	2.03	20.2	35.7	43.8	119
GC05	1999	7	11.15	2.61	2.26	3.59	22.2	39.3	38.4	120
GC06	1999	7	40.28	3.00	2.54	3.88	54.7	17.1	27.9	108
GC07	1999	7	3.46	3.86	2.82	4.32	45.1	20.6	34.3	111
GC08	1999	7	4.40	5.63	3.48	6.44	50.6	11.8	37.0	104
GC09	1999	7	20.04	3.55	2.99	4.37	60.4	14.0	25.5	105
GC10	1999	7	6.99	1.32	0.91	1.91	45.3	21.6	32.2	110
GC11	1999	7	2.47	2.81	1.93	3.57	79.7	5.2	14.5	98

Table 24.1. (*Continued*)

Locality	Year			Nitrate (mg l^{-1})			Percent Land Cover			
		n	Area (km^2)	Mean	Minimum	Maximum	Urban	Forested	Grass	NDVI
GC12	1999	7	4.33	1.69	1.20	2.50	52.2	19.2	28.6	108
GC13	1999	7	1.58	1.83	0.75	2.58	60.7	10.5	28.1	102
GC14	1999	7	4.59	3.09	2.98	3.29	25.9	25.9	48.1	113
Durbin Creek										
DB01	2000	7	131.25	10.65	7.70	15.00	8.4	40.8	50.7	120
DB02	2000	7	48.22	0.99	0.71	1.51	5.5	39.6	54.9	119
DB05	2000	7	77.25	16.92	14.95	19.68	10.5	41.9	47.4	120
DB06	2000	7	48.31	28.66	23.07	37.57	14.8	39.4	45.6	119
DB07	2000	7	18.42	2.08	1.65	2.50	5.3	39.7	55.0	119
DB08	2000	7	41.76	30.86	25.49	33.44	17.1	36.1	46.7	118
DB09	2000	7	3.67	7.60	6.43	8.56	13.2	23.9	63.0	109
DB10	2000	6	13.05	1.19	0.92	1.65	8.4	34.7	56.9	116
DB11	2000	7	16.52	1.44	1.06	1.93	7.8	32.9	59.1	116
DB12	2000	6	2.47	1.42	0.82	2.19	11.0	21.8	67.3	109
DB13	2000	7	4.23	2.12	1.30	2.91	18.1	26.8	55.2	113
DB14	2000	7	1.49	9.50	7.06	10.76	36.3	24.5	39.2	111
DB15	2000	7	33.30	1.79	1.48	2.01	20.5	31.9	47.4	116
DB16	2000	7	16.90	2.06	1.89	2.44	19.4	33.5	46.8	116
DB17	2000	7	6.65	1.91	1.78	2.35	28.2	21.1	50.8	112
DB18	2000	7	9.15	1.91	1.48	2.76	29.4	38.6	31.5	118
Indian Creek										
IR01	2000	7	228.42	0.64	0.42	1.40	1.2	81.5	17.2	130
IR04	2000	7	14.71	0.43	0.37	0.48	0.2	96.9	2.9	131
IR05	2000	7	7.68	0.51	0.26	0.83	0.3	99.0	0.5	132
IR06	2000	7	121.11	0.54	0.29	0.71	1.3	79.5	19.1	130
IR07	2000	7	9.54	0.82	0.67	1.00	0.6	56.9	42.5	125
IR08	2000	7	32.44	0.80	0.49	1.05	3.4	76.1	20.4	128
IR10	2000	7	16.97	0.52	0.38	0.72	0.5	84.7	14.9	131
IR11	2000	7	62.93	0.73	0.59	1.09	1.2	75.8	22.9	
IR13	2000	7	17.47	0.72	0.38	0.88	0.7	66.1	33.0	126
IR14	2000	7	2.09	0.69	0.47	0.96	1.0	65.0	33.9	127
IR15	2000	7	0.52	2.08	0.90	2.69	5.9	17.9	76.4	121
IR16	2000	7	4.42	0.84	0.68	0.98	0.2	54.7	45.1	122
Kings Creek										
KC01	2000	7	45.55	0.90	0.50	1.25	2.8	75.2	21.7	128
KC02	2000	7	8.15	0.72	0.48	1.10	0.2	82.3	17.1	129
KC03	2000	7	12.27	0.75	0.59	0.87	1.3	75.9	23.1	130
KC04	2000	7	4.07	0.84	0.70	1.10	3.6	64.9	31.5	127
KC05	2000	7	17.76	0.91	0.65	1.04	5.9	69.6	23.9	126
KC06	2000	7	1.25	0.78	0.68	0.89	8.9	81.3	9.9	126
KC07	2000	7	2.16	1.03	0.06	2.21	9.7	64.0	24.7	124
KC08	2000	7	2.18	1.35	1.21	1.50	5.9	74.3	18.3	126
Other										
AB01	2000	4	29.39	2.69	1.76	3.09	14.3	26.2	59.4	113
AC01	2000	8	5.80	2.77	2.43	3.68	26.6	20.2	53.2	104

Table 24.1. (*Continued*)

Locality	Year	*n*	Area (km^2)	Nitrate ($mg\,l^{-1}$)			Percent Land Cover			NDVI
				Mean	Minimum	Maximum	Urban	Forested	Grass	
CC01	1999	6	4.24	2.90	2.61	3.29	64.9	13.1	22.1	115
CH01	1999	6	4.83	3.82	3.12	4.69	70.1	12.6	17.4	107
DC01	2000	7	11.15	2.63	1.87	3.10	21.5	39.8	38.7	115
DL01	2000	7	7.24	3.05	1.66	7.84	11.2	53.9	34.7	127
DN01	2000	7	307.03	2.00	0.53	6.38	4.7	67.1	27.8	128
SC01	2000	7	9.21	0.36	0.29	0.40	0.2	82.6	17.0	129
UC01	2000	7	5.15	0.48	0.22	0.72	0.0	90.6	9.4	130
UT01	2000	6	10.35	2.38	1.99	2.83	40.1	18.1	41.2	107

localities; the other ten each had a single sample locality near its confluence with the Enoree River. Samples were collected and processed following the methods of Andersen (2001) and Andersen et al. (2001). Nitrate and nitrite concentrations were measured using a Dionex 120 ion chromatograph. Nitrite concentrations are not reported because 95% of the concentrations were below the detection limit of $0.050\,mg\,l^{-1}$. Although ammonium was not measured, we assumed the concentrations also were negligible based on subsequent analyses of a number of watersheds in the region (G.P. Lewis and C.B. Andersen, unpublished data). Additional major cation and anion results for 116 sample localities are tabulated in Andersen et al. (2001). A complete dataset is available from the authors. Chemical composition reliability was checked using the charge balance method of Freeze and Cherry (1979). In this study, the mean, maximum, and minimum nitrate concentrations for each locality were used for analysis. Mean nitrate concentration data from each sample site were used to analyze the effect of land use and vegetation density on nitrate concentrations in the context of the corresponding drainage area. Because we lacked stream discharge data, we could not determine relationship of nitrate fluxes with land use or vegetation density.

SigmaStat v. 3.1 was used to perform statistical analyses (Systat Software, 2004). Nitrate concentrations in forested watersheds within the Enoree River basin are known to vary significantly from year to year, particularly between drought and non-drought years (Keaton et al., 2005). Because some samples were collected in 1999 and some in 2000, a paired *t*-test was used to determine if the mean nitrate concentrations for seven localities sampled in both years differed significantly. Although mean nitrate concentrations in 1999 were slightly higher than mean concentrations in 2000 (Fig. 24.2), the means did not differ significantly ($p = 0.14$). Therefore, we combined the two datasets for analysis. We

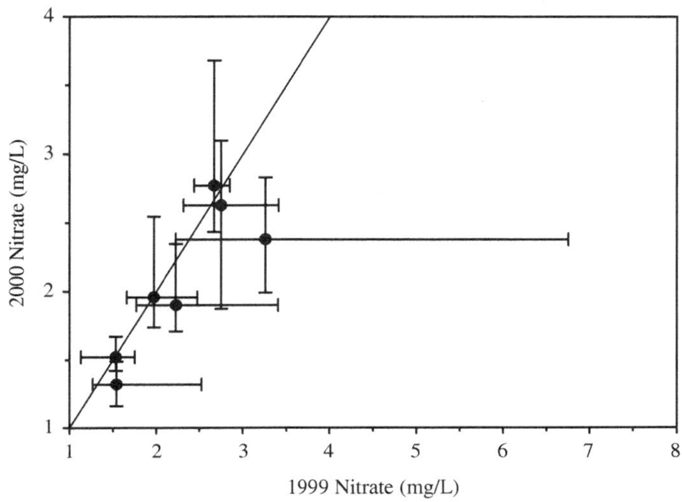

Figure 24.2. Comparison of nitrate concentrations for seven sample localities where water samples were collected in both 1999 and 2000. Filled circles represent the mean nitrate concentrations and bars represent the range of concentrations.

used Kruskal-Wallis One Way Analysis of Variance on Ranks tests to determine if the median nitrate concentrations, land cover percentages, and NDVI values differed significantly among the nine tributary watersheds with multiple sample localities. This non-parametric test was used because the data failed either the Kolmogorov-Smirnov test for normality or a test for equal variance. We used standard linear regression analysis on log-transformed data to test for significant relations among nitrate concentrations, land covers, and NDVI.

24.3.2. Watershed delineation

A digital elevation model (DEM) was used to delineate the drainage area corresponding to each of the 134 sample localities (Fig. 24.1, Table 24.1). An "Hydrology Analysis" extension (ESRI, 2005) was used with ArcGIS system to process the DEM, identify flow directions and sinks, fill the sinks, and then calculate flow accumulations for each grid cell. Once flow direction and flow accumulation layers were generated, the contributing drainage area for any specific point within the watershed was calculated by clicking on the points of interest, which in our case were the sampling localities. Once the watershed was delineated, total area of each basin was calculated by multiplying the total number of grid cells within the watershed by the grid dimension, which was $30\,m \times 30\,m$, or $0.009\,km^2$.

24.3.3. Land cover

A Landsat 7 ETM+, 30-m spatial resolution imagery (path/row/year-month-date are 17/36/2000-06-01 and 18/36/2000-06-10) covering the area of the Enoree River basin was obtained. A combination of supervised and unsupervised classification (ISODATA algorithm with 10 iterations and 0.95 convergence threshold with the initial clustering means derived from training dataset collected using supervised sampling method) methods, along with post-classification clumping and sieving were used to develop land cover data for the entire basin. Final classification for the entire Enoree basin included the following land cover types: water, commercial/industrial, agricultural, high-density residential, low-density residential, grass/pasture, and forest. Ground truth information and aerial photographs taken in 1999 were used to verify the classification results. Also, on November 16, 2003, each sample locality in the Brushy Creek watershed, which had the greatest diversity in land use, was visited and field-checked for agreement with land cover data. At each sample locality, agreement was found between qualitative field assessment and digital land cover data.

24.3.4. Normalized difference vegetation index

The spectral reflectance values from the Landsat ETM+ image were used to compute the normalized difference vegetation index (NDVI), one of more than twenty existing vegetation indexes that gauge small differences between the amount, distribution, and vitality of green biomass (Chen and Brutsaert, 1998; Boone et al., 2000; Jensen, 2000). We selected NDVI because it is widely used and has been shown to detect changes in the amount of green biomass and chlorophyll content (Bannari et al., 1995; Jensen, 2000). We used bands 4 (infrared, 0.75–0.90 μm) and 3 (red, 0.63–0.69 μm) of the Landsat ETM+ data because plant reflection and absorption of radiation are greatest in these wavelengths. The NDVI values for each pixel were calculated using the relationship

$$\mathrm{NDVI} = \left(\frac{(\mathrm{ETM\ Band4} - \mathrm{ETM\ Band3})}{(\mathrm{ETM\ Band4} + \mathrm{ETM\ Band3})}\right)$$

The computed NDVI values ranged from -1.0 to 1.0, where vegetated areas will typically have values higher than zero and other non-vegetated areas, such as water, snow, or barren areas, will have negative values.

These values then were scaled (which is a simple way to amplify small difference between two NDVI values) in order to enhance the visual differences in density of vegetation in the given area using the following relationship:

$$\text{ScaledNDVI} = 100(\text{NDVI} + 1)$$

After scaling, NDVI values ranged from 0 to 200, in which values 100 and below indicated a lack of vegetation, while values above 100 indicated the presence of vegetation, with higher numbers representing denser vegetation cover.

24.4. Results

The sample localities drained watersheds ranging in area from 0.2 to 307 km^2 (Table 24.1) with considerable variation in land cover. In the nine tributary watersheds with multiple sample localities, the drainages were hierarchical and nested. As such, the chemical composition of the downstream localities and their relations to land cover represent an integration of the upstream localities.

24.4.1. Nitrate concentrations

Median nitrate concentrations differed significantly ($p < 0.001$) among the nine tributary watersheds with multiple sample localities. Nitrate concentrations ranged from below the detection limit of 0.05 $mg\,l^{-1}$ to greater than 30 $mg\,l^{-1}$. The highest concentrations occurred downstream of a wastewater treatment plant discharging into Durbin Creek. The highest nitrate concentration unaffected by a known point-source was about 13 $mg\,l^{-1}$. Brushy Creek, Rocky Creek, Gilder Creek, and Durbin Creek watersheds had the highest mean and median concentrations of nitrate (Fig. 24.3). Durbin Creek had relatively high mean and median nitrate concentrations even when the sample localities downstream of the wastewater treatment plant were removed from analysis. The mean concentration of nitrate in Durbin Creek was above the 75th percentile value because of two headwater sample localities that had mean nitrate concentrations greater than 7 $mg\,l^{-1}$. The lowest mean and median nitrate concentrations were found in the mostly forested Indian Creek and Kings Creek. The Upper Enoree, Beaverdam Creek, and Mountain Creek watersheds had intermediate nitrate concentrations, typically between 1 and 3 $mg\,l^{-1}$. The ten watersheds with a single sample locality each (grouped as "other" in Fig. 24.3) had mean nitrate concentrations ranging from

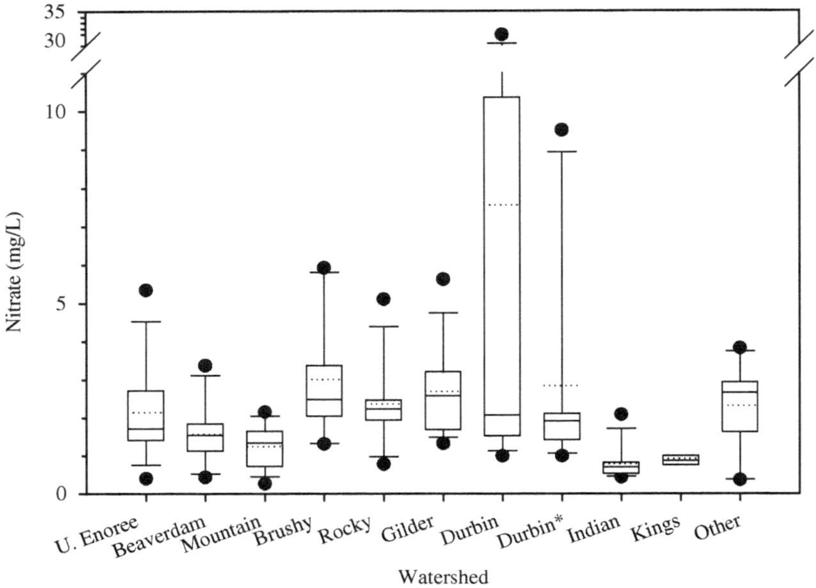

Figure 24.3. Mean nitrate concentrations in streams of tributary watersheds of the Enoree River basin. Watersheds are listed in the downstream direction. "Other" refers to ten watersheds with only a single sample locality. "Durbin" includes four sample localities downstream of a wastewater treatment plant; "Durbin*" removes those four localities. In the box plot, the median is represented by a solid horizontal line, the mean by a dotted horizontal line, the limits of the box represents the 25th and 75th percentiles, the bars represent the 5th and 95th percentiles, and the filled circles represent outliers.

less than $0.50\,mg\,l^{-1}$ to greater than $3.0\,mg\,l^{-1}$. In Brushy Creek, Rocky Creek, Gilder Creek, and Durbin Creek (excluding WWTP-affected samples), the highest nitrate concentrations occurred in headwater streams with small drainages, although the overall correlation between nitrate concentrations and drainage areas for all sample localities was very weak and not statistically significant ($p = 0.076$).

24.4.2. Land cover

The Enoree River basin as a whole had 9% ($170\,km^2$ out of $1893\,km^2$ area) of its drainage area classified as urban, 58% as forested, and 33% as grass/pasture land covers. The majority of the urban land cover was concentrated along the headwater regions of the Enoree River basin and was strongly influenced by the expansion of the Cities of Greenville and Spartanburg. Qualitative analysis of aerial photographs and ground truthing indicates that the vast majority of the grass/pasture land cover is grass in urban areas and pasture or hayfields in rural areas. Row crops constitute only a small percentage of the grass/pasture land cover, and intensive feedlot operations are absent.

For urban, forested, and grass/pasture land covers, the percentage of cover differed significantly ($p<0.001$) among the nine tributary watersheds with multiple sample localities (Fig. 24.4). Proportionally, the Brushy, Rocky, and Gilder Creek watersheds were the most urbanized and the least forested (Fig. 24.5). The Beaverdam, Indian, and Kings Creek watersheds were the most forested watersheds. The Upper Enoree, Mountain, and Durbin Creek watersheds were dominated by mixtures of forest and grass/pasture with lesser amounts of developed land cover.

Proportionally, watershed land cover formed either an urban to grass/pasture gradient or a grass/pasture to forested gradient (Fig. 24.5). The relations among the various land covers were complex and non-linear (Fig. 24.6). In particular, the relation between forested and urban land cover was inverse but non-linear.

24.4.3. NDVI of drainage areas

The NDVI indicates the density of vegetation cover, or "greenness" of a watershed (Fig. 24.7). The weighted mean NDVI differed significantly ($p<0.001$) among the nine tributary watersheds with multiple sample localities (Fig. 24.4). Among all drainage basins, the weighted mean NDVI values ranged from 97 to 144. Watersheds with NDVI value less than 100 were completely developed with little or no vegetation cover. Although no significant relationship ($p = 0.47$) was observed between weighted mean NDVI of a watershed and the watershed area, the NDVI values for watersheds smaller than 20 km^2 showed maximum variation. NDVI showed a significant, negative correlation ($p<0.001$) with percentage of urban land cover and a significant, positive correlation ($p<0.001$) with percentage of forested land cover but showed no significant relationship ($p = 0.07$) with percentage of grass/pasture land cover (Fig. 24.8). Spatially, highest NDVI values were observed alongside the streams where minimally disturbed riparian vegetation was present (Fig. 24.9). The lowest NDVI values were observed in the headwater regions of the watersheds where intense development resulted in loss of vegetative cover around the streams. The Upper Enoree, Beaverdam, Indian and Kings Creeks had more areas with contiguous and densely forested vegetation cover (darker shades of gray in Fig. 24.9) compared to Brushy, Rocky, Gilder, and Durbin Creeks (lighter shades of gray in Fig. 24.9).

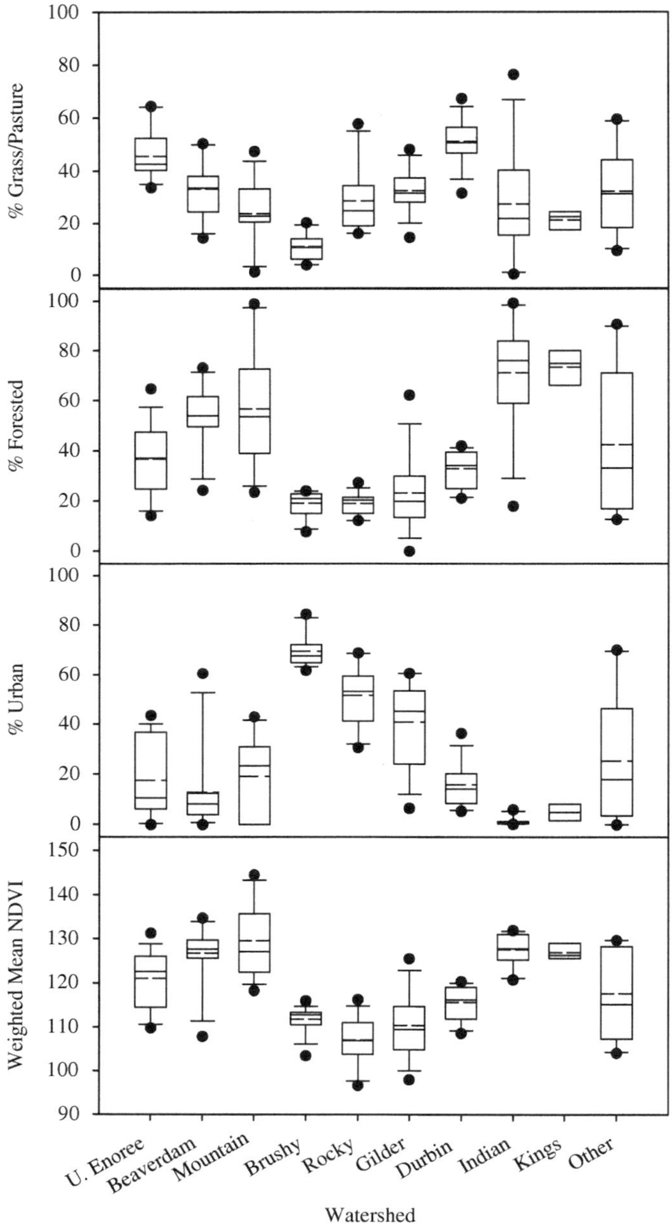

Figure 24.4. Land cover of sample locality drainages in the tributary watersheds of the Enoree River basin. Note that weighted NDVI values are most similar to the percent forested land cover.

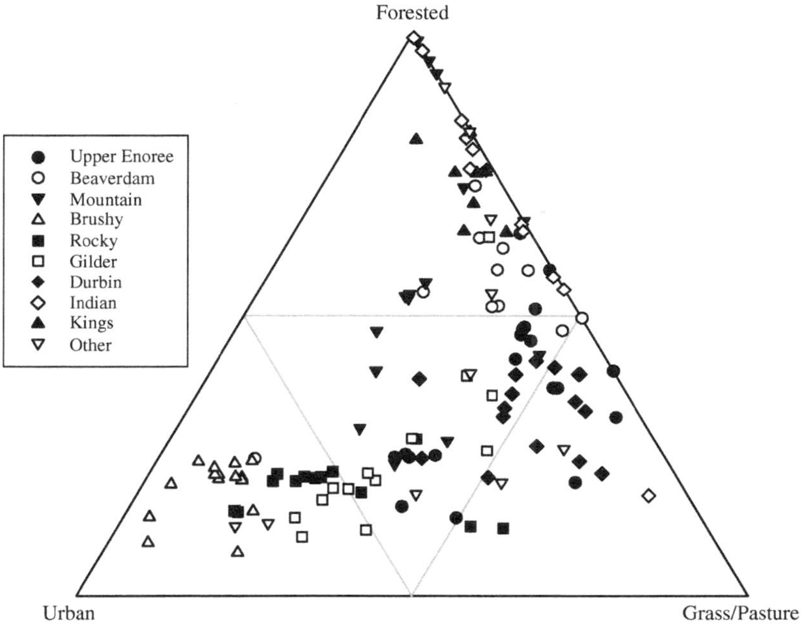

Figure 24.5. Ternary plot showing proportions of land cover for sample locality drainages in the Enoree River basin.

24.4.4. Relations of nitrate with land cover and NDVI

Mean nitrate concentrations showed a significant, positive correlation ($p < 0.001$) with percentage of urban land cover and a significant, negative correlation ($p < 0.001$) with percentage of forested land cover (Fig. 24.10). Mean nitrate concentrations also showed a significant, negative correlation ($p < 0.001$) with mean NDVI (Fig. 24.10). However, mean nitrate concentrations were not significantly correlated with percentage of grass/pasture land cover ($p = 0.17$). Nitrate concentrations were most variable in watersheds with about 5–20% urban land cover (Fig. 24.10).

24.5. Discussion

We interpret land cover in the Enoree River basin to represent the urbanization or reforestation of an originally agricultural (grass/pasture) land cover (Figs. 24.5 and 24.6). Urbanization is correlated with increased

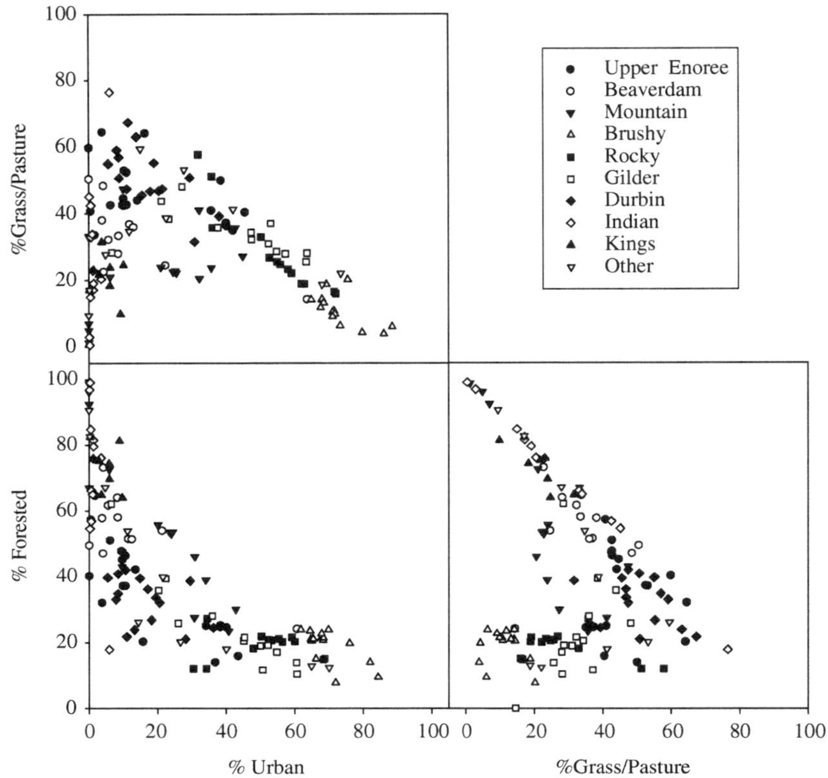

Figure 24.6. Relations between land cover types for sample locality drainages in the Enoree River basin.

nitrate concentrations, whereas reforestation is correlated with decreased nitrate concentrations. Previous studies have shown that nitrate concentrations in forested watersheds are low, especially watersheds undergoing reforestation (Bormann and Likens, 1979; Swank, 1988; Clark et al., 2000). In our discussion, we focus on the relation between urbanization and increased nitrate concentrations.

24.5.1. Potential urban sources of nitrate

In tributary watersheds of the Enoree River basin, other than effluent discharged by a single WWTP, point-source inputs of nitrogen appear insignificant. As a result, land cover is the most important factor controlling stream nitrate concentrations, which were positively correlated

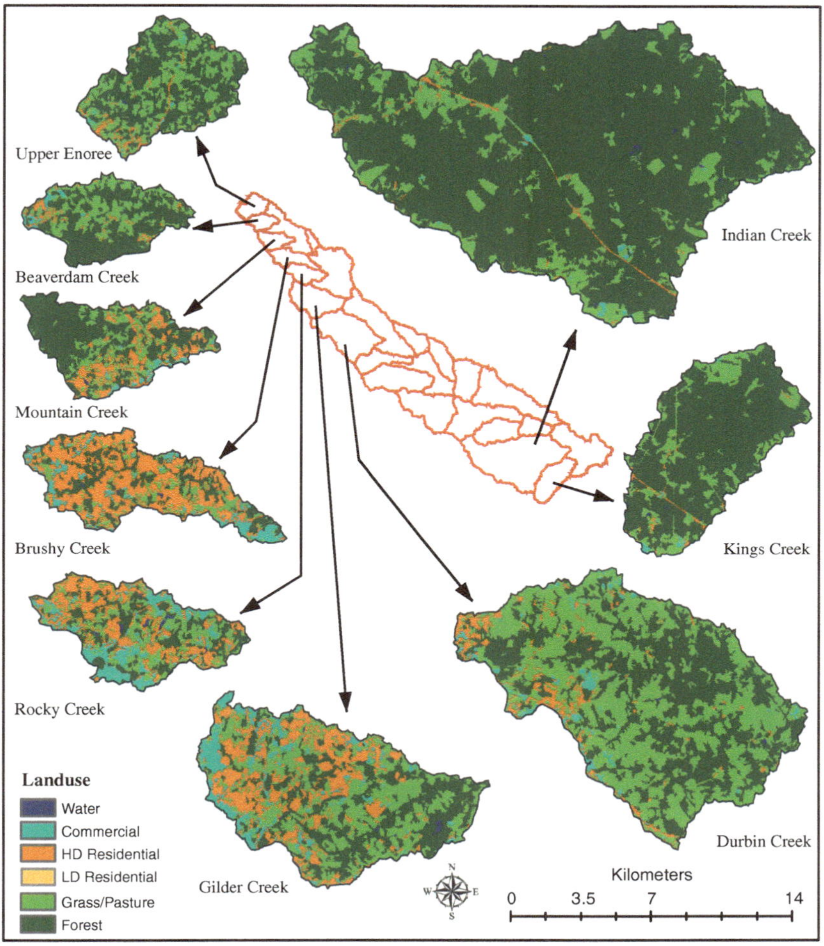

Figure 24.7. Land cover maps for tributary watersheds in the Enoree River basin. Land cover classification is based on satellite images for 2000. See methods for details. For analytical purposes, commercial and high-density (HD) residential land covers are combined as "urban" land cover. The scale in the figure applies to the land cover maps of different watersheds.

with urban land cover in the Enoree River basin. For example, based on the linear regression of nitrate concentration and percentage of urban land cover, watersheds with about 0.02% urban cover would have nitrate concentrations around 0.2–0.4 $mg\,l^{-1}$, whereas watersheds with >90% urban cover would be expected to have concentrations around 3 $mg\,l^{-1}$.

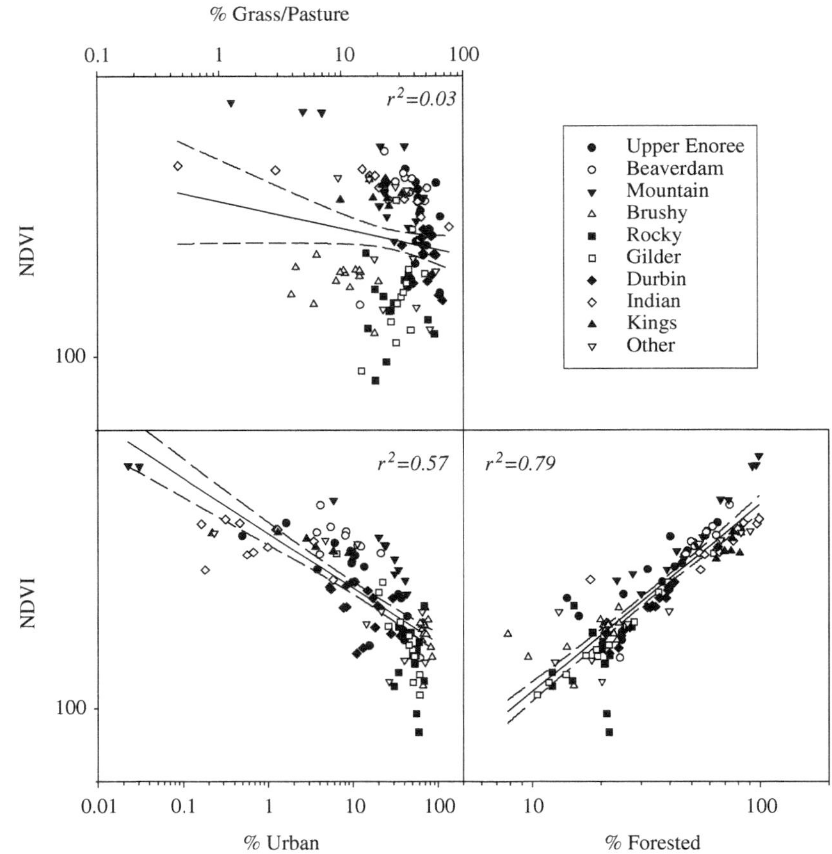

Figure 24.8. Relations between NDVI and land cover. Regression lines are solid and 95% confidence intervals are represented by dashed lines. The NDVI scale is logarithmic. The relationships are statistically significant ($p < 0.001$) except for NDVI versus grass/pasture ($p = 0.07$).

However, we have observed that watersheds with only 5–20% urban land cover had the most variable nitrate concentrations and some of the highest overall concentrations.

High nitrate concentrations in urbanized watersheds could result from several factors, including nitrogen from wet and dry atmospheric deposition, fertilizers added to lawns, septic systems, leaking sewer pipes, and lower nitrogen uptake by vegetation. At present, we lack data to determine the relative contributions of these factors to nitrate concentrations in our study area. Although septic systems are important sources of

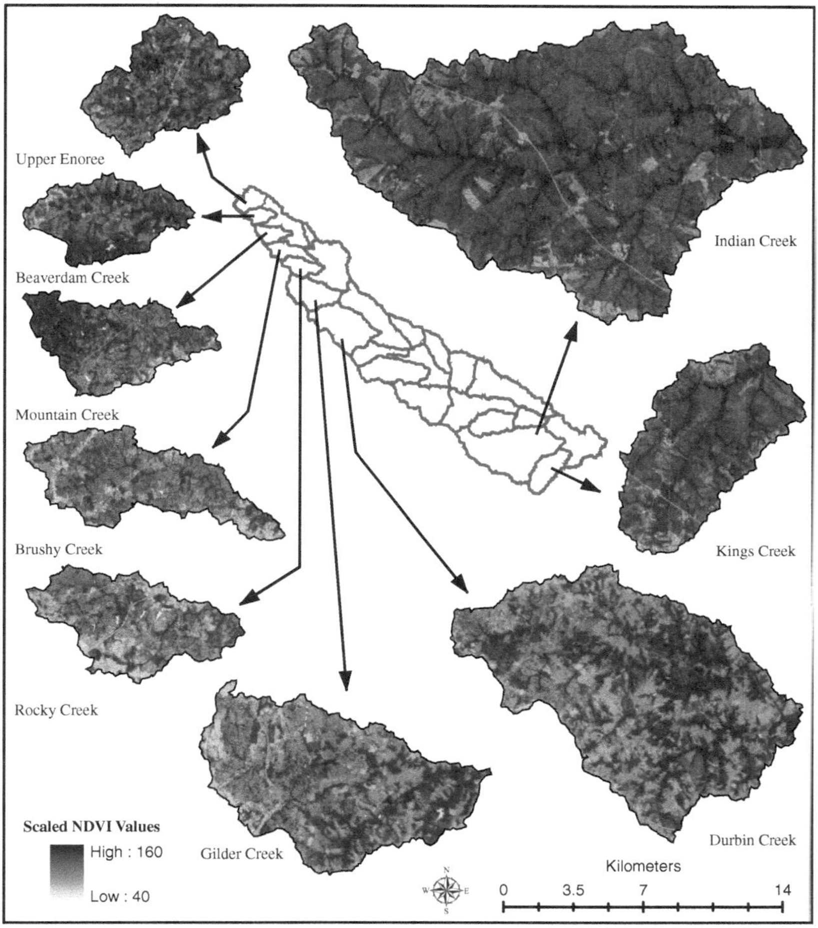

Figure 24.9. Normalized difference vegetation index (NDVI) maps for tributary watersheds in the Enoree River basin. The scale in the figure applies to the landuse map of different watersheds.

nitrates in urban areas of other regions (e.g., Hoare, 1984), our observations suggest that houses in the more urbanized watersheds which are within or adjacent to the city of Greenville (Brushy Creek, Rocky Creek, and Gilder Creek) are served by sewer rather than septic systems. Houses in the less urbanized (more rural) and less densely populated watersheds are more likely to have septic systems, although these watersheds tended to have lower nitrate concentrations.

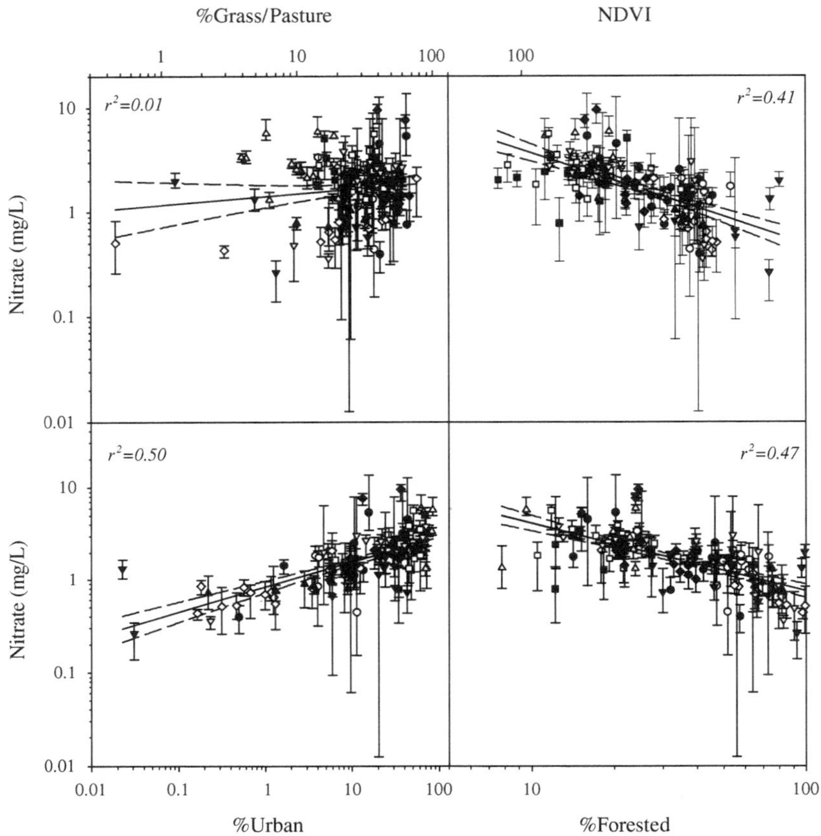

Figure 24.10. Relations between mean nitrate concentrations and land cover for each sample locality. Watersheds are represented by the same symbols as in previous diagrams. Regression lines are solid and 95% confidence intervals are represented by dashed lines. Relationships are statistically significant ($p < 0.001$) except for nitrate versus grass/pasture ($p = 0.17$).

The altered hydrology of urban areas also may contribute to high nitrate concentrations in urban streams. Because urban areas have more impervious surface cover than do forests and grasslands, storm flows are greater in urban areas (Paul and Meyer, 2001). As a result, storm flows erode streambeds, accelerating stream channel incision. For example, in the Maryland coastal plain, channel incision of urban streams results in lowered water tables and more deeply oxidized riparian soils (Groffman et al., 2002). These soils thus have lower denitrification potentials and higher nitrification potentials, which contributes to higher nitrate concentrations in soil water moving into the stream channel (Groffman et al., 2002). Urbanization of headwater streams, in particular, may exert an important control on nitrogen transport in streams because small headwater streams make up a large portion of the total river drainage length of a watershed, collecting a majority of the water and dissolved nutrients from the surrounding terrestrial ecosystem (Peterson et al., 2001). We have observed channel incision in streams in the Enoree River basin, as well, but have not yet quantified the degree of incision or its effect on nitrogen biogeochemistry.

24.5.2. Sources of variability in nitrate concentrations

Although we have demonstrated for the Enoree River basin that there are statistically significant, moderately strong relations among nitrate concentrations, land cover, and NDVI, there clearly is much unexplained variation in nitrate concentrations. Variability in the hydrology and biogeochemical functioning of riparian zones may account for some of the nitrate variability. There is abundant evidence from the literature that vegetated riparian zones can reduce the flux of nitrates into stream channels through processes such as uptake by vegetation and denitrification (see review by Hill, 1996). However, if nitrate-laden water moves below the rooting zone, vegetative uptake will not occur (Hill, 1996). Likewise, lack of sufficient organic carbon in anoxic zones in riparian soils may limit the activity of denitrifying bacteria (Hill et al., 2000). As mentioned previously, increased depth of water tables along incised urban stream channels may also limit nitrate removal via denitrification. In our study, at least some of the urban stream reaches were bordered by forested riparian zones, although the widths of these zones varied.

Although NDVI provides a measure of the density of plant biomass in watersheds, it may overestimate plant biomass in the case of urban areas (e.g., residential areas), which have many trees (and thus some canopy cover which would be detected by remote sensing) but little under-story vegetation. For example, in older neighborhoods with larger trees, impervious surfaces such as streets, roofs of houses, or lawns may promote runoff of rainwater in spite of the tree cover. This may explain the strong relation between NDVI and forested land cover, but a relatively weaker relation between NDVI and urban land cover (Fig. 24.8). The overestimation of plant biomass by NDVI, and correspondingly lower assimilation and denitrification potentials, may have contributed to the variability in the nitrate–NDVI relationship.

At present, we are unable to explain the particularly high variability in nitrate concentrations in watersheds with 5–20% urban land cover.

However, among these watersheds is a great variety of spatial arrangements of land cover. For example, in some of these watersheds, the base of the watershed is more densely vegetated and the headwaters are more urbanized. In others, the opposite is true, even though the percentage of urban and forested land covers are similar in these two cases. Thus, the distribution and exact location of vegetation within these watersheds is probably important to the movement of nitrates to the stream channels.

One additional source of variation in nitrate concentrations is the presence of artificial ponds formed behind earthen dams, which are common in the urbanized watersheds (e.g., ponds in residential areas, parks, and golf courses). Such ponds have a major impact on the hydrology of stream systems by increasing the residence time of water, and may affect nutrient cycling (Smith et al., 2002). Nitrate concentrations in streams draining artificial ponds are lower than at other sampling localities in a given watershed. For example, in the Upper Enoree and Rocky Creek watersheds, the lowest nitrate concentrations occurred in streams draining ponds (Table 24.1, sites UE08, UE13, RC08, and RC12). The locality with the lowest nitrate concentration in the Mountain Creek watershed (MC17) also occurred downstream of a small artificial lake, although the lake's inlet stream drained a heavily forested watershed in which nitrate concentrations are typically low even without the influence of ponds (G.P. Lewis and C.B. Andersen, unpublished data). The ability of ponds and lakes to retain nitrates has been documented in other regions and has been attributed to processes such as algal and microbial nitrate assimilation and denitrification (Saunders and Kalff, 2001). In the present study, sample localities downstream of ponds in our study represent a small fraction of all sample localities, and nitrate concentrations increase further downstream of the ponds, possibly due to soil water or groundwater inputs of nitrates. However, ponds are prevalent in the headwater regions of the majority of the watersheds, so the total impact on nutrient cycling remains to be documented.

24.6. Conclusions

Humans have modified the global biogeochemical cycle of nitrogen substantially. In part, this modification has resulted in a higher flux of nitrogen to the ocean through river systems. In comparison to the contribution of nitrogen by agricultural land use and point-sources to rivers, the impact of urbanization is not well understood. Our study clearly shows that urban land cover is associated with higher concentrations of stream nitrate than forested land cover, well above what is considered normal for undeveloped, forested landscapes. The relation between land cover and stream nitrate concentrations, however, is only moderately strong, in part because of the complex spatial distribution of land cover types within individual watersheds and variation in riparian function and stream morphology. Future research will examine the processes by which urbanization increases stream nitrate concentrations in more detail.

ACKNOWLEDGMENTS

We thank the many students and faculty participants in Furman University's River Basins Research Initiative, especially John Wheeler, Sandy Wheeler, Kenneth Sargent, Christy Gullikson, Neal Gandy, and Tyler Bax. Reviews by Steve Coghlan and Timothy Warner improved the clarity of the manuscript. Funding for this work was provided by the NSF-REU Program (EAR-9820605), South Carolina Department of Health and Environmental Control NPS Program (EQ-9-461), the Associated Colleges of the South, and Furman University.

REFERENCES

Andersen, C.B., 2001. The problem of sample contamination in a fluvial geochemistry research experience for undergraduates. J. Geosci. Educ. 49, 351–357.

Andersen, C.B., Lewis, G.P., Sargent, K.A., 2004. Influence of wastewater treatment effluent on concentrations and fluxes of solutes in the Bush River, South Carolina, during extreme drought conditions. J. Environ. Geosci. 11, 28–41.

Andersen, C.B., Sargent, K.A., Wheeler, J., Wheeler, S., 2001. Fluvial geochemistry of selected tributary watersheds in the Enoree River Basin, SC. South Carolina Geol. 43, 57–71.

Ballester, M.V.R., Victoria, D.C., Krusche, A.V., Coburn, R., Victoria, R.L., Richey, J.E., Logsdon, M.G., Mayorga, E., Matricardi, E., 2003. A remote sensing/GIS-based physical template to understand the biogeochemistry of the Ji-Parana river basin (Western Amazonia). Remote Sens. Environ. 87, 429–445.

Bannari, A., Morin, D., Bonn, F., 1995. A review of vegetation indices. Remote Sens. Rev. 13, 95–120.

Boone, R.B., Galvin, K.A., Smith, N.M., Lynn, S.J., 2000. Generalizing El Nino effects upon Maasai livestock using hierarchical clusters of vegetation patterns. Photogramm. Eng. Rem. Sens. 66(6), 737–744.

Bormann, F.H., Likens, G.E., 1979. Pattern and Process in a Forested Ecosystem. Springer-Verlag, New York, p. 253.

Boyle, C.A., Lavkulich, L., Schreier, H., Kiss, E., 1997. Changes in land cover and subsequent effects on Lower Fraser basin ecosystems from 1827 to 1900. Environ. Manage. 21, 185–196.

Camp, W., 1960. Soil survey of Spartanburg County, SC, U.S. Department of Agriculture, p. 83, and maps.

Camp, W., 1975. Soil survey of Greenville County, SC, U.S. Department of Agriculture, p. 71, and maps.
Camp, W., Jones, W.E., Miford, P.R., Hearn, S.H., Aull, L.E., 1960. Soil survey of Newberry County, SC, U.S. Department of Agriculture, p. 66, and maps.
Camp, W., Meltzer, J.C., Fleming, W.H., Andrew, L.E., 1975. Soil survey of Laurens and Union Counties, SC, U.S. Department of Agriculture, p. 66, and maps.
Chen, D., Brutsaert, W., 1998. Satellite-sensed distribution and spatial patterns of vegetation parameters over a tallgrass prairie. J. Atmos. Sci. 55(7), 1225–1238.
Clark, G.M., Mueller, D.K., Mast, M.A., 2000. Nutrient concentrations and yields in undeveloped stream basins of the United States. J. Am. Water Res. Assoc. 36(4), 849–860.
Douglas, T.A., Chamberlain, C.P., Blum, J.D., 2002. Land use and geologic controls on the major elemental and isotopic (^{15}N and $^{87}Sr/^{86}Sr$) geochemistry of the Connecticut River watershed, USA. Chem. Geol. 189, 19–34.
ESRI, 2005. ArcObjects Online: Hydrology Modeling, Last accessed 30 June 2005. Available at http://arcobjectsonline.esri.com/ArcObjectsOnline/Samples/Spatial%20Analyst/Hydrology%20Modeling/HydrologyModeling.htm.
Filoso, S., Martinelli, L.A., Williams, M.R., Lara, L.B., Krusche, A., Ballester, M.V., Victoria, R.L., Camargo, P.B., 2003. Land use and nitrogen export in the Piracicaba River Basin, Southeast Brazil. Biogeochemistry 65, 275–294.
Freeze, R.A., Cherry, J.A., 1979. Groundwater. Prentice-Hall, Englewood Cliffs, NJ, p. 604.
Galloway, J.N., Cowling, E.B., 2002. Reactive nitrogen and the world: 200 years of change. Ambio 31(2), 64–71.
Goolsby, D.A., 2000. Mississippi Basin nitrogen flux believed to cause Gulf hypoxia. EOS T. Am. Geophys. Un. 81(29), 321–327.
Groffman, P.M., Boulware, N.J., Zipperer, W.C., Pouyat, R.V., Band, L.E., Colosimo, M.F., 2002. Soil nitrogen cycle processes in urban riparian zones. Environ. Sci. Technol. 36(21), 4547–4552.
Hill, A.R., 1996. Nitrate removal in stream riparian zones. J. Environ. Qual. 25, 743–755.
Hill, A.R., Devito, K.J., Campagnolo, S., Sanmugadas, K., 2000. Subsurface denitrification in a forest riparian zone: Interactions between hydrology and supplies of nitrate and organic carbon. Biogeochemistry 51, 193–223.
Hoare, R.A., 1984. Nitrogen and phosphorus in Rotorua urban streams. New Zealand J. Mar. Freshwater Res. 18, 451–454.
Hunsaker, C.T., Levine, D.A., 1995. Hierarchical approaches to the study of water quality in rivers. BioScience 45(3), 193–203.
Jaworski, N.A., Howarth, R.W., Hetling, L.J., 1997. Atmospheric deposition of nitrogen oxides onto the landscape contributes to coastal eutrophication in the northeast United States. Environ. Sci. Technol. 31(7), 1995–2004.
Jensen, J.R., 2000. Remote Sensing of the Environment: An Earth Resource Perspective. Prentice-Hall, Saddle River, N.J, p. 544.
Ji, L., Peters, A.J., 2003. Assessing vegetation response to drought in the northern Great Plains using vegetation and drought indices. Remote Sens. Environ. 87, 85–98.
Keaton, M., Haney, D., Andersen, C.B., 2005. Impact of drought upon fish assemblage structure in two South Carolina piedmont streams. Hydrobiologia 545, 209–223.
Likens, G.E., Bormann, F.H., Johnson, N.M., Fisher, D.W., Pierce, R.S., 1970. Effects of forest cutting and herbicide treatment on nutrient budgets in the Hubbard Brook watershed-ecosystem. Ecol. Monogr. 40(1), 23–47.
Meybeck, M., 1982. Carbon, nitrogen, and phosphorus transport by world rivers. Am. J. Sci. 282, 401–450.

Meybeck, M., Helmer, R., 1989. The quality of rivers: From pristine stage to global pollution. Palaeogeogr. Palaeoecol. 75, 283–309.

Moreau, S., Bertru, G., Buson, C., 1998. Seasonal and spatial trends of nitrogen and phosphorus loads to the upper catchment of the river Vilaine (Brittany): Relationships with land use. Hydrobiologia 373/374, 247–258.

Paul, M.J., Meyer, J.L., 2001. Streams in the urban landscape. Ann. Rev. Ecol. Syst. 32, 333–365.

Peierls, B.L., Caraco, N.F., Pace, M.L., Cole, J.J., 1991. Human influence on river nitrogen. Nature 350, 386–387.

Peterson, B.J., Wollheim, W.M., Mulholland, P.J., Webster, J.R., Meyer, J.L., Tank, J.L., Martí, E., Bowden, W.B., Valett, H.M., Hershey, A.E., McDowell, W.H., Dodds, W.K., Hamilton, S.K., Gregory, S., Morrall, D.D., 2001. Control of nitrogen export from watersheds by headwater streams. Science 292, 86–90.

Rabalais, N.N., Turner, R.E., Scavia, D., 2002. Beyond science into policy: Gulf of Mexico Hypoxia and the Mississippi River. BioScience 52(2), 129–142.

Roy, S., Gaillardet, J., Allegre, C.J., 1999. Geochemistry of dissolved and suspended loads of the Seine River, France: Anthropogenic impact, carbonate, and silicate weathering. Geochem. Cosmochim. Ac. 63, 923–938.

Saunders, D.L., Kalff, J., 2001. Nitrogen retention in wetlands, lakes, and rivers. Hydrobiologia 443, 205–2112.

Smith, S.V., Renwick, W.H., Bartley, J.D., Buddemeier, R.W., 2002. Distribution and significance of small, artificial water bodies across the United States landscape. Sci. Total Environ. 299, 21–36.

Stoms, D.M., Estes, J.E., 1993. A remote sensing research agenda for mapping and monitoring biodiversity. Int. J. Rem. Sens. 14, 1839–1860.

Strahler, A.N., 1952. Dynamic basis of geomorphology. Geol. Soc. Am. Bull. 63, 923–938.

Swank, W.T., 1988. Stream chemistry responses to disturbance. In: Swank, W.T., Crossley, D.A. Jr. (Eds.), Forest Hydrology and Ecology at Coweeta. Springer-Verlag, New York, pp. 339–357.

Systat Software, 2004. SigmaStat 3.1 User's Manual. Systat Software Inc., Point Richmond, CA, p. 848.

Tucker, C.J., Sellers, P.J., 1986. Satellite remote sensing of primary production. Int. J. Remote Sens. 7, 1396–1416.

Vitousek, P.M., Aber, J.D., Howarth, R.W., Likens, G.E., Matson, P.M., Schindler, D.W., Schlesinger, W.H., Tilman, D.G., 1997. Human alteration of the global nitrogen cycle: Sources and consequences. Ecol. Appl. 7(3), 737–750.

Zhiliang, S., Qun, L., Shumel, Z., Hui, M., Ping, Z., 2003. A nitrogen budget of the Changjiang River catchment. Ambio 32, 65–69.

Section VI: Application of Geochemical Principles in Environmental Quality and Remediation Research

Developments in Environmental Science, Volume 5
D. Sarkar, R. Datta and R. Hannigan (Editors)

ISSN: 1474-8177/DOI:10.1016/S1474-8177(07)05025-5

Chapter 25

Remediation of arsenical pesticide applied soils using water treatment residuals: Preliminary greenhouse results

Rupali Datta, Dibyendu Sarkar, Hussein Hussein and Chacharee Therapong

Abstract

Long-term application of arsenical pesticides in agricultural fields has resulted in increased background concentration of this toxic metalloid in soils. In addition, leaching of arsenic (As) from chromated copper arsenate (CCA)-treated wood found in decks, docks, playground equipment, and garden construction has contributed further to soil As contamination. Arsenic contamination is particularly severe in cattle dipping vat sites, where soil As concentrations range between 700 and 2000 $mg\,kg^{-1}$. Plants growing in As-contaminated soils could pose a potential health hazard on human and livestock populations. Several studies have successfully used water treatment residuals (WTRs) to bind phosphorus (P) in soils, resulting in reduced P in runoff from fields, which would otherwise cause P-enrichment, and eutrophication of water bodies. Since P and As exhibit very similar chemical properties, it is logical to assume that As, like P, will be similarly and significantly retained by WTRs. The main objective of this study was to evaluate the use of various WTRs as low-cost chemical amendments to bind As in pesticide-applied soils, reducing As availability to plants. Rice was used as the test crop. Soils contaminated with arsenical pesticides and amended with two types of WTRs (Al and Fe) were used for a greenhouse column study. Rice was grown in the columns for a period of six months. The results obtained in this study show that both Al- and Fe-WTRs are very effective in reducing phytoavailable As in the soils. Addition of WTRs resulted in considerable improvement in the growth of rice plants in comparison to control plants with no WTR added, in which growth of rice was severely impaired. While Al-WTR was more effective in reducing As concentration in rice plants, Fe-WTR was more effective in improving the growth of rice in As-contaminated soils.

25.1. Introduction

Arsenic is an element of great concern in terrestrial as well as aquatic environments because of the high toxicity and carcinogenicity of some As species. Arsenic has been classified as a Group A human carcinogen (Southworth, 1995). Organic forms of As are considered non-carcinogenic; however, transformation of As from organic to inorganic forms is possible in soil environments (Rodriguez, 1998). USEPA adopted a new drinking water standard in 2001, lowering the maximum contaminant level (MCL) of As from 50 to 10 ppb (USEPA, 2001). Arsenic contamination of the environment can occur from natural as well as anthropogenic sources. The primary natural sources of arsenic include hot springs, igneous rock, sedimentary rock, metamorphic rock, seawater, mineral deposits, and volcaniclastic materials (Bowen, 1979; Cai and Ma, 2003). Anthropogenic sources include indiscriminate disposal of wastes from mining, milling, and smelting of ores (Lindau, 1977), raw and spent oil shale (Shendrikar and Faudel, 1978) and coal fly ash amendments (Hansen et al., 1984). Long-term application of arsenical pesticides in agricultural lands has also resulted in high levels of As residues in certain soils (Murphy and Aucott, 1998). The soil As problem is particularly critical in cattle dipping vat sites, which are almost ubiquitously present in ranches or former ranching areas in various parts of the country, particularly in the southern United States. Ng (1997) assessed soil As to range between 700–2000 mg kg^{-1} in cattle dip sites. Although the use of arsenical pesticides as dipping solutions has long been banned, there exists a large number of abandoned vat sites in the southern United States that are seriously contaminated by As.

Plants (either wild or cultivated) growing in As-contaminated soils or water could pose a potential health hazard on human and livestock populations. Arsenic is translocated to most plant organs, although highest concentrations are often found in roots and older leaves (Tamaki and Frankenberger, 1992). Pickering et al. (2000) reported that most of As was locked up in plant roots while relatively small quantities translocated into the aboveground tissues. In rice plants, As concentration was reported to be 4–8-fold higher in root tissue when compared to the shoot (Liu et al., 2004). In addition to the As species, toxicity of As to the plants is affected by several biogeochemical factors such as soil texture, organic matter, nature and constituents of minerals, pH, redox potential, and competing ions (Adriano, 2001). Our knowledge on As metabolism in plants is limited. There are reports of As metabolism in mammals and bacteria involving methylation and biotransformation to As-lipid compounds, but there are no reports of these processes playing a role in As detoxification in higher plants (Nissen and Benson, 1982). While arsenate is a chemical analogue of phosphate and may interfere with oxidative phosphorylation (Terwelle and Slater, 1967), arsenite is a neutral species at natural pH values and inhibits the activity of enzymes by binding to thiol groups. Methylarsonic acid (MMA) and dimethylarsenic acid (DMA) also form anions in soils but are much less toxic than inorganic species (Sohrin et al., 1997) and may block protein synthesis (Sckerl and Frans, 1969).

Water treatment residuals (WTRs) are a by-product of drinking water treatment plants. Metal salts, such as ferric chloride and aluminum sulfate (alum), are commonly used in the municipal water treatment processes to destabilize colloids for subsequent flocculation and water clarification (Butkus et al., 1998; Ippolito et al., 1999). WTRs are generally composed of Fe/Al-oxides, activated C, and high molecular weight, long-chain, water-soluble polymers (Elliott and Dempsey, 1991). Land-filling is a common practice for the disposal of this material; however, recent studies have shown that benefits associated with the reuse of WTR as a soil amendment include improved soil structure (El-Swaify and Emerson, 1975), increased moisture-holding capacity (Bugbee and Frink, 1985), and increased availability of nutrients for various plants (Heil and Barbarick, 1989).

A significant amount of research work has been performed to investigate the advantages of applying WTR to reduce P in surface runoff for P-enriched soils, which could otherwise end up causing eutrophication of surface water bodies. Moore and Miller (1994) used alum sludge (Al-WTR) to reduce P concentration in runoff from land to which poultry litter was applied. Maurice et al. (1998) reported that partially dried alum sludge used as a component of poultry litter also reduced movement of P. Gallimore et al. (1999) attributed the ability of Al-WTR to reduce P in runoff from poultry litter-amended fields to the P-fixation potential of amorphous Al-oxides in the WTR. Adding WTR at higher ratios, on the other hand, could immobilize bioavailable P and induce P deficiency in crops (Ippolito et al., 1999). However, Jacobs and Teppen (2000) found no yield reductions in corn or decreases in extractable P levels even when Al-WTR was applied at rates as high 135 $Mg\,ha^{-1}$ to soils highly enriched in P from repeated fertilizer and manure applications. Butkus et al. (1998) used Fe-WTR as a soil amendment and showed that the reduction in available P is due to binding of P to both amorphous ferric hydroxide and the cationic polyelectrolyte, a quarternary polyamine.

Since P and As exhibit very similar chemical properties, it is logical to assume that As, like P, will be similarly and significantly retained by WTR components, such as amorphous Fe/Al-oxides and cationic polymers. Both phosphates and arsenates undergo similar types of retention in soil minerals via primarily inner-sphere surface complexation. Thus, P has demonstrated a strong ability to compete with As for sorption sites in environmentally important pH ranges (Matera and Le Hecho, 2001). The retention of P by WTR is strongly hysteretic (Butkus et al., 1998; Ippolito et al., 1999). Apparently, if Al-WTR and/or Fe-WTR are capable of retaining As in a similar irreversible fashion as P, it may be possible to use WTR as an in-situ amendment in As-contaminated environmental systems. However, there are very few published studies on As retention by WTR.

The objective of this study was to evaluate the use of various WTRs as low-cost chemical amendments in As-contaminated soils. Immokalee series soil from Florida, with minimal As retention capacity was used for the study. The soil was spiked with either sodium arsenate or DMA. Al- and Fe-WTRs were used as amendments to bind As in the soils. Rice (*Oryza sativa* L) plants were used as a test crop to determine the efficacy of WTRs in reducing phytoavailable As in the contaminated soils.

25.2. Materials and methods

To evaluate the role of WTRs in As-phytoavailability, it is important to use a soil that has minimal As retention capacity. Immokalee series soil from Florida was selected to conduct the greenhouse study. Immokalee series soil is sandy with low pH, low extractable Fe/Al, Ca/Mg, and low organic matter content. Being sandy, acidic and lacking positively charged adsorptive surfaces (e.g. amorphous Fe + Al oxides), it is likely to have minimal As retention capacity (Pierce and Moore, 1980; Oscarson et al., 1981), thereby increasing potentially phytoavailability of As. Immokalee series soil was collected from Southwest Florida Research and Education Center, Immokalee, Florida.

The soil was characterized for its texture, pH, EC, cation exchange capacity (CEC), total Fe and Al, oxalate-extractable Fe and Al, total and extractable Ca and Mg, total and extractable P, soil organic matter, and total As (Table 25.1). Standard protocols outlined in the Soil Science Society of America Handbooks for Chemical and Mineralogical Analysis (Klute, 1996; Sparks, 1996) were followed for soil characterization.

The soil was spiked with two As compounds, sodium arsenate and DMA at two rates, 675 and 1500 mg kg^{-1}. This high-soil As concentrations were chosen following the assessment of Ng (1997) who estimated soil As concentrations in cattle dip sites to range between 700 and 2000 mg As per kg. Contaminated soils as well as control samples (soil

Table 25.1. General chemical properties of Immokalee soil

pH		6.0
EC ($\mu S\,cm^{-1}$)		59
CEC ($cmol\,kg^{-1}$)		777
Soil organic matter ($g\,kg^{-1}$)		8.4
Total-recoverable As ($mg\,kg^{-1}$)		0.8
P ($mg\,kg^{-1}$)	Mehlich 3	4.0
	Total	210
Ca + Mg ($mg\,kg^{-1}$)	Mehlich 3	270
	Total	1200
Fe + Al ($mg\,kg^{-1}$)	Oxalate	66
	Total	212

EC, electrical conductivity; CEC, cation exchange capacity.

with no pesticide application) were loaded in polyvinyl chloride (PVC) columns 15″ high × 6″ diameters in three replicates.

Rice, a fast growing, high-biomass plant was used as the test crop. Rice seeds (thirty seeds per column) were sown into the columns and allowed to grow for a period of six months. The columns were arranged in a randomized block design and rotated periodically to account for variances in temperature and sunlight within the greenhouse and maintained at 80% pot holding capacity. Once every month, the pots were over-watered in order to induce leaching.

To evaluate the role of WTR-amendment in reducing As phytoavailability in contaminated soils, two types of WTRs were used. Al-WTR was obtained from the Manatee County Water Treatment Plant in Bradenton, FL. Fe-WTR obtained from the Hillsboro River Water Treatment Plant in Tampa, FL. The soils, wetted to 80% of their water holding capacity, were mixed with 1500 $mg\,kg^{-1}$ of Sodium arsenate and DMA separately. The WTRs were applied at 2 rates of 5% and 10% following the field study of Gallimore et al. (1999). Soils were loaded in PVC columns and rice was grown as described above. All the above treatments were done in triplicates.

Plant samples were digested using HNO_3/H_2O_2 digestion procedures (Carbonell et al., 1998). The acid digests were analyzed for total As using a graphite furnace atomic absorption spectrophotometer (GFAAS).

25.3. Results and discussion

25.3.1. Effect of arsenical pesticides on plant growth

The quantitative effect of organic and inorganic arsenical pesticides on plant growth is shown in Fig. 25.1. The results showed that the growth of rice was more severely impaired by the addition of sodium arsenate

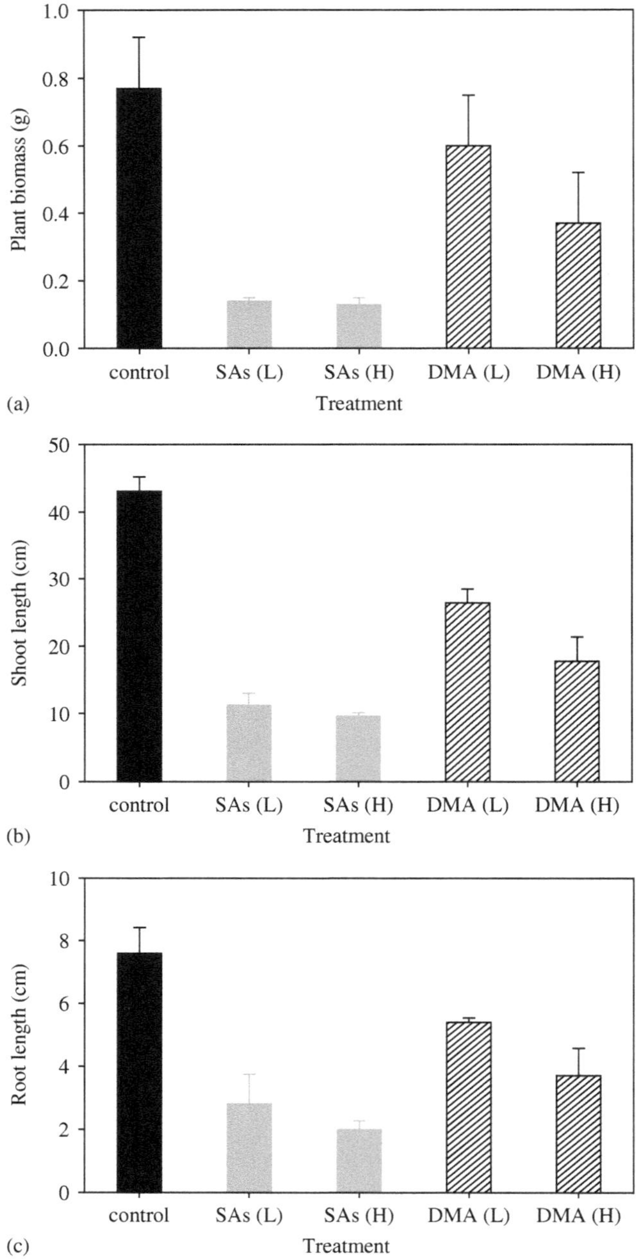

1.0
0.8
0.6
0.4
0.2
0.0
Plant biomass (g)
control
SAs (L)
SAs (H)
DMA (L)
DMA (H)
(a)
Treatment
50
40
30
20
10
0
Shoot length (cm)
control
SAs (L)
SAs (H)
DMA (L)
DMA (H)
(b)
Treatment
10
8
6
4
2
0
Root length (cm)
control
SAs (L)
SAs (H)
DMA (L)
DMA (H)
(c)
Treatment

compared to DMA. The average total biomass of the plants grown in either 675 or 1500 mg kg^{-1} sodium arsenate was 0.13 g in dry weight ($\pm$0.02) compared to 0.6 g ($\pm$0.2) in DMA at the same concentrations (Fig. 25.1a). As expected, the average total biomass of the control plants (grown in soil with no pesticide added) was higher (0.8 g) than the plants grown in pesticide-contaminated soils. Increasing DMA concentration to 1500 mg kg^{-1} significantly reduced the plant biomass ($p < 0.001$) by about 38% when compared to the plants grown in 675 ppm of DMA and by about 52% when compared to the control plants (Fig. 25.1a). Fig. 25.1b shows that the addition of sodium arsenate caused a significant reduction ($p < 0.0001$) in shoot length by an average of 74% compared to the control plants with no significant difference between low and high concentrations of sodium arsenate applied. On the other hand, addition of 675 and 1500 mg kg^{-1} DMA to the soils caused a reduction of 39% and 60% respectively, in shoot lengths compared to the control plants. Unlike shoots, roots were less affected by addition of As in soils (Fig. 25.1c). However, the effect of inorganic and organic As on root length followed a similar trend as total biomass and shoot lengths. The root lengths were not very much affected by the lower concentration of sodium arsenate and DMA. At the higher concentration of 1500 mg kg^{-1} of sodium arsenate and DMA, root lengths decreased by 75% and 53% respectively, compared to control plants (Fig. 25.1c).

Our results indicate that phytotoxicity of As is influenced by the species of As applied. Rice plants treated with sodium arsenate produced lower biomass and reduced plant shoot and root lengths compared to plants treated with DMA with no notable differences between low and high concentrations applied. Milam et al. (1988) and Marin et al. (1993) reported a reduction of biomass/plant growth of rice plants subjected to arsenate. Tsutsumi (1982), on the other hand, observed no reduction in rice plant height at 125 ppm of arsenate applied to the soils. However, increasing arsenate concentration up to 312.5 ppm decreased the plant height by 63% compared to control plants. On the other hand, there are several contradictory reports on the effect of DMA on plant growth. While several reports of varied effects of As on plant growth have been published (Tang and Miller, 1991; Marin et al., 1992; Sneller et al., 1999; Abdein et al., 2002), mechanism of As toxicity in plants remain unclear. The toxicity of inorganic arsenicals such as arsenate is due to interference

Figure 25.1. Effect of arsenical pesticide-contaminated soil on the growth of rice plants. Immokalee soil was contaminated using either sodium arsenate (SAs) or dimethylarsenic acid (DMA) either at 675 mg kg^{-1} (L) or 1500 mg kg^{-1} (H) of As. Effect on total biomass (a), shoot length (b) and root length (c) were recorded. Error bars represent $\pm$SE.

with oxidative phosphorylation (Terwelle and Slater, 1967). On the other hand, organic arsenicals, such as DMA, are less toxic to plants and are likely to block protein synthesis (Sckerl and Frans, 1969).

25.3.2. Arsenic accumulation in plant tissues

The concentration of As in the plant tissues is presented in Fig. 25.2a. The results show that As concentration ranged from 25 $\mu g\, g^{-1}$ to 271 $\mu g\, g^{-1}$ in plants treated with DMA and sodium arsenate, respectively (Fig. 25.2a). In general, As concentration in plants treated with 1500 $mg\, kg^{-1}$ sodium arsenate was 10-fold higher than in the plants treated with DMA at the same concentration. However, at the lower rate of 675 $mg\, kg^{-1}$, As concentrations in the plants treated with sodium arsenate were only 2-fold higher than those treated with DMA at the same rate.

Although As concentrations were higher in plants treated with sodium arsenate, the plants accumulated more As when treated with the same concentration DMA in the soil (Fig. 25.2b). The highest As accumulation in plant tissues was recorded in the plants treated with 675 $mg\, kg^{-1}$ of DMA (57.8 μg/plant dry weight) followed by 36 and 25 μg/plant dry weight when sodium arsenate was added at 1500 and 675 $mg\, kg^{-1}$ respectively. The lowest accumulation of As in plant tissues was recorded when plants were treated with 1500 mg kg of DMA. Since inorganic arsenicals are more toxic than organic arsenicals (Fowler, 1977; Adriano, 2001), plants grew better in soil contaminated with DMA, hence were able to accumulate more As in their tissues over a time period of six months. Earlier studies by Tlustoš et al. (2002) showed that DMA was adsorbed to a much lesser extent than As(V) in the soils, making it more phytoavailable. These findings are in agreement with the results of Marin et al. (1992) who found that phytoavailability of As is higher when supplied as DMA than As(V) for two rice cultivars grown in solution culture. However, in this study, we observed that As concentrations in plant tissues were higher when plants were treated with sodium arsenate, when compared to DMA (Fig. 25.2a).

25.3.3. Effect of WTRs on As phytoavailability

WTRs were added to the soils contaminated with 1500 $mg\, kg^{-1}$ of As to evaluate their capacity of irreversibly retain As, thereby reducing its phytoavailability. Soils spiked with 1500 $mg\, kg^{-1}$ of either sodium arsenate or DMA were amended using two types of WTRs (Fe-WTR

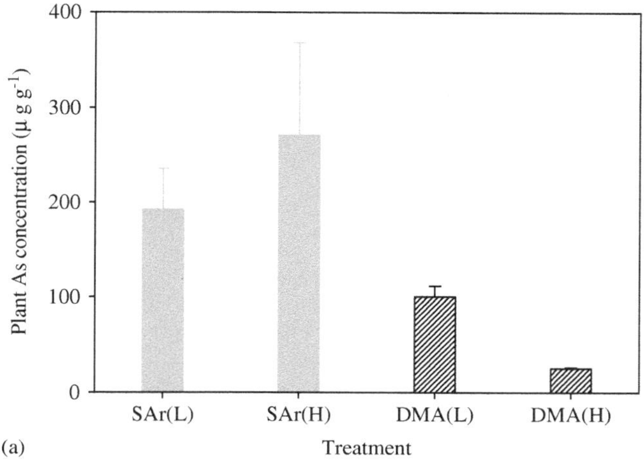

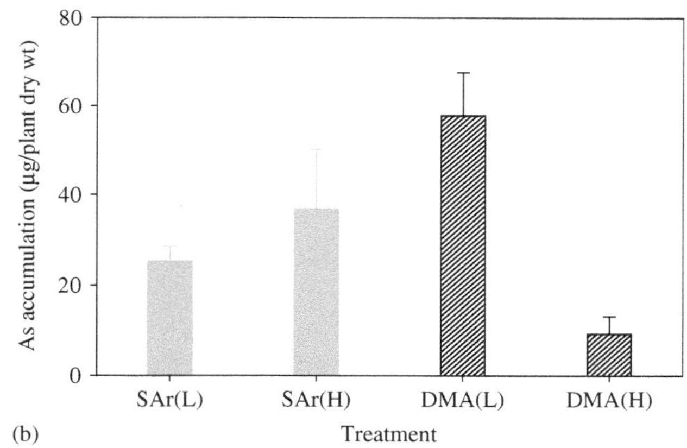

Figure 25.2. Arsenic uptake and accumulation in rice plants grown in arsenical pesticide-contaminated soils. Immokalee soil was contaminated using either sodium arsenate (SAs) or dimethylarsenic acid (DMA) either at 675 mg kg^{-1} (L) or 1500 mg kg^{-1} (H) of As. Arsenic concentration in plant tissues (a) and As accumulation per plant (b) was determined after acid digestion of the plant tissues and analyzing the digests using a graphite furnace atomic absorption spectrophotometer. Error bars represent $\pm$SE.

and Al-WTR) at two rates (5% and 10%). The addition of Fe-WTRs had a marked effect on the growth of rice plants. It was observed that plant biomass increased significantly ($p<0.001$) when Fe-WTR was added at the rate of 10% to the soils contaminated with 1500 mg kg^{-1} sodium arsenate (Fig. 25.3a). Compared to the control

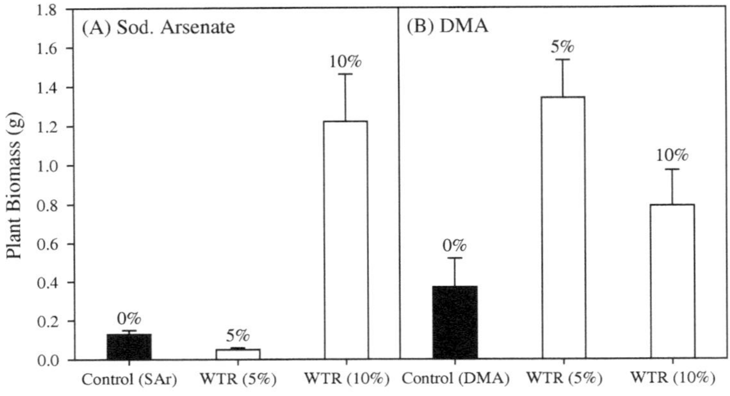
(A) Sod. Arsenate
(B) DMA
Plant Biomass (g)
1.8
1.6
1.4
1.2
1.0
0.8
0.6
0.4
0.2
0.0
0%
5%
10%
Control (SAr)
WTR (5%)
WTR (10%)
Control (DMA)
Treatment (Fe-WTR)
(a)

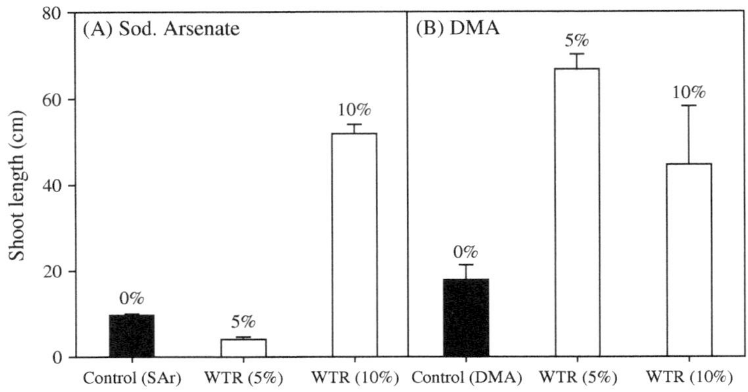
(A) Sod. Arsenate
(B) DMA
Shoot length (cm)
80
60
40
20
0
0%
5%
10%
Control (SAr)
WTR (5%)
WTR (10%)
Control (DMA)
Treatment (Fe-WTR)
(b)

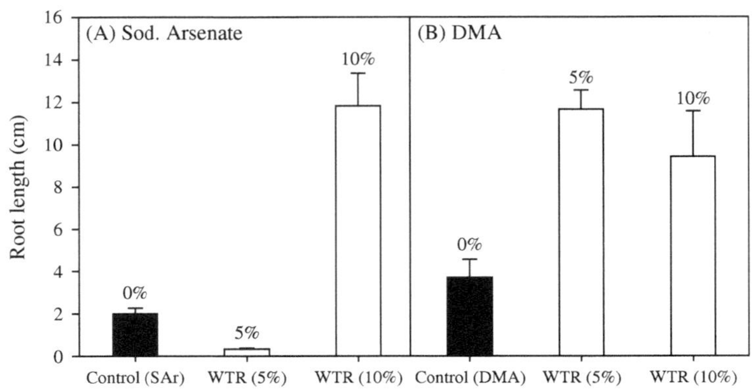
(A) Sod. Arsenate
(B) DMA
Root length (cm)
16
14
12
10
8
6
4
2
0
0%
5%
10%
Control (SAr)
WTR (5%)
WTR (10%)
Control (DMA)
Treatment (Fe-WTR)
(c)

(0% of Fe-WTR), the total plant biomass increased by 90% while shoot and root lengths increased by the average of 82% (Fig. 25.3b, 3c). Addition of 5% Fe-WTR did not improve the growth of rice plants in soils contaminated with sodium arsenate. In case of DMA, the plant growth was significantly increased ($p<0.001$) when Fe-WTR was applied compared the control (0% WTR). Addition of 5% Fe-WTR resulted in an increase in total plant biomass by 96% (Fig. 25.3a) when compared to the control plants. Shoot and root lengths were increased by an average of 71% compared to the controls (Fig. 25.3b, 3c). At 10% Fe-WTR amendment, the total plant biomass increased by an average of 50% (Fig. 25.3a), and shoot and root lengths increased by an average of 60% when compared to control (Fig. 25.3b, 3c). Increased iron concentration in the soils due to Fe-WTR amendment was found to result in improved plant growth. Iron is essential for the functioning of a number of redox proteins in the electron transport chains of respiration and photosynthesis; it is a component of antioxidant enzymes such as catalases and peroxidases; it is essential also in iron–sulfur proteins such as ferredoxin, superoxide dismutase and aconitase (Marschner, 1995). Despite its abundance on the earth surface, soil iron is almost exclusively present in its oxidized form, which has very low solubility in water. Transgenic lettuce plants accumulating higher levels of iron exhibited enhanced growth resulting in higher biomass compared to control plants (Goto et al., 2000).

Application of Al-WTR to the soils amended with 1500 mg kg^{-1} of sodium arsenate or DMA caused no major improvement in plant growth (Fig. 25.4). Generally, when Al-WTR was applied to soils amended with 1500 mg kg^{-1} of sodium arsenate, there was no major effect on plant biomass, shoot, and root lengths. On the other hand, when Al-WTR was applied to the soils amended with 1500 mg kg^{-1} of DMA, while there was no significant effect on plant biomass (Fig. 25.4a), addition of 5% Al-WTR to the DMA-contaminated soil increased shoot length by 17% (Fig. 25.4b) and root length by 30% (Fig. 25.4c). Application of 10% Al-WTR to DMA-contaminated soils caused a significant increase in shoot and root lengths by an average of 56% (Fig. 25.4b,c).

Figure 25.3. Effect of water treatment residual (WTR)-amendment on growth of rice plants on arsenical pesticide-contaminated soil. Immokalee soil was contaminated using either sodium arsenate (A) or DMA (B) at 1500 mg kg^{-1} of As and amended with Fe-WTR at two rates (5% or 10%). Control plants were grown in soils spiked with 1500 mg kg^{-1} of As using either sodium arsenate or DMA with no WTR amendment. Effect on total biomass (a), shoot length (b), and root length (c) were recorded. Error bars represent $\pm$SE.

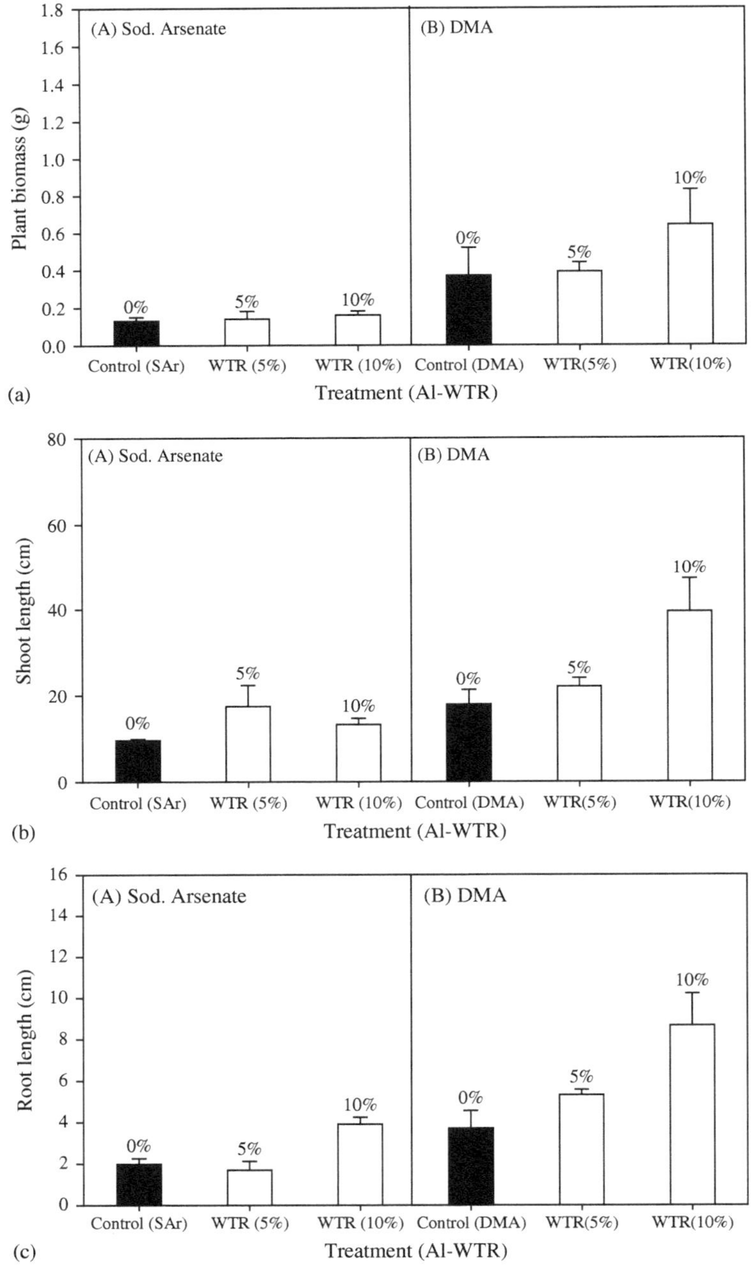

(a)
Plant biomass (g)
1.8
1.6
1.4
1.2
1.0
0.8
0.6
0.4
0.2
0.0
(A) Sod. Arsenate
(B) DMA
0%
5%
10%
0%
5%
10%
Control (SAr)
WTR (5%)
WTR (10%)
Control (DMA)
WTR(5%)
WTR(10%)
Treatment (Al-WTR)
(b)
Shoot length (cm)
80
60
40
20
0
(A) Sod. Arsenate
(B) DMA
0%
5%
10%
0%
5%
10%
Control (SAr)
WTR (5%)
WTR (10%)
Control (DMA)
WTR(5%)
WTR(10%)
Treatment (Al-WTR)
(c)
Root length (cm)
16
14
12
10
8
6
4
2
0
(A) Sod. Arsenate
(B) DMA
0%
5%
10%
0%
5%
10%
Control (SAr)
WTR (5%)
WTR (10%)
Control (DMA)
WTR(5%)
WTR(10%)
Treatment (Al-WTR)

Application of WTRs into highly As-contaminated soils had a significant effect in reduction of As concentration in plant tissues compared to control plants. Arsenic concentration was substantially decreased by an average of 90% and 98% when Fe- or Al-WTR was added into soils amended with 1500 mg kg^{-1} of sodium arsenate respectively (Fig. 25.5a) with no significant difference between low and high rate of WTR applied. In case of DMA-contaminated soils, As concentration in plant tissues increased from 25 μg g^{-1} (in the control plants) up to 39 μg g^{-1} when 5% of Fe-WTR was applied (Fig. 25.5a). Increasing the percentage of Fe-WTR applied up to 10% caused a reduction of As concentration to 14 μg g^{-1} (Fig. 25.5a). Al-WTR had an opposite effect on As concentration in plant tissues compared to the effect of Fe-WTR. At 5% of Al-WTR, As concentration in plants decreased significantly from 25 μg g^{-1} (in control plants) to 5.7 μg g^{-1} in plants treated with 5% Al-WTR, while there was no significant reduction observed when Al-WTR rate increased up to 10% (Fig. 25.5b).

The results obtained in this study indicate the beneficial effects of WTR amendment on growth of rice plants in As-contaminated soils. Both Al and Fe-WTRs were very effective in reducing phytoavailable As in the soils, and as a result, the growth of the plants improved. In general, Al-WTR was slightly more effective in reducing As concentration in rice plants. However, Fe-WTR was more effective in increasing the growth of rice plants when compared to Al-WTRs. This effect could be due to the toxicity of Al to plants. Aluminum toxicity has been reported to result in a reduced and damaged root system, which in turn causes the affected plants to be susceptible to mineral nutrient deficiencies (Foy, 1988). Aluminum toxicity is one of the most important yield-limiting factors for rice grown on acid upland and lowland acid sulphate soils (IRRI, 1978). On the other hand, Fe is an essential micro-nutrient in plants, and crop plants widely suffer from Fe deficiency (Marschner, 1995). Excess iron can also cause loss in crop yield, especially in acid soils (Fageria and Rabelo, 1987). Hence, it is essential to carefully calibrate the rate of WTR amendment to the As-contaminated soil for optimum plant growth and reduced phytoavailability of As.

Figure 25.4. Effect of water treatment residual (WTR)-amendment on growth of rice plants on arsenical pesticide-contaminated soil. Immokalee soil was contaminated using either sodium arsenate (A) or DMA (B) at 1500 mg kg^{-1} of As and amended with Al-WTR at two rates (5% or 10%). Control plants were grown in soils spiked with 1500 mg kg^{-1} of As using either sodium arsenate or DMA with no WTR amendment. Effect on total biomass (a), shoot length (b), and root length (c) were recorded. Error bars represent ±SE.

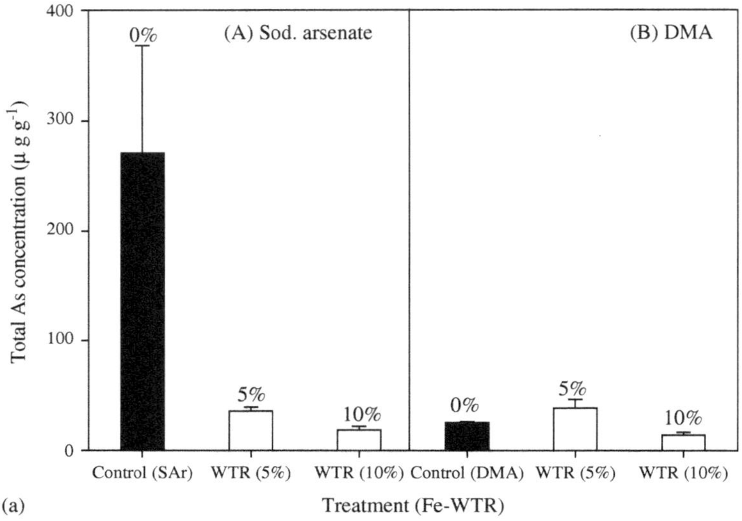

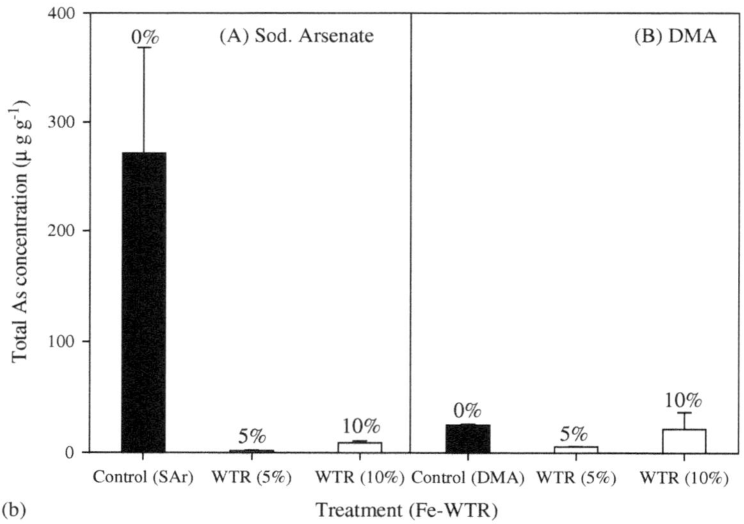

Figure 25.5. Effect of water treatment residual (WTR)-amendment on As uptake in rice plants grown on arsenical pesticide-contaminated soil. Arsenic concentration was determined in the tissues of rice plants grown in soils spiked with 1500 mg kg^{-1} of sodium arsenate (A) or DMA (B) and amended with Fe-WTR (a) or Al-WTR (b) at two rates (5% or 10%). Control plants were grown in soils spiked with 1500 mg kg^{-1} of sodium arsenate or DMA with no WTR-amendment. Arsenic concentration in plant tissues was determined after acid digestion of the plant tissues and analyzing the digests using a graphite furnace atomic absorption spectrophotometer. Error bars represent ±SE.

REFERENCES

Abdein, M.J., Cresser, M.S., Meharg, A.A., Feldmann, J., Cotter-Howells, J., 2002. Arsenic accumulation and metabolism in rice (*Oryza sativa* L.). Environ. Sci. Technol. 36, 962–968.

Adriano, D.C., 2001. Trace elements in terrestrial environments: Biogeochemistry, bioavailability, and risks of metals. Springer, New York, NY.

Bowen, H.J.M., 1979. Environmental chemistry of the elements. Academic Press, New York, NY.

Bugbee, G.J., Frink, C.R., 1985. Alum sludge as soil amendment: Effects on soil properties and plant growth. New Haven Connecticut Agric. Exp. Stn. Bull. 827, .

Butkus, M.A., Grasso, D., Schulthess, C.P., Wijnja, H., 1998. Surface complexation modeling of phosphate adsorption by water treatment residual. J. Environ. Qual. 27, 1055–1063.

Cai, Y., Ma, L.Q., 2003. Metal tolerance, accumulation, and detoxification in plant with emphasis on arsenic in terrestrial plants. In: Cai, Y., Braids, O.C. (Eds.), Biogeochemistry of Environmentally Important Trace Elements. ACS symposium series 835. Washington DC, American Chemical Society, pp. 95–114.

Carbonell, A., Aarabi, A.M.A., DeLaune, R.D., Gambrell, R.P., Patrick, W.H. Jr., 1998. Arsenic in wetland vegetation: Availability, phytotoxicity, uptake and effects on plant growth and nutrition. Sci. Total Environ. 217, 189–199.

Elliott, H.A., Dempsey, B.A., 1991. Agronomic effects of land application of water treatment sludges. J AWWA 83, 126.

El-Swaify, S.A., Emerson, W.W., 1975. Changes in the physical properties of soil clays due to precipitated aluminum and iron hydroxides: I. Swelling and aggregate stability after drying. Soil Sci. Soc. Am. J. 39, 1056–1063.

Fageria, N.K., Rabelo, N.A., 1987. Tolerance of rice cultivars to iron toxicity. J. Plant Nutr. 10, 653–661.

Fowler, B.A., 1977. Toxicology of environmental arsenic. In: Goyer, R.A. and Mehlan, M.A. (Eds.), Toxicology of Trace Elements. Wiely, New York, pp. 79–122.

Foy, C.D., 1988. Plant adaptation to acid aluminum-toxic soils. Commun. Soil Sci. Plant Anal. 19, 959–987.

Gallimore, L.E., Basta, N.T., Storm, D.E., Payton, M.E., Huhnke, R.H., Smolen, M.D., 1999. Water treatment residuals to reduce nutrients in surface runoff from agricultural land. J. Environ. Qual. 28, 1474–1478.

Goto, F., Yoshihara, T., Saiki, H., 2000. Iron accumulation and enhanced growth in transgenic lettuce plants expressing the iron-binding protein ferritin. Theor. Appl. Genet. 100, 658–664.

Hansen, L.D., Gilberman, D., Fisher, G.L., Eatough, D.J., 1984. Chemical speciation of elements in stack-collected, respirable size, coal fly ash. Environ. Sci. Technol. 18, 181–186.

Heil, D.M., Barbarick, K.A., 1989. Water treatment sludge influence on the growth of Sorghum-Sudan grass. J. Environ. Qual. 18, 292–298.

International Rice Research Institute, 1978. Annual Report for 1977. Los Bavos, The Philippines.

Ippolito, J.A., Barbarick, K.M., Redente, E.F., 1999. Co-application effects of water treatment residuals and biosolids on two range grasses. J. Environ. Qual. 28, 1644–1650.

Jacobs, L.W., Teppen, B.J., 2000. WTR as a soil amendment to reduce non-point source pollution from phosphorus-enriched soils. Fourteenth Annual Residuals and Biosolids Management Conference. Water Environment Federation. February 27–March 1, Boston, MA.

Klute, A., 1996. Methods of Soil Analysis. Part 1: Physical and Mineralogical Methods. SSSA Publications, Madison, WI.

Lindau, L., 1977. Emissions of As in Sweden and their reduction. Environ. Health Perspect. 19, 25–30.

Liu, W.J., Zhu, Y.G., Smith, F.A., Smith, S.E., 2004. Do phosphorus nutrition and iron plaque alter arsenate (As) uptake by rice seedlings in hydroponic culture? New Phytol. 162, 481–488.

Marin, A.R., Masscheleyn, P.H., Patrick, W.R., Jr., 1992. The influence of chemical form and concentration of arsenic on rice growth and tissue arsenic concentration. Plant Soil 139, 175–183.

Marin, A.R., Masschenlyn, P.H., Patrick, W.H., Jr., 1993. Soil redox-potential stability of arsenic species and its influence on arsenic uptake by rice. Plant Soil 152, 245–253.

Marschner, H., 1995. Mineral Nutrition of Higher Plants, second ed.. Academic Press, London, UK.

Matera, V. and Le Hecho, I., 2001. Arsenic behavior in contaminated soils: Mobility and speciation. In: Selim, H.M. and Sparks, D.L. (Eds.), Heavy metal release in soils. Lewis Publishers, Boca Raton.

Maurice, D.V., Lightsey, S.F., Hamrick, E., Cox, J., 1998. Alum sludge and zeolite as components of broiler litter. J. Appl. Poultry Res. 7, 263–267.

Milam, M.R., Marin, A.R., Sedberry, J.E., Jr., Bligh, D.P., Sheppard, R., 1988. Effect of water management, arsenic and zinc on selected agronomic traits and rice grain yield. In: Annual Progress Report. Northeast Ridge Research Station: Baton Rouge, LA, pp. 105–108.

Moore, P.A. Jr., Miller, D.M., 1994. Decreased phosphorus solubility in poultry litter with aluminum, calcium, and iron amendments. J. Environ. Qual. 23, 325–330.

Murphy, E.A., Aucott, M., 1998. An assessment of the amounts of arsenical pesticides used historically in a geographic area. Sci. Total. Environ. 218, 89–101.

Ng, J.C., 1997. Speciation, bioavailability, and toxicology of arsenic in the environment. Ph.D. Dissertation, University of Australia, Queensland.

Nissen, P., Benson, A.A., 1982. Arsenic metabolism in freshwater and terrestrial plants. Physiol. Plant 54, 446.

Oscarson, D.W., Huang, P.M., Defosse, C., Herbillon, A., 1981. Oxidative power of Mn(IV) and Fe(III) oxides with respect to As(III) in terrestrial and aquatic environments. Nature 291, 50–51.

Pickering, I.J., Prince, R.C., George, M.J., Smith, R.D., George, G.M., Salt, D.E., 2000. Reduction and coordination of arsenic in Indian mustard. Plant Physiol. 122, 1171–1178.

Pierce, M.L., Moore, C.B., 1980. Adsorption of arsenite on amorphous iron hydroxide from dilute aqueous solution. Environ. Sci. Technol. 14, 214–216.

Rodriguez, R.R., 1998. Bioavailability and biomethylation of arsenic in contaminated soils and solid wastes. Ph.D. Dissertation, Oklahoma State University, Stillwater, OK.

Sckerl, M.M., Frans, R.E., 1969. Translocation and metabolism of MAA-14C in johnsongrass and cotton. Weed Sci. 17, 421–427.

Shendrikar, A.D., Faudel, G.B., 1978. Distribution of trace metals during oil shale retorting. Environ. Sci. Technol. 12, 332–334.

Sneller, F.E.C., Van Heerwaarden, L.M., Kraaijeveld-Smit, F.J.L., Ten Bookum, W.M., Koevoets, P.L.M., Schat, H., Verkleij, J.A.C., 1999. Toxicity of arsenate in *Silene vulgaris*, accumulation and degradation of arsenate-induced phytochelatins. New Phytol. 144, 223–232.

Sohrin, Y., Matsui, M., Kawashima, M., Hojo, M., Hasegawa, H., 1997. Arsenic biogeochemistry affected by eutrophication in Lake Biwa, Japan. Environ. Sci. Technol. 31, 2712–2720.

Southworth, R.M., 1995. Land Application Pollutant Limit for Arsenic. Part 503 USEPA, Washington, DC.

Sparks, D.L., 1996. Methods of Soil Analysis. Part 3: Chemical Methods. SSSA Publications, Madison, WI.

Tamaki, S., Frankenberger, W.T.J., 1992. Environmental biochemistry of arsenic. Rev. Environ. Contam. Toxicol. 124, 79–110.

Tang, T., Miller, M., 1991. Growth and tissue composition of rice grown in soil treated inorganic copper, nickel and arsenic. Commun. Soil Sci. Plant Anal. 22, 2037–2045.

Terwelle, H.F., Slater, E.C., 1967. Uncoupling of respiratory-chain phosphorylation by arsenate. Biochim. Biohphys. Acta 143, 1–17.

Tlustoš, P., Goessler, W., Száková, J., Balik, J., 2002. Arsenic compounds in leaves and roots of radish grown in soil treated by arsenite, arsenate and dimethylarsenic acid. Appl. Organometal. Chem. 16, 216–220.

Tsutsumi, M., 1982. Compartive toxicity of arsenite and arsenate to rice seedlings under various levels of phosphate supply. Soil Sci. Plant Nutr. 26, 561–569.

USEPA, 2001. National Primary Drinking Water Regulations; Arsenic and Clarifications to Compliance and New Source Contaminants Monitoring. Final Rule Federal Register/ Vol. 66, No. 14 (40 CFR Parts 9, 141, and 142). Available at http://www.epa.gov/ safewater/ars/arsenic_finalrule.html

Developments in Environmental Science, Volume 5
D. Sarkar, R. Datta and R. Hannigan (Editors)

ISSN: 1474-8177/DOI:10.1016/S1474-8177(07)05026-7

Chapter 26

Water quality issues in the outer coastal plains: New Jersey

Tait Chirenje, Claude Epstein and Raymond Mueller

Abstract

New Jersey is the most densely populated state in the US and yet it is home to one of the most pristine groundwater resources in the country, the Kirkwood–Cohansey Aquifer System. This chapter discusses some of the activities that affect water quality in New Jersey's coastal plains. The most significant activities are human induced and they include excessive groundwater withdrawal and the addition of non-point and point source contaminants. The effects of those human activities that alter surface soils in ways that promote the movement of chemicals to streams and groundwater is included. The chemicals addressed include point source contaminants coming from superfund sites, landfills, underground storage tanks (USTs), sewage facilities, stormwater runoff facilities, storage lagoons, and other "state permitted" discharges. The contaminants include nitrates (NO_3^-), phosphates, pesticides, some chlorides and trace metals, and many volatile organic compounds (VOCs). Some of these come directly from their natural or human-induced deposition on the surface or come about because of the secondary impact of these deposited chemicals on surface soils and sediments. Well water withdrawals have also altered the pre-development groundwater and surface water flow patterns. Chlorides from brackish bays and tidal rivers and from the ocean itself have recharged coastal plain aquifers in Raritan Bay, Delaware and along the Atlantic Ocean. These pumping centers are also drawing chlorides and other ions from saline groundwater coming from deeper aquifers. In addition, VOCs and Fe have recharged underlying aquifers from contaminated rivers and point sources toward these pumping centers.

26.1. Introduction

Water supply and quality are among the most important environmental issues confronting us today, with most observers ranking them second only to overpopulation. Water touches every aspect of our lives including sanitation, health, agricultural production, and economic growth. It even poses national security risks in many countries (Lawford and Fort, 2003). Therefore, it should be no surprise that the use (or misuse) of water has long been the subject of passionate debate among various sections of civil society.

Lakes, rivers, and streams have long been viewed as cheap dependable sources of water. However, human activities have had significant impacts on them. For example, water bodies have been used for the disposal of waste materials, and where waste is not discharged directly, it eventually finds its way into waterways. Draining, channelization, paving land surfaces, agricultural, industrial and construction have resulted in the loading of many rivers and streams with sediment and other materials, thus altering their hydrology. Before it makes its way back into rivers, streams and lakes, water undergoes considerable changes in its chemistry, temperature, microbial populations and physical attributes due to both anthropogenic and natural causes. This chapter discusses water quality issues in the coastal plain, with specific emphasis on New Jersey. It does not go into water supply issues or riparian rights and/or the identification of specific culprits for contamination.

26.1.1. Water quality criteria

Through the years, the phrase "water quality" has been used to refer to those characteristics of water that make it suitable for a specific use. Overall water quality is determined by its thermal, physical, chemical, and biological integrity. However, rarely do we ever need water that meets the standards for all four criteria. The Clean Water Act (1972), which sets water quality standards for all contaminants in surface waters, takes the specific uses of a water body into account in regulating the quality of the nation's surface waters. The discrimination of water according to specific uses enables us to deem certain water suitable for use for a specific purpose, e.g., irrigation, while it may not be safe for other uses such as drinking or industrial purposes. In order to distinguish between water that is suitable for particular uses, e.g., irrigation and not drinking, the United States Environmental Protection Agency (USEPA), in conjunction with several industries, educational and private institutions, has developed measurable water quality indicators for each specific use. For example, salinity, specific ion toxicity, and low infiltration (from high sodium [Na] and low calcium [Ca] in soil), make otherwise good quality water useless as irrigation water for agriculture (Ayers and Westcot, 1994).

The surface and groundwater quality of the New Jersey coastal plain has been the focus of much attention recently (Ayers et al., 2000; Fischer et al., 2004). Studies of specific chemical components have been undertaken in various counties of the New Jersey coastal plain to address issues of public health. These include chlorides (Cl^-), nitrates (NO_3^-), iron (Fe), mercury (Hg), radium (Ra), volatile organic compounds (VOCs) and pesticides.

In our discussion of water quality issues in the coastal plain, reference will be made to the United States Geological Survey (USGS) New Jersey well monitoring program which makes use of about 150 randomly sited shallow wells in the state (Fig. 26.1; USGS, 2004a, b). These wells are located in urban (60), agricultural (60), and undeveloped (30) areas. The parameters measured by these wells differ depending on their location (input factors). For example, the urban/suburban wells measure non-point source pollution coming from fertilizers and pesticides, septic tank discharges, leaky sewers, automobiles, VOCs, and other urban contaminants. The agricultural wells generally measure non-point source pollutants from agricultural fertilizers and pesticides, liming agents, mobilized metals, and others. The wells in undeveloped areas measure atmospheric discharges, e.g., Hg, acid deposition, NO_3^-, pesticides, VOCs, etc. Before we engage in detailed discussions about this study, we will discuss watersheds and their roles in surface and groundwater quality.

26.1.2. Watersheds

As far back as the late 1800s, John Wesley Powell described a watershed as "that area of land, a bounded hydrologic system, within which all living things are inextricably linked by their common water course and where, as humans settled, simple logic demanded that they become part of a community." In other words, a watershed is the area of land where all of the water that is under it or drains off of it goes into the same place (USEPA, 1995a). The USEPA adopted a watershed approach to address the problems associated with locating both point and non-point source discharges (Younos et al., 2003; USEPA, 1995a, b, c) because of challenges they were facing in the enforcement of the Clean Water Act in large watersheds.

The most fundamental processes and components in watersheds include (i) the chemical, mineralogical, and physical nature of the source

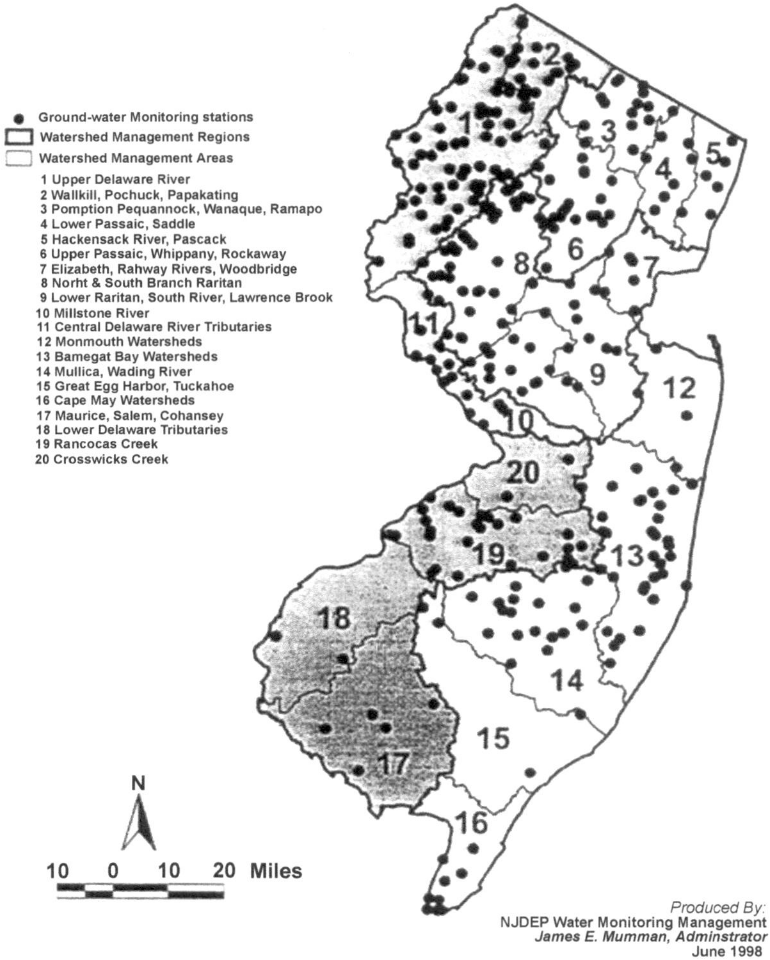

Figure 26.1. New Jersey groundwater monitoring program network.

regions of chemicals and sediment, (ii) the nature of their transport and the processes occurring along pathways, and, (iii) the identification of, and chemical processes occurring within, sinks. While understanding all these enables us to link all the processes that are occurring in New Jersey's watersheds and come up with a full understanding of water issues in this region, this discussion focuses on the source regions of chemicals and sediment and select processes occurring in the sink regions (Fig. 26.2).

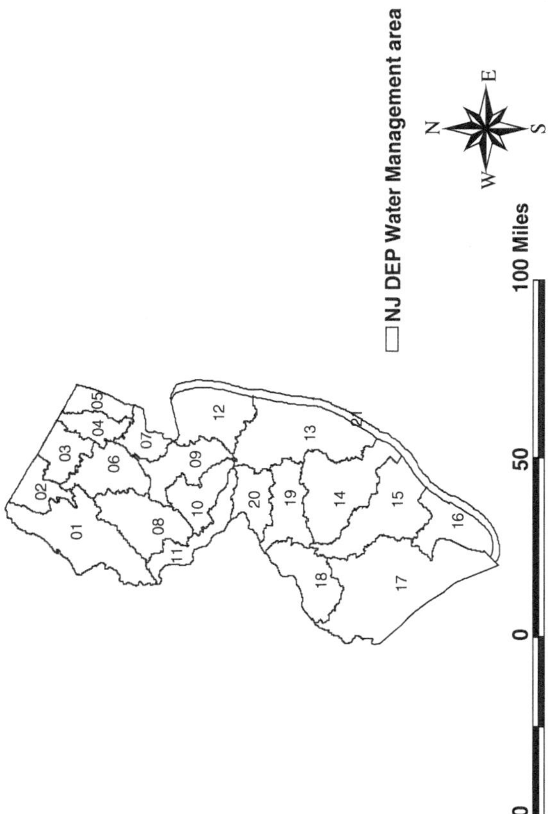

Figure 26.2. New Jersey water management areas.

26.2. Factors affecting water quality

Both surface and groundwater quality are affected by a multitude of factors. Although both natural and anthropogenic factors are equally important, in this discussion emphasis will be placed on how anthropogenic factors alter the natural environment. Two such activities are (i) the introduction of contaminants to the ground surface, leading to polluted runoff into surface water bodies, and (ii) well water withdrawals, which induce the infiltration of surface water and the intrusion of brackish and saltwater. Well water withdrawals also draw in contaminated groundwater from greater depths and distances within the aquifer or adjacent aquifers. A few factors that are of most significance in the coastal plain have been identified for this discussion. They include, but are not limited to: land use changes (agriculture to urban) and, development (increased urbanization and industrial activity), stormwater management, and treated wastewater disposal.

26.2.1. Application of contaminants to the surface of the New Jersey coastal plain

Both point and non-point source pollution is significant in the coastal plain. Point source discharges are carefully monitored by both the NJDEP and USEPA. These include superfund sites (sites containing contaminated groundwater or other releases that are a direct threat to health and safety; USEPA, 2004a), landfills (NJDEP, 2004a), underground storage tanks (USTs; NJDEP, 2004b, d), and sites of permitted discharge (NJDEP, 2004c). Table 26.1 shows the distribution of such sites in select counties in New Jersey.

The kinds of facilities that have been given discharge permits by the state include stormwater runoff systems, municipal solid waste plants, infiltration lagoons, spray irrigation fields, dredge spoils, injection wells, biosolid land applications, thermal waste water, sewer overflow, etc. Table 26.2 presents the relative percentages of the areas where permitted discharges are issued. In order to understand the full impact of these discharges, a discussion on the hydrologic and geochemical setting of the New Jersey coastal plain is necessary.

26.2.1.1. The hydrologic and geochemical setting of the New Jersey coastal plain

The New Jersey coastal plain is made up of a wedge of unconsolidated sediments that thickened from the Delaware River Valley, which lies in the western part of New Jersey, toward the Atlantic Ocean, which borders the eastern side of the state. The northern and southern margins of the

Table 26.1. NJ point-source inventory

County	Landfills	Superfund sites	Registered underground storage tanks	NJPDES "discharge" permits
Atlantic	37	10	476	242
Burlington	37	14	800	420
Camden	33	10	801	707
Cape May	18	1	245	197
Cumberland	26	5	283	192
Gloucester	29	7	359	243
Monmouth	32	11	982	441
Ocean	36	12	825	313
Salem	16	1	164	131
Total	264	71	4935	2886

Table 26.2. Relative proportions of costal plain discharges

Discharged to:	% of Coastal plain permits
Surface runoff	31.3
Groundwater recharge	29.9
Surface water discharge	20.3
Storage	5.5
Rejuvenated and recycled	4.7
Combination	8.3

Atlantic coastal plain are bounded by the brackish waters of the Raritan and Delaware Bays, respectively. The coastal plain can be further divided into an inner and outer coastal plain. The inner coastal plain lies beneath the Delaware River Valley below Trenton, New Jersey, forming a northeast to southwest trending restricted belt of sand, clay, and greensand units whose geologic ages range from Lower Cretaceous to the lowermost Miocene. These units make up the base of the wedge overlying pre-Cretaceous bedrock. Except along the outcrop belts of these units, the aquifers within the inner coastal plain (the Potomac–Raritan–Magothy, the Englishtown, the Wenonah–Mt. Laurel, the Vincentown and the Piney Point Aquifer Systems) are confined. The outer coastal plain occupies most of the coastal plain by far and is commonly referred to as the "Pine Barrens." The outer coastal plain includes generally coarse textured sand and gravel units whose geologic age is younger (Miocene) than the inner coastal plain. The units of the outer coastal plain, overlying the units of the inner coastal plain, constitute an unconfined aquifer (the Kirkwood–Cohansey Aquifer System; Zapecza, 1989). This aquifer lies in the New

Jersey pine barrens and it contains about 17 trillion gallons of water. This water source is rated one of the cleanest in the United States (USEPA, 1995c).

26.2.1.2. Impact of development on the New Jersey coastal plain hydrology/ geochemistry

Rhodehamel (1970) defined the hydrologic cycle of the outer coastal plain as consisting of a single input (precipitation) and three outputs (evapotranspiration, groundwater recharge, and surface runoff). This also applies to the inner coastal plain although further refinement of the definition is necessary to account for the hydrologic and geochemical impact from human activities. Figures 26.3 and 26.4 are simplified schematics representing the pre-developed (without significant human settlement and construction) and developed hydrologic cycles of the New Jersey coastal plain.

The increase in well water withdrawals reversed the relationship between surface water and groundwater. In the pre-development condition, groundwater discharged from aquifers into the Raritan and Delaware Bays, the Delaware River, and the Atlantic Ocean. The increase in population and well water withdrawals formed massive cones of depression in some parts of the coastal plains. As a result, surface water recharges some parts of the coastal plain aquifers resulting in saltwater intrusion.

In addition, the cones of depression along the margins of the Delaware River have induced infiltration of river water into portions of the underlying aquifers. Moreover, since the rate of withdrawal from the New

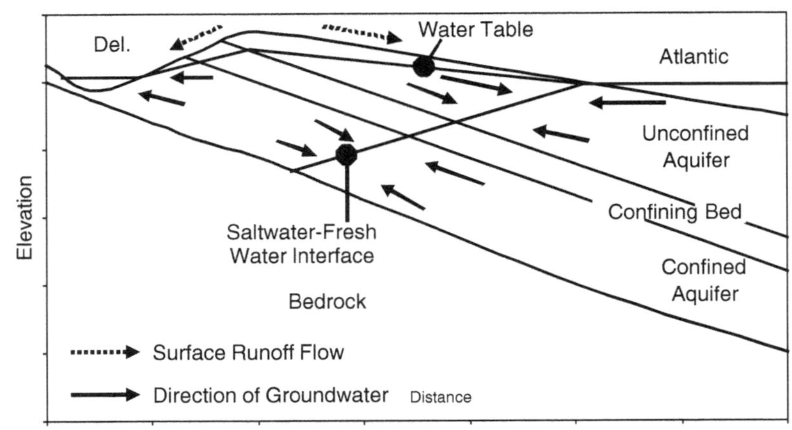

Figure 26.3. Pre-developed hydrologic cycles of the New Jersey coastal plain.

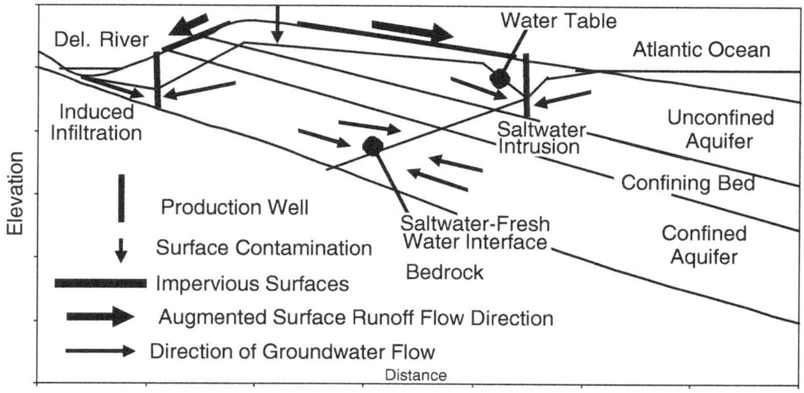

Figure 26.4. VOC concentrations in public supply wells (USGS, 2004b).

Jersey side of the coastal plain increased while that on the Pennsylvania side decreased, groundwater from Pennsylvania has been drawn under the river into New Jersey (Sloto, 2003).

Deeper within coastal plain aquifers, saline groundwater has been formed by the combination of saltwater recharge from the geologic past and the long-term accumulation of dissolved solids (Meisler, 1989). This deeper saline groundwater has been drawn westward toward pumping centers on the New Jersey side of the Delaware Valley.

Transfer of organic and inorganic materials from the atmosphere and the surface of the coastal plain, to the unconfined aquifers, adjacent streams, and the nearby coast has also increased with development. Superfund sites, landfills, septic systems, storage lagoons, leaky USTs, land application facilities, increased volumes of applied fertilizers and pesticides, deposition of air pollutants such as acid rain, heavy metals and organics, as well as accidental chemical spills and illegal dumping are all sources of these materials. Superfund sites are of particular interest. Although New Jersey is the fourth smallest state in the US, it is home to the largest number of superfund sites (114 in 2003), with Pennsylvania being the only other state to top 100 sites (NJDEP, 2003).

Finally, the construction of impervious surfaces (i.e., roads, buildings, industrial and commercial centers) and stormwater runoff facilities have altered two elements of the hydrologic cycle: sediment movement and chemical loss in runoff. Since most of the coastal plain surface is made up of unconsolidated sediments covered by forest or agriculture, groundwater recharge constituted a far greater percentage of hydrologic output than surface runoff during pre-development times (Rhodehamel, 1970).

The emplacement of impervious surface cover and the construction of stormwater runoff networks have increased surface runoff while decreasing groundwater recharge at the same time. Consequently, more organic and inorganic materials make their way into streams and bays.

26.2.1.3. Sources of chemicals of concern to water quality

The chemicals affecting water quality by natural and human activities include chlorides, NO_3^-, Fe, Hg, Ra, VOCs, fertilizers, and pesticides. Their sources vary (Ervin et al., 1994).

Chlorides: Coastal plain streams and aquifers are generally not contaminated with chlorides (Table 26.3; Watt and Johnson, 1992; Watt et al., 1994, 2003; Lacombe and Rosman, 1995; Johnson and Watt, 1996; Johnson and Charles, 1997). Stream concentrations, however, may be elevated briefly in winter during periods of snow when salt is applied to roadways. More long-term elevation results when contamination is by other sources, e.g., saline groundwater drawn toward large cones of depression, intrusion of brackish water from bays and ocean (and up streams during times of drought; DeLuca et al., 2003).

Saline groundwater underlies fresh groundwater in coastal plain aquifers. A transition zone, called the saltwater–fresh water interface, marks the boundary between these two water bodies. The depth of this boundary can extend from sea level by the shore to a depth between a few hundred to over a thousand feet beneath the surface of New Jersey. The

Table 26.3. Chloride, nitrate & ammonia concentrations in coastal plain watershed

	Cl^-		NO_3+NO_2		NH_3+Org N	
	Min	Max	Min	Max	Min	Max
Streams						
Mullica	4.1	23	<0.05	4.18	<0.20	2.0
Great Egg Harbor	1.1	180	0.49	5.91	0.11	18
Upper Maurice	1.1	36	<0.05	3.70	<0.20	5.10
Toms, Meteconk. Kettle	6.4	12	<0.01	0.94	0.30	0.80
Rancocas, Assunpink, Blacks, Crafts	3.0	35	<0.01	3.10	0.30	4.80
Salem, Raccoon, Oldmans, Alloway, Stow	2.8	24	0.70	1.31	0.30	1.5
Aquifers						
Mullica	<0.10	16	<0.10	10	<0.20	0.90
Great Egg Harbor	2.3	16	<0.10	11	<0.20	0.80
Upper Maurice	2.2	35	<0.01	3.60	<0.20	1.00
Toms, Meteconk. Kettle	2.0	55	0.02	5.75	<0.05	0.52
Rancocas, Assunpink, Blacks, Crafts	2.8	35	<0.10	9.80	<0.20	2.10
Salem, Raccoon, Oldmans, Alloway, Stow	3.5	39	0.08	24	<0.20	2.60

distance inland to which this layer of salt water extends beneath the overlying freshwater increases with aquifer depth, and extends to within approximately 10 miles of the Delaware River for the deepest aquifer (the Potomac–Raritan–Magothy [PRM] Aquifer System, Meisler, 1989). Furthermore, the pumping center in the Camden vicinity has drawn the fresh water–saltwater interface to the northeast, thereby elevating chloride levels in parts of Gloucester and Salem counties. Saltwater contamination from surface saltwater bodies also induced by pumping centers has raised chloride levels in different aquifers in different parts of the coastal plain (Schaefer, 1983; Spitz and Barringer, 1992).

Iron: Iron enters surface and groundwater by the dissolution of iron-bearing minerals under anaerobic and low pH conditions. These conditions exist where the decomposition of organic matter (OM) consumes oxygen and generates acids in such settings such as muck soils, organic-rich sediments, accumulated organic wastes, leaky sewers and septic systems, landfills, and wastewater disposal systems. In addition, these conditions may occur under impervious surfaces such as clay or asphalt (Ervin et al., 1994).

Iron is common in both coastal plain surface water and groundwater. Near wetlands, Fe occurs in its lower oxidation state [Fe(II)] in the low pH, anaerobic conditions associated with elevated dissolved organic carbon (DOC) concentrations, especially during the summer (Johnson and Barringer, 1993). Once Fe(II) is recharged to underlying aquifers, it encounters oxygenated groundwater or atmospheric oxygen when it discharges at the surface. It then transforms into insoluble Fe(III) and precipitates as $Fe(OH)_3$ through processes often facilitated by iron bacteria (Zampella et al., 2001).

Many groundwater discharge sites on the coastal plain possess soils and surface deposits enriched in an Fe precipitate (i.e., limonite), locally called *bog iron*. Bog iron was heavily used in the late 18th and early 19th centuries for the production of Fe in the numerous iron furnaces built throughout the coastal plain (Pierce, 1957).

Iron concentrations above the National Secondary Drinking Water Standard (NSDWS, $300\,\mu g\,l^{-1}$) occur in most coastal plain streams. The NSDWS is also exceeded in groundwater within and just down flow of the outcrop of the Potomac–Raritan–Magothy Aquifer System along the east side of the Delaware River. This region is highly urbanized (Lewis et al., 1991; Ervin et al., 1994). Iron also occurs in elevated concentrations in the other coastal plain aquifers (the Englishtown, Wenonah–Mt. Laurel, Vincentown, Piney Point, and Kirkwood–Cohansey Aquifer Systems, Watt and Johnson, 1992; Watt et al., 1994, 2003; Lacombe and Rosman, 1995; Johnson and Watt, 1996; Johnson and Charles, 1997; DeLuca et al., 2003).

The aforementioned situations represent generally "natural" groundwater flow conditions, but the direction of groundwater flow has been greatly altered by the massive pumping center near Camden. Groundwater, heavily enriched in Fe, has been drawn from old industrial sites in Philadelphia, under the Delaware River, to New Jersey by (a) the cessation of pumping on the Philadelphia side of the river, and (b) the progressive decrease in hydraulic head in the pumping center near Camden (Sloto, 2003).

Mercury: Mercury is a minor trace element in the sediments of the New Jersey coastal plain. The only mineral known to contain traces of Hg in this area is glauconite and it is mainly found in the inner coastal plain (Dooley, 1992). By far, the preponderant sources of Hg are aerial deposits. Other sources include geologic deposits, oceanic volatilization, and human sources (i.e., the processing of alkalis and metals and the incineration of medical wastes and fossil fuel combustion at electric plants). Once deposited, it may occur as the organic soluble complexes (e.g., methyl mercury, CH_3Hg) in ground and surface water. It is rarely found at toxic levels in surface and groundwater but it does become concentrated as CH_3Hg in fish and wildlife (USGS, 2000) through bioaccumulation and biomagnification, reaching toxic levels in some fish (NJDEP, 2002).

New Jersey DEP and USGS studies show mercury in the Kirkwood–Cohansey aquifer at levels above the national primary drinking water standard (NPDWS, $2\,\mu g\,l^{-1}$) in 265 of 2239 wells sampled (USGS, 1997). They identified four possible sources of the Hg, which include: atmospheric deposition from regional sources, historical land-applied mercurial pesticides, household contaminants, and point sources such as landfills and industries. Historical use of pesticides and Hg-based exterior paint were believed to be the more significant sources.

In terms of air emissions, NJDEP boasts the most stringent Hg regulations for municipal solid waste incinerators in the US. As a result, great strides have been made in reducing Hg emissions from municipal waste combustors, which includes a reduction from 2000 kg in 1993 to about 360 kg in 1996. This was achieved using both emission controls and source reduction. Approximately 90% of small facilities (notably, apartment complexes, which add about 65 kg per year) shut down their incinerators and are using alternate waste disposal options, resulting in cleaner air (NJDEP, 2002).

Radium: Radium reaches groundwater through dissolution from radium-bearing minerals in soils and aquifers under similar conditions. The source of the Ra found in the surface and groundwater of the coastal plain is probably from the decay products of impurities in minerals present in coastal plain sediments. These minerals, making up minute percentages of coastal plain sediments, include ilmenite, monazite, rutile, sphene, tourmaline, zircon, glauconite, and muscovite (Dooley, 1992).

Radium concentrations exceeding the USEPA NPDWS have been reported from Kirkwood–Cohansey wells in the outer coastal plain (Szabo et al., 1997). Three isotopes of Ra (Ra-224, Ra-226, and Ra-228) have been identified in the coastal plain by the USGS (Szabo et al., 1997; Szabo and DePaul, 1998). These studies report that 33% of the 170 Kirkwood–Cohansey wells exceeded the USEPA maximum contamination level (MCL, 5pCi/L) for the combined concentration of Ra-226 and Ra-228. But this elevation was most common in agricultural areas situated on the Bridgeton Formation. The Bridgeton Formation is made up of radioactive minerals transported from northern New Jersey, Pennsylvania and New York. The normally low pH of the outer coastal plain is made more acidic by the application of fertilizers, thereby leaching Ra from these minerals into the underlying unconfined aquifer.

Zapecza and Szabo (1986) reported elevated Ra concentrations in parts of the coastal plain adjacent to the fall line in the southeastern United States. This would be equivalent to the concentration level in the inner coastal plain of New Jersey. Table 26.4 shows the metal results of the New Jersey surface water monitoring program.

Volatile Organic Compounds: Volatile organic compounds are considered the third major threat to groundwater quality behind pesticides and nitrates (USEPA, 1988). They come from various point sources such as landfills, spills, leaky USTs and above-ground storage tanks (ASTs),

Table 26.4. Trace metal results from the New Jersey surface water monitoring study

Detectable trace elements	N	Number of wells in which trace elements detected by land use			Maximum value detected ($\mu g\,l^{-1}$)	NJ drinking water MCL ($\mu g\,l^{-1}$) 1996
		Agricultural ($N = 31$)	Urban ($N = 22$)	Undeveloped ($N = 22$)		
Arsenic	70	9	5	7	112	501
Barium	70	30	22	18	1180	2000
Cadmium	69	6	7	1	16	5
Chromium	69	13	10	4	3.6	100
Copper	70	20	14	7	38	1300AL
Lead	70	8	8	2	11	15AL
Mercury	70	2	1	0	1.7	2
Selenium	70	13	10	4	13.1	50
Total Detections	–	101	77	43	–	–

After Serfes (2004).

and waste lagoons, landfills and contaminated runoff and recharge. They occur in highest concentrations in urbanized areas. In the inner coastal plain, they occur largely in the outcrop area of the Potomac–Raritan–Magothy (PRM) aquifer system (Ervin et al., 1994), with higher occurrence in the more urbanized lower aquifer. They are also found in the outer coastal plain in the Kirkwood–Cohansey aquifer system. Withdrawals from public supply wells, having large radii of influence, tend to intensify VOC concentrations by intercepting recharge from many point sources (Ervin et al., 1994). In such cases, monitoring wells could under-represent VOC exposure in drinking water due to their small radii of influence. Common organic chemicals found in the New Jersey surface water monitoring study (Serfes, 2004) were methyl tertiary butyl ether (MTBE, detected in 4 and 10 out of 31 and 22 agricultural and urban wells, respectively), chloroform (detected in 7, 8, and 9 out of 31, 22, and 18 wells in agricultural, urban, and undeveloped land, respectively), perchloroethylene (detected in 4, 3, and 0 out of 31, 22, and 18 wells in agricultural, urban, and undeveloped land, respectively), and toluene (detected in 4, 2, and 1 out of 31, 22, and 18 wells in agricultural, urban and undeveloped land, respectively).

Nitrates reach surface and groundwater largely from precipitation, fossil fuel combustion, agricultural and residential fertilizers, and wastewater. Pesticides and herbicides generally come from agriculture, but also from residential neighborhoods, gardens, lawns, golf courses, parks, and landfills. Both nitrates and pesticides are discussed in more detail in later sections.

26.2.1.3.1. Impact of well water withdrawal on water quality
The population of the New Jersey coastal plain has increased rapidly over the last century, with the state having a population of 8,700,000 in 2005 (US Census Bureau, 2005). One consequence has been the increase in groundwater withdrawal through production wells. Pumping centers of various sizes in aquifers have altered the pre-development groundwater flow pattern, drawing contamination from water overlying unconfined aquifers and from nearby streams, bays, and coastal waters (Ervin et al., 1994).

26.2.2. Retention basins, stream hydrology and pollutant load

Retention or detention basins are used to delay, capture, store, treat, or infiltrate stormwater runoff so it does not discharge directly into a surface water body. Under perfect conditions, the retained or detained water is only discharged from the basin either by infiltration or evaporation. In reality, stormwater basins are only effective in reducing on-site and downstream flooding by controlling the rate of stormwater discharge. Good stormwater management practices (SMPs) take the watershed approach, i.e., stormwater management in sub-watersheds must take into account the changes caused by new and existing development in the whole watershed.

Other benefits from stormwater basins include water quality improvement such as sediment removal, aesthetics and recreational opportunities. Although stormwater basins also remove sediments, most of them are not designed for that purpose. Specific sedimentation basins, used during construction, are used for this purpose, i.e., to control off-site migration of sediment. This helps preserve the capacity of downstream reservoirs, ditches, diversions, waterways, and streams. Sometimes sedimentation basins are converted to stormwater basins at the end of the construction period.

26.2.2.1. Stormwater management and coastal water quality

Increased development and urbanization has led to an increase in paved and impervious areas leading to an increase in stormwater runoff volume and non-point source pollution in New Jersey. Contrary to the popular belief that stormwater runoff is natural and harmless, urban stormwater is the second largest source of water quality damage in estuaries (USEPA, 1995a, b). It is also a substantial contributor to the damage of other water bodies. The decrease in trees, grasses, and other plants that normally serve as filters has led to less filtration of the even higher pollutant loads in many urban areas. The long list of potential pollutants in urban stormwater includes nutrients (e.g., NO_3^- and phosphates), heavy metals (e.g., chromium [Cr], copper [Cu], lead [Pb], and zinc [Zn]), microbial contamination (bacteria and fungi), organic materials (oil, grease, pesticides, plasticizers, and flame retardants) and sediment and litter (Table 26.5). Sources of these pollutants include streets, parking lots, driveways, golf courses and lawns, leaking sewers and septic tanks, pet wastes, and construction sites (Oberts and Osgood, 1991).

Urban stormwater quality issues are particularly important in New Jersey because it is the most urbanized state in the United States. Currently, 15% of the total land area in New Jersey is considered urbanized (Nizeyimana et al., 2001) and this urbanization trend continues to increase (national urbanization average ~3%; Obarska-Pempkowiak, 2001). Because stormwater basins are not designed to handle 100-year floods, stormwater quality is often closely related to surface water quality. Increased urbanization and paving has led many stormwater basins to fail

Table 26.5. Common pollutant load in stormwater

Pollutant	Typical concentration
Total suspended solids	$80\,mg\,l^{-1}$
Total phosphorus	$0.30\,mg\,l^{-1}$
Total nitrogen	$2.0\,mg\,l^{-1}$
Total organic carbon	$12.7\,mg\,l^{-1}$
Fecal coliform bacteria	3600 MPN/100 ml
E. Coli bacteria	1450 MPN/100 ml
Petroleum hydrocarbons	$3.5\,mg\,l^{-1}$
Cadmium	$2\,\mu g\,l^{-1}$
Copper	$10\,\mu g\,l^{-1}$
Lead	$18\,\mu g\,l^{-1}$
Zinc	$140\,\mu g\,l^{-1}$
Chlorides (winter only)	$230\,mg\,l^{-1}$
Insecticides	$0.1\text{–}0.2\,\mu g\,l^{-1}$
Herbicides	$\leqslant 5\,\mu g\,l^{-1}$

From Schueler (1987, 1995, 1996, 1997); Rabbanal and Grizzard (1996), USEPA (1983) and Oberts and Osgood (1991).

even with rain events below the 10-year floods, exacerbating the situation. A national survey of waterways in 1998 found that 65% of New Jersey's assessed waterways failed to meet water quality standards (USEPA, 1998). Cahill Associates (1993) demonstrated that a typical suburban-density development with 23% impervious cover would deprive groundwater aquifers of over 40 million gallons of recharge per square mile annually. Even more alarming is the fact that it is estimated that 100-year floods would occur at a frequency of 5 years if impervious cover increases beyond 65% (Hollis, 1999). Barringer et al. (1994) observed that as development occurred, there were increases in flow and flow variability in two streams in New Jersey, although it was difficult to attribute the increase to specific causes. Table 26.6 shows the effect of development on the area of impervious surface.

Although efforts to control soil erosion and sediment control date as far back as 1975, when the New Jersey Soil and Sediment Control Act was enacted (N.J.S.A. 4:24–3 and 4:24–42) and the Standards for Soil Erosion and Sediment Control in New Jersey (N.J.A.C. 2:90–1.1), municipalities were only required to condition development approvals from the local Soil Conservation District for projects disturbing more than 5000 square feet of land. The recently published New Jersey stormwater rules (NJDEP, 2004 c, d) require counties, through municipalities, to develop stormwater plans in their jurisdiction and submit them to the state for approval. This will go a long way in addressing stormwater quality in the state.

Table 26.6. Typical impervious surface coverages for common land uses

Surface cover	Impervious surface (% of total site)
Forest	5
Parks	15
Lawns/pastures	20
Playgrounds	25
Residential (units/acre)	
1	15
2	25
3	34
4	42
5	48
6	52
7	56
Business district or surgical center	95

Adapted from Wells (1994).

26.2.3. Treated wastewater disposal and coastal water quality

Our success in protecting and improving coastal water quality largely depends on our ability to treat urban wastewater and manage stormwater. Under provisions of the Clean Water Act, the USEPA has the authority to implement pollution control programs such as setting wastewater standards for industry, fund the construction of sewage treatment plants under the construction grants program, and address the critical problems posed by non-point source pollution.

The main objectives of wastewater treatment are to convert sewage into suitable end products that can be discharged into any environmental media (e.g., waterways) economically and efficiently. This is done through a set of treatment procedures that include: (a) preliminary treatment (screening for large objects), (b) primary treatment (sedimentation), (c) secondary (biological) treatment, and (d) tertiary treatment. The extent of each treatment is dependent on the quality requirements of the receiving streams.

Since 1972, the federal government has invested over $84 billion dollars in improving wastewater treatment plants, with the private sector investing more than double that amount (Association of State and Interstate Water Pollution Control Administrators, AWPCA, 1992). This has led to a significant reduction of pollution in discharge zones of treated sewage effluent. For example, the number of people served by municipal systems that provided no wastewater treatment or had only primary treatment decreased from 34 million to 1.15 million from 1972 to 1992. At the same time, the number of people on systems with advanced treatment increased from about 6 million to more than 60 million. New Jersey benefited immensely from these investments by improving the water quality on its coastal areas substantially, and beach and bay closings have been dramatically reduced since 1988.

26.2.3.1. Septic tanks, USTs and landfills

The Barnegat Bay watershed is one of New Jersey's 96 watersheds. It comprises a 1700 square kilometer area encompassing all of the land and water in Ocean County, as well as parts of Monmouth County. More than 500,000 people live within the Barnegat Bay Watershed, and that population doubles in the summer with an influx of visitors from Philadelphia, New York, and other parts of New Jersey. Studies by the Barnegat Bay National Estuary Program (BBNEP) Scientific and Technical Advisory Committee (2001) identified nutrient and pathogen pollution as high priority management areas in the watershed. Most of the nutrient flow and pathogens were attributed to urban land uses, especially the use of septic tanks and other effluent discharges. For example, the BBNEP estimated that the amount of septic effluent flow that recharged the groundwater table in Ocean County amounted to 1,330,000 gal day^{-1} (5,030,000 l day^{-1}). They arrived at this value by multiplying the total of 8437 septic systems within the Barnegat Bay watershed by 3.5 persons per system, and 45 gal (170 l) per person per day (USEPA, 1980). Nitrate loadings from all septic systems were estimated to range from 216,930 to 295,300 kg year^{-1}.

Septic tanks do considerable damage due to the relatively short distance between the bottom of the septic system and the sandy texture of the Cohansey sands, which comprise most of the Barnegat Bay watershed. Sometimes homeowners, unaware of the harmful effect of their actions, flush paint thinners, degreasers, cleaners, and detergents into septic systems, exacerbating the problem.

Underground storage tanks are another source of pollutants in the Barnegat Bay watershed. They include USTs for storing gasoline, cleaning fluids, farm chemicals, and/or home heating oil. In New Jersey, tanks with a capacity of 2000 gal (7570 l) must be periodically tested and to have leak detection measures, but there are no such requirements for the homeowner (NJDEP, 2004c). Homeowners only have to worry about USTs when there is change of ownership. The NJDEP now maintains a database of USTs in every county.

Landfills are another significant source of pollutants in this watershed. Landfills that were constructed in the past were a little more than dump sites (Hill, 2004). They were constructed without liners or caps so that rainwater percolated through the garbage, concentrated sewage and chemical wastes and, ultimately drained down into underlying soils and groundwater. This problem has since been solved for new landfills by requiring liners, leachate collection systems, and methane capturing devices. Although old landfills still contribute to the problem (Hill, 2004), many of these sites have found their way on the national priority list (NPL) to accelerate remediation.

26.2.4. Agricultural nutrients and pesticides

Agricultural runoff (non-point source pollution) carries sediment, nutrients, salts, pesticides, and fertilizers, which threaten both groundwater and surface water. The discussion of agricultural input in New Jersey should be concentrated in the agricultural south, which is, coincidentally, located in the New Jersey Pinelands. A New Jersey Pinelands Commission study by Zampella (1994) demonstrated that land-use related disturbance had a substantial effect on the natural water chemistry of Pinelands streams. He demonstrated a gradient of increasing pH, specific conductance, and concentrations of soluble Ca, Mg, total nitrite plus NO_3^-, ammonia (NH_3) and phosphate that correlated with watershed disturbance (increasing urban and agricultural land-use intensity and wastewater flow). However, NO_3^- and phosphates in many of the 14 sites sampled were traced back to wastewater treatment plants rather than agricultural areas. Nonetheless, agricultural areas still contribute a considerable amount of nutrients, salts, pesticides, and sediment to waterways.

26.2.4.1. Sediment and salts

Sediments have been discussed under urban stormwater. This section discusses agricultural stormwater and runoff. In New Jersey, agricultural stormwater is regulated under Tier 2 municipalities (NJDEP, 2004c, d).

Agricultural activities contribute much of the global sediment supply to rivers, lakes, estuaries, and the world's oceans (Milliman and Syvitski, 1992). Sediment pollution affects water bodies in many ways such as: through excessive turbidity, reducing photosynthesis, increasing temperature in the water column, and deposition in river and lake beds, leading to ecological and physical changes of the water bodies. Sediments may also cover fish eggs and prevent spawning in some lakes. The fine fraction of eroded soils also carries adsorbed chemicals, e.g. NO_3^-, phosphates, chlorinated pesticides, VOCs, and metals, which alter stream or lake chemistry, hence affecting aquatic biota. The presence of NO_3^- and phosphates in waterways leads to eutrophication (the excessive promulgation of aquatic vegetation, notably algae, due to nutrient enrichment in waterways), which leads to decreased dissolved oxygen (DO) levels as the same aquatic vegetation decompose.

Livestock acreage in New Jersey predominantly consists of small equestrian enterprises. The often barren exercise lots and paddocks possess compacted soils. These impervious soils generate sediment and excess runoff that picks up animal wastes releasing not only sediment, but also NO_3^-, phosphates, OM, and pathogens to surface waters. This is a significant problem in some counties where horse farms comprise half the acreage of all agricultural land, e.g., Ocean County.

Zampella et al. (2001) showed that developed land and upland agriculture explained 84–89% of the variability in pH, EC, Ca, Mg, and Cl^-, with agriculture accounting for a greater percentage of the variation in Ca and Mg concentrations. Agriculture also accounted for 53% of the variability in sulfate concentrations. Developed urban land explained 84% of the variability in Cl^- concentrations, possibly from deicing salts (Zampella et al., 2001). The elevated levels of Ca and Mg were associated with land-use related watershed disturbance, with liming being cited as a possible culprit (Johnson and Watt, 1996).

26.2.4.2. Nitrates and phosphates

Nutrients from fertilizers play a significant role in eutrophication of water bodies in New Jersey. Although adults are less susceptible, NO_3^- in excess of $10\,mg\,l^{-1}$ can cause human infant methemoglobinemia (blue baby syndrome), which can be fatal (USEPA, 1988). Although excessive phosphates have not been linked to any physiological problems in humans, they are often the limiting nutrient for eutrophication to occur in water bodies. For this reason, activities that release phosphates tend to correlate very well with eutrophication.

Due to the diverse land uses in New Jersey, it is difficult to attribute the sources of nutrients to one specific land use activity. Nitrates and phosphates may originate from a variety of activities including: inorganic fertilizers used in agricultural, residential or commercial areas, wastewater discharges, septic tanks, pet wastes, other land-applied animal wastes, or from other organic residuals such as sludge.

Because New Jersey is highly urbanized, it is prudent to also consider the contribution of urban areas to nutrient pollution. Applying loading rates given by Palone and Todd (1997) to the Barnegat Bay watershed, lawn fertilizers contribute 5–10 $kg\,ha^{-1}\,year^{-1}$ of nitrates from typical residential lots. This means that there is the potential for 286,580–573,200 $kg\,year^{-1}$ of

NO_3^- wash off from lawns in the Barnegat Bay watershed if you take the 1997 residential population and divide it by 3.5 persons per household and multiplying by application rates for NO_3^- on lawn fertilizers.

Less than 25% of New Jersey residents test their soils before applying fertilizers and even fewer use slow release fertilizers (Grandin, 1993). Even professional lawn service companies surveyed at the same time did not regularly test lawn before applying fertilizer. This contributes to increased NO_3^- and phosphate losses through leaching and erosion. In fact, studies done by the BBNEP Scientific and Technical Advisory Committee showed that urban areas contributed more NO_3^- and phosphates to the watershed than agricultural areas (BBNEP Scientific and Technical Advisory Committee, 2001).

Generally NO_3^- is not a significant aquifer contaminant, but some trends in NO_3^- concentrations have been reported. Nitrate concentrations in aquifers decrease with depth (MacLeod et al., 1995). In the outer coastal plain, NO_3^- concentrations are higher in the Kirkwood–Cohansey aquifer system beneath agricultural land than other types of land use (MacLeod et al., 1995; Szabo et al., 1997). In the inner coastal plain, NO_3^- is generally low in aquifers but in the outcrop of the Middle and Upper Aquifers (PRM), both NO_3^- and NH_3 are elevated especially beneath the urban areas around the City of Camden and adjacent parts of Gloucester County (Ervin et al., 1994). Ammonia has been reported at or above the New Jersey Secondary Standard (i.e., $0.5\,mg\,l^{-1}$) in several urbanized streams in the inner coastal plain (DeLuca et al., 2003). Figure 26.5 shows the distribution of NO_3^- in groundwater (U.S. Geological Survey, 2004a).

26.2.4.3. Pesticides

Pesticides tend to be applied to large areas, thus they are considered non-point source contaminants. Both the United States Department of Agriculture (USDA) and the New Jersey Experimental Station have been recommending different types of pesticides since the late 1800s. This long-term use has resulted in pesticide residues of certain persistent pesticides in soil at concentrations that exceed the NJDEP residential soil cleanup criteria and may pose risk especially to children who ingest these soils. For example, up to 5% (240,000 acres) of the total area of New Jersey (5 million acres) may be impacted by the historical use of arsenical pesticides. Cheaper, relatively safer and more effective pesticides, e.g., organochlorines, were only developed in the late 1960s (Hayes and Laws, 1991; NJDEP, 1999). Lead arsenate was the pesticide of choice for apple (and other fruit) orchards, golf courses, turf farms and vegetable fields, and was usually applied at several pounds per acre. Calcium arsenate was

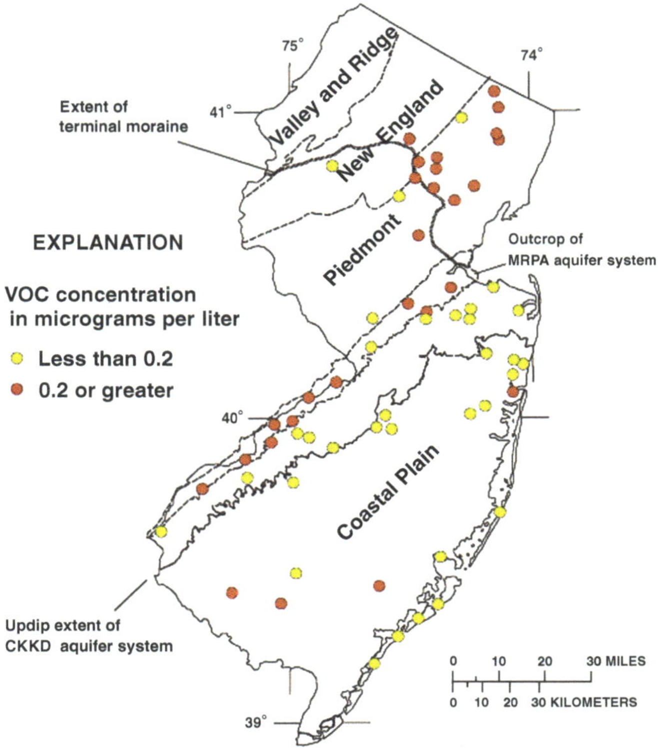

Figure 26.5. Nitrates in water from wells sampled by the USGS.

used on white potato farms (Murphy and Aucott, 1998). Lead arsenate contamination is suspected predominantly in six counties (Burlington, Cumberland, Gloucester, Hunterdon, Monmouth and Salem) because they account for most of the fruit production over the last 90 years, with Burlington and Gloucester being the largest producers.

Interestingly, arsenical pesticides use was reduced mainly due to its cost and effectiveness rather than health and safety considerations. The organochlorines, dichlorodiphenyl trichloroethane (DDT), Aldrin and Dieldrin, were applied in varying proportions on different crops. Hence, it is difficult to predict where they are likely to be found in the state. However, the USEPA banned the use of DDT in 1972 and that of Dieldrin and Aldrin on foodstuffs in 1974 and on all products in 1987 (USEPA, 1990). All three chemicals have made the dirty dozen list, a list of 12 of the most dangerous organic chemicals to be phased out from use internationally, developed at the Stockholm Convention in 1998.

As agricultural land is converted to other land uses, developers, municipal officials, homebuyers, and even lending agencies are inquiring and even testing for pesticide residues in soils. It is not unusual for some municipalities to require environmental assessments of land as part of their site approval process.

Ayers et al. (2000) studied the nature of pesticides in coastal plain surface water and shallow groundwater. In almost all streams and 63% of the groundwater samples tested pesticides were detected below NPDWS standards. Four herbicides (atrazine, metachlor, prometon, simazine) and some of their metabolites occur commonly. Atrazine and metolachlor are found in streams and shallow groundwater in agricultural areas, and are applied largely by licensed applicators. Prometon and simazine, also found in streams and shallow groundwater, are commonly used by non-licensed applicators such as homeowners and occur in urban areas. Prometon is used to clear roadways, railroad beds, and other right of ways. Consequently, it may also occur in areas of mixed land use. Simazine is found in streams and shallow groundwater in both agricultural and urban areas.

Ayers et al. (2000) reported that the insecticide carbaryl was rarely found in groundwater samples in spite of its common usage. Carbonates generally have a short half life and are easily metabolized by plants. They further reported that many organochlorine insecticides were commonly detected in low concentrations. Dieldrin, used in residential areas for termite control and in agricultural areas for insect control, has been detected in stream sediments. It was also detected in 24–27% of the monitoring and production wells sampled in their study and in 20% of the streams draining agricultural areas. The use of dieldrin was discontinued for agricultural use in the 1970s and for termite control in the 1980s. The fact that it is still being detected is testimony to the stability of organochlorines in the environment.

Results from the New Jersey surface water monitoring study showed a preponderance of pesticides and herbicides in wells in agricultural areas. Wells in undeveloped land, typically had concentrations below detection level (Serfes, 2004). Common pesticides and herbicides were atrazine (detected in 14 and 5 out of 31 and 22 agricultural and urban wells, respectively), deethylatrazine (detected in 15 and 4 out of 31 and 22 wells in agricultural and urban land, respectively), dieldrin (detected in 2 wells in both agricultural and urban land, respectively). Others that were detected include metolachlor (detected in 16 and 3 wells in agricultural and urban land, respectively), DDE (detected in 4 and 1 well in agricultural and urban land, respectively), prometone (detected in 4 and 7 wells in agricultural and urban land, respectively), and simazine (detected in 10 and 3 wells out of 31 and 22 wells in agricultural and urban land, respectively; Serfes, 2004).

26.2.5. Trace metals and their effects on stream geochemistry

Iron occurs in its soluble ferrous state in low pH, anaerobic conditions. However, streams in the New Jersey Pinelands are characterized by an unusual reddish brown color, as a result of decomposed plant material and dissolved Fe(III). Due to the underlying parent material, groundwater in the Coastal Plain area contains large concentrations of dissolved Fe(II) and Mn(II). These metals are oxidized under oxic environments such as grottos, springs, and swamps, a process often facilitated by Fe bacteria (e.g., Lepothrix), causing the chemical reactions that precipitate the ore masses. A thin film of rust forms on its surface and floats. The iron oxide combines with sand and gravel to form a low-grade iron ore that includes siderite (iron carbonate, a white ore) and other iron oxides (red ores) that are deposited along stream banks.

Both hydrous ions of Fe and Mn affect chemical reactions in surface waters due to their ion exchange capacities derived from pH-dependent surface charge (Parks, 1967). This means that the hydrous metal oxides are anion and cation exchangers in acid and alkaline solutions, respectively. There is limited exchange under neutral conditions. Manganese is generally oxidized to Mn(II) oxide in oxic and alkaline conditions, leading to greatest adsorption under oxidized alkaline conditions when it changes to Mn(IV) [insoluble] (Parks, 1967). On the other hand, Stumm and Lee (1960, 1961) found that Fe(III) was the only species found in slightly acidic to alkaline conditions under oxidizing conditions. They noted that Fe(II) was stable in oxic conditions only under highly acidic conditions, e.g., acid mine drainage. Thus Fe(II) would precipitate out as a hydroxide or carbonate under conditions of moderate to high carbonate alkalinity. Ferric iron hydroxide precipitates in an amorphous form, giving it unique absorption properties. Freshly precipitated Fe(III) hydroxide exhibits high sorption compared to aged Fe(III) hydroxide, possibly due to the rearrangement of the structure, tending to a more crystalline form (Morgan and Stumm, 1965). It is important to understand boundary conditions that lead to formation of precipitates, e.g., change in pH and redox, because metal-facilitated transport and sequestration takes place at the same time.

Apart from their economic importance, both dissolved and suspended Fe are important in colloid generation and transport (in conjunction with suspended organics). These colloids play a significant role in the retention, mobility, and availability of various trace elements and nutrients, as well as the degradability of various xenobiotics (e.g., pesticides). For example, several workers have reported that adsorption (or co-precipitation) of As on Fe oxides is a major factor in determining the concentration of dissolved and suspended As in groundwater (Welch and Lico, 1998), rivers (Johnson and Thornton, 1987; Fuller and Davis, 1989), pore waters (Masscheleyn et al., 1991), lake water (Belzile and Tessier, 1990; DeVitre et al., 1991; Moncure et al., 1992), and in estuaries. Work done by Welch et al. (2000) provided evidence of Fe oxides affecting As in water, including: (1) correlation between Fe and As in sediments and sediment extracts, (2) observations that As concentrations vary with diel pH cycles, and (3) laboratory column studies.

The adsorption of trace elements, e.g., As, by $Fe(OH)_3$ depends on pH, the concentration of $Fe(OH)_3$ and competing ions. Co-precipitated As is released from iron oxide as the crystalline growth of the oxide proceeds. Although As(III) is adsorbed faster than As(V), adsorption proceeds very quickly for both species, with 90% of total As being adsorbed within $24\,h^{-1}$ (Raven et al., 1998).

Iron compounds can also release As into the environment. For example, Lovley (1991) showed a biologically mediated reaction (dissimilatory Fe reduction) that releases As from Fe oxide using OC. In fact, the high concentrations of As in some groundwater can be attributed solely to the dissolution of Fe compounds (Nagorski and Moore, 1999). Shapiro (1964) found that some natural water organics reduced the size of ferric hydrous oxide precipitates to between 0.1–0.45 μm in the presence of natural OM. This enhanced colloid-facilitated transport of other metals through the peptization of the Fe species. In the absence of natural OM, ferric hydrous precipitates were larger than 0.45 μm and were filtered out by standard membrane filters. Since organics by themselves exhibit ion exchange, their interaction with hydrous ferric oxides may affect their adsorption of and/or complexation with metals such as Cu, Cd, and Cr. This has profound effects on the aquatic chemistry of streams.

26.3. Conclusion

We discussed water quality issues in the Coastal Plains of New Jersey. Human activity considerably influences water quality in the coastal plain by applying chemicals or altering surface soils in ways that promote the movement of chemicals to streams and groundwater. Some of these are point-source contaminants coming from superfund sites, landfills, USTs, sewage facilities, stormwater runoff facilities, storage lagoons, and other "state permitted" discharges. The contaminants include NO_3^-, phosphates, pesticides, some chlorides and trace metals, and many VOCs. Some of these are non-point source contaminants coming directly from their natural or human-induced deposition on the surface or come about because of the secondary impact of these deposited chemicals on surface soils and sediments.

However, a second human activity that has had a great deal of impact indirectly comes from well water withdrawals, which have progressively risen with population growth. Pumping centers in various parts of the coastal plain have altered the pre-development groundwater and surface water flow patterns. Chlorides from brackish bays and tidal rivers and from the ocean itself have recharged coastal plain aquifers in Raritan Bay, Delaware, and along the Atlantic Ocean. These pumping centers are also drawing chlorides and other ions from saline groundwater coming from deeper aquifers. In addition, VOCs and Fe have recharged underlying aquifers from contaminated rivers and point sources toward these pumping centers.

REFERENCES

Association of State and Interstate Water Pollution Control Administrators, 1992. America's Clean Water: The States' Evaluation of Progress 1972–1992.

Ayers, R.S., Westcot., D.W., 1994. Water Quality for Agriculture. UNFAO Publication, pp. 24–64.

Ayers, M.A., Kennen, J. G., Stackelberg, P.E.,. 2000. Water Quality in the Long Island-New Jersey Coastal Drainages, New York and New Jersey, 1996–98: U.S. Geological Survey Circular 1201, p. 40.

Barnegat Bay National Estuary Program (BBNEP) Scientific and Technical Advisory Committee, 2001. Barnegat Bay National Estuary Program Characterization Report.

Barringer, T.H., Reiser, R.G., Price., C.V., 1994. Potential effects of development of flow characteristics of two New Jersey streams. Water Resour. Bull. 30(2), 6–40.

Belzile, N., Tessier, A., 1990. Interactions between arsenic and iron oxyhydroxides in lacustrine sediments. Geochim. Cosmochim. Acta 54, 103–109.

Cahill Associates, 1993. Stormwater Management Systems, Porous Pavement System with Underground Recharge Beds, Engineering Design Report, Cahill Associates.

DeLuca, M.J., Hoppe, H.L., Heckathorn, H.A., Riskin, M.L., Gray, B.J., Melvin, E., Liu, N.A., 2003. Water Resources Data New Jersey 2003. Volume 3. Water-Quality Data: U.S. Geological Survey Water-Data Report NJ-03-3, p. 668.

DeVitre, R., Belzile, N., Tessier, A., 1991. Speciation and adsorption of arsenic on diagenetic iron oxides. Limnol. Oceanogr. 36(7), 1480–1485.

Dooley, J.H., 1992. Natural Sources of Mercury in the Kirkwood-Cohansey Aquifer System of the New Jersey Coastal Plain. New Jersey Geological Survey Report 27, p. 17.

Ervin, E.M., Voronin, L.M., Fusillo, T.V., 1994. Water Quality of the Potomac-Raritan-Magothy Aquifer System in the Coastal Plain, West-Central New Jersey. U.S. Geological Survey Water-Resources Investigations Report 94–4113, p. 114.

Fischer, J.M., Riva-Murray, K., Hickman, R.E., Chichester, D.C., Brightbill, R.A., Romanok, K.M., Bilger, M.D., 2004. Water Quality in the Delaware Basin, Pennsylvania, New Jersey, New York & Delaware, 1998–2001. U.S. Geological Survey Circular 1227, p. 38.

Fuller, C.C., Davis, J.A., 1989. Influence of coupling of sorption and photosynthetic processes on trace element cycles in natural waters. Nature 340, 152–154.

Grandin, B. A., 1993. Exploring householder's use of lawn and garden chemicals in a subwatershed of the Great Swamp: Preliminary findings. Technical Report, Department of Human Ecology, Cook College, Rutgers University, New Brunswick, New Jersey.

Hayes, Laws, 1991. Handbook of Pesticide Toxicology. Academic Press, San Diego, CA.

Hill, M.K., 2004. Understanding Environmental Pollution, second ed.. Cambridge University Press, Cambridge, UK.

Hollis, G.E., 1999. The effect of urbanization on floods of different recurrence interval. Water Resour. Res. 11(3), 434.

Johnson, C.A., Thornton, I., 1987. Hydrological and chemical factors controlling the concentrations of Fe, Cu, Zn and As in a river system contaminated by acid mine drainage. Water Res. 21(3), 359–365.

Johnson, M.L., Charles, E.G., 1997. Hydrology of the unconfined aquifer system, Salem River area: Salem River and Raccoon, Oldmans, Alloway, and Stow Creek basins, New Jersey 1993–94. U.S. Geological Survey Water-Resources Investigations Report 96–4195, 5 sheets.

Johnson, M.L., Watt, M.K., 1996. Hydrology of unconfined aquifer system, Mullica River basin, New Jersey, 1991–92. U.S. Geological Survey Water-Resources Investigations Report 94–4234, 6 sheets.

Lacombe, P.J., Rosman, R., 1995. Hydrology of the unconfined aquifer system in the Upper Maurice River basin and adjacent areas in Gloucester County, New Jersey 1986–87. U.S. Geological Survey Water-Resources Investigations Report 92–4128, 5 sheets.

Lawford, R., Fort, D., 2003. Introduction. In: Water: Science, Policy and Management. Water Resources Monograph 16. American Geophysical Union. 10.1029/016WM01. p. 1.

Lewis, J.C., Hochreiter Jr., J.J., Barton, G.J., Kozinski, J., Spitz, F.J., 1991. Hydrogeology of, and Ground-Water Quality in, the Potomac-Raritan-Magothy Aquifer System in the Logan Township Region, Gloucester and Salem Counties, New Jersey. U.S. Geological Survey: Water-Resources Investigations Report 90–4142, p. 92.

MacLeod, C.L., Barringer, T.H., Vowinkel, E.F., Price, C.V., 1995. Relation of nitrate concentrations in groundwater to well depth, well use, and land use in Franklin Township, Gloucester County, New Jersey, 1970–85. U.S. Geological Survey Water-Resources Investigations Report 94–4174, p. 29.

Masscheleyn, P.H., Delaune, R.D., Patrick, W.H. Jr., 1991. Arsenic and selenium chemistry as affected by sediment redox potential and pH. J. Environ. Qual. 20(4), 446–456.

Meisler, H., 1989. The occurrence and geochemistry of salty groundwater in the northern Atlantic coastal plain. U.S. Geological Survey Professional Paper-1404-D.

Milliman, J.D., Syvitski, 1992. Geomorphic/tectonic control of sediment discharge to the ocean: The importance of small mountain rivers. J. Geol. 100, 525–544.

Moncure, G., Jankowski, P.A., Drever, J.I., 1992. The hydrochemistry of arsenic in reservoir sediments, Milltown, Montana. In: Kharaka, Y.K., Maest, A.S. (Eds.), Water-Rock Interaction. Rotterdam, Balkema, pp. 513–516.

Morgan, J., Stumm, W., 1965. Role of multivalent hydrous metal oxides on limnological transformations. Proceedings 2nd International Conference of Water Pollution Research, Pergamon Press, New York, NY.

New Jersey Department of Environmental Protection (NJDEP), 1999. Findings and recommendations for the remediation of historic pesticide contamination. http://www.state.nj.us/dep/special/hpctf/final/hpctf99.pdf

New Jersey Department of Environmental Protection (NJDEP), 2002. New Jersey Mercury Task Force, Volume III: Sources of Mercury in New Jersey, p. 21.

New Jersey Department of Environmental Protection (NJDEP), 2003. Publicly funded clean-up site status report. http://www.nj.gov/dep/srp/publications/site_status/2003/pdf
New Jersey Department of Environmental Protection (NJDEP). Division of Surface Water, 2004a. Landfill Information: http://www.state.nj.us/dep/dshw/lrm/landinfo.htm (updated 5/29/02).
New Jersey Department of Environmental Protection, (NJDEP) Division of Water Quality, 2004b. Database: http://www.state.nj.us/dep/dwq/database.htm (updated 6/2/04).
New Jersey Department of Environmental Protection (NJDEP). Bureau of Underground Storage Tanks, 2004c. Underground Storage Tank Inventory: http://www.nj.gov/dep/srp/bust/ustri.htm (updated 3/5/04).
New Jersey Department of Environmental Protection (NJDEP), 2004d. Municipal Stormwater General Permit Guidance Materials. http://www.nj.gov/dep/stormwater/sw_guidance.htm (updated 5/3/05).
Obarska-Pempkowiak, H., 2001. Retention of selected heavy metals: Cd, Cu, and Pb in a hybrid wetland system. Water Sci. Technol. 44(11-12), 463–468.
Oberts, G.L., Osgood, R.A., 1991. Water quality effectiveness of a detention/wetland treatment system and its effect on an urban lake. Environ. Manage. 15(1), 131–138.
Palone, R., Todd, A., 1997. Chesapeake Bay riparian handbook: A guide for establishing and maintaining riparian forest buffers. USDA, Forest Service, NA-TP-02.97, Radnor, Pennsylvania.
Parks, G.A., 1967. Aqueous surface chemistry of oxides and complex oxides minerals, in Equilibrium Concepts in Natural Water Systems, Advances in Chemistry Series 67. American Chemical Society, Washington, DC, pp. 121–160.
Pierce, A.D., 1957. Iron in the Pines: The Story of New Jersey's Ghost Towns and Bog Iron. Rutgers University Press, New Brunswick, NJ.
Rhodehamel, E.C., 1970. A hydrologic analysis of the New Jersey Pine Barrens Region. New Jersey Department of Environmental Protection Water-Resources Circular 22, p. 35.
Schaefer, F.L., 1983. Distribution of chloride concentrations in the principal aquifers of the New Jersey coastal plain, 1977–1981. U.S. Geological Survey Water-Resources Investigations Report 83–4061, p. 56.
Serfes, M., 2004. New Jersey's Ambient Groundwater Monitoring Network . New Jersey Water Monitoring Council Meeting Presentation (09/24/04).
Shapiro, J., 1964. The effect of yellow organic acids on iron and other metals in water. J. Am. Wat. Wks. Ass. 56, 1062–1082.
Sloto, R.A., 2003. Historical ground-water-flow patterns and trends in iron concentrations in the Potomac-Rarirtan-Magothy aquifer system, in parts of Philadelphia, Pennsylvania and Camden and Gloucester Counties, New Jersey: U.S. Geological Survey Water-Resources Investigations Report 03–4255, p. 37.
Spitz, F.J., Barringer, T.H., 1992. Ground-water hydrology and simulation of saltwater encroachment, shallow aquifer system of southern Cape May County. New Jersey. U.S. Geological Survey Water-Resources Investigations Report 91–4191, p. 87.
Stumm, W., Lee, G.F., 1960. The chemistry of aqueous iron. Schweizerische Zeitschruft fur Hydrology XXI, 295–319.
Stumm, W., Lee, G.F., 1961. Oxygenation of ferrous iron. Ind. Eng. Chem. 53, 143–146.
Szabo, Z., DePaul, V., 1998. Radium-226 and radium-228 in shallow groundwater. U.S. Geological Survey Fact Sheet FS-062-98, Southern New Jersey, p. 6.
Szabo, Z., Rice, D.E., McLeod, C.L., Barringer, T.H., 1997. Relation of distribution of radium, nitrates and pesticides to agricultural land use and depth. U.S. Geological Survey Water-Resources Investigations Report 96-4165A, p. 107.

United States Census Bureau, 2005. New Jersey Quick Facts. http://quickfacts.census.gov/qfd/states/34000.html (updated 11/16/05).

U.S. Environmental Protection Agency (USEPA), 1988. National Water Quality Inventory: 1988 Report to Congress.

U.S. Environmental Protection Agency (USEPA), 1990. Chemical Information Fact Sheets. U.S. Environmental Protection Agency, Office of Pesticide Programs, Special Review Branch, Washington, D.C.

United States Environmental Protection Agency (USEPA), 1995a. Watershed protection. A project focus. EPA841-R-95-003.

United States Environmental Protection Agency (USEPA), 1995b. Watershed protection. A statewide approach. EPA841-R-95-004.

United States Environmental Protection Agency (USEPA), 1995c. Watershed protection. A watershed approach framework: People working together to protect public health and the environment—community by community, watershed by watershed. http://www.epa.gov/owow/watershed/framework

USEPA Office of Water, National Water Quality Inventory, 1998. Report to Congress, www.epa.gov/305b/98report/nj.html; Widener Environmental and Natural Resources Law Clinic analysis, 1999.

U.S. Geological Survey, 1997. Mercury in groundwater, soils and sediments of the Kirkwood-Cohansey Aquifer in the New Jersey Coastal Plain. USGS Open File 95–475.

U.S. Geological Survey, 2000. Mercury in the Environment. U.S. Geological Survey Fact Sheet 146–00.

U.S. Geological Survey, 2004a. Nitrate in water from wells sampled by USGS, http://nj.usgs.gov/gw/maps/nitrate.gif

U.S. Geological Survey, 2004b. Total VOCs in water from wells sampled by USGS, http://nj.usgs.gov/gw/maps/voc.gif

Watt, M.K., Johnson, M.L., 1992. Water resources of the unconfined aquifer system of the Great Egg Harbor River Basin, New Jersey 1989–90. U.S. Geological Survey Water-Resources Investigations Report 91-4126, 5 sheets.

Watt, M.K., Johnson, M.L., Lacombe, P.J., 1994. Hydrology of the unconfined aquifer system, Toms River, Metedeconk River, and Kettle Creek Basins, New Jersey, 1987–90, 5 sheets.

Watt, M.K., Kane, A.C., Charles, E.G., Storck, D.A., 2003. Hydrology of unconfined aquifer system, Rancocas Creek area: Rancocas, Assunpink, Blacks, and Crafts Creek basins, New Jersey 1996, 5 sheets.

Welch, A.H., Lico, M.S., 1998. Factors controlling As and U in shallow groundwater, southern Carson Desert, Nevada. Appl. Geochem. 13(4), 521–539.

Welch, A.H., Westjohn, D.B., Helsel, D.R., Wanty, R.B., 2000. Arsenic in groundwater of the United States: Occurrence and geochemistry. Groundwater 38(4), 589–604.

Wells, C., 1994. Skinny streets and one sided sidewalks: A strategy for not paving paradise. Technical Note 38: Watershed Protection Techniques 1(3), 135–137.

Younos, T., de Leon, R., Lewicki, C., 2003. Integrating service learning into watershed management programs: Opportunities and challenges. J. Am. Water Res. Ass. Paper # 01243.

Zampella, R.A., 1994. Characterization of surface water quality along a watershed disturbance gradient. JAWRA 30(4), 605–611.

Zampella, R.A., Bunnell, J.F., Laidig, K.J., Dow, C., 2001. The Mullica River Basin. A report to the Pinelands Commission on the status of the landscape and selected aquatic and wetland resources. Pinelands Commission, New Lisbon, New Jersey.

Zapecza, O.S., 1989. Hydrogeologic framework of the New Jersey coastal plain. U.S. Geological Survey Professional Paper1404-B, p. 49.

Developments in Environmental Science, Volume 5
D. Sarkar, R. Datta and R. Hannigan (Editors)

ISSN: 1474-8177/DOI:10.1016/S1474-8177(07)05027-9

Chapter 27

Spatial and Temporal Trends in Surface Water Quality in a Segment of the San Antonio River, Texas

Andrea Anderson, Rachana Nagar and Dibyendu Sarkar

Abstract

Water quality as a function of land use was examined in the upper San Antonio River in the city of San Antonio, Texas. Five water sampling sites representative of different point and nonpoint pollution sources were spread over a distance of 3.2 km. Surface grab sampling was performed on a monthly basis between November 2004 and April 2005 excluding December. Water quality parameters that were evaluated include pH, dissolved oxygen, temperature, total dissolved solids (TDS), total nitrate–nitrogen, total orthophosphate, turbidity, alkalinity and hardness using standard analytical protocols. Results were statistically analyzed by MANOVA. Findings were compared to state (Texas Environmental Quality Commission) and/or federal (U.S. Environmental Protection Agency) limits to establish whether or not parameters were in compliance with those standards or guidelines. Of the routine water quality parameters examined, only turbidity and nitrate–nitrogen exceeded specific standards or guidelines in the segment of the San Antonio River that was sampled. Turbidity and nitrate–nitrogen also showed spatial and temporal trends, which were possibly, affected by land use and local precipitation patterns. Overall, this segment of the upper San Antonio River was considered to be relatively unpolluted and in most parts, unaffected by the land use.

27.1. Introduction

Uncontaminated water resources are declining in a global scale due to human overpopulation and a concomitant increase in the industry and agricultural activities. The pollution that enters the waterways can threaten the health of human drinking water and aquatic ecosystems.

Land use has a profound effect on water quality and surface water is particularly vulnerable to such pollution (Sekhar and Raj, 1995). Specifically, land use is known to affect the amount and quality of runoff during and following rainfall (Richards and Host, 1994). The U.S. Environmental Protection Agency (EPA) has identified some of the leading sources of pollution in rivers and streams of the United States as agriculture, urban runoff, storm sewers and municipal point sources (EPA, 2000).

The San Antonio River originates in central Texas and flows southeast for 346 km where it merges with the Guadalupe River before finally emptying into the Gulf of Mexico (Fig. 27.1a). Overall, the length of the San Antonio River is 362 km. The San Antonio River Basin consists of six major watersheds: Leon, Medio, Salado, Cibolo Creeks, and the Medina and San Antonio Rivers (Fig. 27.1a). Land use along the river is characterized by agriculture, ranching, industry, forestland and urban environments (San Antonio River Authority, 2003). Associated with the urban environment are numerous municipal discharge outflows as well as urban runoff.

The San Antonio River passes through five ecoregions that vary in terms of elevation, soil types and vegetation (SARA, 2003). For instance, in the area where sampling took place, elevation ranges from 200 to

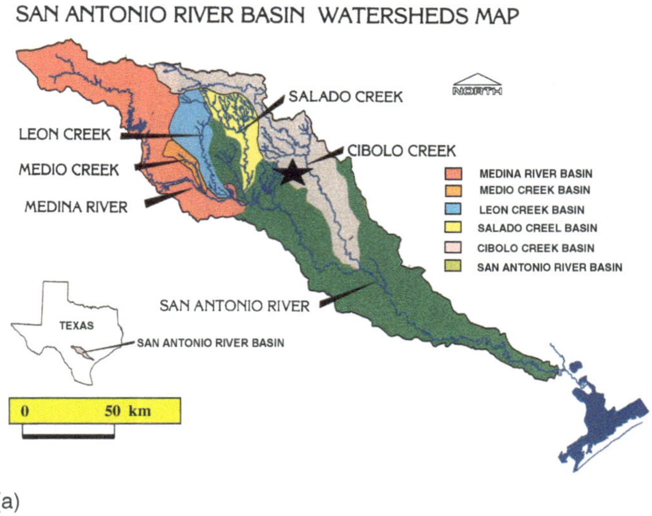

Figure 27.1a. Map shows the location of the San Antonio River Basin in Texas and that the San Antonio River flows to the Gulf of Mexico. (Map courtesy of the San Antonio River Authority). The star on the watershed map indicates the city of San Antonio.

670 m, the soil is generally shallow and underlain by limestone and the vegetation is predominantly oak and juniper woodlands. Meanwhile, the most downstream portion of the river is characterized by elevations of 2–45 m, soils are acidic sand to sandy loams with clay at the river bottom, and vegetation includes mesquite, acacia and tall grass prairies. There are four major aquifers beneath the San Antonio River: the Trinity, Carrizo, Edwards and Gulf Coast aquifers. In the study area, only the Trinity and Edwards aquifers are present (SARA, 2003). These aquifers are hydraulically interconnected and characterized by limestone and dolomite in the upper portions and sand in the lower portions (Boghici, 2004).

Although it is not a heavily polluted river (SARA, 2003), there are potential point sources of pollution (such as municipal effluent pipes) and nonpoint pollution sources (such as surface water runoff from golf courses, residential zones and urban roadways). Pollution from these sources may have negative effects on the freshwater ecosystem of the San Antonio River, which includes at least 73 species of fish (SARA, 2003).

Historically, nitrogen and phosphorous have been problem nutrients in almost all tributaries of the San Antonio River Basin (Ging, 1999; Bush et al., 2000; SARA, 2003). The San Antonio River Authority (SARA) speculated that wastewater discharge is a primary factor contributing to high amounts of these nutrients (SARA, 2003). The Lower San Antonio River can be expected to have high levels of nitrogen and phosphorous since it flows through the agricultural regions of Texas. Agricultural activities can have a major influence on nutrient input into rivers (Puckett, 1995). Meanwhile, the Upper San Antonio River passes adjacent to golf courses and residential zones, both of which may have fertilizer washed into the river during rain events. In high quantities, both of these nutrients can cause water systems to become eutrophic and can contribute to increase suspended sediments (Beránková and Ungerman, 1996; Drolc and Koncan, 1996). In a eutrophic system, excess nutrients promote algal blooms. When the algae die, bacteria decompose the organic matter via aerobic respiration, a process that depletes oxygen in the water system.

Periodic assessment of water quality in a river is essential to discover the appropriate water resource protection plan (Belic and Belic, 1996). The overall health of a waterway can be evaluated by routine chemical tests that include pH, dissolved oxygen (DO), temperature, conductivity (which can be used to calculate TDS), nitrate–nitrogen, phosphorous, turbidity, alkalinity and water hardness. The reported study constituted a six-month assessment program in a 3.2-km segment of the San Antonio River, which measured all these typical water quality parameters and compared the results to the standards established by the Texas Commission on Environmental Quality (TCEQ) and the EPA.

Selected sampling sites in this study were representative of various land uses that potentially contribute to water pollution along a small segment of the San Antonio River. High levels of pollutants originating from these points and nonpoint sources can degrade water quality and threaten the local habitat of fish and aquatic macroinvertebrates. The purpose of this study is to establish if the levels of pollutants in a short segment of the San Antonio River are compliant with state and federal standards and/or guidelines. Data will be useful in determining the water quality of the output from this segment and the effect of this output on the remaining portions of the river.

27.2. Sampling locations

Water samples were collected at a depth of less than 0.46 m one time a month from the upper San Antonio River at five different locations (Fig. 27.1b) five times during a six-month period (November 2004 to April 2005). Sites 1 (extreme upstream) and 5 (extreme downstream) were approximately 3.2 km apart. Site 1 was located within 5 m of the origin of the San Antonio River at a well, called the Blue Hole in the University of Incarnate Word campus. Situated 0.4 km downstream from Site 1, Site 2 was 15 m downstream from a pipe discharging water from the San Antonio Zoo. Located 0.4 km downstream from Site 2, Site 3 was next to a wastewater discharge pipe owned by the local municipal wastewater company (San Antonio Water System, SAWS). Approximately 0.8 km downstream from Site 3, Site 4 was located between a golf course on the one side of the river and a residential zone on the other side. Site 5 was 1.6 km downstream from Site 4 and occurred within an industrial zone of San Antonio immediately downstream from a discharge facility. In the confines of this study, Site 1 was most upstream while Site 5 was the most downstream. The location of this study was approximately 29.45° latitude north and 98.48° longitude west.

27.3. Methods and materials

The water parameters that were measured at each site included pH, dissolved oxygen (DO), temperature, conductivity (which was used to calculate TDS), nitrate–nitrogen, total orthophosphate, turbidity, total alkalinity and water hardness. Methods for collecting, storing and analyzing the samples followed standard EPA protocols (EPA, 1997). Surface grab samples from a depth of less than 0.46 m were collected at the midpoint of the portion of the stream width containing 50% of the total flow

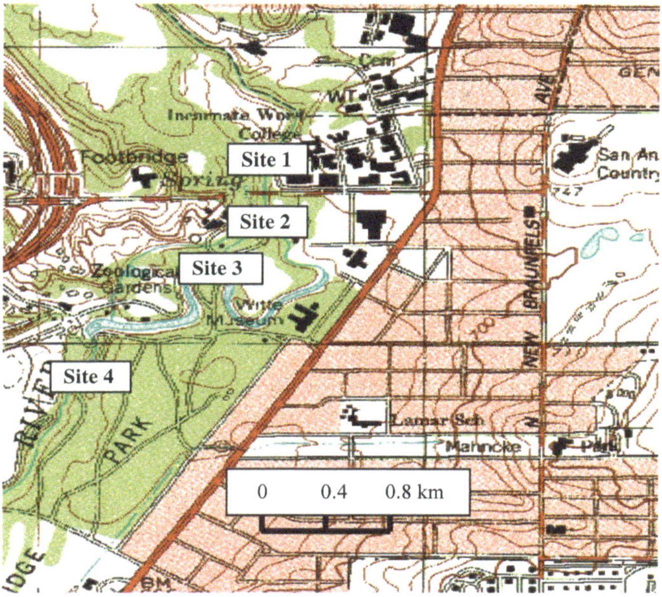

(b)

Figure 27.1b. Sampling sites in the San Antonio River. Site 1 is where the San Antonio River originates. Site 2 is in the vicinity of a discharge pipe from the Zoological Gardens. Site 3 is immediately downstream from a municipal discharge pipe. Site 4 is situated between a golf course on one side of the river and a residential zone on the other. Site 5 is not shown but is approximately 1.6 km downstream from Site 4 and is situated in a light industrial zone of the city of San Antonio. (This figures was taken from http://www.topozone.com.)

(EPA, 1997). Three replicate samples were collected in 125-ml Nalgene HDPE bottles. All samples was stored on ice in the field and immediately transferred to a refrigerator at 4°C upon return to the laboratory.

All instruments for measuring water parameters were calibrated. Dissolved oxygen and temperature were measured with a field meter (850081, SPER Scientific Ltd). The pH, electrical conductivity (EC) and nitrate–nitrogen were measured with an Oakton 510 series meter. Immediately upon return to the lab, pH was measured after a three-point calibration. Using EC, TDS was calculated with the equation

$$EC * 0.564 = TDS \ \ (\text{mg l}^{-1})$$

expressed at 25°C, wherein 0.564 is a conversion factor for the chemical composition of dissolved solids (Boehnke and Delumyea, 2000). Nitrate–nitrogen levels were measured with an ion specific probe and calibration was performed with certified standard solutions ($1\,mg\,l^{-1}$, $10\,mg\,l^{-1}$ and $100\,mg\,l^{-1}$).

Turbidity was measured using a turbidity meter (Cole Parmer Model #8391-40) following a two-point calibration with the certified standards 0.5 NTU and 10 NTU. Total orthophosphate levels were analyzed with the Molybdenum Blue Ascorbic Acid Method (Sparks, 1996) using a Cary 50 Bio UV-Visible Spectrophotometer with a wavelength set at 880 nm. A reagent solution was made by mixing 2.5 M sulfuric acid, 0.03 M ammonium molybdate, 0.004 M antimony potassium tartrate and 0.1 M ascorbic acid. Samples and standards were prepared using this reagent.

Water hardness was measured using the 0.01 M EDTA titration method in which samples were first treated with five drops of a certified standard pH 10 buffer solution and 50 mg Eriochrome black T. Total alkalinity was analyzed with a semi-automatic titrator (Titronic 96) using 0.02 N sulfuric acid. Phenolphthalein was not added to any sample since the pH never went over 8.3. Four drops of methyl orange were first added followed by a titrated amount of 0.02 N sulfuric acid until an approximate pH of 4.3 was achieved as noted by change in color from orange to pink (Sparks, 1996).

Measured water parameters were compared to federal (EPA) and state (TCEQ) maximum or minimum standards or informal guidelines in order to determine compliance. Specifically, criteria for federal standards originated from EPA (1986) while state criteria originated from Chapter 307 of the Texas Surface Water Quality Standards that are specifically designed for the upper San Antonio River (Texas Surface Water Quality Standards, 2000; SARA, 2003). For turbidity and total alkalinity, EPA (1986) served as the guideline since TCEQ does not specify a limit for these parameters.

27.4. Statistical analysis

MANOVA was used to determine the presence of significant spatial and temporal differences in the water quality data. The two factors of the model were site location (Sites 1–5) and the date of collection (November 2004 and January–April 2005). The interaction of these factors was included. Tukey's multiple comparison test was conducted to locate significant differences between the sites and dates of collection. Various regression analyses were also performed.

27.5. Results

The statistical interaction of the site and date of sampling was significant for pH ($p < 0.001$). Generally, pH tended to increase slightly downstream

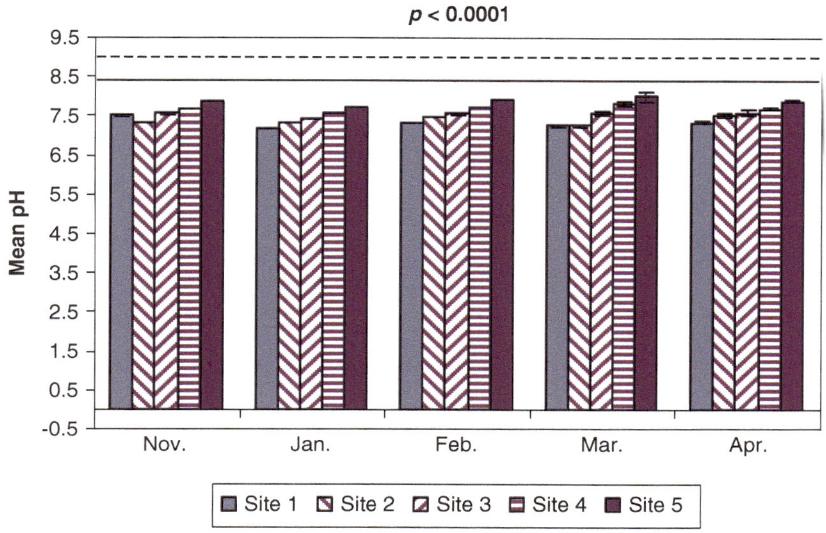

Figure 27.2. Comparisons of the mean (±SD) pH of each site from November 2004 through April 2005, exclusive of December, to the maximum federal and Texas state standards. Refer to Fig. 27.1b for Site descriptions. The *p*-value indicates the interaction across sites and dates of sampling.

from Site 1 to Site 5 (Fig. 27.2). However, the effect of this increase on organisms and water quality is not detrimental since the pH range at all sites for all collection dates ranged between 7.2 and 7.9. In fact, pH of upper San Antonio River complied with both state and federal maximum standards (Fig. 27.2) and is not judged to be problematic.

The statistical interaction of the site and date of sampling was significant for DO ($p<0.001$). Generally, the DO data exhibited a tendency to increase slightly in a downstream manner (Fig. 27.3). The lowest DO levels were observed during the month of November 2004; however, this level (5.4 $\mathrm{mg\,l^{-1}}$, Site 3) was in compliance with both state and federal minimum standards (Fig. 27.4). The highest DO measurement (11.5 $\mathrm{mg\,l^{-1}}$, Site 2) was present during January 2005 when the water sampled had generally lower temperatures (Fig. 27.4). Since DO values ranged from 5.4 to 11.5 $\mathrm{mg\,l^{-1}}$, there were adequate amounts of oxygen for sustaining life.

The statistical interaction of site and date of sampling was significant for temperature ($p<0.001$). An examination of the temperature at each site during each month of collection revealed that not only is the temperature range small (22.6–25.7°C); it also complies with both state and

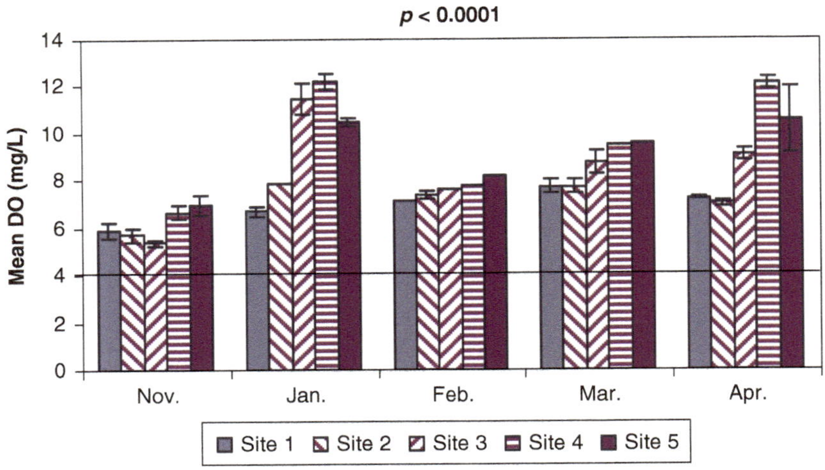

Figure 27.3. Comparisons of the mean ($\pm$SD) dissolved oxygen (DO) concentrations ($mg\,l^{-1}$) at each site from November 2004 through April 2005, exclusive of December, to the federal and Texas state minimum standards. Refer to Fig. 27.1b for Site descriptions. The p-value indicates the interaction across sites and dates of sampling.

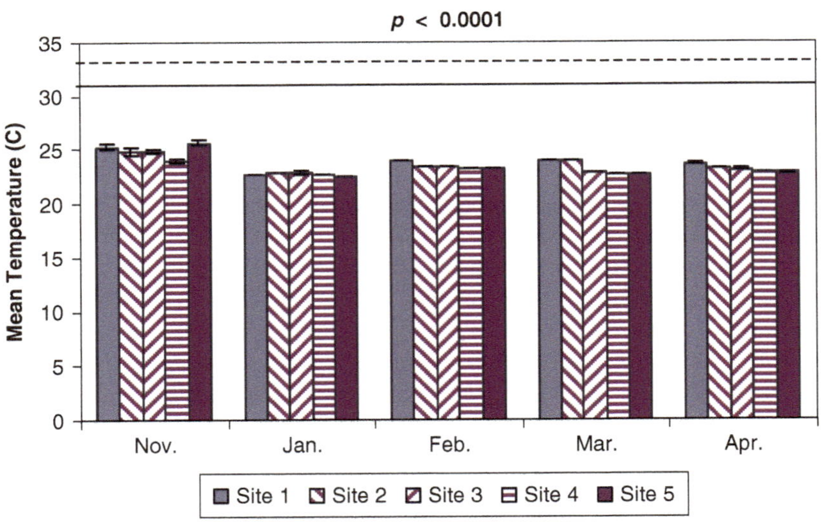

Figure 27.4. Comparisons of the mean ($\pm$SD) temperature at each site from November 2004 through April 2005, exclusive of December, to the federal (solid line) and Texas state (dashed line) maximum standards. Refer to Fig. 27.1b for Site descriptions. The p-value indicates the interaction across sites and dates of sampling.

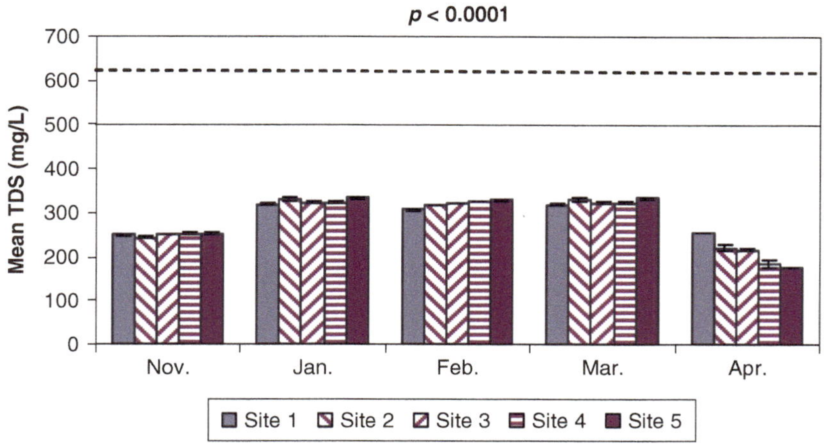

Figure 27.5. Comparisons of the mean (±SD) concentration of total dissolved solids (TDS) at each site from November 2004 through April 2005, exclusive of December, to the federal and the Texas state maximum standards. Refer to Fig. 27.1b for Site descriptions. The *p*-value indicates the interaction across sites and dates of sampling.

federal maximum standards (Fig. 27.4). Hence, water temperature was not considered a detriment to organisms or water quality for this portion of the San Antonio River.

Total dissolved solids complied with both federal and state maximum limits (Fig. 27.5). Highest TDS values occurred from January to March 2005 while the lowest level of TDS was present at Site 5 during April 2005. In addition, during April, TDS concentrations decreased slightly in a downstream manner. Generally, TDS levels were almost two times lower than the federal limit and 1.5 times lower than the state limit. Based on these low-TDS values, this parameter was deemed not one of concern for this segment of the San Antonio River.

The state and federal maximum standards for nitrate–nitrogen were exceeded for all dates and sites of sampling, excluding Site 1 during the month of March (Fig. 27.6). During November 2004, nitrate levels at all sites were higher than those during any other month of sampling. Also during the November sampling, there appeared to be a trend in which nitrates were decreasing in a downstream manner from 33 mg nitrate per liter at Site 1 to 19 mg nitrate per liter at Site 5. This decrease suggests dilution of nitrates in this segment of the river. Regression analysis revealed a weak inverse relationship between nitrate and DO ($R^2 = 0.55$, $p < 0.0001$).

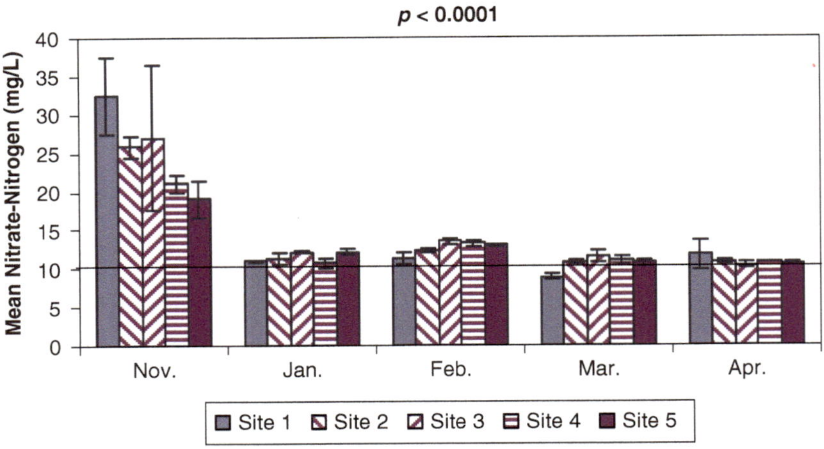

Figure 27.6. Comparisons of the mean (±SD) nitrate–nitrogen concentration ($mg\,l^{-1}$) at each site from November 2004 through April 2005, exclusive of December, to the federal and the Texas state maximum standards. Refer to Fig. 27.1b for Site descriptions. The *p*-value indicates the interaction across sites and dates of sampling.

Total orthophosphate levels in this portion of the San Antonio River ranged from 0.0004 to 0.08 $mg\,l^{-1}$. The statistical interaction of site and date of sampling was not significant for orthophosphate, nor was site of sampling, when considered individually. The only significant factor was the date of sampling ($p<0.0001$). Total orthophosphate concentration was high during November, but still was beneath the federal maximum standard (Fig. 27.7). Despite past studies, which have indicated that phosphorous can be a nuisance nutrient in the San Antonio River (SARA, 2003), it was not judged to be problematic for this segment of the San Antonio River during the study period. Generally, the observed dissolved oxygen concentrations further support the lack of negative effects resulting from phosphorous – a known eutrophic agent.

Turbidity exceeded EPA guidelines at all dates and sites of sampling (Fig. 27.8). Spatially, turbidity tended to increase downstream. Consequently, Site 5 had the highest levels of turbidity at all sampling times. Temporally, turbidity decreased from November 2004 to April 2005. Regression analysis revealed a weak inverse correlation between DO and turbidity ($R^2 = 0.38$, $p<0.0001$).

There appears to be a sufficient buffering capacity of the water sampled at Sites 1 to 5 in the San Antonio River at all sampling dates. Alkalinity values were well above the federal minimum guideline (Fig. 27.9).

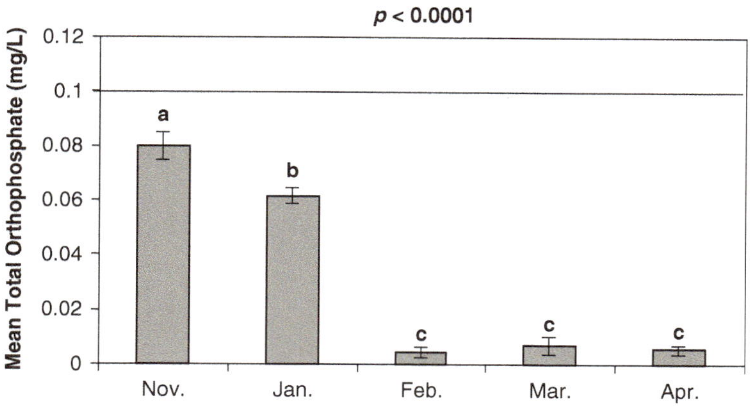

Figure 27.7. Comparisons of the mean (±SD) total orthophosphate concentrations ($mg\,l^{-1}$) from November 2004 through April 2005, exclusive of December, to the federal maximum standard. Refer to Fig. 27.1b for Site descriptions. The *p*-value indicates the interaction across sites and dates of sampling. Different letters indicate significant differences in means (Tukey's multiple comparison test, $p = 0.05$).

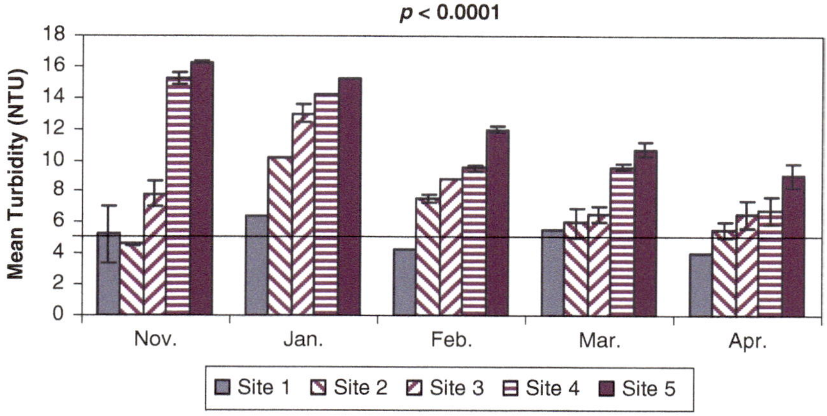

Figure 27.8. Comparisons of the mean (±SD) turbidity level (NTU) at each site from November 2004 through April 2005, exclusive of December, to the federal informal guideline. Refer to Fig. 27.1b for Site descriptions. The *p*-value indicates the interaction across sites and dates of sampling.

Alkalinity values above 20 mg $CaCO_3$ per liter are considered healthy and even 400 mg $CaCO_3$ per liter is not considered dangerous to humans (EPA, 1986). Calculations also suggest that bicarbonate was the predominant ion contributing to total alkalinity (data not shown).

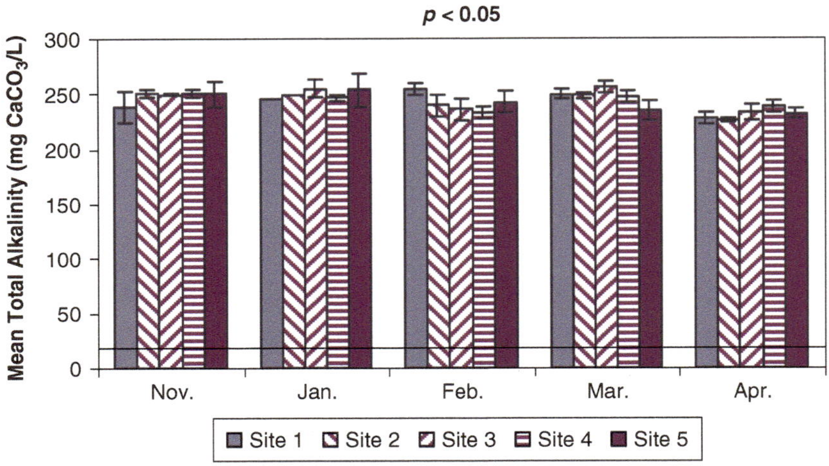

Figure 27.9. Comparisons of the mean ($\pm$SD) total alkalinity at each site from November 2004 through April 2005, exclusive of December, to the informal federal minimal guideline. Refer to Fig. 27.1b for Site descriptions. The *p*-value indicates the interaction across sites and dates of sampling.

Considering that part of the San Antonio River Basin overlies a limestone aquifer (Edwards Aquifer), water hardness values for the sites sampled were lower than expected, ranging from 36 to 52 mg $CaCO_3$ per liter (Fig. 27.10). Water in upper San Antonio River is considered "soft" since the $CaCO_3$ concentration is $<75\,mg\,l^{-1}$ (EPA, 1986).

27.6. Discussion

The majority of the water parameters that were measured complied with state and federal standards and is of no concern for deterioration of water quality. Specifically, these parameters were pH, DO, temperature, TDS, orthophosphate, alkalinity, and hardness. The parameters of potential concern were turbidity and nitrate–nitrogen, both of which exceeded specific state and/or federal limits. Since water quality can change drastically within a short-period time, all of these results should be viewed as short-term findings. For instance, high rainfall can dilute nutrient levels, increase nutrient levels by causing runoff and/or increase turbidity. To ensure consistent compliance and healthy water quality, routine water quality monitoring should be ongoing over a prolonged period. As part of Texas Clean Rivers Program, the San Antonio River Authority oversees the monitoring of water quality in the San Antonio River by measuring parameters bimonthly (SARA, 2003).

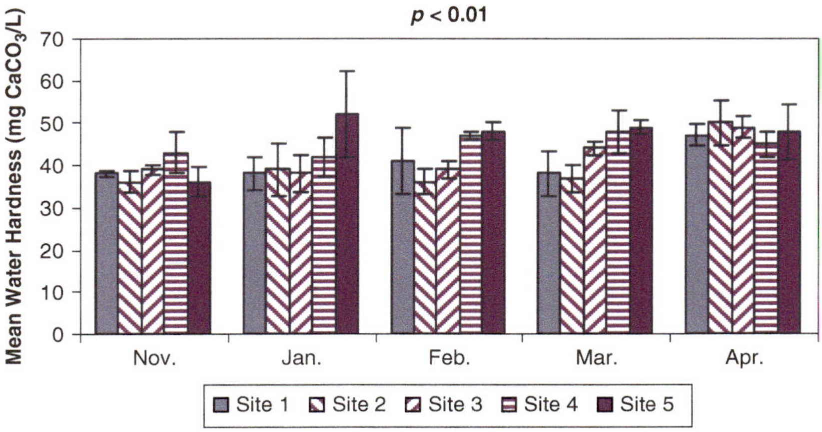

Figure 27.10. Comparisons of the mean (±SD) water hardness levels (mg $CaCO_3$ per liter) at each site from November 2004 through April 2005, exclusive of December. No federal or state standards are established. Refer to Fig. 27.1b for Site descriptions. The *p*-value indicates the interaction across sites and dates of sampling.

The overall healthy condition of this segment of the San Antonio River suggests that upstream portions of the river were not overtly polluted during the duration of the study or that if they were polluted; the pollutants were diluted before reaching this segment of the river. Land use along the upper parts of the river includes ranching, occasional landfills and municipal discharge facilities associated with the San Antonio River Basin (SARA, 2003). Likewise, for most months of sampling results suggest that water quality downstream from the study area would not dramatically degrade by inflow from this segment. An exception to this finding was noticeable during November 2004 when nitrate levels were 1.9–3.3 times greater than the state and federal standards for nitrate–nitrogen (Fig. 27.6). These findings for nitrate are ecologically significant in light of the lower San Antonio River already possessing relatively high-nitrate levels due to being predominantly agricultural land use (SARA, 2003).

Distinct spatial and temporal patterns exist for nitrate–nitrogen in the sampled segment of the river. During November 2004, when rainfall was highest (Fig. 27.11), nitrates decreased in a downstream manner. This pattern suggests a diluting of this nutrient in the study segment. There is also a possible seasonal variation for nitrate. However, our sampling only reflects five months of data and determining such a trend accurately would require additional months of data over the course of at least one year. Nitrate levels were higher during November as compared to the January through April sampling times. Brooker and Johnson (1984)

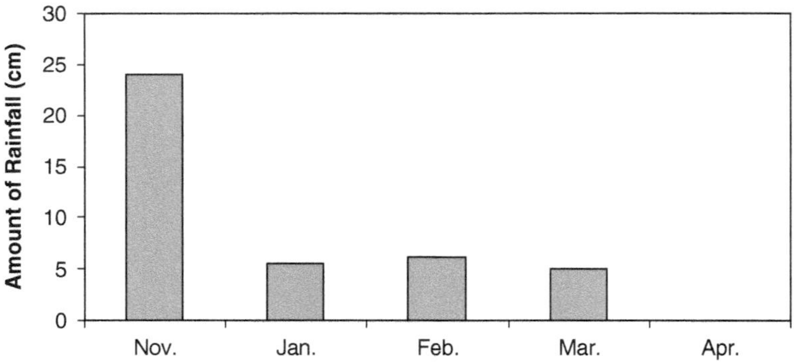

Figure 27.11. Recorded monthly rainfall (cm) for the San Antonio area during the sampling period. (Data are from the National Weather Service, 2005)

noted changes in the concentrations of nutrients in rivers as a function of sampling time and related the variabilities to seasons and climate.

Rainfall recorded in the San Antonio area during the duration of the study was highest in November compared to January through April (Fig. 27.11). Runoff-containing nitrates may have contributed to the concentrations measured. Upstream sources that may have contained nitrate include a golf course and an outdoor public sport fields in the Olmos Park area. During this period of high rain, fertilizer for maintaining these grounds may have been carried into Olmos Creek, which drains into the San Antonio River upstream from Sites 2 to 5. Studies in the United States have found that nitrate levels in creeks increased while passing through golf courses (Mallin and Wheeler, 2000) and that nutrient concentrations of golf course waterways were higher than those in reference locations (Lewis et al., 2002). In the suburban area upstream from the collection sites, residential and public lawns that are maintained with fertilizer could have contributed to elevated nitrate levels. Urban sprawl can increase dissolved solutes such as nitrates in a river and this increase is proportional to the quantity of the developed land area in suburban portions of the watershed (Interlandi and Crockett, 2003).

The San Antonio River Basin is known to have nitrate–nitrogen problems and wastewater has been implicated as a main source (SARA, 2003). In this study, Sites 2 and 3 were close pipes releasing water with high amounts of nitrates. Site 2 was located in the vicinity of an effluent pipe from the San Antonio Zoological Society. Animal feedlots are analogous to nonpoint pollution sources, which have animal wastes; some animal feedlots are known to have contributed high-nitrate levels to nearby waterways (Fryar et al., 2000; Harter et al., 2002). In addition, widespread-ranching activities that occur in the San Antonio River Basin upstream from the sampling sites could contribute to the presence of animal wastes in the river (SARA, 2003). Site 3 was located close to a pipe releasing recycled water by the municipal wastewater treatment organization. A recent study on water quality in the Central Texas region indicated that discharge from the same municipal wastewater treatment plant had the greatest effect on nutrient concentrations compared to any other identifiable source (Ging, 1999). The effluent pipe at Site 3 was removed sometime following the November sampling. This pipe was not present during the January-to-April sampling dates, when nitrate levels were lower at Site 3 as well as the sites downstream from it.

There has been a noticeable trend of increasing nitrate fluxes into waterways of the United States (Bollinger et al., 2000). This trend is a concern because high-nitrate levels can contribute to eutrophication, which is a serious problem in many rivers, and has severe detrimental effects on riverine ecosystems (Dijk et al., 1994). Although this trend appears to exist in the portion of the San Antonio River that was examined, the degree of noncompliance varies for each site and time of sampling. Our nitrate–nitrogen findings are consistent with the San Antonio River Basin Report (SARA, 2003) which also identifies nitrate–nitrogen as a problem associated with both point and nonpoint sources. Nevertheless, the dissolved oxygen concentrations that were measured comply with state and federal limits and are values that support life. Prior to the construction of an upgraded municipal wastewater treatment station (1980s) for the city of San Antonio, nitrate–nitrogen levels in the San Antonio River were higher than current levels (SARA, 2003).

Temporal and spatial trends exist for turbidity in the upper San Antonio River Basin. Generally, turbidity decreased from November 2004 to April 2005. The high amount of precipitation during November (Fig. 27.11) probably caused the high values that were measured during this month. Meanwhile, the lowest amount of rainfall occurred during April, when turbidity values were also the lowest. Spatially, turbidity increased downstream; this trend may be related to water flow at the time of sampling (EPA, 1997). Higher flow rate may cause sediments to be suspended over a longer distance.

The urban location of all sampling sites may have been a factor contributing to high turbidity. Benaabidate (2004) indicated that sites located close to urbanized and industrialized areas tend to contain high turbidity. Surface water bodies, such as rivers passing through cities, are considered to be vulnerable to transient pollution events arising from accidents and contaminated runoff events (Beck, 2005).

Turbidity interferes with recreational swimming because submerged hazards cannot be seen (EPA, 1986). Turbid water affects fish by reducing their growth rate and resistance to disease, preventing the development of fish eggs and larvae, altering migration and reducing food availability (EPA, 1986; Anderson, 2003; Richardson, 2003). If the suspended matter is organic, photosynthesis and the concentration of dissolved oxygen will be reduced and algal growth may ensue (EPA, 1986). Siltation from urban runoff is another nonpoint pollution source that is capable of altering aquatic habitat, suffocating fish eggs and benthic organisms and interfering with both drinking water treatment and recreational use of a river (EPA, 1997).

Speculation on the sources of turbidity is ambiguous in this segment of the San Antonio River. Siltation, urban runoff, wastewater discharge, eroding stream banks, fertilizer from golf course or residential zones, and animal wastes from the San Antonio Zoo discharge are potential sources. The temporal pattern for turbidity (Fig. 27.8) closely followed the rainfall pattern during the study period (Fig. 27.11). High rainfall close to the date of sampling in November possibly contributed to the relatively high-turbidity values in the downstream sites.

27.7. Conclusions

In the segment of the San Antonio River that was sampled, the majority of the routine water quality parameters were at levels lower than that would be considered detrimental to the health of the waterway. Such parameters include pH, DO, temperature, TDS, orthophosphate, alkalinity and hardness. Only turbidity and nitrate–nitrogen values exceeded specific limits established by state and/or federal agencies. In addition, both these water quality parameters revealed spatial and temporal trends. High turbidity can have negative effects on fish, macroinvertebrates, and plants and poses a recreational hazard to swimmers, whereas nitrate–nitrogen can cause eutrophication, which degrades the quality of water in an ecosystem. Results from this study indicate that the upper San Antonio River Basin can be considered relatively unpolluted and in most parts, unaffected by the land use. Future studies regarding the effects of seasonal variation on water quality of the San Antonio River are necessary.

ACKNOWLEDGMENTS

We wish to acknowledge the Environmental Geochemistry Laboratory at the University of Texas at San Antonio for providing the analytical facilities and the means to conduct this study. We are thankful to Mr. Mario Quispe and Mr. Eloy Flores for their technical assistance during sampling and analysis.

REFERENCES

Anderson, B.S., 2003. Ecotoxicologic impacts of agricultural drain water in the Salinas River, California, USA. Environ. Toxicol. Chem. 22, 2375–2384.

Beck, M.B., 2005. Vulnerability of water quality in intensively developing urban watersheds. Environ. Modell. Softw. 20, 381–400.

Belic, S.S., Belic, A.V., 1996. Water quality changes in a small plain agricultural catchment area. Water Sci. Technol. 33, 205–210.

Benaabidate, L., 2004. Assessment of water quality and toxicity of polluted Rivers Fez and Sebou in the region of Fez (Morocco). Environ. Pollut. 131, 163–172.

Beránková, D., Ungerman, J., 1996. Nonpoint sources of pollution in the Morava River basin. Water Sci. Technol. 33, 127–135.

Boehnke, D.N., Delumyea, R.D., 2000. Laboratory Experiments in Environmental Chemistry. Prentice Hall, New Jersey, pp. 81–89.

Boghici, R., 2004. Hydrogeology of Edwards–Trinity Aquifer of Texas and Coahuila in the border region. Texas Water Development Board. Texas Documents: W600.7 R299 NO.360, pp. 91–114. Accessed October 9, 2005, at http://www.twdb.state.tx.us/publications/reports/GroundWaterReports/GWReports/R360AEPC/Ch04.pdf

Bollinger, J.E., Steinburg, L., Englande, A.J., Crews, J.P., Hughes, J.M., Velasco, C., Watanabe, K.H., Swalm, C.M., Mendler, J.M., White, L.E., George, W.J., 2000. Nutrient load characterization from integrated source data for the lower Mississippi River. J. Am. Water Resour. Assoc. 36, 1375–1390.

Brooker, M.P., Johnson, P.C., 1984. The behavior of phosphate, nitrate chloride and hardness in twelve welsh rivers. Water Res. 18, 1155–1164.

Bush, P.W., Ardis, A.F., Lynne, F., Ging, P.B., Hornig, C.E., Jennifer, L.R., 2000. Water quality in south-central Texas, 1996–98. U.S. Geological Survey Circular 1212, p. 32.

Dijk, G.M., Liere, L., Admiraal, W., Bannink, B.A., Cappon, J.J., 1994. Present state of the water quality of European rivers and implications for management. Sci. Total Environ. 145, 187–195.

Drolc, Á., Koncan, Ź., 1996. Water quality modeling of the river Sava, Slovenia. Water Resour 30, 2587–2592.

Environmental Protection Agency, 1986. Criteria for Water. EPA 440/5-86-001, retrieved on March 3, 2005, at http://www.epa.gov/waterscience/criteria/wqcriteria.html

Environmental Protection Agency, 1997. Volunteer Stream Monitoring: A Methods Manual. EPA 841-B-97-003, retrieved on November 20, 2004, at http://www.epa.gov/volunteer/stream/stream.pdf

Environmental Protection Agency, 2000. Water Quality and River Flow Interim Environmental Objectives: Water Objectives. NSW Environment Protection Authority, Sydney, retrieved on March 23, 2005, at http://www.epa.nsw.gov.au/ieo

Fryar, A.E., Stephen, A.M., Mullican, W.F. III, Katherine, D.R., Bennett, P.C., 2000. Nitrate reduction during ground-water recharge, Southern High Plains, Texas. J. Contam. Hydrol. 40, 335–363.

Ging, P.B., 1999. Water quality assessment of south central Texas—Descriptions and comparisons of nutrients, pesticides, and volatile organic compounds at three intensive fixed sites, 1996–98. U.S. Geological Survey Water Resources Investigations Report 99-4155, p. 24.

Harter, T., Davis, H., Mathews, M.C., Meyer, R.D., 2002. Shallow groundwater quality on dairy farms with irrigated forage crops. J. Contam. Hydrol. 55, 287–315.

Interlandi, S.J., Crockett, C.S., 2003. Recent water quality trends in the Schuylkill river, Pennsylvania, USA: A preliminary assessment of the relative influences of climate, river discharge and suburb development. Water Res. 37, 1737–1748.

Lewis, M.A., Boustany, R.G., Dantin, D.D., Quarles, R.L., Moore, J.C., Stanley, R.S., 2002. Effect of a coastal golf complex on water quality, periphyton and seagrass. Ecotox. Environ. Safe. 53, 154–162.

Mallin, M., Wheeler, T., 2000. Nutrient and fecal coliform discharge from coastal North Carolina golf courses. J. Environ. Qual. 29, 979–986.

National Weather Service, 2005. National Weather Service Forecast Office: Climate Information, retrieved on June 05, 2005, at http://www.srh.noaa.gov/ewx/html/climate.htm

Puckett, L.J., 1995. Identifying the major sources of nutrient water pollution. Environ. Sci. Technol. 29, 408–414.

Richards, C., Host, G., 1994. Examining land-use influences on stream habitats and macroinvertebrates: A GIS approach. Water Resour. 30, 729–738.

Richardson, K., 2003. Anthropogenically- induced changes in the environment: Effect on fisheries. In: Sinclair, M., Valdimarsson, G. (Eds.), Responsible Fisheries in the Marine Ecosystem. CAB International, Wallingford, England, pp. 275–289.

Sekhar, M.C., Raj, P.A., 1995. Landuse-water quality modeling: A case study. Water Sci. Technol. 31, 383–386.

Sparks, D.L., 1996. Methods of soil analysis: Chemical methods. Soil Sci. Soc. Am. Book Series (Part 3). Soil Sci. Soc. Am., Inc, Medison, WI.

San Antonio River Authority, 2003. Basin Summary Report for the San Antonio River Basin. Texas Commission on Environmental Quality. Retrieved from the worldwide web on June 5, 2005, at http://www.sara-tx.org/site/water_quality/water_qual_mon/clean_rivers/assessment_report.html

Texas Surface Water Quality Standards, 2000. Texas Natural Resource Conservation Commission, Chapter 307, retrieved on January 23, 2005, at http://www.tnrcc.state.tx.us/oprd/rules/pdflib/307%60.pdf

Developments in Environmental Science, Volume 5
D. Sarkar, R. Datta and R. Hannigan (Editors)

ISSN: 1474-8177/DOI:10.1016/S1474-8177(07)05028-0

Chapter 28

Beneficial utilization of drinking-water treatment residuals as contaminant-mitigating agents

K.C. Makris and G.A. O'Connor

Abstract

Waste recycling and re-utilization are two energy-efficient processes that have gained popularity due to environment-friendly and cost-reductive advantages. This chapter deals with an example of such processes: the beneficial utilization of the waste by-product generated from the drinking-water treatment process as cost-effective sorbents for contaminants, such as arsenic (As) and phosphorus (P). Drinking-water treatment residuals (WTRs) are primarily amorphous masses of iron (Fe), aluminum (Al) hydroxides or $CaCO_3$ that also contain sediment and humic substances removed from the raw water, as well as coagulating agents added to raw water, such as activated carbon and polymers. WTRs are produced in large quantities at the drinking-water treatment plants and are typically disposed of in landfills. This chapter begins with WTR-relevant definitions and types of WTRs produced, and explains how WTRs are produced. Regulations and alternatives for WTR disposal are discussed. General physical and chemical properties of different types of WTRs are presented. Most of the WTR-pertinent literature focuses on P immobilization by WTRs in poorly P-sorbing soils, and the long-term stability of sorbed P. The data suggest that land application of WTR can be a best management practice to reduce the environmental effects of P. We also discuss the potential for using WTRs as sorbents for arsenic (As). Potential limitations associated with WTR use are discussed, and future research needs addressed.

Advances in science and technology have provided steadily increasing energy and material efficiencies in the production of goods and services (Batterham, 2003). Waste recycling and re-utilization are two energy-efficient processes that have gained popularity due to environment-friendly and cost-reductive advantages. An example of such processes is the beneficial utilization of the waste by-product generated from the drinking-water treatment process as cost-effective sorbents for prospective contaminants. Drinking-water treatment residuals (WTRs) are primarily amorphous masses of iron (Fe), aluminum (Al) hydroxides or $CaCO_3$ that also contain sediment and humic substances removed from the raw water, as well as coagulating agents added to raw water, such as activated carbon and polymers (Elliott and Dempsey, 1991; Dayton and Basta, 2001). More than 2 million Mg of WTRs are generated from drinking-water treatment facilities in the U.S. every day (Prakash and Sengupta, 2003). Several authors (Elliott et al., 2002a, b; Dayton et al., 2003; Lind, 2003; Makris, 2004; Novak and Watts, 2004) have suggested that WTRs can serve as low-cost soil amendments to reduce environmental impacts of various oxyanions, notably phosphorus (P). Our purpose is to review the properties of WTRs and to demonstrate how WTRs can serve as contaminant-remediation agents.

28.1. What are WTRs?

New drinking-water treatment technologies (membrane separation, ion exchange, precipitative softening, granular activated carbon, and disinfection/UV) have emerged, but the majority of drinking-water treatment plants are still based upon coagulation/filtration principles. Addition of Fe, Al, or Ca salts in conjunction with polyelectrolytes, such as polymers and/or granulated activated carbon to raw water removes colloids, color, sediment and common contaminants from surface and groundwater supplies intended for potable water use. In the basic pH environment of a typical drinking-water treatment facility, Fe or Al salts added in the raw water hydrolyze to form Fe or Al hydroxides. These remove phosphorus (P), arsenic (As), and other contaminants from solution through adsorption and/or co-precipitation reactions similar to the removal reactions in soils (Livesey and Huang, 1981). Iron and/or Al salts combined with small quantities of polyelectrolytes (e.g., polymers) or activated carbon are typically used by the waste- and drinking-water treatment industry to remove different inorganic and organic contaminants (Maurer and Boller, 1999). For example, coagulation using ferric chloride can reduce As concentrations by as much as 90% with minimal cost (McNeil and Edwards, 1995; Scott et al., 1995). Elliott and Dempsey (1991) reported that 56% of 32 facilities used alum as the coagulant, 28% used ferric chloride, 9% used polyaluminum chloride, and 6% used only polymer. A more recent survey in 100 FL drinking-water treatment plants showed that 49, 20, and 6 plants used lime, ferric chloride, and alum, respectively (Townsend et al., 2001). Depending on the metal salt used, the resulting WTRs are referred to as Al-WTRs (use of Al salt) or Fe-WTRs (Fe salt) or Ca-WTRs ($CaCO_3$).

28.1.1. Disposal of WTRs

Drinking-WTRs can be disposed: (a) directly to a receiving stream; (b) to sanitary sewers; (c) to a landfill, assuming that the residual contains no free-draining water and does not have toxic characteristics as defined by the TCLP test; and (d) by land application (Chwirka et al., 2001). A 1991 survey of 612 utilities serving populations of > 50,000 showed that landfill disposal was the predominant disposal method followed by land application > sanitary sewer disposal > direct stream discharge > lagooning (Kawczyinski and Achtermann, 1991). In the state of Florida, the most acceptable long-term disposal method of WTRs is landfilling (Townsend et al., 2001). The disposal cost of non-hazardous WTRs is low; estimated at < \$50 Mg^{-1} (Meng et al., 2001).

Environmental concerns over the direct disposal of WTRs to streams or landfilling have shifted the attention to more environmentally benign disposal methods, such as land-application. Disposal of WTRs is currently regulated at the state and not at the federal level, thus, different regulations exist for WTR disposal at different US states. Drinking-WTRs are specifically exempt from the 40 CFR Part 503 land disposal regulations for biosolids (USEPA, 1996). Thus, WTRs can be land-applied without having to meet metal (including As) limitations of the Part 503 regulation. However, land applications of mixtures of drinking-water and waste-water residuals (biosolids) are subject to the Part 503 rule. Combining biosolids with WTRs typically increases Fe/Al concentrations in the mixed biosolids, and subsequent land-application of the mixtures can significantly increase P sorption in amended soils (Elliott et al., 2002a).

28.1.2. Contaminants in WTRs

Contaminant leachability from land-applied WTRs is of major concern for regulatory agencies. Total mean arsenic (As) concentrations for 5 Al-WTRs (11.3 $mg\,kg^{-1}$) and 3 Fe-WTRs (7.0 $mg\,kg^{-1}$) were above the industrial direct exposure limit for As in FL (3.7 $mg\,kg^{-1}$), but leachable As concentrations using the SPLP test were well below FL groundwater guidance concentration (FGGC) (0.05 $mg\,l^{-1}$), posing little threat for groundwater contamination with As (Jain et al., 2005). Total mean As concentrations for 20 Ca-WTRs were well below the residential direct exposure limit for As in FL (0.8 mg kg^{-1}) (Jain et al., 2005). Total volatile and semi-volatile organic compounds, as well as nitrogen-, or phosphorus-, or chlorine-based pesticide concentrations, are usually well below detection limit (Jain et al., 2005). Mean total Al concentrations in 5 Al-WTRs were above the residential soil clean-up target level (SCTL) of 72,000 mg kg^{-1}, but not for Fe- or Ca-based WTRs (Jain et al., 2005). In 3 out of 5 Al-WTRs tested, SPLP-based leachable Al concentrations were above the FGGC value of 0.2 mg l^{-1} (Jain et al., 2005). Mean SPLP-based leachable Ba, Cd, Cr, Cu, Pb, Hg, Ni, and even Mo concentrations in Fe-, Al-, and Ca-based WTRs collected in FL were below their respective FGGC values (Jain et al., 2005). Special attention has been given to Mo because relatively small forage-Mo concentrations can induce Cu deficiency (molybdenosis) in grazing animals, if the forage Cu concentration is also low ($<$10 mg kg^{-1}) (Elliott and Taylor, 2000). State regulations in Pennsylvania mandate a ceiling limit of 18 mg kg^{-1} Mo in WTRs for land application (Elliott and Taylor, 2000). Other Fe-, and Al-WTRs tested by Elliott and Dempsey (1991) showed that mean total Cr, Ni, Pb, and Zn concentrations were within the common range for soils, implying that total metal soil concentrations will remain largely unaffected by WTR application at typical loading rates. In addition, air-dried Fe- and Al-based WTRs usually contain small numbers of coliforms ($<$20 coliforms g^{-1}) arising from air-drying, long-term storage, and chlorine addition during the drinking-water purification process (Elliott and Dempsey, 1991).

28.1.3. Macroscopic properties of WTRs

Drinking-water treatment plant facilities use different water sources and different chemicals. Thus, the WTRs produced can have widely different elemental compositions and sorption capacities. Drinking-WTRs are commonly characterized by broad particle size distributions. Very coarse fragments ($>$2 mm) exist in small numbers, and the greatest number of WTR particles is usually found in the micrometer (1–5 μm) size range. The broad size distribution is evidenced by the predominance of large amounts (% number distribution) of small particles (0.1–10 μm) (Makris, 2004). Depending on the residuals' organic carbon (C) content, Al- or Fe-WTR particles are dark, or light in color. Calcium-WTRs are typically white, reflecting the high $CaCO_3$ content. Drinking-WTR particles normally resist dissolution in aqueous solutions, unless exposed to strongly acidic (pH $<$ 3) and/or chemically reducing conditions (applicable only for Fe-WTRs).

Table 28.1. Selected properties of WTRs

Properties	Al-based	Fe-based	Ca-based	Reference
pH	5.0–8.2	4–9.2	8.4–11.0	1, 2, 3
KCl-P ($mg\,kg^{-1}$)	2.2–5.6	6.2–6.3	0.8	1, 4
Total C ($g\,kg^{-1}$)	8.5–225	94–206	114	1, 2, 3, 4
Total N ($g\,kg^{-1}$)	3.0–10.0	5.0–11.0	0.3	1, 4
Total Al ($g\,kg^{-1}$)	15–300	2.2–10.0	0.5	1, 2, 3
Total Fe ($g\,kg^{-1}$)	5.0–66	109–251	0.4	1, 2, 3, 4
Total Ca ($g\,kg^{-1}$)	3.0–50	16.4	310–520	2, 4
Total P ($g\,kg^{-1}$)	0.2–4.4	0.3–3.2	0.04–0.2	1, 2, 3, 4
Oxalate Al ($g\,kg^{-1}$)	1.3–91	0.2–9.8	0.03	1, 2, 4
Oxalate Fe ($g\,kg^{-1}$)	2.3–5.8	108–195	0.35	1, 2, 4
Oxalate P ($g\,kg^{-1}$)	0.05–3.0	0.15–2.6	0.05	1, 2, 4

1: Makris, 2004; 2: Dayton et al., 2003; 3: Elliott and Dempsey, 1991; 4: Elliott et al., 2002a.

The pH of Al-WTRs is commonly circumneutral (5.0–7.0), although some exhibit alkaline pH values as great as 8.2 (Table 28.1). Typically, Fe-based WTRs are alkaline (pH values 7.2–9.2) although the pH values for two Fe-WTRs investigated by Makris (2004) were 4 and 5.6. Calcium-WTRs are highly alkaline with pH values in the order of 8.2 to 11 (Elliott and Dempsey, 1991).

Phosphorus solubility (KCl-extractable P) in WTRs is typically low and represents only a small fraction (<0.2–0.7%) of total P for all three WTR classes. The KCl-extractable P is considered the most available (labile) pool of P (Elliott et al., 2002b). The very low KCl-extractable P levels in WTRs implies that WTRs can serve as additional sinks for P immobilization in poorly P-sorbing soils.

Total C values for the Al- and Fe-WTRs vary widely depending on the raw water source (surface- or ground-water). Often, however, total C values are high, reaching levels on the order of 200 $g\,kg^{-1}$ regardless of the WTR class. Total C determinations could overestimate organic C content because the combustion method normally used measures both organic and inorganic C. However, organic C determination with the Walkley–Black method gave similar results to total C analysis data for 7 WTRs, suggesting minimal inorganic C (Makris et al., 2005a). Iron- and Al-WTRs are also characterized by fair amounts (up to 10 $g\,kg^{-1}$) of total N that could serve as N sources for plants when WTRs are land applied. Calcium-WTRs typically contain negligible amounts of N, as N is volatized at high pH (Table 28.1).

Total Al concentrations range from 15 to 300 g $Al\,kg^{-1}$ in Al-WTRs (Table 28.1). Small amounts of Al can also occur in Fe- and Ca-WTRs. Total Fe concentrations range from 109 to 251 g $Fe\,kg^{-1}$ in Fe-WTRs;

small amounts of Fe are often found in Al-WTRs, but minimal amounts in Ca-WTRs. Total Ca concentrations are commonly high in Ca-WTRs (310–520 $g\,kg^{-1}$), and minimal in Fe- or Al-WTRs. The variability in total Al, Fe, and Ca concentrations in the WTRs can partially be explained by the fact that different drinking-water treatment facilities add different amounts of Al/Fe/Ca salts in the treatment of raw waters, and by differences in the chemical composition of inorganic colloids suspended in the raw water.

Total P concentrations of Fe- and Al-WTRs typically range from 0.2 to 4.4 $g\ P\,kg^{-1}$, and about 10-fold less in Ca-WTRs. Phosphorus in WTRs originates from the raw water treated in drinking-water treatment plants, and ultimately becomes part of WTR structure. As the KCl-extractable P data show, most of the P in WTRs is insoluble in aqueous suspensions. Other data (Makris et al., 2004b) showed that the P is occluded in the WTR structure and is minimally released over time.

X-ray diffraction analysis of Fe- and Al-WTRs (data not shown) suggests that amorphous Al or Fe hydroxides dominate the Al- and the Fe-WTRs, respectively, with no apparent crystalline components (Makris, 2004). Oxalate-extractable P, Fe, and Al are usually associated with the amorphous phase of the particles. Oxalate-extractable Al values are highly variable, but typically represent 30–90% of total Al of Al-WTRs (Dayton et al., 2003; Makris, 2004; Dayton and Basta, 2005), which supports the amorphous model of Al-WTRs. Typically, Fe-WTRs have lower (45–64%) oxalate-extractable Fe values as a fraction of total Fe (Makris et al., 2005a). Phosphorus retention is strongly related to amorphous Fe and Al concentrations. Gallimore et al. (1999) and Dayton et al. (2003) concluded that the amorphous, as determined with the oxalate extraction, rather than the total Al content of WTRs determines their effectiveness in reducing runoff-P. Thus, the lower oxalate-extractable metal percentage in Fe-WTRs than Al-WTRs suggests less P sorption effectiveness.

28.1.4. Microscopic properties of WTRs

Limited information exists on microscopic properties of WTRs, such as specific surface area (SSA) and porosity (Ippolito et al., 2003; Makris et al., 2004b). Scanning-electron microscope secondary images revealed the irregular shapes and variable sizes of Fe- and Al-WTR particles (Makris et al., 2004b). Particle surfaces range from rough to fairly smooth. Elemental spectra, obtained with scanning electron microscopy coupled with energy-dispersive spectroscopy (SEM-EDS), verified the presence of P, Al, and Fe, as well as Si, Ca, and Na. No clustering of P (zones of obvious P precipitation) was evident, proving that P originally in the WTRs existed as an integral part of the particles (Makris et al., 2004b).

Isothermal (70°C, 10 h), thermogravimetric plots of weight losses for an Fe-WTR revealed an initially fast release of water, followed by a kinetically driven stage where hysteretic water was slowly released from the interior of the WTR particles (Makris and Harris, 2006). The data suggest that sufficient amounts of water exist in the internal surfaces of air-dried WTRs (Makris and Harris, 2006). Hyde and Morris (2000) incubated raw, dry and frozen Al- and Fe-WTR particles (up to 120 $Mg\,ha^{-1}$) with two soils having above-optimum soil test extractable P concentrations at 25°C for 21 days. Soil test P levels were reduced by 64%, 28%, and 23%, respectively, suggesting a potential WTR drying effect on P extractability. Drinking-WTRs usually contain significant amounts of carbon (17–149 $g\,kg^{-1}$; Dayton and Basta, 2005). The large amount of organic C in WTRs can cause WTRs to deviate from ideal metal hydroxide physicochemical behavior. Hydrophobic and hydrogen bonding forces (rather than electrostatic) between organic molecules and mineral surfaces may influence solute retention by WTRs. Such interactions can also affect the physicochemical nature of WTRs and differentiate them from ideal metal hydroxides. For example, cationic polyelectrolytes added during the raw water treatment process accounted for a significant portion of sorbed P by Fe-WTRs (Butkus et al., 1998). Steric effects and hydrophobicities imposed by organic compounds present in WTRs may influence P (and other oxyanions) sorption kinetics and diffusivities. Clearly, WTRs are complex and simple electrostatic attraction for anions is insufficient to fully explain solute retention on WTRs.

Traditional specific surface area (SSA) measurements utilizing N_2 gas (BET-N_2) revealed large differences in SSAs among different Fe- and Al-WTRs (Fig. 28.1) (Makris et al., 2005a). Phosphorus loaded to WTRs reduced N_2-based SSAs, except for the Lowell and Holland materials. However, BET-N_2 SSAs did not correlate significantly with the P sorption capacities of the materials (maximum initial P load 10 g $P\,kg^{-1}$). Makris et al. (2005a) hypothesized that N_2 molecules may not have reached all sorption sites due to diffusional restrictions. De Jonge and Mittelmeijer-Hazeleger (1996) showed that SSAs of three soil organic matter samples were underestimated by BET-N_2 measurements. The significant C contents (34 to 210 $g\,kg^{-1}$) of WTRs studied by Makris et al. (2005a) were hypothesized to affect BET-N_2 SSA measurements.

Makris et al. (2005a) also found that micropore volume distributions determined by the Saito-Foley (SF) method (Saito and Foley, 1991) increased with increasing N_2-based SSAs for all WTRs. Micropore volumes decreased when P was added, except for the two materials (Lowell and

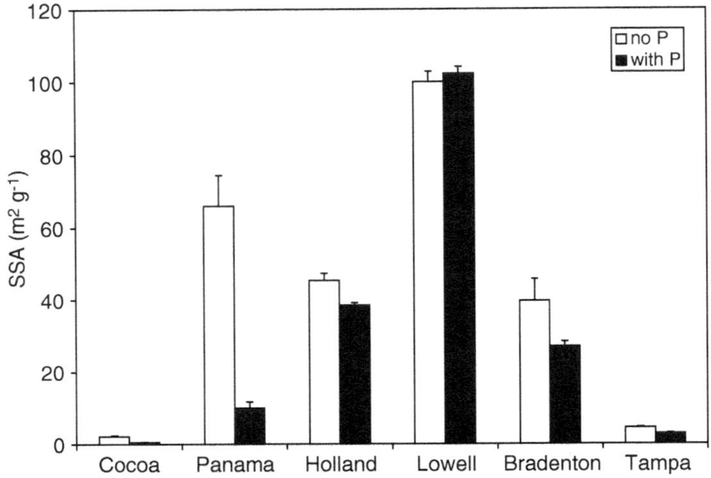

Figure 28.1. BET-N_2 SSA measurements for untreated and P treated WTRs (10 g P kg^{-1} initial load) after 40 days. Error bars denote one standard deviation of two replicated runs (Makris et al., 2005a).

Holland) with the smallest C contents. Figure 28.2 represents micropore distributions for two materials: one from the group that showed micropore volume reduction when P was added (Bradenton), and one from the group that did not (Lowell). The contrasting behaviors may be related to differences in microporosity (Table 28.2). The Lowell and Holland Al-WTRs had the smallest proportion of micropore volume (18 and 9%, respectively), whereas the Bradenton Al-WTR had the greatest (46%).

Materials with relatively large micropore volumes experienced significant pore volume reductions following P addition. Materials with relatively small meso- and micro-pore volumes were unaffected by P addition. The data were interpreted as being consistent with restricted micropore access imposed by sorbed phosphate molecules.

The CO_2- and N_2-based SSA data suggest that organic compounds trapped in WTR-micropores regulate the diffusion of gas molecules in and out of micropores. A strong linear negative correlation ($r^2 = 0.86$) was observed between total C and the N_2/CO_2 SSA ratios (Fig. 28.3). Materials with low total C contents (Holland and Lowell, WTRs) showed little difference in the amounts of N_2 and CO_2 sorbed (N_2/CO_2 SSA ratio close to 1). However, as total C content of WTRs increased, so did the difference in SSAs measured by CO_2 and N_2 (CO_2N_2). Apparently, organic C restricted diffusion and sorption of N_2 to a much greater extent than for CO_2. Similar use of the N_2/CO_2 SSA ratio normalized to the

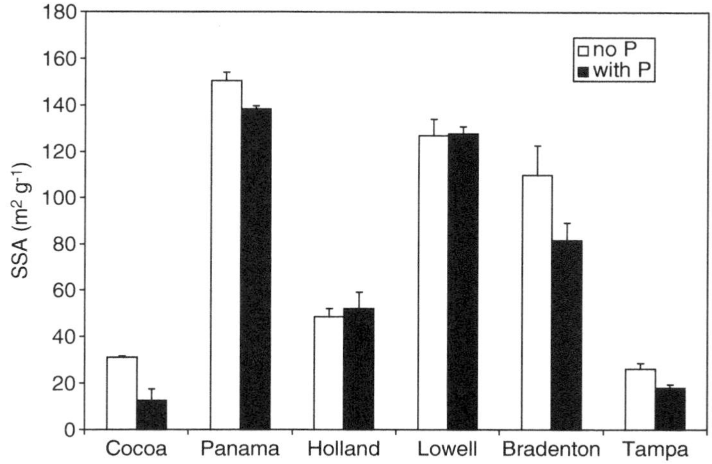

Figure 28.2. Micropore CO_2 SSA measurements for untreated and P treated WTRs (10 g P kg^{-1} initial load) after 40 days (Makris et al., 2005a). Error bars denote one standard deviation of two replicated runs.

Table 28.2. Micro-pore volumes of untreated (no P added) WTRs obtained from BET-N_2 isotherms with the Saito–Foley method and total pore volumes obtained at 0.99 P/P_0 (Makris et al., 2005a).

WTRs	Micro-pore volume ($cm^3 g^{-1}$)	Total pore volume ($cm^3 g^{-1}$)	Micro-pore volume % of total
Al-WTR, Melbourne	8×10^{-2}	2.0×10^{-1}	40
Al-WTR, Lowell	4.3×10^{-2}	2.4×10^{-1}	18
Al-WTR, Bradenton	4.2×10^{-2}	9.2×10^{-2}	46
Al-WTR, Holland	1.05×10^{-2}	1.2×10^{-1}	9
Fe-WTR, Tampa	1.45×10^{-3}	4.3×10^{-3}	34
Fe-WTR, Panama	1.4×10^{-3}	1.2×10^{-2}	12
Fe-WTR, Cocoa	1.2×10^{-3}	7.5×10^{-3}	16

C content was employed by Ravikovitch et al. (2005), studying native grassy and forest Chicago soils. The researchers proposed using N_2/CO_2 SSA ratios to characterize and predict various soils' behavior in sequestration processes involving humic substances (Ravikovitch et al., 2005).

28.2. Beneficial use of WTRs: The case of phosphorus

Intensified agricultural activities have increased P inputs and concentrations in soils. Water bodies can then be contaminated with P via surface

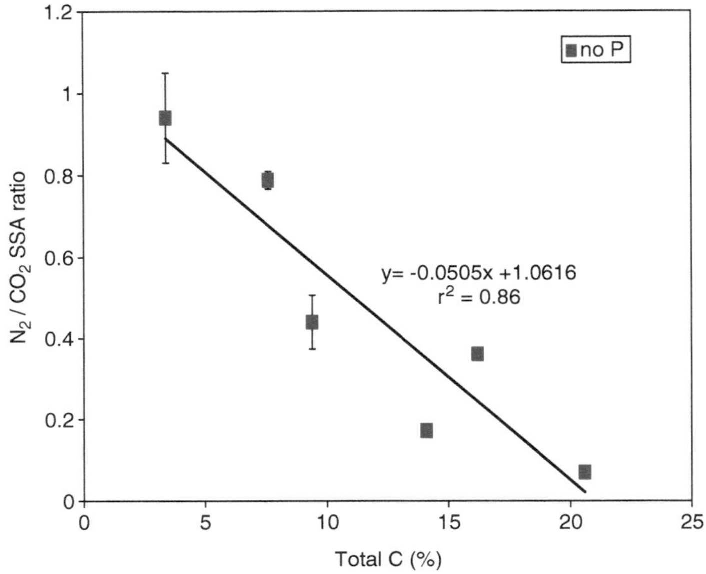

Figure 28.3. Correlation between the ratios of SSA determined with BET-N_2 and CO_2 gas and total C of the untreated (no P added) WTRs examined by Makris et al. (2005a). Error bars denote one standard deviation of two replicates.

runoff, or leaching of soluble and particulate P. Increased P loading of streams, lakes, and rivers can encourage algal blooms and subsequent decreases in water quality. In addition to direct environmental concerns, water quality degradation can increase costs associated with drinking-water treatment.

Poor P-sorbing soils are abundant in southeastern states of U.S.A. These sandy (coarse-textured) soils are characterized by low P-sorbing capacities, and are often accompanied by high water tables. The combination of characteristics makes such soils vulnerable to P losses and negative water quality impacts (He et al., 1999; Novak et al., 2000). Land application of WTRs can be a cost-effective treatment for effectively sorbing excess levels of labile P in soils. The high amorphous Al or Fe content of the WTRs can increase a soil's P sorption capacity (Elliott et al., 1990; Novak and Watts, 2004; Dayton and Basta, 2005; Rhoton and Bigham, 2005).

Phosphorus sorption maxima differ among WTRs with contrasting physicochemical properties. Phosphorus sorption experiments (25 C) with one Al-WTR showed a maximum P sorption capacity of 12,500 mg kg^{-1} (Ippolito et al., 2003). Butkus et al. (1998) were able to load a Fe-WTR with ~20% P by wt. (200,000 mg kg^{-1}). Makris (2004) reacted Al- and Fe-WTR particles with inorganic P solutions at P loads up to 10,000 mg P kg^{-1}. Almost all of the added P was sorbed by most WTRs, although some WTRs sorbed little P. Work by O'Connor et al. (2001) showed that a Ca-WTR was much less effective in sorbing P than a Fe- and an Al-WTR.

Long-term (80 days) P sorption kinetics of WTRs are usually biphasic (Makris et al., 2005a), although Novak and Watts (2005) suggest that first-order kinetics adequately describe the reactions. Both models suggest that P sorption reactions with <2-mm WTR particles may not reach equilibrium in short contact times (up to 1 or 2 days). Phosphorus sorption kinetic data for Al- and Fe-based WTRs (Table 28.3) were best fit to a second-order reaction rate model (Azizian, 2004) except for the Cocoa Beach Fe-WTR, which fit neither a first- or second-order model due to limited P sorption (Makris et al., 2005a). Phosphorus sorption by the other WTRs was initially fast, and was followed by a slow P sorption stage (Makris et al., 2005a). The fast stage of P sorption presumably characterized P retention on highly accessible surfaces like particle exteriors and macropores (Van Riemsdjik and Lyklema, 1980). The slower stage has been characterized as P association, via diffusion, in micropores (Axe and Trivedi, 2002; Makris et al., 2004b).

The magnitude of the slow P sorption stage varies among different WTRs (Table 28.3) (Makris et al., 2005a). The second-order reaction rate coefficients increased with the P sorption capacities of the WTRs. The WTR samples with the greatest final P sorption generally showed significant time-dependency up to 40 or 80 days, whereas WTR samples with low P sorption did not. Apparently, WTRs with relatively low P sorption capacities had minimal rate-limited P surface access (to certain

Table 28.3. Reaction rate constants in Al- and Fe-based WTRs suspensions after a 1000 mg P l^{-1} initial pulse input (Makris et al., 2005a)

Source	Form	First-order rate fit (r^2)	Second-order rate fit (r^2)	Second-order reaction rate k (l s^{-1} mg^{-1})[a]
Holland, MI	Al-based	0.87	0.98	2×10^{-7}
Lowell, AR	Al-based	0.86	0.95	3.4×10^{-5}
Bradenton, FL	Al-based	0.84	0.94	1.3×10^{-4}
Melbourne, FL	Al-based	0.89	0.96	2.4×10^{-4}
Tampa, FL	Fe-based	0.80	0.97	1.07×10^{-7}
Panama City City, FL	Fe-based	0.71	0.77	5.8×10^{-9}
Cocoa Beach City, FL	Fe-based	0.12	0.12	1.24×10^{-9}

[a]Where the slope of a linear fit to a n-order reaction equals $(n\text{–}1)k_n C_o^{n-1}$.

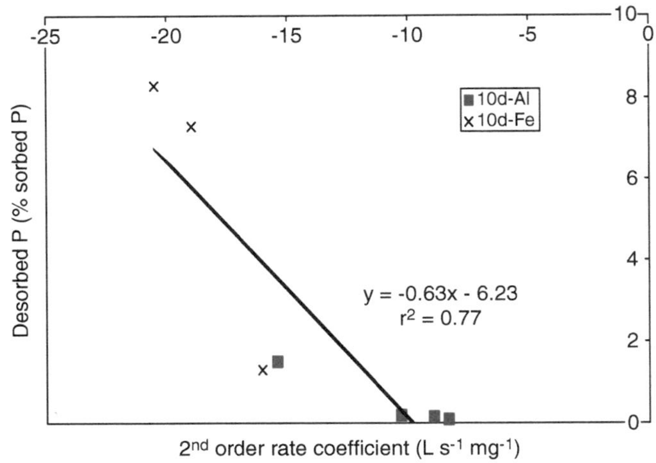

Figure 28.4. Semi (X-axis) log-transformed plot of the second-order rate coefficient changes with the amount of P desorbed after 80 days of reaction for three Fe-WTRs (10 day-Fe) and four Al-WTRs (10 day-Al) (Makris et al., 2005a). The logarithmic transformation was utilized to accommodate large differences in the reaction rate coefficients for the seven WTRs.

micropores) compared with samples with high P sorption capacity. The same trend in rate limitation applies to desorption; the greater the second-order rate coefficient, the smaller the proportion of P desorbed (Fig. 28.4). Despite the large differences (4–5 orders of magnitude) in rate coefficients of Al- and Fe-based WTRs, a log-linear model explained differences in P desorption reasonably well. Limited P desorption suggests great stability of P immobilized by highly-retentive WTRs. Second-order rate coefficients for Fe-based WTRs were generally smaller than those of Al-based WTRs, consistent with less P sorption for the second biphasic (longer-term) sorption stage of the Fe-based WTRs (Makris et al., 2005a). Phosphorus sorption kinetic rates were greater for smaller WTR aggregate sizes ($<500\,\mu m$), than for larger aggregate sizes (500–4000 μm) (Novak and Watts, 2005a), consistent with the effect of increased specific surface area on the smaller size aggregates' P sorption maxima.

As yet unidentified factors apparently limit the ability to adequately predict the highly-variable P sorption capacities of both Fe-, and Al-based WTRs. Traditionally, measurements of oxalate extractable P, Fe and Al in organic wastes, soils, WTRs and their mixtures have been used to explain trends in runoff-P (Gallimore et al., 1999; Dayton et al., 2003), and P leaching losses in soils amended with organic P sources and/or WTRs (Elliott et al., 2002a; Dayton and Basta, 2005). Oxalate extracts non-crystalline forms of Fe and Al and, thus, is expected to release P bound to Fe and/or Al amorphous hydroxides.

Elliott et al. (2002b) explained differences in P-fixing capacities of an Al- and a Fe-WTR by variations in the reactive Fe- and Al-hydroxide contents, as measured by oxalate extraction. Their results indicated that an Al-WTR was more effective in sorbing P than a Fe-WTR. Dayton et al. (2003) found that variations in oxalate-extractable Al concentrations of 20 Al-WTRs explained differences in runoff-P reductions by WTRs ($r^2 = 0.69$, quadratic model). However, Dayton et al. (2003) used short (15 h) equilibration times and modest initial P loads (up to 2500 mg kg^{-1}). A recent attempt to improve P sorption maxima predictions for the same Al-WTRs utilized a different solid:solution ratio (1:100), smaller size aggregates (< 150 μm), and longer equilibration times (6 days), and resulted in a significant ($r^2 = 0.91$) linear relationship between oxalate-extractable Al and P_{max} for Al-WTRs (Dayton and Basta, 2005). Similar work was not conducted for either Fe-, or Ca-based WTRs). Surprisingly, the linear model reported by Dayton and Basta (2005) suggests significant P retention (11.3 g kg^{-1}) at zero oxalate extractable Al. Thus, the empirical regression model predicts that a material with negligible oxalate-extractable Al concentrations would sorb more than 1% soluble P (1.1%). In a long-term (up to 80 days) P sorption study utilizing P loads of 10,000 mg kg^{-1} and 7 Fe-, and Al-based WTRs (<2-mm), no significant correlation existed between oxalate-extractable Fe + Al concentrations with P sorption maxima for the 7 WTRs, despite the high oxalate extractable Fe and Al levels (Makris et al., 2005a). Oxalate-extractable [Fe + Al] accounted for 50–65% of total [Fe + Al] for the Fe-WTRs and 80–90% of total [Fe + Al] for Al-WTRs (Makris et al., 2005a).

A model to accurately predict P sorption capacities for both Fe- and Al-based WTRs of a relatively practical particle size (<2 mm) was developed by Makris et al. (2005a). The model envisioned that organic compounds inherently present in WTRs, impose steric and diffusion restrictions to P diffusion toward the interior of WTR particles. Different CO_2 and N_2 SSA values of WTRs used suggested a significant role of organic compounds trapped in WTR-micropores in regulating the diffusion of gas molecules in and out of micropores. Prediction of the long-term P sorption capacities of WTRs is complex, and seems to require information collected from both N_2- and CO_2-based SSAs. The CO_2-based SSA values better explained P sorption than N_2-based SSA values, but predictions were not perfect. The model utilizes N_2/CO_2 SSA ratios to predict the long-term capacities of the WTRs (Makris et al., 2005a). Organics retard P diffusion toward internal sites, but apparently do not serve as sorption sites for P.

Normalizing P sorption capacities after 40 days to C content of WTRs may provide a means of predicting the long-term P sorption capacities. Makris et al. (2005a) found a significant ($p<0.001$) positive linear relationship between the normalized amount of P sorbed after 1 or 40 days with the N_2/CO_2 SSA ratios of the WTRs (Fig. 28.5). Using measurements of three independent variables, total C, and N_2- and CO_2-based SSAs, Makris et al. (2005a) were able to explain 87% of the variability in the long-term measured P sorption capacities of the WTRs after 40 days. The number (6) of WTRs used in the model was limited, but covered a span of WTRs varying significantly in total C, Fe and Al contents. Accordingly, SSAs of the WTRs varied an order of magnitude. A potential limitation of this model would be the availability of instrumentation required to measure the specific surface areas of the materials.

The model may allow *a priori* predictions of WTR P-sorption capacities and avoidance of having to conduct tedious long-term batch equilibration studies. Model results are encouraging, but further validation using Fe- and Al-based WTRs from different facilities are needed. Clearly, not all WTRs possess the same P retention capacities, and potential users must be aware of these differences before blindly

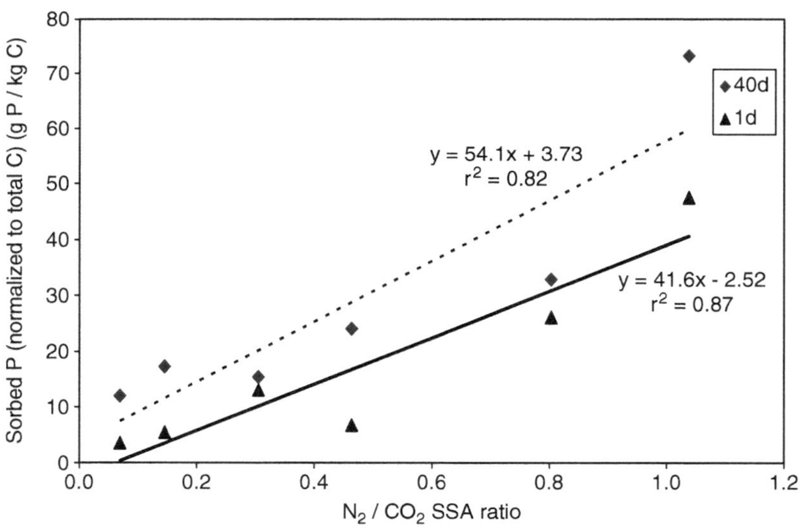

Figure 28.5. Correlation between the SSA ratio of BET-N_2 and CO_2 gas with long-term (40 days) pseudo P sorption capacities of WTRs. Initial P load (2500 mg P kg^{-1}) (Makris et al., 2005a).

depending on any WTR as a best management practice to minimize off-site P losses.

28.2.1. Reduction in runoff-P

Land-application of WTRs can significantly reduce runoff-P from agricultural fields. Haustein et al. (2000) documented decreasing soluble P concentrations in runoff from fields excessively high in soil test P following amendment with an Al-WTR (rates up to 18 $Mg\,ha^{-1}$). Gallimore et al. (1999) applied an Al-WTR to poultry litter-amended soils, and reduced soluble P in surface runoff. Peters and Basta (1996) significantly reduced (~50% of the initial values) soil test-extractable P concentrations of an acidic and calcareous soil incubated with high loading rates of two Al-WTRs ($\cong$60 and 200 $Mg\,ha^{-1}$).

Drinking-WTRs can also reduce runoff-P when used in buffer strips near water bodies (Basta et al., 2003; Dayton et al., 2003). Concentrating the WTRs in strips of land, rather than applying the residuals to an entire watershed, reduces the amount of WTR needed while protecting surface waters from P pollution. Buffer strips are a best management practice to reduce surface water pollution, and the effectiveness of the strips can be greatly improved by amendment with WTRs.

28.2.2. Reduced phosphorus leaching

Land-application of WTRs can also significantly reduce P leaching, but is most beneficial when full contact of soil soluble P with the WTR particles is ensured (Silveira et al., 2006). Surface applications of WTRs reduced runoff-P losses, but P leaching losses continued due to inadequate soil/WTR mixing. Long-term P retention by WTRs is diffusion-controlled (Makris et al., 2004b), so P must be close to WTR particles to be retained. Researchers (Codling et al., 2000) that incorporated either Fe- or Al-based WTRs into poultry litter-amended soils significantly reduced P-leaching. Similarly, surface-applied Al-WTR had little effect on P availability to wheat (*Triticum aestivum* L.) in a greenhouse study, but incorporation into the entire soil significantly decreased P availability (Cox et al., 1997). In a greenhouse column setup, Elliott et al. (2002b) showed that either Fe- or Al-WTRs were able to reduce P leaching in a low P-sorbing FL sand amended with dewatered biosolids and triple superphosphate (TSP) fertilizer. In a greenhouse setting, WTRs reduced P losses to 3.5% (Ca-WTR), 2.5% (Fe-WTR) and <1% (Al-WTR) of applied TSP-P. For the biosolids treatments, all WTRs retarded P vertical movement such that leachate P was not statistically different from the controls.

28.2.3. Stability of sorbed P by WTRs

There is abundant evidence (e.g., Gallimore et al., 1999; Ippolito et al., 1999; Haustein et al., 2000; Dayton et al., 2003; Novak and Watts, 2004, 2005b) that WTRs are effective P sorbents that reduce the environmental lability of P. However, almost all studies have been short-term (<80 days). Insufficient data are available on the long-term stability of P retained by WTRs, or soils amended with WTRs, or metal salts, and the long-term stability of immobilized P is a major concern of state and federal regulators.

To fully understand interactions between P and soil constituents, the effects of time needs to be considered (Scheidegger and Sparks, 1996). Phosphorus sorption kinetics by metal hydroxides and soils are well characterized and generally show a fast sorption phase, followed by a slower reaction rate where sorption may never reach true equilibrium (Bolan et al., 1985). The fast reaction is ascribed to low-energy external surface sites, where ligand exchange is believed to be the main mechanism of adsorption (Bolan et al., 1985). The slow reaction between P and metals with metal hydroxides proceeds for days, months, and even years, and has been attributed to continuing surface precipitation reactions (Van Riemsdijk and Lyklema, 1980; Nooney et al., 1998) or intraparticle diffusion into micropores (Axe and Trivedi, 2002; Makris et al., 2004a). Makris et al. (2004a) identified the latter mechanism as prominent in the retention of P by WTRs, and suggested that P immobilized by WTRs should be essentially irreversibly bound. A 6.5 year field study of WTR effectiveness in reducing water extractable P in two soils with excessively high soil test P levels seems to confirm the contention (Jacobs and Teppen, 2000). Soil samples taken each year for up to 6.5 years after an initial WTR application showed sustained reduction (up to 63% initial values) of water soluble P levels in the WTR-amended plots; the standard soil test for P (Bray I) was unable to detect changes in P lability (preliminary results). Incubation experiments (84 days) conducted by mixing three soils high in Mehlich III-extractable P concentrations with an Al-WTR applied at rates of 1–6% (wt.) showed that WTR incorporation in soils high in soluble P caused larger relative reductions in water-extractable P than Mehlich III-extractable P concentrations (Novak and Watts, 2005b).

Potential WTR particle dissolution, particularly under acidic conditions, is a concern with respect to WTR field applications in humid regions. Particle dissolution in acid soils or aqueous suspensions could release significant quantities of potentially toxic Al and previously immobilized P to the environment. Makris (2004) conducted long-term (80 days) equilibrations of Al-WTRs in unbuffered 0.01 M KCl solutions. Soluble Al concentrations of untreated (no P added) Al-WTRs were below the instrument's (ICP-AES) detection limit (0.03 mg Al l^{-1}). Most P-treated (10,000 mg kg^{-1} initial P load) Al-WTR suspensions released small (<10 mg kg^{-1}) quantities of soluble Al, but one released 46 mg Al kg^{-1}. Overall, the amount of KCl-extractable Al concentrations released from Al-WTRs within 80 days was minimal (<0.1% of oxalate-extractable Al) (Makris, 2004).

Metal (Fe and Al) concentrations were also measured by Makris (2004) during P desorption experiments to a 5 mM oxalate solution. Oxalate (5 mM)-extractable Fe after 160 days (80 days for P sorption and subsequent 80 days for P desorption) was minimal for a highly acidic (pH 3.9) Fe-WTR. Oxalate (200 mM) extractions are commonly performed at pH 3 (McKeague et al., 1971), which dissolves significant amounts of noncrystalline Fe and Al components in WTRs. However, such a large oxalate concentration (200 mM) is rarely encountered in natural systems, so Makris (2004) used 5 mM oxalate to more closely mimic practical conditions (Bhatti et al., 1998). One Fe-WTR released a small amount of oxalate (5 mM)-extractable Fe to solution (~1.0% of the 200 mM-oxalate extractable levels) in the absence of added P. No soluble Fe was detected in supernatants of P-treated WTRs within 1 day. Iron desorbed from WTR surfaces after 20 or 40 days (in the absence of added P) apparently reacted with the added P, resulting in minimum soluble Fe concentrations (<0.02 mg Fe l^{-1}) even at the highest initial P load (10,000 mg kg^{-1}).

Phosphorus desorption from Al-WTRs actually decreased with increasing desorption time, suggesting continuous P sorption (Makris, 2004; Dayton and Basta, 2005). Phosphorus desorption with a 5 mM oxalate solution from different WTRs was minimal (0.2% of sorbed P) (Makris et al., 2005a). Similarly, P desorption with a 0.1 M KCl solution from 5 Al-WTRs after a 6-day equilibration was <10% of the final P sorption maxima (Dayton and Basta, 2005). This phenomenon was also observed in a long-term (211 days) P desorption experiment with another Al-WTR (Ippolito et al., 2003). Maximum percentages of oxalate (5 mM)-desorbable P (% of that previously sorbed) were generally <0.2% for all but one of seven WTRs; one WTR desorbed 1.5% (Makris, 2004). The percentages were observed for desorption experiments conducted either for 10 or 20 days. As desorption time increased to 40 or 80 days, no soluble P concentrations greater than 0.3 mg P l^{-1} were measured for any Al-WTR, suggesting continuous P sorption. Residual

200μm
A
200μm
E
100μm
B
100μm
F
20μm
C
20μm
G
2μm
D
2μm
H

P sorbed during desorption came from the entrained solution after sorption. Apparently, P sorbed by WTRs is chemisorbed on WTR surfaces, and resists desorption to a 5 mM oxalate solution.

Spectroscopic and solid-state characterization of P-loaded WTR particles has been employed to better understand the long-term stability of sorbed P (Ippolito et al., 2003; Makris et al., 2004b). Makris et al. (2004b) reacted Fe- and Al-WTR particles with P for 80 days and then subjected the particles to SEM-EDS (Figs. 28.6 and 28.7) and to electron microprobe analyses. No discrete surficial metal-P phases were detected with SEM-EDS spectroscopy, which led to the hypothesis that P diffuses into particles to reach meso- and micro-pore domains rather than precipitating on external surfaces of the WTRs. Thin cross-sections were prepared that allowed monitoring of the profile depth P distribution in the WTR particles over time. SEM-EDS dot maps of cross-sections from both WTRs qualitatively supported an intraparticle sorption mechanism.

Phosphorus was evenly distributed within the particles, except for some near edge P zonation in Fe-WTR particles after P treatment (Fig. 28.7).

Electron microprobe analysis using wavelength-dispersive spectroscopy (EPMA-WDS) on thin-sections of P-treated WTR particles supported intraparticle P diffusion more quantitatively than SEM-EDS. Cross-sectional P distribution analysis of the P-treated WTRs showed significant ($p < 0.001$) increases in the relative P concentrations in the interior of the particles (approximately 60 μm inside) with time (from 1 to 80 days) (Makris et al., 2004b). Phosphorus concentrations of 80 day-treated particles were significantly greater than 1 day-treated particles, both at the edge and interior (Makris et al., 2004b). Average P concentrations for P-treated particles were slightly greater near the edge, but edge-versus-interior differences were not statistically different at the 95% confidence level.

Ippolito et al. (2003) used EPMA-WDS dot maps to assess P distribution in a P-treated Al-WTR equilibrated for 211 days. Dot maps showed no evidence for P surface precipitation, but a uniform amorphous

Figure 28.6. Scanning electron secondary images of the Al- and Fe-based WTRs. (A) secondary image of representative Al-WTR particles; scale bar = 200 μm. (B) Magnified secondary image of a portion of image (A); scale bar = 100 μm. (C) Secondary image of representative Al-WTR surfaces; rough and smooth surfaces; scale bar = 20 μm. (D) Magnified secondary image of the rough surface of the Al-WTR particle from image (C); scale bar = 2 μm. (E) Secondary image of representative Fe-WTR particles; scale bar = 200 μm. (F) Magnified secondary image of a portion of image (E); scale bar = 100 μm. (G) Secondary image of representative Fe-WTR surfaces; rough and smooth surfaces; scale bar = 20 μm. (H) Magnified secondary image of the rough surface of the Fe-WTR particle from image (G); scale bar = 2 μm. Images D and H show surface porosity, but magnification is not large enough to show microporosity (Makris et al., 2004b).

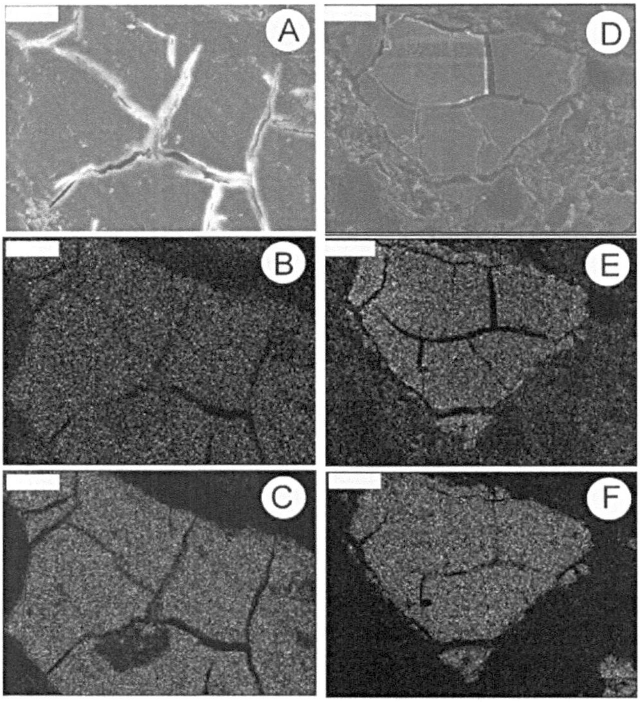

Figure 28.7. Scanning electron secondary images (A, D) and the corresponding P and metal dot maps (B, C, E, and F) of thin cross-sections after 80 days P treatment for both WTRs. (A) Secondary image of a representative Al-WTR cross-section; scale bar = 20 μm. (B) P dot map of the secondary image in (A). (C) Al dot map of the secondary image in (A). (D) Secondary image of a representative Fe-WTR cross-section; scale bar = 20 μm. (E) P dot map of the secondary image in (D). (F) Al dot map of the secondary image in (D). P dot maps of cross-sections for both WTRs show uniform P distribution, with no evidence for surface precipitation. Rarely, and only for the P-treated Fe-WTR, there were indications of zonal P enrichment near the particle edge (Makris et al., 2004b).

Al-P association throughout the particles (Ippolito et al., 2003). EPMA-WDS data support the notion that P moves in a three-dimensional fashion toward the interior of the WTR particles rather than accumulating significantly at the particle surface as by precipitation.

The collective data of Makris et al. (2004b) and Ippolito et al. (2003) have favorable implications for long-term P immobilization. Apparently, once P reaches the WTR microsites, very strong adsorption, and highly hysteretic desorption is likely. Thus, once immobilized by the WTR particles, the P is likely irreversibly bound, barring destruction of the WTR particles.

28.3. Potential limitations of WTR use

28.3.1. Agronomic impacts on phosphorus availability

Studies dealing with land-applied WTRs have identified both negative and positive effects of WTR application to crop yields. Ippolito et al. (1999) conducted a greenhouse study to investigate the efficacy of co-application of Al-WTR and biosolids to two native short grass species (blue gramma-*Bouteloua gracilis*, and western wheatgrass- *Pascopyrum smithii*). Results suggested a linear positive relationship between increasing WTR rate and yield, but a negative relationship between increasing WTR rate and shoot P and Al concentrations. No P deficiency or Al toxicity symptoms were observed, but the trends suggested that very large WTR applications could cause agronomic problems. Other studies have shown that application of WTR10 g WTR kg^{-1} (~20 Mg WTR ha^{-1}) reduced tissue P concentrations, but did not induce other nutrient deficiencies or toxicities (Elliott and Singer, 1988; Heil and Barbarick, 1989; Cox et al., 1997).

The primary concern with the land application of WTRs is the potential for induced plant P deficiencies (Basta et al., 2000). Plant (*maize*) P uptake reductions and germination problems were observed when Al-WTR was applied (up to 40 Mg ha^{-1}) (Rengasamy et al., 1980). Fescue grass yields were decreased in WTR-amended soil columns (up to 80 Mg ha^{-1}), apparently in response to decreasing plant-available P (Lucas et al., 1994).

Others have reported insignificant effects of WTR applications on crop P nutrition. Reductions in exchangeable P concentrations were not accompanied by plant growth limitations in soils amended with biosolids and Al-WTR during a 7-year field experiment (Naylor and Carr, 1997). Harris-Pierce et al. (1993) reported minimal negative effects of WTR (5.6–22.4 Mg ha^{-1}) and biosolids (11.2 Mg ha^{-1}) co-application on native rangeland vegetation. Brown and Sartain (2000) showed that a 2.5% by wt. (56 Mg ha^{-1}) Fe-WTR application rate significantly reduced P leaching from applied fertilizer P, with minimal negative impact on Bermuda grass P uptake.

In addition to agronomic limitations involving P (over-applied WTRs induced plant P deficiencies), there are concerns about potential Al phytotoxicities or elevated soluble Al concentrations in runoff from amended land. Few data exist on the potential for Al-WTR to cause Al phytotoxicity or water quality degradation. Alum applications to agricultural fields had no effect on plant Al concentrations or crop yields (Moore and Miller, 1994; Moore et al., 2000). Land application of

Al-WTR at rates of 11.2 and 44.8 $Mg\,ha^{-1}$ did not increase dissolved solids or Al in surface runoff (Gallimore et al., 1999). Haustein et al. (2000) reported no significant increase of dissolved Al in surface runoff of soils amended with an Al-WTR (2.2–18 $Mg\,ha^{-1}$).

28.3.2. Limited WTR supply

The availability/accessibility of sufficient quantities of WTRs can be a major limitation for projects that require continuous or large WTR applications. Field application rates of WTRs are usually large (25–56 $Mg\,ha^{-1}$), and can out-strip local WTR supplies when hundreds of hectares require treatment. Partial or complete substitution of Al-WTRs for alum may be a cost-effective practice to reduce soluble P in animal waste (Makris et al., 2005b). Complete substitution of Al-WTRs for alum could be practiced in cases of year-around, abundant WTR availability.

28.3.3. Arsenic sorption by WTRs

Another contemporary environmental problem that could be amenable to remediation with WTRs is contamination of soils, waters and wastes with arsenic (As). Arsenic is toxic to man and other living organisms. Arsenic can be found in the waste stream from a variety of industrial processes (petroleum refining, glass melting and smelting of ores). For example, fly ash, a coal combustion residue, can contain As concentrations reaching 180 $mg\,kg^{-1}$ (Qafoku et al., 1999). Arsenic can be also released to the environment through pesticide and fertilizer use (Loebenstein, 1992; Woolson, 1992). Arsenate ($H_2AsO_4^-$ and $HAsO_4^{-2}$) is the predominant As form in well-oxidized soil/water systems, whereas arsenite ($H_3AsO_3^0$ and $H_2AsO_3^-$) occurs mostly in low redox potential environments.

Use of the chromated copper arsenate (CCA)-pressurized lumber is a common practice to preserve wood supplies throughout the U.S. and safe disposal of the As-containing waste lumber can be problematic. Deposition in unlined construction and debris landfills can generate leachate enriched in As, which can endanger local ground waters. Soils associated with playgrounds, abandoned dipping vats, under porches constructed with CCA wood can become As point sources and potentially endanger children playing nearby.

Several As removal techniques exist for As-contaminated soils and waters, but most are cost-prohibitive. Small communities are likely to face expensive and technically imposing challenges to meet the 2006-mandated maximum contaminant level of 10 μg As l^{-1}. Drinking-WTRs are inexpensive sorbents expected to readily sorb As (similar chemistry to

P) and may help address current and future As issues. However, little is known about the extent of As immobilization by WTRs in general and about the retention of various As species [As(V) or As(III)] in particular.

Arsenic(V) (and P) removal in the water treatment industry is based on the well-known sorption of oxyanions by Fe and Al hydroxides (Livesey and Huang, 1981). Sorption of As, like P, on metal hydroxides is initially rapid, but then decreases with increasing equilibration time (O'Reilly et al., 2001). Pierce and Moore (1982) found that for up to 1 $mg\,l^{-1}$ initial arsenate concentrations, the adsorption isotherm of As(V) on ferrihydrite could be described by the Langmuir model. At greater As loads (up to 50 $mg\,l^{-1}$ initial arsenate concentration), data were better described by a two-stage linear model, suggesting a two-site type of As adsorption. Maximum arsenate adsorption occurred at pH 4 and decreased as pH increased. We hypothesize that WTRs can be similarly effective and represent a cost-effective amendment for immobilizing As(V) in As-contaminated sites.

Recent studies showed the strong affinity for soluble As(V) or As(III) by Fe-based or Al-based WTRs (Vandanapu et al., 2005; Makris et al., 2006). After only 2 days reaction, an Al-WTR sorbed nearly all ($>90\%$) of the added As(V) at all initial As loads (up to 20,000 $mg\,kg^{-1}$); an Fe-WTR was less effective (sorbed 50–85%) (Vandanapu et al., 2005). Arsenite sorption experiments with an Fe-, and Al-based WTRs showed that both sorbents exhibited As(III) sorption maxima in the order of $\sim$15,000 $mg\,kg^{-1}$, within a 2-day equilibration period (Makris et al., 2006). Arsenic desorption with a 1 mg P l^{-1} solution was minimal, and decreased with increasing As load (Makris et al., 2006). The data are encouraging in the sense that WTRs apparently behave toward As as they do toward P, with huge As sorption capacities, and minimal As desorption. The desorption study with background P also suggest that WTRs could be effective As sorbents even in systems containing appreciable soluble P. Additional work is necessary to confirm the preliminary results, characterize both As(V) and As(III) sorption by WTRs, and to devise the most practical means to use WTRs to address As contamination concerns.

28.4. Future research needs

1. Field validation of WTR beneficial effects and the lack of negative influences on crop yields is needed to convince farmers of the practicality of using WTRs.
2. Techniques to better characterize WTRs a priori are needed (e.g., wet chemical, and spectroscopic techniques) to identify which WTRs are expected to be effective P and As sorbents and the rates necessary.

3. Studies that promote use of WTRs, independent of as a soil amendment (i.e., using barrels of WTR through which drainage water or potential drinking water is passed to remove contaminants). Field application rates of WTRs are usually large (25–56 $Mg\,ha^{-1}$), and can out-strip local WTR supplies when hundreds of hectares require treatment. Treating the drainage water from the fields (rather than the soil) may be a more efficient means of utilizing WTRs to remove P or As before the pollutants reach water supplies. Use of WTRs in barrels to decontaminate As-polluted water from wells may represent a "low-tech" solution to expensive water treatment plants for small communities faced with meeting more stringent As drinking water.

REFERENCES

Axe, L., Trivedi, P., 2002. Intraparticle surface diffusion of metal contaminants and their attenuation in microporous amorphous Al, Fe, and Mn oxides. J. Colloid Interface Sci. 247, 259–265.

Azizian, S., 2004. Kinetic models of sorption: A theoretical analysis. J. Colloid Interface Sci. 276, 47–52.

Basta, N.T., Dayton, E.A., Storm, D.E., 2003. Advances in WTR research to manage P in soil, runoff, manure, and biosolids. WEF/AWWA/CWEA Joint Residuals and Biosolids Management Conference, "Partnering for a Safe, Sustainable Environment." Baltimore, MD., 19–22 Feb, 2003.

Basta, N.T., Zupancic, R.J., Dayton, E.A., 2000. Evaluating soil tests to predict bermudagrass growth in drinking water treatment residuals with phosphorus fertilizer. J. Environ. Qual. 29, 2007–2012.

Batterham, R.J., 2003. Ten years of sustainability: Where do we go from here. Chem. E. Sci. 58, 2167–2178.

Bhatti, J.S., Comerford, N.B., Johnston, C.T., 1998. Influence of oxalate and soil organic matter on sorption and desorption of phosphate onto a spodic horizon. Soil Sci. Soc. Am. J. 62, 1089–1095.

Bolan, N.S., Barrow, N.J., Posner, A.M.M., 1985. Describing the effect of time on sorption of phosphate by iron and aluminium hydroxides. J. Soil Sci. 36, 187–197.

Brown, E.A., Sartain, J.B., 2000. Phosphorus retention in USGA greens. Soil Crop Sci. Soc. Fl. Proc. 59, 112–117.

Butkus, M.A., Grasso, D., Schulthess, C.P., Wijnja, H., 1998. Surface complexation modeling of phosphate adsorption by water treatment residual. J. Environ. Qual. 27, 1055–1063.

Chwirka, J.D., Narasimhan, R., Scheuer, N., Rousseau, G., 2001. The impact of residuals on the selection of an arsenic treatment process. WEF/AWWA/CWEA Joint Residuals and Biosolids Management Conference. Biosolids 2001: 'Building Public Support'.

Codling, E.E., Chaney, R.L., Mulchi, C.L., 2000. Use of aluminum and iron-rich residues to immobilize phosphorus in poultry litter and litter-amended soils. J. Environ. Qual. 29, 1924–1931.

Cox, A.E., Camberato, J.J., Smith, B.R., 1997. Phosphate availability and inorganic transformation in an alum sludge-affected soil. J. Environ. Qual. 26, 1393–1398.

Dayton, E.A., Basta, N.T., 2001. Characterization of drinking water treatment residuals for use as a soil substitute. Water Environ. Res. 73, 52–57.

Dayton, E.A., Basta, N.T., 2005. A method for determining the phosphorus sorption capacity and amorphous aluminum of aluminum-based drinking water treatment residuals. J. Environ. Qual. 34, 1112–1118.

Dayton, E.A., Basta, N.T., Jakober, C.A., Hattey, J.A., 2003. Using treatment residuals to reduce phosphorus in agricultural runoff. J. Am. Wat. Works Assoc. 95, 151–158.

De Jonge, H., Mittelmeijer-Hazeleger, M.C., 1996. Adsorption of CO_2 and N_2 on soil organic matter: Nature of porosity, surface area and diffusion mechanisms. Environ. Sci. Technol. 30, 408–413.

Elliott, H.A., Dempsey, B.A., 1991. Agronomic effects of land application of water treatment sludges. J. Am.Wat.Works Assoc. 84, 126–131.

Elliott, H.A., Singer, L.M., 1988. Effect of water treatment sludge on growth and elemental composition of tomato (*Lycopersicon esculentum*) shoots. Commun. Soil Sci. Plant Anal. 19, 345–354.

Elliott, H.A., Taylor, M., 2000. Molybdenum content of water treatment residuals. J. Environ. Qual. 29, 1835–1839.

Elliott, H.A., Dempsey, B.A., Hamilton, D.W., DeWolfe, J.R., 1990. Land Application of Water Treatment Sludges: Impacts and Management. AWWA Res. Foundation and American Water Works Association, Denver, CO.

Elliott, H.A., O'Connor, G.A., Brinton, S., 2002a. Phosphorus leaching from biosolids-amended sandy soils. J. Environ. Qual. 31, 1362–1369.

Elliott, H.A., O'Connor, G.A., Lu, P., Brinton, S., 2002b. Influence of water treatment residuals on phosphorus solubility and leaching. J. Environ. Qual. 31, 681–689.

Gallimore, L.E., Basta, N.T., Storm, D.E., Payton, M.E., Huhnke, R.H., Smolen, M.D., 1999. Water treatment residual to reduce nutrients in surface runoff from agricultural land. J. Environ. Qual. 28, 1474–1478.

Harris-Pierce, R., Barbarick, K.A., Redente, E.F., 1993. The effect of sewage sludge application on native rangeland soils and vegetation. Annual Report Meadow Springs Ranch, Fort Collins, CO.

Haustein, G.K., Daniel, T.C., Miller, D.M., Moore, P.A. Jr., Mcnew, R.W., 2000. Aluminum-containing residuals influence high-phosphorus soils and runoff water quality. J. Environ. Qual. 29, 1954–1959.

He, Z.L., Alva, A.K., Li, Y.C., Calvert, D.V., Banks, D.J., 1999. Sorption-desorption and solution concentration of phosphorus in a fertilized sandy soil. J. Environ. Qual. 28, 1804–1810.

Heil, D.M., Barbarick, K.A., 1989. Water treatment sludge influence on the growth of sorghum-sudangrass. J. Environ. Qual. 18, 292–298.

Hyde, J.E., Morris, T.F., 2000. Phosphorus availability in soils amended with dewatered water treatment residual and metal concentrations with time in residual. J. Environ. Qual. 29, 1896–1904.

Ippolito, J.A., Barbarick, K.A., Rendente, E.F., 1999. Co-application of water treatment residuals and biosolids on two range grasses. J. Environ. Qual. 28, 1644–1650.

Ippolito, J.A., Barbarick, K.A., Heil, D.M., Chandler, J.P., Redente, E.F., 2003. Phosphorus retention mechanisms of a water treatment residual. J. Environ. Qual. 32, 1857–1864.

Jacobs, L.W., Teppen, B.J., 2000. WTR as a soil amendment to reduce nonpoint source pollution from phosphorus-enriched soils. p. 9. In: Proc. 14th Annual Residuals and Biosolids Management Conference, Feb. 27–29, 2000, MA. CD-ROM, Water Environment Federation, Alexandria, VA.

Jain, P., Yong-Chul, J., Thabet, T., Witwer, M., Townsend, T., 2005. Recycling of water treatment plant sludge via land application: assessment of risk. J. Res. Sci. Technol. 2, 13–23.

Kawczyinski, E., Achtermann, V., 1991. A water industry database report on residuals handling. pp. 6b-1 to 6b-5. In: Proc. of the AWWA/WEF Joint Residuals Conf. Durham, NC. 11–14 Aug. American Water Works Association, Denver, CO.

Lind, C. 2003. Water treatment residuals for soil P binding. WEF/AWWA/CWEA Joint Residuals and Biosolids Management Conference, "Partnering for a Safe, Sustainable Environmnet." Baltimore, MD. 19–22 Feb, 2003.

Livesey, N.T., Huang, P.M., 1981. Adsorption of arsenate by soils and its relationship to selected chemical properties and anions. Soil Sci. 131, 88–94.

Loebenstein, J.R. 1992. In: Mercury and arsenic wastes: Removal, recovery, treatment and disposal. USEPA Pollution Technology Review; Noyes Data Corporation, NJ, pp. 67–71.

Lucas, J.B., Dillaha, T.A., Reneau, R.B., Novak, J.T., Knocke, W.R., 1994. Alum sludge land application and its effect on plant growth. J. Amer. Wat. Works Assoc. 86, 75–83.

Makris, K.C. 2004. Long-term stability of sorbed phosphorus: Mechanisms and implications. Doctoral dissertation submitted to the University of Florida, Gainesville, FL.

Makris, K.C., El-Shall, H., Harris, W.G., O'Connor, G.A., Obreza, T.A., 2004a. Intraparticle phosphorus diffusion in a drinking-water treatment residual at room temperature. J. Colloid Interface Sci. 277, 417–423.

Makris, K.C., Harris, W.G., O'Connor, G.A., Obreza, T.A., 2004b. Three-dimensional phosphorus sorption by drinking-water treatment residuals: Implications for long-term stability. Environ. Sci. Technol. 38, 6590–6596.

Makris, K.C., Harris, W.G., 2006. Time dependency and irreversibility of water desorption by drinking-water treatment residuals: Implications for sorption mechanisms. J. Colloid Interface Sci. 294, 151–154.

Makris, K.C., Harris, W.G., O'Connor, G.A., Obreza, T.A., Elliott, H.A., 2005a. Physicochemical properties related to long-term phosphorus retention by drinking-water treatment residuals. Environ. Sci. Technol. 39, 4280–4289.

Makris, K.C., O'Connor, G.A., Harris, W.G., Obreza, T.A., 2005b. Relative efficacy of a drinking-water treatment residual and alum in reducing phosphorus release from poultry litter. Comm. Soil Sci. Plant Anal. 36, 2657–2675.

Makris, K.C., Sarkar, D., Datta, R., 2006. Evaluating a drinking-water waste by-product as a novel sorbent for As(V) and As(III). Chemosphere, 64, 730–741.

Maurer, M., Boller, M., 1999. Modeling of P precipitation in wastewater treatment plants with enhanced biological P removal. Water Sci. Tech. 39, 147–163.

McKeague, J.A., Brydon, J.E., Miles, N.M., 1971. Differentiation of forms of extractable iron and aluminum in soils. Soil Sci. Soc. Am. Proc. 35, 33–38.

McNeil, L.S., Edwards, M., 1995. Soluble arsenic removal at water treatment plants. J. Am. Wat. Works Assoc. 87, 105–113.

Meng, X., Korfiatis, G.P., Jing, C., Christodoulatos, C., 2001. Redox transformations of arsenic and iron in water treatment sludge during aging and TCLP extraction. Environ. Sci. Technol. 35, 3476–3481.

Moore, P.A. Jr., Miller, D.M., 1994. Decreasing phosphorus solubility in poultry litter with aluminum, calcium and iron amendments. J. Environ. Qual. 23, 325–330.

Moore, P.A. Jr., Daniel, T.C., Edwards, D.R., 2000. Reducing phosphorus runoff and inhibiting ammonia loss from poultry manure with aluminum sulfate. J. Environ. Qual. 29, 37–49.

Naylor, L.M., Carr, J.S., 1997. Exchangeable phosphorus in soils amended with water treatment residuals, biosolids, and aluminum-rich residues. In: Proc. Conf. Water Residuals Biosolids Management: Approaching the year 2000. Water Environ. Fed., Philadelphia, PA.

Nooney, M.G., Campbell, A., Murrell, T.S., Lin, X.-F., Hossner, L.R., Chusuei, C.C., Goodman, D.W., 1998. Nucleation and growth of phosphate on metal oxide thin films. Langmuir 14, 2750–2755.

Novak, J.M., Watts, D.W., 2004. Increasing the phosphorus sorption capacity of southeastern Coastal Plain soils using water treatment residuals. Soil Sci. 169, 206–214.

Novak, J.M., Watts, D.W., 2005a. Water treatment residuals aggregate size influences phosphorus sorption kinetics and P_{max} values. Soil Sci. 170, 425–432.

Novak, J.M., Watts, D.W., 2005b. An alum-based water treatment residual can reduce extractable phosphorus concentrations in three phosphorus-enriched coastal plain soils. J. Environ. Qual. 34, 1820–1827.

O'Connor, G.A., Elliott, H.A., Lu, P., 2001. Characterizing water treatment residuals phosphorus retention. Soil Crop Sci. Soc. Fl. Proc. 61, 67–73.

O'Reilly, S.E., Strawn, D.G., Sparks, D.L., 2001. Residence time effects on arsenate adsorption/desorption mechanisms on goethite. Soil Sci. Soc. Am. J. 65, 67–77.

Peters, J.M., Basta, N.T., 1996. Reduction of excessive bioavailable phosphorus in soils by using municipal and industrial wastes. J. Environ. Qual. 25, 1236–1241.

Pierce, M.L., Moore, C.B., 1982. Adsorption of arsenite and arsenate on amorphous iron hydroxide. Wat. Res. 16, 1247–1253.

Prakash, P., Sengupta, A.K., 2003. Selective coagulant recovery from water treatment plant residuals using Donnan membrane process. Environ. Sci. Technol. 37, 4468–4474.

Qafoku, N.P., Kukier, U., Sumner, M.E., Miller, W.P., Radkliffe, D.E., 1999. Arsenate displacement from fly ash in amended soils. Water Air Soil Pollut. 114, 185–198.

Ravikovitch, P.I., Bogan, B.W., Neimark, A.V., 2005. Nitrogen and carbon dioxide adsorption by soils. Environ. Sci. Technol. 39, 4990–4995.

Rengasamy, P., Oades, J.M., Hancock, T.W., 1980. Improvement of soil structure and plant growth by addition of alum sludge. Commun. Soil Sci. Plant Anal. 15, 533–545.

Rhoton, F.E., Bigham, J.M., 2005. Phosphate adsorption by ferrihydrite-amended soils. J. Environ. Qual. 34, 890–896.

Saito, A., Foley, H.C., 1991. Curvature and parametric sensitivity in models for adsorption in micropores. AlChE J 37, 429–436.

Scheidegger, A.M., Sparks, D.L., 1996. A critical assessment of sorption-desorption mechanisms at the soil mineral/water interface. Soil Sci 161, 813–831.

Scott, K.N., Green, J.F., Do, H.D., McLean, S.J., 1995. Arsenic removal by coagulation. J. Am. Wat. Works Assoc. 87, 114–126.

Silveira, M.L., Miyattah-Kporgbe, M., O'Connor, G.A., 2006. Phosphorus release from a manure-impacted Spodosol: Effects of a water treatment residual. J. Environ. Qual. 35, 529–535.

Townsend, T.G., Jang, Yong-Chul, Pradeep Jain, Thabet Tolaymat, 2001. Characterization of drinking water sludges for beneficial reuse and disposal. Report submitted to Florida Center for Solid and Hazardous Waste Management, Gainesville, FL, 2001.

U.S. Environmental Protection Agency, 1996. Management of water treatment residuals. EPA/625/R-95/008. Office of research and development, Washington, DC.

Van Riemsdijk, W.H., Lyklema, J., 1980. Reaction of phosphate with gibbsite beyond the adsorption maximum. J. Colloid Interface Sci. 76, 55–66.

Vandanapu, V., Sarkar, D., Datta, R., Makris, K.C., 2005. Arsenic sorption and desorption by drinking-water treatment residuals. EOS Trans. AGU 86 (18) Jt. Assem. Suppl. Abstract B33B-05.

Woolson, E., 1992. Mercury and arsenic wastes: Removal, recovery, treatment and disposal. USEPA Pollution Technology Review; Noyes Data Corporation, NJ, pp. 60–61.

Developments in Environmental Science, Volume 5
D. Sarkar, R. Datta and R. Hannigan (Editors)

ISSN: 1474-8177/DOI:10.1016/S1474-8177(07)05029-2

Chapter 29

Are soils the culprit? Linking natural and anthropogenic watershed processes to the degradation of the Chesapeake Bay

S.M. Lev and B. Brocks

Abstract

Natural weathering processes will tend to enrich the trace metal content of soils relative to the upper continental crust. Resultantly, the erosion of these soils will increase the loading of toxic trace metals to specific watersheds. The trace metal problems of the Chesapeake Bay have been well documented and it is clear that since the industrialization of the United States the flux of anthropogenic contaminants to the bay has steadily increased. However, recent work in the Patuxent River Basin suggests that much of the dissolved metal load is the result of desorption from suspended and bedload sediments and that current estimates of the trace element inputs from the sediment load are inadequate to evaluate their ultimate source. The questions that remain are, what portion of the toxic metal load being transported to the bay is the result of natural processes like chemical weathering and erosion and what portion of soil-derived metals are available to aquatic biota?

An investigation in the Patapsco River watershed, a subordinate drainage system within the Chesapeake Bay watershed, has determined that sediment being transported with in the Patapsco River watershed represents at least two sources. Soil-derived metals appear to dominate the suspended sediment flux during larger storm events for all analytes in this study. During smaller discharge storm events, there appears to be a Zn-enriched source that is a major component of the suspended sediments. This watershed signature is clearly different than the bulk trace metal signature in Baltimore Harbor and the Chesapeake Bay. Instead, the trace metal signature of Harbor sediments displays a pattern that requires an additional local trace metal source(s).

29.1. Introduction

Chesapeake Bay is one of the most studied bodies of water in the world. A primary focus of much of this research has been on the fate and transport of trace metals (e.g. copper, lead, zinc) deposited in the bay (Carpenter et al., 1975; Helz, 1976; Goldberg et al., 1978; Harris et al., 1980; Sinex and Helz, 1981; Nichols et al., 1982; Helz et al., 1985a,b; Kearney et al., 1985; Delfino and Otto, 1986; Marcus et al., 1993). A component of this trace metal problem, that to this point has not been adequately investigated, is the contribution of soil-derived trace metals to the bay. Natural weathering processes will tend to enrich the trace metal content of soils relative to the upper continental crust (Nesbitt and Young, 1982; Nesbitt and Young, 1984; Brimhall and Dietrich, 1987; Nesbitt and Young, 1989; Nesbitt et al., 1990; Maynard, 1992; McLennan, 1995; Nesbitt and Markovics, 1997; Gaillardet et al., 1999). Resultantly, the erosion of these soils will increase the loading of toxic trace metals to specific watersheds. With the increase in industrialization in the U.S. over the past 200 years came an increase in urbanization. The urbanization of large areas of the Chesapeake Bay watershed along with continued agricultural activities has dramatically affected land use and potentially enhanced erosion rates. The changing land use dynamics with in the Chesapeake Bay watershed have the potential to create non-equilibrium denudation conditions in the region. A non-equilibrium denudation region is one in which sediment yield is incongruous with the degree of weathering experienced by soils within the drainage area, which are the primary source of suspended sediment. Meaning, the rate of denudation and the weathering rate in a given region are not in equilibrium, as might be expected, and likely have been upset by anthropogenic influences (McLennan, 1993). These conditions will dramatically affect the flux of soil-derived trace metals being transported to the bay. Stated simply, the anthropogenic flux of trace metals to the bay may mirror and therefore be enhanced by an increase in the input of soil-derived trace metals as a result enhanced erosion brought about by the land use dynamics within the watershed.

The trace metal problems of the Chesapeake Bay have been well documented and it is clear that since the industrialization of the United States the flux of anthropogenic contaminants to the bay has steadily increased (Carpenter et al., 1975; Helz, 1976; Goldberg et al., 1978; Harris et al., 1980; Sinex and Helz, 1981; Nichols et al., 1982; Helz et al., 1985a,b; Kearney et al., 1985; Delfino and Otto, 1986, Marcus et al., 1993). However, recent work in the Patuxent River basin by Riedel et al. (2000) suggests that much of the dissolved metal load is the result of desorption from suspended and bedload sediments and that current estimates of the trace element inputs from the sediment load are inadequate to evaluate their ultimate source. The questions that remain are, what portion of the toxic metal load being transported to the bay is the result of natural processes like chemical weathering and erosion and what portion of soil-derived metals are available to aquatic biota?

The fate of the major elements (aluminum, calcium, sodium, potassium, iron, and magnesium) during weathering is fairly well understood (Nesbitt and Young, 1982; Nesbitt and Young, 1984; Nesbitt and Young, 1989; McLennan, 1995). In fact, the major element composition of soils is commonly used to determine changes in mineralogy and the degree of bedrock alteration associated with weathering and soil-forming processes (Nesbitt and Young, 1982; Nesbitt and Young, 1984; Brimhall and Dietrich, 1987; Nesbitt and Young, 1989; Nesbitt et al., 1990; Maynard, 1992; McLennan, 1995; Nesbitt and Markovics, 1997; Gaillardet et al., 1999). The fate of trace metals is much less understood. The same weathering processes that affect the major element composition of a soil profile and also affect the trace metal composition.

Trace metals can be enriched in soils over the parent material during chemical weathering. The extent of this enrichment can vary significantly based on the bedrock type and the weathering environment. Based on limited data, the trace metal compositions of bulk soils throughout Maryland rarely exceed regulatory limits (MDE, 2001). However, trace metals will tend to be concentrated in the clay size fraction of the soil (e.g., Lev, 2001). While the fine fraction is a minor component (typically $<10\%$), it is more easily transported and the associated metals are typically more readily bio-available (i.e., not bound in the silicate structure). This, coupled with enhanced rates of erosion in both urban and rural settings sets up the possibility that the there is significant metal loading to tributaries of the Chesapeake Bay as a result of natural processes.

This investigation focuses on the Maryland West Shore watershed, known locally as the Patapsco-Back River system, one of the nine subordinate watersheds that feed the bay. In addition to its local significance, the Patapsco-Back River watershed incorporates a wide range of land use and includes the primary drinking water reservoirs for Baltimore County and the City of Baltimore. To investigate the extent of soil-derived trace metal loading in the Patapsco-Back River Watershed, representative soils (bulk and size fractions), suspended sediment and harbor and bay sediments were collected from sampling locations across the rural to urban gradient within the Patapsco-Back River watershed capturing the maximum number of potential sources of trace metals to the bay.

29.2. Sampling locations

The Patapsco-Back River watershed contains eight smaller watersheds including Baltimore Harbor (Fig. 29.1). Samples from this study were collected within four of the eight. These four were selected to maximize the range in bedrock type and land use. In addition, these watersheds drain directly into the main branch of the Patapsco River.

Suspended sediment samples from three separate storm events collected during the Summer and Fall of 2003 were analyzed and compared to integrated soil samples from the same locations and bottom sediments from three locations with in Baltimore Harbor and the Chesapeake Bay (Figs. 29.1 and 29.2). Storm events were sampled during this investigation in order to assess the geochemical characteristics of the sediment flux through the system as a result of storm transport. The watersheds sampled are profiled below with supporting data in Table 29.1.

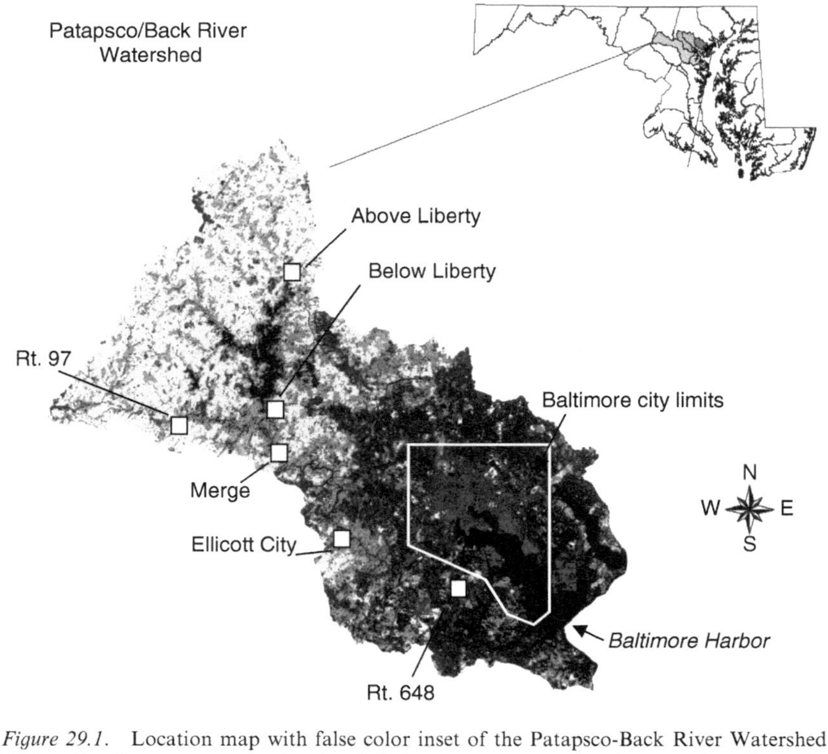

Figure 29.1. Location map with false color inset of the Patapsco-Back River Watershed. Light-colored areas are agricultural and dark-gray areas are sub-urban to urban. Sampling locations with in the watershed are indicated with a square.

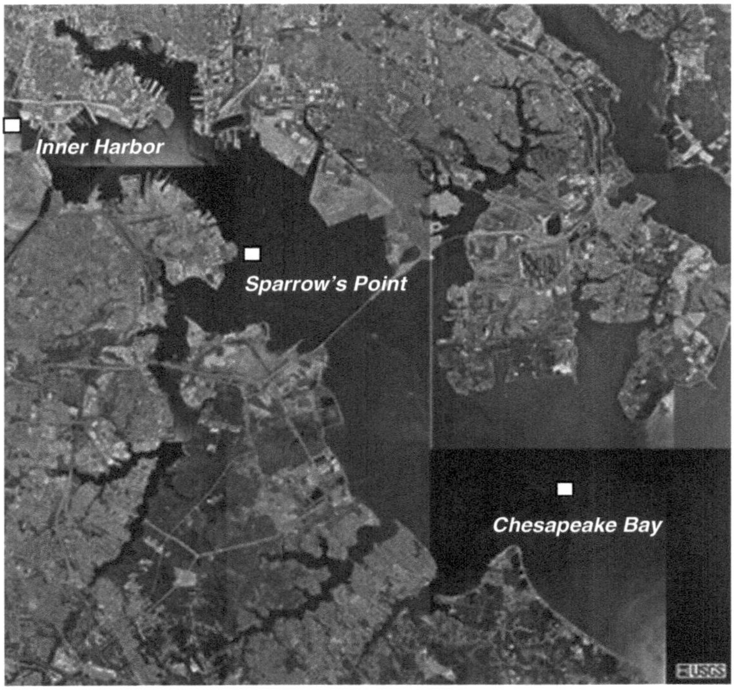

Figure 29.2. Satellite image of Baltimore Harbor and the Chesapeake Bay. Sample locations for this study are indicated with a white square.

Liberty Reservoir. The Liberty Reservoir watershed contains the North Branch of the Pataspco River that empties into and drains Liberty Reservoir, which is one of three major reservoirs that serve the greater Baltimore area. The North Brach was sampled above and below the reservoir (Fig. 29.1) in order to evaluate the impact the reservoir has on sediment transport in this part of the system. More than 80% of the land use in this watershed qualifies as agricultural or forested area with only 6.2% impervious surfaces. This is characteristic of the rural land use areas with in the Patapsco-Back River Watershed and is the rural end-member locale for this investigation.

Soils at the two sample sites have well developed A- and B-horizons and are forming on top of the Lower Pelitic Schist member of the Wissahickon Formation. The Wissahickon Formation is composed of intensively folded and cleaved muscovite-chlorite schist, chloritoid schist, and quartzite. The Lower Pelitic Schist member was formerly mapped

Table 29.1. Land use characteristics for the four subordinate watersheds of the Patapsco-Back River Watershed

Watershed	Characteristics	
Liberty Reservoir (101,404 acres)	Land use (%)	
	Urban	18.92
	Agricultural	48.02
	Forested	32.93
	Wetland	0.05
	Barren	0.08
	% impervous surface	6.3
	Population density (people/acre)	0.7
	Soil erodibility	0.28
	Historic wetland loss (acres)	3987
South Branch Patapsco (54,616 acres)	Land use (%)	
	Urban	18.26
	Agricultural	50.04
	Forested	31.52
	Wetland	0.00
	Barren	0.18
	% impervous surface	6
	Population density (people/acre)	0.79
	Soil erodibility	0.12
	Historic wetland loss (acres)	2745
Patapsco River Lower North Branch (75,513 acres)	Land use (%)	
	Urban	41.99
	Agricultural	12.67
	Forested	43.35
	Wetland	0.78
	Barren	1.21
	% impervous surface	21.9
	Population density (people/acre)	1.95
	Soil erodibility	0.31
	Historic wetland loss (acres)	8422
Baltimore Harbor (55,370 acres)	Land use (%)	
	Urban	75.79
	Agricultural	2.43
	Forested	20.08
	Wetland	1.00
	Barren	0.70
	% impervous surface	35.1
	Population density (people/acre)	3.55
	Soil erodibility	0.14
	Historic wetland loss (acres)	7681

as oligoclase facies of Wissahickon Formation and contains medium to coarse-grained biotite-oligoclase-muscovite-quartz schist with garnet, staurolite, and kyanite; fine- to medium-grained semipelitic schist; and fine-grained granular to weakly schistose psammitic granulite.

South Branch Patapsco. The South Branch watershed contains the second of the two main Patapsco tributaries to the west of Baltimore. The South Branch is roughly half the size of the Liberty Reservoir watershed with a similar land use distribution with a slightly smaller percent of impervious surfaces and a considerably lower soil erodibility factor of 0.12. There was one sampling station (Rt. 97) within this watershed located adjacent to secondary two lane state highway, route 97 (Fig. 29.1).

The soils at this location have the well developed A- and B-horizons. This location falls within the Boulder Gneiss which was formerly mapped as Sykesville and Laurel Formations. Characteristics of this formation include thick-bedded to massive, pebble- and boulder-bearing, arenaceous to pelitic metamorphic rock, typically a medium-grained, garnet-oligoclase-mica-quartz gneiss.

Patapsco River Lower North Branch. The Lower North Branch watershed referred to as the Merge area in this study is the second largest land area of the four watersheds in this study and is a transitional land use area best characterized as suburban. Roughly, 42% of the land use can be characterized as urban, slightly more, 43%, characterized as forested and almost 13% as agricultural. The percent impervious surface exhibits a dramatic increase in this area topping out at nearly 22%. There were two sampling locations within this watershed; the first, on the main branch of the Patapsco just below the merge between the North and South branches and the second, in a suburban area of Baltimore known as Ellicott City (Fig. 29.1).

The sample site below the confluence falls within the Lower Pelitic Schist member of the Wissahickon Formation. This member was formerly mapped as oligoclase facies and is described as medium- to coarse-grained biotite-oligoclase-muscovite-quartz schist with garnet, staurolite, and kyanite; fine- to medium-grained semipelitic schist; and fine-grained granular to weakly schistose psammitic granulite with psammitic beds. The Ellicott City location sits atop the Ellicott City Granodiorite. This unit ranges from biotite granodiorite along the body margin, to quartz monzonite in the core.

Baltimore Harbor. The urban areas directly adjacent to Baltimore's Inner Harbor are all part of the Baltimore Harbor watershed. This highly industrialized portion of the watershed (75.79% urban land use) is more than 35% impervious surface. The suspended sediment sampling location in this watershed was located just inside the Baltimore City limits on the main branch of the Patapsco River (Fig. 29.1). The two sampling locations within Baltimore Harbor also fall within the boundaries of this watershed. The sample site where suspended sediment and soils were collected sits atop the Potomac Group which is comprised of interbedded quartzose gravels; protoquartzitic to orthoquartzitic argillaceous sands; and white, dark gray and multicolored silts and clays, with a thickness of 0–800 feet, within this grouping are three formations, Raritan and Patapsco Formations, Arundel Clay, and the Patuxent Formation.

29.3. Field methods

29.3.1. Suspended sediment and soil sampling

Suspended sediment samples were collected with a depth-integrated sampler with Teflon spout. Approximately two liters of river water were collected during each sampling event. At the time of sampling, velocity measurements were recorded for each site following United States Geological Survey (USGS) methods for estimating discharge (Rantz, 1982). To collect a representative suspended sediment sample, at each site, the sampler was raised and lowered incrementally along a straight-line path across the width of the river. Each site's velocity measurements (Table 29.2) were integrated with their width and depth measurements to give an accurate reading of the discharge rates for each storm event sampled.

At each suspended sediment-sampling site, an area on the floodplain bank was cleared of grass and vegetation to expose 1 m of fresh soil. Approximately 1 kg of soil was collected from each 1-m profile and homogenized prior to analysis.

29.3.2. Bay sediment sampling

Three locations from within the Chesapeake Bay and the mouth of the Patapsco River in Baltimore's Inner Harbor were selected for grab sampling. These sites were selected due to their proximity to the Patapsco River delta (Fig. 29.2). To collect a representative sample at each site, a grab sampler was lowered three times at each sample location and approximately 1 kg of sediment was recovered. All bulk particulate organic matter (e.g. wood and leaf litter) collected from each site during the grab sampling processes was removed after the sample was dried for 72 h and was not included in the analysis.

29.4. Analytical methods

29.4.1. Grain size analysis

Soils that were collected were split for grain size and chemical analysis. Samples for chemical analysis were dried for 72 h, and then powdered to a uniform grain size using an agate mortar and pestle. Moist soil samples

Table 29.2. Storm water data for all events sampled. Sediment yield is presented as an instantaneous values rather than an average. The discharge and suspended sediment were collected at one time during the storm event therefore making the yield calculation valid for the time of sampling only

Sampling location	Date	Instantaneous discharge ($ft.^3 sec^{-1}$)	Suspended sediment ($mg l^{-1}$)	Instantaneous sediment yield ($t km^{-2}$)
Above Liberty Reservior	6.18.03	1977	24.5	0.29
	6.19.03	247	204.9	0.31
	8.7.03	72	5.4	0.002
	9.20.03	381	145.4	0.34
Below Liberty Reservior	6.18.03	525	5.6	0.02
	6.19.03	316	128.3	0.25
	8.7.03	84	12.5	0.01
	9.20.03	1901	2.4	0.03
Rt. 97	6.18.03	174	3.0	0.01
	6.19.03	469	272.4	1.48
	8.7.03	154	3.8	0.01
	9.20.03	1727	5.1	0.10
Merge	6.19.03	2524	1025.6	10.32
	8.7.03	600	2.5	0.01
Ellicott City	6.19.03	3124	963.0	25.42
	8.7.03	1495	146.7	1.85
	9.20.03	1242	385.1	4.04
Rt. 648	6.19.03	4798	633.7	25.69
	8.7.03	10514	25.1	2.23
	9.20.03	1788	130.0	1.96

were sealed within airtight zip bags, labeled, and stored at room temperature for grain size analysis. The moist soil was again split and half was passed through a 63-micron nylon sieve (silt-sized particles) and the remaining half through a 5-micron nylon sieve (clay-sized particles). Deionized water was used in the wet sieving for both the 63- and the 5-micron size fractions. Approximately 5 g of soil was placed on the sieve, and with the aid of deionized water, the soil was agitated within the sieve. What passed the sieve was collected, weighed, dried, and analyzed for trace metals using the methods described below.

29.4.2. Chemical analysis

Bulk samples, size fractions, and acid leachates were analyzed for trace metal concentrations using a Thermo Elemental (VG) PQExcell

Inductively Coupled Plasma-Mass Spectrometer in the Materials Research Laboratory at Towson University. 50 mg sample powder, crushed in an agate mortar and pestle, was weighed into a 30 ml acid cleaned screw top Savilex® Teflon® vial. For bulk metal concentrations, a 3:1 mix of ultrapure 7N HNO_3 and HF was added to each vial, the vial was sealed and placed on a hotplate at 170°C in a metal-free HEPA filtered clean hood overnight. After complete digestion was achieved samples were left uncapped on the hotplate and taken to dryness. Once completely dry, 3 ml of 2% HNO_3 spiked with a 1 ng g^{-1} internal standards (indium and bismuth) was added to the sample and the vial was capped and returned to the hotplate. The sample remained on the hotplate until the remaining solution was clear. The dissolved sample was then transferred to a pre-weighed acid-cleaned 250 ml bottle. The sample was then diluted using the 2% HNO_3 1 ng g^{-1} internal standard solution. Sets of 10 samples were run with a procedural blank, a sample replicate and duplicate as well as at least 1 digestion of National Institute of Standards and Technology (NIST) Soil standard reference material (SRM) 2709. The precision of individual concentrations for all elements is better than 2% based on replicates and NIST standards in addition, recoveries for NIST SRM 2709 for all analytes were consistently between 85 and 105% and the detection limit for all elements measured was below 5 ppb.

Selected samples were split prior to digestion and 50 mg of the powdered sample was used to determine the acid leachable fraction of individual metals using 5 ml of 7N HNO_3. The samples were capped and let stand for 1 h. The leachate was then decanted, taken to dryness on a hotplate in a Teflon vial and prepared for trace element analysis by ICP-MS using the methods described above.

29.5. Results

Suspended sediments. Results are presented in Tables 29.2–29.8. In Table 29.2, the stream and suspended sediment data are presented by site location for each storm event sampled. The discharge and sediment yield calculations are instantaneous values rather than integrated averages since the stream data was collected near peak flow conditions for each storm event. As a result, these should be considered maximum values. Suspended sediment values for all sites and storm events range from a low of 2.4 mg l^{-1} on September 20, 2003, at the BL site to a high of 1025.6 mg l^{-1} on June 19, 2003, at the Merge sampling location (Table 29.2).

Sediment yield. Instantaneous sediment yield (t km^{-2}) values generally increase from the rural portion of the watershed with a low of 0.002 t km^{-2} on August 7, 2003, at the above liberty site to a high of 25.69 t km^{-2} at the urban Rt. 648 location on June 19, 2003 (Table 29.2). While these instantaneous sediment yield values are maximum values some of the values are comparable to the average annual sediment yield of much larger river systems. This observation is considered in more detail within the discussion section of this paper.

Trace metal load. Instantaneous metal load (mg of element/storm sample) calculations are made for all metals in all storms using metal concentration suspended sediment data and are presented in Table 29.3. Zn and Mn consistently dominate the total measured metal load for all storms sampled.

Table 29.3. Instantaneous metal load calculations are reported in mg/sampling event. Land use averages for all storm events are also reported

Sampling location	Date	Suspended sediments					
		Cr	Mn	Ni	Cu	Zn	Pb
Above Liberty Reservior	6.18.03	112.0	2518.3	85.2	105.0	1373.6	12.6
	6.19.03	98.9	1227.0	43.7	32.2	327.9	1.4
	8.7.03	163.8	4757.8	93.4	89.3	1598.0	nd
	9.20.03	4.1	140.6	2.2	2.9	38.2	nd
Below Liberty Reservior	6.18.03	224.1	2688.3	148.0	102.6	1262.7	70.5
	6.19.03	15.7	59.4	6.6	2.7	35.3	2.8
	8.7.03	15.7	2243.7	117.4	64.4	1038.1	33.4
	9.20.03	127.2	3059.2	79.2	85.1	1612.0	nd
Rt. 97	6.18.03	332.6	5423.3	223.8	197.8	2188.9	115.6
	6.19.03	99.6	2014.0	51.8	49.1	228.6	1.5
	8.7.03	215.0	2243.7	117.4	64.4	1038.1	32.0
	9.20.03	193.1	3562.0	108.5	116.1	1322.0	nd
Merge	6.19.03	98.1	1697.0	48.2	49.2	166.6	43.3
	8.7.03	175.7	3584.2	97.7	74.3	2319.1	nd
Ellicott City	6.19.03	98.0	1566.0	47.4	47.2	160.7	43.3
	8.7.03	162.0	2713.0	68.5	55.7	371.7	6.8
	9.20.03	138.3	1941.0	61.0	47.7	242.1	5.9
Rt. 648	6.19.03	136.0	1794.0	66.1	59.7	255.5	1.5
	8.7.03	113.5	2005.2	50.7	59.9	549.3	64.7
	9.20.03	202.3	2689.0	87.1	80.6	446.7	64.2

Table 29.4. Chemical data for suspended sediments collected during storm events. All concentrations are reported in ug g^{-1} and RSDs are generally better than 2%

Sampling location	Date	Instantaneous metal load (mg/sampling event)					
		Cr	Mn	Ni	Cu	Zn	Pb
Above Liberty Reservior	6.18.03	2.7	61.6	2.1	2.6	33.6	0.3
	6.19.03	20.3	251.5	9.0	6.6	67.2	0.3
	8.7.03	0.9	25.7	0.5	0.5	8.6	nd
	9.20.03	0.6	20.4	0.3	0.4	5.6	nd
Below Liberty Reservior	6.18.03	1.3	15.1	0.8	0.6	7.1	0.4
	6.19.03	2.0	7.6	0.8	0.3	4.5	0.4
	8.7.03	0.2	28.0	1.5	0.8	12.9	0.4
	9.20.03	0.3	7.3	0.2	0.2	3.9	nd
Rural average		*3.5*	*52.2*	*1.9*	*1.5*	*17.9*	0.4
Rt. 97	6.18.03	1.0	16.3	0.7	0.6	6.6	0.3
	6.19.03	27.1	548.6	14.1	13.4	62.3	0.4
	8.7.03	0.8	8.5	0.4	0.2	3.9	0.1
	9.20.03	1.0	18.2	0.6	0.6	6.7	nd
Merge	6.19.03	100.6	1740.4	49.4	50.4	170.9	44.4
	8.7.03	0.4	9.0	0.2	0.2	5.8	nd
Transition average		*21.8*	*390.2*	*10.9*	*10.9*	*42.7*	11.3
Ellicott City	6.19.03	94.3	1508.0	45.7	45.5	154.7	41.6
	8.7.03	23.8	398.0	10.0	8.2	54.5	1.0
	9.20.03	53.3	747.5	23.5	18.4	93.2	2.3
Rt. 648	6.19.03	86.2	1136.9	41.9	37.8	161.9	1.0
	8.7.03	2.8	50.2	1.3	1.5	13.8	1.6
	9.20.03	26.3	349.6	11.3	10.5	58.1	8.3
Urban average		*47.8*	*698.4*	*22.3*	*20.3*	*89.4*	*9.3*

29.6. Trace metal concentrations

Cr, Mn, Ni, Cu, Zn, and Pb data for suspended sediments, soils (bulk and size fractions), bay sediments and 7N HNO_3 leachates on Bay samples and soil samples are presented in Tables 29.4–29.8. The results are presented below by sample type.

Suspended sediments. With the exception of two samples, Mn is present at levels between 1200 and 5400 ug g^{-1} for all locations and all storm events. Despite this range there is no overall pattern to this data. Chromium is generally high relative to the upper continental crust (i.e., 35 ug g^{-1}, Taylor and McLennan, 1985). Values are generally greater than 100 ug g^{-1} with a maximum value of 332.6 ug g^{-1} at the Rt. 97 location on June 18, 2003. Nickel and Copper are values are enriched relative to the upper continental crust. Nickel values range from a low

Table 29.5. Chemical data for soils and size fractions collected from the adjacent flood plain at each of the sample locations. All concentrations are reported in ug g^{-1} and RSDs are generally better than 2%

Sampling location		*Soils*					
		Cr	Mn	Ni	Cu	Zn	Pb
Above Liberty Reservior	Bulk	51.4	679.1	22.9	35.0	74.3	11.7
	<63 μm	87.0	1479.0	51.5	75.3	198.8	62.0
	<5 μm	84.6	1405.0	49.3	75.8	201.9	62.6
Below Liberty Reservior	Bulk	123.9	815.1	54.0	23.9	71.2	12.9
	<63 μm	148.7	1381.0	106.3	54.2	141.6	56.1
	<5 μm	185.9	1954.0	141.1	72.2	184.9	41.5
Rt. 97	Bulk	62.6	852.8	29.3	25.2	69.3	16.2
	<63 μm	101.5	1704.0	51.3	48.1	138.3	65.8
	<5 μm	110.1	1929.0	59.6	59.6	171.1	77.2
Merge	Bulk	85.9	785.0	30.8	25.0	77.6	22.2
	<63 μm	94.8	1267.0	48.0	45.0	129.0	43.5
	<5 μm	105.5	1509.0	60.4	54.1	173.6	59.5
Ellicott City	Bulk	121.5	676.8	41.0	22.1	74.4	22.3
	<63 μm	152.9	1302.0	70.5	58.3	164.9	75.2
	<5 μm	158.1	1416.0	87.5	69.0	211.7	93.5
Rt. 648	Bulk	92.3	412.7	41.6	27.3	96.1	64.7
	<63 μm	144.9	513.0	81.0	71.5	216.1	192.7
	<5 μm	115.2	453.2	57.5	46.8	166.5	82.4

of 2.2 ug g^{-1} at the Above Liberty Reservior site on September 20, 2003, to a high of 223.8 ug g^{-1} at the Rt. 97 site on June 18, 2003. Copper values range from a low of 2.7 ug g^{-1} at the Below Liberty Reservior site on June 19, 2003, to a high of 197.8 ug g^{-1} at the Rt. 97 site on June 18, 2003. Lead values are more variable for all samples collected. Zinc values for suspended sediment samples are considerably enriched relative to the upper continental crust for all samples except the June 19, 2003, and September 20, 2003, samples at the Below Liberty and Above Liberty sites, respectively.

Bulk soils and size fractions. Mn values for bulk soil samples are very similar to upper continental crust values. The grain size fractions are enriched to similar levels (i.e., 2–3 times the bulk soil value) relative to the bulk soils. Chromium in the bulk soils is enriched by up to 3.5 times values for the upper continental crust and up to 5.3 times in the grain size separates. Nickel in the bulk soils is enriched relative to crustal values at most of the locations with the lowest concentration of 22.9 ug g^{-1} at the AL site and the highest value of 54 ug g^{-1} at the Below Liberty site. The grain size fractions are enriched relative to the upper continental crust and the bulk soil at all locations. Copper in bulk soils, unlike Nickel, is

Table 29.6. Acid extractable metals data for soils and size fractions collected from the adjacent flood plain at each of the sample locations. All concentrations are reported in $ug\,g^{-1}$ and RSDs are generally better than 2%

Sampling location		7N HNO_3 extractable from soils					
		Cr	Mn	Ni	Cu	Zn	Pb
Above Liberty Reservior	Bulk	5.2	123.0	4.8	7.2	19.2	4.4
	<63 μm	19.2	473.7	15.8	23.0	62.7	15.8
	<5 μm	18.6	520.2	15.0	26.7	75.0	19.8
Below Liberty Reservior	Bulk	16.8	111.0	9.8	4.4	14.0	3.0
	<63 μm	52.5	541.5	39.3	21.6	56.3	19.0
	<5 μm	43.5	527.9	35.7	18.6	49.3	18.2
Rt. 97	Bulk	11.8	255.5	8.0	7.0	23.2	5.0
	<63 μm	22.6	548.3	15.4	15.0	44.7	16.2
	<5 μm	14.8	474.9	12.6	13.4	40.9	16.0
Merge	Bulk	12.8	167.1	7.4	6.8	22.4	5.0
	<63 μm	13.0	258.7	9.8	9.4	28.5	8.4
	<5 μm	19.2	523.6	13.6	16.8	58.3	15.0
Ellicott City	Bulk	20.0	190.8	11.8	10.2	25.9	8.6
	<63 μm	14.6	181.8	8.8	8.0	26.7	11.0
	<5 μm	37.7	394.6	20.0	16.2	56.3	23.0
Rt. 648	Bulk	14.6	78.8	8.8	6.0	22.2	10.0
	<63 μm	32.3	171.5	21.0	16.4	57.3	46.1
	<5 μm	46.3	219.0	27.3	23.4	89.8	39.3

only significantly enriched above crustal values at the Above Liberty site. The copper concentrations in the grain size fractions exhibit a behavior similar to the other metals described and are generally enriched relative to bulk soils and the upper continental crust by as much as three times. Zinc and Pb are not generally enriched in the bulk soils with the exception of the Rt. 648 site which has Zn and Pb concentrations of 96.1 $ug\,g^{-1}$ and 64.7 $ug\,g^{-1}$, respectively. The grain size fractions are generally 2–3 times greater than the bulk soils and upper continental crust concentrations at all sites.

All soil samples and grain size fractions were treated with 7N HNO_3 and the leachates were analyzed for the analytes of interest. This extraction is meant to represent a maximum value for the potentially bioavailable fraction. All samples of bulk soil and associated size fractions yield metals at the $ug\,g^{-1}$ level when treated using the acid extraction method described above. The clay-sized fraction from the Rt. 648 soil sample yields the highest fraction of extractable metals at greater than 40% for all analytes with a high of 54% extractable Zn.

Bay sediments. Surface sediment samples collected from three locations in Baltimore's Inner Harbor and the Chesapeake Bay were analyzed for

Table 29.7. Acid extractable fractions available from soils and size fractions. All concentrations were normalized to individual values for total digestion

Sampling location		Acid extractable fraction from soils					
		Cr	Mn	Ni	Cu	Zn	Pb
Above Liberty Reservior	Bulk	0.10	0.18	0.21	0.21	0.26	0.37
	<63 μm	0.22	0.32	0.31	0.30	0.32	0.25
	<5 μm	0.22	0.37	0.30	0.35	0.37	0.32
Below Liberty Reservior	Bulk	0.14	0.14	0.18	0.18	0.20	0.23
	<63 μm	0.35	0.39	0.37	0.40	0.40	0.34
	<5 μm	0.23	0.27	0.25	0.26	0.27	0.44
Rt. 97	Bulk	0.19	0.30	0.27	0.28	0.33	0.31
	<63 μm	0.22	0.32	0.30	0.31	0.32	0.25
	<5 μm	0.13	0.25	0.21	0.22	0.24	0.21
Merge	Bulk	0.15	0.21	0.24	0.27	0.29	0.23
	<63 μm	0.14	0.20	0.20	0.21	0.22	0.19
	<5 μm	0.18	0.35	0.22	0.31	0.34	0.25
Ellicott City	Bulk	0.16	0.28	0.29	0.46	0.35	0.39
	<63 μm	0.10	0.14	0.12	0.14	0.16	0.15
	<5 μm	0.24	0.28	0.23	0.23	0.27	0.25
Rt. 648	Bulk	0.16	0.19	0.21	0.22	0.23	0.15
	<63 μm	0.22	0.33	0.26	0.23	0.27	0.24
	<5 μm	0.40	0.48	0.48	0.50	0.54	0.48

Table 29.8. Chemical data for Baltimore Harbor and Chesapeake Bay sediments. All concentrations are reported in $ug\,g^{-1}$ and RSDs are generally better than 2%. Acid extractable concentrations were normalized to individual values for total digestion to determine the acid extractable fraction

Sampling location		Harbor and Bay Sediments					
		Cr	Mn	Ni	Cu	Zn	Pb
Inner Harbor	Bulk	172.3	535.4	56.1	106.0	274.6	89.7
	7N HNO_3 extractable	37.0	402.7	14.2	16.6	109.9	19.8
	Acid extractable fraction	0.215	0.752	0.254	0.156	0.400	0.220
Sparrow's Point	Bulk	402.9	879.2	54.7	152.1	665.3	154.0
	7N HNO_3 extractable	33.2	112.6	14.3	28.2	76.2	20.8
	Acid extractable fraction	0.082	0.128	0.261	0.185	0.114	0.135
Chesapeake Bay	Bulk	151.1	1378.0	48.9	56.3	366.0	72.1
	7N HNO_3 extractable	25.7	60.4	3.8	10.6	50.3	11.3
	Acid extractable fraction	0.170	0.044	0.077	0.189	0.138	0.157

their bulk and acid extractable trace metal concentrations. Bulk trace metal concentrations from all Bay sediment samples are generally higher than bulk soil concentrations. Zinc in particular, ranges from 274 to 665 ugg^{-1} and falls between the ranges for bulk soils and suspended

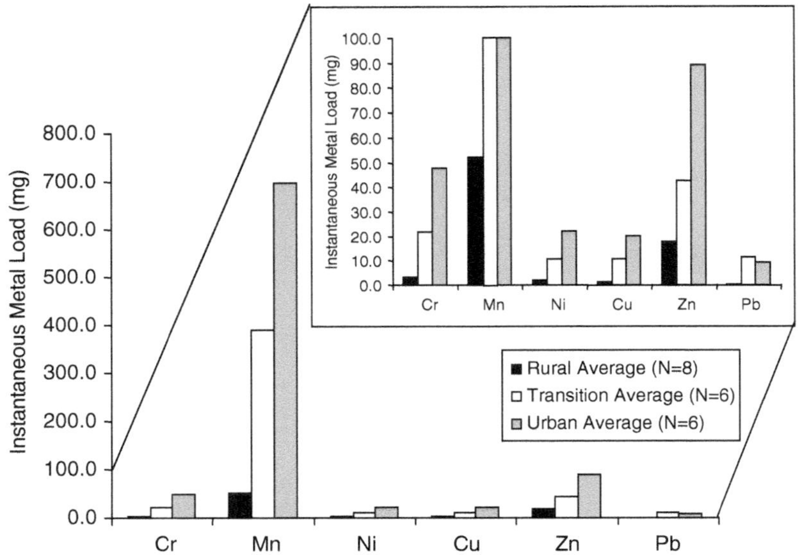

Figure 29.3. Instantaneous metal load values were grouped according to land use and the average values for all elements are plotted by land use. The inset highlights the differences for all elements except Mn. Zn and Mn are the most abundant trace metals for all land uses. Urban sampling sites are making a larger contribution to the total metal load than any of the other land use designations.

sediments. Bulk bay sediments are more comparable to the clay-sized fraction of soil samples although still relatively enriched. With the exception of 75% extractable Mn and 40% extractable Zn at the Inner Harbor sampling location all other extractable metal concentrations from bay sediments are uniformly below 25%.

29.7. Discussion

There are three distinct trends in the data described above. The chemical data suggest a multi-source system comprised of both natural and anthropogenic trace metal sources to both the Patapsco River and the Inner Harbor. The stream data and trace metal load calculations portray a storm water dominated system characterized by disproportionately high-sediment yield through out the Patapsco River watershed, a trace metal flux that is directly related to discharge and a increasing trace metal load during storm events from rural to urban land use zones within the watershed. These relationships are evident in Figs. 29.3, 29.4 and 29.5. In Fig. 29.3, the instantaneous metal load is clearly related to the land use characteristics in the local catchment (see Table 29.1). This suggests, especially for Zn and Mn, that the flux of metal to the Chesapeake Bay is dominated by input from the urbanized portion of the watershed.

In Fig. 29.4, trace metal concentrations are plotted versus suspended sediment concentration for all storm events sampled at all locations. Using Zn as an example, as the amount of sediment being transported during a storm event increases, the Zn concentration decreases. As Zn values decrease, they trend toward values similar to the upper continental crust and Patapsco watershed soils. Therefore, storm events generating larger sediment yields will tend to exhibit a soil-derived trace metal signature while the smaller events trend toward higher, non-soil derived, Zn values indicating the importance of a secondary Zn-enriched non-point source. Despite higher overall concentrations of Zn in the smaller storm events (low yield–low discharge), the low concentration (high yield–high discharge) events are delivering the largest total load of Zn to the bay (Table 29.3 and Fig. 29.5a). In addition, it appears that with increasing storm intensity the trace metal character of the suspended sediment load trends toward the range of concentrations typical for bulk soils within the Patapsco watershed (Fig. 29.5b).

There are two likely scenarios that can account for the elevated Zn concentrations in the suspended sediment of the smaller storm events. The first is that the fine fraction of the soil being eroded and transported through the Patapsco river system is enriched in Zn due to chemical weathering processes. Smaller storm events with lower velocity discharge will tend to carry a smaller average grain size. The fine fraction of all the soils sampled are enriched relative to the bulk soil suggesting that they are a viable source; however, the suspended sediment Zn concentrations are greatly in excess of even the most elevated values in the fine soil fraction (Fig. 29.4). The second scenario is that there is a ubiquitous metal-enriched anthropogenic source of Zn to the Patapsco river system that dominates smaller low-sediment yield events but is diluted by soil-derived metals during larger events when there is more sediment being transported through the system. A plausible non-point source of Zn within the watershed is roadway dust.

In an developed setting, such as the urban portion of the Patapsco Watershed, roadway-derived contaminants have recently been identified as a significant component of urban dust (Kupiainen and Tervahattu, 2005; Omstedt et al., 2005; Samara and Voutsa, 2005). In several Scandinavian and European cities, urban dust in the 2.5–10 μm size range (PM_{10}) is primarily composed of non-tailpipe, roadway-derived particles.

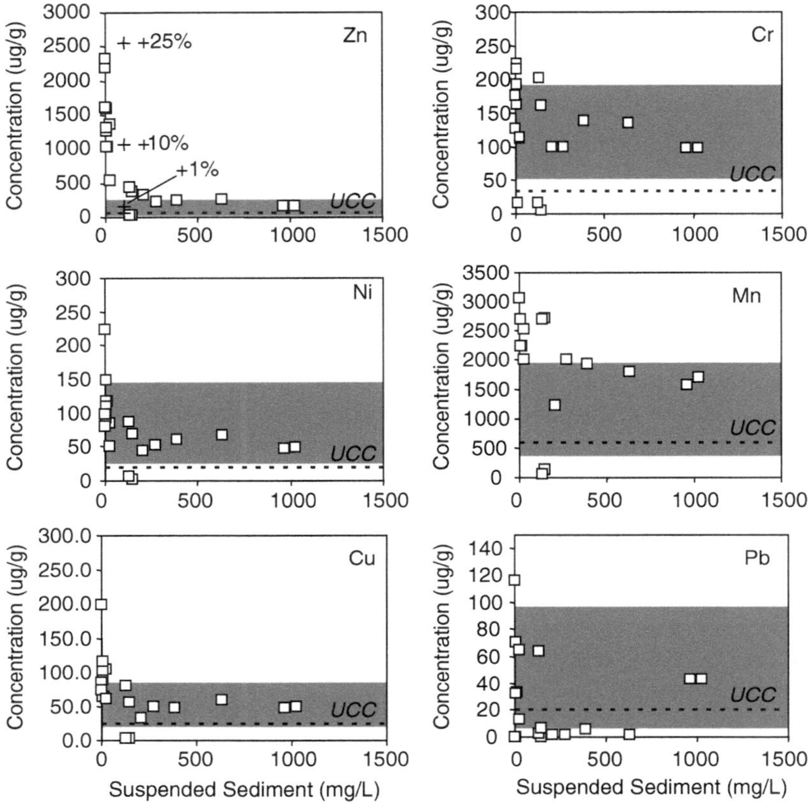

Figure 29.4. Plots of trace metal concentrations in suspended sediments versus suspended sediment concentrations. All metal values are in $\mu g\, g^{-1}$, suspended sediment values are in $mg\, l^{-1}$. The dashed line in each plot represents the average value in the upper continental crust (UCC) as reported in Taylor and McLennan (1985). The gray field in each plot represents the range in concentrations found in the fine soil fraction. The + marks in the Zn plot represent a mixing model between the UCC and tire-derived Zn. The percentages are the total Zn contribution from tires.

A large portion of the roadway particulate load is thought to be geological material (Kupiainen and Tervahattu, 2005). Recent sampling of roadway dust from within the Chesapeake Bay watershed confirms this assertion and documents the relatively low-trace metal load associated with this fraction of the dust (Camponelli et al., 2005). This work also documents the presence of a trace metal enriched (specifically Zn and Cu) component in the roadway dust. One of the major components of roadway dust is thought to be fine rubber fragments derived from tire wear

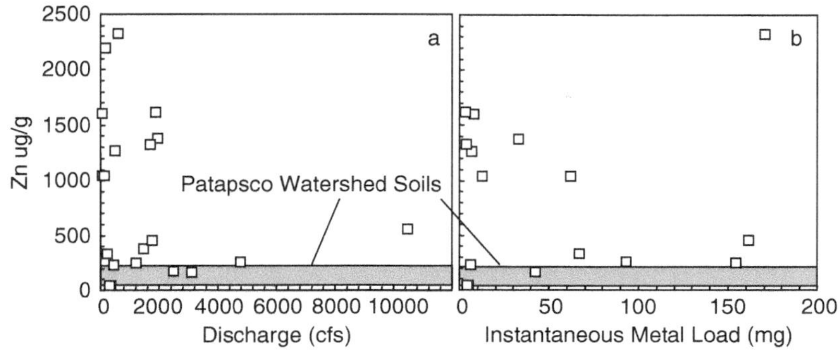

Figure 29.5. (a) Zinc concentrations for suspended sediments are plotted versus discharge. The gray field represents the range of Zn measured in bulk soils analyzed during this investigation. As discharge increases, the Zn concentration of the suspended sediment flux decreases toward soil Zn range indicating the importance of soil-derived Zn in during larger storm events. (b) Zn concentration for suspended sediments is plotted versus instantaneous metal load values calculated for each storm event. There is an inverse trend between Zn concentration of the suspended sediment and the metal load for each event. As in (a) larger storm events have lower overall Zn concentrations in the particulate load; however, the total Zn load is greater. These diagrams suggest that soil-derived Zn is the primary component in the metal load to the Chesapeake Bay.

(Councell et al., 2004). The presence of tire particles in roadway dust from with in the Chesapeake Bay Watershed has been confirmed by Camponelli et al. (2005). These particles tend to be enriched in metals such as Pb, Cd, Cr, Ni, V, Cu, Mn, and Zn and may represent a significant non-point source of metals to roadside environments (Samara and Vousta, 2005; Kupiainen and Tervahattu, 2005 and references therein).

In the case of Zn, tire particles have been documented to contain up to wt.% Zn and are widely distributed on roadway surfaces (Councell et al., 2004). In Fig. 29.4, we present a model for the addition tire-derived Zn to naturally derived suspended sediments. Starting at an upper crustal composition similar to Patapsco soils, the proportion of tire-derived Zn in the suspended sediments is incrementally increased (1%, 10%, and 25%). Based on this simple two-component mixing model, there may be as much as 10% tire-derived Zn in the suspended sediments during the low-sediment yield discharge events (Fig. 29.4). This model is complicated by the fact that there may be other minor roadway-derived sources of Zn in addition to tires and therefore represents a maximum value for tire-derived Zn.

All other metals analyzed exhibit a pattern similar to that of Zn when plotted versus suspended sediment concentrations indicative of storm water controlled grain size effect (Fig. 29.4). However, with the exception of Cu and Mn, the fine fraction of the soils generally overlap the highest metal concentrations in the suspended sediments suggesting that these metals are primarily soil derived within the watershed and by extrapolation, the harbor and the bay. This does not hold up when examining Baltimore Harbor and Chesapeake Bay sediment samples from this study and previous work (Sinex and Helz, 1981; McGee et al., 1999). In Fig. 29.6, harbor, bay and suspended sediment samples from this study and work by Sinex and Helz (1981) and McGee et al. (1999) are plotted in Zn versus Cr concentration space. Also plotted is the upper continental

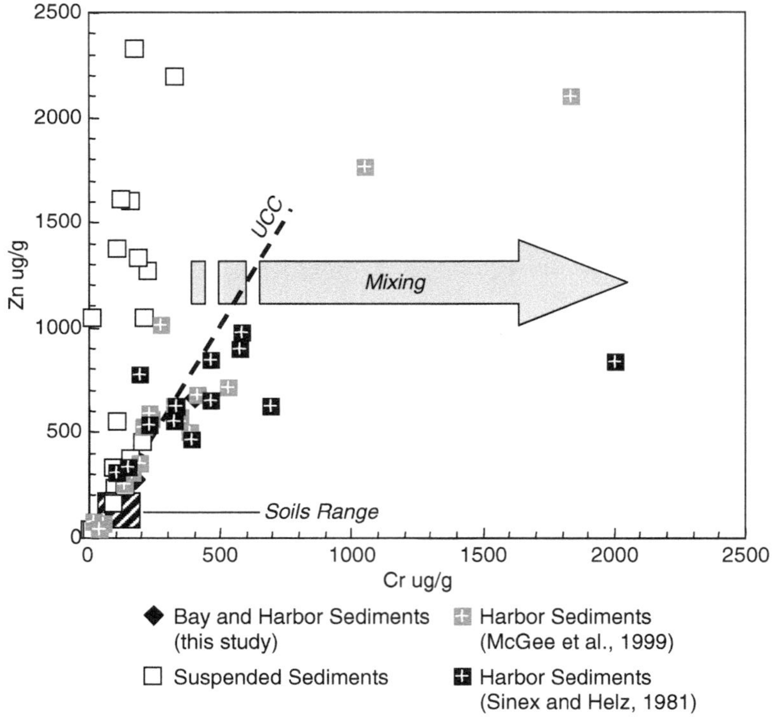

Figure 29.6. Zinc versus Cr concentrations are plotted for suspended sediment from this study (white square), harbor and bay sediments from this study (black diamond) and harbor sediments from two previous investigations (black and gray with white plus). Also plotted for reference in an UCC line that represents the ratio of Zn/Cr in the crust. If soils are the primary source of these metals the point would lie on or near the line. The gray arrow represents a general mixing pathway between the suspended sediments entering Baltimore Harbor from the Patapsco and local point sources adjacent to the Harbor. With no local source addition Harbor sediments should look very similar to the suspended sediment input from the Patapsco River.

crust (UCC) line used previously. If the metals in the Harbor and Bay sediments are soil derived we would expect these samples to lie along the upper continental crust line and within the soils range. If either metal is enriched via an anthropogenic source, the sample(s) would lie off the line in the direction of the enriched source. Based on the chemical characteristics of suspended sediments within the watershed, harbor and bay sediments should be enriched in Zn and essentially crustal with respect to Cr concentrations. This is clearly not the case in for the harbor and bay sediments plotted in Fig. 29.6. The Zn and Cr ratio for the bay sediments from this study and those from Sinex and Helz (1981) and McGee et al. (1999) appear to have a crustal ratio but plot well out of the soils range. This does not appear to be in agreement with the predictions made when considering the chemical composition of the suspended sediments entering the bay. The expectation would be for enriched Zn, but Cr this is comparable to the soils range defined in this study.

The discrepancy between the trace element characteristics of Patapsco River suspended sediments entering Baltimore Harbor and bottom sediments within the Harbor can be explained with a simple two-component mixing model. If the concentration of suspended sediments within the Patapsco River is set as one end member, the Harbor sediments likely represent an intermediate composition between the suspended sediments and a local point source within the Harbor. It is also possible that there is some contribution to Harbor sediments by transported or dredged sediments; however, this does not change the basic premise that there are at least two significant trace metal sources to Baltimore Harbor, suspended sediment and at least one local source. A generalized mixing pathway is plotted in Fig. 29.6. The process of trace metal addition in the Harbor is most clearly demonstrated in Fig. 29.6 with the addition of a Cr-enriched source to watershed-derived Zn-enriched sediments. The suspended sediment concentrations can therefore be considered a lower limit for recently deposited Baltimore Harbor sediments. If average Harbor sediments (McGee et al., 1999) are normalized to average Patapsco suspended sediments (this study) a signature for the non-watershed trace metal source(s) can be discerned. In Fig. 29.7, Cr, Cu, and Pb are enriched by as much as a factor of 2 over Patapsco-derived suspended sediments while Mn, Ni, and Zn are comparable in Harbor and suspended sediments. While these signatures are based on a small number of samples the relative pattern of enrichment in the Harbor is clear.

Regardless of the ultimate source of trace metals to the Harbor, the mobility of these metals remains a question. In an effort to begin to evaluate the availability of the analytes in this investigation bulk soils, size fractions and harbor and bay sediments were treated with 7N HNO_3.

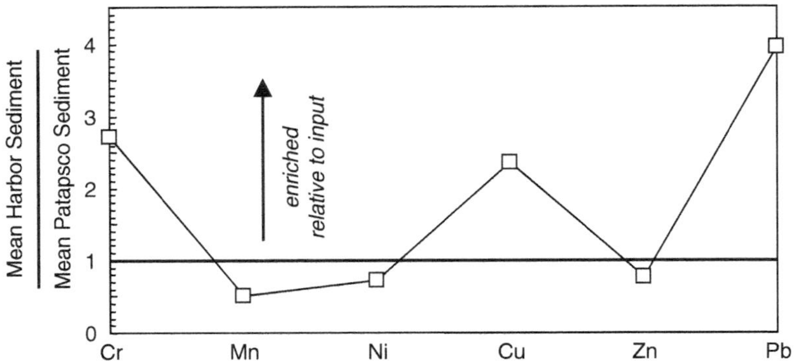

Figure 29.7. Plot of mean Baltimore Harbor sediment concentrations normalized to mean Patapsco suspended sediment input. If the primary source of trace metals to the Harbor is the suspended sediments from the Patapsco River the ratio of Harbor sediment/suspended sediment will be 1. Mn, Ni, and Zn are similar in the suspended sediments and Harbor sediments, while, Cr, Cu, and Pb are highly enriched in the Harbor relative to the suspended sediment input.

While this is an aggressive treatment of these sediments and soils, it will provide an estimate of the upper limit for extractable metals. In Fig. 29.7, average extracted metal patterns are plotted for bulk soils, size fractions and harbor and bay sediments. On average between 20% and 35% of the metals bound in soils are acid extractable. Excluding Mn and Zn, metals bound in harbor and bay sediments are less available at <20% and generally uniform. Zn and Mn are more available in harbor and bay sediments at 21% and 30%, respectively and are enriched within the Patapsco River watershed relative to soils and the upper continental crust (Fig. 29.8). In addition, these metals do not appear to have a significant local source in the Harbor (Fig. 29.7) that may suggest that the watershed trace metal sources may be more reactive than harbor and bay sources. Suspended sediments were not treated due to limited sample size.

29.8. Summary and conclusions

In summary, sediment being transported within the Patapsco River watershed represents at least two distinct trace metal sources. Soil-derived metals appear to dominate the suspended sediment during larger storm events for all analytes in this study. During smaller discharge storm events, there appears to be a Zn-enriched source that is a major

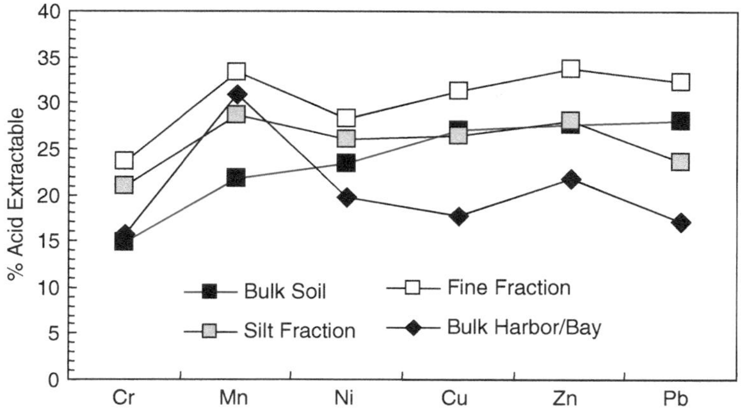

Figure 29.8. Plot of average % acid extractable metals for bulk, silt and the fine fraction of soils from the study area and bulk harbor and bay samples from this study. Trace metals in the soils and hence much of the suspended sediment load are more available than metals bound in harbor and bay sediments.

component of suspended sediment flux. This combined watershed signature is clearly different than the bulk trace metal signature in Baltimore Harbor and the Chesapeake Bay. Instead, the trace metal signature of harbor sediments displays a pattern that requires an additional local trace metal source(s). The local source is enriched in Cr, Cu, and Pb relative to the Patapsco input and is apparently less reactive than the trace metal bearing phases in the suspended sediment load.

One of the primary research questions in this investigation is the importance of soil-derived trace metals to the total flux of trace metals to the Chesapeake Bay. The results of this study point to a storm water controlled system of sediment and therefore trace metal flux. High-discharge storm events carry soil-derived suspended sediments with relatively low-metal concentrations but deliver the greatest metal load to the bay. Smaller storm events in the Patapsco are dominated by a Zn-enriched anthropogenic non-point source that is likely derived from roadway dust. While this non-point source is certainly present during the lager discharge events, its trace metal signature is overwhelmed by the soil signature. By evaluating the background trace metal signature of the sediment flux through the Patapsco River watershed it is possible to identify watershed-derived sources and local point sources to Baltimore Harbor. This is an important step toward characterizing the trace metal flux to the Chesapeake Bay.

There are several more general conclusions that can be drawn from this investigation:

(1) The trace metal concentration of suspended sediments in a mixed land use watershed, typical of the Chesapeake Bay watershed as a whole, do not correlate with the transported trace metal load;
(2) In a mixed land use watershed, soil sources will dominate the trace metal load and therefore the total flux of metals during large storm events;
(3) Urban regions in a mixed watershed will contribute the majority of the trace metal load during most storm water discharge events.

These more broad statements will be useful for future investigations in the Chesapeake Bay watershed and other large at risk regional catchments. The notion that the trace metal concentration of the suspended sediment load may not accurately reflect the total trace metal flux through a system is potentially the most important finding from this investigation.

ACKNOWLEDGMENT

Support for this research was provided by National Science Foundation grant DMR-MRI-011619.

REFERENCES

Brimhall, G.H., Dietrich, W.E., 1987. Constitutive mass balance relations between chemical composition, volume, density, porosity, and strain in metosomatic hydrochemical systems: Results on weathering and pedogenesis. Geochim. Cosmochi. Acta 51, 567–587.

Camponelli, K.M., Casey, R.E., SLev, M., ELanda, R., 2005. Spatial distribution and chemical fractionation of Cu and Zn in a stormwater retention pond. 26th Annual Meeting of the Society of Environmental Toxicology and Chemistry, Baltimore, MD.

Carpenter, J.H., Bradford, W.L., Grant, V., 1975. Processes affecting the composition of estuarine waters (HCO_3, Fe, Mn, Zn, Cu, Ni, Cr, Co, and Cd). In: Cronin, L.E. (Ed.), Estuar. Res. Academic Press, New York, vol. 1, pp. 188–214.

Councell, T.B., Duckenfield, K.U., Landa, E.R., Callender, E., 2004. Tire-wear particles as a source of zinc to the environment. Environ. Sci. Technol. 38, 4206–4214.

Delfino, J.J., Otto, R.G., 1986. Trace metal transport in two tributaries of the upper Chesapeake Bay: The Susquhanna and Bush rivers. Mar. Chem. 20, 29–44.

Gaillardet, J., Dupré, B., Allègre, C.J., 1999. Geochemistry of large river suspended sediments: Silicate weathering or recycling tracer? Geochim. Cosmochim. Acta 63, 4037–4051.

Goldberg, E.D., Hodge, V., Koide, M., Griffin, J., Gamble, E., Bricker, O.P., Matisoff, G., Holdren, G.R. Jr., Braun, R., 1978. A pollution history of the Chesapeake Bay. Geochim. Cosmochim. Acta 42, 1413–1425.

Harris, R., Nichols, M., Thompson, G., Banacki, J., Vadas, G., 1980. Heavy metal inventory of suspended sediment and fluid mud in the Chesapeake Bay. Special scientific report 99, Gloucester Point, VA, Virginia Institute of Marine Science.

Helz, G.R., 1976. Trace element inventory of the northern Chesapeake Bay with emphasis on the influence of man. Geochim. Cosmochim. Acta 40, 573–580.

Helz, G.R., Sinex, S.A., Ferri, K.L., Nichols, M., 1985a. Processes controlling Fe, Mn, and Zn in sediments of the northern Chesapeake Bay. Estuar. Coastal Shelf S. 21, 1–16.

Helz, G.R., Setlock, G.H., Cantillo, A.Y., Moore, W.S., 1985b. Processes controlling the regional distribution of ^{210}Pb, ^{226}Ra and anthropogenic zinc in estuarine sediments. Earth Planet. Sci. Lett. 76, 23–24.

Kearney, M.S., Ward, L.G., Cofta, C.M., Helz, G.R., Church, T.M., 1985. Sedimentology, geochronology, and trace metals in the Nanticole and Choptank rivers, Chesapeake Bay. Technical Report 84, College Park, Maryland, Maryland Water Resources Research Center, University of Maryland.

Kupiainen, K.J., Tervahattu, H., 2005. Size and composition of airborne particles from pavement wear, tires, and traction sanding. Environ. Sci. Technol. 39, 699–706.

Lev, S.M., 2001. Assessing the impact of non-equilibrium denudation on the flux of trace metals to the Chesapeake Bay. Geol. Soc. Am. Abstracts with Programs 33(6), 367.

Marcus, W.A., Nielsen, C.C., JCornwell, C., 1993. Sediment budget-based estimates of trace metal inputs to a Chesapeake estuary. Environ. Geol. 22, 1–9.

Maynard, J.B., 1992. Chemistry of modern soils as a guide to interpreting Precambrian paleosols. J. Geol. 100, 279–289.

McGee, B.L., Fisher, D.J., Yonkos, L.T., Ziegler, G.P., Turley, S., 1999. Assessment of sediment contamination, acute toxicity, and population viability of the estuarine amphipod *Leptocheirus Plumulosus* in Baltimore Harbor, Maryland, USA. Environ. Toxicol. Chem. 18(10), 2151–2160.

McLennan, S.M., 1993. Weathering and global denudation. J. Geol. 101, 295–303.

McLennan, S.M., 1995. Sediment and soil: Chemistry and abundances. In: Rock Physics and Phase Relations, A Handbook of Physical Constants, AGU Reference Shelf 3. Washington, DC, United States, American Geophysical Union, pp. 8–19.

Nesbitt, H.W., MacRae, N.D., Kronberg, B.J., 1990. Amazon deep-sea fan muds: Light REE enriched products of extreme chemical weathering. Earth Planet. Sci. Lett. 100, 118–123.

Nesbitt, H.W., Markovics, G., 1997. Weathering of granodioritic crust, long-term storage of elements in weathering profiles, and petrogenesis of siliciclastic sediments. Geochim. Cosmochim. Acta 61, 1653–1670.

Nesbitt, H.W., Young, G.M., 1982. Early Proterozoic climates and plate motions inferred from major element chemistry of lutites. Nature 299, 715–717.

Nesbitt, H.W., Young, G.M., 1984. Prediction of some weathering trends of plutonic and volcanic rocks based on thermodynamic and kinetic considerations. Geochim. Cosmochim. Acta 48, 1523–1534.

Nesbitt, H.W., Young, G.M., 1989. Formation and diagenesis of weathering profiles. J. Geol. 97, 129–147.

Nichols, M.N., Harris, R., Thompson, G., Nelson, B., 1982. Metal inventory and fate of suspended sediment in the Chesapeake Bay. Special scientific report 106, Gloucester Point, VA, Virginia Institute of Marine Science.

Omstedt, G., Bringfelt, B., Johansson, C., 2005. A model for vehicle-induced nontailpipe emissions of particles along Swedish roads. Atmos. Environ. 39, 6088–6097.

Rantz, S.E., 1982. Measurement and Computation of Streamflow: Volume 1. Measurement of Stage and Discharge. United States Geological Survey Water Supply Paper 2175.

Riedel, G.F., Williams, S.A., Riedel, G.S., Gilmour, C.C., Sanders, J.G., 2000. Temporal and spatial patterns of trace elements in the Patuxent River: A whole watershed approach. Estuaries 23(4), 521–535.
Samara, C., Voutsa, D., 2005. Size distribution of airborne particulate matter and associated heavy metals in the roadside environment. Chemosphere 59, 1179–1206.
Sinex, A., Helz, G.R., 1981. Regional geochemistry of trace elements in Chesapeake Bay sediments. Environ. Geol. 3, 315–323.
State of Maryland, Department of the Environment (MDE), 2001.Cleanup standards for soil and groundwater, Interim Final Guidance Document, p. 97.
Taylor, S.R., McLennan, S.M., 1985. The Continental Crust: Its Composition and Evolution. Blackwell Scientific, Oxford, UK, p. 312.

Section VII: Applications of New Analytical and Quantitative Methods in Environmental Geochemistry Research

Developments in Environmental Science, Volume 5
D. Sarkar, R. Datta and R. Hannigan (Editors)

ISSN: 1474-8177/DOI:10.1016/S1474-8177(07)05030-9

Chapter 30

Characterizing the surface chemistry of oxides with X-ray photoelectron spectroscopy: Assessment regarding surface oxygen valence charge and acid–base properties

M. Ding and B.H.W.S. de Jong

For Titus

Abstract

Variations in valence charge of surficial O (Oxygen) atoms in a compound may be calculated from shifts in the O1s binding energies. Using the National Institute of Standards and Technology (NIST) photoelectron database, a consistent set of charges for O atoms in oxides can be made. It uses as reference frame a theoretical chemical procedure that incorporates a spherical potential around an O atom to account for the field generated by the surrounding atoms in a compound. The complete database indicates no strong correlations between calculated surface O charges and electronegativity, bulk or surface acid–base properties, points of zero charge or refractive index. To account for the large differences in surface acid–base properties, a detailed partitioning of charge on the O atom isolated from the surrounding atoms, a so-called atom in molecule assignment, rather than overall charge transfer must be invoked. It implies that the electron density associated with the O atomic orbitals in a molecular orbital must be partitioned between a bonded and a non-bonded part. The ratio between these two parts of the overall electron density varies, depending on the nature of the interaction between a specific metal atom and the O atoms in its coordination sphere. For silicates, this ratio is reasonably well established. Quantifying it from first principles for other metal–oxide moieties requires additional molecular orbital calculations.

30.1. Introduction

For many decades, structure, reactivity and properties of oxide materials have been studied in a variety of fields, including material science, inorganic chemistry, heterogeneous catalysis, glass and ceramic science, geochemistry, and environmental chemistry. In all these interdisciplinary research activities, much effort has been directed toward characterizing and predicting the surface reactivity of oxides, and in particular their acidic or basic character.

In aqueous solution, the acid–base concept is well defined and is fundamental to all chemistry. However, for solids it is not always obvious how acid–base concepts propagate in the different domains of chemistry. Thus in glass and ceramic science alkali and alkaline earth oxides are basic whereas silica is acidic. In heterogeneous catalysis, an acidic surface is defined as a proton-releasing surface as inferred from colorimetric titration with a basic indicator. In adsorption studies a surplus or deficit of surface protons is associated with a positive or negative surface charge. All these assignments suppose as basic ingredient a charge transfer from an electron source to an electron sink. It is the degree to which this transfer takes place that governs the adhesive and cohesive properties of a substance.

Donating or accepting electrons of elements that make up a substance have been at the root of all chemistry since the introduction of the concept of electronegativity by Berzelius in 1811. Though conceptually easy, it has been difficult to apply quantitatively to chemical processes. This manifests itself by the still in general insurmountable difficulties associated with the prediction of crystal structures from their chemical formulas only, as noted by Maddox (1988). Despite substantial efforts to remedy this, as he calls it, scandal in the physical sciences the issue is by no means settled yet. Sometimes the deficiency is obvious, for instance when using two rather than three body potentials in tetrahedral coordinated solids, more often it is not.

Though already difficult for solids for which accurate crystal structures are known, the problem is compounded for surfaces for which most geometric arrangements have to be inferred indirectly and are always model dependent (Marians and Burdett, 1990). The uneasy feeling prevails that there is a gap in our grasp on the cohesive energy of solids or in the nucleation and growth processes associated with their formation.

Many attempts have been made to systematize charge transfer in solids. Thus, Pauling's second valence rule, in which O atoms function as electron sinks, has in its modern parametric form been applied successfully in systematizing charge transfer between atoms in solids (Brese and

O'Keeffe, 1991; de Jong et al., 2000). Dietzel's cationic field strength and the 15 other different electronegativity scales have been used with varying success in systematizing properties of oxide materials (e.g. Burdett, 1995, 1997). Ketelaar-van Arkel diagrams have been used to assign intuitively the degree of charge transfer in solids (Allen et al., 1993; Jensen, 1995; Mori-Sanchez et al., 2002). Hard-soft acid–base theory and its precursor, Goldschmidt's (1923) atmophile, siderophile, lithophile, and chalcophile classification have rationalized the occurrence of elemental combinations in solids by combining charge transfer with polarizability of the constituent atoms. All these ways of systematizing charge transfer were conceived during the 1920s.

The arrival of computational theoretical chemistry in the 1960s allowed to look with new eyes at charge transfer in solids by examining the atomic orbitals that make up the orthogonal set of molecular orbitals of the system. Isolating characteristic moieties and inspecting the atomic orbital contributions to the overall molecular orbital structure enables testing of the interatomic charge transfer (e.g. Popelier, 2000; Gillespie and Popelier, 2001). It is this atom in a molecule approach, in which one ponders the variation in O atomic orbital coefficients in the molecular orbitals of a molecule, which we shall consider in the following, focusing on O in a surrounding crystalline field. We shall test if, with this notion, at least some part of the gap in our knowledge concerning charge transfer can be bridged.

In some recent studies (Ding et al., 2000, 2005), we looked into the absorbance of atoms and molecules on a ferric oxyhydroxide substrate and attempted to characterize the charge transfer from such substrate to different cations and anions using O1s photoelectron spectroscopy. This spectroscopic technique samples the surficial O layers. Shifts in this O1s spectral line may be correlated with variations of electron density in the valence band as shown in Fig. 30.1. The focus on the O1s core level, with its single or double peak, is advantageous because its spectral interpretation is substantially less complicated and more unequivocal than that of the valence band, which commonly consists of six or more peaks. We concluded from our studies that a solid surface is surrounded by a membrane of O atoms, a two-dimensional O coating on the mineral surface, that varies in charge depending on the nature of the incoming ion. It enabled rationalizations of charge transfer between substrate and adduct in terms of highest occupied (HOMO), which consists primarily of O 2p atomic orbitals, and lowest unoccupied (LUMO) molecular orbitals.

Given the promising nature of these studies we wanted to test if these rationalizations based qualitatively on first principle notions would apply to all oxides and hydroxides for which O1s photoelectron spectra have

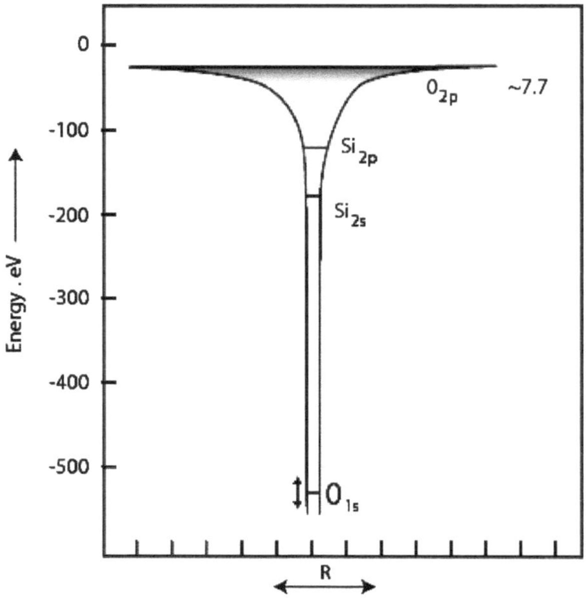

Figure 30.1. The electronic structure of O in silicates. The O1s level varies maximally by about 30 eV according to Bagus and Bauschlicher (1980). The direct gap, i.e. the difference of occupied states on top of the valence band and the continuum is about 7.7 eV for SiO_2 (de Jong, 1989). The top of the valence band consists primarily of O 2p states (de Jong and Brown, 1980).

been recorded; thus that one parameter would suffice to order all acid–base properties of solids and surfaces. We shall show that such parameter does not exist for this close to complete database; that observed surface O charges do not correlate with the standard chemical rationalizations invoking electronegativity variations, and we shall argue that interpretation of surface acid–base properties must include non-bonded electron densities on the O atom. Rather than whole number lone pairs, i.e. two electrons per pair, partially filled lone pairs in Lewis octets coincide with this molecular orbital derived non-bonded electron density concept.

30.2. Method

The overall charge on surface O atom can be ascertained from O1s photoelectron spectroscopy. Bagus and Bauschlicher (1980) calculated the shift in energy of the O1s core level as a function of O valence, i.e. variation in valence electron density, their calculated shifts invalidating

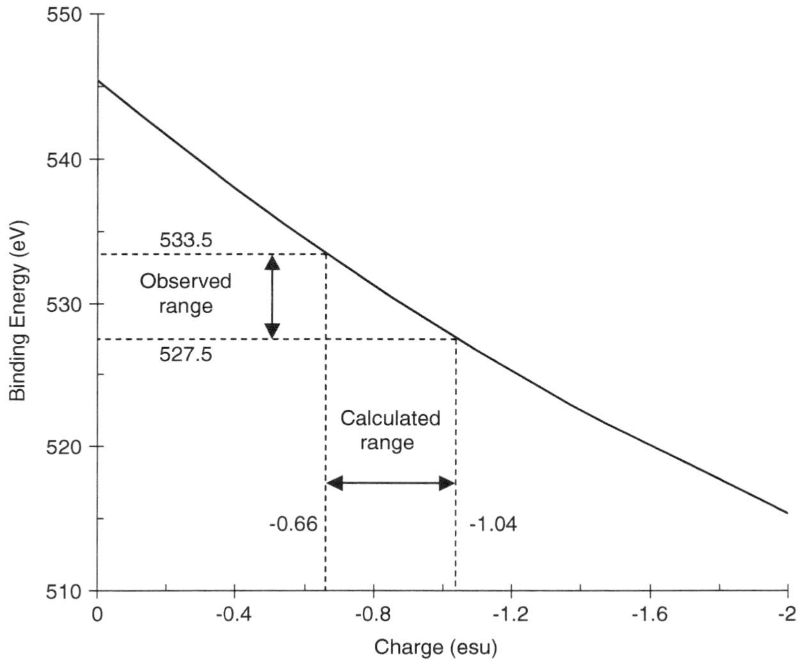

Figure 30.2. O1s core level binding energy versus charge according to Eq. (1) (Ding et al., 2000).

Koopman's theorem which states that the core levels of an atom remain fixed in energy when placed in a compound. Their result is illustrated in Fig. 30.2 and can be couched in the following formula (Ding et al., 2000):

$$Q_O = -4.372 + ((385.023 - 8.976 \times (545.509 - O_{(1s)}BE))^{\frac{1}{2}})/4.488 \qquad (1)$$

where Q_O is the valence charge on the O atom in electrostatic units (esu) and $O_{(1s)}BE$ the experimentally determined binding energy in electron Volts (eV).

Some confidence as to the reality of these charges may be inferred from the close correspondence between the one observed for O in SiO_2 and the observed Mulliken charge density for O in state of the art calculations on the $H_6Si_2O_7$ molecule (Burkhard et al., 1991).

We have used this equation to calculate the O charge for the O1s entries of primarily oxides and hydroxides in the NIST XPS database including also some measurements from elsewhere for comparison (de Jong et al., 1994; Barr et al., 1996; Stoecker, 1996). Our results reveal that the variation in O1s core level for oxide materials vary between Cs_2O at

527.5 eV and $Na_2B_4O_7$ at 533.5 eV corresponding to a valence charge difference of 0.38 esu. The absolute values of these core level shifts are not always well determined and especially insulators as SiO_2 show according to the entries in the NIST XPS database a substantial spread in peak position (for SiO_2: 2.3 eV) which may be attributed among others to the well known charging effects of the sample in the absence of a flood gun as for instance pointed out recently by Gijzeman et al. (2003). We have collected the average O charge values for all tabulated inorganic oxides, and hydroxides in Table 30.1. To highlight some of the results, the least negative O charges in this table are the network-forming oxides and hydroxides SiO_2 (−0.698), B_2O_3 (−0.686), $B(OH)_3$ (−0.668), and P_2O_5 (−0.662); the most negative ones Cs_2O (−1.042), PbO_2 (−0.982), PrO_2 (−0.969), Pb_3O_4 (−0.949), and BaO (−0.949).

30.3. Discussion

Previously, we have used O1s surface charges and their variations upon adsorbance of pollutants from aqueous solutions successfully to delineate the charge transfer on ferric oxyhydroxides (Ding et al., 2000, 2005). Here we want to test if this method can be extended to all oxide and hydroxide substrates, and if inter comparison between oxides is possible despite the, in general, lower coordination of surface O atoms relative to bulk ones and the concomitant damped charge difference (El Shafei, 1996).

We shall start this discussion to test if there exists a relation between XPS-derived surface O charge and two bulk solid properties: electronegativity and the difference between solid acids and bases as used in the ceramics and glass industry. Next we shall look if these O surface charges scale with two surface properties of solids: (i) solid acids and bases as proton donors and acceptors respectively, used in heterogeneous catalysis and (ii) points of zero charge variations of solids in contact with aqueous solutions. Finally, we shall discuss the nature of the O charge in terms of bonding and non-bonding electron density by assessing in detail the electronic structure of the valence band of O.

30.3.1. Oxygen charges and electronegativities

The first issue is if the calculated charges can be correlated with observed electron negativity differences between O and the metal atoms in oxides. Of all the available electronegativity scales we did use the one of Allen (1989, 1992) and Murphy et al., (2000) calculated from the valence shell s and p electrons in the ground-state free atoms. This of course leaves out

Table 30.1. O(1s) Binding energy (eV) and calculated O charge Q (esu) for oxides and hydroxides arranged alphabetically

Oxides	$O_{(1s)}$BE (eV)	Q_o (esu)	Oxides	$O_{(1s)}$BE (eV)	Q_o (esu)	Oxides	$O_{(1s)}$BE (eV)	Q_o (esu)
AgO	529.1	−0.936	Fe_3O_4	529.7	−0.898	Pb_3O_4	528.9	−0.949
Ag_2O	529.6	−0.904	FeOOH	530.8	−0.828	PdO	529.4	−0.917
Al_2O_3	531.0	−0.815	$Fe(OH)_3$	531.3	−0.797	PrO_2	528.6	−0.969
As_2O_3	531.9	−0.759	Ga_2O_3	530.9	−0.822	Pr_2O_3	529.2	−0.930
As_2O_5	531.7	−0.772	Gd_2O_3	531.1	−0.809	PtO_2	531.7	−0.772
B_2O_3	533.1	−0.686	GeO_2	532.1	−0.747	$Pt(OH)_2O$	531.1	−0.809
$B(OH)_3$	533.4	−0.668	H_2MoO_4	530.7	−0.834	$Pt(OH)_4$	531.3	−0.797
BaO	528.9	−0.949	H_2WO_4	530.5	−0.847	ReO_2	530.1	−0.872
$Ba(OH)_2$	531.2	−0.803	HfO_2	530.2	−0.866	ReO_3	531.9	−0.759
BeO	531.7	−0.772	HgO	529.9	−0.885	Rh_2O_3	530.6	−0.841
Bi_2O_3	530.0	−0.879	HoO_3	530.6	−0.841	RuO_2	529.4	−0.917
CaO	530.3	−0.859	In_2O_3	530.6	−0.841	RuO_3	530.7	−0.834
$Ca(OH)_2$	531.2	−0.803	$In(OH)_3$	531.8	−0.766	Sb_2O_3	530.0	−0.879
CdO	529.6	−0.904	K(OH)	531.7	−0.772	Sc_2O_3	529.7	−0.898
CdO_2	530.5	−0.847	La_2O_3	530.0	−0.879	SiO_2	532.9	−0.698
$Cd(OH)_2$	531.7	−0.772	Li_2O	531.3	−0.797	Sm_2O_3	530.4	−0.853
CeO_2	529.2	−0.930	Li(OH)	531.2	−0.803	SnO	529.9	−0.885
Ce_2O_3	529.7	−0.898	Lu_2O_3	530.3	−0.859	SnO_2	530.6	−0.841
Ce_3O_4	530.3	−0.859	MgO	530.7	−0.834	SrO	530.4	−0.853
CoO	529.8	−0.891	$Mg(OH)_2$	530.9	−0.822	Ta_2O_5	530.3	−0.859
Co_2O_3	530.1	−0.872	MnO	529.8	−0.891	Tb_2O_3	529.7	−0.898
Co_3O_4	530.0	−0.879	MnO_2	529.6	−0.904	TeO_2	530.4	−0.853
$Co(OH)_2$	531.3	−0.797	Mn_2O_3	529.5	−0.911	ThO_2	530.3	−0.859
CoOOH	530.5	−0.847	Mn_3O_4	529.5	−0.911	TiO_2	530.2	−0.866
CrO_2	529.3	−0.924	Molec. siev	530.5	−0.847	Ti_2O_3	529.6	−0.904
CrO_3	530.1	−0.872	MoO	530.5	−0.847	Tm_2O_3	531.2	−0.803
Cr_2O_3	530.2	−0.866	MoO_2	530.5	−0.847	UO_2	530.3	−0.859

Table 30.1. (*Continued*)

Oxides	$O_{(1s)}$BE (eV)	Q_o (esu)	Oxides	$O_{(1s)}$BE (eV)	Q_o (esu)	Oxides	$O_{(1s)}$BE (eV)	Q_o (esu)
CrOOH	531.2	−0.803	MoO_3	530.6	−0.841	UO_3	530.4	−0.853
$Cr(OH)_3$	531.2	−0.803	Na_2O	529.7	−0.898	U_3O_7	529.9	−0.885
Cs_2O	527.5	−1.042	NaOH	532.8	−0.704	VO_2	530.0	−0.879
Cs_2O_4	530.5	−0.847	Nb_2O_5	530.4	−0.853	V_2O_3	530.4	−0.853
CuO	529.8	−0.891	$Nb_2O_5\cdots OH$	530.7	−0.834	V_2O_4	530.4	−0.853
Cu_2O	530.5	−0.847	Nd_2O_3	530.1	−0.872	V_2O_5	530.2	−0.866
$Cu(OH)_2$	531.1	−0.809	NiO	529.6	−0.904	WO_2	530.7	−0.834
Dy_2O_3	530.6	−0.841	Ni_2O_3	531.6	−0.778	WO_3	530.5	−0.847
Er_2O_3	530.8	−0.828	$Ni(OH)_2$	531.3	−0.797	Y_2O_3	530.0	−0.879
Eu_2O_3	530.8	−0.828	P_2O_5	533.5	−0.662	Yb_2O_3	530.7	−0.834
FeO	530.0	−0.879	PbO	529.4	−0.917	ZnO	530.4	−0.853
Fe_2O_3	529.9	−0.885	PbO_2	528.4	−0.982	ZrO_2	530.7	−0.834

all the transition metal compounds. Inspection of Table 30.1 shows that electronegativity values do not exist on Allen's scale for compounds with an O charge more negative than -0.9 esu. We have collected the remaining variations in electronegativity in Table 30.2 and ordered them in decreasing metal–O ratio.

Inspection of this table shows that the correlation between charge and electronegativity difference is not very strong. For instance in the XO_2 sequence silica, germania, telluria, all metal atoms are four-fold O coordinated, TeO_2 being somewhat exceptional because of its bipyramidal O coordination with four O atoms and one lone-pair located on Te. The O atom in the Te–O bond is expected to be the least negative based on the O–Te electronegativity difference, whereas of the three oxides it has the most negative calculated charge. Electronegativity or parameters that scale with it such as cationic field strength, reflect bulk properties of elements in solids and apparently tend not to scale with surface properties such as the XPS-derived charges.

Table 30.2. Electronegativity differences, Δ, for selected metal–O linkages using Allen's (1989) electronegativity scale in comparison to calculated O valence charge ordered according to metal–O ratio

Compound	Bond	Δ	Charge
X_2O_5			
P_2O_5	O–P	1.36	−0.662
XO_2			
SiO_2	O–Si	1.69	−0.698
GeO_2	O–Ge	1.62	−0.747
TeO_2	O–Te	1.45	−0.853
X_2O_3			
B_2O_3	O–B	1.56	−0.686
Al_2O_3	O–Al	2.00	−0.815
Ga_2O_3	O–Ga	1.85	−0.822
As_2O_3	O–As	1.40	−0.759
In_2O_3	O–In	1.95	−0.841
Sb_2O_3	O–Sb	1.63	−0.879
XO			
BeO	O–Be	2.03	−0.772
MgO	O–Mg	2.32	−0.834
CaO	O–Ca	2.58	−0.859
SrO	O–Sr	2.65	−0.853
SnO	O–Sn	1.79	−0.885
X_2O			
Li_2O	O–Li	2.69	−0.797
Na_2O	O–Na	2.74	−0.898

30.3.2. Oxygen charge and acid–base characteristics of bulk solids

Electronegativity and acidity and basicity are intimately related to one another (e.g. Huheey et al., 1993). For solids two ways exist to distinguish between acids and bases based on the principal species exchanged in acid–base reactions. This is the proton in the Brönsted (1923) theory focusing on its exchange with solid surfaces, and the O atom in the Flood and Forland (1947) and Flood et al. (1947) theory for non-protonic systems, primarily ceramics (e.g. de Jong et al., 1989), for bulk solids. Though our XPS-derived charges tend to reflect only the first three or four O layers of a surface, we tested if these charges scale with the solid acidity parameters for acidic-basic and amphoteric oxides derived by Smith (1987) based on the Flood and Forland (1947) theory. Our results are collected in Table 30.3; Cs_2O with the most negative value, being the most basic oxide; the intermediate negative values being basic or amphoteric oxides,

Table 30.3. Selected values of basic (most negative) and acidity parameters for selected bulk oxides according to Smith (1987) based on the Flood and Forland (1947) concept

Oxide	Parameter acidity	O charge (esu)
Cs_2O	−15.5	−1.042
Na_2O	−12.5	−0.898
BaO	−10.8	−0.949
SrO	−9.4	−0.853
Li_2O	−9.2	−0.797
CaO	−7.5	−0.859
MnO	−4.8	−0.891
MgO	−4.5	−0.834
ThO_2	−3.8	−0.859
CoO	−3.8	−0.891
FeO	−3.4	−0.879
ZnO	−3.2	−0.853
NiO	−2.4	−0.904
BeO	−2.2	−0.772
Al_2O_3	−2.0	−0.815
Fe_2O_3	−1.7	−0.885
Cu_2O	−1.0	−0.847
H_2O	0.0	
ZrO_2	0.1	−0.834
TiO_2	0.7	−0.866
SiO_2	0.9	−0.698
B_2O_3	1.5	−0.686
V_2O_5	3.0	−0.866
WO_3	4.7	−0.847
MoO_3	5.2	−0.841
P_4O_{10}	7.5	−0.662

and the slight negative or positive values being acidic oxides according to Smith (1987). In solution chemistry clearly not all assignments coincide with this classification as inspection of the paper by Rich (1985) concerning the periodicity in acid–base behavior of oxides and hydroxides shows. For example TiO_2 is only soluble in acid and therefore a basic oxide, ZrO_2 and V_2O_5 show a minimum in their solubility as a function of pH and are therefore amphoteric oxides.

Inspection of this table shows that there is no relation between the bulk basicity of a material and the XPS-derived surface charge on the O atom.

30.3.3. Oxygen charge and acid–base properties of oxide surfaces in heterogeneous catalysis

For solid acids and bases used in heterogeneous catalysis acid/base assignments are based on proton exchange between the surface and basic or acidic indicators (Tanabe, 1970) with concomitant color change. This release of protons and associated acid nature of the surface is of significant industrial importance (Izumi et al., 1992; Tanabe and Holderich, 1999). We tabulated those solids collated in Tanabe's book as acid or base and looked at our XPS-derived surface charges including entries besides oxides and hydroxides in the NIST database, and collected the results in Table 30.4.

Inspection of table indicates some global relation between XPS-derived charges and the acidic (proton donor) and basic nature of the surface. Thus, the O charges on the acidic surface tend to be less negative and thus release adsorbed protons easier or accept electron pairs more readily than basic surfaces. Still among acids there are exceptions, MoO_3, TiO_2, V_2O_5, ZnO, and CeO_2, which are according to Tanabe's criterion acids even though they have an O charge more akin to that observed for basic solid oxides.

30.3.4. Acid–base properties of oxide surfaces in aqueous solutions

Finally, a measure of the acid–base character of a surface in aqueous solutions is the point of zero charge. It can be determined either with a zeta potential meter, potentiometric titration or inferred from the minimum in solubility of a compound in aqueous solutions, or the minimum in rate of dissolution (e.g. Kamiya and Shimokata, 1974) as a function of pH. Three solubility behaviors exist: acidic oxides are soluble in base, basic oxides in acids, and amphoteric oxides show a minimum in solubility usually associated with a change in coordination of the dissolved species as for instance observed for alumina (de Jong et al., 1983). Based

Table 30.4. Acid/base characteristics of solid surfaces and the XPS-derived surface charges

Acids	O charge (esu)	Bases	O charge (esu)
$CaCO_3$	−0.797	$BaCO_3$	−0.803
$Al_2(SO4)_3$	−0.729	$CaCO_3$	−0.797
$BaSO_4$	−0.741	Na_2CO_3	−0.772
$CaSO_4$	−0.735	$SrCO_3$	−0.809
$Cr(SO_4)_3$	−0.747	Al_2O_3	−0.815
$CuSO_4$	−0.772	CaO	−0.859
$FeSO_4$	−0.735	Cr_2O_3	−0.866
$Fe_2(SO_4)_3$	−0.759	MgO	−0.834
K_2SO_4	−0.803	SrO	−0.853
$MnSO_4$	−0.729	ZnO	−0.853
$SrSO_4$	−0.747	BaO	−0.949
$ZnSO_4$	−0.723	BeO	−0.770
$AlPO_4$	−0.704	SiO_2	−0.698
$Ca(NO_3)_2$	−0.656		
MoO_3	−0.841		
TiO_2	−0.866		
V_2O_5	−0.866		
CeO_2	−0.930		
ZnO	−0.853		
As_2O_3	−0.759		
SiO_2	−0.698		

on the data of Baes and Mesmer (1976) and Kragten (1978), Rich (1985) has collated a substantial number of these solubility behaviors from which for the amphoteric elements the point of zero charge can be estimated from their minimum in solubility. The scatter in the PZC data is substantial as inspection of the values quoted by Morrison (1985), de Jong (1989), Blesa et al. (1994), Boldyrev (1996), El Shafei (1996), Kosmulski (1997, 2004), and Sposito (2004) show. Some values, such as in particular that for SiO_2 and even Al_2O_3 (Kosmulski, 2003) are open to continuing debate. In Table 30.5, we have collected the pertinent points of zero charge, designated PPZC by Kosmulski (1997), the pristine point of zero charge, versus O charge, the entries coming from the critically assessed canonical set proposed by him. They are arranged in decreasing metal O ratio in accordance with the already by Parks (1965) observed trends in PPZC as a function of this ratio.

Inspection of these PPZC that fulfill tight criteria with respect to their veracity show that they do not correlate with the calculated XPS charges.

Clearly, there seems to be no or only some tentative relation between the various acidity indices and the XPS calculated O charges. Sources of error in the latter may be at the base of this lack of correlation as mentioned previously, because calculating O charges from XPS requires

Table 30.5. O charge and pristine point of zero charge (PPZC) for various oxides (Kosmulski, 1997, 2002a,b)

Compound	PPZC (pH)	O Charge (esu)
X_2O_5		
Nb_2O_5	4.1	−0.853
Ta_2O_5	5.2	−0.859
XO_2		
RuO_2	5.4	−0.917
TiO_2	5.8	−0.866
ZrO_2	7.8	−0.834
CeO_2	8.1	−0.930
HfO_2	7.6	−0.866
X_2O_3		
Fe_2O_3	9.8	−0.885
Al_2O_3	9.1	−0.815
Y_2O_3	9.1	−0.879
Ga_2O_3	9.0	−0.822
In_2O_3	8.7	−0.841
XO		
ZnO	9.2	−0.853
MgO	12.0	−0.834

Table 30.6. Monoxides with the NaCl structure, their density (gr/cc), refractive index, *n*(sodium D line), and XPS-derived surface charge(esu)

Compound	Structure	Symmetry	Density (gr cc−1)	*n*	O charge (esu)
MgO	NaCl	Cubic	3.56	1.737	−0.834
CaO	NaCl	Cubic	3.32	1.837	−0.859
SrO	NaCl	Cubic	4.75	1.870	−0.853
BaO	NaCl	Cubic	5.72	1.980	−0.949
FeO	NaCl	Cubic	5.50	2.320	−0.879
NiO	NaCl	Cubic	6.80	2.270	−0.904

confidence in the absolute value of the O1s energies. However, one may wonder if the overall charge on O is that what determines the interaction of the electron density of this atom with adducting atoms and molecules.

30.3.5. Oxygen charge and the refractive indices of oxides

To assess if something else affects the interaction of O charge density with adducts we have looked at the refractive indices of oxides. As has been well known since Lorentz these refractive indices scale with the number density and dipole oscillator strength of the O atoms in a compound. In Table 3.6, we have collected some oxides with the NaCl structure, their density, refractive indices (Winchell and Winchell, 1964), and XPS-derived surface charge.

The choice for one structure type presupposes that the surface O atoms in contact with a solvent have a similar O coordination. The XPS-derived O charges are indeed quite similar, even though the bulk refractive indices show large variations, suggesting a very large difference in charge transfer between cation and anion, as may be ascertained from their variation in refractive index, amounting to a maximum variation in light velocity of 43.440 km sec^{-1} between MgO and FeO.

One may wonder why no clear relation exists between variation in overall O charge and any of the material properties of oxides mentioned above. Overall O charge seems not to be the panacea with which to account for a variety of properties, and it may be that the build up of this charge on an O atom in the valence band is what needs to be considered.

30.3.6. Non-bonded electron density on oxygen atoms

The absence of discrete molecules in silicate chemistry in contrast to carbon chemistry was already noted among others by Friedel in the 1870s as one of the principal difficulties in making silicon analogues after carbon-containing molecules. This discreteness of organic molecules made theoretical calculations straightforward in comparison to the extended networks commonly encountered for silicates. It translated in the seventies of the last century in how to calculate an essential silicate moiety embedded in a surrounding, the overall fragment being uncharged. A satisfactory result to account for the spectroscopic properties of silica turned out to be the $H_6Si_2O_7$ molecule (de Jong and Brown, 1980). Isolating the atomic orbitals associated with the Si–O–Si linkage in the overall set of molecular orbitals gave an adequate description of the electronic spectra of quartz, the observed angle bond length variations, as well as charge density on the bridging O atom. Our results indicated that, though the overall O charge remained approximately constant, there was substantial difference in the distribution of this electron density, the top of the valence band containing molecular orbitals that nearly completely consisted of O atomic orbitals only. This charge density was designated the non-bonded O electron density as depicted in Fig. 30.3. It implies that the surface reactivity of oxides hinges on that part of the overall O electron density that is free to react. It is associated with molecular orbitals with the least negative energy, i.e. those orbitals that represent the top of the valence band.

Non-bonded electron density varied as a function of Si–O–Si angle from about four electrons at an Si–O–Si angle of 180 degrees to about 3.5

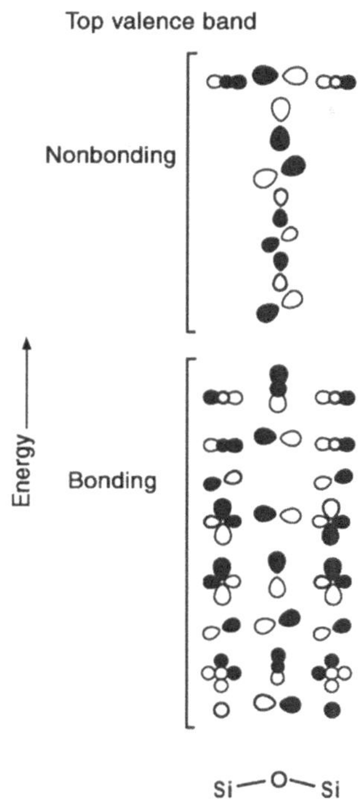

Figure 30.3. Linear combination of atomic orbitals across an Si–O–Si linkage in the $H_6Si_2O_7$ molecule (de Jong, 1989). The overall electron density on the O atom may be partitioned in a bonding and a non-bonding part. Note that the top of the valence band, the highest occupied molecular orbital consists of O 2p states only.

at an angle of 140 degrees, the total number of valence electrons staying about constant around 6.7 electrons, i.e. a valence charge of about −0.7 esu. The four non-bonded electrons on the bridging O correspond to two lone pairs and the rationalization of the virtually always bend Si–O–Si angle in silicates (Liebau, 1985) could readily be rationalized by the Gillespie-Nyholm VSEPR theorem (Gillespie and Popelier, 2001).

Generalizing these results to oxide materials suggests that ionicity or covalency of the constituent bonds in an oxide do not depend on overall charge transfer between O and the atoms in its first coordination sphere. This charge barely varies for surface O as we have demonstrated here. The overall O charge must be partitioned in a part associated with bond formation with atoms in the coordination sphere surrounding the O atoms versus a part on the O atom that is localized on it and free to react and act as an atom in a molecule. It may well be that electronegativity scales might be used to assess this partitioning into non-bonded and bonded electron density but are likely too crude to account for the variation observed in the complete database. To calibrate this charge transfer additional molecular orbital calculations on different oxide moieties must be carried out to test the variation in O non-bonded electron density for characteristic oxides and hydroxides.

The need for such calculations can be illustrated by examining the electronic structure of two SiO_2 polymorphs: Quartz with ^{IV}Si, each O atom connected to two Si atoms, for which a Lewis octet structure can readily be drawn; and Stishovite with ^{VI}Si, each O atom connected to three Si atoms without an obvious octet structure. The valence band structure for Quartz clearly shows a sizable gap between bonded and non-bonded O 2p states as illustrated in Fig. 30.3. For Stishovite such gap does not exist and the O 2p states are continuous with a total width of 11 eV (Ching, 2000). Thus, despite identical electronegativity differences between the constituent atoms in these two phases, a pronounced variation in O electron density distribution occurs presumably with concomitant surface reactivity, disavowing the possibility of simple rationalizations.

30.4. Conclusion

We have demonstrated previously that it is possible to successfully delineate the HOMO–LUMO interaction between ferric oxyhydroxides and various adsorbing atoms and molecules using XPS-derived O charges and that the HOMO consists primarily of O 2p atomic orbitals. Here, we have attempted to extend our procedure to all oxide and hydroxide materials for which reliable XPS data exist. We have shown that no clear correlation exists between these charges for either bulk properties, i.e. electronegativities, bulk acidity or basicity or refractive indices, or surface properties, solid acids and bases and points of zero charge. Clearly, such generalizations are too primitive and do not represent first principle cause and effect relations. This absence of clear correlations leads us to suggest that rather than considering the overall electron density on the O atom, a distinction has to be made between electrons involved in linking with their surroundings versus those free to react. This image is close to Lewis octet assignments at least for tetrahedrally coordinated oxide materials with their lone versus bond pairs, the difference being that the constraints of complete pairs, i.e. a whole, even number of electrons, must be lifted.

ACKNOWLEDGMENTS

Thoughtful comments by Rebecca Sutton and an anonymous reviewer have aided in improving this manuscript.

REFERENCES

Allen, L.C., 1989. Electronegativity is the average one-electron energy of the valence-shell electrons in ground-state free atoms. J. Am. Chem. Soc. 111, 9003–9014.

Allen, L.C., 1992. Extension and completion of the periodic table. J. Am. Chem. Soc. 114, 1510–1511.

Allen, L.C., Capitani, J.F., Kolks, G.A., Sproul, G.D., 1993. Van Arkel-Ketelaar triangles. J. Mol. Struct. 300, 647–655.

Baes, C.F., Mesmer, R.E., 1976. The Hydrolysis of Cations. Wiley, New York, p. 489.

Bagus, P.S., Bauschlicher, C.W., 1980. Core binding energy shifts for free negative ions of oxygen: O^0 to O^{2-}. J. Electron Spectrosc. 20, 183–190.

Barr, T.L., Seal, S., He, H., Klinowski, J., 1996. X-ray photoelectron spectroscopic studies of kaolonite and montmorillonite. Vacuum 46, 1391–1395.

Blesa, M.A., Morando, P.J., Ragzzonis, A.E., 1994. Chemical Dissolution of Metal Oxides. CRC Press, Boca Raton, p. 401.

Boldyrev, V.V., (Ed.), 1996, Reactivity of Solids: Past, Present and Future. Blackwell Science, Oxford.

Brese, N.E., O'Keeffe, M., 1991. Bond-valence parameters for solids. Acta Crystallogr. B 47, 192–197.

Brönsted, J.N., 1923. Einige bemerkungen über den begriff der Säuren und basen. Recl. Trav. Chim. Pays-Bas 42, 718–728.

Burdett, J.K., 1995. Chemical Bonding in Solids. Oxford University Press, New York, p. 319.

Burdett, J.K., 1997. Chemical Bonds a Dialog. Wiley, Chichester, p. 166.

Burkhard, D.J.M., de Jong, B.H.W.S., Meyer, A.J.M.M., van Lenthe, J.H., 1991. $H_2Si_2O_7$: Ab initio molecular orbital calculations show two geometric conformations. Geochim. Cosmochim. Acta 55, 3453–3456.

Ching, W.Y., 2000. First principles calculation of the electronic structures of crystalline and amorphous forms of SiO_2. In: Devine, R.A.B., Duraud, J.-P., Dooryhee, E. (Eds.), Structure and Imperfections in Amorphous and Crystalline Silicon Dioxide. Wiley, pp. 168–179.

De Jong, B.H.W.S., 1989. Glass, Ullmann's encyclopedia of chemistry and physics. VCH Verlag Weinheim, Ser. A 12, 365–432.

De Jong, B.H.W.S., Adams, J.W., Aitken, B.A., Dickinson, J.E., Fine, G.J., 1989. Glass ceramics, Ullmann's encyclopedia of chemistry and physics. VCH Verlag Weinheim, Ser. A 12, 433–448.

De Jong, B.H.W.S., Brown, G.E., 1980. Polymerization of silicate and aluminate tetrahedra in glasses, melts, and aqueous solutions. I. Electronic structure of H_6Si_2O7, $H_6AlSiO7^{1-}$, and $H_6Al_2O_7^{2-}$. Geochim. Cosmochim. Acta 44, 491–511.

De Jong, B.H.W.S., Ellerbroek, D., Spek, A.T., 1994. Low temperature structure of lithium nesosilicate, Li_4SiO_4, and its Li_{1s} and O_{1s} X-ray photoelectron spectrum. Acta Crystallogr. B 50, 511–518.

De Jong, B.H.W.S., Schramm, C.M., Parziale, V.E., 1983. Polymerization of silicate and aluminate tetrahedra in glasses, melts, and aqueous solutions. IV Aluminum coordination in glasses and aqueous solution and comments on the aluminum avoidance principle. Geochim. Cosmochim. Acta 47, 1223–1236.

De Jong, B.H.W.S., Super, H.T.J., Frijhoff, R.M., Spek, A.L., Nachtegaal, G., 2000. Mixed alkali systems: Dietzel's theorem, X-ray structure, hygroscopicity, and ^{29}Si MASNMR of $NaRbSi_2O_5$ and $NaCaSi_2O_5$. Zeitschrift fuer Kristallographie 215, 397–405.

Ding, M., de Jong, B.H.W.S., Niessen, S.J.B., van der Meer, J.P.M., van Hinsberg, V.J., 2005. Black ferric oxyhydroxide and its sorption of toxic substances. Geochimica 34, 41–50.

Ding, M., de Jong, B.H.W.S., Roosendaal, S.J., Vredenberg, A., 2000. XPS studies on the electronic structure of bonding between solids and solutes: Adsorption of arsenate, chromate, phosphate, Pb^{2+} and Zn^{2+} ions on amorphous black ferric oxyhydroxide. Geochim. Cosmochim. Acta 64, 209–219.

El Shafei, G.M.S., 1996. The polarizing power of metal cations in (hydr)oxides. J. Colloid Interf. Sci. 182, 249–253.

Flood, H., Forland, T., 1947. The acidic and basic properties of oxides. Acta Chem. Scand. 1, 592–604.

Flood, H., Forland, T., Roald, B., 1947. The acidic and basic properties of oxides III relative acid–base strengths of some polyacids. Acta Chem. Scand. 1, 790–798.

Gijzeman, O.L.J., Mens, A.J.M., van Lenthe, J.H., Mortier, W.J., Weckhuysen, B.M., 2003. The effect of chemical composition and structure on XPS binding energies in zeolites. J. Phys. Chem. B 107, 678–684.

Gillespie, R.J., Popelier, P.L.A., 2001. Chemical Bonding and Molecular Geometry. Oxford University Press, New York, p. 268.

Goldschmidt, V.M., 1923. Geochemische Verteilungsgesetze der Elemente. Videnskapsselskapets Skrifter. I. Mat.-naturv. Klasse 3, 3–17.

Huheey, J.E., Keiter, E.A., Keiter, R.L., 1993. Inorganic Chemistry, fourth ed.. HarperCollins, New York, p. 964.

Izumi, Y., Urabe, K., Onaka, M., 1992. Zeolite, Clay, and Heteropoly Acid in Organic Reactions. VCH, Weinheim, p. 166.

Jensen, W.B., 1995. A quantitative van Arkel diagram. J. Chem. Educ. 72, 395–398.

Kamiya, H., Shimokata, A., 1974. The role of salts in the dissolution of powdered quartz. In: Prague Cadek, J., Paces, T. (Eds.), Proc. Int. Symp. Water–Rock Interact.. Czech Geological Survey, pp. 426–429.

Kosmulski, M., 1997. Attempt to determine pristine points of zero charge of Nb_2O_5, Ta_2O_5, and HfO_2. Langmuir 13, 6315–6320.

Kosmulski, M., 2002a. The pH-dependent surface charging and the point of zero charge. J. Colloid Interf. Sci. 253, 77–86.

Kosmulski, M., 2002b. Electric charge density of silica, alumina, and related surfaces. Encyclopedia of Surface and Colloid Science, Ser A, Hubbard ed. Dekker, New York, pp. 1627–1636.

Kosmulski, M., 2003. Point of zero charge of a corundum-water interface probed with optical second harmonic generation and atomic force microscopy: new approaches to oxide surface charge, Comment. Geochim. Cosmochim. Acta 67, 319–320.

Kosmulski, M., 2004. pH dependent surface charging and points of zero charge II Update. J. Colloid Interf. Sci. 275, 214–224.

Kragten, J., 1978. Atlas of Metal–Ligand Equilibria in Aqueous Solution. Ellis Horwood, New York, p. 781.

Liebau, F., 1985. Structural Chemistry of Silicates—Structure, Bonding, and Classification. Springer-Verlag, Heidelberg, p. 347.
Maddox, J., 1988. Crystals from first principles. Nature 335, 201.
Marians, C.S., Burdett, J.K., 1990. Geometric constraints: A refined model for the structure of silica glass. J. Non-Cryst. Solids 124, 1–21.
Mori-Sanchez, P., Martin Pendas, A., Luana, V., 2002. A classification of covalent, ionic, and metallic solids based on the electron density. J. Am. Chem. Soc. 124, 14721–14723.
Morrison, W.H., 1985. Stabilization of aqueous oxide pigment dispersions. J. Coat. Technol. 57, 55–65.
Murphy, L.R., Meek, T.L., Allred, A.L., Allen, L.C., 2000. Evaluation and testing of Pauling's electronegativity scale. J. Phys. Chem. 104, 5867–5871.
Parks, G., 1965. The isoelectric point of solid oxides, solid hydroxides, and aqueous hydroxo complex systems. Chem. Rev. 65, 177–198.
Popelier, P., 2000. Atoms in Molecules: An Introduction. Prentice Hall, New Jersey, p. 164.
Rich, R.L., 1985. Periodicity in the acid–base behavior of oxides and hydroxides. J. Chem. Educ. 62, 44–46.
Smith, D.W.J., 1987. An acidity scale for binary oxides. J. Chem. Educ. 64, 480–481.
Sposito, G., 2004. The Surface Chemistry of Natural Particles. Oxford University Press, New York, p. 242.
Stoecker, M., 1996. X-ray photoelectron spectroscopy on zeolites and related materials. Microporous Mater. 6, 235–257.
Tanabe, K., 1970. Solid Acids and Bases. Academic Press, New York, p. 175.
Tanabe, K., Holderich, W.F., 1999. Industrial application of solid acid–base catalysis. Appl. Catal. A-Gen. 181, 399–434.
Winchell, A.N., Winchell, H., 1964. The microscopical characters of artificial inorganic solid substances: Optical properties of artificial minerals. Academic Press, New York, p. 439.

Developments in Environmental Science, Volume 5
D. Sarkar, R. Datta and R. Hannigan (Editors)

ISSN: 1474-8177/DOI:10.1016/S1474-8177(07)05031-0

Chapter 31

Arsenic speciation in soils: An analytical challenge for understanding arsenic biogeochemistry

Guangliang Liu and Yong Cai

Abstract

The mobility, fate, bioavailability, and overall biogeochemical cycling of arsenic in soil depend on the individual form in which arsenic exists as well as physical and chemical characteristics of the soil. Speciation analysis of arsenic poses a great challenge and is the key to understand the biogeochemistry of arsenic in soils and sediments. The speciation of arsenic in soil may be defined functionally (e.g., bioavailable fraction), operationally, or chemical species-specifically. The extraction of arsenic species from solid matrices generally constitutes a critical step for speciation analysis of arsenic in soils and sediments. Various sequential extraction procedures have been developed to extract operationally defined arsenic species such as water soluble, exchangeable, Fe/Mn oxides associated, organic matter bound, and residual arsenic fractions. Several extractants have been used to leach chemically specific arsenic species, mainly arsenite, arsenate, monomethylarsonic acid (MMA), and dimethylarsinic acid (DMA) from a solid matrix. Efforts have been made to preserve the integrity of arsenic species during extraction. In this paper, extraction procedures with emphasis on the selection of extractants for chemical-species specifically and operationally defined arsenic species were reviewed. In addition, extraction method for evaluating bioavailable arsenic and spectroscopic technique for directly measuring arsenic species in soils were briefly discussed.

31.1. Introduction

31.1.1. Arsenic in the environment

Arsenic is the 20th most abundant element found in the Earth's crust and is ubiquitously distributed in soils, rocks, water, air, plants, and animals

(Cullen and Reimer, 1989; Cotton et al., 1999). It exists in inorganic and organic forms and in different oxidation states (−3, 0, +3, +5) (Frankenberger, 2002; Meng et al., 2003; Watt and Le, 2003). In recent years, there has been a renewed interest in the biogeochemistry of arsenic because of the frequently reported health problems associated with exposure to some toxic arsenic compounds worldwide.

The sources of arsenic in the environment include both natural and anthropogenic. More than 200 different arsenic mineral forms exist in nature, among them about 60% are arsenates, 20% sulfides and sulfosalts, and the remaining 20% include arsenides, arsenites, oxides, silicates, and elemental arsenic (As) (Onishi, 1969). The total arsenic amount in the earth's crust is estimated to be 4.01×10^{16} kg, and the natural arsenic liberation from the lithosphere into the exogenic cycle is about 1.715×10^{7} kg/year, mainly from volcanic exhalations and eruptions. About 4.87×10^{6} kg/year is added due to the submarine volcanism (Matschullat, 2002). Other natural activities that can release arsenic into the environment include erosion of rocks and forest fires (Smedley and Kinniburgh, 2002).

Arsenic compounds have been used in such fields as agriculture, metallurgy, and industry (Zhang et al. 1996; Mandal and Suzuki, 2002). These anthropogenic arsenic sources include insecticides, herbicides, desiccants, wood preservatives, high-temperature combustion (oil and coal burning power plants or waste incineration or cement works), mining process of metals, wastes from intense husbandry (disinfectants), glass ware production (discoloring agent), and electronics industries (admixture in semiconductor production, arsenide as laser material to convert electrical energy into coherent light) (Matschullat, 2002; Smedley and Kinniburgh, 2002). It was estimated that anthropogenic levels of arsenic exceeded natural levels in the environment at a ratio of 3 to 1 (Woolson, 1983).

31.1.2. Arsenic speciation and biogeochemistry

Arsenic is present in environmental and biological systems in a number of forms and species. A number of physiochemical and biological factors can affect arsenic speciation in the environment. It has been found that nearly two dozen arsenic species including organic and inorganic forms are present in the environmental and biological systems (Gong et al., 2002) despite the fact that arsenic is mostly present in its inorganic forms of arsenite (As^{III}) and arsenate (As^{V}) as a component in minerals. The two inorganic arsenic oxyanions, arsenite and arsenate, are widely distributed in the environment and are the major arsenic species found in most water, soil, and sediment samples. Monomethylarsonic acid (MMA), dimethylarsinic acid (DMA), arsenobetaine, arsenocholine, and arsenosugars have been found as common organic arsenic compounds in environmental and biological samples. In many biological systems, arsenic is present mainly in organic forms such as arsenobetaine and arsenocholine. For instance, arsenobetaine exists widely in higher organisms including both invertebrates and vertebrates (Phillips, 1994; Edmonds and Francesconi, 2003; Kuehnelt and Goessler, 2003; 2002) and is by far the major form of arsenic in marine animals, often representing 80% of the extractable arsenic with concentrations ranging from 1–300 $\mu g\,g^{-1}$ dry mass (Francesconi and Edmonds, 1997; Francesconi and Kuehnelt, 2002).

In soils and sediments, arsenic is mainly present in inorganic forms, namely arsenite and arsenate. Although arsenate is the thermodynamically favored form under normal environmental conditions, transformation between arsenite and arsenate can easily occur in the presence of certain microorganisms under both aerobic and anaerobic conditions. As a result, it is likely inaccurate to predict the predominant form of arsenic in soils only based on redox potential in the specific environment. MMA and DMA are major organic arsenic compounds in soils and sediments, but they are generally present in much lower concentrations compared to arsenite and arsenate (Demesmay and Olle, 1997; Kuehnelt and Goessler, 2003). These methylated species are probably produced through microorganisms-mediated oxidation–reduction reactions and can also be demethylated to inorganic arsenic species under specific conditions (Garcia-Manyes et al., 2002; Craig et al., 2003; Kuehnelt and Goessler, 2003). These two organic species, as well as arsenite and arsenate, play important roles in the biogeochemical cycling of arsenic in soil and sediment environment.

The environmental fate, bioavailability, and toxicity of arsenic vary dramatically with the chemical forms of arsenic. The two inorganic forms of arsenic were previously known to be much more toxic than their organic forms and arsenite has higher toxicity than arsenate (Jain and Ali, 2000). Recent studies, however, suggested that monomethylarsinous acid (MMA^{III}) and dimethylarsinous acid (DMA^{III}) were at least as toxic as, if not so more than, arsenite (Lin et al., 2001; Mass et al., 2001; Petrick et al., 2001; Le et al., 2000, 2004). Dimethylarsinic acid (DMA^{V}), once viewed simply as a detoxification product of inorganic arsenic, has been proven to be a complete carcinogen in the rat urinary bladder (Wei et al., 1999; Kenyon and Hughes, 2001). It is believed that arsenite is more mobile than arsenate in most environmental conditions, suggesting that arsenite in soils is subject to leaching into soil solution and thus is able to move longer distances both in horizontal and vertical directions compared with arsenate. The knowledge of the chemical forms of arsenic is critical for the understanding of arsenic mobilization to the aqueous phase in equilibrium with the soils or sediments and for the assessment of environmental impacts of arsenic. Frequently the lack of the speciation information is the major limitation to our understanding of the biogeochemical cycling of arsenic. Therefore, it has become very important and challenging to accurately quantify the amount of each form of arsenic that is present in the environment through speciation analysis.

31.1.3. Arsenic speciation analysis

Speciation analysis of an element has been defined by Florence (1982) as the determination of the concentrations of the individual physico-chemical forms of the element in a sample that together, constitute its total concentration. Speciation analysis involves a complex scheme of operations. Numerous publications concerned with speciation analysis of an element have appeared. Depending on the research interests, this term has been used in a number of different ways in such fields as environmental chemistry, biogeochemistry, biochemistry, and environmental toxicology. In a recent attempt to end the present confusion regarding the usage of the term speciation (analysis), IUPAC has defined speciation analysis as analytical activities of identifying and/or measuring the quantities of one or more individual chemical species in a particular system (Templeton et al., 2000). This definition confines speciation to chemical species that refer to isotopic composition, electronic or oxidation state, and/or complex or molecular structure. Such activities as classification of an analyte or a group of analytes from a certain sample according to physical (e.g., size, solubility) or chemical (e.g., bonding, reactivity) properties are defined as fractionation. The fractionation was often called speciation and regarded as a type of speciation analysis (e.g., Van Herreweghe et al., 2003). The following discussion adopted this approach and defined fractionation as a part of speciation analysis.

Arsenic speciation analysis of soils has been carried out in three different ways which can be defined as: chemical species-specifically, operationally, and functionally. The chemical species-specifically defined speciation analysis involves various levels of structure in the IUPAC definition including electronic or oxidation state, inorganic compounds and complexes, organoarsenic compounds, and organic and macromolecular complexes. This is the most common type of arsenic speciation. Various inorganic and organic arsenic compounds may be further combined with inorganic (e.g., Fe and Mn oxides) or organic ligands (e.g., thiol and humic substances) to produce coordination complexes with different degrees of thermodynamic stability and lability (Templeton et al., 2000; Gong et al., 2002). The operationally defined speciation analysis differentiates these arsenic species according to the extractants used. For example, arsenic species can be categorized into water soluble, exchangeable, Fe/Mn oxides associated, organic matter bound, and residual fractions or mobile and mobilizable fractions. The functionally defined speciation analysis has recently been used to extract and evaluate the bioaccessible arsenic fraction, i.e., the fraction that dissolves in the gastrointestinal tract and is available to be absorbed into the blood (bioavailable). From the point of view of biogeochemistry, the most important structure levels of arsenic speciation are arsenic compounds with different oxidation states and chemical bonds, those physically distinguished forms present with different molecular sizes (particulate, colloidal, and dissolved phases), and those with different operationally defined forms (Tessier et al., 1979; Cai, 2003).

31.2. Arsenic speciation analysis in soil

The operationally defined as well as chemical species-specifically defined speciation analyses of arsenic in soil have been extensively studied whereas few studies have dealt with functionally defined arsenic species. Different sequential extraction procedures (SEP) have been developed for the operationally defined speciation analysis of arsenic for the purpose of assessing mobility of arsenic. Meanwhile, several sets of extraction and analysis procedures for chemical specifically defined arsenic species in soil, mainly arsenite, arsenate, MMA and DMA, have also been studied. All these studies, however, had limited success. For operationally defined speciation analysis, the specificity and the reproducibility of SEP were usually questionable. For chemical specifically defined speciation analysis, analytical protocols are not available to date to achieve satisfactory recoveries as well as avoidance of species transformation during extraction steps.

Accurate measurement of arsenic species in soil is determined by a number of factors (Loeppert et al., 2003). Most important of these is the selection of suitable extraction procedures that can selectively extract/release target arsenic species (different phase association forms or oxidation states) from the solid matrix while at the same time protecting the original integrity of arsenic species. There is a large number of suitable methods for the identification of the dissolved arsenic compounds; these include: hydride generation (HG) atomic fluorescence or absorbance spectrometry (AFS or AAS) and high performance liquid chromatography (HPLC) coupled with inductively coupled plasma mass spectrometry (ICPMS) or AFS. Combinations of other chromatographic separation with spectrometric detection have also been utilized to arsenic speciation analysis although to a lesser extent. Details in instrumentation for arsenic speciation analysis were previously reviewed (Guerin et al., 1999; Vilanó et al., 2000; Gong et al., 2002) and therefore are not discussed further in this paper.

The extraction of arsenic species constitutes a critical step for speciation analysis of arsenic in soils and sediments. This is true for chemical species-specifically defined as well as operationally defined speciation analyses. There are some exceptions in cases where direct spectroscopic techniques have been applied to determine arsenic speciation. Sequential extraction procedures are usually adopted to differentiate operationally defined arsenic species while simple extraction using a selective extractant for arsenite or arsenate is applied to distinguish arsenic species with different oxidation states. Both extraction procedures are generally done by mechanical shaking. Recently, these extraction procedures have been simplified by introducing ultrasonic or microwave-assisted extraction techniques (Thomas et al., 1997). In the following text, direct spectroscopic speciation techniques are briefly introduced and then extraction procedures with the emphasis on the selection of extractants for chemical species-specifically and operationally defined arsenic species are reviewed in detail.

31.2.1. Direct arsenic speciation in soil using spectroscopic techniques

The identification of arsenic-bearing phases in solid samples, especially in minerals or mineral waste, has been attempted using spectroscopic techniques without sample extraction. The commonly used instruments include extended X-ray absorption fine-structure spectroscopy (EXAFS) (Foster et al., 1998; Savage et al., 2000), scanning electron microscope-energy dispersive spectrometer (SEM-EDS), X-ray powder diffraction (XRD) (Jones et al., 1997; Juillot et al., 1999), and electron microprobe analysis (Davis et al., 1996). These direct spectroscopic techniques, however, are limited by their sensitivity for many real environmental samples.

By using EXAFS for arsenic analysis in mine tailings, Foster et al. (1998) observed the formation of scorodite ($FeAsO_4 \cdot 2H_2O$) in mining wastes and also suggested that arsenates could be substituted by sulfates in the crystalline structure of jarosite ($KFe_3(SO_4)_2(OH)_6$). Savage et al. (2000) recently confirmed this substitution by using the same technique. By using electron microprobe analysis, Davis et al. (1996) observed the formation of metal arsenic oxides, iron–arsenic oxide similar to scorodite ($FeAsO_4 \cdot 2H_2O$) and arsenic phosphates such as $FeAlPbAs(PO_4)(SO_4)$ in smelter-impacted soils.

In addition to studying arsenic-bearing phase in minerals, direct spectroscopic techniques have also been used to determine arsenic oxidation state and coordination environment of arsenic-metal oxides complex in soils. A shift in the X-ray absorption near edge structure (XANES) spectra to a higher energy (e.g., from ~11,871 to ~11,874 eV) can be an indication of oxidation of As^{III} to As^{V} (Manning et al., 2003; Smith et al., 2005). EXAFS is another method to differentiate As^{V} from As^{III}. This can be done by comparing the coordination number (N) and interatomic distance (R_{As-O}) of the As–O shell, a shell of oxygen atoms surrounding the As atom which is formed when As is associated with metal oxides in soils. In theory, there are three O atoms surrounding As for As^{III} ($N = 3$) while four O atoms for As^{V} ($N = 4$). It was known by EXAFS that R_{As-O} values for As^{III} and As^{V} were approximately 1.79 Å and 1.69 Å, respectively (Fendorf et al., 1997; Foster et al., 1998; Manning et al., 1998). By fitting adjustable parameters of the theoretical EXAFS function to the experimental data, Manning et al. (2003) obtained an N value of 3.54 at a distance (R_{As-O}) of 1.71–1.72 Å for Aiken clay and Wyo loam soil samples fortified with 20 ml of 67 μM As^{III} to 2 g soil. The obtained N and R_{As-O} values provided evidences that As^{III} initially spiked has been partially oxidized to As^{V} so that a mixture of As^{III} and As^{V} was observed in these soils. Also in this study, a second shell of Fe atoms (As–Fe shell) further away from the As–O shell was observed, indicating that arsenic was predominantly combined to Fe oxide surface. In addition to X-ray absorption spectroscopy (XANES and EXAFS), the energy dispersive X-ray microanalysis (EDXMA) was also used to investigate association forms of arsenic in solid phase. Using EDXMA technique combined with SEP, Lombi et al. (2000) confirmed that 0.2 M NH_4-oxalate and 0.2 M NH_4-oxalate buffer + ascorbic acid could extract amorphous and poorly crystalline hydrous oxides of Fe and Al associated and well-crystallized hydrous oxides of Fe and Al bound arsenic fractions from five Cambisol soils in Austria, respectively.

31.2.2. Sequential extraction

The sequential extraction technique used for studying solid-phase association of elements, originally proposed by Tessier et al. (1979) and modified by many authors (e.g., Lum and Edgar, 1983; Voigt et al., 1996; Roussel et al., 2000), has been extensively applied to metals and metalloids. This procedure involves sequential chemical extractions separating trace metals into five fractions: exchangeable, bound to carbonates, bound to Fe–Mn oxides, bound to organic matter, and residual. Simplified sequential extraction methods have also been developed for practical applications (Sahuquillo et al., 1999, 2003; Quevauviller, 2002). For example, the three-level evaluation system, which separates metals into mobile, mobilizable, and pseudo total metal fractions, seems to be attractive because of its simplicity and potential application in risk assessment and risk management (Gupta, 1996; Cai et al., 2002).

Sequential extraction procedures provide operational definition of the metal phases one is trying to measure. Strictly speaking, the measurements performed by sequential extractions correspond to fractionation studies and not to real speciation measurements. Sequential extraction was developed to evaluate the lability and availability of trace metal in real environment, but the physicochemical conditions used during extraction procedures (strong reagents and short extraction time) did not exactly coincide the naturally occurring processes (weak reagents and slow kinetics). Additionally, the extractants used in each extraction step are not specifically selective for the target phase of trace metal. It has been observed that same extractants could extract different phase of metal species when applied to different samples. The validation of sequential extraction protocols are hard to achieve, though several attempts have been made to do so (Rendell et al., 1980; Tipping et al., 1985; Kheboian and Bauer, 1987; Martin et al., 1987). The specificity and reproducibility of the method greatly depend upon the chemical properties of the element and the chemical composition of the samples. Accordingly, these methods provide, at best, a gradient for the physicochemical association strength between trace elements and solid particles rather than their actual speciation (Nirel and Morel, 1990).

Despite these inherent limitations, sequential extraction procedures have been extensively applied to assess the mobility, environmental fate and behavior, and environmental risk of trace metals in soils and sediments. Some useful information has been obtained in the fields of environmental chemistry and geochemistry through extensive individual validation and consideration of particular conditions. Most of the sequential extraction procedures are developed for metal cations (Martin et al., 1987; Hirner, 1992; Shan and Chen, 1993; Ngiam and Lim, 2001; Gleyzes et al., 2002), but recently speciation studies dealing with oxyanions such as arsenate (Gruebel et al., 1988; Keon et al., 2001; Fedotov et al., 2005) have been considerably extended. Table 31.1 summarizes some commonly used sequential procedures for arsenic speciation in soils and sediments.

Table 31.1. Sequential extraction protocols used for speciation analysis of arsenic in soil

Extractants	As fraction	Reference
1 M $MgCl_2$	Ionically bound As	Keon et al. (2001)
1 M NaH_2PO_4	Strongly adsorbed As	
1 N HCl	As coprecipitated with AVS, carbonates, Mn oxides, and very amorphous Fe oxyhydroxides	
0.2 M ammonium oxalate/ oxalic acid	As coprecipitated with amorphous Fe oxyhydroxides	
0.05 M Ti^{III}-citrate-EDTA-bicarbonate	As coprecipitated with crystalline Fe oxyhydroxides	
10 M HF	As oxides and As coprecipitated with silicates	
16 N HNO_3	As coprecipitated with pyrite and amorphous As_2S_3	
16 N HNO_3 + 30% H_2O_2	Orpiment and remaining recalcitrant As minerals	
$(NH_4)_2SO_4$ (0.05 M)	Non-specifically sorbed	Wenzel, et al. (2001); Fedotov, et al. (2005)
$(NH_4)H_2PO_4$ (0.05 M)	Specifically sorbed	
NH_4-oxalate buffer (0.2 M)	Amorphous Fe and Al oxides bound	
NH_4-oxalate buffer (0.2 M) + ascorbic acid (0.1 M)	Crystalline Fe and Al oxides bound	
HNO_3/H_2O_2	Residual	
NH_4Cl (1 M)	Soluble	Jones et al. (1997)
NaOH (0.1 M)	Non-occluded Fe/Al oxides	
Citrate-bicarbonate (0.3 M)	Amorphous Fe	
Citrate-dithionite-bicarbonate (0.3 M)	Reducible Fe	
HCl (1 M)	Ca-associated	
30% H_2O_2 + 0.8 M NH_4OAC	Organic matter	
Acids	Residual	
Anion-exchange resin	Exchangeable	McLaren et al. (1998)
$NaHCO_3$ (0.5 M)	Easily labile	
NaOH (0.1 M)	Adsorbed on Al/Fe	
HCl (1 M)	Ca-associated	
Aqua regia	Recalcitrant As	
Ultrapure water	Water soluble	Shiowatana et al. (2001)
$NaHCO_3$ (0.5 M)	Surface adsorbed	
NaOH (0.1 M)	Fe- and Al-associated	
HCl (1 M)	Carbonate-bound	
HNO_3-HF (1:1)	Residual	
Water	Water soluble	Van Herreweghe et al. (2003)
Anion exchange membrane strips	Exchangeable anions	

Table 31.1. (Continued)

Extractants	As fraction	Reference
NH_4F (0.5 M)	NH_4F-extractable	
NaOH (0.1 M)	NaOH-extractable	
Sodium citrate (0.5 M) + $NaHCO_3$ (1 M) + $Na_2S_2O_4$	Reducible	
HNO_3/H_2O_2	Oxidizable	
HCl + HNO_3 + HF	Residual	
HCH_3COO (0.1 M)	Exchangeable	Sahuquillo et al. (1999, 2003); Quevauviller (2002)
$NH_2OH \cdot HCl$ (0.5 M)	Reducible	
NH_4CH_3COO/H_2O_2	Oxidizable	
$NaNO_3$ (0.1 M)	Mobile	Cai et al. (2002)
KH_2PO_4 (0.1 M)	Mobilizable	
HNO_3/H_2O_2	Residual	

The sequential extraction procedures listed in Table 31.1 can be divided into three categories: traditional phase-association scheme, BCR (the Community Bureau of Reference) scheme, and three-level differential system (mobile, mobilizable, and residual). Traditional phase-association sequential extraction scheme differentiates arsenic species based on the classification system of Tessier et al. (1979) by choosing extraction reagents commonly used for metals, Se or P. For instance, based on the chemical similarity of P and As, modified versions of the sequential procedures for P (Chang and Jackson, 1957) have been adopted for As (Woolson et al., 1973). The extraction reagents in this scheme include NH_4Cl, NH_4F, NaOH, and H_2SO_4, corresponding to easily exchangeable, and Al-, Fe-, and Ca-associated As. However, the adaptability of extraction steps from SEP for P and Se is questionable when they are used for As fractionation. Some extractants that have been used in SEPs for P and Se, including NH_4NO_3, NaOAc, $NH_2OH \cdot HCl$, EDTA, NH_4OH, and NH_4F, were shown to either have only low extraction efficiency for As, or to be insufficiently selective or specific for the phases targeted. This was possibly caused by re-adsorption on other mineral phases during reductive and oxidative dissolution of As from a certain mineral phase as well as subsequent desorption of As in the next extraction step (Gruebel et al., 1988).

In order to overcome these limitations, Wenzel et al. (2001) integrated the Chang and Jackson (1957) procedure for P, the Saeki and Matsumoto (1994) procedure for Se, and the Han and Banin (1995) approach to extract carbonates-associated metal fractions into a five-step SEP for As. In this scheme, 0.05 M $(NH_4)_2SO_4$, 0.05 M $NH_4H_2PO_4$, 0.2 M NH_4-oxalate, 0.2 M NH_4-oxalate buffer + ascorbic acid, and HNO_3/H_2O_2, were sequentially used to extract non-specifically sorbed, specifically sorbed, amorphous and poorly crystalline hydrous oxides of Fe and Al associated, well-crystallized hydrous oxides of Fe and Al bound, and residual arsenic fractions. The results of repeatability and recovery tests suggested that this SEP was dependable. The application of this SEP in 20 soils showed a partitioning pattern of As among these five fractions as follows: amorphous and poorly crystalline hydrous oxides of Fe and Al associated (42.3%) > well-crystallized hydrous oxides of Fe and Al bound (29.2%) > residual (17.5%) > specifically sorbed (9.5%) > non-specifically sorbed (0.24%) (Wenzel et al, 2001).

The lack of uniformity in a number of SEP protocols resulted in little comparability amongst the data obtained with different schemes. In order to achieve comparability among various studies and among different laboratories, the European Commission attempted to establish well-defined and standardized sequential extraction schemes for assessing lability of trace metals, including arsenic in soil (Quevauviller, 1998a, 1998b, 2002). A three-step extraction scheme (acetic acid—step 1; hydroxylammonium chloride—step 2; and, hydrogen peroxide/ammonium acetate—step 3) has been developed and improved by the addition of a fixed volume of dilute nitric acid into hydroxylamine hydrochloride to adjust pH to 1.5 in Step 2. Corresponding with these three steps, exchangeable (associated mainly with carbonates), reducible (associated with Fe/Mn oxides), and oxidizable (associated with organic matter) metal fractions were extracted, respectively. Speciation of these three metal fractions was known to be a good estimation of extractable metal in soils (Sahuquillo et al., 2003). This three-step sequential extraction protocol has been used for developing the sediment certified reference material (CRM 601) (Quevauviller et al., 1997; Sahuquillo et al., 1999). The so-called BCR scheme now appears to be internationally recognized as a reference method for soil and sediment studies (Davidson et al., 1998; Martin et al., 1998; Gómez Ariza et al., 2000; Tokalioglu et al., 2000; Pérez Cid et al., 2001), and this is supported by the increasing number of scientific papers referring to this protocol.

In order to simplify the sequential extraction method and make it easy to use for practical application, the protocol for separating metals into mobile, mobilizable, and pseudo total metal fractions, has also been widely used (Gupta et al., 1996). In this scheme, the mobile fraction was generally extracted by neutral unbuffered salt solutions, such as $NaNO_3$, $CaCl_2$, and NH_4NO_3, whereas buffered and unbuffered complexing and chelating reagents like EDTA, DTPA + $CaCl_2$, and acetic acid have been proposed as extraction media for the mobilizable fraction. This simple sequential extraction procedure has been adopted to evaluate the mobility of arsenic in South Florida golf course soils by extracting mobile arsenic fraction with 0.1 M $NaNO_3$ and mobilizable arsenic with 0.1 M KH_2PO_4, respectively (Cai et al., 2002).

31.2.3. Extraction of arsenate and arsenite

Similar to arsenic speciation using SEP, difficulties in chemical specifically defined speciation analysis of arsenic in soils lie in the choice of the extraction media. A desirable extraction method has to quantitatively extract all arsenic species and to preserve the identity of the native species in the samples during extraction process. Moreover, the arsenic species in the soil extracts should be stable until analysis and the solvent used to extract samples should not interfere with the species analysis. The stability of arsenic species depends on a number of factors, such as the pH value and chemical composition of the extraction media, and soil properties (Wang et al., 1995). For example, Pantsar-Kallio and Manninen (1997) investigated the stability of dissolved As^{III} and As^{V} in soil extracts obtained using $NaNO_3$, K_2SO_4, $NaHCO_3$, and Na_2CO_3, in the pH range of 1 to 12. It was found that As^{V} was stable in the whole pH range whereas As^{III} was converted to As^{V} at a pH value of 1 and 12.

A number of studies have investigated the arsenic speciation involving As^{III} and As^{V} in soil and sediment samples using a variety of solvent extraction techniques (Chappell et al., 1995; Manning and Martens, 1997; Thomas et al., 1997; Pongratz, 1998; Ellwood and Maher, 2003; Pizarro et al., 2003). However, the results do not always reflect the original arsenic speciation status from the samples studied due to possible conversion between arsenic species, especially oxidation of As^{III} to As^{V} during the extraction process (Wang et al., 1995; Pantsar-Kallio and Manninen, 1997; Ellwood and Maher, 2003; Georgiadis, 2004). A few studies have shown limited success in preserving the original arsenic redox integrity during extraction of different arsenic species. Commonly used extraction methods for speciation analysis of As^{III}, As^{V}, MMA, and DMA are illustrated in Table 31.2 and some representative procedures were taken as examples and further discussed in detail in the following text.

It is worthy noting that no suitable CRMs of the soils and sediments are nowadays available for validating As speciation in such materials (Pongratz, 1998). Reference materials certified for total As content were therefore used for speciation analysis to validate extraction methods and to intercompare the results with other laboratories. The validation and

Table 31.2. Extractants used for speciation analysis of arsenic in soil

Extractants	Extracting method	Sample type	Arsenic species targeted	Recovery for total arsenic (%)	Reference
Hydroxylammonium hydrochloride	Occasional shaking at 95°C for 8 h	Estuarine sediments	As^{V}, As^{III}	98.2[a]	Gómez-Ariza et al. (1998)
		CRM 320	As^{V}, As^{III}	67.0[b]	
Orthophosphoric acid	Microwave irradiation at 40 W for 20 min	CRM 320	As^{V}, As^{III},	93.6–96.5[b]	Gallardo et al. (2001)
		SRM 2709	As^{V}, As^{III},	37.1–61.6[b]	
		CRM 007-040	As^{V}, As^{III}, MMAA, DMAA	92.2–99.1[b]	
		Indonesian soil/ sediment	As^{V}, As^{III}, MMAA[c], DMAA[c]	72.5–96.6[a]	
		CRM 320	As^{V}, As^{III},	81.9[b]	Thomas et al. (1997)
		CRM 141	As^{V}	50.5[b]	
		IAEA Soil-7	As^{V}	86.6[b]	
		In-house soil	As^{V}	81.0[a]	
		In-house sediment	As^{V}, As^{III}	99.8[a]	
	Heated at 150 °C for 3 h	Contaminated soil	As^{V}, As^{III}, MMA, DMA	82	Pizarro et al., 2003
Phosphoric acid and ascorbic acid	Microwave irradiation at 20–60 W for 10 min	GBW07405	As^{V}, As^{III}, MMA	81.7[b]	Garcia-Manyes et al. (2002)
		GBW07311	As^{V}, As^{III},	101[b]	
		BCR 320	As^{V}, As^{III},	95.8[b]	
		Soil	As^{V}, As^{III}, MMA[c], DMA[c]	56.5–101[a]	

Table 31.2. (*Continued*)

Extractants	Extracting method	Sample type	Arsenic species targeted	Recovery for total arsenic (%)	Reference
Phosphoric acid and hydroxylamine hydrochloride	Agitated on a mixing wheel for 1 h	BCSS-1	As^{V}, As^{III},	32–72	Ellwood and Maher (2003)
		NIST 1646			
		PACS-1			
		NIST 1941			
KH_2PO_4 and K_2HPO_4	Shaken on a reciprocating shaker for 2 h	MESS-1	As^{V}, As^{III},	<5	Manning and Martens (1997)
		PACS-1			
		NIST 1646			
		NIST 1633b			
Ammonium oxalate	Agitating at room temperature for 1 h	Contaminated soil	As^{V}	0.04–20[a]	Bissen and Frimmel (2000)
Sodium carbonate			As^{V}	0.04–20[a]	
Sodium bicarbonate			As^{V}	0.04–20[a]	
hydrochloric acid—chloroform water	Shaken vigorously for 30 min	Soil	As^{V}, As^{III}, organic As	85	Chappell et al. (1995)
10 mM phosphate + 0.5% NaDDC	Shaken at 300 rpm on a orbital shaker for 1–24 h	PACS-2	As^{V}, As^{III}	7	Georgiadis (2004); Georgiadis et al. (2006)
		Everglades soil			

[a]Recoveries are calculated against total arsenic concentration obtained through strong acid (e.g., nitric acid) digestion procedure.
[b]Recoveries are calculated against certified value of total arsenic concentration.
[c]MMA or DMA were detected only in few cases for real soil/sediment samples.

intercomparisons are questionable, at least in some cases, because the speciation of As in reference materials may have been modified by the rather energetic treatments to which these samples have been submitted during their preparation (Gallardo et al., 2001).

Hydrochloric acid was a reagent used for the extraction of As^{III} and As^{V} from soil in earlier studies for speciation (Chappell et al., 1995). The strong acidic condition used is probably too aggressive and is suspected to lead to the alternation of arsenic species (Gong et al., 2000; Gallardo et al., 2001). In order to avoid species changing, some weak extractants such as phosphate, acetate, citrate, and oxalate buffers have been proposed for the selective extraction of As^{III}, As^{V}, MMA, and DMA (Bissen and Frimmel, 2000; Shi et al., 2003; Georgiadis et al., 2006).

In a study of evaluating the mobility and differentiating the chemical species of arsenic, an extraction procedure at different pH values similar to the natural environment has been developed (Bissen and Frimmel, 2000). Stepwise extractions were performed with 0.3 M ammonium oxalate ($pH = 3$), milli-Q water ($pH = 5.8$), 0.3 M sodium carbonate ($pH = 8$), and 0.3 M sodium bicarbonate ($pH = 11$). As^{III}, As^{V}, MMA, and DMA were monitored by HPLC-ICPMS in each extract. As^{V} was the predominant form while neither MMA nor DMA was found in the extracts of two contaminated soils sampled at a former tannery site and a former paint production site (Bissen and Frimmel, 2000). The concentrations of arsenic species in different extracts varied remarkably and the overall extraction yields were fairly low (0.04% for tannery site and 20% for paint production site, respectively), indicating that the mobilization of arsenic depended on the pH value of the extraction solution and the kind of extracted soil (Bissen and Frimmel, 2000). Similarly, an analytical procedure for determination of As^{III} and As^{V} in soil extracts obtained by sequentially applying water, 0.6 M KH_2PO_4 solution, 1% (v/v) HCl solution and 1% (w/v) NaOH solution was developed (Shi et al., 2003).

Phosphate was another frequently used mild extractant for leachable As species in solid samples. The similarity of chemical properties between phosphate and arsenate, such as ion size and acid dissociation constants, made the ion-exchange reaction between phosphate and arsenate occur easily. Through this kind of ion-exchange reaction, phosphate is able to release extractable As fractions from soils. However, it was found that As^{III} could be oxidized to As^{V} during phosphate extraction process. Georgiadis et al. (2005) attempted to add complexing reagents such as ethylenediaminetetraacetic acid (EDTA) and sodium diethyldithiocarbamate trihydrate (NaDDC) or reductant hydroxylamine hydrochloride ($NH_2OH \cdot HCl$) into phosphate to preserve As^{III} integrity. The results for a Florida Everglades peat soil with high content of organic matter and low content of Fe and Mn and marine sediment CRM PACS-2 with high content of Fe and Mn suggested that the addition of EDTA in the phosphate solution did not prevent As^{III} from oxidation in the extracts despite the fact that EDTA was an effective preservative for As^{III} in water samples. When 1% $NH_2OH \cdot HCl$ was added, it was effective to preserve As^{III} in a limited time (< 12 h). The addition of 0.5% NaDDC, however, minimized As^{III} oxidation and the recoveries of spiked As^{III} into these samples ranged from 80 to 120%. The tests on several other types of soil and sediment samples indicated that the combined phosphate and NaDDC solution seemed to be a promising extraction procedure for the speciation analysis of phosphate-extractable As^{III} and As^{V} (Georgiadis et al., 2006).

In addition to these mild extractants mentioned above, a simple extraction for arsenic species using certain particular reagent could be feasible for soils and sediments with specific characteristics. For instance, reducible reagents are likely efficient extraction media for solid samples with high content of Fe oxides because As species are predominantly associated with this phase. Gómez-Ariza et al. (1998) reported a procedure for the extraction and analysis of arsenic species (As^{III}, As^{V}, MMA, and DMA) from estuarine iron oxide rich sediments using $NH_2OH \cdot HCl$ solution as leaching agent. When samples were extracted with 40 ml of 0.4 M $NH_2OH \cdot HCl$ at 95°C for 8 h with occasional shaking, no changes in the arsenate oxidation state were observed despite the reducing character of the extractant. As^{V} was found to be the predictably predominant form while As^{III} was present as a minor constituent. A good recovery higher than 90% ($As^{III} + As^{V}$ versus total As concentration determined after *aqua regia* digestion) was obtained for this oxic estuary sediment sample after three repetitive extractions. However, low recovery (67.0%) was observed for CRM 320 river sediment, indicating the influence of the matrix composition on the extraction.

More recently, the use of phosphoric acid for the extraction of arsenic from soils has been studied. Several studies have investigated the influence of phosphoric acid concentration, extraction time, extraction method (shaking, heating, or microwave irradiation), and the addition of reducible agents such as $NH_2OH \cdot HCl$ and ascorbic acid on arsenic extraction efficiency and species integrity (Demesmay and Ollé, 1997; Thomas et al., 1997; Hegesen and Larsen, 1998; Martin et al., 2001; Garcia-Manyes et al., 2002; Ellwood and Maher, 2003; Pizarro et al., 2003). The efficiency of orthophosphoric acid extraction was shown to be related with acid concentration. High concentrations of phosphoric acid could extract more arsenic species from the soil, but the oxidation of As^{III} to As^{V} was also observed at high concentration (Hegesen and Larsen, 1998; Martin et al., 2001). Ascorbic acid as well as $NH_2OH \cdot HCl$ was observed to enhance the protection of As^{III} from oxidation. The recovery of spiked As^{III} increased up to 98.9% with addition of ascorbic acid from 93.7 without the addition of ascorbic acid (Garcia-Manyes et al., 2002) while recoveries between 89 and 104% were observed for As^{III} spiked into four oxic sediment CRMs and an anoxic sediment when $NH_2OH \cdot HCl$ was used as a sacrificial oxidant (Ellwood et al., 2003). Regarding extraction method, microwave assisted extraction seemed to be more efficient compared with mechanical shaking (Thomas et al., 1997).

Applying microwave-assisted phosphoric acid extraction method to an arsenic-contaminated soil sample, Garcia-Manyes et al. (2002) evaluated the stability of As species in the soil extracts. The extraction was conducted using a solution containing 1 M phosphoric acid and 0.1 M ascorbic acid and 60 W microwave power during 10 min extraction. The extract was then analyzed for the concentration of As^{III}, As^{V}, DMA, and MMA at different periods of time (0, 3, 6, and 7 days after extraction). The results showed that As^{III} remained practically the same as the original, whereas both DMA and MMA had decreased more than 50% after the extract was kept for 6 days at 4°C under darkness. After 7 days, As^{III}, DMA, and MMA were not detectable in the soil extract. The authors suspected that the original species As^{III}, DMA, and MMA were probably transformed into As^{V} (Garcia-Manyes et al., 2002).

From these studies, it seems that phosphoric acid combined with appropriate reducible agents was an efficient extraction method for the speciation analysis of As species in soil. Meanwhile, it should be noted that the efficiency of phosphoric acid extraction significantly depended upon the nature of the material analyzed. For example, excellent (90–100%) recoveries of total As were obtained for the sediment and the sludge reference materials (CRM 320 and CRM 007-040) whereas yield did not exceed 62% for the soil reference materials (SRM 2709) (Gallardo et al., 2001).

It should be kept in mind that most extraction procedures for arsenite and arsenate discussed above were developed to evaluate arsenic mobility or leachability in soils, which suggests that the extraction of arsenic species is incomplete and only labile or selected solvent extractable arsenic fractions in soils are extracted. When using 1 mM phosphate buffer to extract As from several sediment CRMs, very low extraction yields (at most 2.2%) were obtained (Manning and Martens, 1997). Georgiadis et al. (2006) extracted 7% of total arsenic from PACS-2 using 10 mM phosphate combined with 0.5% NaDDC, whereas Bissen and Frimmel (2000) extracted 0.04–20% of arsenic from contaminated soils using four different extractants, namely 0.3 M ammonium oxalate, Milli-Q water,

0.3 M sodium carbonate, and 0.3 M sodium bicarbonate. These extractions with low yields could provide an estimate for labile arsenic pool. However, information provided by these procedures may not be sufficient for overall evaluation of arsenic biogeochemistry. With the assistance of microwave, orthophosphoric acid extraction seemed to be efficient to some kinds of soils and sediments (Demesmay and Ollé, 1997; Thomas et al., 1997).

31.2.4. Extraction of functionally defined arsenic species

The functionally defined speciation analysis of As in soil was generally used to evaluate the bioavailability of As. In this extraction scheme, synthetic leaching fluids closely analogous to those of the human stomach and small intestine was utilized to release bioavailable As fraction from mining waste or contaminated soil at pH, temperature, and duration which mimics human gastro-intestinal tract environment. For example, Ruby et al. (1996) developed a physiologically based extraction test (PBET) to assess in vitro bioavailability of metals. The simulated gastric fluid was prepared by adding 1.25 g pepsin (activity of 800–2500 units mg^{-1}), 0.5 g sodium citrate, 0.5 g malate (DL-Malic acid, disodium salt 97%), 420 μl lactic acid syrup (free acid 85%) and 500 μl acetic acid into 1 l of de-ionized water, and adjusting pH to 2.5 with HCl. This method has been used for different types of samples (Williams et al., 1998; Rodriguez et al., 1999; Reimer et al., 2003). Williams et al. (1998), using a modified version of PBET, evaluated the post-ingestion bioavailability of arsenic in alluvial soil and mineral beneficiation waste in Thailand. It was found that an average stomach absorption of 11.2% of total soil As could occur, which provided a direct basis for assessment of As bioavailability following juvenile hand–mouth ingestion. Reimer et al. (2003) applied this test to a mining area at Yellowknife, Canada and found that As in crushed rock samples ($<12\%$ of total As extracted) was less bioavailable than As in lawn soils ($>30\%$ of total As extracted). The PBET provides a potentially valuable mechanism for refining risk assessments of sites subject to natural or anthropogenic As contamination.

31.3. Concluding remarks

Speciation analysis of arsenic in soils and sediments can provide crucial information in our understanding the biogeochemical cycle of arsenic and meanwhile is a challenge in the field of environmental and analytical chemistry. The speciation of arsenic in soil may be defined functionally, operationally, or chemical specifically. The extraction of arsenic species from solid matrices generally constitutes a critical step for speciation analysis of arsenic in soils and sediments. Sequential extraction procedures were used to extract operationally defined arsenic species (phase association forms) while simple extraction with selective extractants were applied to leach chemically specific arsenic species (As^{III}, As^{V}, MMA, and DMA) from a solid matrix. Simulated gastric fluid was usually used to release functionally defined As species (bioavailable As fraction). Additionally, direct spectroscopic technologies without an extraction step were also useful tools for speciation analysis of As, which could provide information for As phase association as well as As oxidation state. The combined application of different types of speciation analysis seems to be an encouraging area and will serve to give a better view of As biogeochemical cycling.

REFERENCES

Bissen, M., Frimmel, F.H., 2000. Speciation of $As(III)^{III}$, $As(V)^{V}$, MMA and DMA in contaminated soil extracts by HPLC-ICP/MS. Fresenius J. Anal. Chem. 367, 51–55.

Cai, Y., 2003. Biogeochemistry of environmentally important trace elements: Overview. In: Cai, Y., Braids, O.C. (Eds.), Biogeochemistry of Environmentally Important Trace Elements. American Chemical Society, Washington, DC, pp. 1–10.

Cai, Y., Cabrera, J.C., Georgiadis, M., Jayachandran, K., 2002. Assessment of arsenic mobility in the soils of some golf courses in South Florida. Sci. Total Environ. 291, 123–134.

Chang, S.C., Jackson, M.L., 1957. Fractionation of soil phosphorus. Soil Sci. 84, 286–293.

Chappell, J., Chiswell, B., Olszowy, H., 1995. Speciation of arsenic in a contaminated soil by solvent extraction. Talanta 42, 323–329.

Cotton, F.A., Wilkinson, G., Murillo, C.A., Bochmann, M., 1999. Advanced Inorganic Chemistry, sixth ed. Wiley, New York.

Craig, P.J., George, Eng., Jenkins, R.O., 2003. Occurrence and pathways of organometallic compounds in the environment: General consideration. In: Craig, P. (Ed.), Organometallic Compounds in the Environment. second ed. Wiley, pp. 1–55.

Cullen, W.R., Reimer, K.J., 1989. Arsenic speciation in the environment. Chem. Rev. 89, 713–764.

Davidson, C.M., Duncan, A.L., Littlejohn, D., Ure, A.M., Garden, L.M., 1998. A critical evaluation of the three-stage BCR sequential extraction procedure to assess the potential mobility and toxicity of heavy metals in industrially contaminated land. Anal. Chim. Acta 363, 45–55.

Davis, A., Ruby, M.V., Bloom, M., Schoof, R., Freeman, G., Bergstrom, P.D., 1996. Mineralogic constraints on the bioavailability of arsenic in smelter-impacted soils. Environ. Sci. Technol. 30, 392–399.

Demesmay, C., Ollé, M., 1997. Application of microwave digestion to the preparation of sediment samples for arsenic speciation. Fresenius' J. Anal. Chem. 357, 1116–1121.

Edmonds, J.S., Francesconi, K.A., 2003. Organoarsenic compounds in the marine environment. In: Craig, P. (Ed.), Organometallic Compounds in the Environment. second ed. Wiley, pp. 195–222.
Ellwood, M.J., Maher, W.A., 2003. Measurement of arsenic species in marine sediments by high-performance liquid chromatography-inductively coupled plasma mass spectrometry. Anal. Chim. Acta 477, 279–291.
Fedotov, P.S., Fitz, W.J., Wennrich, R., Morgenstern, P., Wenzel, W.W., 2005. Fractionation of arsenic in soil and sludge samples: Continuous-flow extraction using rotating coiled columns versus batch sequential extraction. Anal. Chim. Acta 538, 93–98.
Fendorf, S., Eick, M.J., Grossl, P., Sparks, D.L., 1997. Arsenate and chromate retention mechanisms on goethite. 1. Surface structure. Environ. Sci. Technol. 31, 315–320.
Florence, T.M., 1982. The speciation of trace elements in waters. Talanta 29, 354–373.
Foster, A.L., Brown, G.E., Tingle, T.N., Parks, G.A., 1998. Quantitative arsenic speciation in mine tailings using X-ray absorption spectroscopy. Am. Mineral. 83, 553–568.
Francesconi, K.A., Edmonds, J.S., 1997. Arsenic and marine organisms. In: Sykes, A.G. (Ed.), Advances in Inorganic Chemistry. Academic Press.
Francesconi, K.A., Kuehnelt, D., 2002. Arsenic compounds in the environment. In: Frankenberger, W.T. Jr. (Ed.), Environmental Chemistry of Arsenic. Marcel Dekker, New York, pp. 51–94.
Frankenberger, W.T.Jr., (Ed.), 2002, Environmental Chemistry of Arsenic. Marcel Dekker, New York.
Gallardo, M.V., Bohari, Y., Astruc, A., Potin-Gautier, M., Astruc, M., 2001. Speciation analysis of arsenic in environmental solids reference materials by high-performance liquid chromatography–hydride generation–atomic fluorescence spectrometry following orthophosphoric acid extraction. Anal. Chim. Acta 441, 257–268.
Garcia-Manyes, S., Jimenez, G., Padro, A., Rubio, R., Rauret, G., 2002. Arsenic speciation in contaminated soils. Talanta 58, 97–109.
Georgiadis, M., 2004. Aarsenic speciation in soils and sediments. M.S. Thesis. Florida International University, Miami.
Georgiadis, M., Cai, Y., Solo-Gabriele, H.M., 2006. Extraction of arsenate and arsenite species from soils and sediments. Environ. Pollut. 141, 22–29.
Gleyzes, C., Tellier, S., Astruc, M., 2002. Fractionation studies of trace elements in contaminated soils and sediments: A review of sequential extraction procedures. Trends Anal. Chem. 21, 451–467.
Gómez Ariza, J.L., Giraldez, I., Sánchez Roads, D., Morales, E., 2000. Comparison of the feasibility of three extraction procedures for trace metal partitioning in sediments from south–west Spain. Sci. Total Environ. 246, 271–283.
Gómez-Ariza, J.L., Sánchez-Rodas, D., Giráldez, I., 1998. Selective extraction of iron oxide associated arsenic species from sediments for speciation with coupled HPLC-HG-AAS. J. Anal. Atom. Spectrom. 13, 1375–1379.
Gong, Z., Lu, X., Ma, M., Watt, C., Le, X.C., 2002. Arsenic speciation analysis. Talanta 58, 77–96.
Gruebel, K.A., Davis, J.A., Leckie, J.O., 1988. The feasibility of using sequential extraction techniques for arsenic and selenium in soils and sediments. Soil Sci. Soc. Am. J. 52, 390–397.
Guerin, T., Astruc, A., Astruc, M., 1999. Speciation of arsenic and selenium compounds by HPLC hyphenated to specific detectors: A review of the main separation techniques. Talanta 50, 1–24.

Gupta, S.K., Vollmer, M.K., VM, R., Krebs, R., 1996. The importance of mobile, mobilizable and pseudo total heavy metal fractions in soil for three-level risk assessment and risk management. Sci. Total Environ. 178, 11–20.

Han, F.X., Banin, A., 1995. Selective sequential dissolution techniques for trace metals in arid-zone soils: The carbonate dissolution step. Commun. Soil Sci. Plant Anal. 26, 553–576.

Hegesen, H., Larsen, E.H., 1998. Bioavailability and speciation of arsenic in carrots grown in contaminated soil. Analyst 23, 791–796.

Hirner, A.V., 1992. Trace element speciation in soils and sediments using sequential chemical extraction methods. Int. J. Environ. Anal. Chem. 46, 77–85.

Jain, C.K., Ali, I., 2000. Arsenic: Occurrence, toxicity, and speciation techniques. Water Res. 34, 4304–4312.

Jones, C.A., Inskeep, W.P., Neuman, D.R., 1997. Arsenic transport in contaminated mine tailings following liming. J. Environ. Quality 26, 433–439.

Juillot, F., Ildefonse, Ph., Morin, G., Calas, G., De Kersabiec, A.M., Benedetti, M., 1999. Remobilization of arsenic from buried wastes at an industrial site: Mineralogical and geochemical control. Appl. Geochem. 14, 1031–1048.

Kenyon, E.M., Hughes, M.F., 2001. A concise review of the toxicity and carnogenicity of dimethylarsinic acid. Toxicology 160, 227–236.

Keon, N.E., Swartz, C.H., Brabander, D.J., Harvey, C., Hemond, H.F., 2001. Validation of an arsenic sequential extraction method for evaluating mobility in sediments. Environ. Sci. Technol. 35, 2778–2784.

Kheboian, C., Bauer, C.F., 1987. Accuracy of selective extraction procedures for metal speciation in model aquatic sediments. Anal. Chem. 59, 1417–1423.

Kuehnelt, D., Goessler, W., 2003. Organoarsenic compounds in the terrestrial environment. In: Craig, P. (Ed.), Organometallic Compounds in the Environment. second ed. Wiley, pp. 223–275.

Le, X.C., Lu, X.F., Li, X.F., 2004. Arsenic speciation. Anal. Chem. 76, 27A–33A.

Le, X.C., Lu, X.F., Ma, M., Cullen, W.R., Aposhian, H.V., Zheng, B., 2000. Speciation of key arsenic metabolic intermediates in human urine. Anal. Chem. 72, 5172–5177.

Lin, S., Del Razo, L.M., Styblo, M., Wang, C., Cullen, W.R., Thomas, D.J., 2001. Arsenicals inhibit thioredoxin reductase in cultured rat hepatocytes. Chem. Res. Toxicol. 14, 305–311.

Loeppert, R.H., Jain, A., El-Haleem, M.A.A., Biswas, B.K., 2003. Quantity and speciation of arsenic in soils by chemical extraction. In: Cai, Y., Braids, O.C. (Eds.), Biogeochemistry of Environmentally Important Trace Elements. American Chemical Society, Washington, DC, pp. 42–56.

Lombi, E., Sletten, R.S., Wenzel, W.W., 2000. Sequentially extracted arsenic from different size fractions of contaminated soils. Water Air Soil Pollut. 124, 319–332.

Lum, K.R., Edgar, D.G., 1983. The determination of arsenic by flame AAS using the Zeeman effect and its application to the analysis of sediments extracts. Int. J. Environ. Anal. Chem. 15, 241–248.

Mandal, B.K., Suzuki, K.T., 2002. Arsenic round the world: A review. Talanta 58, 201–235.

Manning, B.A., Fendorf, S., Goldberg, S., 1998. Surface structures and stability of arsenic(III) on goethite: Spectroscopic evidence for inner-sphere complexes. Environ. Sci. Technol. 32, 2383–2388.

Manning, B.A., Fendorf, S.E., Suarez, D.L., 2003. Arsenic(III) complexation and oxidation reactions on soil. In: Cai, Y., Braids, OC. (Eds.), Biogeochemistry of Environmentally Important Trace Elements. American Chemical Society, Washington, DC, pp. 57–69.

Manning, B.A., Martens, D.A., 1997. Speciation of As(III) and As(V) in sediment extracts by high-performance liquid chromatography-hydride generation atomic absorption spectrophotometry. Environ. Sci. Technol. 31, 171–177.

Martin, A., Lopez-gonzalvez, M.A., Barbas, C., 2001. Development and validation of extraction methods for determination of zinc and arsenic speciation in soils using focused ultrasound: Application to heavy metal study in mud and soils. Anal. Chim. Acta 442, 305–318.

Martin, J.M., Nirel, P.M.V., Thomas, A.J., 1987. Sequential extraction techniques: promises and problems. Mar. Chem. 22, 313–341.

Martin, R., Sánchez, D.M., Gutierrez, A.M., 1998. Sequential extraction of U, Th, Ce, La and some heavy metals in sediments from Ortigas river, Spain. Talanta 46, 1115–1121.

Mass, M.J., Tennant, A., Roop, B.C., Cullen, W.R., Styblo, M., Thomas, D.J., Kligerma, A.D., 2001. Methylated trivalent arsenic species are genotoxic. Chem. Res. Toxicol. 14, 355–361.

Matschullat, J., 2002. Arsenic in the geosphere—a review. Sci. Total Environ. 249, 297–312.

McLaren, R.G., Naidu, R., Smith, J., Tiller, K.G., 1998. Fractionation and distribution of arsenic in soils contaminated by cattle dip. J. Environ. Qual. 27, 348–354.

Meng, X., Jing, C., Korfiatis, G.P., 2003. A review of redox transformation of arsenic in aquatic environments. In: Cai, Y., Braids, OC. (Eds.), Biogeochemistry of Environmentally Important Trace Elements. American Chemical Society, Washington DC, pp. 70–83.

Ngiam, L.S., Lim, P.E., 2001. Speciation patterns of heavy metals in tropical estuarine anoxic and oxidized sediments by different sequential extraction schemes. Sci. Total Environ. 275, 53–61.

Nirel, P.M.V., Morel, F.M.M., 1990. Pitfalls of sequential extractions. Water Res. 24, 1055–1056.

Onishi, H., 1969. Arsenic. In: Wedepohl, K.H. (Ed.), Handbook of Geochemistry. Springer-Verlag, New York Vol. II-2, Chapter 33.

Pantsar-Kallio, M., Manninen, P.K.G., 1997. Speciation of mobile arsenic in soil samples as a function of pH. Sci. Total Environ. 204, 193–200.

Pérez Cid, B., Fernández Alborés, A., Fernández Gómez, E., Falqué López, E., 2001. Use of microwave single extractions for metal fractionation in sewage sludge samples. Anal. Chim. Acta 431, 209–218.

Petrick, J.S., Jagadish, B., Mash, E.A., Vasken Aposhian, H., 2001. Monomethylarsonous acid (MMA^{III}) and arsenite: LD_{50} in hamsters and in vitro inhibition of pyruvate dehydrogenase. Chem. Res. Toxicol. 14, 651–656.

Phillips, D.J.H., 1994. The chemical forms of arsenic in aquatic organisms and their interrelationships. In: Nriagu, J.O. (Ed.), Arsenic in the Environment, Part I. Cycling and Characterization. Wiley, New York, pp. 263–288.

Pizarro, I., Gómez, M., Cámara, C., Palacios, M.A., 2003. Arsenic speciation in environmental and biological samples: Extraction and stability studies. Anal. Chim. Acta 495, 85–98.

Pongratz, R., 1998. Arsenic speciation in environmental samples of contaminated soil. Sci. Total Environ. 224, 133–141.

Quevauviller, Ph., 1998a. Operationally-defined extraction procedure for soil and sediment analysis. Part 2: Certified reference materials. Trend Anal. Chem. 17, 632–642.

Quevauviller, Ph., 1998b. Operationally-defined extraction procedure for soil and sediment analysis. Part 1: Standardization. Trend Anal. Chem. 17, 289–298.

Quevauviller, Ph., 2002. Operationally-defined extraction procedure for soil and sediment analysis. Part 3: New CRMs for trace-element extractable contents. Trend Anal. Chem. 21, 774–785.

Quevauviller, Ph., Rauret, G., López-Sánchez, J.F., Rubio, R., Ure, A., Muntau, H., 1997. Certification of trace metal extractable contents in a sediment reference material (CRM 601) following a three-step sequential extraction procedure. Sci. Total Environ. 205, 223–234.

Reimer, K.J., Ollson, C.A., Koch, I., 2003. An approach for characterizing arsenic sources and risk at contaminated sites: Application to gold mining sites in Yellowknife, NWT, Canada. In: Cai, Y., Braids, O.C. (Eds.), Biogeochemistry of Environmentally Important Trace Elements. American Chemical Society, Washington, DC, pp. 166–180.

Rendell, P.S., Batley, G.E., Cameron, A.J., 1980. Adsorption as a control of metal concentration in sediments extracts. Environ. Sci. Technol. 14, 314–318.

Rodriguez, R.R., Basta, N.T., Casteel, S.W., Pace, L.W., 1999. An in vitro gastrointestinal method to estimate bioavailable arsenic in contaminated soils and solid media. Environ. Sci. Technol. 33, 642–649.

Roussel, C., Bril, H., Fernandez, A., 2000. Arsenic speciation: Involvement in evaluation of environmental impact caused by mine wastes. J. Environ. Qual. 29, 182–188.

Ruby, M.V., Davis, A., Schoof, R., Eberle, S., Sellstone, C.M., 1996. Estimation of lead and arsenic bioavailability using a physiologically based extraction test. Environ. Sci. Technol. 30, 422–430.

Saeki, K., Matsumoto, S., 1994. Selenite adsorption by a variety of oxides. Commun. Soil Sci. Plant Anal. 25, 2147–2158.

Sahuquillo, A., Lopez-Sanchez, J.F., Rubio, R., Rauret, G., Thomas, R.P., Davidson, C.M., Ure, A.M., 1999. Use of a certified reference material for extractable trace metals to assess sources of uncertainty in the BCR three-stage sequential extraction procedure. Anal. Chim. Acta 382, 317–327.

Sahuquillo, A., Rauret, G., Rehnert, A., Muntau, H., 2003. Solid sample graphite furnace atomic absorption spectroscopy for supporting arsenic determination in sediments following a sequential extraction procedure. Anal. Chim. Acta 476, 15–24.

Savage, K.S., Tingle, T.N., O'Day, P.A., Waychunas, G.A., Bird, D.K., 2000. Arsenic speciation in pyrite and secondary weathering phases, mother Lode Gold District, Tuolumne County, California. Appl. Geochem. 15, 1219–1244.

Shan, X.Q., Chen, B., 1993. Evaluation of sequential extraction for speciation of trace metals in model soil containing natural minerals and humic acid. Anal. Chem. 65, 802–807.

Shi, J.B., Tang, Z.Y., Jin, Z.X., Chi, Q., He, B., Jiang, G.B., 2003. Determination of As(III) and As(V) in soils using sequential extraction combined with flow injection hydride generation atomic fluorescence detection. Anal. Chim. Acta 477, 139–147.

Shiowatana, J., McLaren, R.G., Chanmekha, N., Samphao, 2001. A fractionation of arsenic in soil by a continuous-flow sequential extraction method. J. Environ. Qual. 30, 1940–1949.

Smedley, P.L., Kinniburgh, D.G., 2002. A review of the source, behavior and distribution of arsenic in natural waters. Appl. Geochem. 17, 517–568.

Smith, P.G., Koch, I., Gordon, R.A., Mandoli, D.F., Chapman, B.D., Reimer, K.J., 2005. X-ray absorption near-edge structure analysis of arsenic species for application to biological environmental samples. Environ. Sci. Technol. 39, 248–254.

Templeton, D.M., Ariese, F., Cornelis, R., Danielsson, L-.G., Muntau, H., Leeuwen, H.P.V., Lobinski, R., 2000. Guidelines for terms related to chemical speciation and fractionation of elements. Definitions, structural aspects, and methodological approaches (IUPAC Recommendations 2000). Pure Appl. Chem. 72, 1453–1470.

Tessier, A., Campbell, P.G., Bisson, M., 1979. Sequential extraction procedure for the speciation of particulate trace metals. Anal. Chem. 51, 844–850.

Thomas, P., Finnie, J.K., Williams, J.G., 1997. Feasibility of identification and monitoring of arsenic species in soil and sediment samples by coupled high performance liquid chromatography-inductively coupled plasma mass spectrometry. J. Anal. Atom. Spectrom. 12, 1367–1372.

Tipping, E., Hetherington, N.B., Hilton, J., Thompson, D.W., Howles, E., Hamilton-Taylor, J., 1985. Artifacts in the use of selective chemical extraction to determine the distribution of metals between oxides of manganese and iron. Anal. Chem. 57, 1944–1946.

Tokalioglu, S., Kartal, S., Elçi, L., 2000. Determination of heavy metals and their speciation in lake sediments by flame atomic absorption spectrometry after a four-stage sequential extraction procedure. Anal. Chim. Acta 413, 33–40.

Van Herreweghe, S., Swennen, R., Vandecasteele, C., Cappuyns, V., 2003. Solid phase speciation of arsenic by sequential extraction in standard reference materials and industrially contaminated soil samples. Environ. Pollut. 122, 323–342.

Vilanó, M., Padró, A., Rubio, R., 2000. Coupled techniques based on liquid chromatography and atomic fluorescence detection for arsenic speciation. Anal. Chim. Acta 411, 71–79.

Voigt, D.E., Brantley, S.L., Hennet, R.J.C., 1996. Chemical fixation of arsenic in contaminated soils. Appl. Geochem. 11, 633–643.

Wang, J., Tomlinson, M.J., Caruso, J.A., 1995. Extraction of trace elements in coal fly ash and subsequent speciation by high-performance liquid chromatography with inductively coupled plasma mass spectrometry, Plenary lecture. J. Anal. Atom. Spectrom. 10, 601–610.

Watt, C., Le, X.C., 2003. Arsenic Speciation in Natural Waters. In: Cai, Y., Braids, OC. (Eds.), Biogeochemistry of Environmentally Important Trace Elements. American Chemical Society, Washington, DC, pp. 11–32.

Wei, M., Wanibuchi, H., Yamamoto, S., Li, W., Fukushima, S., 1999. Urinary bladder carcinogenicity of dimethylarsinic acid in male F344 rats. Carcinogenesis 20, 1873–1876.

Wenzel, W.W., Kirchbaumer, N., Prohaska, T., Stingeder, G., Lombi, E., Adriano, D.C., 2001. Arsenic fractionation in soils using an improved sequential extraction procedure. Anal. Chim. Acta 436, 309–323.

Williams, T.M., Rawlins, B.G., Smith, B., Breward, N., 1998. In-vitro determination of arsenic bioavailability in contaminated soil and mineral beneficiation waste from Ron Phibun, Southern Thailand: A basis for improved human risk assessment. Environ. Geochem. Health 20, 169–177.

Woolson, E.A., 1983. Industrial, Biomedical and Environmental Perspective. Van Nostrand Reinhold Company, New York.

Woolson, E.A., Axley, J.H., Kearney, P.C., 1973. The chemistry and phytotoxicity of arsenic in soils: II. Effects of time and phosphorus. Soil Sci. Soc. Am. Proc. 37, 254–259.

Zhang, X., Cornelis, R., Kimpe, J.D., Mees, L., 1996. Speciation of toxicologically important arsenic species in human serum by liquid chromatography-hydride generation atomic absorption spectrometry. J. Anal. Atom. Spectrom. 11, 1075–1079.

Developments in Environmental Science, Volume 5
D. Sarkar, R. Datta and R. Hannigan (Editors)

ISSN: 1474-8177/DOI:10.1016/S1474-8177(07)05032-2

Chapter 32

Surficial characterization of dioxin in Midland, Michigan, using non-euclidean geostatistics

Patrick Kinnicutt

Abstract

Much research has been done characterizing surficial contaminants using geostatistics or other spatial estimation technique. This chapter examines the use of a non-Euclidean distance metric combined with geostatistical techniques to model the surficial distribution of dioxins in Imerman Park near Midland, Michigan. This chapter also examines the applicability of geostatistics to small data sets.

An overview of the dioxin sampling in Midland, MI, will be examined, followed by a brief overview of geostatistical theory, variogram modeling, and the use of non-Euclidean distance metrics to capture the geologic processes. Preliminary results of a case study evaluating the surficial dioxin distribution in Imerman Park downstream from the Dow Midland plants will then be presented, comparing the use of a flood plain non-Euclidean distance norm versus a Euclidean distance norm.

32.1. Background

Dow Chemical Corporation has emitted dioxins into the Tittabawassee River from their Midland, Michigan, plants since 1897 (Dow, 2005). Figure 32.1 shows the location of Midland, Michigan, in relation to the state of Michigan. Midland is located approximately 160 km north of Lansing, Michigan's state capitol. As shown in Fig. 32.2, the Tittabawassee River flows through downtown Midland, with the Dow Chemical plants located adjacent to the river, downstream from where the Tittabawassee and Chippewa Rivers join. The Michigan Department of Environmental Quality (MDEQ) performed some initial tests over the past five years along the Tittabawassee River, finding higher than normal dioxin levels in the Tittabawassee River flood plain, downstream from the Dow Chemical plants. Upstream dioxin levels in the Pine, Chippewa, and

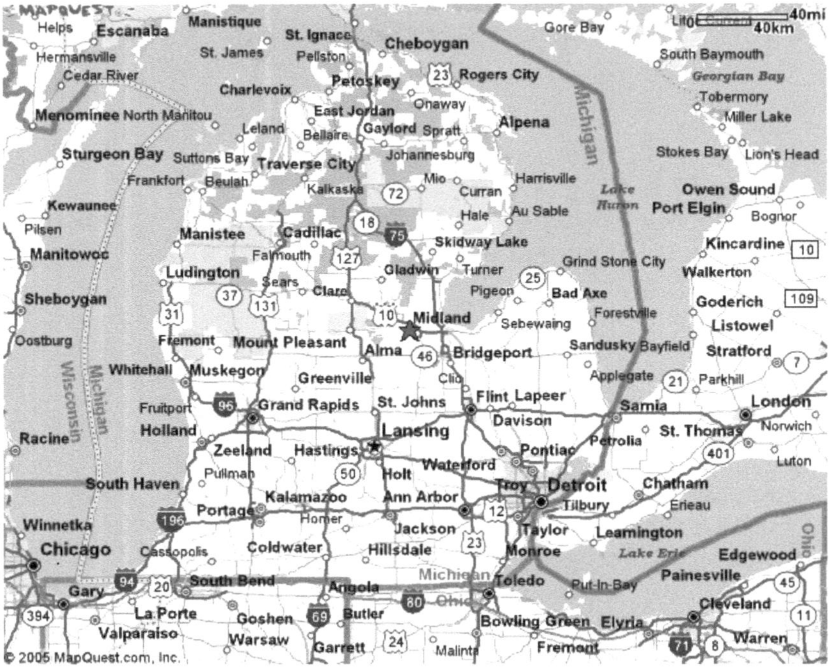

Figure 32.1. Figure showing the location of Midland, Michigan (represented as a star) in relation to the state of Michigan [from Mapquest].

Tittabawassee Rivers had background levels of dioxin (MDEQ, 2003). These elevated dioxin levels were found downstream of the Dow plants, all the way to where the Tittabawassee River joins the Saginaw River and further until the Saginaw River empties into the Saginaw Bay. Further downstream, other plants exist that are dioxin sources. No other plants or point sources of dioxin exist immediately downstream of the Dow plants in Midland, however, so the dioxin found in the Tittabawassee River between Midland and Freeland is all believed to have originated from the Dow plants.

32.1.1. Dioxin overview

The term dioxin is a general term used to refer to approximately 210 chemicals with similar chemical properties (MDEQ, 2003). These chemicals are generally formed as a by-product of burning chlorine-based chemical products with hydrocarbons. Several sources of dioxin include

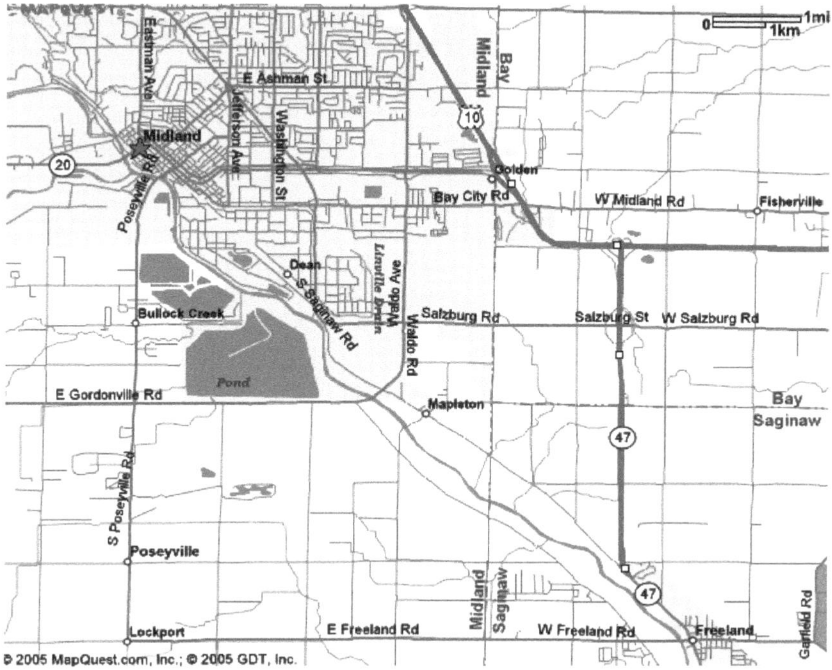

Figure 32.2. Figure showing a street map of downtown Midland, Michigan, with Dow Chemical's location (represented as a star) [from Mapquest].

burning activities from incinerators, pesticides, herbicides, pulp mills, and PVC (polyvinyl chloride). These chemicals may be found pervasively in air, water, soil, and sediments. Agent Orange is one example of an herbicide used by the military during the Vietnam War that led to many health problems in veterans.

Health effects from overexposure to dioxins are an issue of current debate and controversy between Dow and the MDEQ. One universally accepted health effect is the development of chloracne (MDEQ, 2003). Chloracne is a skin condition that may look similar to teenage acne, and is caused by exposure to dioxins. Other possible health effects have been linked to overexposure to dioxins, although these effects have not been universally accepted. These effects include a lowered immune response, hormonal changes, developmental problems in children, reproductive problems in pregnant mothers, and cancer.

Exposure to dioxins is most effectively accomplished by direct ingestion, through eating various foods in ones diet (Clark, 2004). Beef is a common source of dioxin in the diet, as well as dairy, chicken, pork, fish, and eggs. Vegetables can also be a source of dioxins, although to a lesser extent than meats. Dioxin exposure can also be obtained through direct contact with soil or sediments contaminated with dioxins.

The 210 chemicals that are classified as dioxins all have differing toxicity levels; some chemicals are more toxic than others (Clark, 2004). The most toxic form of dioxin is 2,3,7,8-tetrachlorodibenzo-para-dioxin (2,3,7,8-TCDD). Since all dioxins are usually characterized as a group in environmental remediation studies, a toxic equivalency (TEQ) was developed converting concentrations of other dioxin chemicals to the equivalent concentration of 2,3,7,8-TCDD, with this number representing the concentration of all dioxins at a site.

The EPA has an established human exposure limit of dioxins set at 1000 parts per trillion (ppt) for mandatory cleanup. The MDEQ calculated a different exposure limit of 90 ppt as the threshold for cleanup (MDEQ, 2003). The MDEQ calculated this threshold based on an increased cancer risk of 1 additional cancer per 100,000 individuals above the normal risk. This threshold discrepancy has been a major point of disagreement between Dow and the MDEQ in their negotiations for cleanup in the Tittabawassee River flood plain. Recently, the MDEQ agreed to use the EPA standard 1000 ppt residential contact criteria as the threshold for Dow to mitigate the area, due to complaints from Dow and from the Midland residents (Midland, 2005).

32.1.2. Michigan DEQ dioxin testing summary

The MDEQ tested the sediments under the Tittabawassee River for dioxins and other contaminants (MDEQ, 2002). Dredging the river at various locations within each transect was performed; these samples were then combined to create composite samples by mixing the various dredge samples. A representative sample was then tested for dioxin within each transect's composite sample. Several shallow cores were also taken adjacent to the river. Figure 32.3 shows a part of the Tittabawassee flood plain and a summary of results. Nine transects were sampled over a 22 mile river distance. Each transect is represented in Fig. 32.3 as a rectangle, and each transect has a value next to the rectangle displaying the TEQ for dioxins in ppt. The transect dioxin levels downstream from Dow showed consistent readings well above the 90 ppt threshold set by the MDEQ, but less than the 1000 ppt threshold mandated by the EPA. However, several of the shallow cores did show TEQ concentrations above 1000 ppt adjacent to the river.

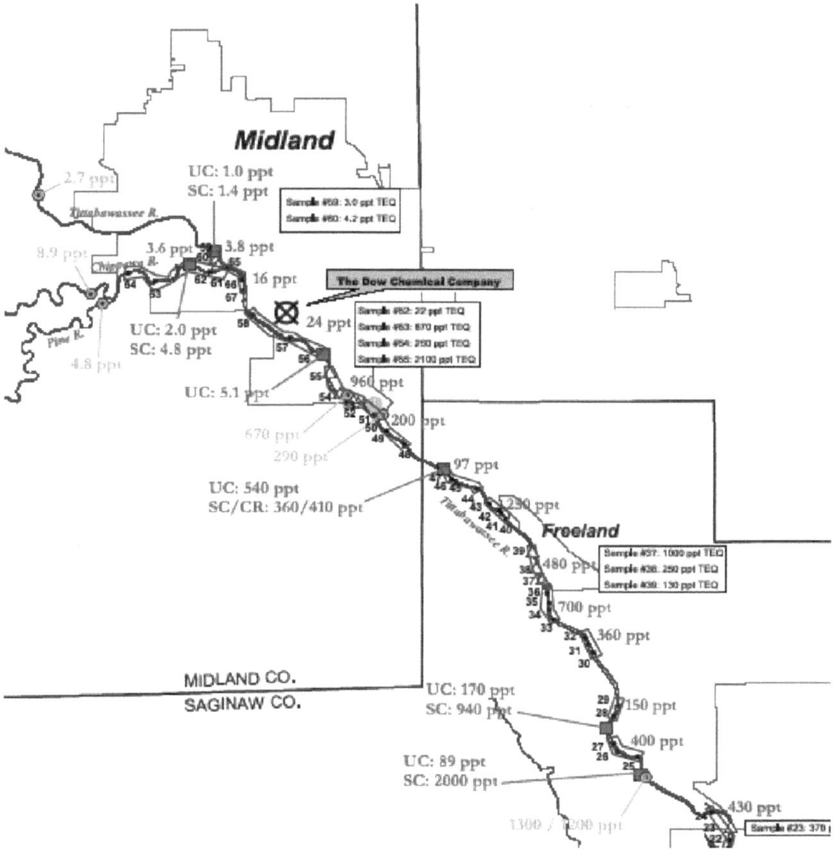

Figure 32.3. Tittabawassee River dioxin sample results between Midland and Saginaw, Michigan (MDEQ, 2002).

During phase 2 of testing, the MDEQ took over 200 shallow soil samples within the Tittabawassee River flood plain (MDEQ, 2003). These samples were taken over a 22 mile length at various locations. The sample depths for the shallow cores were taken at 1″, 3″, 6″, and 15″ below ground level (bgl). These samples also showed many locations where the TEQ was greater than the 90 ppt MDEQ threshold. Several samples were also found to be greater then the 1000 ppt EPA threshold. Dow and the MDEQ have been negotiating the extent of mitigation; recently, the MDEQ agreed to mitigation at the 1000 ppt level (Midland, 2005).

During phase 2, three deep soil borings were taken using a hydraulic probe sampler. Samples were taken at these three borings 24″, 36″, 48″, and 60″ below ground level. Of these samples, several of the samples at the 36″–48″ depth were found to be above the MDEQ threshold. All the samples tested from the 60″ depth tested at or near the background levels. Figure 32.4 shows the dioxin background levels for various locations in Michigan (MDEQ, 1999). As shown in Fig. 32.4, the background dioxin concentrations in Michigan's Lower Peninsula average well less than 10 ppt, with the area near Saginaw measuring around 13 ppt.

Figure 32.4. Background dioxin levels in Michigan (MDEQ, 1999).

The MDEQ lastly tested 22 drinking wells for dioxins (MDEQ, 2003). In these tests, none of the drinking well samples tested above the 33 ppt level set by the EPA. Current studies are focusing on bioaccumulation in flora and fauna in the flood plain, and blood monitoring of Midland residents that live in the flood plain is being discussed.

32.1.3. Geostatistical characterization of surficial contaminants

Characterization of environmental contaminants is usually done by taking samples at various spatial locations at a particular site and then using spatial interpolation methods to infer the spatial distribution of contaminants. Many different techniques have been used to do this site characterization; several will be described in the following paragraphs. For many contaminants, these spatial interpolation methods extend to three dimensions (x, y, and z) since the contaminant plume can migrate into the groundwater and into aquifers. Much of this discussion will focus on surficial characterization, as studies have shown that dioxins will normally stick to the soil near the surface (MDEQ, 2003). Also, no data samples exist from the MDEQ tests at sufficient depths to do any 3D modeling; the case study will focus on 2D characterization of surficial contaminants.

When looking at natural phenomena in earth science data sets such as with contaminant characterization, the sample data does not exhibit statistical independence; rather, a spatial dependence in the random variable(s) of interest exists (Isaaks and Srivastava, 1989). This spatial dependence is known as spatial autocorrelation. A positive spatial autocorrelation tells us that as two data locations get closer together, their random variable values will be more similar than two data locations further apart. Figure 32.5 shows an example of this correlation. As the separation distance $|\mathbf{u} - \mathbf{v}|$ increases, the covariance $c_{ij}(|\mathbf{u} - \mathbf{v}|)$ decreases

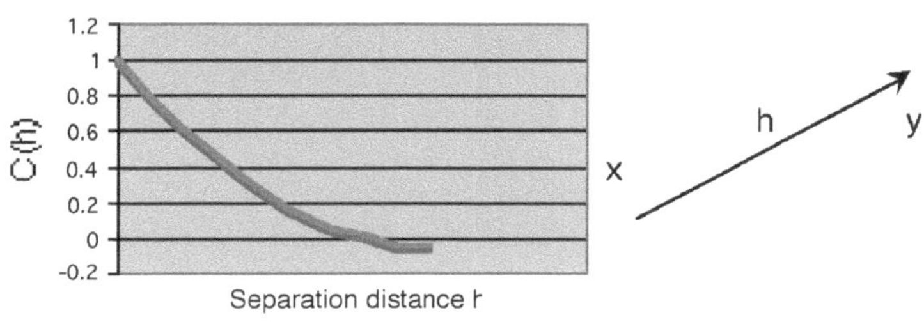

Figure 32.5. As the separation distance between two data points increases, the covariance decreases.

(Deutsch and Journel, 1998), until a certain range is reached. Once two data locations become further apart than this range, the data values no longer exhibit any correlation so become independent. The sample covariance can be calculated from Eq. (1), where **u** and **v** are two data location vectors, $|\mathbf{u}-\mathbf{v}|$ is the separation distance vector, $c^*_{ij}(|\mathbf{x}-\mathbf{y}|)$ is the covariance as a function of separation distance, $Z_i(\mathbf{u})$ and $Z_i(\mathbf{v})$ are the data values at the tail and head of the data pair, respectively, and $\bar{Z}_i$ and $\bar{Z}_j$ are the means at the tail and head respectively.

$$c^*_{ij}(|\mathbf{u}-\mathbf{v}|) = \left\langle \left(Z_i(\mathbf{u}) - \bar{Z}_i\right) \cdot \left(Z_j(\mathbf{v}) - \bar{Z}_j\right) \right\rangle \tag{1}$$

The vector **h**, as shown in Eq. (2), usually denotes the separation distance between two data locations.

$$\mathbf{h} = |\mathbf{u}-\mathbf{v}| \tag{2}$$

Geostatistics, as defined by Matheron (Journel and Huijbregts, 1978), is the application of the formalism of random functions to the estimation of natural phenomena. Geostatistics studies natural phenomena that change over space and/or time (Deutsch and Journel, 1998). In geostatistics, spatial variability is commonly used instead of spatial correlation. With spatial variability, as two data points become further apart, the spatial variability increases, until the distance becomes large enough where the two data values are independent. The sample (semi-)variogram provides a measure of the spatial variability, and is computed in Eq. (3),

$$\gamma^*(\mathbf{h}) = \tfrac{1}{2}E\left[\left\{Z(\mathbf{u}) - Z(\mathbf{u}+\mathbf{h})\right\}^2\right] \tag{3}$$

In Eq. (3), $\gamma^*(\mathbf{h})$ is the experimental or sample variogram measure for all data pairs within the separation distance **h**, while $Z(\mathbf{u})$ and $Z(\mathbf{u}+\mathbf{h})$ are the data values at two locations separated by the separation distance **h**. The sample variogram is one half of the second-order moment between these data samples. One advantage of this variogram measure over the covariance measure shown in Eq. (1) is that the means of the corresponding head and tail values are not required. Other spatial measures (Deutsch and Journel, 1998) have been defined to capture the spatial variability in sampled data.

Other than looking at the spatial description of the data samples via the covariance and/or variogram measures, non-spatial statistical measures such as the mean, median, interquartile range, range, variance, skew, and kurtosis are also typically computed in order to obtain a good understanding of the data. In the case of multivariate data, measures such as covariance and correlation coefficient are also computed. Graphs such as histograms and scattergrams provide excellent visual tools for analyzing univariate and multivariate data.

Once the data analysis has been completed examining the spatial and non-spatial sample statistics, spatial estimation or simulation techniques can be applied to create a map or volume of predicted values (Isaaks and Srivastava, 1989). In the case of surficial contaminant characterization, the relevant predicted values would be either a map of contaminant concentrations or a map of probabilities, showing the probability of the concentration exceeding the desired threshold. All these spatial estimation techniques share a fundamental equation form, as shown in Eq. (4) (Isaaks and Srivastava, 1989),

$$Z^*(\mathbf{u}) = \sum_{i=1}^{N} w_i \cdot Z(\mathbf{u}_i) \tag{4}$$

In Eq. (4), $Z^*(\mathbf{u})$ represents the estimate of the random variable at a particular location, N is the sample size, and w_i are weights assigned to each data value $Z(\mathbf{u}_i)$. Spatial estimation methods differ basically in how the weights assigned to each data value are determined. For instance, with the inverse distance squared method, each weight w_i is a function of $1/d_i^2$, where d_i is the separation distance between the known data sample location and the estimation location.

Kriging methods are a set of spatial interpolation methods that provide a variety of ways to compute the weights, w_i, from Eq. (4) (Isaaks and Srivastava, 1989). These methods are often referred to as BLUE methods; BLUE is an acronym for "best linear unbiased estimator" (Isaaks and Srivastava, 1989). Kriging estimates are linear because their general form follows Eq. (4), which is a linear weighted combination of data. They are unbiased since they attempt to set the mean residual or error to 0, and they are best because they try to minimize the variance of the errors. Kriging methods rely on the use of a variogram model to model the spatial variability and compute weights that are "best" and "unbiased".

Several different variogram models exist in the literature; Deutsch describes five models, although many more exist (Deutsch and Journel, 1998). A variogram model provides a model of variability, describing how the data varies as a function of both the distance and direction. The variogram model replaces the Euclidean distance as used in an algorithm like inverse distance with a structural distance; the variogram model is a two-point statistic, since it only considers two locations at a time. The variogram model usually takes the Euclidean distance between one data value and the estimation location, converting this distance into a variogram value, which is then used to solve for the weights. A typical variogram model used in surficial contaminant characterization can capture geometric and zonal anisotropies (Deutsch and Journel, 1998). Equation (5) shows an example of an exponential variogram model (Deutsch and Journel, 1998).

$$\gamma(\mathbf{h}) = c \cdot \left[1 - \exp\left(-\frac{3\mathbf{h}}{a}\right)\right] \tag{5}$$

In Eq. (5), $\gamma(\mathbf{h})$ is the (semi-)variogram model, c is the sill or maximum variogram value, $\mathbf{h}$ is the separation distance between a data point and the estimation point, and α is the effective range. A variogram ellipse, with a major and minor axis as well as an azimuth direction, models the change in the range a as a function of the direction.

The variogram model does not have to use the Euclidean distance. A non-Euclidean distance norm can be developed to capture objective and/or subjective spatial and/or temporal information about the processes that caused the contaminant to be distributed as it is. In this approach, the spatial coordinates are transformed into a "flattened" geographical space, then classical geostatistics is performed in the transformed space; after modeling in the "flattened" space, the results are resampled back into the original coordinate space. For example, Fig. 32.6 shows an example of a contaminant characterization model, where the contaminant was deposited via stream processes. In this figure, the use of a Euclidean distance metric may actually lead to the two ovals being too highly correlated when predicting, since the Euclidean distance is less than the non-Euclidean distance. This non-Euclidean distance, referred to as the water distance (Rathbun, 1998), takes account of the contaminant's depositional process. Figure 32.6-B displays the same geologic area as Fig. 32.6-A, although this figure shows the flattened logical grid skeleton generated by using the water distance to model the contaminant distribution. This approach was used by Barbaras et al. (2001) to model uncertainty

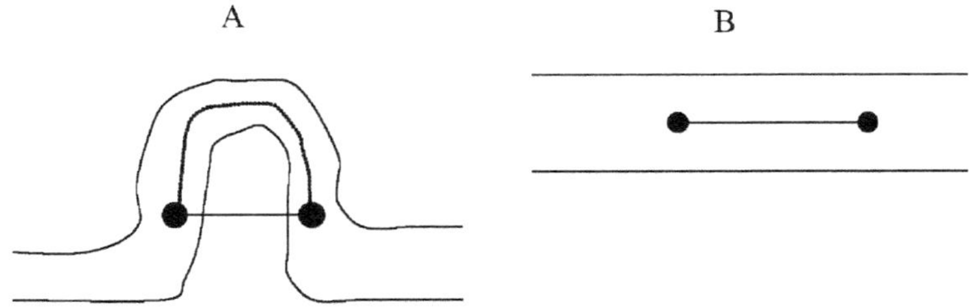

Figure 32.6. (A) Two points appear closer in distance using a Euclidean distance metric than when using a non-Euclidean metric that accounts for the oxbow in a meandering stream; (B) view of the non-Euclidean metric after flattening the model, removing the fluvial meandering process.

for three-dimensional dioxin data in the Passaic River sediments; in Barbaras's study, they used a software package called *Gridgen* to transform the coordinates, using a water distance metric, since their study examined the dioxin distribution in the river sediments.

The selection of a non-Euclidean distance norm must be based on both objective and subjective knowledge regarding the processes that are known to have contributed to the contamination. In the case study presented later, the dominating process leading to the surficial dioxin contamination involved flooding events, where the flood caused contaminated sediments to wash ashore in the flood plain.

The choice of norm, whether Euclidean or non-Euclidean, should not be done a priori at the expense of excluding alternative measures of distance. These alternative norms may provide yet unrealized insights into the nature of the random process. Departures from the Euclidean distance have been suggested based on the geology of the domain of interest, for example non-convex sampling regions. Rathbun (1998) and Rendu and Readdy (1992) discuss such situations. The great-arc distance has also found some use in geostatistics, for example Cressie et al. (1990). The goal is to model spatial variability as well as possible and the model that accomplishes this may not have to correspond perfectly to reality (Curriero, 1996). Much research and development has been done to incorporate non-Euclidean distances in commercial GIS software, in particular ESRI's ArcGIS (Krivoruchko and Gotway, 2003). One method Krivoruchko describes for determining the non-Euclidean metric involves the use of cost-weighted distance to produce a cost surface.

Several flavors of kriging have been developed over the years to account for the use of subjective and objective data, and to address limitations of the variogram model approach (Deutsch and Journel, 1998). The use of a non-Euclidean distance norm as input to the variogram model provides a mechanism to take geologic processes into account in the structural model, as in Fig. 32.5. In Fig. 32.5, the contaminant in most likelihood flow through the river as a dissolved load or suspended load attached to sediments, finally getting deposited downstream in the river. Thus, at any one time it makes sense that the distance norm should follow the water distance (Rathbun, 1998). It also makes sense that when looking at contaminant concentration values on the surface transported by flood processes, the concentrations in the river and in the flood plain should be segregated and modeled separately. In the river, the contaminant is constantly being subjected to fluvial processes, causing temporal variations in the contaminant concentrations. However, in the flood plain the contaminants (in the case of dioxins) will cling to surficial soil and not be transported, until the next flood event occurs. In this scenario, a non-Euclidean distance norm based on flood plain maps and DEM maps can be used to model the variability in contaminants. If one is examining a contaminant that flows readily in the groundwater, the hydrogeology can be taken into account if information is available to obtain better models of spatial variability.

Many different spatial interpolation algorithms may be used to solve a particular application. Saito and Goovaerts (2000) evaluated several algorithms to characterize dioxins in a well-sampled (over 600 data samples) EPA Superfund site at Piazza Road in Missouri, examining the use of ordinary kriging, log-normal kriging, multi-Gaussian kriging and indicator kriging. During this study, they found that log-normal kriging consistently gave the best results in terms of smallest prediction errors and false positives, with indicator kriging producing good results. Saito and Goovaerts (2000) later examined the use of indicator kriging and p-field simulations at the same site. In this study, indicator kriging was used to provide a probability distribution that the dioxin levels did not exceed the threshold, then p-field simulations were done to see if the average pollutant concentrations in a remediation unit exceed the regulatory threshold. Saito and Goovaerts used Euclidean distance norms in both studies.

32.2. Case study: Midland, Michigan dioxin characterization

This case study shows an application of non-Euclidean distance norms in the estimation of dioxin concentrations in Imerman Park. Many studies concentrate on the use of the water metric (Saito and Goovaerts, 2002; Rendu, 1998) to model contaminant concentrations in streams. This study examines the use of a flood plain distance metric applied on land, with non-ideal data. The data in this study is similar to that found in many sites; there are few data samples and the data samples are clustered in various locations, with large areas left unsampled. Inverse distance and sequential gaussian simulation algorithms were chosen for comparison with this study. GSLIB programs (Deutsch and Journel, 1998) were used for data analysis, simulation, and presentation of results. The inverse distance squared algorithm used a custom program designed by the author, and the coordinate transformation from and to Euclidean space from the non-Euclidean norm was done as pre-processing and post-processing steps to using sgsim and gamv (Deutsch and Journel, 1998), using a coordinate conversion program written by the author.

The MDEQ collected shallow samples at various locations along the Tittabawassee River flood plain, as described earlier (MDEQ, 2003).

One site was examined in a pilot study to see what effect the use of a non-Euclidean distance metric as opposed to a Euclidean distance metric would have on the interpreted results. The site chosen was Imerman Park, located approximately 11 ½ miles downstream from the Dow plants. This site was chosen because it had the largest number of shallow data sample locations, 14. Imerman Park is a site where Dow Chemical has an established interim response activities (IRA) plan in place with the MDEQ. Figure 32.7 shows an aerial photograph of Imerman Park, adjacent to the Tittabawassee River in Saginaw Township, Michigan.

Figure 32.7. Areal photograph of Imerman Park in Saginaw Township, Michigan.

The blue lines show the extent of the 100-year flood plain, and the stars denote the sample locations for the 14 locations tested in the park. Imerman Park encompasses approximately 96 acres, and it includes public playgrounds, a baseball field, boat launch areas, picnic sites, hiking trails and pet exercise areas. The park entrance is shown in the upper right corner of Fig. 32.7. As shown by the blue lines, most of the park is located within the 100-year flood plain.

Figure 32.8 displays a contour map of the flood plain in Imerman Park that was used in the study. This contour map shows approximate boundaries for flooding events, using a 20-year contour interval. The author converted these contours into a "flattened" coordinate space, using a preprocessing program written by the author to transform the coordinates.

Figure 32.9 displays the cumulative frequency plot of surficial dioxin concentrations measured in ppt from the 14 sample locations. As shown in Fig. 32.9, the minimum measured dioxin concentration was 5.2 ppt, with a maximum data sample of 1400 ppt and a median of 580 ppt. From the interquartile range, it is shown that a little over 25% of the samples tested less than the MDEQ limit of 99 ppt, while a little over 25% of the samples tested greater than the EPA threshold of 1000 ppt.

Figure 32.10 displays the same 14 data samples as shown in Fig. 32.7, color-coded to display the dioxin concentrations. Figure 32.10 shows that the high concentration dioxin values are clustered in the southwest section of the park, near the boat launch and the Tittabawassee River.

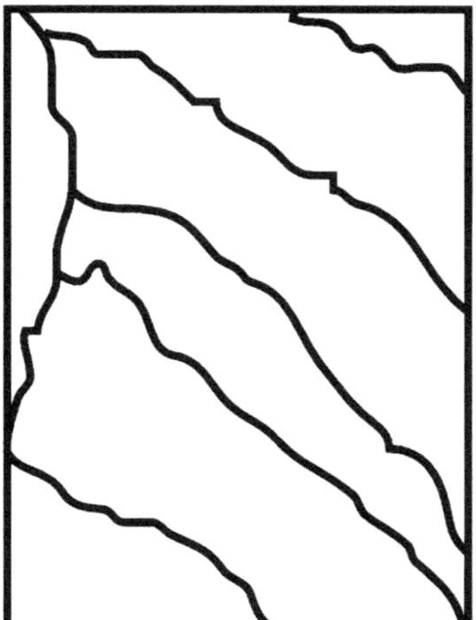

Figure 32.8. Figure showing flood plain contours at a 20 year interval.

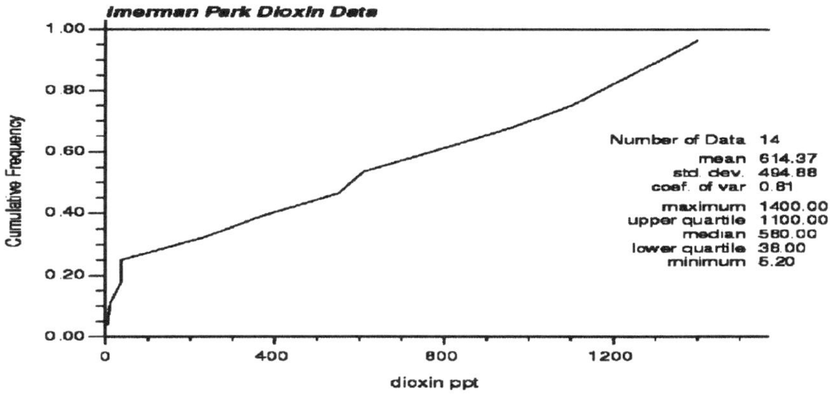

Figure 32.9. Univariate non-spatial statistical measures of the surficial dioxin concentrations in Imerman Park.

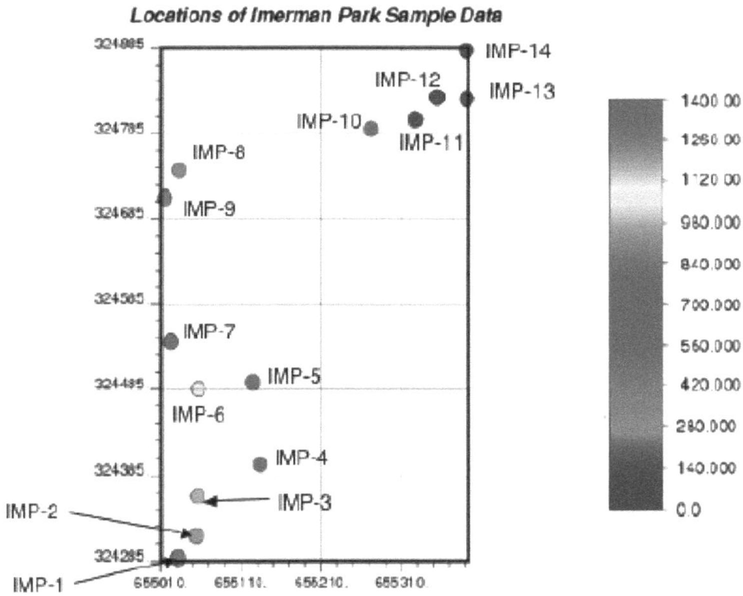

Figure 32.10. Data sample locations shown in a grid outlining Imerman Park.

Low dioxin levels are found in the northeast park section, near the park entrance and the 100-year flood plain contour. Testing was done primarily at the southwest and the northeast corners of Imerman Park, with no sampling performed in the center or southeast sections of the park.

Figure 32.11 shows the dioxin concentrations estimated using an inverse distance squared algorithm and a Euclidean distance norm. The data values are honored at the sampled locations, with the estimated values far from the sampled locations converging on the mean of 614 ppt. This data convergence occurred because that area of the park was equidistant to the two data clusters. The inverse distance algorithm is an estimation technique; this algorithm produces deterministic results and does not produce any uncertainty measurements.

Figure 32.12 shows the spatial dioxin distribution modeled using sequential Gaussian simulation, with a Euclidean distance norm. In all the simulations, 100 realizations were run. As with the inverse distance squared algorithm, the sequential Gaussian simulation algorithm also

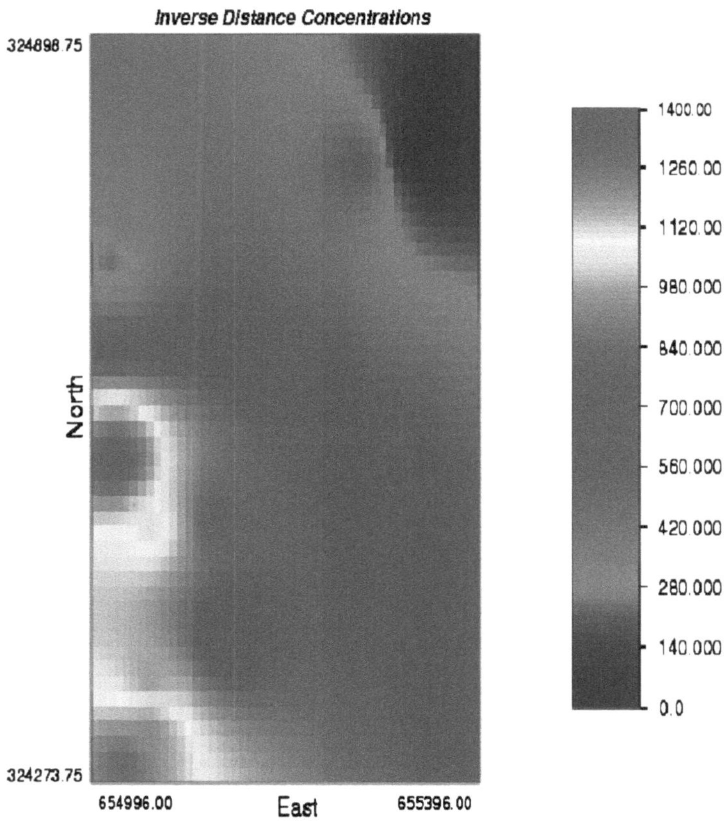

Figure 32.11. Estimated dioxin concentrations, estimated using the inverse distance squared algorithm. This estimation was performed using Euclidean distance norms.

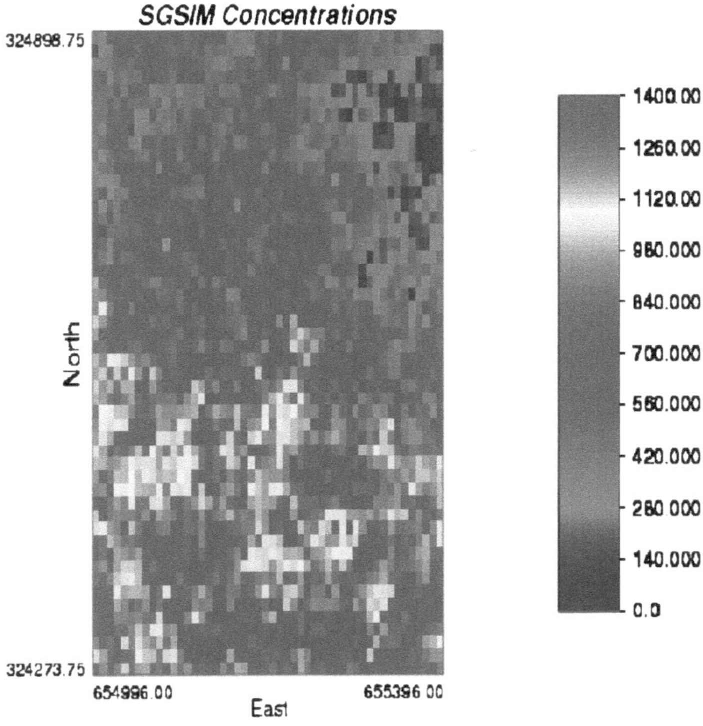

Figure 32.12. Simulation results using the sequential Gaussian simulation algorithm. This simulation was performed using Euclidean distance norms, with 100 realizations.

honors the sampled data values, but as shown generates a "salt-and-pepper", non-smoothed concentration map. The sequential Gaussian simulation algorithm is useful for obtaining uncertainty measurements and introducing randomness in the simulated results from multiple realizations, while honoring the data samples and the sample data histogram. In the sequential Gaussian simulations, the data were declustered prior to performing a normal score transform of the data. A normal score transform is performed to transform the data histogram into a Gaussian distribution. Simulation is done in normal space, then the results are back transformed, yielding results that match the data histogram. The models generated were stored in a rectangular geometry; this is a limitation of the GSLIB software.

With both the inverse distance and the sequential Gaussian algorithms, hot spots of high and low dioxin levels are shown. The remediation areas vary, however; with the sequential Gaussian simulation, hot spots can be seen where no data has been collected, indicating that perhaps further testing may be desired in these areas where the dioxin levels are uncertain.

Figures 32.13 and 32.14 display spatially the probabilities that the dioxin concentration exceeds the MDEQ's 90 ppt threshold for cleanup. In the inverse distance map, no uncertainty is captured in the algorithm. The entire park would have to be remediated with the exception of the far northeast corner, if this map were used for making the remediation decision. The sequential Gaussian simulation algorithm enables multiple realizations to be done, each using a different random walk in a Monte-Carlo simulation; using this map, a probability threshold can be set by the

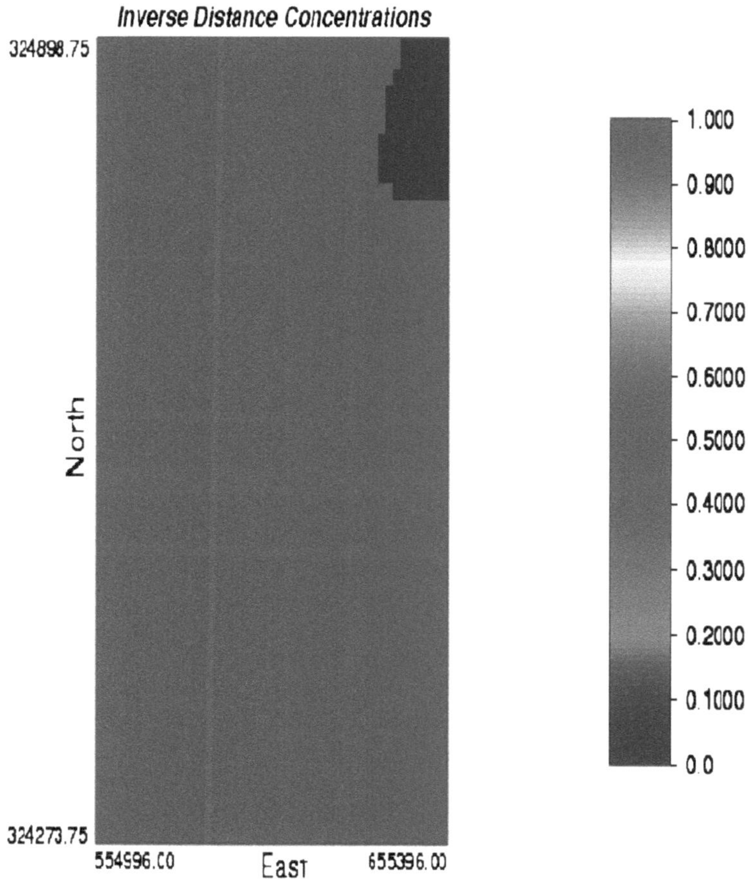

Figure 32.13. Inverse distance probability map, showing the probability that the dioxin concentration is higher than the MDEQ threshold of 90 ppt.

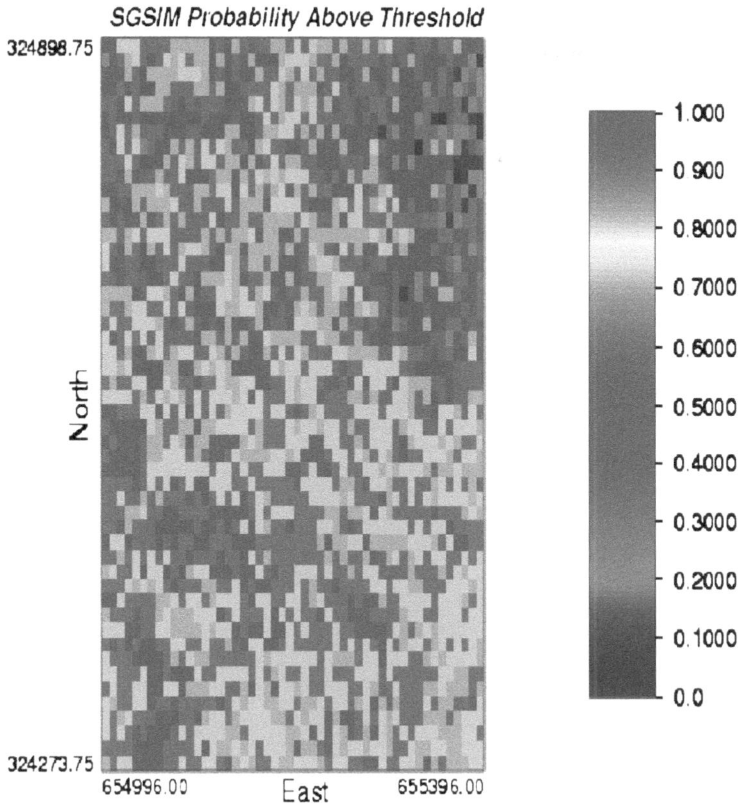

Figure 32.14. Sequential Gaussian simulation probability map, showing the probability that the dioxin concentration is higher than the MDEQ threshold of 90 ppt.

policy makers to determine what extent of the park needs to be remediated. To calculate the probability map for the simulations, the percentage of realizations for each pixel where the simulated value exceeded the target threshold was calculated.

A non-Euclidean distance norm was next used along with sequential Gaussian simulation to perform simulations to characterize the dioxin concentrations in Imerman Park. Figure 32.15 displays the simulated concentrations using a non-Euclidean distance metric. In the non-Euclidean simulation, first a transformed grid was generated as a preprocessing step to sgsim. Gamv was then run to examine the experimental variogram measure using the transformed grid space, then sgsim was run using a variogram model specified using the experimental variogram

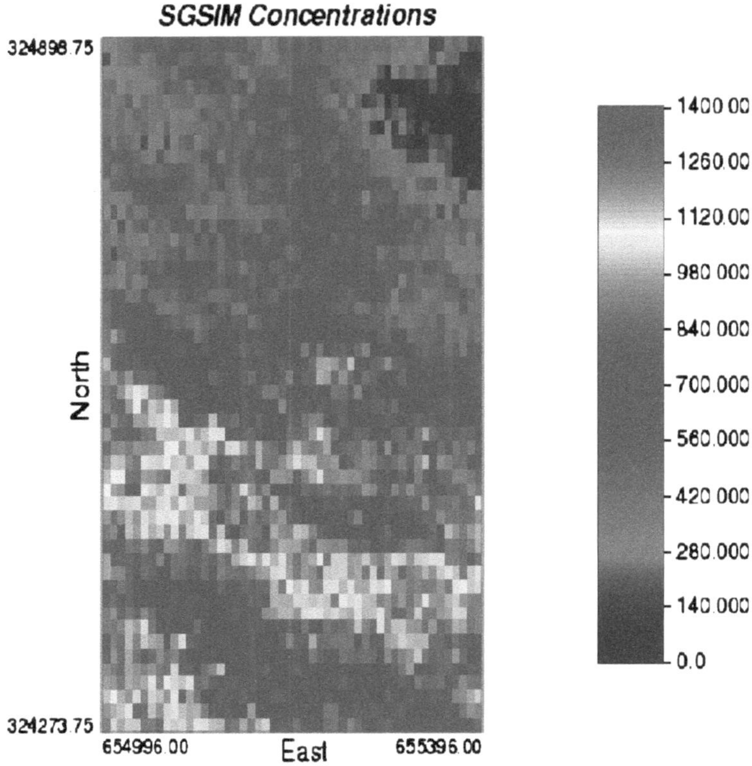

Figure 32.15. Sequential Gaussian simulation results using a non-Euclidean distance norm.

results. The results were then resampled as a post-processing step into the Euclidean space, for display and comparison. Comparing Figs. 32.12 and 32.15, it can be seen that the dioxin concentrations have more structure in the distribution, trending in the direction of the flood plain distance metric. The univariate data distribution still matches that of the sample data, but the spatial distribution has taken into account the flood stages would affect how the dioxins get transported and deposited.

When examining the probabilities of dioxin existing in concentrations above the MDEQ threshold of 90 ppt, the non-Euclidean results again show more structure, trending along the direction of movement of water in the flood plain. Figure 32.16 shows non-Euclidean sequential Gaussian's probability maps. Based on the non-Euclidean results and based on the acceptable probability threshold value as set by the decision makers, Fig. 32.16 shows that it looks more certain that the southwest

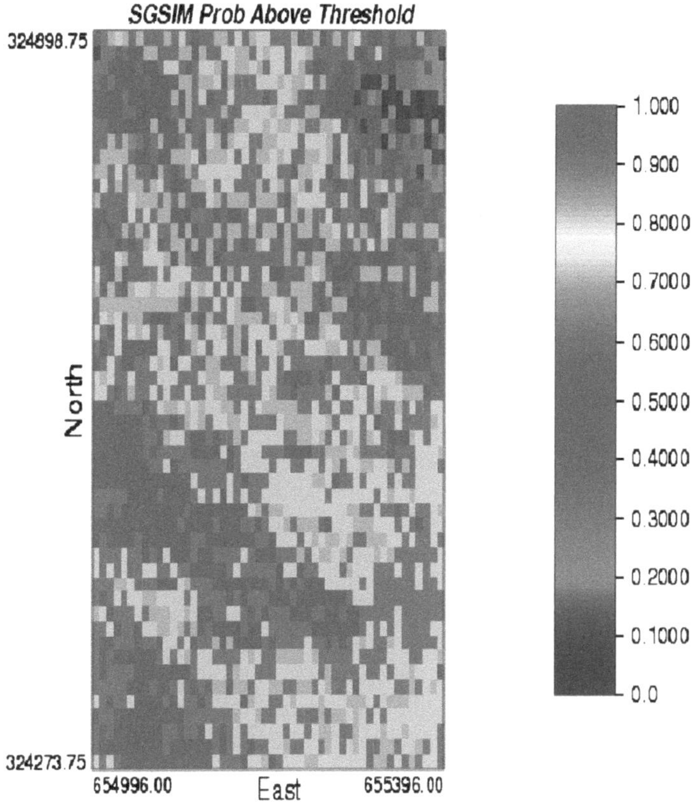

Figure 32.16. Sequential Gaussian simulation probability map using a non-Euclidean distance norm, showing the probability that the dioxin concentration is higher than the MDEQ threshold of 90 ppt.

half of the park needs remediation, with no remediation action required for the northeast corner of the park. This is in contrast to Fig. 32.14, where the simulated results show more randomness in the dioxin distribution.

When performing an actual remediation, usually the remediation unit is much larger than the model grid's resolution. Saito (Saito and Goovaerts, 2002) describes this in more detail. In general, once a geostatistical model is performed, the model is resampled into a lower resolution grid, by taking the arithmetic average of all the model grid cell values that fall within a reservoir unit. A probability threshold is determined by the regulatory agency, and if this probability is exceeded, cleanup will be performed in the reservoir unit.

One major problem with modeling this site arises because of the lack of sufficient data support. There were only 14 samples taken in Imerman Park, and this data was clustered in three primary areas. Given these data limitations, the data histogram, and hence the variogram model chosen and the normal score data transform, contain high degrees of uncertainty. In a remediation study, the author suggests using the resulting model as a guide to aid in developing a more detailed sampling plan, then updating the model as more data is collected. Deutsch (Deutsch and Journel, 1998) describes how to simply calculate statistical measures on multiple simulations. In general, statistics such as variances, medians, etc. can be calculated on a pixel-by-pixel basis using the multiple realizations.

32.3. Conclusions

The use of non-Euclidean distance norms for modeling the spatial variability in a random variable may lead to more realistic results and more informed decision-making when determining the remedial action in a contaminated site. This non-Euclidean norm used must be site and domain specific, to capture the objective and subjective knowledge regarding the processes that led to the distribution of the natural phenomena.

Ideally, more exploratory samples should be taken and the model iteratively updated and refined. There were not enough data samples to perform error analysis on this data set; one area of future work involves obtaining more exhaustive data sets to use with error analyses, and trying to get further samples taken at Imerman Park to validate the model results.

ACKNOWLEDGMENTS

I would like to thank the Michigan Department of Environmental Quality for providing me with the geo-referenced dioxin test results. Thanks also go to the editors and anonymous reviewers for their valuable and constructive comments. Lastly, I would also like to thank my wife and Central Michigan University's Geology Department for their support in this project.

REFERENCES

Barabas, N., Goovaerts, P., Adriens, P., 2001. Geostatistical assessment and validation of uncertainty for three-dimensional dioxin data from sediments in an estuarine river. Environ. Sci. Technol. 35(16), 3294–3301.

Clark, J.M., 2004. Health Risk Analysis of Tittabawassee Fish with Dioxin, EPA Open-File Report, July 30, 2004.
Cressie, N., Gotway, C.A., Grondona, M.O., 1990. Spatial prediction from networks. Chemometrics and Intelligent Laboratory Systems 7, 251–271.
Curriero, F.C., 1996. The use of non-Euclidean Distances in Geostatistics [Ph.D. thesis]. Manhattan, Kansas State University.
Deutsch, C.V., Journel, A.G., 1998. GSLIB: Geostatistical Software Library and User's Guide, second ed.. Oxford University Press, New York, p. 369.
Dow, 2005. History of dioxins and furans in the Tittabawassee River, 2005. From http://www.dow.com/facilities/namerica/michigan/dioxin/what/history.htm web site.
Isaaks, E.H., Srivastava, R.M., 1989. An Introduction to Applied Geostatistics. Oxford University Press, New York, p. 592.
Journel, A.G., Huijbregts, Ch.J., 1978. Mining Geostatistics. Academic Press, New York.
Krivoruchko, K., Gotway, C.A., 2003. Using Spatial Statistics in GIS. In: Proceedings, Modsim 2003 International Congress on Modeling and Simulation, Townsville, Queensland, Australia, pp. 713–718.
MDEQ, 1999. Michigan Department of Environmental Quality, 1999, Michigan soil background Data, from http://www.michigan.gov/deq/ web site.
MDEQ, 2002. Taylor, Allan B. and McCabe, John M., 2002, Baseline chemical characterization of Saginaw Bay Watershed sediments, from http://www.michigan.gov/deq/ web site.
MDEQ, 2005. Michigan Department of Environmental Quality, 2003, FINAL REPORT: Phase II Tittabawassee/Saginaw River Dioxin Flood Plain Sampling Study, from http://www.michigan.gov/deq/ web site.
Midland [2005] 2005. Interim Response Activities Decided: Dioxin Contamination Issues in Midland, from http://www.midland-mi.org/government/Dioxin/Updates.htm#Framework web site.
Rathbun, S.L., 1998. Spatial modeling in irregularly shaped regions: Kriging estuaries. Environmetrics 9, 109–129.
Rendu, J.M., Readdy, L., 1992. Geology and the Semivariogram—A critical relationship. In: T.B. Johnson, R.J. Barnes (Eds.), 17th Application of Computers and Operations Research in the Mineral Industry, American Institute of Mining, Metallurgical, and Petroleum Engineers, Inc., pp. 771–783.
Saito, H., Goovaerts, P., 2000. Geostatistical interpolation of positively skewed and censored data in a dioxin contaminated site. Environ. Sci. Technol. 34(19), 4228–4235.
Saito, H., Goovaerts, P., 2002. Accounting for measurement error in uncertainty modeling and decision making using indicator kriging and p-field simulation: Application to a dioxin contaminated site. Environmetrics 13, 555–567.

Developments in Environmental Science, Volume 5
D. Sarkar, R. Datta and R. Hannigan (Editors)

ISSN: 1474-8177/DOI:10.1016/S1474-8177(07)05033-4

Chapter 33

Black shale weathering contribution to stream chemistry using end-member mixing analysis

Leonette Cox, George Morara Ogendi and Robyn Hannigan

Abstract

A mathematical approach known as end-member mixing analysis (EMMA) was applied to determine the end-member components (overland/shallow subsurface, runoff, and groundwater) contributing to the surface water chemistry of two first-order headwater tributaries of the Little Red River in Arkansas. Binary mixing between two end-members explained only a portion of the variance in major anion and cation chemistry of the surface waters. Therefore, ternary mixing was investigated using the EMMA method. Previous studies using EMMA have involved first- and second-order streams of various bedrock types. This study is unique in that it is the first, to our knowledge, to apply this approach to the study of first-order streams draining black shales. The tributaries we investigated (Cove Creek and Begley Creek) are underlain mainly by metal-rich Mississippian Fayetteville Shale. Weathering of this bedrock could affect water quality. Where surface waters are in intimate contact with these shales, it is necessary to determine the extent to which these waters contribute to the chemistry of the stream as a first approximation to the significance of black shale weathering to the catchment.

The mixing analysis results suggest that the contributions from overland/shallow subsurface flow, runoff from the black shale outcrop, and groundwater play a major role in making up the surface water chemistry of Begley and Cove Creeks under both storm and base flow conditions. The relative contribution of these end-members varies with flow. As expected, the mixing results showed that groundwater dominates during base flow. During storm flow the surface water chemistry represents a mix of overland/shallow subsurface flow and black shale runoff. We found that during storm flow groundwater and overland/shallow subsurface flow were, in general, indistinguishable across storm hydrographs resulting in storm flow chemistries represented by binary mixtures of two chemically distinct end-members. Ion concentration ratios are similar for overland/shallow subsurface flow and groundwater because of the lithologic similarities of the rocks in contact with these waters. Runoff from the black shale outcrop does contribute to the surface water chemistry.

33.1. Introduction

Water flowing through a stream channel is composed of water from a variety of sources also known as end-members. These end-members include precipitation, runoff, shallow subsurface (e.g., bank storage), and groundwater that may be further influenced by atmospheric deposition, biological activity, chemical weathering, and the geology of the area. As a result, the chemistry of streams is highly variable. Geochemical hydrograph separation and modeling of hydrochemistry can provide insights into the temporal variability in stream chemistry and the relative contributions of end-member components.

This study focused on black shale weathering as a non-point source of dissolved metals to surface waters of headwater tributaries of the Little Red River. Major ion concentrations were used to identify the major contributing end-members to the surface waters. K^+, Ca^{2+}, Mg^{2+}, Na^+, Cl^-, NO_3^-, alkalinity, acid neutralizing capacity, and silica are among the most commonly used parameters (Pinder and Jones, 1969; DeWalle and Poinke, 1994; Elsenbeer et al., 1995; Jarvie et al., 2000; LaDouche et al., 2001). Many models have been developed to assess the interactions between physical and chemical hydrology. However, when conservative binary mixing fails to explain the observed variation in surface water chemistry, a least-squares mathematical approach called end-member mixing analysis (EMMA) can be used to explain variation under ternary-mixing conditions. Using this approach, we identified the dominant end-member components influencing surface water chemistry in the headwaters of the Little Red River. If waters of the end-members determined by EMMA are in contact with shales, we can use this as our first approximation of the significance of black shale weathering and the potential of weathering of this metal-rich bedrock as a non-point source of dissolved metals to the system. Further analysis of black shale weathering products and trace metal chemistry can then be used to more precisely assess the significance of black shale weathering in influencing the metal chemistry of the surface waters and sediments in black shale draining systems.

EMMA has the advantage of using directly measured solute concentrations to generate verifiable information about the mixing patterns making up the stream chemistry (Christophersen et al., 1990; Hooper et al., 1990). This method simultaneously uses multiple solutes as tracers and relies on binary or ternary component mixing. It utilizes conservation of mass to explain the variation in solute concentration based on the relative composition of the end-members that contribute to the waterway. The end-members chosen must have solute concentrations that are distinct from the stream water and they should bind the stream water data in a mixing diagram. The model assumes that the source solutions are invariant in time and space but it is often difficult to sufficiently sample stream component compositions either spatially or temporally. However, the model can be validated by the linear fit of EMMA predicted versus observed solute concentration and an evaluation of the hydrological patterns predicted. Even if EMMA fails to sufficiently predict the observed stream chemistry the information obtained can be used as a guide for future studies.

This study used EMMA to explore the dissolved chemistry of surface waters draining Mississippian Fayetteville Shale bedrock of first-order streams (Cove Creek and Begley Creek) in the headwaters of the Little Red River in central Arkansas. Previous studies utilizing EMMA have focused on both first- and second-order streams that drain various common bedrock types (e.g., granite, sandstone, limestone) but none have studied surface water draining black shale bedrock (Christophersen et al., 1990; Christophersen and Hooper, 1992; Hooper et al., 1990; Jarvie et al., 2000; Burns et al., 2001; Hannigan and Bickford, 2003). Black shales are metal-rich and sedimentary rocks that are easily weathered. In watersheds dominated by these shales, their presence is often reflected in increased metal loading of waterways and sediments/soils (Chon et al., 1996; Loukola-Ruskeeniemi et al., 1996, 1998; Petsch et al., 2000; Tuttle et al., 2001; Ogendi et al., 2004a). Regions of central and southeastern USA are particularly affected as weathering of shales may be a significant source of pollution in small creeks and streams (Foley et al., 2001). Increases in dissolved metal concentrations may have a negative impact on living organisms including deficiency and toxicity symptoms in plants and livestock such as restricted growth, incomplete cell wall development and gastrointestinal disorders (Kim and Thorton, 1993; Loukola-Ruskeeniemi, 1994, 1996; Ogendi et al., 2004a,b).

In areas where surface or groundwaters are in intimate contact with metal-rich bedrock, it is necessary to determine the extent to which these waters contribute to the chemistry of the stream. Using EMMA to identify the major end-member components contributing to the surface water chemistry of the headwater tributaries of the Little Red River as a guide to the significance of black shale weathering to the chemistry of the stream is a novel approach.

33.2. Methods

33.2.1. Study site

Sampling sites were located on Devil's Backbone, Cove Creek, Begley Creek, and Middle Fork. The streams are first-order streams that are part

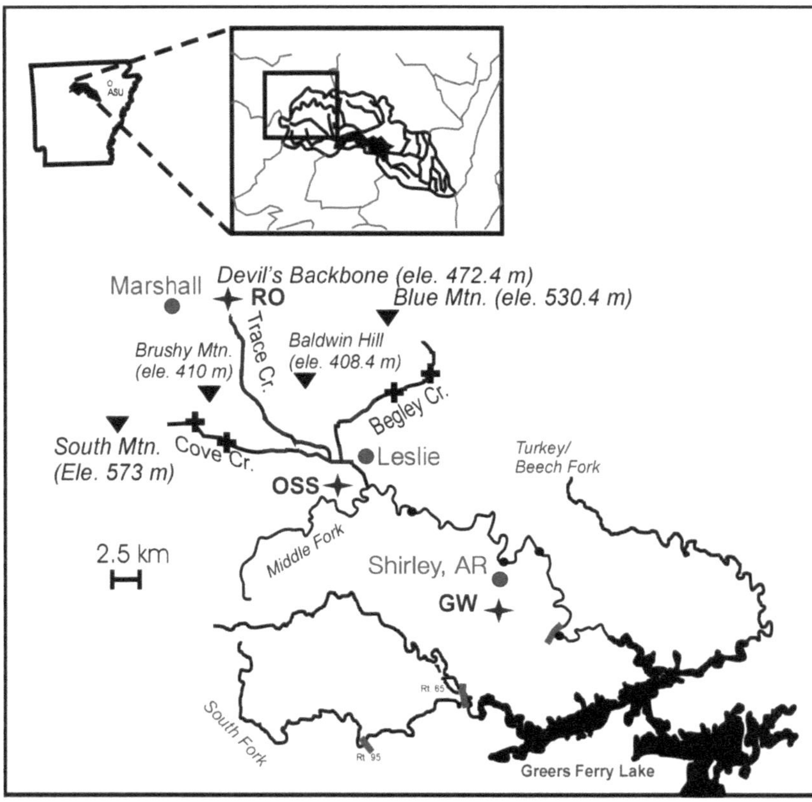

Figure 33.1. Schematic diagram of the study site showing sampling locations. Surface water was collected at Begley Creek and Cove Creek. Runoff from black shales was collected from outcrop and shale seeps at Devil's Backbone during a storm event. Overland/shallow subsurface samples were collected from bank soils along the Middle Fork. Groundwater was collected from a well near Shirley AR. Surface water sampling subsites are indicated by black crosses. End-members are noted as stars (GW, groundwater; OSS, overland/shallow subsurface; RO, runoff). Adapted from Ogendi et al. (2004a).

of the Little Red River watershed (HUC 11010014) in Arkansas, USA (Fig. 33.1). The Little Red River has a total watershed area of 2901.75 km^2 and a total river length of 4882.67 km. The streams drain the northern portion of the Little Red River watershed that is >75% forest riparian and ~20% agricultural/urban riparian. The study focused on the unpopulated upper reaches of the streams. Details on the geology of the study area are given elsewhere in this book (Ogendi et al., 2007) as well as Renken (1998). In summary, the Devil's Backbone, Cove Creek, and Begley Creek flow over the Mississippian Fayetteville Shale (~350 mya), which according to Hudson et al. (2001) rests conformably on the Batesville Sandstone. The Fayetteville Shale is in intimate contact with surface as well as groundwaters of Devil's Backbone, Cove Creek, and Begley Creek (Ogendi et al., 2004b). The Middle Fork drains over the Mississippian Pitkin Limestone that rests conformably on the Mississippian Fayetteville Shale. The mineralogy of the Fayetteville Shale as well as that of the Pitkin Limestone is described in Murthy (2003). The most abundant clay minerals in the Fayetteville Shale are kaolinite, illite, and smectite, whereas micas, quartz, and feldspars form the main mineral constituents of this rock. Calcite and illite dominate the mineral composition of the Pitkin Limestone. The Fayetteville Shale contains a significantly higher organic carbon (5–8wt% than the Pitkin Limestone (⩽1wt %) (Murthy, 2003; Ogendi et al., 2004b).

Water flowing off the face of the Devil's Backbone shale outcrop was collected during a storm event and represents the runoff end-member. Surface waters drain southward from the Blue Mountain catchment into Begley Creek. Begley and Cove Creeks have surface water in direct contact with Fayetteville Shale in the streambed; surface water was collected from both these streams to make up the stream water dataset for EMMA. We did not attempt to model the chemistry of Trace Creek because chloride was non-conservative suggesting an additional end-member in this region such as shallow or deep saline groundwater or saline soil water (Bethke and Marshak, 1990; Winter, 1999; Davis et al., 2003). Middle Fork near Marshall drains an area of limestone. Surface water collected here serves as our overland flow end-member because the major ion chemistry of the overland flow in this region was found to be indistinguishable from the soil water and bank storage fluids of Cove and Begley Creeks. Given that it is easier to collect sufficient quantities of surface waters at Middle Fork than the shallow subsurface fluids at Cove or Begley we selected this option, although less than ideal. Surface waters from Cove and Begley Creeks mix with Middle Fork. Groundwater data from Middle Fork near Shirley represents our groundwater end-member because the region shares a common aquifer. Groundwater data were obtained from Earthinfo Inc. USGS Quality of Water for Central USA (http://www.earthinfo.com/).

The end-members were determined by the availability of data and the likelihood of them contributing to the stream. The runoff from Devil's Backbone, overland flow/shallow subsurface (Middle Fork) and groundwater are potential end-members because of the geology and hydrology of the area. The shales of the northern Little Red River watershed are non-permeable but they are highly fractured allowing the upflow of groundwater through the more permeable limestone aquifer into the streambeds.

33.2.2. Sample collection

Stream water samples were collected during storm and base flow at various times during the year in nitric acid-washed high-density polyethylene bottles at randomly determined subsites. Samples were filtered using a 0.45 μm Gelman in-line filter with the fraction passing through the filter operationally defined as dissolved. 125 ml was acidified with nitric acid for analysis of cations, and 125 ml was left unacidified for analysis of anions. All samples were stored at ~4°C prior to analysis.

Water quality parameters such as temperature, pH, electrical conductivity, dissolved oxygen, and stream flow were measured in the field using various meters. Discharge was calculated from stream velocity. Major ion concentrations were measured by ion-chromatography (Dionex DX 120). Major anion concentrations were measured within 24 h of collection and major cation concentrations were measured within 48 h of collection. Alkalinity was determined by titration.

33.2.3. End-member mixing analysis

Alkalinity/HCO_3^-, Cl^-, K^+, Mg^{2+}, Ca^{2+}, Na^+, and SO_4^{2-} concentrations and water quality parameters for Begley and Cove Creeks stream water are summarized in Tables 33.1 and 33.2. Binary mixing plots of the various solutes were generated to show variation in solute concentrations and to perform a simple two-component mixing analysis as in Jarvie et al. (2000). A principal component analysis (PCA) was performed on a standardized correlation matrix of the stream water dataset and a model was generated that accounted for the most variance (~97%) in the dataset. The number of end-members needed to complete the model is equal to the number of principal components plus one. The end-member solute compositions were then standardized against the stream water data and multiplied by the eigenvalues of the matrix for the stream water values then projected onto a mixing diagram. The mixing diagram consisted of

Table 33.1. Physical water quality parameters for the stream samples used in this study

Date	Site	T (°C)	pH	DO	Conductivity (μS)	Discharge ($m^3 s^{-1}$)
May-04	BC	24.0	6.05	6.30	260.50	0.118
May-04	BC	24.0	6.04	6.38	258.50	0.130
May-04	CC	24.1	6.80	6.80	224.50	0.115
May-04	CC	24.2	6.8	6.70	221.50	0.0975
Jan-04	BC	4.0	7.12	11.40	237.75	0.213
Jan-04	BC	4.1	7.09	10.40	233.00	0.190
Jan-04	CC	7.9	7.60	10.80	245.50	0.180
Jan-04	CC	8.5	7.50	10.60	249.50	0.145
Nov03	BC	22.5	5.80	6.40	212.00	0.298
Nov-03	BC	22.4	5.78	6.30	211.00	0.343
Nov-03	CC	22.1	6.80	6.70	203.25	0.290
Nov-03	CC	22.0	6.75	6.80	205.50	0.313
Sep-03	BC	25.1	7.89	8.16	372.17	0.0550
Sep-03	BC	25.1	7.75	8.24	369.17	0.0550
Sep-03	CC	24.9	8.11	8.17	361.67	0.0425
Sep-03	CC	25.0	8.22	8.13	366.67	0.0450
Aug-03	BC	24.0	6.30	6.40	205.75	0.325
Aug-03	BC	23.9	6.10	6.30	197.50	0.333
Aug-03	CC	25.2	7.10	6.70	195.50	0.303
Aug-03	CC	25.0	7.00	6.70	196.25	0.315
June-03	BC	21.3	6.52	9.54	165.50	0.798
June-03	BC[a]	–	–	–	169.50	0.828
June-03	CC	20.8	6.57	9.24	161.25	0.768
June-03	CC[a]	–	–	–	156.00	0.775

BC, Begley Creek; CC, Cove Creek.
[a]Sample split.

the principal components of stream water determined from EMMA and the end-member values. The end-member proportions were tested within the model to derive the fraction of each end-member contributing to stream chemistry. The model resulted in the predicted percentage frequency of various end-member contribution combinations that best explain stream chemistry and provided details about the relative contributions of the end-members used.

33.3. Results

Alkalinity/HCO_3^-, Cl^-, K^+, Mg^{2+}, Ca^{2+}, and SO_4^{2-} were chosen for EMMA because the concentration of these solutes were available from USGS Quality of Water Data for the end-members being tested. The solute concentrations for overland/shallow subsurface, runoff, and

Table 33.2. Solute concentrations in stream water samples from the Little Red River. Concentrations of bicarbonate in mg l^{-1} all others in meq l^{-1}

Date	Site	HCO_3^- (mg l^{-1})	Cl^-	K^+	Mg^{2+}	Ca^{2+}	Na^+	$SO4^{2-}$
May-04	BC	220.00	0.11	0.14	0.89	3.95	0.04	1.07
May-04	BC	224.00	0.11	0.14	0.89	3.95	0.05	1.09
May-04	CC	230.00	0.18	0.16	1.09	3.28	0.10	0.27
May-04	CC	235.00	0.18	0.15	1.08	3.36	0.09	0.28
Jan-04	BC	296.59	0.15	0.19	1.20	5.33	0.06	1.44
Jan-04	BC	301.99	0.14	0.19	1.20	5.34	0.06	1.47
Jan-04	CC	336.77	0.27	0.23	1.58	4.75	0.15	0.40
Jan-04	CC	344.09	0.26	0.22	1.56	4.87	0.14	0.41
Nov03	BC	177.54	0.09	0.16	0.99	4.39	0.05	0.86
Nov-03	BC	180.77	0.09	0.16	0.99	4.40	0.05	0.88
Nov-03	CC	206.20	0.16	0.16	1.08	3.24	0.10	0.25
Nov-03	CC	210.68	0.16	0.15	1.07	3.32	0.09	0.25
Sep-03	BC	459.34	0.23	0.28	1.76	7.84	0.09	2.23
Sep-03	BC	467.69	0.22	0.28	1.76	7.85	0.10	2.27
Sep-03	CC	480.22	0.38	0.32	2.16	6.50	0.20	0.57
Sep-03	CC	490.66	0.37	0.30	2.14	6.67	0.19	0.58
Aug-03	BC	182.09	0.09	0.15	0.97	4.33	0.05	0.88
Aug-03	BC	185.40	0.09	0.16	0.97	4.33	0.05	0.90
Aug-03	CC	201.23	0.16	0.20	1.35	4.05	0.13	0.24
Aug-03	CC	205.61	0.15	0.19	1.33	4.15	0.12	0.24
June-03	BC	117.87	0.06	0.07	0.44	1.97	0.02	0.57
June-03	BC	120.01	0.06	0.07	0.44	1.97	0.02	0.58
June-03	CC	153.67	0.12	0.08	0.52	1.56	0.05	0.18
June-03	CC	157.01	0.12	0.07	0.51	1.60	0.05	0.19

BC, Begley Creek; CC, Cove Creek.

groundwater are summarized in Table 33.3. A PCA reduced the six major ions used in two principal components that explained 98% of the variability of the dataset. The first component, U1, explained 79% of the variability and the second component, U2, explained 19%. The mixing diagram of these two principal components along with the end-member components is shown in Fig. 33.2. The surface water samples did not fall within the end-member mixing triangle. Although the graphic fit to the stream data was poor, the EMMA model did not fail to identify the end-member contributions. The relation between observed solute concentration and EMMA predicted concentration showed strong correlation (Fig. 33.3; Alkalinity/HCO_3^- $r^2 = 0.90$; Mg^{2+} $r^2 = 0.91$; Ca^{2+} $r^2 = 0.98$; Cl^- $r^2 = 0.75$; SO_4^{2-} $r^2 = 0.83$; K^+ $r^2 = 0.93$). Since the predictions were linearly correlated with the observed concentrations we examined which combinations of the sampled end-members EMMA predicted as comprising the surface water chemistry.

Table 33.3. End-member solute concentrations from USGS water quality data used in EMMA in meq l^{-1} unless otherwise indicated

	HCO_3^- (mg l^{-1})	Cl^-	K^+	Mg^{2+}	Ca^{2+}	SO_4^{2-}
Overland/shallow subsurface	517	0.231	0.285	1.8	8.1	2.44
Black shale runoff	345	0.02	0.054	0.197	0.6	0.1
Groundwater	29.00	0.275	0.203	1.241	2.74	0.05

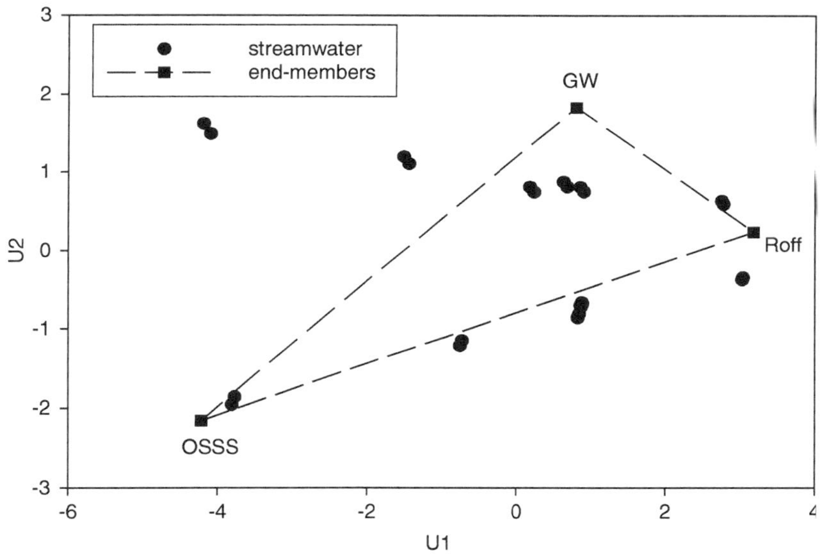

Figure 33.2. Mixing diagram of the principal components U1 and U2 of the stream water and end-member data for the Little Red River. The end-members do not adequately bind the stream water chemistry. The model did yield insights into the presence of three end-members but variance related to discharge needs to be isolated from the variance associated with end-member mixing.

The most frequent combination of end-members that explained the majority of the variance in stream water chemistry was a combination of the overland/subsurface flow and runoff (Fig. 33.4; 41.67%). The combination of the overland/subsurface flow, runoff, and groundwater end-members had a frequency of 33.33%, overland/subsurface flow and groundwater 16.67%, and the runoff and groundwater combination 8.33%. No single end-member accounted for the mixing pattern is

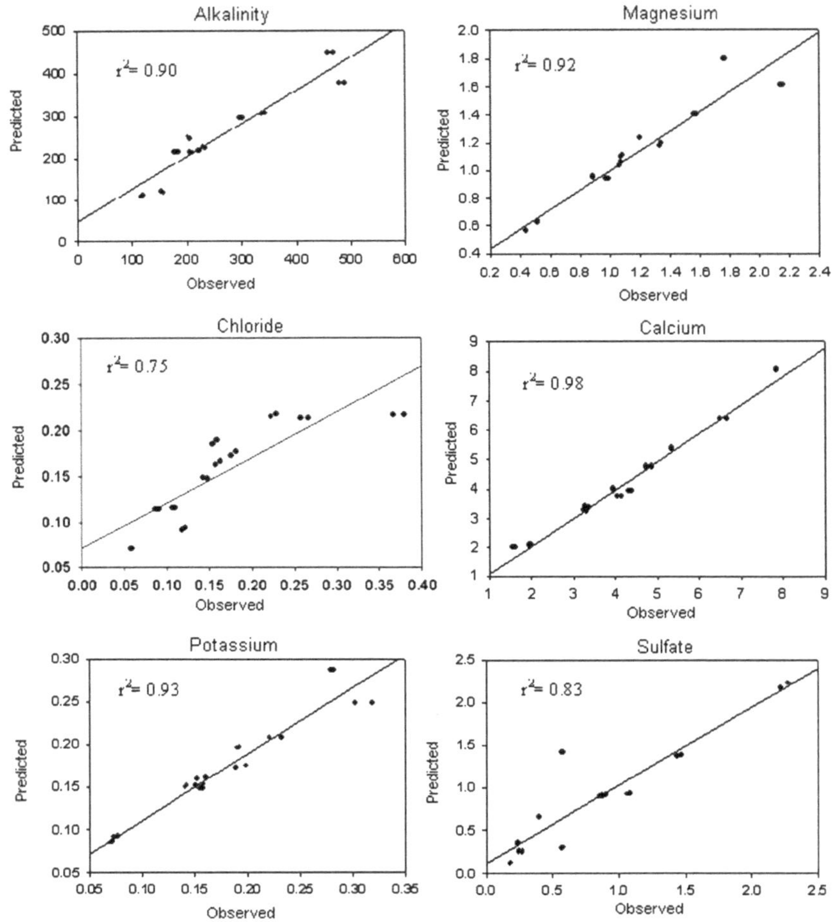

Figure 33.3. Observed stream water chemistry versus EMMA predicted solute concentrations. These results indicate that EMMA has not failed in identifying the contributing end-members to stream chemistry.

observed. As expected, major solute concentrations were lowest during storm flow (Fig. 33.5).

Binary mixing plots of the various solutes show two distinct mixing patterns. Simple binary mixing best describes the relation between surface water concentrations of Cl^- and Mg^{2+}, and HCO_3^- and Mg^{2+} (Fig. 33.6a). Ternary mixing is suggested by the relation between other solutes (Fig. 33.6b).

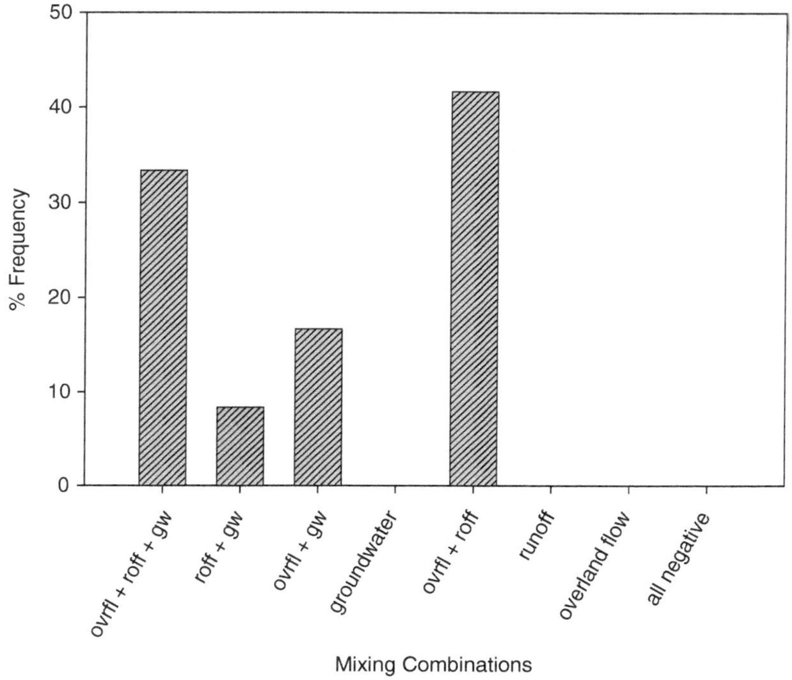

Figure 33.4. Histogram shows the relative contributions of various end-members to stream water chemistry. The percentage frequency at which the variation in the stream water dataset could be explained by a particular end-member or end-member combinations is plotted. A combination of overland/shallow subsurface water and black shale runoff was determined to be the dominant combination of end-member sources (41.67%). A combination of the three end-members, groundwater, overland/shallow subsurface, and black shale runoff had a frequency of 33.33%. This suggests stream chemistry is mainly controlled by these two end-members (ovrfl, overland/shallow subsurface flow; roff, runoff; gw, groundwater).

33.4. Discussion

The stream water sample data did not fit with the mixing triangle of the end-members. Although the graphic fit to the stream data was poor, the EMMA model did not fail to identify the end-member contributions. The poor graphic fit may have been due to seasonal variability within the end-member dataset. The end-member solute concentrations were obtained from USGS data and the seasonal variability could not be adequately captured. The EMMA model assumes that the source solutions are invariant in time and space. Although this assumption is often violated, the

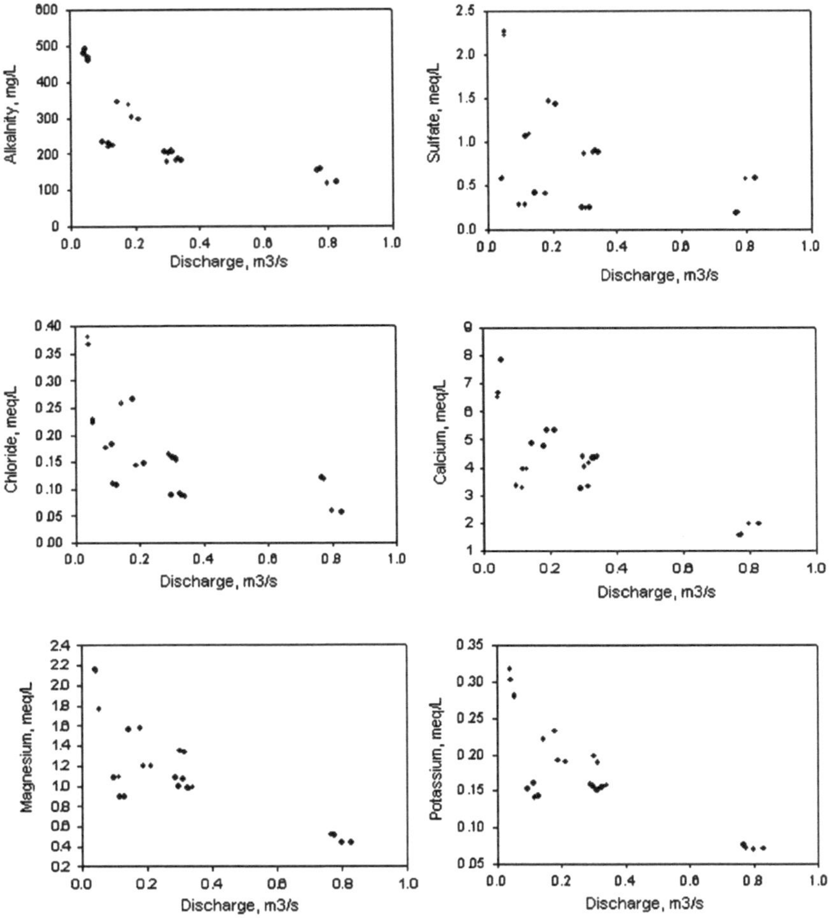

Figure 33.5. Total stream discharge versus major solute concentration used in EMMA. Solute concentrations dilute with increase flow suggesting changes in source waters making up stream chemistry during high flow or dilution of solute concentration.

linear fit of EMMA predicted versus observed stream water data validates the model. The seasonal variability in the dataset is also apparent in the stream water data but the sampling frequency gives a seasonal average of solute concentrations.

The PCA determined that mixing of two components accounts for the majority of variation in the surface water chemistry data (~98%). Given the results of the EMMA model a combination of the two dominant

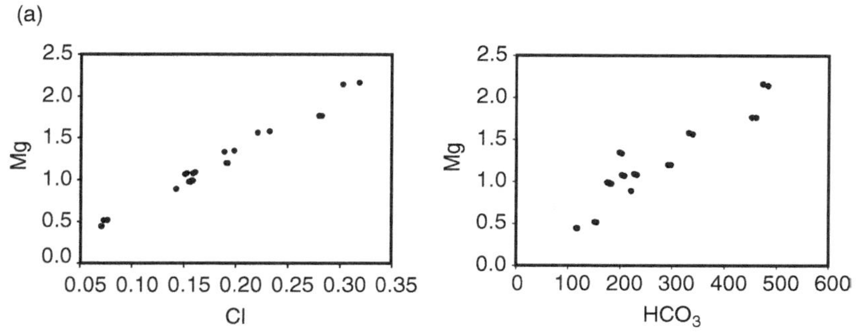

Figure 33.6. (a) Binary mixing plots of the solutes employed in EMMA. The relations between Mg^{2+} and Cl^- and Mg^{2+} and HCO_3^- are simple binary mixing between two end-members as shown by the straight-line relationship. The primary source of these solutes remains the same across the hydrograph. (b) The relation between the other solutes is not binary as shown by the lack of linearity between the solute pairs with several showing two distinct mixing patterns. The dilute effect of storm flow is evidenced at lower stream concentrations such that one of the end-members is present throughout the hydrograph. This is supported by the EMMA model.

end-members, overland/shallow subsurface flow and runoff, were determined to be the dominant end-members accounting for the majority of the variance in stream water chemistry (Fig. 33.4). The solute-mixing relations show that the dissolved surface water Cl^- and Mg^{2+}, and HCO_3^- and Mg^{2+} chemistries are best explained by mixing of two discrete end-members (Fig. 33.6a). Major ion chemistry across the hydrograph is best explained by binary mixing of end-member components with different base flow and storm flow chemistries (Fig. 33.5). Since ion concentrations decrease as discharge increases, a mixing pattern where ion concentrations are low most likely occurs at storm flow and vice versa. The differences in chemistries may result from changes in the flowpaths of water during precipitation events or from dilution of source water inputs. EMMA determined these end-member components to be overland/subsurface flow that includes both soil water and pre-event bank storage water with a chemical signature similar to the surface waters in the limestone draining Middle Fork. The second end-member, runoff from black shale, mixes with the overland/subsurface waters during and just after precipitation events leading to the development of a dissolved chemistry distinct from base flow.

Two end-member components mixing explain 98% of stream water chemistry variability in the dataset, but to explain 100% of the variance a third end-member must be included. Binary mixing plots of HCO_3^- versus

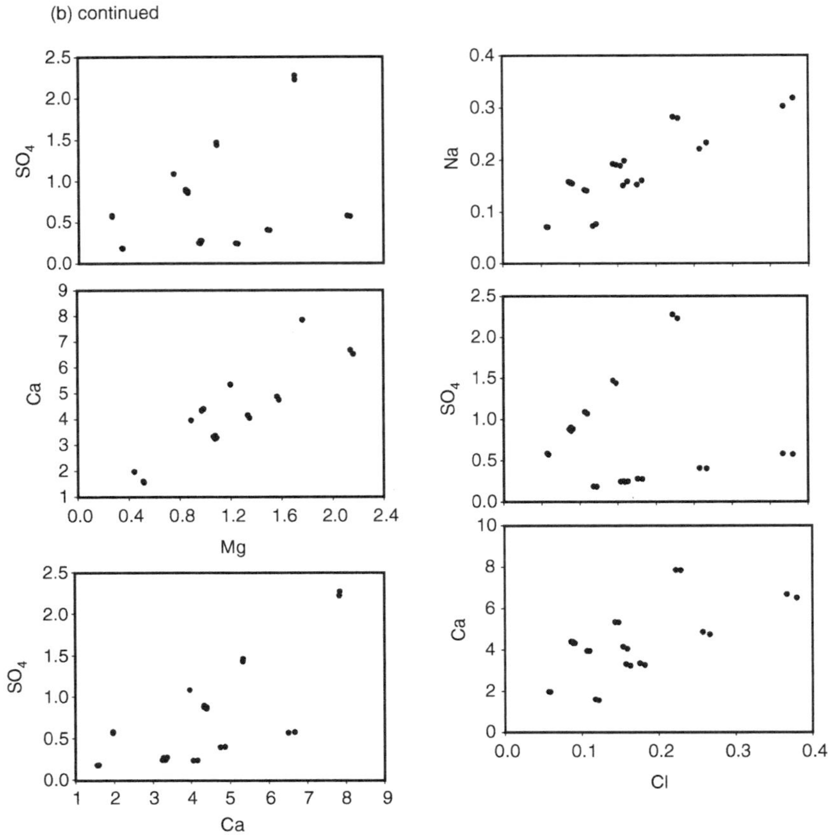

Figure 33.6. Continued

SO_4^{2-}, Ca^{2+}, Na^+, Cl^-; Na^+ versus SO_4^{2-}, Ca^{2+}, Mg^{2+}; Mg^{2+} versus Ca^{2+}, SO_4^{2-}; Cl^- versus Na^+, SO_4^{2-}, Ca^{2+}, and Ca^{2+} versus SO_4^{2-} show a ternary-mixing pattern (Fig. 33.6b). Mixing lines converge at low-ion concentrations pointing to a common storm flow end-member component where the ion concentrations are more dilute but two distinct base flow end-members where the ion concentrations are higher.

Single straight-line binary-mixing plots for Mg^{2+} and Cl^- and Mg^{2+} and HCO_3^- show that there is one base flow and one storm flow end-member components; binary-mixing plots for the other solutes show that there are two base flow components and one storm flow component. This is best explained by the flushing of pre-event (old) water into the stream after a rain event. The binary mixing relations are best explained as two

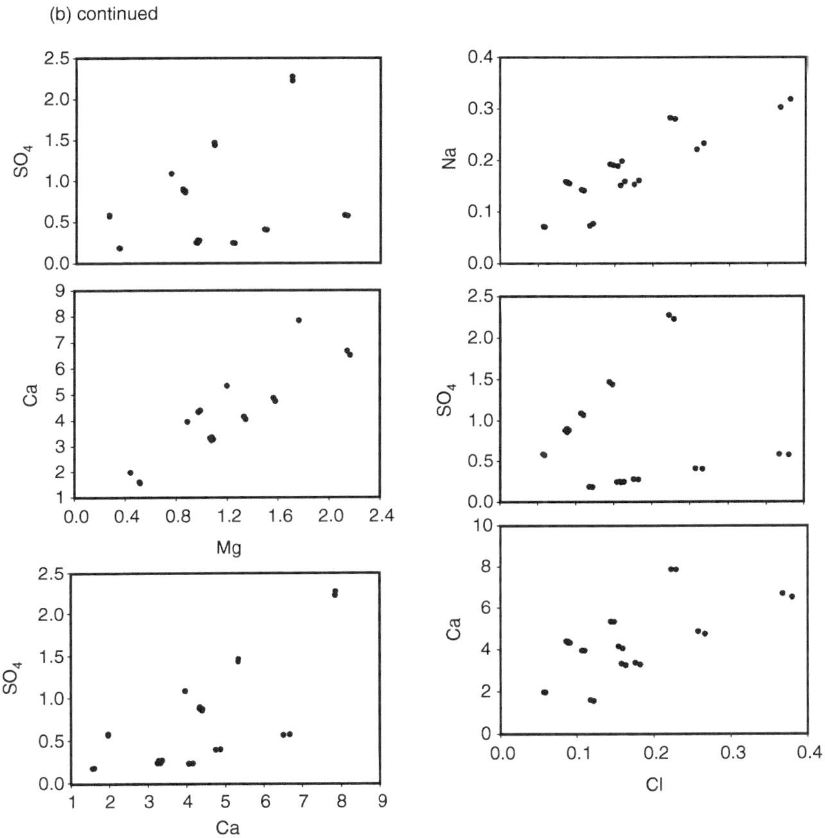

Figure 33.6. Continued

end-members (overland/shallow subsurface and runoff) being present during both storm and base flows with a third end-member, groundwater, only influencing stream water chemistry during true base flow. Ion concentration ratios are similar for overland/shallow subsurface flow and groundwater because of the lithologic similarities where these waters flow. This would also explain why the frequency percentage of the three end-members combination (33.33%) was close to the percentage of the overland/shallow subsurface flow and runoff end-member combinations (41.67%). Although the spring from which groundwater collected was not as close to the headwaters as the other end-members, its chemistry must be similar to the overland/shallow subsurface flow closer to the headwaters for it be identified by EMMA as a discrete end-member.

Runoff from the shales is a significant contributor to the chemistry of the surface water during storm flow. This means that weathering products from black shales are being contributed to the surface waters of the Begley Creek and Cove Creek.

33.5. Conclusion

The PCA determined that two principal components can account for the majority but not all of the variation in the surface water chemistry data. Given that the number of end-members is equal to the number of principal components plus one means that mixing of three discrete end-member sources are required to explain the mixing patterns observed (ternary mixing). Major ion relationships and discharge data suggest that the majority of variation in stream water chemistry of the black shale draining streams can be explained by binary conservative mixing of two end-member components; however, the identity of these end-members change between storm and base flow. During storm flow, the surface water chemistry represents mixing of runoff and overland/shallow subsurface waters. During base flow, groundwater and overland/shallow subsurface waters mix resulting in chemistries distinct from storm flow. In this way, three end-members are required to explain the mixing pattern observed. The individual contributions of these end-members vary between storm and base flow conditions. Runoff from the black shale outcrop does contribute to the surface water chemistry but it is lessened during base flow. The contribution of black shale runoff to surface water chemistry through the hydrograph has implications regarding the potential loading of metals and organic matter into the streams from mechanical and chemical weathering.

The end-members used in this study are not ideal in that runoff water samples contain precipitation as well black shale runoff. Overland/shallow subsurface waters included a mixture of pre-event soil water and hyporheic water. Since waters in intimate contact with shales have been determined to be significant contributors to surface water chemistry future studies will elucidate the role of the hyporheic zone and soil water in the delivery of metals to the streams. Future studies to determine how rock-weathering products, especially metals, are transformed in the hydrograph are warranted. This involves resolving the end-member components determined here and quantifying the contributions of increased precipitation, changes in storm runoff and how metal species are transformed as they move through the stream.

ACKNOWLEDGMENTS

This project was funded thanks to a grant by the National Science Foundation (EAR 0233611). We are also grateful to the Water Rock Life Lab and the Arkansas Bioscience Institute at Arkansas State University for field and analytical support. Finally, we acknowledge Jane M. Hammarstrom and an anonymous reviewer for their critiques and comments that significantly improved this manuscript.

REFERENCES

Bethke, C.M., Marshak, S., 1990. Brine migrations across North America—The plate tectonics of groundwater. Ann. Rev. Earth Planet. Sci. 18, 287–315.

Burns, D.A., McDonnell, J.J., Hooper, R.P., Peters, N.E., Freer, J.E., Kendall, C., Bowen, K., 2001. Quantifying contributions to storm runoff through end-member mixing analysis and hydrologic measurements at the Panola Mountain Research Watershed (Georgia, USA). Hydrol. Process. 15, 1903–1924.

Christophersen, N., Hooper, R.P., 1992. Multivariate analysis of stream water chemical data: The use of principal components analysis for the end-member mixing problem. Water Resour. Res. 28, 99–107.

Christophersen, N., Neal, C., Hooper, R.P., Vogt, R.D., Andersen, S., 1990. Modeling streamwater chemistry as a mixture of soilwater end-members—A step towards second-generation acidification models. J. Hydrol. 116, 307–320.

Chon, H.T., Cho, C.H., Kim, K.W., Moon, H.S., 1996. The occurrence and dispersion of potentially toxic elements in areas covered with black shales and slates in Korea. App. Geochem. 11, 69–76.

Davis, S.N., Moysey, S., DeWayne, L.C., Zreda, M., 2003. Chlorine-36 in groundwater of the United States: Empirical data. Hydrogeol. J. 11, 217–227.

DeWalle, D.R., Poinke, H.B., 1994. Stream flow generation on a small agricultural catchment during autumn recharge: II. Stormflow periods. J. Hydrol. 163, 23–42.

Elsenbeer, H., Lorieri, D., Bonnell, M., 1995. Mixing model approaches to estimate storm flow sources in an overland flow-dominated tropical rainforest catchment. Water Resour. Res. 31, 2267–2278.

Foley, N., Southworth, S., Schultz, A.P., Ayuso, R.A., Robinson, G.R., Seal, R.R., 2001. Geochemical, mineralogical, and environmental characteristics of metamorphosed black shales of the Central Appalachians, with comparisons to metalliferous shales of the Northern Appalachians. In: Adams, D.B., Burke, K., Hemingway, B., Keay, J., Yurewicz, M., comps., U.S. Geological Survey Appalachian Region Integrated Science Workshop Proceedings, Gatlinburg, Tennessee, October 22–26, U.S. Geological Survey Open-File Report 01-406, pp. 111–114 (abstract).

Hannigan, R.E., Bickford, N., 2003. Hydrochemical variation in a spring-fed river, Spring River, Arkansas. Environ. Geosci. 10(4), 167–188.

Hooper, R.P., Christophersen, N., Peters, N.E., 1990. Modelling streamwater chemistry as a mixture of soilwater end-members—An application to the Panola Mountain catchment, Georgia, U.S.A.. J. Hydrol. 116, 321–343.

Hudson, M.R., Kyle, E.M., Pezzutti, D., 2001. Geologic map of the Jasper Quadrangle, Newton and Boone Counties, Arkansas. Miscellaneous field studies map MF-2356. U.S. Geological Survey, Denver, CO.

Jarvie, H.P., Neal, C., Tappin, A.D., Burton, J.D., Hill, L., Neal, M., Harrow, M., Hopkins, R., Watts, C., Wickman, H., 2000. Riverine inputs of major ions and trace elements to the tidal reaches of the River Tweed, UK. Sci. Total Environ. 251/252, 55–81.

Kim, K.W., Thorton, I., 1993. Influence of uraniferous black shales on cadmium, molybdenum and selenium in soils and crop plants in the Deog-Pyong area of Korea. Environ. Geochem. Health 15, 119–133.

LaDouche, B., Probst, A., Viville, D., Idir, S., Baque, D., Loubet, M., Probst, J.L., Bariac, T., 2001. Hydrograph separation using isotopic, chemical and hydrological approaches (Strengbach catchment, France). J. Hydrol. 242, 255–274.

Loukola-Ruskeeniemi, K., 1994. Environmental geochemistry of proterozoic black shales in eastern Finland: A pilot study. In: Bornhorst, T.J., McDowell, S.D. (Eds.), Institute on Lake Superior Geology. Proceedings of the 40th annual meeting, May 11–14, Michigan Technological University, Houghton MI, 37 p. 37.

Loukola-Ruskeeniemi, K., Tenhola, M., Paukola, T., Uutela, A., 1996. Mustaliuskeiden vaikutus Sotkamon Talvivaarassa (English summary: Environmental impact of black shales on watercourses at Talvivaara, Sotkamo, eastern Finland). Vouriteollisus-Berghanteringen 54, 49–53.

Loukola-Ruskeeniemi, K., Uutela, A., Tenhola, M., Paukola, T., 1998. Environmental impact of metalliferous shales at Talvivaara in Finland, with indication of lake acidification 9000 years ago. J. Geochem. Explor. 64, 395–407.

Murthy, R., 2003. Trace and major element fractionation during diagenesis of phosphate nodules. M.S. Thesis, Arkansas State University.

Ogendi, G.M., Farris, J.L, Hannigan, R.E., 2004a. Black shale trace metal concentrations and toxicity: Preliminary findings. In: Wanty, R.B., Seal II, R.R. (Eds.), Proceedings of the 11th International Symposium on Water Rock Interaction, Vol. 2. Taylor and Francis Group plc, London, UK, pp. 1359–1362.

Ogendi, G.M., Hannigan, R.E., Farris, J.L., Smith, D., 2004b. The impact of black shale weathering on sediment quality. J. Ark Acad. Sci. 58, 84–90.

Ogendi, G.M., Hannigan, R.E., Farris, J.L., 2007. Association of dissolved organic carbon with stream discharge and dissolved metals concentrations in black shale-draining streams. In: Sarkar, D., Datta, R., Hannigan, R. (Eds.), Development in Environmental Science. Elsevier Ltd., Vol. 5, pp. 245–272.

Petsch, S.T., Berner, R.A., Eglinton, T.I., 2000. A field study of the chemical weathering of ancient sedimentary organic matter. Org. Geochem. 31, 475–487.

Pinder, G.F., Jones, J.F., 1969. Determination of the groundwater component of peak discharge from the chemistry of total runoff. Water Resour. Res. 5, 438–445.

Renken, R., 1998. Groundwater atlas of the United States; Segment 5; Arkansas, Louisiana and Mississippi. Hydrologic Investigations Atlas, Report: HA-730-F: F1-F28.

Tuttle, M.L.W., Goldhaber, M.B., Breit, G.N., 2001. Mobility of metals from weathered black shale: The role of salt efflorescences. Geol. Soc. Am. Abstracts with Programs 33, A–191.

Winter, T.C., 1999. Relation of streams, lakes, and wetlands to groundwater flow systems. Hydrogeol. J. 7, 28–45.

Section VIII:
Conclusion

Developments in Environmental Science, Volume 5
D. Sarkar, R. Datta and R. Hannigan (Editors)

ISSN: 1474-8177/DOI:10.1016/S1474-8177(07)05034-6

Chapter 34

Current trends and future directions in environmental geochemistry research

Dibyendu Sarkar, Konstantinos C. Makris and Rupali Datta

Abstract

Increasing population growth and rapid industrialization with passing time have augmented the rate of deterioration of the environment, which is having a clear negative impact on human and ecological health. Although considerable research is now being undertaken in critical areas such as global change, environmental quality, and ecological protection, more collaborative, inter-disciplinary research is needed in order to ensure long-term environmental sustainability. A thorough understanding of the geochemistry of the earth's intrinsic processes is needed to develop a comprehensive, holistic model for the protection of the environment to ensure long-term sustainability in a global scale. This chapter briefly discusses some of the more salient current trends in environmental geochemistry research and a few of the many potential future directions of research in this area, often in close collaboration with experts in other scientific disciplines and policy makers geared towards maintaining earth system sustainability.

Technological and scientific advances aiming at improving the quality of our lives often come with a substantial pressure on the environment's resources and sustainability. Sustainability is intrinsically related to environmental quality. Maintaining a clean environment is of utmost importance to the public because of its impact on human and ecological health. Understanding the chemistry of the environment is necessary to better understanding environmental quality issues. Hence, environmental geochemistry as a discipline has been steadily gaining prominence for the last few decades due to increasing public awareness of major environmental issues relating to human and ecological health. The emergence of new concepts and innovative practices in environmental geochemistry has inspired us to assemble original scientific works from a group of experts from various areas of geochemistry in this book. This chapter briefly discusses certain current trends a few of the many potential future directions of innovative research in this fast-emerging field.

One of the areas where geochemistry plays a very important role in nuclear waste management. Advances in nuclear waste management are critical in accurately predicting nuclear waste stability in depository areas (e.g., Yucca Mountain) under a variety of harsh soil environments. Accurate modeling of long-term (hundreds of years) kinetics of nuclear waste degradation in soil environments is of utmost importance for the regulatory agencies.

Phytoremediation of toxic metals in contaminated soils with the use of specific plants that actively or passively absorb large quantities of toxic compounds from Rhizosphere has also become an active field of geochemical research due to its environment-friendly character and its relatively low-implementation cost. Mechanistic studies that underline the major factors influencing contaminant absorption capacity of plants are of great importance. Still unresolved are the mechanisms by which low molecular organic plant exudates interact with soil organic carbon during contaminant uptake by the plant.

A comprehensive understanding of aqueous speciation of toxic metalloids in environmental systems is still incomplete. Although lately much research has been focused in this area, this is one emerging research field that will continue to develop as a result of recent advances in the use of hyphenated analytical techniques such as IC-ICPMS or HPLC-ICPMS in environmental research in the future. Speciation of metalloids, such arsenic, chromium, selenium, etc., in soil and water environments are prime examples of innovative concepts and practices that are currently developed for use in environmental geochemical research. Future research on this subject will most likely to focus on better understanding the association of different metalloid species with dissolved organic carbon, and how this complexation influences the overall metalloid availability in soil solution environments.

Characterization and monitoring of surface- and groundwater geochemistry is another area that is continuously evolving due to vast improvements in sensitivity of the new state-of-the-science analytical instrumentation. Research in soil and sediment geochemistry is also an active field, and attempts are being made to better understand and predict fate and transport of contaminants in soils both qualitatively and semiquantitatively. This generation has experienced tremendous advances in analytical methodology that are capable of measuring micro- and nanogram quantities of chemicals and their corresponding toxicities on human and ecological health. A significant number of findings have been reported on new and emerging contaminants in water and soil systems, which were previously unknown to the scientific and regulatory communities. Accordingly, maximum contaminant levels for contaminants in water and soil have been adjusted (mostly lowered) to comply with updated federal regulations.

Such developments in the analytical sciences and better understanding of the toxicology of the emergent contaminants have played a tremendous role in countering the negative effects of anthropogenic input to the environment, which has increased exponentially over the years with increased modernization of the agricultural and industrial practices. Increased toxicant input in the form of excess fertilizers, manures, biosolids, and industrial wastes has resulted in rapid deterioration of soil and water quality. A more comprehensive understanding of all the physical, chemical, and biological factors influencing chemical fate and transport in both unsaturated and saturated media is needed prior to developing the ability to accurately model such processes in a wide range of spatial and temporal scales. Development of such all-encompassing, holistic models will tremendously improve the predictive capabilities that are so needed to maintain a sustainable environment.

High-tech molecular instrumentation techniques, such as XANES (X-ray absorption near edge structure) and EXAFS (extended X-ray absorption fine structure), will continue to serve as robust analytical tools in elucidating sorption/desorption mechanisms of chemicals at the solid/liquid and solid/air interface. In the future, more advanced molecular techniques will allow us to conduct such studies at the micrometer scale (e.g., X-ray spectromicroscopy). The increase in spatial resolution will be accompanied by concomitant advances in understanding complex reactions at the true microscopic level of micron-based and sub-micron particles.

Risk assessment and management provides a new frontier in environmental geochemistry research that generates useful information on the health effects of contaminated environment on humans via a wide range of exposure routes, including direct ingestion of soils, especially by children in playgrounds and residential yards. This has become a more serious problem in recent years because of the "urban sprawl" that the world is experiencing with passing time. Yesterday's farmlands that experienced continuous fertilizer and pesticide application for decades are becoming today's backyards as urban development continues to encroach upon rural landscape. Potential bioaccessibility of metals and metalloids in contaminated soils and their health effects due to prolonged exposure in low doses is one area that has not yet been adequately studied. More comprehensive models are needed to predict metal bioaccessibility in a wide range of environmental conditions. New in-vitro bioaccessibility tests with better predictive capabilities are required to simulate the trends offered by more dependable, yet extremely expensive animal trials. Because estimated risk is a direct function of the input parameters, a better understanding of bioaccessibility is needed to more accurately predict the health risk associated with chronic human exposure to contaminated media.

Future research in environmental geochemistry should focus on the complex interactions of the chemicals and the environmental media at the solid/solution interface in both macro- and microscales. Characterization and monitoring, microbe–contaminant interactions, coupled with modeling, high-performance computing, as well as molecular characterization of submicron colloidal particles will quite emerge as major themes driving the research in environmental geochemistry for the upcoming decades. There is a great need for innovative, effective, yet inexpensive contaminant remediation technologies, which demands more focused research in this area. Inter-disciplinary studies coupling the fundamental biological, chemical, and physical processes are inevitable prerequisites in gaining further insight into such complex reactive processes that control contaminant mobility and transport in the environment.

The need to provide mechanistic answers to unsolved environmental problems related to geochemical processes will possibly shift the current trend of emphasis on short-term (1–2 years) studies to longer-term (>5–10 years) monitoring and analysis of environmental processes. With the progress of time, economics will play a more crucial role in determining the applicability of the emerging environmental remediation technologies. Decisions on the feasibility of applicable environmental technologies will be made more and more not only on the basis of science, but on the cost-effectiveness of the methods. Striking the perfect balance between science and will be absolutely necessary to make educated decisions on the appropriate environmental technologies. If financial viability takes precedence over scientific soundness without exercising proper judgment and discrimination, it can turn out to be a slippery slope causing irreversible damage to the environment, thereby hurting human and ecological health.

Inter-disciplinary studies addressing complex organo-mineral processes ranging from the pore scale to the field-scale should be given more attention in environmental geochemical research. The obvious knowledge gap that currently exists between the variety of microbial processes in aquatic environments and their ultimate effects on contaminant fate and transport will be the driving force behind developing more fruitful research collaborations between geochemists and microbiologists. There is a great need for unraveling the heterogeneous and complex interactions of the microbes with the chemicals and the porous media responsible for contaminant immobilization in soils, and/or plume cleanup in contaminated groundwater. Microbial genetics and proteomics should become more commonly used tools in recognizing the specific strains responsible for such biodegradation processes; applying these techniques to genetically modify strains to further accelerate and/or enhance contaminant cleanup need to be a research priority.

Global climate change currently is a major focus area of the scientific community at large. Collaborative efforts in an international scale are already in progress in order to understand the rates and processes that are responsible for causing climate change. For example, a lot of research is currently being done and some significant policy implementation has been made in recent years to reduce greenhouse gas emissions. Geochemists and atmospheric scientists will continue to combine their skills and knowledge to investigate the mechanisms behind greenhouse gas emissions in a wide range of spatial scales, from microbial activities in the minute soil pores to the role of rapid industrialization in continuous generation of Sox, Nox, and acid rain.

New inter-disciplinary collaboration pathways will open up between the environmental geochemists and biomedical scientists as more linkages continue to be established between pathogenic compounds and infectious agents in the environment and their potential human exposure. One such infectious agent that has received serious attention in recent years is the prions, which have proved to be rather persistent in soils. Prions cause transmissible spongiform encephalopathies (TSE) in cattle (the "mad cow" disease). One of the major sources of cattle exposure to prions is the contaminated soil, and the leading cause of human exposure to prions is via consumption of the contaminated meat. The soil can be contaminated by causative agent as a result of (i) accidental dispersion from storage plants of meat and bone meal, (ii) incorporation of bone meal into fertilizers, (iii) spreading of effluents of slaughterhouses, and (iv) burial of carcasses of contaminated animals. A thorough understanding of the geochemical behavior of prions in various environmental systems and their subsequent toxicological effects on bovine or humans is urgently needed.

There are many other areas of environmental and biomedical research where the input of geochemists are not only advantageous but absolutely required. A lot of research is currently being done, but a lot more needs to be done with passing time. Overall, the future seems promising for environmental geochemists as there are major unresolved issues and challenges that need to be met in order to ensure environmental sustainability, protection of natural ecosystems, and for the protection of human health caused by exposure to toxic compounds in soil, plants, water, and air.

Index

Papier | Fördert gute Waldnutzung
FSC® C083411

Kolibri 360 GmbH
Ferdinand-Jühlke-Straße 7
99095 Erfurt, Deutschland
produktsicherheit@kolibri360.de